Quasiconformal Mappings in the Plane and Complex Dynamics

Luis T. Magalhães

Quasiconformal Mappings in the Plane and Complex Dynamics

Springer

Luis T. Magalhães
Instituto Superior Técnico
University of Lisbon
Lisboa, Portugal

ISBN 978-3-031-80117-4 ISBN 978-3-031-80115-0 (eBook)
https://doi.org/10.1007/978-3-031-80115-0

Mathematics Subject Classification: 30-01, 37F10, 37F30, 30C62

This Springer imprint is published by the registered company Springer Nature Switzerland AG
The registered company address is: Gewerbestrasse 11, 6330 Cham, Switzerland

To

Lia
Eric
Miguel

Preface

This book is a natural continuation of the author's book *Complex Analysis and Dynamics in One Variable with Applications*,[1] Springer, 2025, designed to support the initial steps of students of Mathematics, Physics, or Engineering in their initial years at a university in studying Complex Analysis and to provide a coherent and solid foundation for subsequent studies, while delving into the fascinating Complex Dynamics.

Similarly to that book, the intention now is to facilitate the access to the bases of quasiconformal mappings in the complex plane, especially in view of applications to Complex Dynamics, without the need of relying on dispersed bibliographical sources. This led to include extensive, but simplified, appendices with the necessary real analysis that usually appears to students in more advanced studies. The book is also intended to be useful to graduate students concentrating in Dynamical Systems, Complex Analysis, or related fields, since the important topics included are fertile for future research, are not found together in other textbooks and are presented as much as possible in a self-contained fashion.

For the convenience of readers, references to background material needed throughout the text are mostly to the author's book mentioned above. However, most prospective readers do not need to go through that book, since the required background may have been acquired with one of several widely used beginning One Variable Complex Analysis textbooks, such as those with the scope of the excellent classic L.V. Ahlfors'[2] *Complex Analysis*, or of the E.M. Stein and R. Shakarchi's *Complex Analysis*, the chapters on Complex Analysis of the W.Rudin's *Real and Complex Analysis*, B.Chabat's *Introduction à L'Analyse Complexe, Tome*

[1] In the present text, the equivalent to the content of that book is assumed as a prerequisite and explicit references to it are by its acronym *CADOVA*.

[2] Lars Ahlfors (1907–1966) received the *Fields Medal* in 1936, when it was first awarded, for contributions to Riemann surfaces and the opening of new fields in Mathematical Analysis, and he later received the *John von Neumann Lecture Prize* of the *SIAM—Society of Industrial and Applied Mathematics* of USA in 1960, and the *Leroy P. Steele Prize* of the *AMS—American Mathematical Society* in 1982.

1: Fonctions d'Une Variable, R. Remmert's *Theory of Complex Functions*, or J. Conway's[3] *Functions of One Complex Variable* (2 volumes), to name a few as examples.

These books do not deal with Complex Dynamics, the topic of the last chapter of the author's book mentioned above, nor even with the uniformization of Riemann surfaces, considered in its preceding chapter, which usually appears only in more advanced textbooks. Readers not acquainted with it may either take its validity for granted or study this chapter of the author's book or an alternative source. For Complex Dynamics, widely used alternative textbooks on what is needed here and is covered in the author's book final chapter are, for instance, parts of J. Milnor's[4] *Dynamics in One Complex Variable* or L. Carleson and T.W. Gamelin's *Complex Dynamics*.[5]

The selection of topics in this book is not found in other existing textbooks, because in addition to the intention of describing the bases of quasiconformal mappings in an accessible way even for students in an initial phase of university studies, it is oriented toward application in Complex Dynamics. The seminal idea of applying quasiconformal mappings in Complex Dynamics was of D. Sullivan in 1982, in analogy with Klein groups.[6] It was the most remarkable development in recent decades in the study of the dynamics of holomorphic functions. The flexibility achieved with perturbations of holomorphic functions by quasiconformal mappings has been particularly fertile by allowing the analysis of aspects of the dynamics defined by iteration of complex functions inaccessible within the strict scope of holomorphic functions. In addition to Complex Dynamics and Klein groups, quasiconformal mappings have been applied to Geometry, Topology, Potential Theory and Partial Differential Equations, Functional Analysis and Calculus of Variations, Harmonic Analysis, Electrostatics, Fluid Dynamics, and Nonlinear Elasticity, but these applications are not considered here.

We re-encounter the fertility of the study of functions in the complex plane as a starting point for concepts in more general contexts and for diverse applications, highlighted in the author's book mentioned at the beginning of this Preface. In that book, I defended the benefits of preceding the study of areas such as Algebraic Topology, Algebraic Geometry, Riemannian Geometry, Dynamical Systems, Harmonic Analysis, and Elliptic Partial Differential Equations by the study of Complex Analysis in these directions, as it was the context where many of their basic concepts naturally emerged in a relatively simple framework. Similarly, I also consider

---

[3] Elias Stein (1931–1918) received the *Leroy P. Steele Prize for Lifetime Achievement* of the *AMS* in 2002. Shakarchi, Rami (1975-). Walter Rudin (1921–2010) received the *Leroy P. Steele Prize* of the *AMS* in 1993. Chabat, Boris (1917–1987). Remmert, Reinhold (1930–2016). John Conway received the *Leroy P. Steele Prize* of the *AMS* in 2000 (1937–2020).

[4] John Milnor (1931-) received the *Fields Medal* in 1962, the *Leroy P. Steele Prize for Lifetime Achievement* of the *AMS* in 2011 and the *Abel Prize* in the same year.

[5] Carleson, Lennart (1928-). Gamelin, Theodore William (1939-).

[6] Klein groups are subgroups of the matrices group of complex 2×2 matrices with determinant 1 modulo the center, represented by the conformal homeomorphisms in $\mathbb{C}_\infty$. Klein, Felix (1849–1925). Denis Sullivan (1941-) was awarded the *Oswald Veblen Prize in Geometry* of the *AMS* in 1971, the *Steele Prize for Lifetime Achievement* of the *AMS* in 2006, and the *Abel Prize* of 2022.

very useful to precede the general study of quasiconformal mappings and their applications, focusing initially on functions in the complex plane, the framework where they originated, an idea that was instrumental to carve the scope of the book.

There are several excellent books on quasiconformal mappings in the plane, but without Complex Dynamics. Some of them are listed in the final Bibliography. Books including the bases of quasiconformal mappings and considering substantial aspects of Complex Dynamics are scarce and concentrate in restricted aspects of some topics considered here, as, for example, the books of E. Faria and W. Melo and of B. Branner and N. Fagella[7] listed in the final Bibliography.

There are several important topics appearing here for the first time in a self-contained (except for contents covered in the author's book mentioned at the beginning) presentation in a general textbook. Since referees of this book manuscript suggested it would be useful for prospective readers to include here a description of how its contents differ from existing books, a not exhaustive list of selected examples of such topics is given next, just by themselves accounting for more than half of the book:

- The simpler proof of the measurable Riemann mapping theorem, with Hilbert space and Fourier-Plancherel transform, obtained in 2000 by A. Douady and X. Buff.[8]
- The 2008 proof of K. Astala, T. Iwaniec and G.J. Martin, simplifying the 2007 proof of L. Kovalev, that δ-monotone functions in a Hilbert space of dimension greater than 1 are quasiconformal mappings, providing a bit of additional light on quasiconformity.[9]
- The proof by R. Mañé[10] in 1985 that complex rational functions with Herman[11] ring cycles are not structurally stable by proving they are not infinitesimally stable.
- The 1998 work of C. McMullen[12] and D. Sullivan considering several notions of stability and bifurcation and completing the study initiated in 1983 by R. Mañé, P. Sad,[13] and D. Sullivan on structural stability and bifurcation and related hyperbolicity and genericity.

[7] de Faria, Edson (1962-).de Melo, Wellington (1946–2016).Branner, Bodil (1943-).Fagella, Núria (1967-).

[8] Riemann, Bernhard (1826–1866). Hilbert, David (1862–1943). Fourier, Jean-Baptiste Joseph (1768–1830). Plancherel, Michel (1885–1967). Douady, Adrien (1935–2006). Buff, Xavier (1971-).

[9] Kovalev, Leonid (1977-). Astala, Kari (1953-). Iwaniec, Tadeusz (1947-). Martin, Gaven John (1958-).

[10] Mañé, Ricardo (1948–1995).

[11] Herman, Michael-Robert (1942–2000).

[12] Curtis McMullen (1958-) was awarded the *Fields Medal* in 1988 for work in Complex Dynamics and hyperbolic geometry, and the *Humboldt Research Award* of the German *Alexander von Humboldt Foundation* in 2011.

[13] Sad, Paulo (1952-).

- The concept and general properties of the dynamics of polynomial-like functions introduced in 1985 by A. Douady and J. Hubbard, motivated by observations on the dynamics of the Newton method for computing zeros of functions applied to complex cubic polynomials in 1983 by J. Curry, L. Garret[14] and D. Sullivan.
- A simplified, but detailed, description of the fundamental Écalle cylinders theory for the study of perturbations of parabolic fixed points, initiated in 1984–1985 by A. Douady and J. Hubbard and continued in 1989 by P. Lavaurs and in 1998 by M. Shishikura.[15]
- The important work of M. Shishikura published in 1987 on the optimal upper bounds of the number of attracting or neutral periodic orbits, the number of Herman rings cycles, and the number of cycles of rational functions in the complex plane with quasiconformal surgery, and his work published in 1988 proving that the Hausdorff dimension of the Mandelbrot set boundary and generically of the Julia set of quadratic maps in the complex plane with parameters in that boundary is 2, as conjectured 13 years before by B. Mandelbrot, with holomorphic motions, Écalle cylinders (involving the measurable Riemann mapping theorem) and stability and bifurcation.[16]
- The proof of the slight extension of the Thurston[17] theorem described in 2011 by X. Buff, G. Cui, and L. Tan of the characterization of postcritically finite rational maps in the complex plane that are not Lattès functions by the Thurston obstructions, found by W. Thurston in 1982 and presented by him at a conference in 1983, with proof published in 1993 by A. Douady and J. Hubbard.[18]
- The proof of C. McMullen in 1994 that when Thurston obstructions fail marginally and the map is not a Lattès function, there are rotation domains (i.e., Siegel disks or Herman rings), using results on meromorphic quadratic differentials with almost all trajectories closed due to several authors, mainly J. Jenkins in 1957, H. Renelt in 1976, and K. Strebel[19] in 1967 and 1976.
- The application of the above-mentioned Thurston theorem to maximize the number of zeros of harmonic polynomials of a complex variable with Complex Dynamics, based on works of D. Khavinson and G. Świątek in 2003, D. Khavinson and G. Neumann in 2006, and L. Geyer in 2008, with application to Gravitational Lensing.[20]

---

[14] Hubbard, John (1945-). Newton, Isaac Newton (1642–1726). Curry, James (1948-). Garrett, Lucy.

[15] Écalle, Jean (1950-). Lavaurs, Pierre (1963-). Shishikura, Mitsuhiro (1960-).

[16] Hausdorff, Felix (1868–1942). Mandelbrot, Benoit (1924–2010). Julia, Gaston (1893–1978).

[17] William Thurston (1946–2012) was awarded the *Fields Medal* in 1982 for having *"revolutionized study of topology in* 2 and 3 dimensions, showing interplay between analysis, topology, and geometry", the *Oswald Veblen Prize in Geometry* of the *AMS* in 1976 and the *Steele Prize for Seminal Contribution to Research* of the *AMS* in 2012.

[18] Cui, Guizhen (1966-). Tan, Lei (1963–2016).

[19] Lattès, Samuel (1873–1918). Siegel, Carl Ludwig (1896–1981). Jenkins, James (1923–2012) Renelt, Heinrich (1942-). Strebel, Kurt (1921–2013).

[20] Khavinson, Dmitry (1956-). Świątek, Grzegorz (1964-). Neumann, Geneva.

- The notions and several results of Teichmüller space of holomorphic and rational functions in the plane and the corresponding space of deformations due to C. McMullen and D. Sullivan in 1998, with characterizations of quasiconformal mappings in a hyperbolic Riemann surface isotopic relative to the ideal boundary through quasiconformal conjugacies, obtained by C. Earle and C. McMullen in 1988 using the notions and properties of conformal barycenter and conformal barycentric extension introduced in 1986 by A. Douady and C. Earle.[21]

Some other topics are presented very differently from the few other books that address them, as, for instance, holomorphic motions, in one whole book chapter, a most important subject initiated in 1983 by R. Mañé, P. Sad, and D. Sullivan, and pursued by several others, especially in 1986 by D. Sullivan and W. Thurston, and by L. Bers and H. Royden, in 1991 and 1995 by Z. Slodkowski, in 1994 by C.J. Earle, I. Kra, and S.L. Krushkal, in 2001 by K. Astala and G.J. Martin, in 2004 by E.M. Chirka, and in 2015 by F. Gardiner, Y. Jiang, and Z. Wang.[22] But even more classic topics like the geometric and analytic definitions of quasiconformal mappings in the plane in the first chapter and their possible extensions, in the third chapter, are presented in a different fashion, hopefully simpler for students of these fascinating subjects, than alternative books.

The notion of quasiconformal mapping in the complex plane stems from weakening that of conformal homeomorphism by allowing at each point bounded distortion of angles by homeomorphisms. As such, they also preserve the topology and have geometric properties of bounded distortion of circles and ellipses, and of affine geometry. The flexibility they provide allows using deformations of holomorphic functions that proved to be most useful in several areas and have been the basis for the most important contributions to the dynamics of iterations of holomorphic functions since D. Sullivan had in 1982 the brilliant idea of applying them in this context.

Quasiconformal mappings appeared in 1859–1860 at the hands of N.-A. Tissot in cartography (the same context where conformal mappings appeared) about three quarters of a century before the name quasiconformal was created. N.-A. Tissot expressed the local distortion of angles, distance, and area by a field of ellipses, called *ellipse indicatrix*, a representation routinely used afterwards by cartographers for more than one century. The detailed study of quasiconformal mappings was initiated in 1928 by H. Grötzsch, who considered them as C^1 functions of two real variables, but their wide usefulness is revealed when this restriction is dropped. This led L. Ahlfors to adopt in 1935 a geometric definition in one of his works judged

[21] Teichmüller, Oswald (1913–1943).

[22] Lipman Bers (1914–1993) was awarded the *Leroy P. Steele Prize* of the *AMS* in 1975. Royden, Halsey (1928–1993). Slodkowski, Zbigniew. Earle, Clifford (1935–2017). Kra, Irwin (1937-). Krushkal, Samuel Leibovich (1938-). Chirka, Evgeniĭ Mikhaĭlovich (1942-). Gardiner, Cyril Frederick (1930-). Jiang, Yunping. Wang, Zhe.

as determinant for awarding him the *Fields Medal* in 1936, where he developed a geometric version of Nevanlinna theory and introduced the name quasiconformal.[23]

Functions with quasiconformal properties had been considered in the context of solutions of certain partial differential equations in works of several authors in 1916, 1919, 1935, and 1938, referred to in the introduction to Chap. 1, but the connection with the geometric notion of quasiconformal mapping was only noticed several decades later, around 1955.The analytic definition of quasiconformal mappings in the plane is centered in relation of the functions partial derivatives by the Beltrami[24] partial differential equation with a Beltrami coefficient establishing ellipses eccentricity and orientation in the tangent space of the domain which are transformed by the derivative of the function considered with real variables to circles in the tangent space of the function range. So, this definition requires an appropriate notion of solutions of linear partial differential equations, in terms of distributions and relying on real analysis.

The quasiconformal mappings theory was notably developed in 1937 and 1939 by O. Teichmüller, who introduced equivalence classes of quasiconformal mappings modulo conformal homeomorphisms in Riemann surfaces with a metric defined by the maximal distortion of circles to ellipses in the surfaces tangent spaces.

The clarification of the relationship of geometric and analytic definitions of quasiconformal mappings in the plane led, around 1955, to an analytic definition as a solution of the Beltrami partial differential equation with bounded coefficient and the associated measurable Riemann mapping theorem, which is the central object of Chap. 2. This theorem had been proved in particular cases and with alternative proofs by several people, and in general conditions by C. Morrey in 1938. In 1960 L. Ahlfors and L. Bers gave a proof for measurable Beltrami coefficient depending continuously or holomorphically on parameters with solutions with similar regularity on the parameters, paving the way to applications of perturbations of holomorphic functions made possible by the flexibility allowed by quasiconformal mappings when compared to conformal homeomorphisms.[25]

In Chap. 4, the extension of the notion of holomorphic motion is considered as specified by D. Sullivan and W. Thurston in 1986 following the use of this concept by R. Mañé, P. Sad, and D. Sullivan in 1983 in the context of structural stability and bifurcation of dynamical systems defined by iteration of complex rational functions considered in Chap. 5.

The work of R. Mañé, P. Sad, and D. Sullivan in 1983 included establishing relationships with hyperbolicity and genericity properties, and was completed in 1998 by W. McMullen and D. Sullivan. The chapter is centered in the result of R. Mañé, P. Sad, and D. Sullivan on holomorphic motions with the important contribution of L. Bers and H. Royden in 1986 allowing to have uniqueness of the extension, which is necessary for certain results of dynamical systems structural

---

[23] Tissot, Nicolas-Auguste (1824–1897). Grötzsch, Herbert (1902–1993). Nevanlinna, Rolf (1895–1980).

[24] Beltrami, Eugenio (1835–1900).

[25] Morrey, Charles (1907–1984).

Stability. They obtained a simple upper bound of the Beltrami coefficient of each quasiconformal mapping of a holomorphic motion of a region in the Riemann sphere $\mathbb{C}_\infty$ with parameter in an open disk or in the open upper complex half plane $\mathbb{H}$ in terms of the parameter value. To get this result they used Teichmüller spaces, in particular, the Bers embedding proved in 1960 by L. Bers, establishing that the Teichmüller space of such a region can be embedded in the space of quadratic differentials of the set in a complex half plane through Schwarzian derivatives. So, Chap. 4 includes an introduction to Teichmüller spaces of regions in $\mathbb{C}_\infty$ with a covering by the open upper complex half plane $\mathbb{H}$, in a simple and concrete context and with the handy side benefit of one of these spaces being the universal Teichmüller space which contains all Teichmüller spaces of Riemann surfaces with a covering by $\mathbb{H}$, as observed by L. Bers in[26] 1965.

Chapter 4 also includes the remarkable 1985 proof by R.Mañé that complex rational functions with Herman ring cycles are not structurally stable, by proving that they are not infinitesimally stable, a notion he introduced then in Complex Dynamics This notion had been used for diffeomorphisms in differential manifolds by J. Robbin in 1971 (inspired by a J.Moser article of 1969) as an intermediate property to prove that diffeomorphisms hyperbolicity and transversality implies structural stability under C^1 perturbations. R. Mañé extended it in 1971 to endomorphisms in differential manifolds and in 1987 he proved the equivalence of infinitesimal stability and structural stability for C^1 diffeomorphisms. In 1988 and 1989, H.Ikeda obtained for endomorphisms properties of infinitesimal stability similar to those for diffeomorphisms.

Chapter 6 is dedicated to the quasiconformal surgery proposed by D. Sullivan, A. Douady, and J. Hubbard in 1982 and 1983, although L. Bers had introduced two decades earlier, in 1960, the pioneering idea underlying quasiconformal surgery of using with remarkable fertility extensions to $\mathbb{C}$ of quasiconformal mappings in $\mathbb{H}$ defined to be conformal homeomorphisms in $-\mathbb{H}$. Quasiconformal surgery was particularly fertile for the study of the dynamics of complex rational functions in the most recent 4 decades.

Chapter 7 presents the interesting proofs, published in 1998 by M. Shishikura, that the Mandelbrot set boundary has Hausdorff dimension 2, as conjectured by B. Mandelbrot in 1985, as well as the Julia sets of complex quadratic polynomials z^2+c for a generic subset of points c of the Mandelbrot set boundary, hence for a dense subset of this boundary. The proofs are based on the geometry of orbits in neighborhoods of parabolic fixed points using Écalle cylinders theory. This is a good opportunity to introduce this important topic of general interest, initiated in 1984–1985 by A. Douady and J. Hubbard and continued in 1989 by P. Lavaurs and

[26] Robbin, Joel (1941-). Jürgen Moser (1928–1999) was awarded the *George Birkhoff Prize* of the *AMS* and the *SIAM* in 1968 *"for outstanding contributions to applied mathematics in the highest and broadest sense,"* the *James Craig Watson Medal* in 1969 by the *U.S. National Academy of Sciences* for contributions to Astronomy, and the *John von Neumann Lecture Prize* of the *SIAM* in 1984. Ikeda, Hiroshi. Schwarz, Hermann (1843–1921).

in 1998 by M. Shishikura. The name was given by A. Douady due to its connection with the 1975 J. Écalle work on holomorphic functions tangent to the identity.

Chapter 8 is about postcritically finite rational functions and combinatorial equivalence. These functions are important because several aspects of the dynamics they define can be obtained from information about postcritical orbits, but they have a reinforced importance because, excluding Lattès functions (which do not occur in polynomial functions), the set of postcritical finite rational functions is small, since as attracting basins of attracting periodic orbits contain critical points, if there were a non-postcritical attracting periodic orbit, there would exist an infinite postcritical orbit, and consequently postcritically finite rational functions do not have non-postcritical attracting orbits. The complex rational functions can be identified with branched coverings of $\mathbb{C}_\infty$ with functions defined in $\mathbb{C}_\infty$. In 1982 W. Thurston presented a topological characterization of postcritically finite rational functions that are not Lattès functions relating them bijectively, with Teichmüller spaces, to the homotopy classes of branched coverings of $\mathbb{C}_\infty$ with a finite set of branching points, provided they do not have Thurston obstructions, expressed in terms of the spectral radius of matrices.[27] The result was presented in 1983 by W. Thurston at a conference and the proof was published in 1993 by A. Douady and J. Hubbard based on notes they took at that conference. In this chapter we have the proof of a slightly more general result published in 2011 by X. Buff, G. Cui, and L. Tan.

In Chap. 9, we consider basic aspects of Teichmüller spaces of Riemann surfaces as a preliminary step to the final chapter, where Teichmüller spaces of holomorphic dynamical systems are introduced and to enhance the contact with the general Teichmüller spaces theory. The links of Teichmüller spaces with quasiconformal mappings are strengthened in this chapter, including descriptions of the universal Teichmüller space from many different perspectives. This space is the Teichmüller space of $\mathbb{H}$, used in Chap. 4 for holomorphic motions, and it contains the Teichmüller space of any Riemann surface with a covering by $\mathbb{H}$. Besides clarifying the notion of Teichmüller space, those descriptions allow to obtain some general properties of Teichmüller spaces from those of the universal Teichmüller space.

The last chapter is on aspects of Teichmüller spaces of holomorphic dynamical systems, due to C. McMullen and D. Sullivan in 1998, which are a fundamental basis of such Dynamical Systems Theory. For this, the study of Teichmüller spaces, initiated in Chap. 4 for holomorphic motions, and continued in Chap. 8 for a characterization of postcritically finite rational functions and in Chap. 9 with Teichmüller spaces of Riemann surfaces, is further pursued.

Thus, the last six chapters, approximately half of the book, considerably expand the study of the dynamics of iteration of rational functions in the final chapter of the author's book mentioned at the beginning of this Preface.

---

[27] In 2020 Dylan Paul Thurston (1972-) obtained a necessary and sufficient condition expressed in terms of extremal length and elastic graphs. D.P. Thurston is son of William Thurston and was PhD student of Vaughan Jones (1952–2020), who was awarded the *Fields Medal* in 1990 for having *"discovered an astonishing relationship between von Neumann algebras and geometric topology"* and as a result having *"found a new polynomial invariant for knots and links in 3-space."*

The book has a substantial introduction to Teichmüller spaces theory, of fundamental importance since they allow to characterize the possible conformal structures of Riemann surfaces by considering classes of quasiconformal mappings of Riemann surfaces homotopic modulo conformal homeomorphisms and, therefore, they can be seen as the set of equivalence classes by deformation of conformal structures in a Riemann surface, which was why O. Teichmüller introduced them (for compact Riemann surfaces), but this important topic is not pursued here since that would deviate from the main objectives of this book. This restricted, but we hope useful, introduction can be pursued by studying some of the books listed in the final Bibliography.

The study of quasiconformal mappings in the complex plane, although with contributions from Topology and Geometry, is at the frontier of complex and real analysis. As this book is intended to be accessible to students with a good background in Mathematical Analysis at the usual level of the two initial university years, minimally resorting to additional bibliography, it includes eight appendices (together a substantial part of the book) to facilitate the access to certain aspects of real analysis, namely elements of Lebesgue[28] integral and measure, distributions, Fourier transform, partially absolutely continuous functions, relationships of differentiability and properties of partial derivatives, properties of continuous invertible functions, and a global inverse function theorem. Two other appendices give the bases of projections and separating hyperplanes of convex sets and an extension of the Schwarz lemma used in one of the chapters.

The scope is strictly restricted to quasiconformal mappings in the complex plane, not considering them in $\mathbb{R}^n$ or more general spaces (Euclidean or metric). Such extensions were initiated in 1938 by M.A. Lavrentiev, in 1940 by A. Markushevich, and in 1941 by M. Kreines, but reached maturity only after an interregnum of almost two decades, with contributions from 1959 onward by C. Loewner, F. Gehring, and J. Väisälä, among others.[29]

I had the fortunate opportunity of having been introduced to quasiconformal mappings by O. Lehto[30] in the elegant lectures of an engaging course he offered at the University of Minnesota in 1982 when he was writing a "good part" of his book *Univalent Functions and Teichmüller Spaces* listed in the final Bibliography and I was a Research Fellow at the *IMA—Institute for Mathematical Analysis and Applications* created that year. I would also like to thank several colleagues with whom I shared my enthusiasm in writing parts of this book. In particular, Manuel Ricou, who shares office with me and had the patience to listen attentively on the many occasions I abruptly interrupted what he was doing for chatting about these fascinating topics. To the others, I address a collective acknowledgment,

---

[28] Lebesgue, Henri (1875–1941).

[29] Lavrentiev, Mikhail (1900–1980). Markushevich, Alexei (1908–1979). Kreines, Mikhail (1903-). Löwner, Karl (1893–1968), changed name to Charles Loewner. Frederick Gehring (1925–2012) was awarded the *Steele Prize for Lifetime Achievement* of the AMS in 2006. Väisälä, Jussi (1935-)

[30] Lehto, Olli (1925–2020).

regrettably anonymous. I am also grateful to the Springer senior and production editors Robinson dos Santos and Kirithiga Nandini, and to Sinduja Ravi and Aravajy Meenahkumary of Straive, for their friendly cooperation.

Lisboa, Portugal
April 2025

Luis T. Magalhães

Contents

About the Author

Luis T. Magalhães was Professor of Mathematics at IST—*Instituto Superior Técnico*—of University of Lisbon (formerly of Technical University of Lisbon) from 1993 to December 2021 at his jubilee.

He was awarded the degrees of Electrical Engineer—Telecommunications and Electronics—by IST in 1975, and MSc (1980) and PhD (1982) in Applied Mathematics by Brown University, USA.

He worked at IST (1972–1997, 2002–2005, 2012–), IMA—Institute for Mathematics and Its Applications, University of Minnesota, USA (1982–1983, 1985), Division of Applied Mathematics, Brown University, USA (1978–1983), and Biology Center of the Gulbenkian Institute of Science, Portugal (1972–1978).

In Portugal: President of the General Council of the University of Algarve (2013–2017); President of the Knowledge Society Agency (UMIC), the national agency with the mission of coordinating the National Policy on Information Society and Information and Communication Technology (ICT) (Jul 2005–Jan 2012); President of the Science and Technology Foundation (FCT), the national agency with the mission of furthering the scientific and technological knowledge through evaluation and funding of people, projects and institutions (1997–2002).

Member of the Scientific Council of *Universidade Aberta* (2019–20); Member of the General Council of the Foundation for National Scientific Computing (FCCN), the agency responsible for the National Research and Education Network (2009–2012); Member of the Consultative Council of the Luso-American Foundation for Development (FLAD) (1997–2011); Coordinator and Founder of the Center of Mathematical Analysis, Dynamical Systems and Applications to Engineering (1991–1997), and proponent of changing its name to Center of Mathematical Analysis, Geometry and Dynamical Systems in 1997; Chair of the Mathematics Department of IST (1993–1997), Co-Founder with João Sentieiro of the Institute of Systems and Robotics (ISR), and Vice-Director of its Lisbon branch (1991–1996).

At International Research Organizations and Partnerships: President of the Council of the INL—International Iberian Nanotechnology Laboratory (2008-2012), President of the General Council of the INL Installation Committee (2007–2011); Member of the Council of EGI—European Grid Initiative (2009–

2012), Member of the Governing Board of ESF—European Science Foundation (2000–2002); Member of the Governing Boards of the International MIT—Portugal (2006–2011), Carnegie Mellon University—Portugal (2006–2011), University of Texas at Austin—Portugal (2007–2011), Harvard Medical School—Portugal (2009–2011); Member of the Steering Committee of the *Fraunhofer Gesellscahft*—Portugal Partnership (2007–2012).

At ICANN: Member of the Strategy Panel on ICANN Role in the Internet Governance Ecosystem, chaired by Vint Cerf (2013–2014), Member of GAC—Governmental Advisory Committee (2009–2012).

At the European Union (EU): Chair (2016–2018), Acting Chair (2015), and Vice-Chair (2013–2016) of the HORIZON 2020 Advisory Board on Research Infrastructures Including e-Infrastructures, European Co-Chair of the EU—Africa 8th Partnership, on Science, Information Society and Space (2011, 2012), Member of the Working Group of the National ICT Research Directors Forum on the FET—Future and Emerging Technologies Flagships (2010–2012).

At OECD: Member of the OECD Committee on Digital Economy Policy (formerly Committee on Information, Computers and Communication Policy) (2005–2022), its Vice-Chair in 2009, 2010, 2017, 2018, and Member of its Extended Bureau in 2009–2019; Member of the Steering Group of the OECD Horizontal Project GOING DIGITAL— Digital Transformation of the Economy and Society (Dec2016–2020); Member of the Steering Group of the 2016 Ministerial Meeting on Digital Economy, Innovation, Growth and Social Prosperity (2014–2016); Chair of the OECD Working Party on Measurement and Analysis of the Digital Economy (MADE) (formerly Working Party on Indicators for the Information Society (WPIIS)) (2011–2016).

At United Nations (UN): Member of the UN Commission on Science and Technology for Development (CSTD) (2009–2012), Member of the MAG—Multistakeholder Advisory Group of the IGF—Internet Governance Forum, appointed by the UN Secretary-General (Set-Dec 2007).

He is co-author, with Jack K. Hale and Waldyr M. Oliva, of the books *An Introduction to Infinite Dimensional Dynamical Systems—Geometric Theory,* Springer-Verlag, 1984, and *Dynamics in Infinite Dimensions,* Springer-Verlag, 2002, an the author of *Álgebra Linear como Introdução a Matemática Aplicada,* Texto Editora, 1989, *Integrais Múltiplos,* Texto Editora, 1993, and *Integrais em Variedades e Aplicações,* Texto Editora, 1993, and *Complex Analysis and Dynamics in One Variable with Applications*, Springer, 2025.

Chapter 1
Definitions of Quasiconformal Mapping in the Plane

1.1 Introduction

As the name suggests, the notion of quasiconformal mapping corresponds to weakening that of a conformal homeomorphism. The condition of being a homeomorphism is maintained, but bounded distortion of angles at each point is allowed. These functions do not have a complex derivative and their derivatives as linear transformations in $\mathbb{R}^2$ map circles to ellipses with bounded eccentricity in the whole domain, which can vary from point to point as well as the orientation of the major ellipses axes. As they are homeomorphisms, the relationship of circles in the domain tangent space with ellipses in the range tangent space corresponds to a similar relationship of circles in the range tangent space with ellipses in the domain tangent space which are preimages by the derivative as a linear transformation in $\mathbb{R}^2$.

Quasiconformal mappings are regular enough for validity of the derivatives chain rule and the integral formulas for length and area, but not sufficiently for conjugacy with them not to change derivatives. As homeomorphisms, they preserve the topology, and by the geometry of bounded distortion of circles and ellipses, they have affine geometry properties.

As mentioned in the Preface, the idea of quasiconformal mappings arose in 1859–1860, by the hand of N.-A. Tissot who introduced the use of ellipse fields, called *ellipse indicatrix* (Fig. 1.1[1]). The detailed study of quasiconformal mappings started in 1928 by H. Grötzsch, about three quarters of a century before the name quasiconformal appeared. H. Grötzsch observed that a square cannot be transformed to a nonsquare rectangle by a conformal homeomorphism preserving the vertices and sought to obtain mappings preserving them as close to conformal

[1] Author and date of Figure : Stefan Kühn, 09.12.2004, under license *Creative Commons Attribution-Share Alike 3.0 Unported.* https://commons.wikimedia.org/wiki/File:Tissot_behrmann.png

L. T. Magalhães, *Quasiconformal Mappings in the Plane and Complex Dynamics*,
https://doi.org/10.1007/978-3-031-80115-0_1

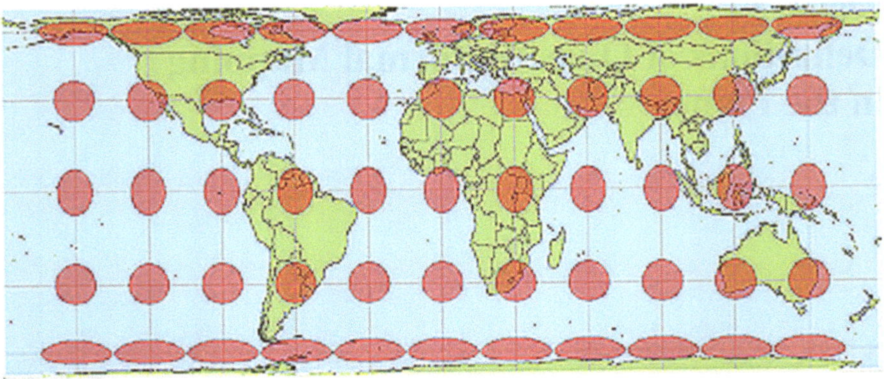

Fig. 1.1 Cartography of the Earth surface with Behrmann cylindrical equal area projection with superimposed *ellipse indicatrix* of N.-A. Tissot

homeomorphisms as possible. At the time, he only considered C^1 quasiconformal mappings when considered from subsets of[2] $\mathbb{R}^2$ to $\mathbb{R}^2$.

The name quasiconformal was given by L. Ahlfors in a paper published in 1935 where he developed a geometric version of Nevanlinna theory.

In 1937, O. Teichmüller introduced spaces of equivalence classes of quasiconformal mappings modulo conformal homeomorphisms in Riemann surfaces with a metric defined by the maximal distortion of circles to ellipses. However, functions with quasiconformal properties had appeared as solutions of partial differential equations in the works of L. Liechtenstein (in 1916), A. Korn (in 1919), M.A. Lavrentiev (in 1935), and C. Morrey (in 1938), but the relationship with the geometric notion of quasiconformal mappings only began to be noticed decades later, around[3] 1955.

To benefit from the flexibility provided by quasiconformal mappings, they cannot be restricted to be C^1 when considered as functions of subsets of $\mathbb{R}^2$ to $\mathbb{R}^2$. There are several equivalent ways of defining them, some more analytic and others more geometric or topological, the latter developed mainly by A. Pfluger in 1951 and L. Ahlfors in 1953. It is not easy to prove the equivalence of some of the definitions, but they illuminate the notion of quasiconformal mappings from different angles. In 1955–1957, A. Mori, Z. Yûjôbô and L. Bers (independently) contributed to the unification of geometric and analytic notions by showing that quasiconformal mappings in the geometric sense are differentiable almost everywhere.[4]

The analytic definitions of quasiconformal mappings focus on relations of partial derivatives of the functions, in particular of being solutions of the Beltrami partial

[2] Behrmann, Walter (1882–1955).

[3] Liechtenstein, Leon (1878–1933). Korn, Arthur (1870–1945). Lavrentiev, Mikhail Alekseevich (1900–1980).

[4] Pfluger, Albert (1907–1993). Mori, Akira. Yûjôbô, Zuiman.

differential equation with Beltrami coefficient that specifies the eccentricity and orientation of the ellipses in the tangent space of the domain mapped by the derivative of the function, considered from $\mathbb{R}^2$ in $\mathbb{R}^2$, to circles in the tangent space of the range, called complex dilatation. Therefore, these definitions depend on an appropriate definition of solutions of linear partial differential equations, involving distributions and real analysis. The Beltrami equation was considered for the first time in 1822 by C.F. Gauss when he proved that every surface with real analytic metric admits isothermal local coordinates, i.e., local coordinates such that the Riemannian metric is conformal to the Euclidean metric.[5]

1.2 Regular Quasiconformal Mapping

H. Grötzsch considered in 1928 C^1 homeomorphisms preserving orientation $f=(u,v)$ from regions of $\mathbb{R}^2$ to $\mathbb{R}^2$. The preimages of circles with center at 0 by the derivative of one of these homeomorphisms are ellipses with center at 0. To evaluate the ellipses axes directions and the axes lengths ratio, observe that, with

$$z=x+iy\,,\qquad f=u+iv\,,\qquad x=\tfrac{1}{2}(z+\overline{z})\,,\qquad y=i\tfrac{1}{2}(\overline{z}-z)\,,$$
$$F\big(z,\overline{z}\big)=u(x,y)+iv(x,y)=u\big(\tfrac{1}{2}(z+\overline{z}),i\tfrac{1}{2}(\overline{z}-z)\big)+iv\big(\tfrac{1}{2}(z+\overline{z}),i\tfrac{1}{2}(\overline{z}-z)\big)\,,$$

it is

$$\frac{\partial F}{\partial z}=\frac{1}{2}\Big(\frac{\partial u}{\partial x}+\frac{\partial v}{\partial y}\Big)+i\frac{1}{2}\Big(\frac{\partial v}{\partial x}-\frac{\partial u}{\partial y}\Big)=\frac{1}{2}\Big(\frac{\partial f}{\partial x}-i\frac{\partial f}{\partial y}\Big)\,,$$
$$\frac{\partial F}{\partial \overline{z}}=\frac{1}{2}\Big(\frac{\partial u}{\partial x}-\frac{\partial v}{\partial y}\Big)+i\frac{1}{2}\Big(\frac{\partial v}{\partial x}+\frac{\partial u}{\partial y}\Big)=\frac{1}{2}\Big(\frac{\partial f}{\partial x}+i\frac{\partial f}{\partial y}\Big)\,,$$

where $\frac{\partial F}{\partial z}$, $\frac{\partial F}{\partial \overline{z}}$ are evaluated at $(z,\overline{z})$ and the partial derivatives of u,v at (x,y), with $z=x+iy$. To simplify the notation, denote

$$f_z=\tfrac{1}{2}\Big(\frac{\partial f}{\partial x}-i\frac{\partial f}{\partial y}\Big)\,,\qquad\qquad f_{\overline{z}}=\tfrac{1}{2}\Big(\frac{\partial f}{\partial x}+i\frac{\partial f}{\partial y}\Big)\,.$$

The Jacobian matrix of f satisfies

$$Df(x_0,y_0)\begin{bmatrix}x\\y\end{bmatrix}=f_z z+f_{\overline{z}}\overline{z}\,.$$

As for $a,b,z\in\mathbb{C}$ with $T(z)=az+b\overline{z}$, the preimage of the circle with center 0 and radius 1 by the linear transformation T is the set of points z solutions of $|T(z)|=1$. With $z=|z|e^{i\theta}$, $a=|a|e^{i\alpha}$, and $b=|b|e^{i\beta}$, as

[5] Gauss, Carl Friedrich (1777–1855).

$$
\begin{aligned}
|az+b\bar{z}|^2 &= |z|^2\big[(|a|^2+|b|^2)+2|a||b|\cos(\alpha-\beta+2\theta)\big] \\
&= |z|^2\big[(|a|^2+|b|^2)\big(\cos^2\tfrac{\alpha-\beta+2\theta}{2}+\sin^2\tfrac{\alpha-\beta+2\theta}{2}\big) \\
&\quad +2|a||b|\big(\cos^2\tfrac{\alpha-\beta+2\theta}{2}-\sin^2\tfrac{\alpha-\beta+2\theta}{2}\big)\big] \\
&= |z|^2\big[(|a|+|b|)^2\cos^2\tfrac{\alpha-\beta+2\theta}{2}+(|a|-|b|)^2\sin^2\tfrac{\alpha-\beta+2\theta}{2}\big] \\
&= |z|^2\,\big|(|a|+|b|)\cos\tfrac{\alpha-\beta+2\theta}{2}+i\big||a|-|b|\big|\sin\tfrac{\alpha-\beta+2\theta}{2}\big|^2,
\end{aligned}
$$

the equation $|T(z)|=1$ is equivalent to

$$
|z|\big[(|a|+|b|)\cos\tfrac{\alpha-\beta+2\theta}{2}+i\big||a|-|b|\big|\sin\tfrac{\alpha-\beta+2\theta}{2}\big]=1\,.
$$

This is a Cartesian equation of an ellipse with center 0, smaller half axis of length $\frac{1}{|a|+|b|}$ making the angle $\frac{\beta-\alpha}{2}$ with the real axis and larger half axis with length $\frac{1}{||a|-|b||}$ making an angle with the real axis $\frac{\beta-\alpha}{2}+\frac{\pi}{2}$ (Fig. 1.2). The operator norm of T is the reciprocal of the length of the ellipse smaller half axis, and the absolute value of the determinant of T is the quotient of the areas of the disk of area 1 and of the region bounded by the ellipses, and its is positive or negative according to T to preserve or invert the plane orientation, i.e.,

$$
\|T\|=|a|+|b|\,, \qquad \det T=(|a|+|b|)\,(|a|-|b|)=|a|^2-|b|^2.
$$

Thus, the operator norm of the Jacobian matrix and the Jacobian of f are

$$
\|Df\|=|f_z|+|f_{\bar{z}}|\,, \qquad \det Df=|f_z|^2-|f_{\bar{z}}|^2,
$$

and the Jacobian is positive for homeomorphisms preserving the plane orientation, i.e., such that $|f_z|>|f_{\bar{z}}|$. The quotient of the length of the ellipses larger axis by that of the smaller axis is at each point

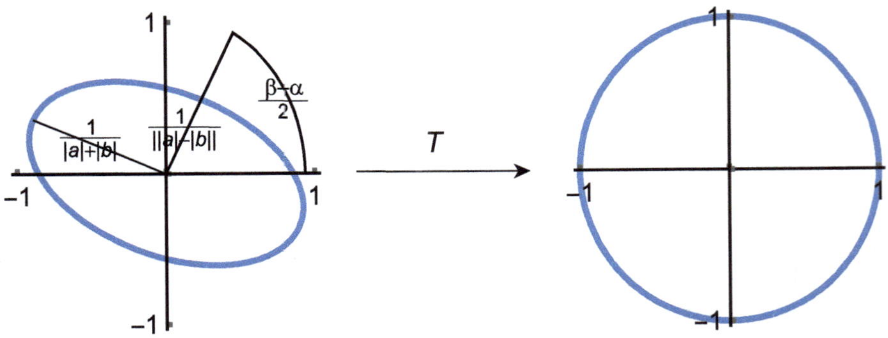

Fig. 1.2 Relationship of ellipses in the domain with circle images by the derivative of a C^1 function $f=(u,v)$ of $\mathbb{R}^2$ to $\mathbb{R}^2$, with $a=|a|e^{i\alpha}=f_z$, $b=|b|e^{i\beta}=f_{\bar{z}}$

$$K_f = \frac{|f_z|+|f_{\bar z}|}{|f_z|-|f_{\bar z}|} = \frac{\|Df\|^2}{\det Df} \geq 1\,,$$

and is called the **dilatation** at the point where the derivatives are evaluated.

The notion of quasiconformal mapping does not require that it be a C^1 function when considered from a subset of $\mathbb{R}^2$ to $\mathbb{R}^2$, but when it is C^1, it corresponds exactly to having dilatation, in the previous sense, uniformly bounded throughout the domain. If K is an upper bound of the dilatation, it is said to be a **regular K-quasiconformal mapping** and f is said to be a **regular quasiconformal mapping** if it is a K-regular quasiconformal mapping for some $K \geq 1$.

We have the following direct relationship of dilatation with the sum of squares of the Jacobian matrix components and the Jacobian of f, whose proof is left as exercise:

$$K_f(x+iy) + \frac{1}{K_f(x+iy)} = \frac{1}{\det Df(x,y)}\left[\left(\frac{\partial u}{\partial x}\right)^2+\left(\frac{\partial u}{\partial y}\right)^2+\left(\frac{\partial v}{\partial x}\right)^2+\left(\frac{\partial v}{\partial y}\right)^2\right],$$

with the partial derivatives evaluated at (x, y).

We call $\kappa_f = \frac{f_{\bar z}}{f_z}$ the **Beltrami coefficient** or **complex dilatation** of f at a given point, with the derivatives evaluated at the point.

The relationship of dilatation with the Beltrami coefficient at a point is

$$K_f = \frac{1+|\kappa_f|}{1-|\kappa_f|}\,, \qquad |\kappa_f| = \frac{K_f-1}{K_f+1}\,.$$

f preserves the plane orientation if and only if $|\kappa_f| < 1$ or, equivalently, $K_f > 1$, and f is a conformal homeomorphism if and only if $\kappa_f = 0$, or equivalently $K_f = 1$.

(1.1) Examples:

1. The function $f(x, y) = (x, 2y)$ defined for $(x, y) \in \mathbb{C}$ is not holomorphic. As a function of $\mathbb{R}^2$ to $\mathbb{R}^2$, it is a linear transformation mapping circles with center at 0 to ellipses with center at 0, horizontal smaller axis and vertical major axis of lengths, resp.,[6] equal to and twice the circles diameters. Therefore, f is a regular 2-quasiconformal mapping in $\mathbb{C}$. The dilatation is a constant $K_f = 2$ and the complex dilatation κ_f has absolute value $|\kappa_f| = \frac{K_f-1}{K_f+1} = \frac{1}{3}$ in the whole complex plane.

   Since $\frac{\partial f}{\partial x} = (1, 0)$ and $\frac{\partial f}{\partial y} = (0, 2)$,

   $$f_z = \tfrac{1}{2}\left(\tfrac{\partial f}{\partial x} - i\tfrac{\partial f}{\partial y}\right) = \tfrac{1}{2}(3, 0)\,, \qquad f_{\bar z} = \tfrac{1}{2}\left(\tfrac{\partial f}{\partial x} + i\tfrac{\partial f}{\partial y}\right) = -\tfrac{1}{2}(1, 0)\,,$$

   and the complex dilatation is $\kappa_f(z) = \frac{f_{\bar z}}{f_z} = -\frac{1}{3}$ in $\mathbb{C}$.
2. The function defined in the complex plane by $f(z) = |z|^{K-1}z$, for $K > 0$, is a radial elongation or contraction of the plane relative to 0 according to $K > 1$ or

[6] "Respectively" or "respective" are abbreviated "resp." throughout the text.

$K < 1$, and in the latter case or if $K = 1$, it is not defined at 0. As a function of two real variables in $\mathbb{R}^2$ $f(x, y) = (x^2+y^2)^{\frac{K-1}{2}}(x, y)$ is differentiable in the whole plane if $K > 1$ and in the whole plane except 0 if $K \leq 1$. It is

$$f(z) = (z\overline{z})^{\frac{K-1}{2}} z = z^{\frac{K+1}{2}} \overline{z}^{\frac{K-1}{2}} ;$$

thus, $f_z(z) = \frac{K+1}{2} z^{\frac{K-1}{2}} \overline{z}^{\frac{K-1}{2}}$ and $f_{\overline{z}}(z) = \frac{K-1}{2} z^{\frac{K+1}{2}} \overline{z}^{\frac{K-3}{2}}$ in the whole plane if $K > 1$ and in the whole plane except 0 if $K < 1$. f does not satisfy the Cauchy-Riemann equations[7] $f_{\overline{z}}=0$, except if $K=1$; thus, f is not holomorphic except for $K = 1$, when it is the identity in $\mathbb{C} \setminus \{0\}$. Since $\frac{f_{\overline{z}}}{f_z}(z) = \frac{K-1}{K+1}\frac{z}{\overline{z}}$, with $\kappa_f(z)=\frac{K-1}{K+1}\frac{z}{\overline{z}}$ it is $|\kappa_f(z)|=\frac{K-1}{K+1}<1$, f is a regular K-quasiconformal mapping in $\mathbb{C} \setminus \{0\}$ with complex dilatation $\kappa_f(z)=\frac{K-1}{K+1}\frac{z}{\overline{z}}$ at each $z\neq 0$.

3. Every C^1 function f of an open set $\Omega \subset \mathbb{R}^2$ to $\mathbb{R}^2$ with derivative at all points invertible in each region C with compact closure included in Ω, by the Weierstrass theorem[8] of extrema of continuous functions, has bounded partial derivatives. Thus, $\|Df\|$ and $\det Df$ are also bounded. $\det Df$ does not change sign and $|\det Df|$ is lower bounded by a $m > 0$; thus, the dilatation of f, $K_f = \frac{\|Df\|^2}{|\det Df|}$, is bounded in C. When considered as a complex function, if orientation is preserved, it is a regular quasiconformal mapping in each region with closure a compact subset of Ω.

   Similar examples of quasiconformal mappings can be obtained under more general conditions. Every diffeomorphism f defined from an open set $\Omega \subset \mathbb{R}^2$ to $\mathbb{R}^2$ (i.e., a differentiable function with differentiable inverse at all points) preserving orientation is quasiconformal as a complex function of complex variable in any region C with compact closure included in Ω, since $\det Df$ and $\|Df\|$ are locally, resp., lower bounded and upper bounded by some numbers > 0. Thus, considering an open cover of C by sets where such properties hold, there is a finite subcover of C and, therefore, there is $M > 0$ such that $\|Df\|^2 \leq M \det Df$ in C. Therefore, the dilatation of f satisfies $K_f = \frac{\|Df\|^2}{\det Df} \leq M$, and f is a regular M-quasiconformal mapping.

   In conclusion, *every orientation preserving diffeomorphism defined in a region $\Omega \subset \mathbb{C}$ is a regular quasiconformal mapping in each region with closure a compact subset of Ω.*

If f is a regular quasiconformal mapping, $\left|\kappa_{f^{-1}}\big(f(z)\big)\right| = |\kappa_f(z)|$, and the preimage by the derivative of f^{-1} at $f(z)$ of the circle with radius 1 and center 0 is the ellipse corresponding to f at z rotated by $\frac{\pi}{2}$.

The Beltrami coefficients are invariant under compositions at the right by conformal homeomorphisms h defined in the range of f, i.e., $\kappa_{f\circ h}=\kappa_f$, as expected since a conformal homeomorphism does not change the mapping of ellipses to circles.

[7] Cauchy, Augustin-Louis (1789–1857).

[8] Weierstrass, Karl (1815–1897).

However, the Beltrami coefficients are not invariant under compositions at the left by conformal homeomorphisms g with range including the domain of f, because, although eccentricity is not changed, the orientation of the ellipses axes can change. In this case, *if g is a conformal homeomorphism in a set where a Beltrami coefficient κ_f is defined*

$$\kappa_{f\circ g}(z) = \frac{\overline{g'(z)}}{g'(z)}\, \kappa_f\big(g(z)\big)\,.$$

It is useful to have formulas for the Beltrami coefficients of compositions and inverses of functions at points of regularity. By the chain rule,

$$(f\circ g)_{\overline{z}}(z) = f_z\big(g(z)\big)\, g_{\overline{z}}(z) + f_{\overline{z}}\big(g(z)\big)\, \overline{g}_{\overline{z}}(z)\,,$$
$$(f\circ g)_z(z) = f_z\big(g(z)\big)\, g_z(z) + f_{\overline{z}}\big(g(z)\big)\, \overline{g}_z(z)\,.$$

Thus

$$\kappa_{f\circ g} = \frac{(f\circ g)_{\overline{z}}}{(f\circ g)_z} = \frac{\kappa_g + (\kappa_f\circ g)\,[\overline{g}_{\overline{z}}/g_z]}{1+(\kappa_f\circ g)\,[\overline{g}_z/g_z]} = \frac{\kappa_g + (\kappa_f\circ g)\,[\overline{g}_{\overline{z}}/g_z]}{1+(\kappa_f\circ g)\,\kappa_g\,[\overline{g}_z/g_{\overline{z}}]}$$

Since $g_z = \frac{1}{2}\big(\frac{\partial g}{\partial x} + i\frac{\partial g}{\partial y}\big)$ and $g_{\overline{z}} = \frac{1}{2}\big(\frac{\partial g}{\partial x} - i\frac{\partial g}{\partial y}\big)$, it is $\overline{g}_{\overline{z}} = \overline{g_z}$ and $\overline{g}_z = \overline{g_{\overline{z}}}$. Therefore, $\frac{\overline{g}_{\overline{z}}}{g_z} = \frac{\overline{g_z}}{g_z} = e^{-i2\operatorname{Arg} g_z}$ and $\frac{\overline{g}_z}{g_{\overline{z}}} = \frac{\overline{g_{\overline{z}}}}{g_{\overline{z}}} = e^{-i2\operatorname{Arg} g_{\overline{z}}}$. Consequently, we have the following formula for the Beltrami coefficient of functions at points of regularity:

$$\kappa_{f\circ g} = \frac{(f\circ g)_{\overline{z}}}{(f\circ g)_z} = \frac{\kappa_g + (\kappa_f\circ g)\, e^{-i2\operatorname{Arg} g_z}}{1+(\kappa_f\circ g)\,\kappa_g\, e^{-i2\operatorname{Arg} g_{\overline{z}}}}\,.$$

With $f = g^{-1}$, this formula gives $0 = (g^{-1}\circ g)_{\overline{z}} = \kappa_g + (\kappa_{g^{-1}}\circ g)\, e^{-i2\operatorname{Arg} g_z}$. Thus, at points of regularity,

$$\kappa_{g^{-1}} = (\kappa_g\circ g^{-1})\, e^{i2\operatorname{Arg} g_z}\,.$$

It is also useful to have a formula for the composition of a function with the inverse of another at points of regularity. Replacing f by $f\circ g^{-1}$ in the above formula for $\kappa_{f\circ g}$ gives

$$\kappa_f = \kappa_{f\circ g^{-1}\circ g} = \frac{\kappa_g + (\kappa_{f\circ g^{-1}}\circ g)\, e^{-i2\operatorname{Arg} g_z}}{1+(\kappa_{f\circ g^{-1}}\circ g)\,\kappa_g\, e^{-i2\operatorname{Arg} g_{\overline{z}}}}\,.$$

Therefore,

$$\kappa_{f\circ g^{-1}}\circ g = \frac{(\kappa_f - \kappa_g)\, e^{i2\operatorname{Arg} g_z}}{1-\kappa_f\kappa_g\, e^{i2(\operatorname{Arg} g_z - \operatorname{Arg} g_{\overline{z}})}} = \frac{(\kappa_f - \kappa_g)\, e^{i2\operatorname{Arg} g_z}}{1-\kappa_f\kappa_g\, e^{-i2(\operatorname{Arg} \kappa_g)}} = \frac{\kappa_f - \kappa_g}{1-\kappa_f\overline{\kappa_g}}\, e^{i2\operatorname{Arg} g_z}\,,$$

using in the last equality $\overline{z} = z e^{-i2\operatorname{Arg} z}$ for all $z\in\mathbb{C}$. Consequently, the formula for the Beltrami coefficient of the composition of a function with the inverse of another at points of regularity is with $w = g(z)$

$$\kappa_{f\circ g^{-1}}(w) = \frac{\kappa_f(z) - \kappa_g(z)}{1-\kappa_f(z)\,\overline{\kappa_g(z)}}\, e^{i2\operatorname{Arg} g_z(z)}\,.$$

The function $d(f, g) = \sup d_P(\kappa_f, \kappa_g)$, with d_P the distance in the Poincaré metric[9] of[10] B_1, for which the length element is $ds = \frac{2\,|dw|}{1-|w|^2}$, is a distance of quasiconformal mappings f, g, identifying those that differ by composition with a conformal homeomorphism, and it is called **Teichmüller distance**.

The differential equation with unknown f and Beltrami coefficient $\kappa_f(z)$ given at each point z of the considered set is called **Beltrami equation**:

$$f_{\bar{z}} = \kappa_f(z)\, f_z\,.$$

The Beltrami coefficient κ_f defines a **field of ellipses** specified by eccentricity and direction of the larger axis at each point. Thus, to solve the Beltrami equation corresponds to finding functions with a field of ellipses specified at each point by the Beltrami coefficient.

1.3 Geometric Definition of Quasiconformal Mapping

As the requirement of quasiconformal mappings to be C^1 is excessive, it is convenient to define them without this condition. The geometric definition given below follows the path initiated by H. Grötzsch in 1928 when he took the first steps toward the notion of quasiconformal mapping by posing the problem: *Since a conformal homeomorphism of a square onto a (nonsquare) rectangle mapping vertices to vertices does not exist, what are the functions closer to conformal homeomorphisms with this property?* The closeness condition is made precise by requiring the smallest dilatation K_f possible. This is why a quasiconformal mapping is defined as a homeomorphism preserving orientation with bounded dilatation of quadrilaterals.

(1.2) Solution of Grötzsch problem: *If $f = (u, v)$ is a C^1 homeomorphism with dilatation K_f defined in an open set containing a rectangle $R \subset \mathbb{R}^2$ with $R' = f(R)$ a rectangle in $\mathbb{R}^2$ with sides images of the sides of R, a, b are the lengths of two consecutive sides of R and a', b' those of the resp. images, then*

$$\frac{a'/b'}{a/b} \leq \sup_R K_f\,,$$

called **Grötzsch inequality**, *with equality if f is the affine function*

$$f(z) = \tfrac{1}{2}\big[\big(\tfrac{a'}{a}+\tfrac{b'}{b}\big)z+\big(\tfrac{a'}{a}-\tfrac{b'}{b}\big)\bar{z}\,\big].$$

---

[9] See Section 5 of Chapter 11 of *CADOVA*. Poincaré, Henri (1854–1912).

[10] $B_r(z)$ denotes the disk with radius $r > 0$ and center z, and $B_r = B_r(0)$.

Proof Without loss of generality, the coordinate system is such that R has vertices $(0,0)$, $(a,0)$, $(0,b)$, (a,b). Then,

$$a'b = \int_0^a \|f'(x,y)\|\,dx \int_0^b 1\,dy = \int_0^a\int_0^b (|f_z|+|f_{\bar z}|)\,dxdy$$

$$= \int_0^a\int_0^b \sqrt{\frac{|f_z|+|f_{\bar z}|}{|f_z|-|f_{\bar z}|}}\sqrt{(|f_z|+|f_{\bar z}|)\,(|f_z|-|f_{\bar z}|)}\,dxdy\,.$$

By Cauchy-Schwarz inequality,

$$(a'b)^2 \le \int_0^a\int_0^b \frac{|f_z|+|f_{\bar z}|}{|f_z|-|f_{\bar z}|}\,dxdy \int_0^a\int_0^b (|f_z|+|f_{\bar z}|)(|f_z|-|f_{\bar z}|)\,dxdy$$

$$= \iint_R K_f \int_0^a (|f_z|+|f_{\bar z}|)\,dx \int_0^b (|f_z|-|f_{\bar z}|)\,dy = a'b'\iint_R K_f\,.$$

Thus, $\frac{a'/b'}{a/b} \le \frac{1}{ab}\iint_R K_f \le \sup_R K_f$.

If f is the affine function in the statement, $K_f = \frac{|f_z|+|f_{\bar z}|}{|f_z|-|f_{\bar z}|} = \frac{a'/a}{b'/b}$. □

A **quadrilateral** $Q = Q(z_1, z_2, z_3, z_4)$ is a Jordan curve[11] in the Riemann sphere $\mathbb{C}_\infty$ with 4 of its points z_1, z_2, z_3, z_4, called **vertices**, ordered consistently with a simple path describing the Jordan curve. The arc in Q with end points at vertices z_j and z_{j+1} (identifying $z_5 = z_1$) without any of the other vertices is called the **side** j of Q. By the Jordan curve theorem,[12] a Jordan curve in $\mathbb{C}_\infty$ separates it in two regions; the **region bounded by a Jordan curve** described by a simple Jordan path is defined to be the region relative to which the direction of this path is positive. For a quadrilateral $Q = Q(z_1, z_2, z_3, z_4)$, the region bounded by the Jordan curve Q for a simple path describing it in the direction agreeing with the order of the vertices is also denoted $Q = Q(z_1, z_2, z_3, z_4)$, an ambiguity easily resolved by context.

A **homeomorphism of quadrilaterals** is a homeomorphism of the closure of the region bounded by one of them onto the closure of the region bounded by other mapping vertices to vertices and preserving their orders.

It is not possible that all homeomorphisms of quadrilaterals be conformal because once the images of 3 points are fixed by a conformal homeomorphism, the image of the 4th point is also fixed. So, the set of all quadrilaterals is partitioned in equivalence classes of conformal quadrilaterals.

By the Riemann mapping theorem,[13] a quadrilateral can be mapped by a conformal homeomorphism h_Q onto a quadrilateral $Q'\big(-\frac{1}{k}, -1, 1, \frac{1}{k}\big)$, $0 < k < 1$, bounding the open upper complex half plane $\mathbb{H}$. The elliptic function

$$T(z) = \int_0^z \frac{1}{\sqrt{(1-w^2)\,(1-k^2w^2)}}\,dw\,, \qquad z \in Q'\big(-\tfrac{1}{k}, -1, 1, \tfrac{1}{k}\big)\,,$$

[11] Jordan, Camille (1838–1922).

[12] See Section 8 of Appendix A of *CADOVA*.

[13] See Section 5 of Chapter 10 of *CADOVA*.

defines a conformal homeomorphism of $\mathbb{H}$ (the region bounded by the quadrilateral Q') onto the interior of the quadrilateral $\left(-A+iB, -A, A, A+iB\right)$, which is a rectangle[14] R with vertices the images of the vertices of the quadrilateral Q', where $A=\int_0^1 g$, $B=\int_1^{1/k} g$, $g(t)=\frac{1}{\sqrt{|1-t^2|\,(1-k^2t^2)}}$. By the Carathéodory theorem of boundary correspondence,[15] this homeomorphism can be extended by continuity to the boundary.

The **canonical mapping of a quadrilateral** Q is the composition $T\circ h_Q$ of the conformal homeomorphism in the preceding paragraph, and the **canonical rectangle of a quadrilateral** Q is the image R of Q by the canonical mapping of Q (Fig. 1.3).

Each equivalence class of conformal quadrilaterals contains a rectangle and all similar rectangles, i.e., with the same quotient of the length a of the side connecting the 1st and 2nd vertices by the length b of the side connecting the 2nd and 3rd vertices. $\boldsymbol{M(Q)}=\frac{a}{b}$ is called the **quadrilateral Q module**.

There is a bijection of the equivalence classes of conformal quadrilaterals to each possible value $M>0$ of the resp. modules. The effect of cyclic permutations of a quadrilateral vertices on the resp. modules is

$$M\big(Q(z_1,z_2,z_3,z_4)\big)=\frac{1}{M\big(Q(z_2,z_3,z_4,z_1)\big)}=M\big(Q(z_3,z_4,z_1,z_2)\big)=\frac{1}{M\big(Q(z_4,z_1,z_2,z_3)\big)}.$$

A homeomorphism f defined in a region $\Omega\subset\mathbb{C}$ **preserves orientation** if the image $f(C)$ of positive (resp., negative) orientation of Jordan curves $C\subset\Omega$ relative to the region they bound is also positive (resp., negative).

The **dilatation of a quadrilateral** Q with closure in a region $\Omega\subset\mathbb{C}$ by a homeomorphism preserving orientation f defined in Ω is the quotient of the quadrilateral modules $K(Q)=\frac{M(f(Q))}{M(Q)}$, and the **dilatation of quadrilaterals** by f is the supremum K_f of the dilatations of all quadrilaterals Q, $\operatorname{cl} Q\subset\Omega$.

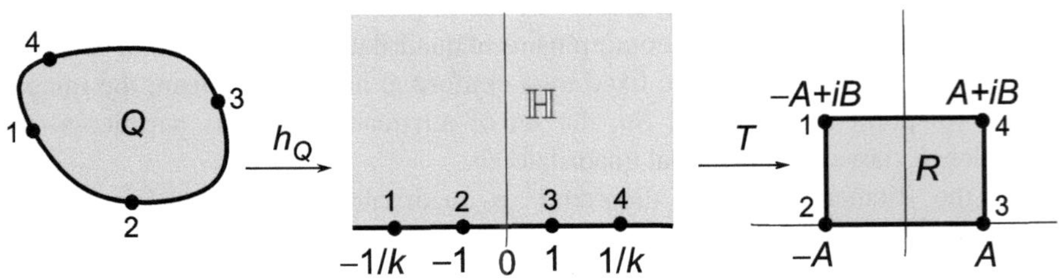

Fig. 1.3 Canonical mapping of quadrilateral Q onto canonical rectangle R (h_Q conformal homeomorphism onto $\mathbb{H}$, T elliptic function, $k\in]0,1[$, $A, B>0$)

[14] See the first of the exercises on elliptic functions and the exercise on the Schwarz-Christoffel formula in the first group of exercises at the end of Chapter 10 of *CADOVA*. Christoffel, Elwin Bruno (1829–1900).

[15] See Section 5 of Chapter 10 of *CADOVA*. Carathéodory, Constantin (1873–1950).

By (1.2), this definition of dilatation extends that given at the beginning of the section on regular quasiconformal mappings.

Since the dilatations of quadrilaterals with the vertices cyclically permuted so that the two first vertices become the second and third initially considered are the reciprocal of each other, the dilatation of a homeomorphism f preserving orientation is $K_f \geq 1$. If f is a conformal homeomorphism preserving orientation, it maps quadrilaterals in the domain of f to quadrilaterals with the same module, and the dilatation is $K_f = 1$. Therefore, conformal homeomorphisms preserving orientation are 1-quasiconformal mappings. For C^1 homeomorphisms when considered as functions of real variables, the converse also holds because $K_f = 1$ if and only if $|\frac{\partial F}{\partial \bar{z}}| = 0$, what is equivalent to the Cauchy-Riemann equations, where F is the function defined from the function f as at the section beginning. Therefore, the dilatation is a measure of deviation of orientation preserving homeomorphisms from conformity.

Geometric Definition of Quasiconformity
A **quasiconformal mapping** in a region $\Omega_\infty \subset \mathbb{C}$ is a homeomorphism $f : \Omega \rightarrow \mathbb{C}$ preserving orientation with finite dilatation of quadrilaterals K_f; it is called a *K*-**quasiconformal mapping** if $K_f \leq K$.

(1.3) *Inverses of K-quasiconformal mappings are K-quasiconformal. Compositions of K_1-quasiconformal mappings with K_2-quasiconformal mappings are $K_1 K_2$-quasiconformal.*

Proof It is immediate from the definitions. □

1.4 Quasiconformal Mapping and Extremal Length

The geometric definition of quasiconformal mapping just given is based on quadrilateral module, using the concept of conformal homeomorphism. It is convenient to have a characterization of quasiconformal mappings independent of the concept of conformal homeomorphism, what is achieved with the extremal length considered by L. Ahlfors and A. Beurling[16] in 1946, which gives a method for estimation of quadrilateral modules.[17]

---

[16] See Section 8 of Chapter 10 of *CADOVA*. Beurling, Arne (1905–1986).

[17] A function from a measurable subset of $\mathbb{R}^n$ or $\mathbb{C}$ to a topological space is said to be a **Borel function** if its preimages of open sets are Borel sets. A **Borel set** in X ($\mathbb{R}^n$ or $\mathbb{C}$) is an element of the smallest set $\mathscr{B}$ of subsets of X containing open and closed subsets of X and all countable unions of its elements.

(1.4) *The quadrilateral module of $Q \subset \mathbb{C}$ is the reciprocal of the extremal length $\lambda_\Omega(\Gamma)$ of the set Γ of rectifiable curves in Q with one end in side 1 and the other in side 3 of Q:*

$$M(Q) = \frac{1}{\lambda_Q(\Gamma)} = \left[\sup_\rho \frac{[L_\rho(\Gamma)]^2}{A_\rho(Q)}\right]^{-1} = \inf_\rho \frac{A_\rho(Q)}{[L_\rho(\Gamma)]^2},$$

where $\sup_\rho$ is for all metrics with element of length $\rho(z)\,|dz|$ and with ρ Borel measurable in Q, conformal to the Euclidean metric and such that $0 < A_\rho(Q) < \infty$, with $L_\rho(\Gamma)$ the infimum of the lengths of the curves belonging to Γ, $A_\rho(Q)$ the area of Q in the metric defined by ρ, and the $\inf_{\{\gamma:\,\gamma^*\in\Gamma\}} \rho$ in the last term of the above equalities is achieved with $\rho = |f'|$ for f the canonical mapping of Q onto a rectangle.

Proof Immediate by Example (10.30.1) of Chapter 10 of *CADOVA*. □

As a consequence, since extremal length is a conformal invariant[16], it also is a **quasiconformal invariant** in the sense that

$$\tfrac{1}{K}\lambda_\Omega(\Gamma) \le \lambda_{f(\Omega)}\big(f(\Gamma)\big) \le K\lambda_\Omega(\Gamma),$$

for K-quasiconformal mappings f in a region $\Omega \subset \mathbb{C}$ and Γ the set of rectifiable curves in Ω.

The Rengel inequality[18] gives useful bounds of a quadrilateral module.

(1.5) Rengel inequality: *If $M(Q)$ is the quadrilateral module of Q*

$$\frac{[L_{24}(Q)]^2}{A(Q)} \le M(Q) \le \frac{A(Q)}{[L_{13}(Q)]^2},$$

where $A(Q)$ is the area of Q and $L_{jk}(Q)$ the distance of the sides j and k of Q in the Euclidean metric, with equality if and only if Q is a rectangle.

Proof The inequalities are straightforward, as are the equalities if Q is a rectangle. To prove that equality in one of the inequalities implies that Q is a rectangle, with f the inverse of the canonical mapping of Q onto $R\big(0, M(Q), M(Q)+i, i\big)$, notice that

$$M(Q)[L_{13}(Q)]^2 = M(Q) \inf_{\gamma^*\in\Gamma}\left(\int_{f^{-1}(\gamma^*)} |f'(w)||dw|\right)^2 \le \int_0^{M(Q)}\left(\int_0^1 |f'(u+iv)|dv\right)^2 du$$
$$\le \int_0^{M(Q)}\int_0^1 |f'(u+iv)|^2 dv\,du = A(Q),$$

[18] Rengel, Ewald.

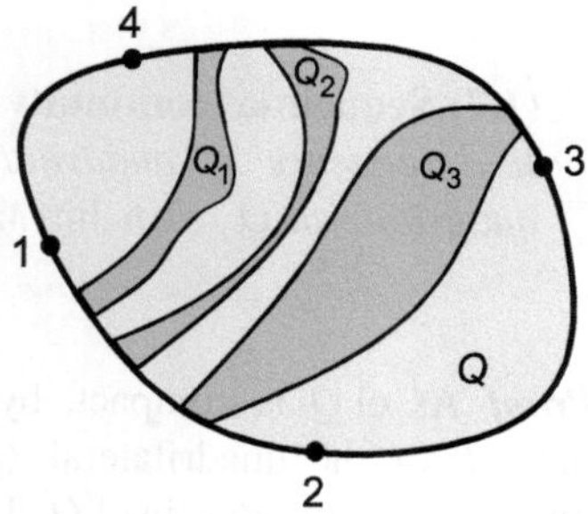

Fig. 1.4 Superadditivity and monotonicity of quadrilateral module: $M(Q_1)+M(Q_2)+M(Q_3) \leq M(Q)$

the last inequality by the Cauchy-Schwarz inequality with one of the factors 1. Thus, if $M(Q)[L_{13}(Q)]^2 = A(Q)$, then

$$\int_0^1 |f'(u+iv)|^2 dv = \left(\int_0^1 |f'(u+iv)| dv\right)^2$$
$$= \inf_{\gamma^* \in \Gamma} \int_{f^{-1}(\gamma^*)} |f'(w)||dw|, \quad u \in [0, M(Q)].$$

The first of these equalities is an equality in a Cauchy-Schwarz inequality with one of the factors 1; therefore, $f'(u+iv)$ is independent of v, and due to the second equality, it is constant. So, f is a similarity mapping and Q is a rectangle. Equality in the first Rengel inequality leads to an analogous conclusion. □

(1.6) Superadditivity and monotonicity of quadrilateral module: *If $Q_1, Q_2, \ldots \subset \mathbb{C}$ are quadrilaterals bounding disjoint regions with closures in the closure of the region bounded by a quadrilateral Q and the sides 1 and 3 of Q_n are subsets of the resp. sides of Q, then* (Fig. 1.4)

$$\sum_n M(Q_n) \leq M(Q);$$

if Q is a rectangle, equality holds in this inequality if and only if all Q_n are rectangles and $\sum_n A(Q_n) = A(Q)$.

Proof Since the quadrilateral module is a conformal invariant, Q can be assumed to be the rectangle $R(0, M(Q), M(Q)+i, i)$ and, then, $L_{13}(Q) \geq 1$; by Rengel inequality, $\sum_n M(Q_n) \leq \sum_n A(Q_n) \leq A(Q) = M(Q)$. For equality in the first inequality, the Rengel inequality has to be equality for every Q_n, and this happens if and only if each Q_n is a rectangle. □

A sequence of quadrilaterals $\{Q_n\}$ is said to **converge from inside** to a quadrilateral Q if $\mathrm{cl}\, Q_n \subset \mathrm{cl}\, Q$ and the distance of each point on each side of Q_n to the correspondent side of Q tends 0 when $n \to +\infty$.

(1.7) Sequential continuity of quadrilateral module:
If a sequence of quadrilaterals $\{Q_n\} \subset \mathbb{C}$ *converges from inside to a quadrilateral* Q, *then* $\lim M(Q_n) = M(Q)$.

Proof As $\mathrm{cl}\, Q$ is compact, by the Heine-Cantor theorem,[19] the canonical mapping f of the quadrilateral Q onto the rectangle $R\big(0, M(Q), M(Q)+i, i\big)$ is uniformly continuous in $\mathrm{cl}\, Q$. Thus, $\{f(Q_n)\}$ converges from inside to R. For every $0 < \varepsilon < \frac{1}{2}\min\{M(Q), 1\}$, there is $N \in \mathbb{N}$ such that $n > N$ implies $f(Q_n) \geq M(Q) - 2\varepsilon$ and $f(Q_n) \geq 1 - 2\varepsilon$, and by the Rengel inequality,

$$\frac{(M(Q)-2\varepsilon)^2}{M(Q)} \leq M\big(f(Q_n)\big) \leq \frac{M(Q)}{(1-2\varepsilon)^2},$$

and, as the canonical mapping is conformal, $M\big(f(Q_n)\big) = M(Q_n)$. Since $\varepsilon > 0$ is arbitrary, $M(Q_n) \to M(Q)$. □

Despite being defined globally in a region, the property of an orientation preserving homeomorphism to be K-quasiconformal is local, as follows.

(1.8) *A homeomorphism* f *of a region* $\Omega \subset \mathbb{C}$ *onto* $f(\Omega) \subset \mathbb{C}$ *is* K*-quasiconformal if and only if each point of* Ω *has a neighborhood where* f *is* K*-quasiconformal.*

Proof It is immediate from the definition that if f is K-quasiconformal in Ω, it is K-quasiconformal in every open subset of Ω.

Conversely, if each point of Ω has a neighborhood where f is K-quasiconformal and Q is a quadrilateral such that $\mathrm{cl}\, Q \subset \mathbb{C}$, since $\mathrm{cl}\, Q$ is compact, there is a finite cover of $\mathrm{cl}\, Q$ by open sets where f is K-quasiconformal. Consider canonical mappings of quadrilaterals Q and $Q' = f(Q)$ onto the rectangles, resp., R and R' and subdivide R in a finite number of vertical bands, consider the correspondent subdivision of Q in bands Q_j, and after this, subdivide the image of each band $Q'_j = f(Q_j)$ in a way corresponding to a subdivision of R' in horizontal bands. These subdivisions define a finite set of disjoint quadrilaterals Q_{jk} and the union of their closures is $\mathrm{cl}\, Q$. Then,

$$M(Q) = \sum_j M(Q_j), \quad \frac{1}{M(Q_j)} \geq \sum_k M(Q_{jk}), \quad M(Q') \geq \sum_j M(Q'_j), \quad \frac{1}{M(Q'_j)} = \sum_k \frac{1}{M(Q_{jk})}.$$

With subdivisions in smaller bands, each Q_{jk} is included in some element of the cover of $\mathrm{cl}\, Q$, where f is K-quasiconformal and $M(Q_{jk}) \leq KM(Q'_{jk})$,

[19] Heine, Heinrich Eduard (1821–1881). Cantor, Georg (1845–1918).

$M(Q) \le KM(Q')$. Since Q is an arbitrary quadrilateral with $\mathrm{cl}\, Q \subset \Omega$, f is a K-quasiconformal mapping in Ω. □

If f is an orientation preserving homeomorphism defined in a region $\Omega \subset \mathbb{C}$, the **dilatation of** f **at a point** $z \in \Omega$ is $F_f(z) = \inf_{U_z} K_f(U_z)$ with the infimum over all connected neighborhoods of z included in Ω and $K_f(U_z)$ the dilatation of f in U_z.

By the preceding result, *an orientation preserving homeomorphism defined in a region* $\Omega \subset \mathbb{C}$ *is a K-quasiconformal mapping if and only if the dilatation of* f *at each point* $z \in \Omega$ *satisfies* $F_f(z) \le K$ *and* $K_f(\Omega) = \sup_{z\in\Omega} F_f(z)$.

(1.9) *A quasiconformal mapping is a conformal homeomorphism if and only if it is* 1*-quasiconformal.*

Proof Immediately from the definition, a conformal homeomorphism is 1-quasiconformal. The converse requires equality in the inequalities of the proof of the preceding result, and this holds if and only if the quadrilaterals are rectangles and f is the identity. □

It was seen that K-quasiconformal mappings can be defined without reference to conformal homeomorphisms using extremal length, and, with the preceding result, *conformal homeomorphisms can be defined without reference to differentiability as being* 1*-quasiconformal mappings.*

The following result is that the limit of a sequence of uniformly convergent quasiconformal mappings on compact subsets of a region is a quasiconformal mapping if it is a homeomorphism.[20]

(1.10) *If* $\{f_n\}$ *is a sequence of K-quasiconformal mappings defined in a region* $\Omega \subset \mathbb{C}$ *uniformly convergent in compact subsets to a homeomorphism* f *defined in* Ω, *then* f *is K-quasiconformal.*

Proof Let Q be a quadrilateral such that $\mathrm{cl}\, Q \subset \Omega$. A sequence $\{Q_n\}$ of quadrilaterals converging from inside to Q can be constructed considering a conformal homeomorphism h mapping Q onto B_1 with the images of the vertices of Q denoted z_j, $j = 1, 2, 3, 4$, and extending fh to a conformal homeomorphism of $\mathrm{cl}\, Q$ onto $\mathrm{cl}\, B_1$. h^{-1} is uniformly continuous in $\mathrm{cl}\, B_1$, and the quadrilaterals $Q_n = h^{-1}(B_{1-\frac{1}{n}}(0))$ with vertices $h^{-1}(1-\frac{1}{n})z_j$, $j = 1, 2, 3, 4$, converge from inside to Q, and each ∂Q_n is an analytic curve.

[20] It is possible to prove the result without requiring a priori that the limit is a homeomorphism.

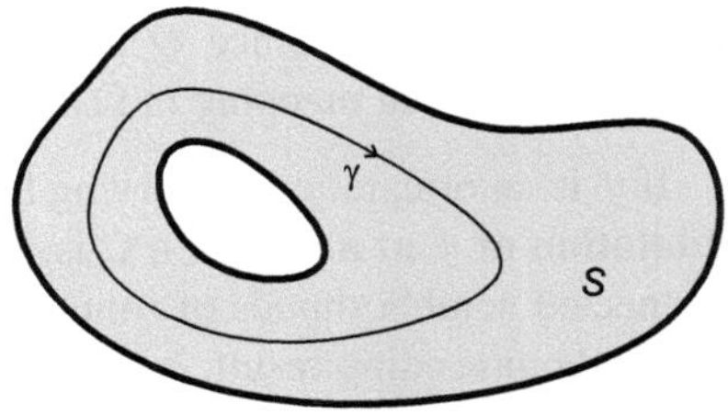

Fig. 1.5 Ring S with a Jordan path γ not homotopic to a constant path in S

For large m, $f_m(Q_n) \subset f(Q)$ and there exists a subsequence with $f_n(Q_n) \subset f(Q)$. Since $f_n \to f$ uniformly in cl Q and f is uniformly continuous in cl Q, the quadrilaterals $f_n(Q_n)$ converge from inside to $f(Q)$. Thus, by the sequential continuity of the quadrilateral module, $M\big(f_n(Q_n)\big) \to f(Q)$ and $M(Q_n) \to M(Q)$ when $n \to +\infty$. Since $M\big(f_n(Q_n)\big) \leq K M(Q_n)$ for all $n \in \mathbb{N}$, it is $M\big(f(Q)\big) \leq K M(Q)$. □

1.5 Quasiconformal Mapping and Ring Module

The geometric definition of quasiconformal mapping can be with rings instead of quadrilaterals.

A **ring** in $\mathbb{C}_\infty$ is a doubly connected region $S \subset \mathbb{C}_\infty$, i.e., such that there is a simple closed path in S not homotopic to a constant path and simple closed paths in S not homotopic in S to such a path are homotopic to a constant path (Fig. 1.5). An **annulus** is a ring with boundary the union of two circles with the same center.

Every ring $S \subset \mathbb{C}_\infty$ *is conformal to an annulus (possibly degenerated to a point or the complex plane),* $B_{R_2} \backslash \text{cl}\, B_{R_1}$ *with* $0 \leq R_1 \leq R_2 \leq +\infty$. A conformal homeomorphism of S onto the annulus $C = B_{R_2} \backslash \text{cl}\, B_{R_1}$ is called a **canonical mapping of the ring** S. The range C of a canonical mapping of a ring is called a **canonical image of the ring** S.

Two annuli are conformal if and only if the quotient of the larger to the smaller radius of the pair of circles bounding them is equal.[21] Hence, all canonical images of a ring are annuli with the same quotient of larger to smaller radius.

If a ring S is conformal to an annulus $B_{R_2} \backslash \text{cl}\, B_{R_1}$, with $0 \leq R_1 \leq R_2 \leq +\infty$, then $M(S) = \frac{1}{2\pi} \log \frac{R_2}{R_1}$, is[22] called its **ring module**.

For each $m > 0$, there is only one conformal class of rings with module m, and there are two conformal classes of rings with module $+\infty$, one if $R_1 = 0$, $R_2 > 0$ or $R_1 > 0$, $R_2 = +\infty$ and another if $R_1 = 0$ and $R_2 = +\infty$.

Like quadrilateral module, the ring module can be obtained without consideration of conformal homeomorphisms using extremal length.

---

[21] See Section 8 of Chapter 10 of *CADOVA*.

[22] Some authors define the ring module without dividing by 2π.

(1.11) *The module of a ring $S \subset \mathbb{C}_\infty$ is the reciprocal of the extremal length $\lambda_S(\Gamma)$ of the set Γ of rectifiable Jordan curves in S separating the connected components of ∂S:*

$$M(S) = \frac{1}{\lambda_S(\Gamma)} = \left[\sup_\rho \frac{[L_\rho(\Gamma)]^2}{A_\rho(S)} \right]^{-1} = \inf_\rho \frac{A_\rho(S)}{[L_\rho(\Gamma)]^2} ,$$

where $\sup_\rho$ *is for all metrics with element of length* $\rho(z)\,|dz|$ *and* ρ *Borel measurable in* S*, conformal to the Euclidean metric and such that* $0 < A_\rho(S) < \infty$*, with* $L_\rho(\Gamma)$ *the infimum of the lengths of the curves belonging to* Γ*,* $A_\rho(S)$ *the area of* S *in the same metric, and* $\inf_\rho$ *in the last term of the above equalities attained with* $\rho = \left|\frac{f'}{f}\right|$ *for* f *the canonical mapping of the ring* S *onto an annulus.*

Analogously, if Γ^* *is the set of rectifiable curves in* S *with each end point in one of the two connected components of* ∂S

$$M(S) = \lambda_S(\Gamma^*) = \sup_\rho \frac{[L_\rho(\Gamma^*)]^2}{A_\rho(S)} = \inf_\rho \frac{[L_\rho(\Gamma^*)]^2}{A_\rho(S)} .$$

Proof Immediate with Example (10.30.2) in Chapter 10 of *CADOVA*. □

(1.12) Rengel inequality for ring module:
If $M(S)$ is the module of a bounded ring S with 0 in the region bounded by one of the connected components of ∂S not intersecting ∂S

$$M(S) \le \frac{1}{(2\pi)^2} \iint_S \frac{1}{x^2+y^2}\, dxdy ,$$

with equality if and only if B is a disk with center 0 or $M(S) = +\infty$.

Proof By the Example (10.30.2) of *CADOVA*, the length of each curve in the set Γ considered in the preceding result is $\ge 2\pi$. With the metric with element of length $\rho(z)|dz|$, where $\rho(z) = \frac{1}{|z|}$, the length of each curve in the set Γ considered in the preceding result is $\ge 2\pi$. With this metric,

$$M(S) = \frac{1}{\lambda_S(\Gamma)} = \inf_\rho \frac{A_\rho(S)}{[L_\rho(\Gamma)]^2} \le \frac{1}{(2\pi)^2} \iint_S \rho^2 = \frac{1}{(2\pi)^2} \iint_S \frac{1}{x^2+y^2}\, dxdy .$$

For the annulus $C_{R_1,R_2} = B_{R_2} \backslash \mathrm{cl}\, B_{R_1}$

$$\frac{1}{(2\pi)^2} \iint_{C_{R_1,R_2}} \frac{1}{x^2+y^2}\, dxdy = \frac{1}{(2\pi)^2} \int_0^{2\pi} \int_{R_1}^{R_2} \frac{1}{r^2} r\, dr d\theta = \frac{1}{2\pi} \log \frac{R_2}{R_1} ,$$

what, by the same example of *CADOVA*, is the reciprocal of the extremal length $\lambda_S(\Gamma)$ of Γ in C in the Euclidean metric, and we have equality in the inequality in the statement. In the case of equality with finite $M(S)$, considering the inverse f of the canonical mapping of the ring S onto the annulus C_{R_1,R_2} with $0<R_1<R_2<+\infty$,

$$\frac{1}{2\pi}\log\frac{R_2}{R_1}=M(S)=\frac{1}{(2\pi)^2}\iint_S\frac{1}{x^2+y^2}\,dxdy=\frac{1}{(2\pi)^2}\int_{R_1}^{R_2}\int_0^{2\pi}\frac{1}{|f(re^{i\theta})|^2}|f'(re^{i\theta})|^2r\,d\theta dr\,.$$

Since, for $\gamma_r:[0,2\pi]\to\mathbb{C}$ such that $\gamma_r(\theta)=re^{i\theta}$, with $r\in]R_1,R_2[$,

$$\int_0^{2\pi}\frac{f'(re^{i\theta})}{f(re^{i\theta})}\,ire^{i\theta}d\theta=\int_{\gamma_r}\frac{f'(z)}{f(z)}\,dz=\pm i2\pi\,,$$

and, by the Cauchy-Schwarz inequality,

$$(2\pi)^2=\left|\int_0^{2\pi}\frac{f'(re^{i\theta})}{f(re^{i\theta})}\,ire^{i\theta}d\theta\right|^2\le 2\pi r^2\int_0^{2\pi}\left|\frac{f'(re^{i\theta})}{f(re^{i\theta})}\right|^2d\theta\,,$$

with equality if and only if $\frac{f'(z)}{f(z)}z$ is constant on every circle γ_r^* with center 0 and radius $r\in]R_1,R_2[$, which, with the equality in the preceding formula, means that in each of these circles $\frac{f'(z)}{f(z)}z$ is 1 or -1. With this inequality,

$$\frac{1}{(2\pi)^2}\int_{R_1}^{R_2}\int_0^{2\pi}\frac{1}{|f(re^{i\theta})|^2}\,|f'(re^{i\theta})|^2r\,d\theta dr\ge\frac{1}{2\pi}\int_{R_1}^{R_2}\frac{1}{r}\,dr=\frac{1}{2\pi}\log\frac{R_2}{R_1}\,.$$

Due to the first formula of this paragraph, we have equality in this inequality, what is equivalent to equality in the mentioned Cauchy-Schwarz inequality and also equivalent to $z\mapsto\frac{f'(z)}{f(z)}z$ to be constant and equal to 1 or -1 in S, i.e., to one of the functions $\frac{f(z)}{z}$ or $zf(z)$ to be constant in S. Thus, $f(z)$ or $\frac{1}{f(z)}$ is the product of a constant with the identity function, and since f is the inverse of a canonical mapping of the ring S onto the annulus C_{R_1,R_2}, it is equivalent to S being an annulus with center 0.

If $M(S)=+\infty$, the ring S is conformal to $B_{R_2}\setminus \mathrm{cl}\,B_{R_1}$ with $R_1=0$ or $R_2=+\infty$. In any of the three possible cases

$$\iint_S\frac{1}{x^2+y^2}\,dxdy=\int_0^{2\pi}\int_{R_1}^{R_2}\frac{1}{r}\,dr=2\pi\int_{R_1}^{R_2}\frac{1}{r}\,dr=+\infty\,,$$

and we have equality in the inequality in the statement. □

(1.13) Superadditivity and monotonicity of ring module:
If $S_1,S_2,\ldots\subset\mathbb{C}$ are disjoint rings, $\mathrm{cl}\,S_1,\mathrm{cl}\,S_2,\ldots\subset S$*, S is a ring, and each S_n separates the connected components of $\mathbb{C}\setminus S$, then* (Fig. 1.6)

(continued)

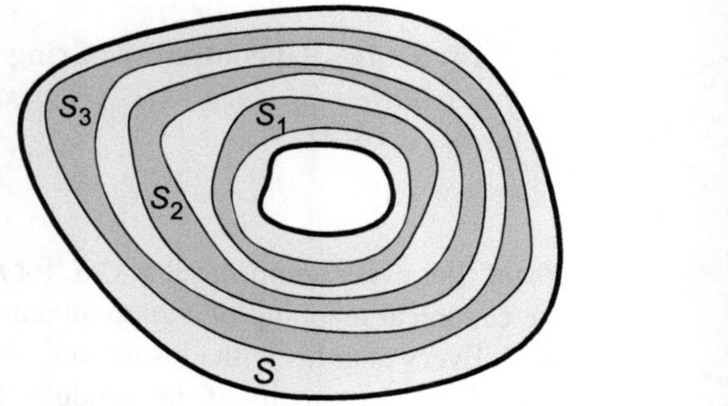

Fig. 1.6 Superadditivity and monotonicity of ring module: $M(S_1)+M(S_2)+M(S_3)\leq M(S)$

(1.13) (continued)

$$\sum_n M(S_n) \leq M(S)\,.$$

If $S\subset\mathbb{C}$ is an annulus and $M(S)$ is finite, the inequality holds with equality if and only if all S_n are annuli and $\sum_n A(S_n)=A(S)$.

Proof Without loss, S is an annulus centered at 0, $S=B_{R_2}\backslash \mathrm{cl}\, B_{R_1}$ with $0<R_1<R_2<+\infty$, as this can be ensured if $M(S)$ is finite by application of a conformal homeomorphism, and if $M(S)$ is infinite, the inequality holds trivially. Each $M(S_n)$ is finite and, by the Rengel inequality of ring module,

$$\sum_n M(S_n) \leq \frac{1}{(2\pi)^2}\iint_{\bigcup_n S_n}\frac{1}{x^2+y^2}\,dxdy \leq \frac{1}{(2\pi)^2}\iint_S \frac{1}{x^2+y^2}\,dxdy = \frac{1}{2\pi}\log\frac{R_2}{R_1} = M(S)\,.$$

Equality in the Rengel inequality holds if and only if the ring is an annulus centered at 0. Thus, the inequality in the statement is equality if and only if all S_n are annuli centered at 0 and, denoting the radii of the circles bounding them by, resp., $0<R_{1,n}<R_{2,n}$,

$$\iint_{\bigcup_n S_n}\frac{1}{x^2+y^2}\,dxdy = \iint_S \frac{1}{x^2+y^2}\,dxdy\,,$$

i.e., $\sum_n \log\frac{R_{n,2}}{R_{n,1}} = \log\frac{R_2}{R_1}$, equivalent to $\prod_n \frac{R_{n,2}}{R_{n,1}} = \frac{R_2}{R_1}$. Since the annuli S_n are disjoint, they can be ordered so that $R_{1,n}<R_{2,n}\leq R_{1,n+1}$ for all n, and for $\prod_n \frac{R_{n,2}}{R_{n,1}}=\frac{R_2}{R_1}$ to hold, it is necessary that $\cup_n S_n = S$ and, consequently, $\sum_n A(S_n)=A(S)$. □

A sequence of rings $\{S_n\}$ is said to **converge from inside** to a ring S if $\mathrm{cl}\, S_n\subset \mathrm{cl}\, S$ and the spherical distance of the connected component of ∂S_n bounding the smallest of the regions bounded by its two connected components to the connected component of ∂S that bounds the smallest of those regions tends to 0 when $n\to+\infty$ and analogously for the other connected component of ∂S_n and ∂S.

(1.14) Sequential continuity of ring module:
If a sequence of rings $\{S_n\} \subset \mathbb{C}_\infty$ *converges from inside to a ring* S, *then* $\lim M(S_n) = M(S)$.

Proof For every compact set $K \subset S$, for n large, $\mathbb{C}\backslash S_n \subset \mathbb{C}\backslash K$. The images of S_n by the canonical mapping of S onto an annulus C centered at 0 converge from inside to C. Every annulus with closure in C is a subset of such images for n large. Then, apply the monotonicity of ring module (1.13). □

There is an obvious relationship of the ring module of an annulus $S = B_{R_2}\backslash \mathrm{cl}\, B_{R_1}$ with the quadrilateral module of a rectangle Q obtained cutting out of S a radius, e.g. the intersection of S with the real positive semiaxis, and applying the logarithm, since the lengths of the sides L_{12}, image of the arc of a circle with radius R_1, and L_{23}, image of the radial segment in the positive real semiaxis, of Q are, resp., 2π and $\log \frac{R_2}{R_1}$, and $M(Q) = \frac{2\pi}{\log R_2/R_1}$, which is equal to $\frac{1}{M(S)}$ (Fig. 1.7). The following result establishes a more general relationship of ring and quadrilateral modules including this and, essentially, resulting from it.

(1.15) Relationship of ring and quadrilateral modules:
If $Q_1, Q_2, \ldots \subset \mathbb{C}$ *are quadrilaterals bounding disjoint regions and* S *is a ring separating the sides* L_{12} *from the sides* L_{34} *of each one of the quadrilaterals* Q_n (Fig. 1.8), *then*

$$\sum_n M(Q_n) \leq \frac{1}{M(S)}.$$

If S *is an annulus, equality holds if and only if the sides* L_{12} *and* L_{34} *of each* Q_n *are subsets of* ∂S *and the sides* L_{23}, L_{14} *of each* Q_n *are radial segments of the annulus* S *and* $\sum_n A(Q_n) = A(S)$.

Proof Without loss of generality, the two Jordan curves connected components of ∂S are analytic, since if they were not, they could be uniformly approximated by analytic curves and (1.14) applies. Consider the function $\rho = \left|\frac{h'}{h}\right|$ in S and 0 outside

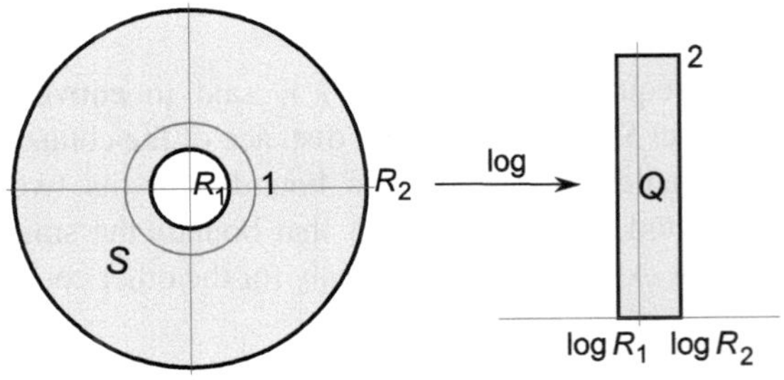

Fig. 1.7 Equality of ring module of an annulus S and a quadrilateral module of Q defined by the vertices of the rectangle image of S minus a radius of the annulus by the logarithm: $M(S) = M(Q)$

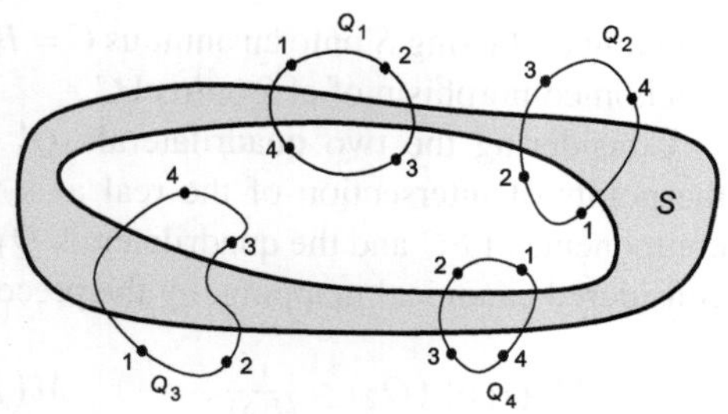

Fig. 1.8 Ring S and quadrilaterals $Q_1, Q_2, \ldots$ considered in (1.15)

of S, where h is a canonical mapping of the ring S onto an annulus $B_{R_2} \setminus \mathrm{cl}\, B_{R_1}$. If Γ_n is the set of rectifiable curves with end points in each one of the sides of Q_n included in the, resp., bounded and unbounded region of $\mathbb{C}$ separated by S, then the intersection of each curve in Γ_n with ∂S is a finite set, and ρ is continuous on each curve belonging to Γ except at a finite number of points. Thus, $\rho(z)\,|dz|$ is the element of length of a metric, and, by (1.4),

$$M(Q_n) = \frac{1}{\lambda_{Q_n}(\Gamma_n)} \leq \left[\frac{[L_\rho(\Gamma_n)]^2}{A_\rho(Q_n)}\right]^{-1} = \frac{A_\rho(Q_n)}{[L_\rho(\Gamma_n)]^2},$$

for each curve $\gamma^* \in \Gamma_n$,

$$\sum_n M(Q_n) \leq \sum_n \left[\int_{\gamma^* \cap S} \left|\frac{f'(z)}{f(z)}\right| |dz|\right]^{-2} \iint_{\Omega_n \cap S} \left|\frac{h'}{h}(x, y)\right|^2 dxdy$$

$$\leq \left[\int_{R_1}^{R_2} \frac{1}{r}\, dr\right]^{-2} \int_0^{2\pi} \int_{R_1}^{R_2} \frac{1}{r^2} r\, dr d\theta = 2\pi \left[\int_{R_1}^{R_2} \frac{1}{r}\, dr\right]^{-1} = 2\pi \left[\log \frac{R_2}{R_1}\right]^{-1} = (2\pi)^2 M(S).$$

If S is an annulus, in the above inequalities, the first equality holds if and only if Q_n satisfies the condition in the last sentence of the statement. If yes, $M(Q_n) = \frac{A(Q_n)}{M(S)A(S)}$ and $\sum_n M(Q_n) = \frac{\sum_n A(Q_n)}{M(S)A(S)}$. Thus, equality in the inequality in the statement holds if and only if, additionally, $\sum_n A(Q_n) = A(S)$, i.e., the sum of the areas of the regions bounded by the sides of the quadrilaterals is equal to the area of the ring. □

(1.16) Ring module is a quasiconformal invariant:
If $S \subset \mathbb{C}$ is a ring and $f : S \to \mathbb{C}$ is K-quasiconformal

$$\tfrac{1}{K} M(S) \leq M\big(f(S)\big) \leq K M(S).$$

Proof It suffices to prove one of the inequalities, as applying it to f^{-1} gives the other.

As in the proof of the preceding result, it can be assumed without loss of generality that the connected components of ∂S are analytic curves. A canonical mapping of the ring S onto an annulus $C = B_{R_2} \backslash \mathrm{cl}\, B_{R_1}$ can be extended by continuity to a homeomorphism of cl S onto cl C.

Considering the two quadrilaterals $Q'_1 \subset \mathbb{H}$ and $Q'_2 \subset -\mathbb{H}$ with vertices at the points of intersection of the real axis with the circles that are the connected components of ∂C and the quadrilaterals Q_1, Q_2 preimages of, resp., Q'_1, Q'_2 of the considered canonical mapping, by the preceding result,

$$M(Q_1)+(Q_2) \le \frac{1}{M(S)}, \qquad M\big(f(Q_1)\big)+M\big(f(Q_2)\big) \le \frac{1}{M(f(S))},$$

Since f is a K-quasiconformal mapping,

$$M\big(f(S)\big) \le \frac{1}{M(f(Q_1))+M(f(Q_2))} \le \frac{K}{M(f(Q_1))+M(f(Q_2))} = K\,M(S)$$

□

The characterization of quasiconformal mappings by the ring module was established in 1962 by F. Gehring and J. Väisälä based on an analytic definition of conformal homeomorphism. The following proof, which does not use the analytic definition, is of E. Reich,[23] also in 1962.

(1.17) Characterization of quasiconformal mappings by ring module: *If $\Omega \subset \mathbb{C}$ is a region, a homeomorphism $f:\Omega \to \mathbb{C}$ preserving orientation is a K-quasiconformal mapping if and only if $M\big(f(S)\big) \le K\,M(S)$ for every ring S with* cl $S \subset \Omega$.

Proof The necessity of the condition for K-quasiconformal mapping follows from the preceding result.

To prove sufficiency, it is necessary to prove that if $M\big(f(S)\big) \le K\,M(S)$ for every ring S with cl $S \subset \Omega$, then $M\big(f(Q)\big) \le K\,M(Q)$ for every quadrilateral Q with the closure of the region it bounds included in Ω. By the invariance of quadrilateral module under conformal homeomorphisms, it can be assumed without loss of generality that Q, $f(Q)$ are the rectangles, resp., $R = (0, M, M+i, i)$, $R' = (0, M', M'+i, i)$. Denote $R'_n \subset R'$ the rectangles with parallel sides to those of R' and at a distance $\frac{1}{n}$ of the corresponding sides of R', for $n > \frac{2}{\min(1,M')}$ and $R_n = f^{-1}(R'_n) \subset R$. The quadrilaterals R_n converge from inside to $Q = R$ when $n \to +\infty$ and, by (1.7), $M(Q_n) \to M$ when $n \to +\infty$. Divide R'_n by n^3-1 equally spaced vertical line segments in n^3 rectangles with bases of length $\frac{M'-2/n}{n^3}$ and height $1-2/n$, denoted R'_{nk}, $k=1,\ldots,n^3$. Denote $R_{nk} = f^{-1}\big(R'_{nk}\big)$, $k=1,\ldots,n^3$. By the superadditivity of quadrilateral module (1.6), $\sum_{n=1}^{n^3} M(R_{nk}) \le M(R_n)$. In particular, there exists $p \in \{1,\ldots,n^3\}$ such that $M(Q_{np}) \le \frac{M(R_n)}{n^3}$.

[23] Reich, Edgar (1927–2009).

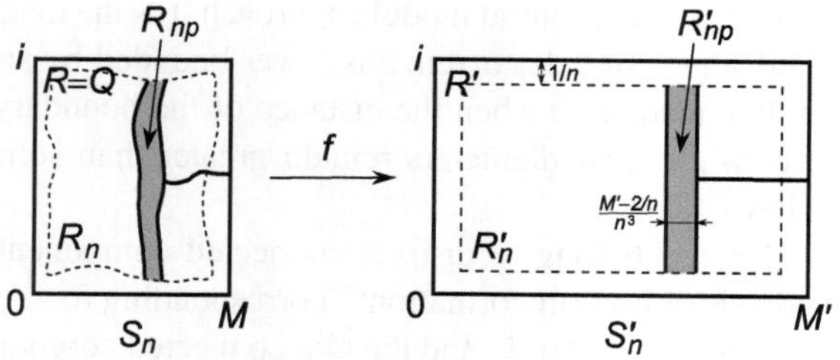

Fig. 1.9 Rings S_n and S'_n (with boundaries the continuous lines) considered in the proof of (1.17)

Consider now the rings $S'_n = R_n \setminus [(\partial R'_{np} \setminus \partial R'_n) \cup L_n]$, where L_n is the line segment $[c_{np}, M'] \times \{\frac{1}{2}\}$ with $c_{np} + i\frac{1}{2} \in \partial R'_{np}$, $]c_{np}, M'] \times \{\frac{1}{2}\} \cap R'_{np} = \emptyset$, and $S_n = f^{-1}(S_{n'})$ (Fig. 1.9). The module of the quadrilateral with the same vertices as R_{np} and included in S_n is the reciprocal of the module of R_{np}; so, by the relationship of ring and quadrilateral module, (1.11),

$$M(S_n) \le M(R_{np}) \le \frac{M(R_n)}{n^3}.$$

Considering for $n \in \mathbb{N}$ with $n^2 > M'$ the metric with element of length $\rho(z)\,|dz|$ and $\rho(z) = \frac{n^3}{M'-2/n}$, the rectangle obtained from R'_{np} by extending the vertical sides to have height 1 and $\rho(z) = n$ if $z \in S'_n \setminus R'_{np}$, the length in this metric of each curve $\gamma^* \subset S'_n$ with end points in each one of the connected components of $\partial S'_n$ is $L_\rho(\gamma^*) \ge 1$, and, by the formula for ring module in terms of extremal length relative to the set Γ^*_n of curves in S'_n with end points in each one of the components of $\partial S'_n$ in (1.11),

$$M(S'_n) = \lambda_{S'_n}(\Gamma^*_n) = \inf_\rho \frac{[L_\rho(\Gamma^*_n)]^2}{A_\rho(S'_n)} \ge \left[A_\rho(S'_n)\right]^{-1} > \left[n^3\left(\frac{M'}{n} + \frac{1}{M'-2/n}\right)\right]^{-1};$$

thus,

$$\frac{M(S'_n)}{M(S_n)} > \left[\left(\frac{M'}{n} + \frac{1}{M'-2/n}\right) M(R_n)\right]^{-1}.$$

Therefore, $K \ge \frac{M(S'_n)}{M(S_n)} \to \frac{M'}{M}$ when $n \to +\infty$ and, consequently, $M' \le KM$, i.e., $M\big(f(Q)\big) \le KM(Q)$. □

The module of a ring S can be upper bounded in terms of the length of curves in the spherical metric, what is used at the end of Chap. 3. As in this case $A_\rho(S) \le \pi$, by (1.4), $M(S) \le \frac{\pi}{k_0^2}$, where k_0 is the infimum of the lengths of curves separating the components of $\mathbb{C} \setminus S$ in the spherical distance. If C is one of these curves and (a_1, b_1), (a_2, b_2) are pairs of points in each one of the connected components of $\mathbb{C} \setminus S$ such that $d_S(a_j, b_j) \ge \delta > 0$ for $j \in \{1, 2\}$, with d_S the spherical distance, then $k_0 \ge 2\delta$, and $M(S) \le \frac{\pi}{4\delta^2}$.

Just as quadrilateral module approach 0 as the distance of sides b tends to 0 while the distance of sides a remains lower bounded by some positive constant, the ring module tends to 0 when the distance of the boundary connected components tends to 0, while their diameters remain greater than some positive constant, as shown below.

If z_1, z_2 belong to distinct connected components of $\mathbb{C}\backslash S$ and $\varepsilon = d_S(z_1, z_2)$, with a Möbius transformation[24] corresponding to a rotation of $\mathbb{C}_\infty$, these points can be mapped to $\pm\tan\frac{\varepsilon}{2}$, and the two connected components of the complement of the annulus $\mathscr{C} = B_{\tan\frac{\delta}{2}} \backslash \mathrm{cl}\, B_{\tan\frac{\varepsilon}{2}}$ contain points of each of the connected components of $\mathbb{C}\backslash S$. Thus, each curve C separating those components has points in the two connected components of $\mathbb{C}\backslash\mathscr{C}$. With the metric with element of length $\frac{1}{|z|}|dz|$ if $z \in \mathscr{C}$ and 0 if $z \notin \mathscr{C}$, it is $A(S) = 2\pi \log \frac{\tan(\delta/2)}{\tan(\epsilon/2)}$ and $L(C) = 2\log\frac{\tan(\delta/2)}{\tan(\epsilon/2)}$. Therefore, taking into account (1.4), we have the following result.

(1.18) *If $S \subset \mathbb{C}$ is a ring with the diameters of the connected components of $\partial S > \delta > 0$ and the spherical distance of them smaller than $\varepsilon > 0$, then*

$$M(S) \leq \left[\frac{2}{\pi} \log \frac{\tan(\delta/2)}{\tan(\epsilon/2)}\right]^{-1}.$$

The following property, established with the Koebe one quarter theorem for *schlicht* functions,[25] is used in the next chapter in the proof of the measurable Riemann mapping theorem given in 2000 by A. Douady and X. Buff, who established it to be used for that proof.

(1.19) *If $0 < R_1 < R_2$ and $h : B_{R_2} \to \mathbb{C}$ is K-quasiconformal, $h(0) = 0$, $h'(0) = 1$, and h is holomorphic in B_{R_1}, then $h(B_{R_2}) \supset B_{R_0}$ with $R_0 = \frac{1}{4} R_1^{1-\frac{1}{K}} R_2^{\frac{1}{K}}$ and $|h(z)| \geq R_0$ for $|z| \geq R_2$, $z \in \mathbb{C}$* (Fig. 1.10).

Proof Let H_U be a conformal homeomorphism of $U = h(B_{R_2})$ onto B_R such that $H_U(0) = 0$ and $H_U'(0) = 1$. For $\varepsilon > 0$, let $r(\varepsilon) = \varepsilon\big(1 + o(1)\big)$ when $\varepsilon \to 0$ be such that $h(B_\varepsilon) \subset B_{r(\varepsilon)}$. The restriction of $G = H_U \circ h$ to $B_{R_2}\backslash \mathrm{cl}\, B_\varepsilon$ is a K-quasiconformal mapping with values in $B_R\backslash$, cl $B_{r(\varepsilon)}$ whose restriction to $B_{R_1}\backslash \mathrm{cl}\, B_\varepsilon$ is holomorphic, i.e., 1-quasiconformal. By the superadditivity and quasiconformal invariance of ring module,

[24] Möbius, August Ferdinand (1790–1868).

[25] See Section "Univalent Functions" in Chapter 10 of *CADOVA*. *Schlicht*, which in German means simple, is usually used today for injective holomorphic functions $h : B_1 \to \mathbb{C}$ normalized to $h(0) = 0$, $h'(0) = 1$. Koebe, Paul (1882–1945).

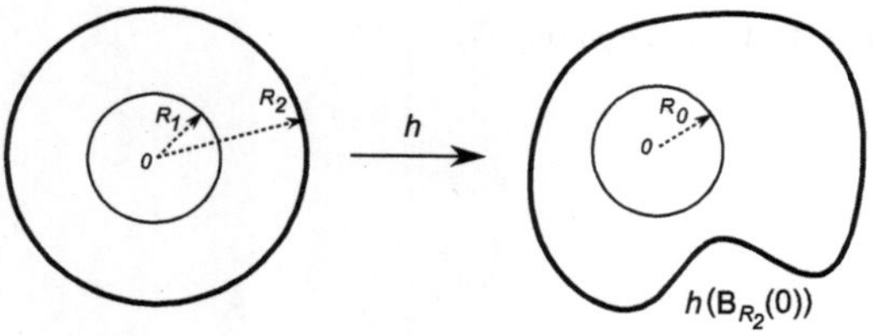

Fig. 1.10 Ranges of K-quasiconformal mappings $h: B_{R_2} \to \mathbb{C}, h=0, h'=1,$ holomorphic in B_{R_1}, $0<R_1<R_2$, contain B_{R_0}, $R_0=\frac{1}{4}R_1^{1-1/K}R_2^{1/K}$

$$\frac{1}{2\pi}\log\frac{R}{r(\varepsilon)}=M\big(B_R\backslash \text{cl}\, B_{r(\varepsilon)}\big)\geq M\big(B_{R_1}\backslash \text{cl}\, B_\varepsilon\big)+\frac{1}{K}M\big(B_{R_2}\backslash \text{cl}\, B_{R_1}\big)$$
$$=\frac{1}{2\pi}\log\frac{R_1}{\varepsilon}+\frac{1}{K}\frac{1}{2\pi}\log\frac{R_2}{R_1}=\frac{1}{2\pi}\log\big(\frac{1}{\varepsilon}R_1^{1-\frac{1}{K}}R_2^{\frac{1}{K}}\big).$$

Therefore, $R\geq\frac{r(\varepsilon)}{\varepsilon}R_1^{1-\frac{1}{K}}R_2^{\frac{1}{K}}$ for $\varepsilon>0$ small and $R\geq R_1^{1-\frac{1}{K}}R_2^{\frac{1}{K}}$. By the Koebe one quarter theorem, U contains $B_{\frac{R}{4}}$. □

1.6 Extremals of Ring Modules

In this section, we consider the question of how to find a ring with maximal module separating two given disjoint closed sets $A_1, A_2 \subset \mathbb{C}$. If A_1, A_2 are connected, the answer is immediate, since a connected component of $\mathbb{C}\setminus(A_1 \cup A_2)$ is a ring, and by the monotonicity of the ring module, it is an extremal module of separating rings.

The next easiest case is a set A_1 with two points and a Jordan curve A_2 not separating them. By the conformal invariance of the ring module, without loss of generality, we can consider $A_1=\{0, r\}$ with $0<r<1$ and $A_2=\partial B_1$.

The open disk B_1 without a slit in the real axis from 0 to r, with $0<r<1$, is called **Grötzsch ring**, $G_r=B_1\backslash\{x+i0: 0\leq x\leq r\}$, and its module is denoted $\boldsymbol{M_G(r)}$ (Fig. 1.11).

In 1928, H. Grötzsch obtained the following result.

(1.20) Grötzsch ring module theorem: *If S is a ring separating $A_1=\{0, r\}$ with $0<r<1$ and $A_2=\partial B_1$, then $M(S)\leq M_G(r)$.*

Proof Considering the annulus $C_r=B_{e^{2\pi M_G(r)}}\backslash \text{cl}\, B_1$ and extending a conformal homeomorphism f from $G_r\cap\mathbb{H}$ onto $C_r\cap\mathbb{H}$ to G_r by reflection relative to the real axis, we obtain a conformal homeomorphism from G_r onto C_r, also denoted f. As the function $\rho=|\frac{f'}{f}|$ is symmetric relative to the real axis, it can be extended by continuity to a continuous function defined in $B_1\backslash\{0, r\}$. With Γ the set of rectifiable curves separating the two connected components of the complement of S and the metric with element of length $\rho(z)\,|dz|$, by (1.11), and as

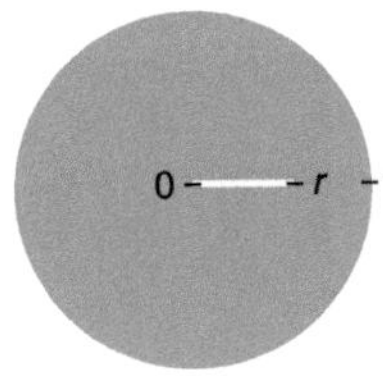

Fig. 1.11 Grötzsch ring G_r

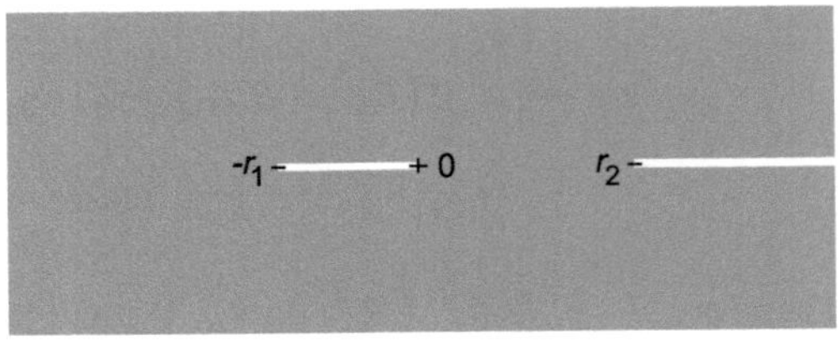

Fig. 1.12 Teichmüller ring T_{r_1,r_2} with $r_1, r_2 > 0$

$A_\rho(S) \le A_\rho(G_r) = (2\pi)^2 M_G(r)$, it is $M(S) \le \frac{(2\pi)^2 M_G(r)}{[L_\rho(\Gamma)]^2}$. Each curve $\gamma^* \in \Gamma$ can be decomposed as a disjoint union of subarcs γ_1^*, γ_2^*, both with end points in the real axis, one in $\{x + i0 : -1 < x < 0\}$ and the other in $\{x + i0 : r < x < 1\}$. It is $L_\rho(\gamma_j^*) \ge \pi$, $j = 1, 2$. As $L_\rho(\gamma^*) = L_\rho(\gamma_1^*) \cup L_\rho(\gamma_2^*)$, $M(S) \le \frac{(2\pi)^2 M_G(r)}{(2\pi)^2} = M_G(r)$. □

We consider now the case of $A_1 = \{a_1, b_1\}$, $A_2 = \{a_2, b_2\} \subset \mathbb{C}$ with $a_j \ne b_j$, for $j = 1, 2$, and a_1, b_1, a_2, b_2 in a straight line or a circle in that order. There is a Möbius transformation mapping these points to points on the extended real axis $-r_1, 0, r_2, \infty$, with $r_1, r_2 > 0$. The following result establishes that the complex plane minus the slits in the real axis from $-r_1$ to 0 and from r_2 to ∞, $\boldsymbol{T_{r_1,r_2}} = \mathbb{C} \setminus \{x + i0 : -r_1 \le x \le 0 \text{ or } x \ge r_2\}$, called **Teichmüller ring**, has the extremal ring module separating $A_1 = \{-r_1, 0\}$ and $A_2 = \{r_2, \infty\}$ with $r_1, r_2 > 0$ (Fig. 1.12).

(1.21) Teichmüller ring module extremal: *If S is a ring separating $A_1 = \{-r_1, 0\}$ and $A_2 = \{r_2, \infty\}$ with $r_1, r_2 > 0$, then*

$$M(S) \le 2\, M_G\left(\sqrt{\frac{r_1}{r_1 + r_2}}\right),$$

with equality if S is the Teichmüller ring T_{r_1,r_2}.

Proof As points in the complex plane symmetrical relative to a circle are collinear with the circle center and the product of their distances to the center is equal to the square of the circle radius,[26] 0 and r_2 are symmetric relative to the circle $\partial B_R(-r_1)$ with $R = \sqrt{r_1(r_1 + r_2)}$, and the two slits in the real axis are

[26] See the exercise on the subject at the end of Chapter 3 of *CADOVA*.

symmetric relative to this circle. The set $\mathbb{C}\setminus\partial B_R(-r_1)$ is a union of two regions conformal to the Grötzsch ring G_r with $r=\sqrt{\frac{r_1}{r_1+r_2}}$. We consider the conformal homeomorphism to a canonical region of one of the two regions,[27] which can be extended by reflection relative to the circle to a conformal homeomorphism f from the Teichmüller ring S to a canonical region. By the superadditivity of ring module, $M(S)=2M(G_r)=2M_G(r)$. Therefore, if S is the Teichmüller ring T_{r_1,r_2}, there is equality in the inequality in the statement.

The function $\rho=|\frac{f'}{f}|$ is symmetric relative to the real axis. Thus, it can be extended by continuity to $\mathbb{C}\setminus\{-r_1,0,r_2\}$. As in the proof of (1.20), in the metric with element of length $\rho(z)|dz|$, the length of a rectifiable curve separating $\{-r_1,0\}$ and $\{r_2,\infty\}$ is $\geq 2\pi$. Therefore, for any ring S separating these sets, $M(S)=\frac{1}{\lambda_S(\Gamma)}$, where $\lambda_S(\Gamma)$ is the extremal length of the set Γ of rectifiable Jordan curves in S separating A_1 and A_2, and

$$M(S)\leq\frac{1}{(2\pi)^2}A_\rho(S)\leq\frac{1}{(2\pi)^2}\iint_{\mathbb{R}^2}\Big|\frac{f'}{f}\Big|^2=2M_G(r).$$

□

For points $a_1,b_1,a_2\in\mathbb{C}$ and $b_2=\infty$ not in a same straight line or circle, O. Teichmüller obtained in 1938 the following result.

(1.22) Teichmüller ring module theorem: *If S is a ring separating $A_1=\{0,z_1\}$ and $A_2=\{z_2,\infty\}$ with $z_1,z_2\in\mathbb{C}$, then*

$$M(S)\leq 2\,M_G\Big(\sqrt{\tfrac{|z_1|}{|z_1|+|z_2|}}\Big).$$

Proof Let F_2 be the connected component of $\mathbb{C}\setminus S$ containing A_2. The set $\Omega=\mathbb{C}\setminus F_2\subsetneq\mathbb{C}$ is a simply connected region. By the Riemann mapping theorem,[28] there is a conformal homeomorphism h of Ω onto B_1 with $h(0)=0$, $w_1=h(z_1)>0$. The function $f(z)=-\frac{4|z_2|h(z)}{[1-h(z)]^2}$ is a conformal homeomorphism from Ω onto $\mathbb{C}\setminus[\,|z_2|,+\infty[$ with $-f(z_1)=\frac{4|z_2|w_1}{(1-w_1)^2}>0$, and $f(S)$ is a ring separating $\{|z_2|,\infty\}$ and $\{0,f(z_1)\}$. By the preceding result,

$$M(S)=M\big(f(S)\big)\leq 2\,M_G\Bigg(\sqrt{\frac{-f(z_1)}{-f(z_1)+|z_2|}}\Bigg),$$

with equality if $f(S)$ is the Teichmüller ring $\mathbb{C}\setminus([f(z_1),0]\cup[|z_2|,+\infty[)\times\{0\}$. By the monotonicity of ring module, $r\mapsto M_G(r)$ is a decreasing function. To finish the proof, it suffices to show that $-f(z_1)\geq|z_1|$.

[27] See the end of Section 5 of Chapter 10 of *CADOVA* for remarks on canonical regions of multiply connected regions.

[28] See Section 5 of Chapter 10 of *CADOVA*.

The function $g=\frac{h^{-1}}{h^{-1\prime}(0)}$ is *schlicht*. By the distortion theorem for *schlicht* functions,[29] $|g(z)|\leq\frac{|z|}{(1-|z|)^2}$. On the other hand, by the Koebe one quarter theorem[29], $g(B_1)\supset B_{\frac{1}{4}}$, and, consequently, $\left|\frac{h^{-1}}{(h^{-1})'(0)}\right|\leq\frac{1}{4}$. Thus, $|(h^{-1})'(0)|\leq 4h^{-1}(z)$ for $z\in B_1$ and, in particular, $|(h^{-1})'(0)|\leq 4|z_2|$. By the preceding two periods, $-f(z_1)=\frac{4|z_2|w_1}{(1-w_1)^2}\geq h^{-1}(w_1)=z_1$. □

The upper bound in the preceding result may fail to be the maximum of the ring module separating $A_1=\{0,z_1\}$ and $A_2=\{z_2,\infty\}$ with $z_1,z_2\in\mathbb{C}$.

For rings separating $A_1=\{z_1,z_2\}\subset\mathbb{C}\setminus\{0\}$ and $A_2=\{0,\infty\}$:

(1.23) Variant of the Teichmüller ring module theorem:
If S is a ring separating $A_1=\{z_1,z_2\}\subset\mathbb{C}\setminus\{0\}$ and $A_2=\{0,\infty\}$, then

$$M(S)\leq M_G\left(\frac{|\sqrt{z_1}-\sqrt{z_2}|}{\sqrt{2(|z_1|+|z_2|)}}\right),$$

where $\sqrt{z_1}$, $\sqrt{z_2}$ are evaluated in the same branch of the square root in S.

Proof Let F_1, F_2 be the connected components of $\mathbb{C}\setminus S$ and F_2 that containing 0. $\Omega=\mathbb{C}\setminus F_2\subsetneqq\mathbb{C}$ is a simply connected region and does not contain 0. Hence, there are two holomorphic branches of the square root in Ω, $\pm\sqrt{\;}$. These functions define conformal homeomorphisms from Ω onto disjoint regions. Denote the images of S by these mappings, resp., S', S'', and those of F_1 by, resp., F_1', F_1'', and $A=\mathbb{C}\setminus(F_1'\cup F_1'')$. Since S' and S'' separate F_1' and F_1'', by the superadditivity of ring module, $M(S')+M(S'')\leq M(A)$, and as $M(S')=M(S'')=M(S)$, it is $M(S)\leq\frac{1}{2}M(A)$. The ring A separates $\left\{\sqrt{z_1},\sqrt{z_2}\right\}$ and $\left\{-\sqrt{z_1},-\sqrt{z_2}\right\}$. The Möbius transformation $h(z)=\frac{z-\sqrt{z_1}}{z+\sqrt{z_1}}$ maps A onto the ring A' which separates $\{0,w_1\}$ and $\{w_1^{-1},\infty\}$, where $w_1=\frac{\sqrt{z_2}-\sqrt{z_1}}{\sqrt{z_2}+\sqrt{z_1}}$. By the preceding result,

$$M(S)\leq\tfrac{1}{2}M(A)=\tfrac{1}{2}M(A')\leq\tfrac{1}{2}\,2\,M_G\left(\sqrt{\frac{|w_1|}{|w_1|+|(w_1)^{-1}|}}\right)=M_G\left(\frac{|\sqrt{z_1}-\sqrt{z_2}|}{\sqrt{2(|z_1|+|z_2|)}}\right).$$

□

The inequality in the statement holds with equality if the inequalities in the last formula of the proof are equalities. The first is equality if and only if S', S'' are disjoint annuli with the area of their union equal to the area of A and the second if it is equality in the inequality of the Teichmüller ring module theorem used in the proof, i.e., if and only if

$$M(A')=2\,M_G\left(\sqrt{\frac{|\sqrt{z_1}-\sqrt{z_2}|}{\sqrt{2(|z_1|+|z_2|)}}}\right).$$

[29] See Section "Univalent Functions" in Chapter 10 of *CADOVA*.

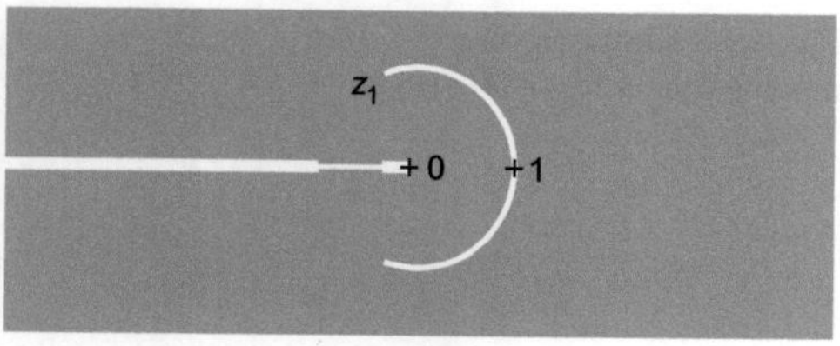

Fig. 1.13 Mori ring

The **Mori ring** is the complex plane without two slits, one the whole negative real semiaxis and the other an arc of circle with center at 0, radius 1, and end points at z_0, $\overline{z_0} \neq 0$ with $|z_0 - 1| \leq \sqrt{2}$, i.e., the set

$$S_{z_0} = \mathbb{C} \setminus \big(\{x + i0 \colon x \geq 0\} \cup \{z \in \mathbb{C} \colon |z| = 1,\ |z-1| \leq |z_0 - 1|\}\big).$$

The following result was obtained in 1956 by A. Mori (Fig. 1.13).

(1.24) Mori ring module theorem:
If S is a ring separating $A_1 = \{z_1, z_2\} \subset \operatorname{cl} B_1$ and $A_2 = \{0, \infty\}$

$$M(S) \leq M_G\Big(\tfrac{1}{2}\sqrt{2 - \sqrt{4 - |z_1 - z_2|^2}}\Big),$$

with equality if and only if S is, up to rotation around 0, a Mori ring S_{z_0} with $z_0 \in \partial B_1$ and $|z_0 - 1| = \max\{|z_1 - 1|, |z_2 - 1|\}$.

Proof With $w_j = \sqrt{z_j}$, it is $|w_j| \leq 1$, for $j = 1, 2$, and $\frac{|\sqrt{z_1} - \sqrt{z_2}\,|}{\sqrt{2(|z_1| + |z_2|)}} \geq \frac{|w_1 - w_2\,|}{2}$. By (1.23) and as $r \mapsto M_G(r)$ is a decreasing function, $M(S) \leq M_G\big(\frac{1}{2}|w_1 - w_2|\big)$.

Since

$$\begin{aligned}|w_1 - w_2\,|^4 + |w_1^2 - w_2^2|^2 &= |w_1 - w_2\,|^2\big(|w_1 - w_2\,|^2 + |w_1 + w_2\,|^2\big)\\ &= |w_1 - w_2\,|^2\, 2\,\big(|w_1|^2 + |w_2|^2\big) \leq 4|w_1 - w_2\,|^2,\end{aligned}$$

it is $|w_1 - w_2\,|^4 - 4|w_1 - w_2\,|^2 + |z_1 - z_2|^2 \leq 0$, i.e., $(|w_1 - w_2\,|^2 - 2)^2 \leq 4 - |z_1 - z_2|^2$,

$$2 - |w_1 - w_2\,|^2 \leq \big|\,|w_1 - w_2\,|^2 - 2\,\big| \leq \sqrt{4 - |z_1 - z_2|^2}\,.$$

Thus, $|w_1 - w_2\,| \geq \sqrt{2 - \sqrt{4 - |z_1 - z_2|^2}}$. With this inequality applied to the result of the first paragraph, since M_G is a decreasing function, we have the upper bound in the statement.

The inequality in the first line of this proof holds with equality if and only if $|z_1| = |z_2| = 1$. If yes, it is equality in the inequality with two square roots at the end of the preceding paragraph. So, the inequality in the statement holds with equality

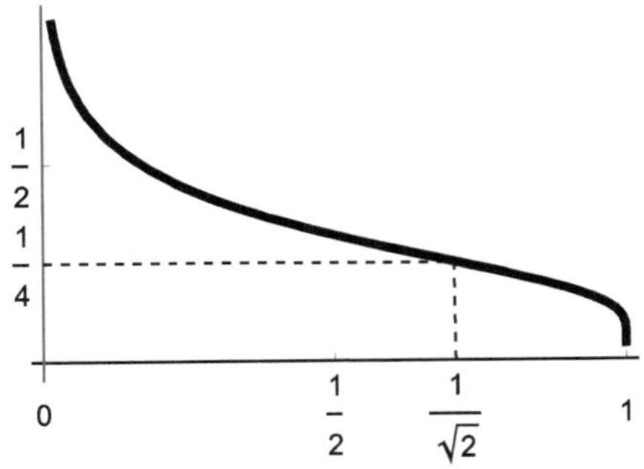

Fig. 1.14 Grötzsch ring module function $r \mapsto M_G(r)$

if and only if $|z_1| = |z_2| = 1$ and the inequality of the preceding result used at the beginning of this proof holds with equality, i.e.,

$$M(S) = M_G\left(\frac{|\sqrt{z_1} - \sqrt{z_2}\,|}{\sqrt{2(|z_1| + |z_2|)}}\right).$$

If C is a smallest arc of ∂B_1 with end points z_1, z_2 and includes them, the ring A' considered in the proof of the preceding result is conformal to a Teichmüller ring. Thus, the 2nd inequality in the last formula of the proof of the preceding result and the 1st inequality of that formula if $\mathbb{C} \setminus S = C \cup L$, where L is a ray symmetric relative to C and containing its starting point, hold with equality. Hence, we have equality in the inequality of the statement if and only if S is the Mori ring S_{z_0} or one of its rotations around 0. □

The upper bounds of the ring module obtained in the last five results are expressed in terms of the Grötzsch ring module $M_G(r)$. So, it is convenient to have basic properties of the function $r \mapsto M_G(r)$ (Fig. 1.14).

(1.25) Basic properties of the Grötzsch ring module:

1. $M_G(r) = \frac{1}{4}\frac{E(\sqrt{1-r^2})}{E(r)}$, *where* $E(r) = \int_0^1 \frac{1}{\sqrt{(1-t^2)(1-r^2t^2)}}\,dt$.
2. $M_G(r) = 2\,M_G\left(\frac{2\sqrt{r}}{1+r}\right) = \frac{1}{2}M_G\left(\left(\frac{1-\sqrt{1-r^2}}{r}\right)^2\right)$.
3. $M_G(r)\,M_G\left(\sqrt{1-r^2}\right) = \frac{1}{16}$.
4. $r \mapsto M_G(r)$ *is a strictly decreasing homeomorphism of* $]0, 1[$ *onto* $]0, +\infty[$.
5. $M_G\left(\frac{1}{\sqrt{2}}\right) = \frac{1}{4}$.
6. $M_G(r)\,M_G\left(\frac{1-r}{1+r}\right) = \frac{1}{8}$.
7. $\frac{1}{2\pi}\log\frac{(1+\sqrt{1-r^2})^2}{r} < M_G(r) < \frac{1}{2\pi}\log\frac{4}{r}$.
8. $M_G(r) - \frac{1}{2\pi}\log\frac{4}{r} \to 0$ *when* $r \to 0$.
9. $M_G(r) = \frac{1}{2\pi}\log\left(\frac{4}{r} - r - \delta_1(r)\right)$ *for* $r \in]0, 1[$, *where* δ_1 *is a function defined in* $]0, 1[$ *with values in* $\left]\frac{1}{4}r^3, 2r^3\right[$.
10. $M_G(r) + \frac{\pi}{4}\frac{1}{\log(1-r)} \to 0$, *and* $M'_G(r) \to -\infty$, *when* $r \to 1$.

Proof

1. There is a conformal homeomorphism of $G_r \cap \mathbb{H}$ to $(B_R \setminus \mathrm{cl}\, B_1) \cap \mathbb{H}$, with $R = e^{2\pi M_G(r)}$. This conformal homeomorphism can be extended to the whole Grötzsch ring G_r by symmetry relative to the circle ∂B_1, giving a conformal homeomorphism of G_r onto $(B_R \setminus \mathrm{cl}\, B_{R^{-1}}) \cap \mathbb{H}$, which can be mapped by a conformal homeomorphism onto the quadrilateral $Q_r(\infty, 0, r, \frac{1}{r}) \subset \mathbb{H}$ with module $M(Q_r) = 4\, M_G(r)$. The function

$$h(z) = \frac{1}{i\, 2\sqrt{r}} \int_{\infty}^{z} \frac{1}{\sqrt{z(z-r)(z-\frac{1}{r})}}\, dz$$

is a conformal homeomorphism of Q_r onto the rectangle

$$R_r\Big(0\,,\, E\big(\sqrt{1-r^2}\,\big)\,,\, E\big(\sqrt{1-r^2}\,\big) + i\,E(r)\,,\, i\,E(r)\Big)\,,$$

where

$$E(r) = \int_0^1 \frac{1}{\sqrt{(1-t^2)(1-r^2t^2)}}\, dt\,.$$

Thus, $M_G(r) = \frac{1}{4} M(Q_r) = \frac{1}{4}\, \frac{E(\sqrt{1-r^2})}{E(r)}$.

2. $f(z) = -\frac{4z}{(1-z)^2}$ is a conformal homeomorphism of the Grötzsch ring G_r onto the Teichmüller ring $T_{r_1,1}$ with $r_1 = \frac{4r}{(1-r)^2}$, which, by (1.21), has module $2M_G\Big(\sqrt{\frac{r_1}{1+r_1}}\,\Big)$. Thus,

$$M_G(r) = 2\, M_G\Big(\sqrt{\tfrac{r_1}{1+r_1}}\,\Big) = 2\, M_G\Big(\tfrac{2\sqrt{r}}{1+r}\Big)\,.$$

Using this formula $R = \frac{2\sqrt{r}}{1+r}$, which, since $0 < r < 1$, is equivalent to $r = \big(\frac{1-\sqrt{1-R^2}}{R}\big)^2$, we get $M_G\big(\big(\frac{1-\sqrt{1-R^2}}{R}\big)^2\big) = 2\, M_G(R)$.

3. The Teichmüller ring $T_{r_1,1}$ and also $-T_{1,r_1}$, by (1.21), have modules

$$M\big(T_{r_1,1}\big) = 2\, M_G\Big(\sqrt{\tfrac{r_1}{r_1+1}}\,\Big)\,, \qquad M\big(-T_{1,r_1}\big) = 2\, M_G\Big(\sqrt{\tfrac{1}{1+r_1}}\,\Big)\,.$$

Thus,

$$4\, M_G\Big(\sqrt{\tfrac{r_1}{r_1+1}}\,\Big)\, M_G\Big(\sqrt{\tfrac{1}{1+r_1}}\,\Big) = M\big(T_{r_1,1}\big)\, M\big(-T_{1,r_1}\,\big)\,.$$

The intersections of the rings $T_{r_1,1}\,, -T_{1,r_1}$ with $\mathbb{H}$ are conformal to the quadrilaterals, resp., $Q_1(\infty, -r_1, 0, 1)\,,\, Q_2(-r_1, 0, 1, \infty) \subset \mathbb{H}$ which, as seen in 1, have modules $M(Q_1) = 2\, M(T_{r_1,1})$ and $M(Q_2) = 2\, M(-T_{1,r_1})$. Since Q_1 and Q_2 have the same sides and differ by cyclic permutation of vertices, $M(Q_1)\, M(Q_2) = 1$, and

$$M\big(T_{r_1,1}\big) M\big(-T_{1,r_1}\big) = \tfrac{1}{4}\,, \qquad M_G\Big(\sqrt{\tfrac{r_1}{r_1+1}}\,\Big) M_G\Big(\sqrt{\tfrac{1}{1+r_1}}\,\Big) = \tfrac{1}{16}\,.$$

With $r=\sqrt{\frac{1}{1+r_1}}$, it is $\sqrt{\frac{r_1}{r_1+1}}=\sqrt{1-r^2}$. Thus, $M_G(r)\,M_G\big(\sqrt{1-r^2}\big)=\frac{1}{16}$.

4. It is immediate from the definition of $M_G(r)$ that it is strictly decreasing. By 1 and by the Leibniz[30] rule, the function M_G is continuous. By 2, $\lim\limits_{r\to 0} M_G(r)=2\lim\limits_{r\to 0} M_G(r)$, which can only be 0 or $+\infty$, and as M_G is a decreasing function, $\lim\limits_{r\to 0} M_G(r)= +\infty$. By 3,

$$\lim_{r\to 1} M_G(r) = \frac{1}{16}\lim_{r\to 1}\frac{1}{M_G(\sqrt{1-r^2})} = \frac{1}{16}\lim_{r\to 0}\frac{1}{M_G(r)} = 0\,.$$

Therefore, $r\mapsto M_G(r)$ is strictly decreasing from $]0,1[$ onto $]0,+\infty[$, and consequently, it is a homeomorphism of these intervals.

5. Using $r=\frac{1}{\sqrt{2}}$ in 3 gives $M_G\big(\frac{1}{\sqrt{2}}\big)^2=\frac{1}{16}$. Thus, $M_G\big(\frac{1}{\sqrt{2}}\big)=\frac{1}{4}$.

6. By 2, $M_G\big(\sqrt{1-r^2}\big) = \frac{1}{2}M_G\big(\frac{1-r}{1+r}\big)$, and, by 3, $M_G(r)\,\frac{1}{2}M_G\big(\frac{1-r}{1+r}\big) = \frac{1}{16}$. Therefore, $M_G(r)\,M_G\big(\frac{1-r}{1+r}\big)=\frac{1}{8}$.

7. We apply a conformal homeomorphism of the Grötzsch ring G_r onto a ring S with one of the connected components of the boundary the set $\{x+i0:\ -1\le x\le 1\}$ and the other a circle with center 0 and radius R, which must be $R=\frac{1+\sqrt{1-r^2}}{r}$. $h(z)=\frac{1}{2}\big(z+\frac{1}{z}\big)$ is a conformal homeomorphism of an annulus $B_\rho\backslash \mathrm{cl}\,B_1$ onto the region bounded by the ellipse E_ρ with axes $\rho\pm\frac{1}{\rho}$ without a slit $\{x+i0:\ -1\le x\le 1\}$. If $\rho+\frac{1}{\rho}\le 2R$, then $E_\rho\subset S$ and, by the monotonicity of ring module, $\frac{1}{2\pi}\log\rho=M(E_\rho)<M(S)=M_G(r)$. Analogously, if $\rho-\frac{1}{\rho}\ge 2R$, then $S\subset E_\rho$ and $\frac{1}{2\pi}\log\rho>M_G(r)$. As $\frac{1}{2R-r}<r$, if $\rho=2R-r=\frac{1}{r}(1+\sqrt{1-r^2})^2$ or $\rho=\frac{4}{r}$, then $\rho+\frac{1}{\rho}\le 2R$. Therefore,

$$\frac{1}{2\pi}\log\frac{(1+\sqrt{1-r^2})^2}{r} < M_G(r) < \frac{1}{2\pi}\log\frac{4}{r}\,.$$

8. By 7, $\frac{1}{2\pi}\log\frac{(1+\sqrt{1-r^2})^2}{4} < M_G(r)-\frac{1}{2\pi}\log\frac{4}{r} < 0$ and, as the 1st term in these inequalities $\to 0$ when $r\to 0$, we have $M_G(r) - \frac{1}{2\pi}\log\frac{4}{r}\to 0$ when $r\to 0$.

9. Define recursively a sequence by $r_0=r$, $r_{n+1}=\Big(\frac{1-\sqrt{1-r_n^2}}{r_n}\Big)^2$. By 2, $M_G(r_n)=2^n M_G(r)$, and by 7,

$$\frac{1}{2^n 2\pi}\log\frac{(1-\sqrt{1-r_n^2})^2}{r_n} < M_G(r) < \frac{1}{2^n 2\pi}\log\frac{4}{r_n}\,.$$

Since $\lim r_n=0$, it is $M_G(r)=\lim\frac{1}{2^n 2\pi}\log\frac{4}{r_n}$. By 7, with $n=0$,

$$\frac{1}{2\pi}\log\frac{(1-\sqrt{1-r^2})^2}{r} < M_G(r) < \frac{1}{2\pi}\log\frac{4}{r}\,.$$

Since $(1+x)^2+(1-x)^2=2(1+x^2)$,

[30] Leibniz, Gottfried (1646–1716).

$$2\big(1+\sqrt{1-r^2}\big)-\big(1+\sqrt[4]{1-r^2}\big)^2=\big(1-\sqrt[4]{1-r^2}\big)^2<4\,.$$

Thus, for $r\in\,]0,1[$, it is $M_G(r)=\frac{1}{2\pi}\log\big(\frac{2(1+\sqrt{1-r^2})}{r}-\delta(r)\big)$, where δ is a function from $]0,1[$ to $]0,r^3[$, which, as

$$2\big(1+\sqrt{1-r^2}\big)=4-r^2-\frac{r^4}{1+\sqrt{1-r^2}}\,,$$

can also be written $M_G(r)=\frac{1}{2\pi}\log\big(\frac{4}{r}-r-\delta_1(r)\big)$, where δ_1 is a function from $]0,1[$ to $\big]\frac{1}{4}r^3,2r^3\big[$.

10. By 3, $M_G(t)\,M_G\big(\sqrt{1-t^2}\big)=\frac{1}{16}$. With $r=\sqrt{1-t^2}$, it is $t=\sqrt{1-r^2}$ and $M_G(r)=\frac{1}{16}\big[M_G\big(\sqrt{1-r^2}\big)\big]^{-1}$. By 7,

$$M_G(r)-\tfrac{1}{16}\Big[\tfrac{1}{2\pi}\log\Big(\frac{4}{\sqrt{1-r^2}}\Big)\Big]^{-1}\to 0\,,\qquad r\to 1\,.$$

Since $\log\frac{4}{\sqrt{1-r^2}}=\log 4-\frac{1}{2}\,[\log(1-r)+\log(1+r)]$, $\log\frac{4}{\sqrt{1-r^2}}+\frac{1}{2}\log(1-r)\to 0$ when $r\to 1$, and $M_G(r)+\frac{\pi}{4}\frac{1}{\log(1-r)}\to 0$ when $r\to 1$; thus, $M'_G(r)\to-\infty$ when $r\to 1$.

□

1.7 Analytic Definition of Quasiconformal Mapping

In the initial part of the preceding section, it was seen that C^1 quasiconformal mappings when considered as functions of real variables are solutions of the Beltrami equation with uniformly bounded Beltrami coefficient. There are important properties that can be obtained for solutions of this partial differential equation which may have solutions not C^1, making clear the convenience of characterizing quasiconformal mappings in the context of distributions,[31] as usually for partial differential equations.

It is well known and simple to establish that the condition $f_{\bar z}=0$ (i.e., the validity of the Cauchy-Riemann equations) suffices for a function C^1 when taken as function of real variables to be holomorphic. With the C^1 requirement weakened to just continuity, the Cauchy-Riemann equations are still sufficient for the function to be holomorphic, as proved by D. Menchoff in 1935, correcting a gap in the attempted proof of H. Looman[32] in 1923; this interesting result, proved with Morera theorem, Baire category,[33] Lebesgue measure, and the Barrow[34] rule for Lebesgue integral, is given at the end of Appendix A. But we need a much weaker condition for sufficiency of the Cauchy-Riemann equations for holomorphy, namely, $f\in L^1_{loc}(U)$

[31] See Appendix B.

[32] Menchoff, Dmitrii (1892–1988). Looman, Herman (1896–1983).

[33] For Morera theorem and Baire category, see, resp., Section 2 of Chapter 6 and Appendix A in *CADOVA*. Morera, Jacinto (1856–1909). Baire, René-Louis (1874–1932).

[34] Barrow, Isaac (1630–1677).

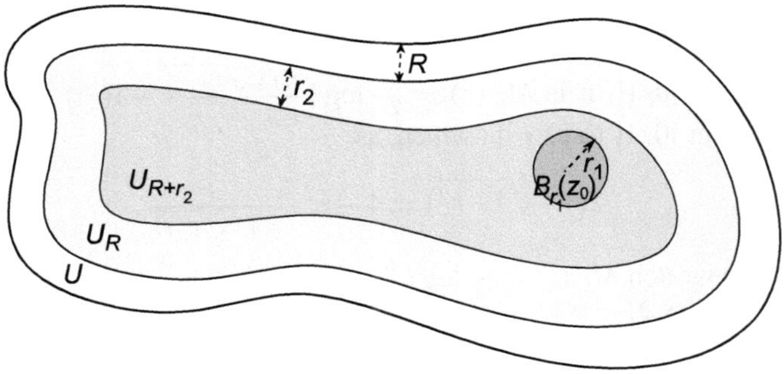

Fig. 1.15 Sets U, U_R, U_{R+r_2}, $B_{r_1}(z_0)$ in the proof of the Weyl lemma

with partial derivatives in the sense of distributions.[35] An immediate consequence is that every 1-quasiconformal mapping is a conformal homeomorphism.

(1.26) Weyl Lemma: *If $U \subset \mathbb{C}$ is open, $f: U \to \mathbb{C}$ with $f \in L^1_{loc}(U)$, and the derivative in the sense of distributions $f_{\overline{z}}=0$, then $f \in H(U)$.*

Proof With the family of mollifiers $\{\Psi_\varepsilon\}_{\varepsilon>0}$ considered in (B.1), if $U_R=\{z \in U : B_R(z) \subset U\}$ with $R>0$ such that $U_R \neq \emptyset$, for every $\varepsilon>0$ small, $f_\varepsilon = f * \Psi_\varepsilon$ is C^∞, $\|f_\varepsilon\|_{L^1} \leq \|f\|_{L^1}$, and $f_\varepsilon \to f$ in $L^1(U_R)$. Since $f_{\overline{z}}=0$ in the sense of distributions, in the classical sense, $(f_\varepsilon)_{\overline{z}}=0$, and, consequently, $f_\varepsilon \in H(U_R)$. If $0<r_1<r<r_2$ and $U_{R+r_2} \neq \emptyset$, for $z_0 \in U_{R+r_2}$ (Fig. 1.15), by the mean value property of holomorphic functions

$$f_\varepsilon(z) = \frac{1}{2\pi}\int_0^{2\pi} f_\varepsilon(z_0+re^{i\theta})\,d\theta\,, \quad z \in B_{r_1}(z_0)\,.$$

If $\eta: \mathbb{R} \to \mathbb{R}$ is C^∞ with support included in $]r_1, r_2[$, $\eta \geq 0$, and $\int_{r_1}^{r_2} \eta = 1$,

$$f_\varepsilon(z) = \frac{1}{2\pi}\int_{r_1}^{r_2}\int_0^{2\pi} f_\varepsilon(z_0+re^{i\theta})\,\eta(r)\,d\theta\,dr\,, \quad z \in B_{r_1}(z_0)\,.$$

Since $\|f_\varepsilon\|_{L^1} \leq \|f\|_{L^1}$ and $f_\varepsilon \to f$ in $L^1(U_R)$, by the Lebesgue dominated convergence theorem, $f_\varepsilon \to f$ when $\varepsilon \to 0$ in $B_{r_1}(z_0)$ and, as the integrand is a continuous function in z, f is continuous in $B_{r_1}(z_0)$, and also in U_{r+r_2}. Hence, $f_\varepsilon \to f$ uniformly on compact subsets of U when $\varepsilon \to 0$, and, by the Weierstrass theorem for sequences of functions,[36] $f \in H(U)$. □

[35] The first result of this type was obtained in 1940 by Hermann Weyl (1885–1955) for Laplace equation. Laplace, Pierre-Simon (1749–1827).

[36] See Section 5 of Chapter 6 of *CADOVA*.

Thus, C^2 solutions of a same Beltrami equation in a region of $\mathbb{C}$ are related by one being the composition of a conformal homeomorphism with the other.

(1.27) *If $\Omega \subset \mathbb{C}$ is a region, $f \in L^1_{loc}(\Omega)$ is a solution of the Beltrami equation in the sense of distributions, and g is a C^2 homeomorphism in Ω as a function of real variables solution of the same Beltrami equation, there is a conformal homeomorphism h of $g(\Omega)$ in $\mathbb{C}$ such that $f = h \circ g$ a.e. in Ω, and f is C^2 in Ω as a function of real variables.*

Proof By (B.2), the derivatives in the sense of the distributions satisfy the chain rule for derivatives, and

$$f_z = (h_z \circ g) g_z + (h_{\overline{z}} \circ g) \overline{g}_z \,, \qquad f_{\overline{z}} = (h_z \circ g) g_{\overline{z}} + (h_{\overline{z}} \circ g) \overline{g}_{\overline{z}} \,,$$

$$0 = f_{\overline{z}} - \kappa f_z = (h_z \circ g)(g_{\overline{z}} - \kappa\, g_z) + (h_{\overline{z}} \circ g)(\overline{g}_{\overline{z}} - \kappa\, \overline{g}_z) = (h_{\overline{z}} \circ g)(\overline{g}_{\overline{z}} - \kappa\, \overline{g}_z) \,.$$

Since $g_{\overline{z}} - \kappa\, g_z = 0$, if it also $\overline{g}_{\overline{z}} - \kappa\, \overline{g}_z = 0$, the derivatives of g and $\overline{g}$ as functions of two real variables would be such that $D\overline{g} Dg^{-1}$ would map circles centered at 0 to circles centered at 0, and consequently, $\varphi = \overline{g} \circ g^{-1}$ would be holomorphic. $\overline{\varphi} = g \circ g^{-1} = 1_\Omega$ would also be holomorphic. Therefore, $\varphi = 0$, and also $g = 0$ in Ω, in contradiction with g being a homeomorphism. Thus, $\overline{g}_{\overline{z}} - \kappa\, \overline{g}_z \neq 0$. By the last equality of the initial period of this proof, $h_{\overline{z}} \circ g = 0$ in Ω and, by the Weyl lemma, h is holomorphic in $g(\Omega)$. □

For example, if f, g are regular quasiconformal mappings in B_1 with the same Beltrami coefficient everywhere, the derivatives of f and g^{-1} as functions of real variables are such that $Df\, Dg^{-1}$ maps circles with center 0 to circles with center 0. Thus, $h = f \circ g^{-1}$ is holomorphic in B_1 and $f = h \circ g$ for some Möbius transformation h.

If $\Omega \subset \mathbb{C}$ is a region, an **absolutely continuous function on lines** in Ω is $f = (u, v) : \Omega \to \mathbb{C}$ such that for each closed rectangle $[a, b] \times [c, d] \subset \Omega$ the functions

$$x \mapsto u(x+iy) \,, \qquad y \mapsto u(x+iy) \,, \qquad x \mapsto v(x+iy) \,, \qquad y \mapsto v(x+iy)$$

are absolutely continuous[37] a.e. in, resp., $[a, b]$ and $[c, d]$.

The following result gives analytic necessary and sufficient conditions for a function to be a K-quasiconformal mapping. The necessity that it be absolutely continuous function on lines was proved in 1955 by K. Strebel. In 1957, A. Mori proved the necessity of 2, but with a different proof than the proof given here, which for absolutely continuous functions on lines uses the proof of A. Pfluger in 1959 with Rengel inequality and for the inequality of the absolute values of the partial derivatives uses the Gehring-Lehto theorem (E.3) proved in 1959. In the same paper

[37] See Appendix A.

where they obtained this theorem, F. Gehring and O. Lehto proposed an analytic definition of quasiconformal mapping, essentially with the necessary condition of 2 in the following result, and proved equivalence to the geometric definition.

(1.28) *If $\Omega \subset \mathbb{C}$ is a region, $f:\Omega \to \mathbb{C}$, $K \geq 1$, and $k = \frac{K-1}{K+1} < 1$, the following conditions are equivalent:*

1. *f is a K-quasiconformal mapping in Ω.*
2. *f is an orientation preserving homeomorphism absolutely continuous on lines in Ω and $|f_{\bar{z}}| \leq k\,|f_z|$ a.e. in Ω.*
3. *f is an orientation preserving homeomorphism, it has locally integrable partial derivatives in the sense of distributions, and $|f_{\bar{z}}| \leq k\,|f_z|$ a.e. in Ω.*

Every quasiconformal mapping f in a region $\Omega \subset \mathbb{C}$ is differentiable a.e. *as a function of real variables absolutely continuous in Ω has Jacobian* $\det Df \in L^1_{loc}(\Omega)$ *and* $f_z, f_{\bar{z}}, \frac{\partial f}{\partial x}, \frac{\partial f}{\partial y} \in L^2_{loc}(\Omega)$.

Proof *(1$\Rightarrow$2)* Let f be a K-quasiconformal mapping, $R =]a,b[\times]c,d[$ such that $\mathrm{cl}\, R \subset \Omega$, $y \in]c,d[$, $I_y =]a,b[\times\{y\}$, $R_y =]a,b[\times]c,y[$, and $A(y) = \mu\big(f(R_y)\big)$ the area of $f(R_y)$ for $y \in]c,d[$, where μ denotes Lebesgue measure in $\mathbb{C}$. As A is an increasing function, A' exists and is finite[38] a.e. in $]c,d[$. If $y_0 \in]c,d[$, $A'(y_0)$ exists and is finite, $]x_n, x'_n[$ for $n = 1, \ldots, N$ are disjoint subintervals of $]a,b[$, $c < y_0 + \delta < d$, $\delta > 0$, and $R_n(\delta) = [x_n, x'_n] \times [y_0, y_0+\delta]$, then the module of the quadrilateral $\partial R_n(\delta)$ with vertices (x_n, y_0), (x'_n, y_0), $(x'_n, y_0+\delta)$, $(x_n, y_0+\delta) \in \mathbb{C}$ is $M\big(\partial R_n(\delta)\big) = \frac{x'_n - x_n}{\delta}$. By the Rengel inequality, the module of the quadrilateral $Q_n(\delta) = \partial f\big(R_n(\delta)\big)$ satisfies $M\big(Q_n(\delta)\big) \geq \frac{[L_{24,n}(\delta)]^2}{\mu(f(R_n(\delta)))}$, where $\mu\big(f\big(R_n(\delta)\big)\big)$ is the area of $f\big(R_n(\delta)\big)$ and $L_{24,n}(\delta)$ is the distance of the sides 2 and 4 of the quadrilateral $Q_n(\delta)$. So, $\lim_{\delta\to 0} L_{24,n}(\delta) = |f(x'_n) - f(x_n)|$ and the dilatation K_f of f satisfies

$$K_f(x'_n - x_n) \geq \frac{M(Q_n(\delta))}{M(\partial R_n(\delta))}(x'_n - x_n) \geq \frac{[L_{24,n}(\delta)]^2}{\mu(f(R_n(\delta)))}\,\delta\,, \quad n = 1, \ldots, N.$$

By the Cauchy-Schwarz inequality, if $z_n, w_n \geq 0$ for $n = 1, \ldots, N$,

$$\left(\sum_{n=1}^{N} z_n\right)^2 = \left(\sum_{n=1}^{N} \frac{z_n}{\sqrt{w_n}}\sqrt{w_n}\right)^2 \leq \left(\sum_{n=1}^{N} \frac{z_n^2}{w_n}\right)\left(\sum_{n=1}^{N} w_n\right),$$

and, as $z_n = L_{24,n}(\delta)$ and $w_n = \mu\big(f\big(R_n(\delta)\big)\big)$,

$$K_f \sum_{n=1}^{N}(x'_n - x_n) \geq \delta \sum_{n=1}^{N} \frac{[L_{24,n}(\delta)]^2}{\mu(f(R_n(\delta)))} \geq \delta\, \frac{\left(\sum_{n=1}^{N} L_{24,n}(\delta)\right)^2}{\sum_{n=1}^{N} \mu(f(R_n(\delta)))}\,.$$

[38] See Appendix A.

Since $A(y_0+\delta)-A(y_0)\geq\sum_{n=1}^{N}\mu\big(f\big(R_n(\delta)\big)\big)$,

$$\left(\sum_{n=1}^{N}L_{24,n}(\delta)\right)^2\leq K_f\,\frac{A(y_0+\delta)-A(y_0)}{\delta}\sum_{n=1}^{N}(x'_n-x_n)\,.$$

Letting $\delta\to 0$,

$$\left(\sum_{n=1}^{N}|f(x'_n)-f(x_n)|\right)^2\leq K_f A'(y_0)\sum_{n=1}^{N}(x'_n-x_n)\,.$$

Thus, $x\mapsto f(x+iy_0)$ is absolutely continuous in I_{y_0} if $y_0\in\,]c,d[$ is such that $A'(y_0)$ exists and is finite, what holds for y_0 a.e. in $]c,d[$. It is proved analogously that $x\mapsto f(x+ix_0)$ is absolutely continuous in $I^*_{x_0}=\{x_0\}\times]c,d[$ if $x_0\in\,]a,b[$ is such that, with $A^*=\mu\big(f(R^*_x)\big)$, where $R^*_x=\,]a,x[\times]c,d[$, the derivative $(A^*)'(x_0)$ exists and is finite, what holds for x_0 a.e. in $]a,b[$. This proves the necessity of f to be absolutely continuous on lines in order that it be a quasiconformal mapping.

(2 ⇒3) Let $f:\Omega\to\mathbb{C}$ be an orientation preserving homeomorphism partially absolutely continuous a.e. satisfying the inequality in 2. As absolutely continuous functions on subsets of $\mathbb{R}$ are differentiable, a.e.,[39] the partial derivatives of f exist a.e. in Ω, and, by the Gehring-Lehto theorem (E.3), f is differentiable a.e. in Ω. As seen at the beginning of the preceding section, the Jacobian of f is $\det Df=|f_z|^2-|f_{\bar z}|^2>0$, and, by (E.4), $\det Df\in L^1_{loc}(\Omega)$. By the inequality in 2, $\det Df\geq(1-k^2)|f_z|$ and

$$|f_{\bar z}|^2<|f_z|^2\leq\frac{1}{1-k^2}\det Df\,.$$

So, $f_z,\,f_{\bar z}\in L^2_{loc}$. Since $\frac{\partial f}{\partial x}=f_z+f_{\bar z}$ and $\frac{\partial f}{\partial y}=i(f_z-f_{\bar z})$, it is $\frac{\partial f}{\partial x},\,\frac{\partial f}{\partial y}\in L^2_{loc}(\Omega)$. If $\varphi\in C_0^\infty(\Omega)$, with Fubini theorem[40] and integration by parts,

$$\iint_\Omega\frac{\partial f}{\partial x}\,\varphi\,dxdy=-\iint_\Omega f\,\frac{\partial\varphi}{\partial x}\,\varphi\,dxdy,\qquad\iint_\Omega\frac{\partial f}{\partial y}\,\varphi\,dxdy=-\iint_\Omega f\,\frac{\partial\varphi}{\partial y}\,\varphi\,dxdy,$$

and $\frac{\partial f}{\partial x}$, $\frac{\partial f}{\partial y}$ are partial derivatives of f in the sense of distributions[39].

(3 ⇒ 1) Let $f:\Omega\to\mathbb{C}$ be an orientation preserving homeomorphism with locally integrable partial derivatives in the sense of distributions satisfying the inequality in the statement. If, in addition, f is C^1 as a function of real variables, the proof that it satisfies the geometric definition of quasiconformal mapping is the solution of Grötzsch problem in (1.2). For the general case, it suffices to ensure that if f is a homeomorphism with locally integrable partial derivatives in the sense of distributions and $|f_{\bar z}|\leq k\,|f_z|$, compositions of f with conformal homeomorphisms also have locally integrable partial derivatives in the sense of

[39] See Appendix D.

[40] Fubini, Guido (1879–1943).

distributions and satisfy the same condition, as guaranteed by (B.2), since with conformal homeomorphisms a quadrilateral can be mapped to a rectangle with the same module.

Hence, *every conformal homeomorphism in a region of $\mathbb{C}$ is absolutely continuous*, since it is a partially absolutely continuous homeomorphism with partial derivatives in L^2_{loc}, and the result follows by (D.4). □

Analytic Definition of Quasiconformity

A **quasiconformal mapping** in a region $\Omega \subset \mathbb{C}_\infty$ is a homeomorphism $f:\Omega \to \mathbb{C}$ preserving orientation with partial derivatives in the sense of distributions in $L^2_{loc}(\Omega)$ and satisfying the Beltrami equation:

$$f_{\bar{z}} = \kappa(z)\, f_z\,,$$

with **Beltrami coefficient**[41] $\kappa \in L^\infty(\Omega)$ such that $\|\kappa\|_{L^\infty} < 1$. We say that f is K-**quasiconformal** if, additionally, $\frac{1+\|\kappa\|_{L^\infty}}{1-\|\kappa\|_{L^\infty}} \leq K$. The infimum of the set of all K such that f is K-quasiconformal is called the **quasiconformity constant** of f in Ω.

Thus, *a function f defined in a region $\Omega \subset \mathbb{C}_\infty$ is quasiconformal if it has a Beltrami coefficient with derivatives in the sense of distributions with absolute values upper bounded by some $k \in [\,0,\,1\,[$, and it is K-quasiconformal with $K \geq 1$ if the Beltrami coefficient is upper bounded by $k = \frac{K-1}{K+1} \in [\,0\,,\,1\,[$.*

An immediate consequence of the analytic definition and the Weyl lemma is that every 1-quasiconformal mapping is a conformal homeomorphism, establishing with the analytic definition the property obtained before with the geometric definition, although there with more work.

The analytic definition is also more convenient for proving the measurable Riemann mapping theorem, which is the subject of the next chapter.

Areas of open measurable subsets of $\mathbb{C}$ are transformed by quasiconformal mappings to the integral of the Jacobian, and, in particular, sets of measure zero are mapped onto sets of measure zero, as in general for injective C^1 changes of variables, as established below using the Green theorem[42] with conditions in the integrand weaker than the classical ones (D.3), along with some immediate consequences of that property.

(1.29) *If $\Omega \subset \mathbb{C}$ is a region, every quasiconformal mapping f defined in Ω maps any measurable set $S \subset \Omega$ onto a measurable set $f(S)$ with area $\iint_S \det Df$. Besides, $f_z \neq 0$* a.e. *in Ω and f is a regular quasiconformal mapping* a.e. *in Ω.*

---

[41] $L^\infty(\Omega)$ is the normed space of functions $f:\Omega \to \mathbb{C}$ with $|f|$ bounded by some $M > 0$ a.e. in Ω with the norm $\|f\|_{L^\infty} = \inf\{M > 0 : |f| \leq M$ a.e. in $\Omega\}$, identifying functions equal a.e. in Ω.

[42] Green, George (1793–1841).

Proof By (1.28), $\frac{\partial f}{\partial x}, \frac{\partial f}{\partial y} \in L^2_{loc}(\Omega)$, and, by (D.3), considering rectangles R with sides parallel to the real and imaginary axes such that the restriction of f to the sides of R is absolutely continuous, denoting the real and imaginary parts of f by, resp., u, v,

$$\iint_R \det DF = \iint_R \left(\frac{\partial u}{\partial x}\frac{\partial v}{\partial y} - \frac{\partial u}{\partial y}\frac{\partial v}{\partial x}\right) = \int_{\partial R} \left(u \frac{\partial v}{\partial x}\, dx + u \frac{\partial v}{\partial y}\, dy\right).$$

Since $\det Df = |f_z|^2 - |f_{\bar{z}}|^2$ and $|f_{\bar{z}}| \leq k\,|f_z|$ a.e. in Ω, if $f_z = 0$ on some set S with positive measure, then $\det Df = 0$ a.e. in S and, by the formula in the preceding paragraph, $f(S)$ has measure 0. Therefore, the quasiconformal mapping f^{-1} would map a set with measure 0 onto a set with positive measure, contradicting the formula of the preceding paragraph applied to f^{-1}. Consequently, $\det Df \neq 0$ and $f_z \neq 0$ a.e. in Ω. Since a quasiconformal mapping is regular at a point if it is differentiable as a function of real variables and has positive Jacobian at that point, f is a regular quasiconformal mapping a.e. in Ω. □

Other consequences of the Green theorem (D.3) are the following formulas, with the first replacing for quasiconformal mappings the Cauchy formula for holomorphic functions or, more generally, for absolutely continuous functions on lines with partial derivatives in L^1_{loc} in a region of $\mathbb{C}$.

(1.30) *If $\Omega \subset \mathbb{C}$ is a region and f is an absolutely continuous function on lines with partial derivatives in $L^p_{loc}(\Omega)$ ($p = 1$ or 2), in particular if f is quasiconformal in Ω and $D \subset \Omega$ is an open set bounded by a rectifiable Jordan curve with* $\mathrm{cl}\, D \subset \Omega$, *then*

$$\int_{\partial D} f(z)\, dz = 2i \iint_D \frac{\partial f}{\partial \bar{z}}, \qquad \int_{\partial D} f(z)\, d\bar{z} = -2i \iint_D \frac{\partial f}{\partial z}.$$

Proof For absolutely continuous functions on lines with partial derivatives in $L^p_{loc}(\Omega)$, it is a consequence of the Green theorem in the form (D.3) with $\mathbf{g}(z) = z$ and $\mathbf{g}(z) = \bar{z}$ for $z \in K$ multiplied by a C^1 function equal to 1 in K and to 0 outside a compact set $K' \supset K$ with $K' \subset \Omega$. For quasiconformal mappings, it is a consequence of the preceding case and of (1.28) to imply that for every compact set $K' \subset \Omega$, $\frac{\partial f}{\partial x}, \frac{\partial f}{\partial y} \in L^2(K') \subset L^1(K')$. □

Chapter 2
Measurable Riemann Mapping Theorem

2.1 Introduction

The measurable Riemann mapping theorem is most important in the theory of quasiconformal mappings. It establishes the existence of solutions of the Beltrami equation and, therefore, of quasiconformal mappings with Beltrami coefficient specified by a function with absolute value <1. It is related to the Riemann mapping theorem as it ensures existence of a quasiconformal mapping of any simply connected region proper subset of $\mathbb{C}$ onto an open disk with radius 1 with Beltrami coefficient given by an arbitrary bounded measurable function except possibly on a set of measure 0.

For Hölder[1] continuous Beltrami coefficients, the local existence of solution of the Beltrami equation was obtained in 1916 by L. Liechtenstein and in 1919 by A. Korn; in 1955, S.-S. Chern gave a simpler proof. M.A. Lavrentiev[2] proved global existence for continuous Beltrami coefficients in 1935. C. Morrey[3] obtained a proof for measurable Beltrami coefficients in 1938. L. Bers and L. Nirenberg[4] in 1954 and B.V. Boyarski[5] in 1957 obtained alternative proofs and additional properties of the solutions. In 1960, L. Ahlfors and L. Bers[6] proved the result for measurable

[1] Hölder, Otto (1859–1937).

[2] Lavrantiev, M.A., Sur une classe de representation continues, *Mat. Sbornik* **42** (1935), 407–423.

[3] Morrey, C., On the solutions of quasilinear elliptic partial differential equations, *Trans. Amer. Math. Soc.* **43** (1938), 126–166.

[4] Bers, L., Nirenberg, L., On a representation theorem for linear elliptic systems with discontinuous coefficients and its applications, *Convegno Internazionale sulle Equazioni Derivate e Parziali* (1954), 111–140.

[5] Boyarski, B.V., Homeomorphic solutions of Beltrami systems, *Doklady* **102** (1955), 661–664 (in Russian).

[6] Ahlfors, L.V., Bers, L., Riemann's mapping theorem for variable metrics, *Ann. of Math.* (2) **72** (1960), 385–404.

L. T. Magalhães, *Quasiconformal Mappings in the Plane and Complex Dynamics*,
https://doi.org/10.1007/978-3-031-80115-0_2

Beltrami coefficients depending continuously or analytically on parameters. In 2000, A. Douady and X. Buff obtained a simpler proof of this result with Hilbert spaces using Fourier-Plancherel transform, which is the proof given in this chapter.[7]

2.2 Existence and Regularity of Solution of Beltrami Equation in $\mathbb{C}$

The proof of the measurable Riemann mapping theorem obtained in 2000 by A. Douady and X. Buff uses the Neumann series, which is simply the extension of numerical geometric series to **Banach spaces**, i.e., complete normed spaces. The sum of a geometric series of ratio r with $|r| < 1$ is $\sum_{n=0}^{\infty} r^n = \frac{1}{1-r}$. Analogously, for a bounded linear transformation $T : X \to X$, with X Banach space and operator norm $\|T\| = \sup\limits_{x\in X\setminus\{0\}} \frac{\|T(x)\|}{\|x\|} < 1$, it is $\sum_{n=0}^{\infty} T^n = (1_X - T)^{-1}$, called **Neumann series**. The proof is the same as for the sum of the geometric series; it boils down to observing that

$$(1_X - T)\sum_{n=0}^{N} T^n = \sum_{n=0}^{N} T^n - \sum_{n=0}^{N} T^{n+1} = 1_X - T^{N+1}, \qquad N \in \mathbb{N};$$

thus, with $r = \|T\| < 1$,

$$\left\| (1_X - T)\sum_{n=0}^{N} T^n - 1_X \right\| = \|T^{N+1}\| \le \|T\|^{N+1} = r^{N+1} \to 0, \text{ when } N \to +\infty,$$

and $(1_X - T)\sum_{n=0}^{\infty} T^n = 1_X$; analogously, $\sum_{n=0}^{\infty} T^n(1_X - T) = 1_X$. Thus, the linear transformation $1_X - T$ has inverse $(1_X - T)^{-1} = \sum_{n=0}^{\infty} T^n$, which is a bounded linear transformation, since

$$\left\| \sum_{n=0}^{\infty} T^n(x) \right\| \le \sum_{n=0}^{\infty} \|T\|^n \|x\| = \sum_{n=0}^{\infty} r^n \|x\| = \tfrac{1}{1-r}\|x\|, \qquad x \in X,$$

and[8] $\|(1_X - T)^{-1}\| \le \frac{1}{1-\|T\|}$.

[7] This chapter is mainly based on the article A. Douady, X. Buff, Le théorème d'intégrabilité des structures presque complexes. *In The Mandelbrot set, theme and variations,* 307–324, London Math. Soc. Lecture Note Ser., 274, Cambridge Univ. Press, Cambridge, 2000. Chern, Shiing-Shen (1911–2004). Louis Nirenberg (1925–2020) was awarded the *Bôcher Memorial Prize* in 1959 and the *Steele Prize for Lifetime Achievement* in 1994 both of the *AMS*, the *Chern Medal* in the *International Congress of Mathematicians* in 2010, the *Steele Prize for Seminal Contribution to Research* of the *AMS* in 2014, and the *Abel Prize* in 2015. Boyarski, Bogdan V. (1931–2018).

[8] Neumann, Carl (1832–1925). Banach, Stefan (1892–1945).

For $U \subset \mathbb{C}$ an open set, $\boldsymbol{H^1_{loc}(U)}$ denotes the set of functions $f \in L^1_{loc}(U)$ with partial derivatives in the sense of distributions in $L^2_{loc}(U)$ such that $f(z) - z \to 0$ when $|z| \to +\infty$. Denote $C^0 = C^0(\mathbb{C})$, $C^\infty = C^\infty(\mathbb{C})$, $C_0^\infty = C_0^\infty(\mathbb{C})$, considering the elements as functions of two real variables, $\mathscr{S} = \mathscr{S}(\mathbb{C})$ the Schwartz space,[9] $L^j = L^j(\mathbb{C})$ for $j = 1, 2, \infty$ and L_0^∞ the subspace of L^∞ of functions with compact support.

The Neumann series and the Fourier-Plancherel transform allow to establish in one stroke existence, uniqueness, and analytic and continuous dependence on parameters of solutions of the Beltrami equation in $\mathbb{C}$.

(2.1) Existence, uniqueness, and analytic and continuous dependence on parameters of solutions of Beltrami equation in $\mathbb{C}$:
If $\kappa : \mathbb{C} \to \mathbb{C}$ is a measurable function with $\|\kappa\|_{L^\infty} < 1$, then there is a unique solution f of the Beltrami equation $f_{\overline{z}} = \kappa f_z$ in H^1_{loc}. If, in addition, κ depends analytically (resp., continuously) on a parameter λ in a Riemann surface (resp., a topological space), then f depends analytically (resp., continuously) on λ in the topology of H^1_{loc}.

Proof If $g(z) = f(z) - z$, the Beltrami equation $f_{\overline{z}} = \kappa f_z$ equals $g_{\overline{z}} = \kappa(1 + g_z)$, which with the Hilbert-Beurling transform[9] $\mathcal{L}$ and $h = g_{\overline{z}}$ can be written $h = \kappa[1 + \mathcal{L}(h)]$ or $(1_{L^2} - \kappa\mathcal{L})(h) = \kappa$. Since $\mathcal{L}$ is an isometry in L^2, the operator norm of $\kappa\mathcal{L} : L^2 \to L^2$ satisfies $\|\kappa\mathcal{L}\| < 1$, and, by the Neumann series, $1_{L^2} - \kappa\mathcal{L} : L^2 \to L^2$ has an inverse $(1_{L^2} - \kappa\mathcal{L})^{-1} = \sum_{n=0}^\infty (\kappa\mathcal{L})^n$. Therefore, the equation $h = \kappa\,[1 + \mathcal{L}(h)]$ has a unique solution $h \in|, L^2, h = (1_{L^2} - \kappa\mathcal{L})^{-1}(\kappa)$. For each $h \in L^2$, the solution of the equation $g_{\overline{z}} = h$ can be obtained with the Fourier-Plancherel transform. Since $\widehat{g_{\overline{z}}} = \widehat{h}$, it is $i\pi\,\zeta\,\widehat{g}(\zeta) = \widehat{h}(\zeta)$; thus, $\widehat{g}(\zeta) = -i\frac{1}{\pi\zeta}\,\widehat{h}(\zeta)$ and, again with Fourier-Plancherel transform, $g = i\widehat{\frac{1}{\pi z}} * h$. By Example (C.3), the function $\frac{1}{\pi z}$ is an eigenvector of the Fourier-Plancherel transform associated with the eigenvalue $-i$, and $g = \frac{1}{\pi z} * h$. Therefore, the Beltrami equation $f_{\overline{z}} = \kappa f_z$ has a unique solution $f \in H^1_{loc}$ given by $f = 1_{\mathbb{C}} + \frac{1}{\pi z} * (1_{L^2} - \kappa\mathcal{L})^{-1}(k)$. If κ depends analytically (resp., continuously) on a parameter λ in a Riemann surface (resp., topological space), then f also depends analytically (resp., continuously) on λ.

By the Cauchy-Schwarz inequality,

$$|f(x+iy) - (x+iy)|^2 = \left| \iint_{\mathbb{R}^2} \frac{1}{\pi(x-s, y-t)} \big[(1_{L^2} - \kappa\mathcal{L})^{-1}(k)\big](s,t)\,dsdt \right|^2$$

$$\leq \left\|(1_{L^2} - \kappa\mathcal{L})^{-1}(k)\right\|_{L^2}^2 \iint_{\mathbb{R}^2} \frac{1}{|\pi(x-s, y-t)|^2}\,dsdt\,, \quad x, y \in \mathbb{R}\,.$$

The last integral converges to 0 when $|(x+iy)| \to +\infty$; thus, $f(z) - z \to 0$ when $|z| \to +\infty$. □

[9] See the end of Appendix C. Laurent Schwartz (1915–2002) was awarded the *Fields Medal* in 1950 for the development of the theory of distributions.

This gives, under the stated conditions, analytic (resp., continuous) dependence of the solutions on the parameters in the topology of L^2, but it is also possible to obtain analytic (resp., continuous) dependence in the topology of uniform convergence in compact sets.

(2.2) *If, with the assumption of* (2.1), f *is the solution of the Beltrami equation in* (2.1) *and* $\lambda \mapsto \kappa_\lambda$ *is an analytic (resp., continuous) function of a parameter* λ *in a Riemann surface (resp., topological space)* Λ, *and for each value of the parameter* $\lambda_0 \in \Lambda$, *there are a neighborhood* $\Lambda' \subset \Lambda$ *containing* λ_0 *and* $k \in]0, 1[$, $R > 0$ *such that for* $\lambda \in \Lambda'$, $\operatorname{supp} \kappa_\lambda \subset B_R \subset \mathbb{C}$, *and* $\|\kappa_\lambda\|_{L^\infty} \leq k$, *then we have analytic (resp., continuous) dependence in the topology of uniform convergence in compact subsets of* $\mathbb{C}$.

Proof *(Continuous dependence)* Each family $\{f_\lambda\}_{\lambda \in \Lambda'}$ of solutions of the Beltrami equation is equicontinuous. Hence, by the Arzelà-Ascoli theorem,[10] it has compact closure in the topology of uniform convergence in compact sets. This topology agrees for the family $\{f_\lambda\}_{\lambda \in \Lambda'}$ with that of H^1_{loc}. Thus, by the preceding result, there is continuous dependence on the topology of uniform convergence in compact sets.

(Analytic dependence) The following general result holds in Fréchet spaces with $E_2 = H^1_{loc}$ and E_1 the space of complex functions with uniform convergence in compact sets: *If* E_1, E_2 *are Fréchet spaces*[11] $\iota : E_1 \to E_2$ *is a continuous linear injection,* Λ *is a Riemann surface,* $F_1 : \Lambda \to E_1$ *is continuous, and* $F_2 = \iota \circ F_1$ *is analytic, then* F_1 *is analytic*. To prove it, assume without loss of generality $\Lambda \subset \mathbb{C}^n$ and $[0, 1]^n$. For $\mathbf{k} = (k_1, \ldots, k_n)$, define

$$c_{\mathbf{k}} = \int_{[0,1]^n} F_1\left(e^{i2\pi(t_1 + \cdots + t_n)}\right) e^{-i2\pi \mathbf{k} \cdot \mathbf{t}}\, d\mathbf{t}.$$

Since F_2 is analytic, $F_2(z) = \sum_{\mathbf{k} \in (\mathbb{N} \cup \{0\})^n} \iota(c_{\mathbf{k}}) z^{\mathbf{k}}$ for $|z| < 1$, and since ι is injective, $F_1(z) = \sum_{\mathbf{k} \in (\mathbb{N} \cup \{0\})^n} c_{\mathbf{k}} z^{\mathbf{k}}$. □

If $\Omega \subset \mathbb{C}$ and $\kappa : \Omega \to \mathbb{C}$ is measurable with $\|\kappa\|_{L^\infty(\Omega)} < 1$, extending κ to $\mathbb{C}$ with value 0 in $\mathbb{C} \setminus \{0\}$, by the preceding result, there is a solution of the Beltrami equation with the extended coefficient κ depending analytically (or continuously) on parameters, and the restriction of this solution to Ω satisfies the Beltrami equation in Ω with the corresponding parameter dependency property. This function is a

[10] See Section 4 of Chapter 10 of *CADOVA*. Arzelà, Cesaro (1847–1912). Ascoli, Giulio (1843–1896).

[11] A **Fréchet space** X is a complete topological linear space with the induced topology for a countable family of seminorms $x \mapsto \|x\|_k$, $k \in \mathbb{N}$, i.e., $U \subset X$ is open if and only if for every $u \in U$ there are $\varepsilon, K > 0$ such that $\{x \in X : \|x - u\|_k < \varepsilon$ to $k \leq K\} \subset U$, such that $\cap_{k \in \mathbb{N}}\{x \in |, X : \|x\|_k = 0\} = \{0\}$. A **seminorm** in a linear space V is defined as a norm except that there can be nonzero vectors with zero seminorms (see Section 6 of Appendix A of *CADOVA*).

quasiconformal mapping in Ω if and only if it is a homeomorphism. Thus, to establish the measurable Riemann mapping theorem for $\Omega \subset \mathbb{C}$, it just remains to prove the solution of the Beltrami equation in Ω is a homeomorphism. For this, it is convenient to obtain regularity properties of solutions of Beltrami equations, what, in addition to allowing to obtain solutions that are homeomorphisms, is also useful to obtain more regularity.

Observe that the solution of the Beltrami equation established in the last previous result is a C^∞ diffeomorphism if the Beltrami coefficient $\kappa \in C_0^\infty$ has Fourier transform with L^1 norm small enough.

(2.3) *If, under the hypothesis of* (2.1), $\kappa \in \mathrm{C}_0^\infty$, *$f$ is the solution of the Beltrami equation in* (2.1) *and additionally:*

1. *If* $\|\widehat{\kappa}\|_{L^1} < 1$, *then* $f \in C^\infty$.
2. *If* $\|\widehat{\kappa}\|_{L^1} < \frac{1}{3}$, *then f is a C^∞ diffeomorphism from $\mathbb{C}$ onto $\mathbb{C}$.*

Proof

(1) By the proof of (2.1), the Beltrami equation for f with coefficient κ is equivalent to $h = \kappa[1 + \mathcal{L}(h)]$, where $h = f_{\bar{z}}$, and, therefore, to

$$\widehat{h}(\zeta) = \widehat{\kappa}(\zeta) + \widehat{\kappa} * \big(\frac{\bar{\zeta}}{\zeta}\,\widehat{h}(\zeta)\big).$$

Since $\kappa \in \mathscr{S}$, we have $\widehat{\kappa} \in \mathscr{S}$. It is proved below with this property that $\widehat{h} \in \mathscr{S}$, implying $h \in \mathcal{S}$. Consequently, h is C^∞.

Suppose $\widehat{h} \notin \mathscr{S}(\mathbb{C})$. The function $g : [0, +\infty[\to [0, +\infty[$ such that $g(t) = \sup_{|\zeta|=t} |\widehat{h}(\zeta)|$ tends to 0 when $t \to +\infty$, but it does not belong to $\mathscr{S}(\mathbb{R})$. Thus, there is $k \in \mathbb{N}$ such that $t^k g(t) \to +\infty$ when $t \to +\infty$. The set $A_m = \{t \geq 0 : t^k g(t) \geq m\} \neq \emptyset$ is closed since it is the preimage of a closed set by a continuous function with domain a closed subset of $\mathbb{R}$, and it is bounded because g is a continuous function with $g(t) \to 0$ when $t \to +\infty$; therefore, A_m is a compact subset of $[0, +\infty[$. By the Weierstrass theorem of extrema of continuous functions in compact sets, g assumes a maximum in A_m at some $t_m \in A_m$, and

$$g(t) \leq g(t_m)\,, \text{ for } t \in A_m\,, \qquad t^k g(t) \leq m \leq (t_m)^k g(t_m)\,, \text{ for } t \notin A_m\,,$$

$$1 \leq \sup_{|t-t_m| \leq \sqrt{t_m}} \frac{g(t)}{g(t_m)} \leq \Big(\frac{t_m}{t_m - \sqrt{t_m}}\Big)^k .$$

With ζ_m such that $|\zeta_m| = t_m$ and $\widehat{\kappa_m} - \widehat{\kappa}\,\chi_{B_{r_m}}$, with χ_B the characteristic function of the set B (equal to 1 in B and 0 outside B), $B_{r_m} \subset \mathbb{C}$ the open disk with center 0 and radius $r_m = \sqrt{t_m}$, and $\|\widehat{\kappa}\|_{L^1} \leq C$,

$$\left|\widehat{\kappa_m} * \left(\tfrac{\overline{\zeta_m}}{\zeta_m}\,\widehat{h}(\zeta_m)\right)\right| \le C \sup_{B_{r_m}} |\widehat{h}| = C\big|\widehat{h}(\zeta_m)\big|\big(1+o(1)\big)\,, \qquad m \to +\infty\,.$$

Since $\kappa \in \mathscr{S}$, also[12] $\widehat{\kappa} \in \mathscr{S}$, and

$$\|\widehat{\kappa} - \widehat{\kappa_m}\|_{L^1} = O\big(\tfrac{1}{(t_m)^j}\big)\,, \quad \widehat{\kappa_m} = O\big(\tfrac{1}{(t_m)^j}\big)\,, \quad m \to +\infty\,, \qquad \text{for } j \in \mathbb{N}\,;$$

thus

$$\begin{aligned}\big|\widehat{h}(\zeta_m)\big| &= \left|\widehat{\kappa_m} * \left(\tfrac{\overline{\zeta_m}}{\zeta_m}\,\widehat{h}(\zeta_m)\right) + \big(\widehat{\kappa} - \widehat{\kappa_m}\big) * \left(\tfrac{\overline{\zeta_m}}{\zeta_m}\,\widehat{h}(\zeta_m)\right) + \widehat{\kappa}(\zeta_m)\right| \\ &\le C\big|\widehat{h}(\zeta_m)\big|\big(1+o(1)\big) + O\big(\tfrac{1}{(t_m)^j}\big)\,, \qquad m \to +\infty\,.\end{aligned}$$

If $\|\widehat{\kappa}\|_{L^1} \le C < 1$, by the preceding formula, $|\widehat{h}(\zeta_m)|\; O\big(\tfrac{1}{(t_m)^j}\big)$ when $m \to +\infty$, for $j \in \mathbb{N}$, contradicting the assumption $\widehat{h} \notin \mathscr{S}$. Therefore, $\widehat{h} \in \mathscr{S}$.

Since in the proof of (2.1), $h = g_{\overline{z}}$ with $g(z) = f(z) - z$, it is $\widehat{g_{\overline{z}}} \in \mathscr{S}$, and since $\widehat{g_z}(\zeta) = \frac{\overline{\zeta}}{\zeta}\,\widehat{g_{\overline{z}}}(\zeta)$ for $\zeta \in \mathbb{C}$, it is $|\widehat{g_z}| = |\widehat{g_{\overline{z}}}|$. Consequently, also $\widehat{g_z} \in \mathscr{S}$. Therefore,[12] $g_z, g_{\overline{z}} \in C^\infty$, and $g \in C^\infty$, $f = g + 1_{\mathbb{C}} \in C^\infty$.

(2) If $\|\widehat{\kappa}\|_{L^1} \le \frac{1}{3}$, by $\widehat{h}(\zeta) = \widehat{\kappa}(\zeta) + \widehat{\kappa} * \big(\frac{\overline{\zeta}}{\zeta}\,\widehat{h}(\zeta)\big)$,

$$\big\|\widehat{h}\big\|_{L^1} \le \|\widehat{\kappa}\|_{L^1} + \|\widehat{\kappa}\|_{L^1}\big\|\widehat{h}\big\|_{L^1}, \quad \|h\|_{L^\infty} \le \big\|\widehat{h}\big\|_{L^1} \le \frac{\|\widehat{\kappa}\|_{L^1}}{1-\|\widehat{\kappa}\|_{L^1}} < \frac{1/3}{2/3} = \frac{1}{2}\,.$$

As $|\widehat{g_z}| = |\widehat{g_{\overline{z}}}| = |\widehat{h}|$, it is $\|g_z\|_{L^\infty} = \|\widehat{g_z}\|_{L^1} = \|\widehat{h}\|_{L^1} < \frac{1}{2}$. Thus

$$|f_z| = |g_z + 1| > 1 - |g_z| > 1 - \tfrac{1}{2} = \tfrac{1}{2}\,, \qquad |f_{\overline{z}}| = |g_{\overline{z}}| < \tfrac{1}{2}\,.$$

By the beginning of Section 2 of the preceding chapter, if Df is the Jacobian matrix of f as a function from a subset of $\mathbb{R}^2$ to $\mathbb{R}^2$, $\det Df = |f_z| - |f_{\overline{z}}| > 0$. So, by the inverse function theorem, f is a local diffeomorphism. Since $f(z) = z + o(1)$ when $|z| \to +\infty$, preimages of bounded sets are bounded, and as f is continuous, preimages of closed sets are closed. Therefore, preimages of compact sets are compact (a function with this property is called a **proper function**). By the global inverse function theorem (F.4) in $\mathbb{R}^2$, f is a diffeomorphism of $\mathbb{C}$ onto $\mathbb{C}$. □

Other conditions of regularity of the solution of Beltrami equation in (2.1) are obtained under corresponding conditions of regularity of the Beltrami coefficient.

---

[12] See Appendix C.

(2.4) *If, in addition to the assumptions of* (2.1), $\kappa \in C^\infty$ *and* $z_0 \in \mathbb{C}$, *then there is a diffeomorphism of* $\mathbb{C}$ *onto* $\mathbb{C}$ *satisfying the Beltrami equation in some neighborhood of* z_0.

If κ *is* C^∞ *with* $|\kappa|<1$ *in a neighborhood of* $z_0 \in \mathbb{C}$, *then there is a neighborhood of* z_0 *where the restriction of the solution of the Beltrami equation in* (2.1) *is a* C^∞ *diffeomorphism onto an open subset of* $\mathbb{C}$.

Proof Assume, without loss of generality, $z_0=0$, as can be ensured by translation of the coordinates origin to z_0. If $k_0 \overset{def}{=} \kappa(0)\neq 0$, consider $T(z)=\frac{1}{1-|k_0|^2}(z-k_0\bar{z})$, which is a linear transformation $T:\mathbb{C}\to\mathbb{C}$. By the beginning of Section 2 of Chap. 1, $\det T=\frac{1-|k_0|^2}{(1-|k_0|^2)^2}=\frac{1}{1-|k_0|^2}>0$. Therefore, T is an invertible linear transformation in $\mathbb{C}$ preserving orientation and with Beltrami coefficient $\kappa_T=\frac{T_{\bar{z}}}{T_z}=-k_0$. By the formulas for Beltrami coefficients of composition of functions at the end of Section 2 of Chap. 1, the Beltrami coefficient of $f\circ T$ is

$$\kappa_{f\circ T}=\frac{\kappa_T+(\kappa_f\circ T)\,e^{-i2\operatorname{Arg}T_z}}{1+(\kappa_f\circ T)\,\kappa_T\,e^{-i2\operatorname{Arg}\overline{T_{\bar{z}}}}}=\frac{\kappa_T+(\kappa_f\circ T)\,e^{-i2\operatorname{Arg}T_z}}{1+(\kappa_f\circ T)\,\kappa_T\,e^{-i2\operatorname{Arg}\overline{T_{\bar{z}}}}}=\frac{k_0+(\kappa_f\circ T)}{1+(\kappa_f\circ T)\,\overline{\kappa_T}}\,.$$

The function $\kappa_{f\circ T}$ is C^∞, satisfies $\kappa_{f\circ T}(0)=0$, and, as $z\mapsto\frac{z-k_0}{1-\bar{k}_0 z}$, is a conformal automorphism of[13] B_1, since $|k_0|\le 1$, also $|\kappa_{f\circ T}|\le 1$. Therefore, without loss of generality, it can be assumed $\kappa_{f\circ T}(0)=0$.

Let $h\in C_0^\infty$ be 1 in a neighborhood U of 0, $h_\varepsilon(z)=h\left(\frac{z}{\varepsilon}\right)$, and $\kappa_\varepsilon=h_\varepsilon\kappa$. By (C.2),

$$\widehat{h_\varepsilon}(\zeta)=\varepsilon^2\,\widehat{h}(\varepsilon\zeta)\,,\quad \frac{\partial\widehat{h_\varepsilon}}{\partial\xi}(\xi,\eta)=\varepsilon^3\frac{\partial\widehat{h_\varepsilon}}{\partial\xi}(\varepsilon\xi,\varepsilon\eta)\,,\quad \frac{\partial\widehat{h_\varepsilon}}{\partial\eta}(\xi,\eta)=\varepsilon^3\frac{\partial\widehat{h_\varepsilon}}{\partial\eta}(\varepsilon\xi,\varepsilon\eta)\,,$$

$$\left\|\widehat{h_\varepsilon}\right\|_{L^1}=\left\|\widehat{h}\right\|_{L^1},\quad \left\|\frac{\partial\widehat{h_\varepsilon}}{\partial\xi}\right\|_{L^1}=\left\|\frac{\partial\widehat{h}}{\partial\xi}\right\|_{L^1},\quad \left\|\frac{\partial\widehat{h_\varepsilon}}{\partial\eta}\right\|_{L^1}=\left\|\frac{\partial\widehat{h_\varepsilon}}{\partial\eta}\right\|_{L^1}.$$

Since $\kappa\in C_0^\infty$, it is $\widehat{\kappa}\in\mathscr{S}$, and $\iint_{\mathbb{R}^2}\widehat{k}(\xi,\eta)\,d\xi d\eta=\kappa(0)=0$. With $\psi:\mathbb{R}\to\mathbb{R}$ C_0^∞ and $\int_{\mathbb{R}}\psi=1$, $K(\xi)=\int_{\mathbb{R}}\widehat{\kappa}(\xi,\eta)\,d\eta$ and $\widehat{\kappa_1}(\xi,\eta)=K(\xi)\,\psi(\eta)$, $\widehat{\kappa_2}=\widehat{\kappa}-\widehat{\kappa_1}$; thus, $\widehat{\kappa}=\widehat{\kappa_1}+\widehat{\kappa_2}$ and, by Fubini theorem, $\int_{\mathbb{R}}K(\xi)\,d\xi=\iint_{\mathbb{R}^2}\widehat{\kappa}(\xi,\eta)\,d\xi d\eta=0$. Consequently,

$$\int_{\mathbb{R}}\widehat{\kappa_1}(\xi,\eta)\,d\xi=0\,,\ \text{for }\eta\in\mathbb{R}\,,\qquad \int_{\mathbb{R}}\widehat{\kappa_2}(\xi,\eta)\,d\eta=0\,,\ \text{for }\xi\in\mathbb{R}\,.$$

With

$$K_1(\xi,\eta)=\int_{-\infty}^{\xi}\widehat{\kappa_1}(t,\eta)\,dt\,,\qquad K_2(\xi,\eta)=\int_{-\infty}^{\eta}\widehat{\kappa_2}(\xi,t)\,dt\,,$$

---

[13] See Section 3 in Chapter 10 of *CADOVA*.

by the fundamental theorem of calculus, $\frac{\partial K_j}{\partial \xi_j}(\xi_1, \xi_2) = \widehat{\kappa}_j(\xi_1, \xi_2)$, $(\xi_1, x_2) \in \mathbb{C}$, and $K_j \in \mathscr{S}$, for $j = 1, 2$; so,

$$\widehat{k}(\xi, \eta) = \widehat{k}_1(\xi, \eta) + \widehat{k}_2(\xi, \eta) = \frac{\partial K_1}{\partial \xi}(\xi, \eta) + \frac{\partial K_2}{\partial \eta}(\xi, \eta) .$$

Therefore,

$$\widehat{\kappa_\varepsilon} = \widehat{h_\varepsilon} * \widehat{\kappa} = \widehat{h_\varepsilon} * \left(\frac{\partial K_1}{\partial \xi} + \frac{\partial K_2}{\partial \eta}\right) = \frac{\partial \widehat{h_\varepsilon}}{\partial \xi} * K_1 + \frac{\partial \widehat{h_\varepsilon}}{\partial \eta} * K_2 ,$$

and, consequently,

$$\|\widehat{\kappa_\varepsilon}\|_{L^1} \le \left\|\frac{\partial \widehat{h_\varepsilon}}{\partial \xi}\right\|_{L^1} \|K_1\|_{L^1} + \left\|\frac{\partial \widehat{h_\varepsilon}}{\partial \eta}\right\|_{L^1} \|K_2\|_{L^1} \le \varepsilon\left(\left\|\frac{\partial \widehat{h_\varepsilon}}{\partial \xi}\right\|_{L^1} \|K_1\|_{L^1} + \left\|\frac{\partial \widehat{h_\varepsilon}}{\partial \eta}\right\|_{L^1} \|K_2\|_{L^1}\right).$$

For $\varepsilon > 0$ small, it is $\|\widehat{\kappa_\varepsilon}\|_{L^1} \le \frac{1}{3}$, and, by the preceding result, the solution of the Beltrami equation in (2.1) is a C^∞ diffeomorphism of $\mathbb{C}$ onto $\mathbb{C}$. The proof ends noting that $\kappa_\varepsilon = \kappa$ in εU. □

The local structure of solutions of the Beltrami equation with coefficient C^∞ with absolute value <1 can be obtained from the local structure of holomorphic functions.

(2.5) *If, additionally to the assumptions in* (2.1), κ *is* C^∞ *in a neighborhood* U *of* $z_0 \in \mathbb{C}$, $|\kappa| < 1$, *and* $f \in H^1_{loc}(U)$ *is a local solution of the Beltrami equation* $f_{\overline{z}} + \kappa f_z$ *in* U, *then there are a* C^∞ *diffeomorphism* g *of* $\mathbb{C}$ *onto* $\mathbb{C}$ *solution of the Beltrami equation* $g_{\overline{z}} = \kappa g_z$ *in some neighborhood* $V \subset U$ *of* z_0, $h \in |, H\big(g(V)\big)$ *such that* $f = h \circ g$, *a* $m \in \mathbb{N} \cup |, \{0\}$ *and a diffeomorphism* $\varphi \in C^\infty(V)$ *defined in* V *with* $\varphi(z_0) = 0$ *such that*

$$f \in C^\infty(V), \qquad f(z) = f(z_0) + \big(\varphi(z)\big)^m, \quad z \in V .$$

Proof By (2.4), there is a C^∞ diffeomorphism g of $\mathbb{C}$ onto $\mathbb{C}$ as a function of 2 real variables solution of the Beltrami equation $g_{\overline{z}} = \kappa g_z$ in some neighborhood $V \subset U$ of z_0. With $W = g(V)$, $w_0 = g(z_0)$, and $h = f \circ g^{-1}$, it is $h \in H^1_{loc}(W)$, $f = h \circ g$, and, by (1.27), h is holomorphic in W and $f \in C^\infty(V)$. By the local structure of holomorphic functions,[14] there are $m \in \mathbb{N} \cup \{0\}$, a neighborhood of w_0 and a holomorphic bijection ψ of this neighborhood onto a neighborhood of 0 with $\psi(w_0) = 0$ and ψ' without zeros such that $h(w) = h(w_0) + \big(\psi(w)\big)^m$ for w in this neighborhood. With $\varphi = \psi \circ g$, we get the formula in the statement, $\varphi(z_0) = 0$, and $\varphi \in C^\infty(V)$ is a diffeomorphism of V onto an open subset of $\mathbb{C}$. □

[14] See Section 4 in Chapter 6 of *CADOVA*.

(2.6) *If additionally to the hypothesis of* (2.1) f *is a solution of the Beltrami equation* (2.1) *and:*

1. *If* $\kappa \in C_0^\infty$, *then* f *is a* C^∞ *diffeomorphism of* $\mathbb{C}$ *onto* $\mathbb{C}$.
2. *If* $\kappa \in L_0^\infty$, *then* f *is a homeomorphism of* $\mathbb{C}$ *onto* $\mathbb{C}$.

Proof

(1) By (2.4), each point of $\mathbb{C}$ has some neighborhood U where the restriction of f is a C^∞ diffeomorphism as a function of two real variables onto an open subset of $\mathbb{C}$. Since $f(z)-z \to 0$ when $z \to +\infty$, f is a proper function, and, by the global inverse function theorem (F.4), $f \in C^\infty$ is a diffeomorphism of $\mathbb{C}$ onto $\mathbb{C}$.

(2) Let $R>0$ be such that $B_R \supset \operatorname{supp}\kappa$ and $\varphi \in C_0^\infty$ such that $\varphi \geq 0$ and $\iint_{\mathbb{R}^2} \varphi = 1$. For each $\varepsilon > 0$, define $\varphi_\varepsilon(z) = \frac{1}{\varepsilon^2}\varphi\left(\frac{z}{\varepsilon}\right)$, which satisfies $\iint_{\mathbb{R}^2} \varphi_\varepsilon = 1$. Consider a sequence $\{\varepsilon_n\} \subset]0,+\infty[$ converging to 0 and the functions $\kappa_n = \varphi_{\varepsilon_n} * \kappa$, which satisfy $\kappa_n \in C_0^\infty$, $\operatorname{supp}\kappa_n \subset B_R$ for $n \in \mathbb{N}$ large, $\|\kappa_n\|_{L^\infty} \leq \|\kappa\|_{L^\infty}$, and $\kappa_n \to \kappa$ in L^2 when $n \to +\infty$. By 1, for large $n \in \mathbb{N}$, the solution f_n of the Beltrami equation (2.1) is a C^∞ diffeomorphism of $\mathbb{C}$ onto $\mathbb{C}$ as a function of two real variables.

Let $r>0$ be arbitrary. For $z_1 \in \operatorname{cl} B_{R+r}$, define the functions $g_{z_1,n} : B_{1/r} \to \mathbb{C}$ by $g_{z_1,n}(s) = \left[f_n\left(\frac{1}{s}+z_1\right) - f_n(z_1)\right]^{-1}$. Each function $g_{z_1,n}$ is K-quasiconformal in $B_{1/r}$ and holomorphic in $B_{1/(2R+r)}$. By (1.19), if $|s| \geq r^{-1}$, then $|g_{z_1,n}(s)| \geq \frac{1}{4} r^{-\frac{1}{K}} (2R+r)^{-1-\frac{1}{K}}$. With $t = \frac{1}{s}$, if $|t| \leq r$, then

$$|f_n(t+z_1) - f_n(z_1)| \leq 4 r^{\frac{1}{K}} (2R+r)^{1-\frac{1}{K}}.$$

Therefore, the function $z \mapsto f_n(t+z) - f_n(z)$ is holomorphic in $\mathbb{C}\backslash B_{R+r}$, with absolute value upper bounded by $4 r^{\frac{1}{K}} (2R+r)^{1-\frac{1}{K}}$ in $\operatorname{cl} B_{R+r}$ and converges to 0 when $|z| \to +\infty$. By the maximum modulus principle[15] for holomorphic functions, $|f(z_2) - f(z_1)| \leq 4 r^{\frac{1}{K}} (2R+r)^{1-\frac{1}{K}}$ for $z_1, z_2 \in \mathbb{C}$ such that $|z_2 - z_1| \leq r$. By the Koebe one quarter theorem,[16] $f_n(B_R) \subset B_{2R}$. Since f_n is univalent[16] in $\mathbb{C}\backslash \operatorname{cl} B_R$, $|f_n(z) - z| \leq 3R$ for $z \in \operatorname{cl} B_R$. The function $z \mapsto f_n(z) - z$ is holomorphic in $\mathbb{C}\backslash B_R$ and converges to 0 when $|z| \to +\infty$. As it is upper bounded by $3R$ on the disk with radius R and center 0, it is also upper bounded by $3R$ in $\mathbb{C}\backslash B_R$. It follows that $\{f_n\}$ is an equicontinuous family of functions. By the Arzelà-Ascoli theorem,[17] it is a normal family. Thus, for any compact subset of $\mathbb{C}$, there is a subsequence of $\{f_n\}$ converging uniformly

[15] See the last Section in Chapter 5 of *CADOVA*.

[16] See the last Section in Chapter 10 of *CADOVA*.

[17] See Section 4 in Chapter 10 of *CADOVA*.

to a function $\widetilde{f}:\mathbb{C}\to\mathbb{C}$. The inverses f_n^{-1} of the considered functions satisfy conditions similar to those of the functions f_n, and $\{f_n^{-1}\}$ is also a normal family. Therefore, there is a subsequence of the subsequence corresponding to the terms of the considered convergent subsequence of $\{f_n\}$ converging to $\widetilde{f}^{-1}$ uniformly on the compact subset of $\mathbb{C}$ considered. Consequently, $\widetilde{f}$ is a homeomorphism of $\mathbb{C}$ onto $\mathbb{C}$. It remains to prove $f=\widetilde{f}$. If $h_n(z)=f_n(z)-z$, then

$$\|(h_n)_z\|_{L^2} = \|(h_n)_{\overline{z}}\|_{L^2} \leq \frac{\|\kappa_n\|_{L^2}}{1-\|\kappa\|_{L^\infty}} .$$

As $\{(h_n)_{\overline{z}}\}$ and $\{(h_n)_z\}$ are bounded in L^2, the former has a subsequence converging weakly to a $\widetilde{h}_1\in L^2$ and the subsequence with the same terms as $\{(h_n)_z\}$ has a weakly convergent subsequence[18] to $\widetilde{h}_2\in L^2$; denote $\widetilde{h}_0=\lim h_n$. It remains to prove that *(1)* $\widetilde{h}_1=(\widetilde{h}_0)_{\overline{z}}$, *(2)* $\widetilde{h}_2=(\widetilde{h}_0)_z$ in the sense of distributions, and *(3)* $\widetilde{h}_1=\kappa\,\widetilde{h}_2+\kappa$. *(1)* is immediate from

$$\begin{aligned}\iint_{\mathbb{R}^2}\widetilde{h}_1\psi &= \lim\iint_{\mathbb{R}^2}(h_n)_{\overline{z}}\,\psi = -\lim\iint_{\mathbb{R}^2}h_n\psi_{\overline{z}}\\ &= -\iint_{\mathbb{R}^2}(\widetilde{h}_0)_{\overline{z}}\,\psi\,, \qquad \psi\in C_0^\infty,\end{aligned}$$

where the sequence is the subsequence of the originally considered sequence obtained above. *(2)* is obtained analogously. To get *(3)*, since

$$\kappa\,\widetilde{h}_2-\kappa_n(h_n)_z = \kappa(\widetilde{h}_2-(h_n)_z)+(\kappa-\kappa_n)(h_n)_z\,,$$

and $\kappa\big(\widetilde{h}_2-(h_n)_z\big)$ converges weakly to 0 in L^2 because

$$\iint_{\mathbb{R}^2}\psi\kappa(\widetilde{h}_2-(h_n)_z) = \iint_{\mathbb{R}^2}(\psi\kappa)(\widetilde{h}_2-(h_n)_z)\to 0\,,$$

as $\|\kappa-\kappa_n\|_{L^2}\to 0$ and $\{\|(h_n)_z\|\}$ is bounded, by Cauchy-Schwarz inequality,

$$\|(\kappa-\kappa_n)(h_n)_z\|_{L^1}\leq\|(\kappa-\kappa_n)\|_{L^2}\|(h_n)_z\|_{L^2}\to 0\,,\qquad n\to+\infty\,;$$

thus, $(h_n)_{\overline{z}}=\kappa_n(h_n)_z+\kappa_n\to\kappa(h_n)_z+\kappa\widetilde{h}_2$ in the sense of distributions. Since $\{(h_n)_{\overline{z}}\}$ converges weakly to $\widetilde{h}_1$ in L^2, it is $\widetilde{h}_1=\kappa\,\widetilde{h}_2+\kappa$, also in the sense of distributions. □

[18] A sequence $\{u_n\}$ in a Hilbert space X is said to **converge weakly** to $u\in X$ if $\langle v,u_n\rangle\to\langle v,u\rangle$ for all $v\in X$.

2.3 Existence and Regularity of Solution of Beltrami Equation in $\Omega \subset \mathbb{C}$

In this section, the results obtained in the previous section are applied to solutions of the Beltrami equation in $\mathbb{C}$ to obtain results for the Beltrami equation in open subsets $\Omega \subset \mathbb{C}$.

(2.7) *If, in addition to the hypothesis of* (2.1), $\kappa \in L_0^\infty$, *f is the solution of the Beltrami equation in* (2.1), *and $\Omega \subset \mathbb{C}$ is open, then $g \in H_{loc}^1(\Omega)$ is a solution in Ω of the same Beltrami equation if and only if there is a function $h \in H\big(f(\Omega)\big)$ such that $g = h \circ f$ in Ω.*

Proof If $\kappa \in C_0^\infty$, this is a simple consequence of the preceding section. By the last result in that section, if f is the solution of (2.1), $f \in C^\infty$ is a diffeomorphism of $\mathbb{C}$ onto $\mathbb{C}$. Therefore, if $g \in H_{loc}^1(\Omega)$ is a solution of the same equation in an open set $\Omega \subset \mathbb{C}$, by (1.27), there is $h \in H\big(f(\Omega)\big)$ such that $g = h \circ f$. Conversely, since $g \in H_{loc}^1(\Omega)$, if $h \in H\big(f(\Omega)\big)$, then $h \in C^\infty\big(f(\Omega)\big)$ and, by (B.2), the derivatives of $g = h \circ f$ in the sense of distributions satisfy the chain rule; as in the beginning of the proof of (1.27), we get that g is a solution in Ω of the same equation.

If $\kappa \in L_0^\infty \backslash C_0^\infty$, the result follows considering sequences of Beltrami equations approximating the given equation by Beltrami equations with coefficients $\kappa_n \in C_0^\infty$ as defined in the beginning of the proof of (2) in the last result of the preceding section and the corresponding sequence of solutions $\{f_n\}$, which are diffeomorphisms of $\mathbb{C}$ onto $\mathbb{C}$. The argument proceeds differently for necessity and sufficiency of the condition in the statement.

Necessity: We saw that $g_n = h \circ f_n \in H_{loc}^1(\Omega)$ is a solution in Ω of the Beltrami equation with coefficient κ_n. We have $\kappa_n \to \kappa$ in every bounded open set V with $\overline{V} \subset \Omega$. For n large $f_n(V) \subset f(\Omega)$, $(f_n)_z \to f_z$ in $L^2(V)$ and $h' \circ f_n \to h' \circ f$ uniformly in V. So, $g = h \circ f \in H_{loc}^1(\Omega)$ and $(h \circ f_n)_z = \big(h' \circ (f_n)\big)(f_n)_z \to (h' \circ f) f_z$. Similarly, $(h \circ f_n)_{\overline{z}} \to (h' \circ f) f_{\overline{z}}$. Since $(h \circ f_n)_{\overline{z}} = \kappa_n (h \circ f_n)_z$, letting $n \to +\infty$, we get $(h \circ f)_{\overline{z}} = \kappa (h \circ f)_z$. Thus, $g = h \circ f$ is a solution in Ω of the Beltrami equation in (2.1).

Sufficiency: Let $g \in H_{loc}^1(\Omega)$ be a solution of the Beltrami equation in (2.1) in an open set $\Omega \subset \mathbb{C}$. If $\varphi \in H_{loc}^1(\Omega)$ is a solution of the Beltrami equation in (2.1) in Ω, define in $f(\Omega)$ the function $h = \varphi \circ f^{-1}$. Let V be a bounded open set with $\mathrm{cl}\, V \subset \Omega$. We would like to have for n large $f(V) \subset f_n(\Omega)$ with $h_n = \varphi \circ f_n^{-1}$ in $f(V)$. It is convenient in this case to consider the linear transformations $(h_n)_{\overline{z}}\, d\overline{z}$, with[19]

[19] It is not necessary for this proof, but a reader familiar with differential forms can recognize that $f_n^*\big((h_n)_{\overline{z}}\, d\overline{z}\big)$ is the value of the *pullback* f_n^* of the 1-form f_n at the 1-form $(h_n)_{\overline{z}}\, d\overline{z}$.

$$f_n^*\big((h_n)_{\overline{z}}\, d\overline{z}\big) = \big((h_n)_{\overline{z}} \circ f_n\big)\big[(\overline{f_n})_z\, dz + (\overline{f_n})_{\overline{z}}\, d\overline{z}\,\big].$$

By (B.2), $(h_n)_{\overline{z}} = (\varphi_z \circ f_n^{-1})(f_n^{-1})_{\overline{z}} + (\varphi_{\overline{z}} \circ f_n^{-1})(\overline{f_n^{-1}})_{\overline{z}}$. Since f_n is a diffeomorphism and $f_n^{-1} \circ f_n = 1_{\mathbb{C}}$, with the chain rule,

$$\big[(f_n^{-1})_z \circ f_n\big](f_n)_z + \big[(f_n^{-1})_{\overline{z}} \circ f_n\big]\big(\overline{f_n}\big)_z = 1\,,$$
$$\big[(f_n^{-1})_z \circ f_n\big](f_n)_{\overline{z}} + \big[(f_n^{-1})_{\overline{z}} \circ f_n\big]\big(\overline{f_n}\big)_{\overline{z}} = 0\,.$$

By Cramer[20] rule,

$$(f_n^{-1})_z \circ f_n = \frac{1}{(f_n)_z\,(1-|k_n|^2)}\,, \qquad (f_n^{-1})_{\overline{z}} \circ f_n = -\frac{k_n}{\overline{(f_n)_z}\,(1-|k_n|^2)}\,.$$

Hence,

$$\begin{aligned} f_n^*\big((h_n)_{\overline{z}}\, d\overline{z}\big) &= \big[\varphi_z (f_n^{-1})_{\overline{z}} \circ f_n + \varphi_{\overline{z}}\,\overline{(f_n^{-1})_z} \circ f_n\big]\,\big[\overline{(f_n)_{\overline{z}}}\, dz + \overline{(f_n)_z}\, d\overline{z}\,\big] \\ &= (-k_n\varphi_z + \varphi_{\overline{z}})\,\frac{\overline{k_n} dz + d\overline{z}}{\overline{(f_n)_z}\,(1-|k_n|^2)}\,\overline{(f_n)_z} = (\varphi_{\overline{z}} - \kappa_n \varphi_z)\,\frac{d\overline{z} + \overline{\kappa_n} dz}{1-|\kappa_n|^2} \\ &= (\kappa\varphi_z - \kappa_n\varphi_z)\,\frac{d\overline{z} + \overline{\kappa_n} dz}{1-|\kappa_n|^2} = (\kappa - \kappa_n)\,\varphi_z\,\frac{d\overline{z} + \overline{\kappa_n} dz}{1-|\kappa_n|^2}\,. \end{aligned}$$

By the beginning of the proof of *(2)* in the last result of the preceding section, for large $n \in \mathbb{N}$, $\|\kappa_n\|_{L^\infty} \le \|\kappa\|_{L^\infty}$ and $\|\kappa - \kappa_n\|_{L^2} \to 0$ for all $n \in \mathbb{N}$. Furthermore, $\varphi_z \in L^2$ and, with $k = \|\kappa\|_{L^\infty}$, it is $\big\|\frac{d\overline{z} + \overline{\kappa_n} dz}{1-|\kappa_n|^2}\big\|_{L^\infty} \le \frac{1+k}{1-k^2} = \frac{1}{1-k}$ for $n \in \mathbb{N}$. So, $u_n \overset{def}{=} (\kappa - \kappa_n)\frac{d\overline{z} + \kappa_n dz}{1-|\kappa_n|^2}$ satisfies $\|u_n\|_{L^\infty} \le \frac{2k}{1-k^2}$ for all $n \in \mathbb{N}$ and $\|u_n\|_{L^2} \to 0$. Since C_0^∞ is dense in L^2, for every $\varepsilon > 0$, there is $\psi \in C_0^\infty$ such that $\|\varphi_z - \psi\|_{L^2} < \varepsilon$, and $\|u_n(\varphi_z - \psi)\|_{L^2} < \frac{2k}{1-k^2}\,\varepsilon$. Since $\|u_n\psi\|_{L^2} \to 0$, for $n \in \mathbb{N}$ large $\|u_n\psi\|_{L^2} \le \varepsilon$. By the triangular inequality, $\|u_n\varphi_z\|_{L^2} \le \big(\frac{2k}{1-k^2} + 1\big)\varepsilon$. Thus, $\|f_n^*\big((h_n)_{\overline{z}}\ d\overline{z}\big)\|_{L^2} \to 0$. By the preceding result, f_n is a diffeomorphism of $\mathbb{C}$ onto $\mathbb{C}$. As it is a solution of the Beltrami equation with coefficient κ_n, it is a K-quasiconformal mapping for some $K \ge 1$ and

$$\tfrac{1}{K}\|f_n^*\big((h_n)_{\overline{z}}\, d\overline{z}\big)\|_{L^2} \le \|(h_n)_{\overline{z}}\, d\overline{z}\|_{L^2} \le K\|f_n^*\big((h_n)_{\overline{z}}\, d\overline{z}\big)\|_{L^2}\,.$$

So, $\big\|(h_n)_{\overline{z}}\, d\overline{z}\big\| \to 0$. Since $h_n \to h$ uniformly in $f(V)$, it is $h_{\overline{z}}\, d\overline{z} = 0$ in $f(V)$ in the sense of distributions. By Weyl lemma (1.26), $h \in H\big(f(V)\big)$. As V is an arbitrary open bounded set with $\operatorname{cl} V \subset \Omega$, we get $h \in H\big(f(\Omega)\big)$ such that $g = h \circ f$. □

(2.8) *If $\Omega \subset \mathbb{C}$ is an open set and $\kappa \in L^\infty$ with $\|\kappa\|_{L^\infty} < 1$, then there is a quasiconformal mapping f in Ω with Beltrami coefficient κ* a.e. *in Ω,*

(continued)

[20] Cramer, Gabriel (1704–1752).

(2.8) (continued)
unique up to left composition by conformal homeomorphisms, i.e., there is a solution $f \in H^1_{loc}(\Omega)$ *of the Beltrami equation* $f_{\bar{z}} = \kappa f_z$ *in* Ω *which is a homeomorphism of* Ω *onto an open set* $U \subset \mathbb{C}$*, and if* $g \in H^1_{loc}(\Omega)$ *is a solution of the same Beltrami equation, then there is a conformal homeomorphism* h *defined in* U *such that* $g = h \circ f$.

Proof Extend κ to $\mathbb{C}$ with value 0 in $\mathbb{C} \setminus \Omega$. By the preceding result, the restriction of the solution f of the Beltrami equation established in (2.1) to Ω satisfies the stated conditions. □

As the equivalence classes by composition with conformal or quasiconformal mappings coincide, the following results hold.

(2.9) Measurable Riemann mapping theorem:
If $\Omega, \Omega' \subset \mathbb{C}$ *are conformal simply connected regions and* $\kappa : \Omega \to \mathbb{C}$ *is measurable with* $\|\kappa\|_{L^\infty} < 1$*, then there is a quasiconformal mapping* f *of* Ω *onto* Ω' *with Beltrami coefficient* κ a.e. *in* Ω*, unique up to left composition by a conformal homeomorphism of* Ω' *onto* Ω'.

Proof It is immediate by the preceding result and, in case $\Omega \subsetneq \mathbb{C}$, by the Riemann mapping theorem.[21] □

(2.10) *Every quasiconformal mapping is a finite composition of quasiconformal mappings with dilatations arbitrarily close to* 1.

Proof By the preceding results, we can assume, without loss of generality, that f_κ is a K-quasiconformal mapping in $\mathbb{C}$ solution of the Beltrami equation with Beltrami coefficient κ. For each $z \in \mathbb{C}$, we divide the geodesic from 0 to $\kappa(z)$ in the spherical metric in n equal length segments in this metric. The subdivision points are denoted $\kappa_j(z)$, $j = 1, \ldots, n+1$, with $\kappa_1(z) = 0$, $\kappa_{n+1}(z) = z$, and $f_k = f_{\kappa_j(z)}$. Then

$$\kappa_{f_{j+1} \circ f_j^{-1}} = \left(\frac{\kappa_{j+1} - \kappa_j}{1 - \kappa_{j+1}\overline{\kappa_j}} \frac{(f_j)_z}{\overline{(f_j)_z}} \right) \circ f_j^{-1} .$$

Every function $g_j = f_{j+1} \circ f_j^{-1}$ is $K^{\frac{1}{n}}$-quasiconformal and $f_\kappa = g_n \circ \cdots \circ g_1$. $K^{\frac{1}{n}}$ can be made arbitrarily close to 1 for large n. □

[21] See Section 5 in Chapter 10 of *CADOVA*.

Chapter 3
Extension and Distortion of Quasiconformal Mapping

3.1 Introduction

An immediate consequence of the measurable Riemann mapping theorem of the previous chapter is that every K-quasiconformal mapping in a region can be extended to a region enlarged by a set of measure zero.

The next simplest case of extension of quasiconformal mappings is the reflection or symmetry principle, which gives extensions preserving dilatation. The extension to the entire complex plane of K-quasiconformal mappings in compact subsets of regions of the complex plane is also established, without determining the optimal dilatation, but obtaining that it depends only on K, the compact set and the region considered.

In 1956, L. Ahlfors and A. Beurling obtained that a necessary and sufficient condition for a real function of a real variable to have a quasiconformal extension in the complex plane is that it be quasisymmetric,[1] i.e., strictly increasing and with quotients of slopes of secants to the left and right of each point in equal length intervals uniformly upper and lower bounded by positive constants.

In 1962, J. Väisälä proved the existence of a quasiconformal mapping g in a (simply connected) region Ω if there is a K-quasiconformal mapping in the intersection of Ω with a region D (possibly multiply connected) having in common with Ω a connected component of the resp. boundaries (with an upper bound of the dilatation of g only dependent on K, D, Ω).

This chapter includes basic properties of quasiconformal Jordan curves or arcs and extensions of quasiconformal mappings defined in regions bounded by quasiconformal Jordan curves and the characterization of quasiconformal Jordan curves obtained in 1962 by M. Tienari[2] as curves separating regions conformal to

[1] Name given in 1963 by John Kelingos (1936–1996).

[2] Tienari, Martti (1935–2013).

L. T. Magalhães, *Quasiconformal Mappings in the Plane and Complex Dynamics*,
https://doi.org/10.1007/978-3-031-80115-0_3

the upper and lower complex half planes with extensions to the real axis that are quasisymmetric functions, as well as the characterization in terms of the notion of Jordan curve of bounded bending obtained by L. Ahlfors in 1963, and other properties related to this notion.

This chapter also includes the important sewing or conformal welding theorem at a Jordan curve, establishing conditions, in terms of quasisymmetric functions on the common boundary of two regions bounded by a Jordan curve, where conformal homeomorphisms can be defined on, resp., the complex upper and lower half plane whose extensions by homeomorphisms to the closures of the regions coincide on the common boundary. This was proved in 1960 by O. Lehto and K. Virtanen[3] and in 1961 by A. Pfluger, who used the measurable Riemann mapping theorem as in the proof here.

Distortion properties by quasiconformal mappings are also considered, including bounded distortion relative to 0 for quasiconformal mappings in circles with fixed center, obtained in 1952 by J. Hersch[4] and A. Pfluger, in terms of the Grötzsch ring module, as well as the Mori theorem establishing that the absolute value of the increases of a K-quasiconformal mapping is upper bounded by 16 times the power $\frac{1}{K}$ of the absolute value of the corresponding upper bound of the independent variable, and this coefficient is optimal as it is the smallest possible for all $K > 0$. An immediate consequence is that such mappings are continuous with Hölder exponent $\frac{1}{K}$ and the family of these mappings is equicontinuous and normal, but there are some more general conditions that guarantee these properties. In particular, it is proved that for families of K-conformal homeomorphisms in a region equicontinuity, Hölder continuity, and normality are equivalent.

In one of the sections, we consider properties of limits of convergent sequences of K-quasiconformal mappings in a region, including that they are, alternatively, constant functions, with only 2 values or K-quasiconformal mappings, and in the latter case the convergence is uniform in compact subsets of the region, as well as the notion of nucleus of a sequence of sets of complex numbers, obtaining the resp. properties for the three types of limits of sequences of K-quasiconformal mappings.

One of the final sections is on the characterization of K-quasiconformal mappings by circular dilatation, as established by F. Gehring in 1960, after being formulated by Z. Yûjôbô in 1955, supplemented with results by A. Mori in 1957 and O. Lehto, K. Virtanen, and J. Väisälä in 1959. Since circular dilatation only involves the notion of distance, it can be used to define quasiconformal mappings in metric spaces.

The chapter ends with a section proving that monotonic functions in the plane are quasiconformal, contributing to another perspective of this type of functions, and a section establishing the Leau-Fatou conjugacy[5] in a neighborhood of a parabolic fixed point of the dynamics defined by iteration of a holomorphic function by application of the measurable Riemann mapping theorem of Chap. 2 and the

---

[3] Virtanen, Kaarlo (1921–2006).

[4] Hersch, Joseph (1925–2012).

[5] Leau, L'eopold (1868–1943). Fatou, Pierre (1878–1929).

Mori theorem mentioned in this introduction, which accounts for a first example of the usefulness of the flexibility provided by quasiconformal mappings and an anticipation of quasiconformal surgery, the central subject of Chap. 6.

3.2 Extension to a Domain Boundary

(3.1) Extension from a region to its union with a set of zero measure: *If $S \subset \mathbb{C}$ has (Lebesgue) measure zero and $\Omega \subset \mathbb{C}$, $\Omega \cup S$ are regions, then every K-quasiconformal mapping f in Ω has a unique extension (by continuity) to a K-quasiconformal mapping in $\Omega \cup S$.*

If S is a regular curve, f has such extension. If $z_0 \in \partial\Omega$ is an isolated point of $\mathbb{C} \setminus \Omega$ and f is a K-quasiconformal mapping in Ω but not in $\Omega \cup S$, we say that f has a ***removable singularity*** *at z_0.*

Proof Let $\kappa = \frac{f_{\bar{z}}}{f_z}$. By the measurable Riemann mapping theorem, there is a unique quasiconformal mapping g solution of the Beltrami equation $g_{\bar{z}} = \kappa g_z$ a.e. in $\Omega \cup S$ with $g_{|\Omega} = f$. The function g is the extension by continuity from f to $\Omega \cup S$. If f is a K-quasiconformal mapping, so is g. The rest is immediate since regular curves have measure zero. □

A consequence is that K-quasiconformity persists under removal of a regular curve from a region in $\mathbb{C}$. Another consequence follows.

(3.2) *The classes of quasiconformal simply connected regions of $\mathbb{C}$ coincide with those of conformal simply connected regions of $\mathbb{C}$.*

Proof By the Riemann mapping theorem, in the set of simply connected regions of $\mathbb{C}$, there are two equivalence classes of conformal homeomorphisms: one with only $\mathbb{C}$ and the set of all other simply connected regions in $\mathbb{C}$. As conformal homeomorphisms are quasiconformal, it remains to prove that a simply connected region $\Omega \subsetneq \mathbb{C}$ is not quasiconformal to $\mathbb{C}$. If it were, as, for example, $\mathbb{C} \setminus [0, +\infty[$ is simply connected, it would be quasiconformal to $\mathbb{C}$, *i.e.*, there would exist a quasiconformal mapping f from $\mathbb{C} \setminus [0, +\infty[$ onto $\mathbb{C}$ and, by the preceding result, f would have extension by continuity to a quasiconformal mapping from $\mathbb{C}$ onto $\mathbb{C}$ with $f(0) \in \mathbb{C}$, what is impossible because such an extension would assume $f(0)$ at 2 distinct points and, therefore, it would not be a homeomorphism. □

If $k > 0$, we call ***k*-quasisymmetric function** on an interval $I \subset \mathbb{R}$ to a continuous strictly increasing function $\varphi : I \to \mathbb{R}$ such that

$$\frac{1}{k} \leq \frac{\varphi(x+t) - \varphi(x)}{\varphi(x) - \varphi(x-t)} \leq k, \qquad x,\ x-t,\ x+t \in I,\ t > 0,$$

and **quasisymmetric function** if it is k-quasisymmetric for some $k > 0$.

(3.3) Beurling-Ahlfors extension:
There is a quasiconformal mapping f from $\mathbb{H}$ onto $\mathbb{H}$ with values given at its boundary by $\varphi:\mathbb{R}\to\mathbb{R}$ if and only if φ is quasisymmetric in $\mathbb{R}$. If yes and f is K-quasiconformal with fixed point ∞, then φ is k-quasisymmetric with $k=1/[M_G^{-1}(\frac{K}{4})]^2$, where $M_G(r)$ is the Grötzsche ring module $G_r=B_1\backslash\{x+i0:0\le x\le r\}$ and f has a K-quasiconformal extension to $\mathbb{C}$ with $f(\overline{z})=\overline{f(z)}$ and dilatation upper bounded by $8k(1+k^2)$.

Proof Let f be a K-quasiconformal mapping from $\mathbb{H}$ onto $\mathbb{H}$. Without loss of generality (up to a conformal homeomorphism), f keeps ∞ fixed. Let $\varphi:\mathbb{R}\to\mathbb{R}$ be the function that gives the values of the continuous extension of f to the boundary of $\mathbb{H}$. One quadrilateral $Q=Q(z_1,z_2,z_3,\infty)$, $z_1,z_2,z_3\in\mathbb{R}$ and $z_1<z_2<z_3$, is conformal to a square if and only if $z_3-z_2=z_2-z_1\overset{def}{=}t$. The module of the quadrilateral $f(Q)=Q'=Q'(w_1,w_2,w_3,\infty)$ with $w_1,w_2,w_3\in\mathbb{R}$ and $w_1<w_2<w_3$ is, by (1.12), $M(Q')=4M_G\left(\sqrt{\frac{w_3-w_2}{w_3-w_1}}\right)$. Since $\frac{1}{K}\le M(Q')\le K$, with $x=z_2$,

$$M_G^{-1}\left(\tfrac{K}{4}\right)\le\sqrt{\frac{\varphi(x+t)-\varphi(x)}{\varphi(x+t)-\varphi(x-t)}}\le M_G^{-1}\left(\tfrac{1}{4K}\right).$$

Defining $\lambda(K)=\frac{1}{[M_G^{-1}(\frac{K}{4})]^2}-1$, it is $M_G\left(\frac{1}{\sqrt{1+\lambda(K)}}\right)=\frac{K}{4}$ and

$$M_G\left(\frac{1}{\sqrt{1+\lambda(K)}}\right)M_G\left(\frac{1}{\sqrt{1+\lambda(1/K)}}\right)=\frac{1}{16}.$$

Replacing in this equation $\lambda\left(\frac{1}{K}\right)$ by $\frac{1}{\lambda(K)}$, by (1.25.3), the equation remains valid; so, $\lambda\left(\frac{1}{K}\right)=\frac{1}{\lambda(K)}$. As $M_G^{-1}\left(\frac{K}{4}\right)=\frac{1}{\sqrt{1+\lambda(K)}}$, the inequalities above are equivalent to

$$\frac{1}{1+\lambda(K)}\le\frac{\varphi(x+t)-\varphi(x)}{\varphi(x+t)-\varphi(x-t)}\le\frac{1}{1+1/\lambda(K)}=\frac{\lambda(K)}{1+\lambda(K)},$$

and also to

$$\frac{1+\lambda(K)}{\lambda(K)}\le\frac{\varphi(x+t)-\varphi(x-t)}{\varphi(x+t)-\varphi(x)}\le 1+\lambda(K).$$

Subtracting 1 from all terms, we get the equivalent inequalities

$$\frac{1}{\lambda(K)}\le\frac{\varphi(x)-\varphi(x-t)}{\varphi(x+t)-\varphi(x)}\le\lambda(K),$$

which are equivalent to

$$\frac{1}{\lambda(K)}\le\frac{\varphi(x+t)-\varphi(x)}{\varphi(x)-\varphi(x-t)}\le\lambda(K),$$

i.e., φ is k-quasisymmetric with $k=\lambda(K)$.

Conversely, if φ is a k-quasisymmetric function, with

$$\alpha(x,y)=\int_0^1 \varphi(x+ty)\,dt\,,\qquad \beta(x,y)=\int_0^1 \varphi(x-ty)\,dt\,,\qquad x,y\in\mathbb{R}\,,$$

$$f(x+iy)=\tfrac{1}{2}\,(1+i)\,[\,\alpha(x,y)-i\beta(x,y)\,]\,,\quad x,y\in\mathbb{R}\,,$$

since

$$f(x+iy)=\tfrac{1}{2}\big[\,\big(\alpha(x,y)+\beta(x,y)\big)+i\big(\alpha(x,y)-\beta(x,y)\big)\,\big]\,,$$

and φ is a strictly increasing real function of one real variable, f is a continuous function mapping $\mathbb{H}$ to $\mathbb{H}$ such that $f(x)=\varphi(x)$ for $x\in\mathbb{R}$ and $f(\overline{z})=\overline{f(z)}$.

Since f is continuous, by (F.2), to prove that f is a homeomorphism of $\mathbb{C}$ onto $f(\mathbb{C})$, it suffices to prove it is injective.

To prove that f is injective, consider $z_1=x_1+iy_1$ and $z_2=x_2+iy_2$, with $x_1,y_1,x_2,y_2\in\mathbb{R}$ and $x_1\le x_2$, such that $f(z_1)=f(z_2)$. Since $f(\overline{z})=\overline{f(z)}$, z_1,z_2 belong to the same complex half plane with the real axis as boundary and, without loss of generality, we can assume $y_1,y_2>0$. Changing the variable of integration, $\alpha(x,y)=\frac{1}{y}\int_x^{x+y}\varphi$ and $\beta(x,y)=\frac{1}{y}\int_{x-y}^{x}\varphi$ for $x,y\in\mathbb{R}$, i.e., the averages of φ in each pair of intervals $\big(\,]x_1,x_1+y_1[\,,\,]x_2,x_2+y_2[\,\big)$ and $\big(\,]x_1-y_1,x_1[\,,\,]x_2-y_2,x_2[\,\big)$ are equal and, as φ is strictly increasing, in each pair one of the intervals is a subset of the other. Therefore, $x_2+y_2\le x_1+y_1$ and $x_2-y_2\le x_1-y_1$. Adding these inequalities, we get $x_2\le x_1$; so, $x_2=x_1$ and, by the same inequalities, $y_2=y_1$; thus, $z_1=z_2$. Therefore, f is injective, and it is a homeomorphism of $\mathbb{C}$ onto $f(\mathbb{C})$.

Now, we prove that f is quasiconformal in $\mathbb{H}$. By the formulas for α and β in the preceding paragraph and the fundamental theorem of calculus for simple integrals, these functions are C^1 and

$$\frac{\partial\alpha}{\partial x}=\frac{1}{y}[\varphi(x+y)-\varphi(x)]\,,\qquad \frac{\partial\beta}{\partial x}=\frac{1}{y}[\varphi(x)-\varphi(x-y)]\,,$$

$$\frac{\partial\alpha}{\partial y}=\frac{1}{y}\big[\varphi(x+y)-\alpha(x,y)\big]\,,\qquad \frac{\partial\beta}{\partial y}=\frac{1}{y}\big[\varphi(x-y)-\beta(x,y)\big]\,.$$

Since φ is k-quasisymmetric, by the definition k-quasisymmetric function, $\frac{1}{k}\frac{\partial\beta}{\partial x}\le\frac{\partial\alpha}{\partial x}\le k\frac{\partial\beta}{\partial x}$. As φ is strictly increasing, the first three partial derivatives of α or β are >0 and the last one <0. Therefore, the Jacobian of the vector field (u,v) defined in $\mathbb{R}\times]0,+\infty[$ corresponding to the complex function $f=u+iv$, as $u=\frac{1}{2}(\alpha+\beta)$ and $v=\frac{1}{2}(\alpha-\beta)$, is $J=\frac{1}{2}\big(\frac{\partial\alpha}{\partial y}\frac{\partial\beta}{\partial x}-\frac{\partial\alpha}{\partial x}\frac{\partial\beta}{\partial y}\big)>0$. By the second section of Chap. 1, the dilatation of f is at each point $K_f=\frac{|f_z|+|f_{\overline{z}}|}{|f_z|-|f_{\overline{z}}|}$ and f is a quasiconformal mapping if K_f is upper bounded. Then

$$K_f=\frac{|f_z|+|f_{\overline{z}}|}{|f_z|-|f_{\overline{z}}|}=\frac{(|f_z|+|f_{\overline{z}}|)^2}{|f_z|^2-|f_{\overline{z}}|^2}\le\frac{(\sqrt{2}\sqrt{|f_z|^2+|f_{\overline{z}}|^2}\,)^2}{|f_z|^2-|f_{\overline{z}}|^2}=2\,\frac{|f_z|^2+|f_{\overline{z}}|^2}{|f_z|^2-|f_{\overline{z}}|^2}$$

$$=\frac{(\frac{\partial\alpha}{\partial x})^2+(\frac{\partial\alpha}{\partial y})^2+(\frac{\partial\beta}{\partial x})^2+(\frac{\partial\beta}{\partial y})^2}{\frac{\partial\alpha}{\partial y}\frac{\partial\beta}{\partial x}-\frac{\partial\alpha}{\partial x}\frac{\partial\beta}{\partial y}}=\frac{\frac{\partial\alpha}{\partial x}\frac{\partial\beta}{\partial x}}{-\frac{\partial\alpha}{\partial y}\frac{\partial\beta}{\partial y}}\;\frac{\frac{\partial\alpha}{\partial x}/\frac{\partial\beta}{\partial x}+(\frac{\partial\alpha}{\partial y})^2/(\frac{\partial\alpha}{\partial x}\frac{\partial\beta}{\partial x})+\frac{\partial\beta}{\partial x}/\frac{\partial\alpha}{\partial x}+(\frac{\partial\beta}{\partial y})^2/(\frac{\partial\alpha}{\partial x}\frac{\partial\beta}{\partial x})}{\frac{\partial\alpha}{\partial y}\frac{\partial\beta}{\partial x}-\frac{\partial\alpha}{\partial x}\frac{\partial\beta}{\partial y}}\,.$$

Since φ is k-quasisymmetric, the left inequality in the definition of k-quasi-symmetric function gives $\varphi(z)-\varphi(z-t)\le k[\varphi(z+t)-\varphi(z)]$, with $z=x+\frac{y}{2}$ and $t=\frac{y}{2}$,

$$\varphi\big(x+\tfrac{y}{2}\big)-\varphi(x)\le k\big[\varphi(x+y)-\varphi\big(x+\tfrac{y}{2}\big)\big],$$

and adding $\varphi(x+y)-\varphi\big(x+\frac{y}{2}\big)$ to both side of the inequality,

$$\varphi(x+y)-\varphi(x)\le(1+k)\big[\varphi(x+y)-\varphi\big(x+\tfrac{y}{2}\big)\big].$$

If $g:[a,b]\to\mathbb{R}$ is a decreasing continuous function with $g(b)>0$, $\int_a^b g>(b-a)g(b)$. Thus, $t\mapsto\varphi(x+y)-\varphi(x+ty)$ is strictly decreasing and vanishes at $t=1$,

$$\int_0^1[\varphi(x+y)-\varphi(x+ty)]dt>\int_0^{\frac{1}{2}}[\varphi(x+y)-\varphi(x+ty)]dt>\tfrac{1}{2}\big[\varphi(x+y)-\varphi\big(x+\tfrac{y}{2}\big)\big].$$

Similarly, $\alpha(x,y)=\int_0^1\varphi(x+ty)\,dt>\varphi(x)$ because φ is strictly increasing, and, therefore, $\frac{\partial\alpha}{\partial x}>\frac{\partial\alpha}{\partial y}$. By the last three periods,

$$\tfrac{\partial\alpha}{\partial x}(x,y)>\tfrac{\partial\alpha}{\partial y}(x,y)=\tfrac{1}{y}\int_0^1[\varphi(x+y)-\varphi(x+ty)]\,dt>\tfrac{1}{2y}\big[\varphi(x+y)-\varphi\big(x+\tfrac{y}{2}\big)\big]\ge\tfrac{1}{2(1+k)}\tfrac{\partial\alpha}{\partial x}(x,y).$$

Analogously, we get

$$\tfrac{\partial\beta}{\partial x}(x,y)>-\tfrac{\partial\beta}{\partial y}(x,y)>\tfrac{1}{2y}\big[\varphi\big(x-\tfrac{y}{2}\big)-\varphi(x-y)\big)\big]\ge\tfrac{1}{2(1+k)}\tfrac{\partial\beta}{\partial x}(x,y).$$

By $\frac{1}{k}\frac{\partial\beta}{\partial x}\le\frac{\partial\alpha}{\partial x}\le k\frac{\partial\beta}{\partial x}$, the estimatives in the last three periods, and the formula for K_f in terms of the partial derivatives of α and β, $K_f\le 8k(1+k)^2$. Hence, f is quasiconformal in the open complex upper half plane $\mathbb{H}$.

Since f is quasiconformal in $\mathbb{C}$ and, by (3.2), only $\mathbb{C}$ is quasiconformal to $\mathbb{C}$, it is $f(\mathbb{C})=\mathbb{C}$, and, as $f(\overline{z})=\overline{f(z)}$, the restriction of f to $\mathbb{H}$ maps it onto itself, and the dilatation of f in $\mathbb{C}$ is bounded by $8k(1+k)^2$. □

This proof gives the dilatation upper bound $8k(1+k)^2$, but in 1956, A. Beurling and L. Ahlfors improved it to k^2, and in 1966, T. Reed[6] obtained the upper bound $8k$ and proved that it is optimal for large k. Combining the upper bounds, we have the upper bound k^2 for $0<k\le 8$ and $8k$ for $k\ge 8$.

(3.4) *If $\Omega,\Omega'\subset\mathbb{C}$ are regions bounded by Jordan curves, φ is a homeomorphism of $\partial\Omega$ onto $\partial\Omega'$ preserving orientation, and $Q=Q(z_1,z_2,z_3,z_4)$ is a quadrilateral in Ω, then there is a K-quasiconformal mapping f of Ω onto Ω' such that*

$$\tfrac{1}{K}M(Q)\le M\big(f(Q)\big)\le KM(Q).$$

(continued)

[6] Reed, Terence (1935–2021).

(3.4) (continued)
Conversely, if $\frac{1}{K} \le M(f(Q)) \le K$ *for quadrilaterals* $Q \subset \Omega$ *with* $M(Q) = 1$ *and vertices* z_4 *fixed, there is a K-quasiconformal mapping* f *of* Ω *onto* Ω' *with extension by continuity to* $\mathrm{cl}\,\Omega$ *satisfying the boundary condition with dilatation upper bounded by a constant depending on* K.

Proof It is a consequence of the previous result due to the invariance of the module with conformal homeomorphisms and the Carathéodory theorem of boundary correspondence,[7] as it can be assumed without loss of generality that Ω and Ω' are complex half planes and $z_4 = \infty$. □

For multiply connected regions, we just deal with existence of quasiconformal mappings with values specified in a connected component of the boundary, for what we have the following result obtained in 1962 by J. Väisälä. It guarantees the existence of a quasiconformal mapping g in a (simply connected) region Ω if there is a K-quasiconformal mapping in the intersection of Ω with a region D (possibly multiply connected), and their boundaries have in common at least one connected component (with the dilatation of g with an upper bound that only depends on K and D).

A **free arc of region boundary** $\Omega \subset \mathbb{C}$ is an arc $C \subset \partial\Omega$ of a Jordan curve with each of its points with a neighborhood union of two connected components, one subset of Ω and the other of $\mathbb{C} \setminus \Omega$. A **free curve of region boundary** $\Omega \subset \mathbb{C}$ is a Jordan curve that is a free arc of the boundary.

(3.5) *If* C, C' *are free curves of regions boundaries, resp.,* $D, D' \subset \mathbb{C}$, $f{:}D \to D'$ *is a K-quasiconformal mapping that can be extended to a homeomorphism of* $D \cup C$ *onto* $D' \cup C'$ *and* Ω, Ω' *are the connected components of, resp.,* $\mathbb{C} \setminus C$, $\mathbb{C} \setminus C'$ *which contain, resp.,* D, D', *then there is a quasiconformal mapping from* Ω *onto* Ω' *with restriction to* C *equal to the restriction of* f *to* C *with an upper bound of the dilatation only depending on* K *and* D.

Proof By the Riemann mapping theorem,[8] we can consider without loss of generality $\Omega = B_1$. Since C is a free curve of the boundary of D, there is $0 \le r < 1$ such that $D \subset B_1 \backslash B_r(0)$ (in the degenerate case with $B_0(0) = \emptyset$). Let $z_1, z_2, z_3, z_4 \in C$ be such that the quadrilateral $Q_1 = Q_1(z_1, z_2, z_3, z_4)$ has Module 1 and z_1, z_2 are the ends of one of the longer sides of Q. Consider in the annulus $B_1 \backslash B_r$ the radial straight line segments passing through z_1 and z_2. These line segments separate the interior of the annulus in two regions with the boundary of one of them containing the points z_3 and z_4. Let Q_2 be the quadrilateral in the boundary

[7] See Section 5 of Chapter 10 of *CADOVA*.
[8] See Section 5 in Chapter 10 of *CADOVA*.

of this region and with the same vertices. By the monotony of the quadrilateral module (1.6), $M\big(f(Q_1)\big) \le M\big(f(Q_2)\big)$ and since f is a K-quasiconformal mapping, $M\big(f(Q_1)\big) < KM\big(Q_2\big)$. To upper bound $M\big(Q_2\big)$, we use the formula in terms of area and extremal length (1.4), $M(Q) = \inf_\rho \frac{A_\rho(Q)}{[L_\rho(\Gamma)]^2}$. Let $s_a(Q_j)$, $s_b(Q_j)$ be the distances of the sides, resp., a and b of Q_j, $j = 1, 2$. Since z_1z_2 is one of the largest sides of Q_1, $s_b(Q_1)$ is the length of the side z_3z_4 and

$$s_b(Q_2) < \tfrac{\pi}{2}\, s_b(Q_1)\,, \qquad s_a(Q_2) \ge \min\left\{1-r\,,\, \tfrac{1}{2}\, s_a(Q_1)\right\} > \tfrac{1-r}{2}\, s_a(Q_1)\,,$$

we have $\frac{s_a(Q_2)}{s_b(Q_2)} > \frac{1-r}{\pi}\,\frac{s_a(Q_1)}{s_b(Q_1)}$ and

$$M(Q_2) < \pi\,\frac{1+2\log(1+(2/\pi)(1-r)\,s_a(Q_1)/s_b(Q_1))}{[\,\log(1+(2/\pi)(1-r)\,s_a(Q_1))/s_b(Q_1)\,]^2}\,.$$

Switching the sides a and b,

$$1 = \frac{1}{M(Q_1)} \le \pi\,\frac{1+2\log(1+2\,s_b(Q_1)/s_a(Q_1))}{[\,\log(1+2\,s_b(Q_1)/s_a(Q_1)\,]^2}\,.$$

Thus,

$$\frac{s_a(Q_1)}{s_b(Q_1)} \ge \frac{2}{e^{\pi+\sqrt{\pi^2+\pi}}-1} > 0.006\,.$$

With this estimate in the preceding inequality, we get an upper bound of $M(Q_2)$ only depending on r, and $M\big(f(Q_1)\big) < KM(Q_2)$ has an upper bound only depending on K and r. By (3.4), there is a quasiconformal mapping from Ω onto Ω' with restriction to C coincident with that of f and an upper bound of the dilatation depending only on K and D. □

3.3 Extensions by Reflection and other Extensions

A region **can be reflected at a free arc of its boundary** C if it is disjoint of the region obtained by reflecting each of its points with respect to C.

(3.6) Reflection principle *If $\Omega, \Omega' \subset \mathbb{C}$ are regions that can be reflected at an arc of a circle or a line segment, resp., C, C' in regions, resp., $\widetilde{\Omega}$, $\widetilde{\Omega}'$ and f is a K-quasiconformal mapping of Ω onto Ω' with $f(C) = C'$, then the extension from f to $\Omega \cup C \cup \widetilde{\Omega}$ with value at each point of $\widetilde{\Omega}$ equal to the value at the symmetrical point relative to C is a K-quasiconformal mapping of $\Omega \cup C \cup \widetilde{\Omega}$ onto $\Omega' \cup C' \cup \widetilde{\Omega}'$.*

Proof By (3.1), f can be extended to a K-quasiconformal mapping in $\Omega \cup C$. To extend this extension to $\widetilde{\Omega}$, denote $\widetilde{z}$ the reflection of $z \in \Omega$ relative to C and $\widetilde{f(z)}$

the reflection of $f(z)$ relative to C'. Extending f to $\Omega \cup C \cup \widetilde{\Omega}$ with $f(\widetilde{z}) = \widetilde{f(z)}$ gives a homeomorphism of $\Omega \cup C \cup \widetilde{\Omega}$ onto $\Omega' \cup C' \cup \widetilde{\Omega}'$ preserving orientation. Quadrilaterals $\widetilde{Q}(\widetilde{z_1}, \widetilde{z_2}, \widetilde{z_3}, \widetilde{z_4}) \subset \widetilde{\Omega}$ and the resp. reflections $Q(z_1, z_2, z_3, z_4) \subset \Omega$ have equal modules, as well as quadrilaterals in $\widetilde{\Omega}'$ and Ω', one a reflection of the other. The dilatation of the extension of f to $\Omega \cup \widetilde{\Omega}$ is upper bounded by K. By (3.1), the extension of f to $\Omega \cup C \cup \widetilde{\Omega}$ is a K-quasiconformal mapping. □

The reflection principle gives extensions of quasiconformal mappings preserving dilatation. The results below are for extensions of quasiconformal mappings under more general conditions, but without determining the optimal dilatation of the resulting extension.

(3.7) *If $\Omega \subset \mathbb{C}$ is a region, $f : \Omega \to \mathbb{C}$ is a K-quasiconformal mapping, and $F \subset \Omega$ is compact, then there is a quasiconformal mapping in $\mathbb{C}$ equal to f in F with an upper bound of the dilatation depending only in Ω, K, F.*

Proof By the compactness of F, there is a finite cover of F by open disks with closure in Ω. The set of the corresponding closed disks is a finite cover of F. Since Ω is an open connected set, it is connected by arcs. So, adding to the union of the closed disks of such a cover of F a finite number of arcs included in Ω gives a connected closed set $F_1 \subset \Omega$. ∂F_1 is a finite union of Jordan curves which are a finite union of arcs of circles included in the boundary of elements of the considered cover of F with a finite number of arcs added to get F_1. The set of connected components of $\mathbb{C}_\infty \setminus F_1$ is an open cover of the compact set $\mathbb{C}_\infty \setminus \Omega$. So, there is a finite open subcover $U_1, \ldots, U_N$ of $\mathbb{C}_\infty \setminus \Omega$. Since the points of $\partial\Omega$ in each U_j belong to $\operatorname{int} U_j$, each $(\mathbb{C}_\infty \setminus \Omega) \cap U_j$ is closed and can be separated by a closed polygonal curve C_j from the closed set $\mathbb{C}_\infty \setminus U_j$. Let D_j denote the region in U_j with boundary C_j. $\widetilde{F} = \mathbb{C}_\infty \setminus \cup_{j=1}^N D_j$ is compact and $F \subset \widetilde{F} \subset \Omega$. Since the possible choices of $\widetilde{F}$ depend only on F and Ω, it suffices to prove the result for $\widetilde{F}$ instead of F. So, it is assumed, without loss of generality, that $\mathbb{C}_\infty \setminus F$ is a union of N regions with closed polygonal curved boundaries. Since $C_j = \partial D_j \subset \Omega$, each $\Omega \cup D_j$ is a region. The sets $\Omega \cap (\mathbb{C}_\infty \setminus \operatorname{cl} D_1)$ are also regions.

C_1 is a free boundary arc of $\Omega \cap D_1$ and $C_1' = f(C_1)$ is a Jordan curve. Let D_1' be the region in $\mathbb{C}_\infty$ with boundary C_1' containing $f(\Omega \cap D_1)$. By (3.5), there is a quasiconformal mapping $g_1 : D_1 \to D_1'$ coinciding in C_1 with f and with dilatation K_1 with an upper bound only depending on K and $\Omega \cap D_1$. The function f_1 equal to g_1 in D_1 and to f in $\Omega \cap (\mathbb{C}_\infty \setminus D_1)$ is a homeomorphism defined in $\Omega \cup D_1$ with dilatation in $(\Omega \cup D_1) \setminus C_1$ upper bounded by $\max\{K, K_1\}$, which, by (3.1), is also upper bounded by the dilatation of f_1 in $\Omega \cup D_1$.

$F_1 = F \cup D_1 = F \cup \operatorname{cl} D_1$ is a compact subset of $\Omega \cup D_1$, and $\mathbb{C}_\infty \setminus (\Omega \cup D_1)$ is a union of regions $D_2, \ldots, D_N$. Repeating the preceding argument, we get a quasiconformal mapping in $\mathbb{C}$ equal to f in F and with dilatation depending only on K, Ω, F. □

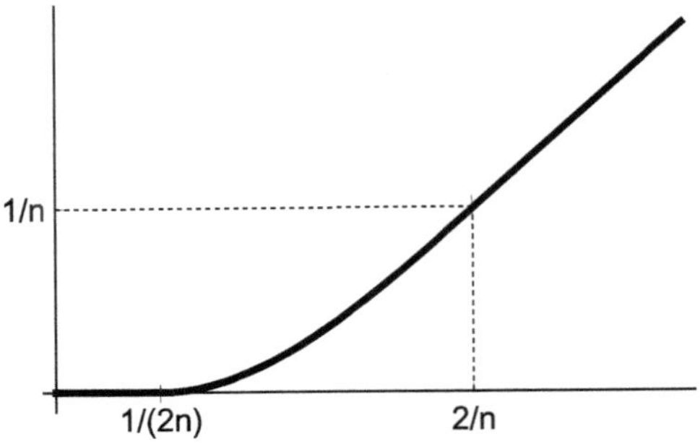

Fig. 3.1 Auxiliary functions $\eta_n\colon\mathbb{R}\to\mathbb{R}$ for the proof of (3.8); they are C^1, odd, $0\leq\eta_n'<1$, $\eta_n(t)=t-\frac{1}{n}$ in $[\frac{2}{n},+\infty[$ and $\eta_n(t)=0$ in $\left[0,\frac{1}{2n}\right]$

The following property was proved by S. Rickman[9] in 1969.

(3.8) *If $U\subset\mathbb{C}$ is open, $K\subset U$ is compact, $\varphi\colon U\to\mathbb{C}$ is a bounded quasiconformal mapping, $\psi\colon U\to\mathbb{C}$ is a homeomorphism onto $\psi(U)$ with $\psi_{|U\setminus K}$ a quasiconformal mapping, and $\psi=\varphi$ in K, then ψ is a quasiconformal mapping in U, and the Jacobian matrices of ψ and φ are equal* a.e. *in K as functions of real variables.*

Proof Since φ is a quasiconformal mapping in U, it belongs to the Sobolev[10] space $H^1(U)$. Consider the sequence $\{\eta_n\}$ of nondecreasing odd functions C^1 from $\mathbb{R}$ to $\mathbb{R}$ with $\eta_n'<$, $\eta_n(t)=t-\frac{1}{n}$ in $[\frac{2}{n},+\infty[$ and $\eta_n(t)=0$ in $\left[0,\frac{1}{2n}\right]$ (Fig. 3.1). $\{\eta_n\circ\mathcal{R}e(\psi-\varphi)\}$ is a Cauchy sequence in $H^1(U)$ converging to $\mathcal{R}e(\psi-\varphi)\in H^1(U)$. Thus, $\mathcal{R}e\,\psi\in H^1(U)$. Similarly, $\mathcal{I}m\,\psi\in H^1(U)$. Hence, $\psi\in H^1(U)$. The differences of corresponding partial derivatives of the real and imaginary parts of ψ and φ are uniformly bounded a.e. in U. As φ is a quasiconformal mapping in U, so is ψ. As in K it is $\eta_n\circ\mathcal{R}e(\psi-\varphi)=0$ and $\eta_n\circ\mathcal{I}m(\psi-\varphi)=0$, the Jacobian matrix of $\psi-\varphi$ is 0 a.e. in K. □

3.4 Quasiconformal Curves

A **quasiconformal Jordan curve (resp., arc)** is a Jordan curve (resp., arc) $C\subset\mathbb{C}$ such that there is a quasiconformal mapping f defined in a region $\Omega\supset C$ with $f(C)$ a circle (resp., line segment).

Therefore, *a closed subarc of a Jordan curve in $\mathbb{C}_\infty$ is a quasiconformal curve if and only if it is a closed subarc of a quasiconformal curve.*

(3.9) *If Ω, $\Omega'\subset\mathbb{C}$ are regions with, resp., C, C' quasiconformal free curves or arcs of regions boundaries and f is a quasiconformal mapping of Ω onto Ω' such that $f(C)=C'$, then f can be extended to a quasiconformal mapping defined in a region containing $\Omega\cup C$.*

[9] Rickman, Seppo (1935–2017).
[10] Sobolev, Sergei (1908–1989).

Proof Let g and h be quasiconformal mappings from a neighborhood U of C to a neighborhood U' of C' such that C and C' are mapped to a line segment or to an arc of circle. There is in $g(U)$ a region D symmetric relative to $g(C)$ containing this set and separated by $g(C)$ in two disjoint sets D_1, D_2 such that $g^{-1}(D_1) \subset \Omega$ and $g^{-1}(D_2) \subset \mathbb{C}\backslash\Omega$. Denote D', D'_1, D'_2 the analogous sets for h. D can be chosen small so that $f\left(g^{-1}(D_1)\right) \subset h^{-1}(D'_1)$. Applying the reflection principle (3.6) to $h \circ f \circ g^{-1}$ in D, f can be extended to a quasiconformal mapping defined in $\Omega_1 = \Omega \cup |, g^{-1}(D)$. □

For multiply connected regions with boundaries quasiconformal curves, G. Springer[11] obtained the following result in 1964.

(3.10) *If $\Omega, \Omega' \subset \mathbb{C}$ are multiply connected regions and $\partial\Omega$, $\partial\Omega'$ are unions of quasiconformal Jordan curves, every quasiconformal mapping f of Ω onto Ω' can be extended to the whole $\mathbb{C}$.*

Proof If Ω is a n-multiply connected region, i.e., $\mathbb{C}\backslash\Omega$ has n connected components bounded by n connected components, as f is a homeomorphism, Ω' is also a n-multiply connected region, and each connected component of $\partial\Omega$ corresponds to a single connected component of $\partial\Omega'$, in the sense that each point of a connected component of $\partial\Omega$ limit point of a sequence in $\{u_n\} \subset \Omega$ corresponds to a limit point of $\left\{f(u_n)\right\} \subset \Omega'$ in a connected component of $\partial\Omega'$. By the preceding result, f can be extended to a quasiconformal mapping in a region Ω_1 containing Ω. By (3.7), f can be extended to a quasiconformal mapping in $\mathbb{C}$. □

In 1962, M. Tienari obtained the following characterization of quasiconformal Jordan curves in terms of conformal homeomorphisms.

(3.11) *A Jordan curve separating $\mathbb{C}_\infty$ in two regions Ω_1, Ω_2 is quasiconformal if and only if for conformal homeomorphisms f_1, f_2 of the open complex half planes, resp., upper $\mathbb{H}$ and lower $-\mathbb{H}$ onto, resp., Ω_1 and Ω_2 coinciding at ∞, and the restriction of $f_2^{-1} \circ f_1$ to the real axis is quasisymmetric.*

Proof If C is a quasiconformal Jordan curve, then f_2 can be extended to a quasiconformal mapping in $\mathbb{C}$. So, there is a quasiconformal mapping $g_1 : \mathbb{H} \to \Omega_1$ with values on the real axis coinciding with those of f_2, and $g_1^{-1} \circ f_1$ is a quasiconformal mapping of $\mathbb{H}$ onto $\mathbb{H}$ with fixed point ∞ and values on the real axis coinciding with those of $f_2^{-1} \circ f_1$. By (3.3), its restriction to the real axis is a quasisymmetric function.

[11] Springer, George (1924–2019).

Conversely, if the restriction of $f_2^{-1} \circ f_1$ to the real axis is a quasisymmetric function, there is a quasiconformal mapping $g_1 : \mathbb{H} \to \mathbb{H}$ with values on the real axis coinciding with those of $f_2^{-1} \circ f_1$ and the restrictions of $f_1 \circ g_1^{-1}$, and f_2 to the real axis coincide. Therefore, these functions define a homeomorphism in $\mathbb{C}$. Since an analytic curve has measure zero, by (3.1), the function is a quasiconformal mapping whose image of the real axis is the curve C and, consequently, C is a quasiconformal curve. □

A **quasiconformal reflection** relative to a Jordan curve J in $\mathbb{C}$ is a quasiconformal mapping that reverses orientation, keeps the points of J fixed, and maps one of the components Ω separated by J in $\mathbb{C}$ to the other $\widetilde{\Omega}$ and vice versa. If such a mapping exists, it is said that the **Jordan curve J admits a quasiconformal reflection**.

(3.12) *A Jordan curve in $\mathbb{C}$ admits a quasiconformal reflection if and only if it is quasiconformal.*

Proof If J is a quasiconformal Jordan curve, there is a quasiconformal mapping $f : \mathbb{C} \to \mathbb{C}$ mapping the region Ω bounded by J onto the open complex upper half plane $\mathbb{H}$. The function $F : \Omega \to \widetilde{\Omega} \overset{def}{=} \mathbb{C} \setminus (J \cup \Omega)$ such that $F(z) = f^{-1}\big(\overline{f(z)}\big)$ is a quasiconformal reflection relative to J.

Conversely, if J admits a quasiconformal reflection F, with T a conformal homeomorphism of $\mathbb{H}$ onto Ω, define f equal to T in $\mathbb{H}$ and to $F \circ T$ in $\mathbb{C} \setminus \mathbb{H}$. This is a quasiconformal mapping whose image of the real axis is J; thus, J is a quasiconformal Jordan curve. □

Given a Jordan curve C separating $\mathbb{C}_\infty$ in two regions Ω_1, Ω_2, quadrilaterals $Q_1 = Q_1(z_1, z_2, z_3, z_4)$ bounding Ω_1 and $Q_2 = Q_2(z_4, z_1, z_2, z_3)$ bounding Ω_2, with z_1, z_2, z_3, z_4 ordered positively relative to Ω_1, are said to be **conjugate quadrilaterals** in the Jordan curve C.

A subarc of a Jordan curve (resp., a Jordan curve) $C \subset \mathbb{C}_\infty$ is said to have **bounded bending** if the set of quotients of the diameter in the spherical distance of the subarcs (resp., subarcs with smaller diameter) of C with ends at any distinct points $z_1, z_2 \in C$ by the spherical distance of these points is bounded.

If C is a subarc of a bounded Jordan curve, spherical distances in this definition can be replaced by Euclidean distances.

(3.13) *A subarc of a Jordan curve $C \subset \mathbb{C}$ has bounded bending if and only if there is $k > 0$ such that $|z_1 - z_2| \le k|z_1 - z_3|$ for all $z_1, z_2, z_3 \in \mathbb{C}$, in this order.*

Consequently, *a sufficient condition for every compact subarc C of a Jordan curve to be a quasiconformal curve is that there is a homeomorphism h of* $]0, 1[$ *onto C such that for some $0 < m < M$,*

$$m\,|t_1-t_2| \le |h(t_1)-h(t_2)| \le M\,|t_1-t_2|\,, \quad t_1, t_2 \in\,]0,1[\,.$$

Every regular closed subarc of a Jordan curve is quasiconformal, as well as every piecewise regular arc that at each common end point of two distinct maximal regular subarcs has tangents to each of the subarcs with angles $\neq 0$.

There are quasiconformal arcs without tangent at any point (i.e., only described by nondifferentiable paths at all points in the domain); they are not rectifiable and have Hausdorff dimension[12] > 1. An example is the **von Koch curve**[13] described by a path defined in $[0, 1]$ limit of a sequence of paths describing curves starting with a line segment of length 1 and with each next term of the sequence obtained from the preceding one by replacing the segment of length $\frac{1}{3}$ of each line segment in the curve centered in the resp.midpoint by other 2 sides of an equilateral triangle with one side coincident with the suppressed line segment and the opposite vertex of the triangle to the same side of the curve as the vertices of triangles previously considered (Fig. 3.2) and with the parameter of the curve representation in each term scaled proportionally onto the interval $[0, 1]$. With the preceding characterization of quasiconformal arcs, it is easy to see that the von Koch curve is quasiconformal. On the other hand, for each parameter $t \in [0, 1]$, the distance of the corresponding points of two successive paths in the sequence is not larger than the height of the equilateral triangle considered to pass from one curve to the next, and the sequence of these heights is $\left\{\sqrt{\frac{1}{3^{2n}} - \frac{1}{3^{2n}2^2}}\right\} = \left\{\frac{1}{3^{n-1/2}}\right\}$. Therefore, the sequence of paths is uniformly convergent, and the limit is a continuous function representing the von Koch curve. It is easy to see that this path is not differentiable at any point, and the sequence of the curves lengths is $\left\{1+\frac{2^{2n-1}}{3^n}\right\}$, which is unbounded. Therefore, the von Koch curve is not rectifiable. It can be proved that it has Hausdorff dimension $\frac{\log 4}{\log 3} \approx 1{,}262$.

Therefore, there are quasiconformal arcs not rectifiable and that even do not have any rectifieable subarcs, what led to consider the question of what are the possible

Fig. 3.2 First four terms of sequence with limit the von Koch curve, which is a quasiconformal despite not having tangent at any point, not rectifiable, and with Hausdorff dimension > 1 ($\frac{\log 4}{\log 3} \approx 1{,}262$)

[12] See the definition at the end of Chap. 5.

[13] von Koch, Niels (1870–1924).

Fig. 3.3 Cardioid described by $\gamma:[0, 2\pi]\to\mathbb{C}$, $\gamma(\theta)=(1-\cos\theta)\, e^{i\theta}$. It is not a quasiconformal curve despite of failing to have tangent at only 1 point and of being rectifiable

Hausdorff dimensions of quasiconformal arcs. In 1973, F. Gehring and J. Väisälä proved that the set of Hausdorff dimensions of quasiconformal arcs is $[1, 2[$.

On the other hand, there are curves not quasiconformal without tangent at just 1 point, (i.e., described by nonclosed paths that just do not have derivative $\neq 0$ at 1 point of the domain $[a, b]$ or closed paths differentiable at all points of $[a, b]$, but with derivatives in a and b symmetric and $\neq 0$). An example is a cardioid (Fig. 3.3).

If $C\subset\mathbb{C}_\infty$ is a Jordan curve or a subarc of a Jordan curve with bounded bending, there is $k>0$ such that the diameter in the spherical distance of the subarc with end points any $z_1, z_2\in C$ is $\leq k\, d_S(z_1, z_2)$, where $d_S(z_1, z_2)$ is the spherical distance of z_1 to z_2. Since the diameter in the spherical distance of a Jordan subarc is $\leq\frac{\pi}{2}$, the preceding inequality is trivially satisfied for z_1, z_2 such that $d_S(z_1, z_2)\geq\frac{\pi}{2k}$.

The quasiconformal Jordan curves[14] can be characterized as follows.

(3.14) Characterizations of quasiconformal Jordan curves:
For $C\subset\mathbb{C}_\infty$ the following conditions are equivalent:

1. *C is a quasiconformal Jordan curve.*
2. *The set of modules of quadrilaterals included in C conjugate to quadrilaterals with Module 1 included in C is upper bounded.*
3. *C is a Jordan curve of bounded bending.*

Proof *(1) $\Leftrightarrow$ (2)* If Q_1, Q_2 are conjugate quadrilaterals in a Jordan curve C as in the definition, there is a quasiconformal mapping f in $\mathbb{C}$ with $f(C)=\mathbb{R}$, and $f(Q_1)$, $f(Q_2)$ are symmetric relative to the real axis and are reflections of each other relative to this axis. So, $M\big(f(Q_1)\big)=M\big(f(Q_2)\big)$. As f is a quasiconformal mapping, the set of quotients $\frac{M(f(Q_2))}{M(f(Q_1))}$ for pairs of conjugate quadrilaterals in C is upper bounded.

Conversely, denoting $\widetilde{\Omega}_2$ the reflection of Ω_2 relative to the real axis, if the set of the modules of quadrilaterals included in C conjugate to quadrilaterals Q_1 with Module 1 included in C is upper bounded, the function $z\mapsto\overline{z}$ defines a bijection from the boundary of Ω_1 to that of $\widetilde{\Omega}_2$ and satisfies $\frac{1}{K}\leq M(Q_2)\leq K$ for some $K>0$. Thus, by (3.4), there is a quasiconformal mapping $f:\Omega_1\to\widetilde{\Omega}_2$ with values

[14] The condition *2* was obtained by A. Pfluger in 1961 and the condition *3* by L. Ahlfors in 1963.

$f(z)=\overline{z}$ for $z\in\partial\Omega_1$. The function $\overline{f}$ is a quasiconformal reflection relative to C. Thus, by (3.12), C is a quasiconformal curve.

(1) ⇔ (3) If C is a bounded Jordan curve and Q_1, Q_2 are conjugate quadrilaterals in C such that $M(Q_1)=1$, with s_a, s_b the distances between the sides, resp., a and b of Q_1 in this quadrilateral and d_a, d_b the corresponding Euclidean distances, then $s_b\geq d_b$ and the inequality obtained in the proof of (3.5) holds

$$\frac{s_a(Q_1)}{s_b(Q_1)}\geq\frac{2}{e^{\pi+\sqrt{\pi^2+\pi}}-1}>0.006\,.$$

Hence, $s_a>0.006\,d_b$, and there is $k>0$ such that for any two points $z_1, z_2\in C$ at least one of the subarcs of C with end points z_1, z_2 is included in a disk with diameter upper bounded by $k|z_1-z_2|$. So, one of the sides b of Q_1 and Q_2 is included in a circle with diameter $k\,d_a$. Thus, $d_a>\frac{0.006}{\pi k}\,d_b$. The module of Q_2 can be estimated with the formula in terms of area and extremal length (1.4), $M(Q_2)=\inf_\rho\frac{A_\rho(Q_2)}{[L_\rho(\Gamma)]^2}$. As above, we obtain an open disk $B_{kd_b/2}(z_0)$ containing a side a of Q_2 and denoting ρ the function equal to 1 in the open disk with center z_0 and radius $\frac{1}{2}kd_b+\frac{0.006}{\pi k}\,d_b$ and to 0 outside it, the ρ-length of an arc connecting the sides a of Q_2 is at least $\frac{0.006}{\pi\,k}\,d_b$. So,

$$M(Q_2)\leq\frac{\pi\,(kd_b/2+0.006\,d_b/\pi\,k)^2}{(0.006\,d_b/\pi k)^2}=\frac{10^6\pi}{36}\left(\tfrac{1}{2}\pi k^2+0.006\right)^2.$$

As *(1) ⇔(2)*, we get that C is a quasiconformal Jordan curve.

Conversely, if C is a quasiconformal Jordan curve, there is a quasiconformal mapping f defined in $\mathbb{C}$ mapping C onto the real axis. If C_1, C_2 are the subarcs of C with distinct end points $z_1, z_2\in C$, denoting p_j a point of $\mathrm{cl}\,C_j$ at a maximal distance from z_1 for $j\in\{1,2\}$, since $|p_j-z_1|\geq|z_1-z_2|$ for $j\in\{1,2\}$, if it is not $|p_1-z_1|, |p_2-z_1|>|z_1-z_2|$, at least one of the two subarcs of C is contained in the disk with center z_1 and radius $|z_1-z_2|$. Otherwise, with $A=B_d(z_1)\backslash\mathrm{cl}\,B_{|z_2-z_1|}(z_1)$, where $d=\min\{|p_1-z_1|, |p_2-z_1|\}$, there is $M(A)=\frac{1}{2\pi}\log\frac{d}{|z_2-z_1|}$, and $f(A)$ separates the points $f(z_1)$, $f(z_2)$ from the points $f(p_1)$, $f(p_2)$. All these points belong to the real axis, and we can choose f to be $f(z_1)=0<f(p_1)<f(z_2)$ and $f(p_2)=\infty$. By (1.22), $M\big(f(A)\big)\leq 2M_G\Big(\sqrt{\frac{f(z_2)}{f(z_2)+f(p_1)}}\Big)$. As M_G is decreasing, $M\big(f(A)\big)<2M_G\big(\frac{1}{\sqrt{2}}\big)$ and, by (1.25. 5), $M\big(f(A)\big)<\frac{1}{2}$. If f is K-quasiconformal

$$\frac{1}{2\pi}\log\frac{d}{|z_2-z_1|}=M(A)\leq KM\big(f(A)\big)<\frac{K}{2}\,.$$

Hence, $d<e^{K\pi}|z_2-z_1|$. If $k=2e^{K\pi}$, at least one of the subarcs C_1, C_2 is included in an open disk with diameter $k|z_2-z_1|$. Therefore, C is a Jordan curve with bounded bending. □

Next, we give local properties of quasiconformal curves. (3.15-3) is excessively weak, since in 1966 S. Rickman established that closed subarcs of Jordan curves are quasiconformal if and only if they have bounded bending, i.e., the implication in 2 can be replaced by equivalence, but we opt here for this weaker statement much simpler to prove.

(3.15) *The following properties hold:*

1. *Closed subarcs of quasiconformal curves have bounded bending.*
2. *Closed quasiconformal subarcs of Jordan curves have bounded bending.*
3. *If an open subarc of a Jordan curve has bounded bending, its closed subarcs are quasiconformal curves.*

Proof

(1) If C' is a closed subarc of a quasiconformal curve C, at least one of the subarcs of C with distinct end points $z_1, z_2 \in C'$ has a diameter in the spherical distance $\leq k\, d_S(z_1, z_2)$. If $k\, d_S(z_1, z_2)$ is smaller than the diameter of the subarc $C \setminus C'$, the considered subarc of C is included in C'. The diameter in the spherical distance of each subarc of C' is upper bounded by k times the spherical distance of the ends of the subarc.

(2) Since every closed quasiconformal subarc of a Jordan curve is a subarc of a quasiconformal curve, by *1*, it has bounded bending.

(3) It is assumed, without loss of generality, that the Jordan curve C is bounded. Every closed subarc C' of C is included in a subarc C'' of C with end points z_1, z_2 at a distance $\geq r > 0$ of C'. As C'' does not separate $\mathbb{C}$, we can choose points $p_j \in B_{r/2}(z_j)$ for $j \in \{1, 2\}$ and a polygonal curve connecting them at a positive distance from C''. Connecting p_1 and p_2 to points of C'' at minimal distance by line segments, we have a closed curve C''' of which C' is a subarc. If $z_1, z_2 \in C'' \cap C'''$ or $z_1, z_2 \in C''' \setminus C''$, then C''' has bounded bending; otherwise, as p_1 and p_2 are linked to C'' by line segments with minimal lengths, by (3.14.3), C''' is a quasiconformal curve and, therefore, so is C'. □

We have the following local characterization of Jordan quasiconformal curves.

(3.16) *A Jordan curve $C \subset \mathbb{C}_\infty$ is quasiconformal if and only if it is a union of the open subarcs of C of bounded bending.*

Proof If C is a quasiconformal Jordan curve, by (3.15), all closed subarcs of C have bounded bending, as do all open subarcs contained in some closed subarc of C.

Conversely, if the Jordan curve C is the union of the open subarcs of C of bounded bending, as it is a compact set, it can be covered by a finite number of open subarcs of C, denoted $C_1, \ldots, C_N$. If $z_1, z_2 \in C$ belong to one of these subarcs, there is $k > 0$ such that the diameter in the spherical distance of the subarc of C with end points z_1, z_2 is $\leq k\, d_S(z_1, z_2)$. If z_1, z_2 do not belong to the same subarc of C considered, $d_s(z_1, z_2) > 0$, and we have $\leq k' d_S(z_1, z_2)$ with $k' = \frac{\pi}{d_s(z_1, z_2)}$. So, the Jordan curve C has bounded bending, and by (3.14.3), the Jordan curve C is quasiconformal. □

3.5 Sewing at a Jordan Curve

The sewing at a Jordan curve theorem gives conditions on the values at a common boundary of two regions bounded by a Jordan curve where conformal Homeomorphisms can be defined onto, resp., the open upper and lower complex half planes whose extensions by homeomorphisms to the closures of the regions coincide on the common boundary. The proof given here uses the measurable Riemann mapping theorem as in the proof published by A. Pfluger in 1961. In 1960, O. Lehto and K. Virtanen had published a proof with approximations of quasiconformal mappings by holomorphic functions and construction of sequences of functions converging to the considered conformal homeomorphisms and used the result to prove the measurable Riemann mapping theorem.

(3.17) Sewing at a Jordan curve theorem *Let* $\Omega_1 \subset \mathbb{C}$ *be a region bounded by a Jordan quasiconformal curve* $C \subset \mathbb{C}$. *There are conformal homeomorphisms* f_1, f_2 *of, resp.,* $\Omega_1, \mathbb{C}\backslash\Omega_1$ *onto the open complex half planes, resp., upper and lower if and only if the extensions to homeomorphisms in the closures of the resp. domains, also denoted* f_1, f_2, *satisfy* $f_2(z) = \varphi\big(f_1(z)\big)$ *for* $z \in C$, *with* φ *a quasisymmetric function defined on the real axis. There are infinitely many pairs of functions* f_1, f_2 *with these properties, but those invertible with fixed points* 0, 1, ∞ *are unique.*

Proof The necessity of $f_2(z) = \varphi\big(f_1(z)\big)$ for $z \in C$, with φ a quasisymmetric function defined on the real axis, is a direct consequence of (3.11).

Conversely, by (3.3), there is a quasiconformal mapping F of the open upper complex half plane $\mathbb{H}$ onto itself extending φ. The Beltrami coefficient κ_F of F is extended to all $\mathbb{C}$ with value 0 outside $\mathbb{H}$, denoted $\kappa_{\widetilde{F}}$. By the measurable Riemann mapping theorem (2.9), there is a quasiconformal mapping $\widetilde{F}$ in $\mathbb{C}$ with Beltrami coefficient $\kappa_{\widetilde{F}}$, unique modulo composition by a conformal homeomorphism. Therefore, there is a conformal homeomorphism g_1 of $\mathbb{H}$ onto $\widetilde{F}(\mathbb{H})$ such that $\widetilde{F} = g_1 \circ F$. The restriction g_2 of $\widetilde{F}$ to $\mathbb{C}\backslash \mathrm{cl}\,\mathbb{H}$ is a conformal homeomorphism of the open lower complex half plane onto $\widetilde{F}(\mathbb{C}\backslash \mathrm{cl}\,\mathbb{H})$. These conformal homeomorphisms extend by continuity to homeomorphisms in the closures of the resp. open complex half planes and, denoting these extensions also, resp., g_1, g_2, on the real axis $g_2 = g_1 \circ F$.

If $\Omega_1 = \widetilde{F}(\mathbb{H})$, $f_1 = g_1^{-1}$ and $f_2 = g_2^{-1}$ satisfy the conditions in the statement. Otherwise, as Ω_1, Ω_2 are simply connected regions, by the Riemann mapping theorem, there are conformal homeomorphisms T_1, T_2, resp., of Ω_1 onto $\widetilde{F}(\mathbb{H})$ and of Ω_2 onto $\widetilde{F}(\mathbb{C}\backslash \mathrm{cl}\,\mathbb{H})$, and $f_1 = g_1^{-1} \circ T_1$, $f_2 = g_1^{-1} \circ T_2$ satisfy the conditions in the statement.

If (f_1, f_2) and (F_1, F_2) are pairs of invertible functions with fixed points 0, 1, ∞ satisfying the conditions of f_1, f_2 in the statement, the function w equal to $F_1^{-1} \circ \widetilde{F}$

in $\operatorname{cl}\mathbb{H}$ and to F_2^{-1} in $\mathbb{C}\setminus\operatorname{cl}\mathbb{H}$ is a homeomorphism defined in $\mathbb{C}$, which is a quasiconformal mapping in the complement of the real axis. As the real axis has measure zero, by the measurable Riemann mapping theorem, w is quasiconformal in $\mathbb{C}$. Since the complex dilatation of w is equal to that of $\widetilde{F}$, these functions can only differ by composition by a conformal homeomorphism, but as they have fixed points 0, 1, ∞, this mapping is the identity, and $w=\widetilde{F}$. Therefore, $F_1=f_1$, $F_2=f_2$.

□

3.6 Distortion by Quasiconformal Mapping in a Disk

In 1952, J. Hersch and A. Pfluger proved the following result on distortion by quasiconformal mappings.

(3.18) Bounded distortion relative to the origin:
For every K-quasiconformal mapping f of B_1 onto $\Omega\subset B_1$ with $f(0)=0$,

$$\varphi_{\frac{1}{K}}(|z|)\le|f(z)|\le\varphi_K(|z|),\quad z\in B_1,$$

where, with M_G the Grötzsch ring module,

$$\varphi_K(r)=(M_G)^{-1}\big(\tfrac{1}{K}M_G(r)\big),$$

and equality is possible in any of the inequalities with $\Omega=B_1$.

Proof The module of the ring B_1 with a slit in the real axis from 0 to $|z|$ for any $z\in B_1$ is $M_G(|z|)$. By the Grötzsch ring module theorem (1.20), its image by f is a ring with module $\le M_G(|f(z)|)$. Since the ring module is a quasiconformal invariant, as established in (1.16), it is $M_G(|z|)\le K\,M_G\big(|f(z)|\big)$; thus, $|f(z)|\le\varphi_K(|z|)$, $z\in B_1$. In order to get a quasiconformal mapping for this inequality to be an equality at a point $z=r>0$, consider a conformal homeomorphism f_1 symmetric relative to the real axis of the Grötzsch ring G_r on the annulus $B_1\setminus\operatorname{cl}B_R$ with $R=e^{-2\pi M_G(r)}$ mapping ∂B_1 onto ∂B_1. The function $f_2(\zeta)=\zeta\,|\zeta|^{1/K-1}$ is an orientation preserving homeomorphism of this annulus onto the annulus $B_1\setminus\operatorname{cl}B_{R/K}$ and differentiable as a function of real variables at all points and has constant dilatation K. Hence, by Grötzsch inequality (1.2), it is a regular K-quasiconformal mapping. There is a conformal homeomorphism f_3 of the last annulus mentioned onto the Grötzsch Ring $G_{\varphi_K(r)}$ which can be chosen so as to be symmetric relative to the real axis and to satisfy $f_3(\partial B_1)=\partial B_1$. The composition $f_3\circ f_2\circ f_1$ is a K-quasiconformal mapping of G_r onto $G_{\varphi_K(r)}$ with fixed point 0. It can be extended by continuity to a quasiconformal mapping defined in B_1. Since $G_{\varphi_K(r)}$ differs from B_1 just by removing an analytic arc, by the measurable Riemann mapping theorem (2.9), that function is a K-quasiconformal mapping of B_1 onto B_1, with equality in the considered inequality, concluding the proof of the second inequality in the formula in the statement and of the possibility of equality.

Applying the result of the preceding paragraph and taking into account that f is injective, $|z| \le \varphi_K(|f(z)|)$ for $z \in B_1$. Since the inverse of φ_K is $\varphi_{\frac{1}{K}}$, the first inequality in the formula in the statement holds, with equality for some K-quasiconformal mapping. □

(3.19) Bounded distortion between pairs of points:
For every K-quasiconformal mapping $f : B_1 \to B_1$

$$\frac{f(z_1)-f(z_2)}{1-\overline{f(z_2)}f(z_1)} \le \varphi_K\left(\left|\frac{z_1-z_2}{1-\overline{z_2}z_1}\right|\right), \quad z_1, z_2 \in B_1\,,$$

and this inequality is optimal.

Proof Apply the preceding result to $T_2 \circ f \circ T_1^{-1}$ with T_1, T_2 Möbius transformations, mapping, resp., z_2, $f(z_2)$ to 0. □

(3.20) Basic properties of the distortion function φ_K:

1. *For* $K > 0$, φ_K *is an increasing homeomorphism of* $]0, 1[$ *onto* $]0, 1[$.
2. $\varphi_K(r) \ge r$ *for* $K \ge 1$ *and* $\varphi_K(r) \le r$ *for* $K \le 1$.
3. $\varphi_{K_1K_2} = \varphi_{K_1} \circ \varphi_{K_2}$.
4. $\varphi_2(r) = \frac{2\sqrt{r}}{1+r}$ *and iterations give formulas for* φ_{2^n}, $n \in \mathbb{N}$.
5. $1 - 12\, r^{\frac{2}{K}} < 4^{-(1-\frac{1}{K})}\, r^{-\frac{1}{K}}\, \varphi_K(r) < 1$.
6. $\lim_{r\to 0} \varphi_K(r)\, r^{-\frac{1}{K}} = 4^{1-\frac{1}{K}}$.

Proof *(1)*, *(2)* and *(3)* are immediate from the definition of φ_K in (3.18) and Property 4 of the Grötzsch ring module M_G in (1.25.4).

(4) Follows from (1.25.2).

(5) With $0 < r < r' < 1$, consider the annulus $B_1 \backslash \mathrm{cl}\, B_{\frac{r}{r'}}$, the ring bounded by the circle $\partial B_{\frac{r}{r'}}$, and the line segment in the real axis $[0, r[$. By the superadditivity of the ring module (1.13), $\frac{1}{2\pi}\log\frac{r'}{r} + M_G(r') \le M_G(r)$. So, $r \mapsto -\frac{1}{2\pi}\log r - M_G(r)$ is increasing and, by (1.25.7), for $K > 1$, it is $\frac{1}{k}\left(\log\frac{4}{r} - M_G(r)\right) < \log\frac{4}{r'} - M_G(r')$. Thus, with $r' = \varphi_K(r)$, $M_G(r') = \frac{1}{K}M_G(r)$ and $\varphi_K(r) < 4^{1-\frac{1}{K}}\, r^{\frac{1}{K}}$. By (1.25.9), for $0 < r < 1$,

$$\frac{\varphi_K(r)}{4-\varphi_K^2(r)-\varphi_K(r)\,\delta_1(\varphi_K(r))} = \left(\frac{r}{4-r^2-r\,\delta_1(r)}\right)^{\frac{1}{K}},$$

and, as $\delta_1(\varphi_K(r)) < 2\varphi_K(r)$, we get $\varphi_K(r) > \left(\frac{r}{4}\right)^{\frac{1}{K}}[4 - 3\varphi_K^2(r)]$. Therefore, $\varphi_K(r) > 4^{1-\frac{1}{K}} r^{\frac{1}{K}} (1 - 12 r^{\frac{2}{K}})$.

6) By 5, $\lim\limits_{r \to 0} 4^{-(1-\frac{1}{K})} r^{-\frac{1}{K}} \varphi_K(r) = 1$. □

The following result is due to A.Mori in a posthumous article of 1956.

(3.21) Mori theorem: *If $f : B_1 \to B_1$ with $f(0) = 0$ is a K-quasiconformal mapping, extending it by continuity to* $\mathrm{cl}\, B_1$

$$|f(z_1) - f(z_2)| \le 16\, |z_1 - z_2|^{\frac{1}{K}}, \quad z_1, z_2 \in \mathrm{cl}\, B_1,$$

and the coefficient 16 *is the smallest valid for all* $K > 0$.

Proof Since $|f(z_1) - f(z_2)| \le 2$, the result holds trivially if $|z_1 - z_2| \ge \frac{1}{8}$.

If $|z_1 + z_2| \le 1$, then $|1 - \overline{z_2} z_1| > \frac{1}{2}$ and, as $\left|1 - \overline{f(z_2)} f(z_1)\right| \le 2$, by (3.19) and 5 of the preceding result,

$$|f(z_1) - f(z_2)| \le 2\, \varphi_K(2\, |z_1 - z_2|) \le 2\, 4^{1-\frac{1}{K}} 2^{\frac{1}{K}} |z_1 - z_2|^{\frac{1}{K}} < 2^3 |z_1 - z_2|^{\frac{1}{K}}.$$

If $|z_1 + z_2| > 1$, extend f to $\mathbb{C}$ by reflection. The module of the annulus $C = B_{\frac{1}{2}} \setminus \mathrm{cl}\, B_{\frac{1}{2}|z_1 - z_2|}$ is $M_G(C) = \frac{1}{2\pi} \log \frac{1}{|z_1 - z_2|}$. The points 0 and ∞ belong to the same connected component of $\mathbb{C} \setminus C$, and, as they are not fixed points of f, $f(C)$ separates the points of $w_1 = f(z_1)$, $w_2 = f(z_2)$ from the points $0, \infty$. By the variant of the Teichmüller ring module theorem (1.23),

$$M(f(C)) \le M_G\left(\frac{|\sqrt{w_1} - \sqrt{w_2}|}{\sqrt{2}\,(|w_1| + |w_2|)}\right) = M_G\left(\frac{|w_1 - w_2|}{(\sqrt{w_1} + \sqrt{w_2})\sqrt{2}\,(|w_1| + |w_2|)}\right) \le M_G\left(\frac{|w_1 - w_2|}{4}\right).$$

By (1.25.7), $M_G(r) < \frac{1}{2\pi} \log \frac{4}{r}$. Thus, $M_G(f(C)) < \frac{1}{2\pi} \log \frac{16}{|w_1 - w_2|}$ and, as $M_G(f(C)) \ge \frac{1}{K} M_G(C)$, the inequality in the statement follows.

It remains to prove the constant 16 in the inequality is the smallest possible for all $K > 0$. Consider the quadrilateral $Q_\alpha = Q_\alpha(0, 1, e^{i\alpha}, -1)$ with interior the upper half disk with radius 1 and center 0 whose image by the conformal homeomorphism $\frac{(1-z)^2}{(1+z)^2}$ is the quadrilateral $Q'_\alpha = Q'_\alpha(1, 0, -\tan^2 \frac{\alpha}{2}, \infty)$ with interior the lower complex half plane. As this quadrilateral is the extremal of the Teichmüller ring module in (1.21),

$$M(Q_\alpha) = M(Q'_\alpha) = 2 M_G\left(\sqrt{\frac{\tan^2(\alpha/2)}{1 + \tan^2(\alpha/2)}}\right) = 2 M_G\left(\sin \tfrac{\alpha}{2}\right).$$

If $\alpha, \beta \in\,]0, \pi[$ and $M_G\left(\sin \frac{\alpha}{2}\right) = K M_G\left(\sin \frac{\beta}{2}\right)$, there is a K-quasiconformal mapping f from the upper half disk with radius 1 and center 0 to itself such that $f(-1) = -1$, $f(0) = 0$, $f(1) = 1$, $f(e^{i\alpha}) = e^{i\beta}$, which can be extended by reflection to a function from B_1 to B_1 with fixed point 0 mapping $e^{i\alpha}$, $e^{i\beta}$ to, resp., $e^{-i\alpha}$, $e^{-i\beta}$. The equation $M_G(\sin\frac{\alpha}{2}) = K M_G(\sin\frac{\beta}{2})$ is $\sin\frac{\beta}{2} = \varphi_K(\sin\frac{\alpha}{2})$ and, with $|z_1 - z_2| = 2\sin\alpha$ and $|f(z_1) - f(z_2)| = 2\sin\beta$, by (3.20.5),

$$|f(z_1)-f(z_2)| = 16^{1-\frac{1}{K}}\big[1+\varepsilon(|z_1-z_2|)\big]\,|z_1-z_2|^{\frac{1}{K}},$$

where $\varepsilon(|z_1-z_2|) \to 0$ when $|z_1-z_2| \to 0$. Therefore, the inequality cannot hold for all $K>0$ with a constant smaller than 16. □

So, *every K-quasiconformal mapping $f: B_1 \to B_1$ is Hölder continuous with exponent*[15] $\frac{1}{K}$, as known before this contribution of A. Mori.[16]

Applying the result to the inverse of f: *Every quasiconformal mapping of a disk onto a disk of $\mathbb{C}$ can be extended to a homeomorphism of the disks.*

Another consequence is: *Every family of quasiconformal mappings $f: B_1 \to B_1$ with $f(0)=0$ is equicontinuous.* By the Arzelà-Ascoli theorem, a family of complex functions defined on an open set is normal if and only if it is pointwise bounded and locally equicontinuous. Therefore, *every family of quasiconformal mappings $f: B_1 \to B_1$ with $f(0)=0$ is normal.*[17]

These results hold under more general conditions, as follows.

(3.22) *A family of K-quasiconformal mappings in a region $\Omega \subset \mathbb{C}$ is equicontinuous and, by the Arzelà-Ascoli theorem, it is a normal family, if for some $d>0$ one of the conditions holds:*

1. *There are $w_1, w_2 \in \mathbb{C} \cup \{\infty\}$ at a spherical distance $>d$ omitted by each mapping in the family.*
2. *There are $w \in \mathbb{C} \cup \{\infty\}$ and distinct $z_1, z_2 \in \Omega \cup \{\infty\}$ such that each mapping in the family omits the value w and this value is at a spherical distance $>d$ of the images of z_1, z_2.*
3. *There are distinct $z_1, z_2, z_3 \in \Omega \cup \{\infty\}$ with images of each pair of them by each mapping of the family at spherical distances $>d$.*

Proof

(1) If $z_0 \in \Omega$, $0<\varepsilon<d$, the circle with center z_0 and radius $r>0$ in the spherical distance is a subset of Ω, $\delta \in]0, r[$ is such that the open annulus in the spherical distance $S_{\delta,r}$ consisting of points at a distance from z_0 between δ and r has ring module $M(S_{\delta,r}) > \frac{K\pi}{4\varepsilon^2}$, and S_δ is the open disk in the spherical

[15] See definition after the proof following this paragraph.

[16] The first estimate of this type with a large constant coefficient C depending on K was by M. Lavrentiev in 1948. In 1953, Z. Yûjôbô got the result with a constant C independent of K sufficiently large. In 1953–54, L. Ahlfors proved it with $C=2^{K^2}$ and in 1955 A. Mori with $C=48$.

[17] For the basic notions and properties of normal families and equicontinuity, including the Arzelà-Ascoli theorem, see Section 4 of Chapter 10 of *CADOVA*.

distance with radius δ and center z_0, as f omits two values a, b at a spherical distance $> d$, the ring $f(S_{\delta,r})$ separates a, b from $f(z_1)$ and $f(z_2)$. By the formula for the ring module in (1.11), $M\big(f(S_{\delta,r})\big) > \frac{\pi}{4\eta^2}$, where η is the minimum of the spherical distances of a to b and of $f(z_1)$ to $f(z_2)$. Since f is a K-quasiconformal mapping, $M\big(f(S_{\delta,r})\big) \geq \frac{1}{K} M(S_{\delta,r})$ and, therefore, $M\big(f(S_{\delta,r})\big) \geq \frac{\pi}{4\varepsilon^2}$. Consequently, $\eta < \varepsilon$ and, since $\varepsilon < d$, η is the spherical distance of $f(z_1)$ to $f(z_2)$, proving equicontinuity of f.

(2) By 1, the family is equicontinuous in the regions $\Omega \setminus \{z_j\}$, $j \in \{1, 2\}$, whose union is Ω, and, therefore, f is equicontinuous in Ω.

(3) Analogously, for the regions $\Omega \setminus \{z_j, z_k\}$, $j, k \in \{1, 2, 3\}$. □

A function f between metric spaces X, Y is **Hölder continuous with exponent** $\alpha > 0$ if there is $C > 0$ such that

$$d_Y\big(f(a), f(b)\big) \leq C\,[d_X(a, b)]^\alpha, \quad a, b \in X.$$

It is **Hölder continuous** if it is Hölder continuous with some exponent $\alpha > 0$. A family of functions between metric spaces X, Y is said to be **uniformly Hölder continuous** if there are constants $C, \alpha > 0$ such that the above inequality holds for all functions of the family.

(3.23) *An equicontinuous family of K-quasiconformal mappings $\mathscr{F}$ in a region $\Omega \subset \mathbb{C}$ is uniformly Hölder continuous with exponent $\frac{1}{K}$ in the spherical distance in each compact subset of Ω. In particular, a K-quasiconformal mapping in a region $\Omega \subset \mathbb{C}$ is Hölder continuous with exponent $\frac{1}{K}$ in the spherical distance at each $F \subset \Omega$ compact; this exponent is optimal.*

Proof It suffices to prove the first statement locally in a neighborhood of each point, which, without loss of generality, is taken to be 0, as this can be achieved with a rotation of $\mathbb{C}_\infty$. U_R denotes the open disk with center 0 and radius $R > 0$ in the spherical distance. Let $U_r \subset \Omega$ with $r \in]0, 1[$ be such that $f(U_r) \subset U_1$ for $f \in \mathscr{F}$. By (3.19) and (3.20.5), if $z_0 \in U_{\frac{r}{2}}$, for each $f \in \mathscr{F}$ and $z \in U_r$,

$$\tfrac{1}{2}|f(z) - f(z_0)| \leq \left|\tfrac{f(z)-f(z_0)}{1-\overline{f(z_0)}\,f(z)}\right| \leq \varphi_K\big(\tfrac{r|z-z_0|}{|r^2-\overline{z_0}z|}\big) \leq 4\big(\tfrac{r|z-z_0|}{|r^2-\overline{z_0}z|}\big)^{\frac{1}{K}} \leq 4\big(\tfrac{2|z-z_0|}{r}\big)^{\frac{1}{K}} \leq \tfrac{8}{r}|z - z_0|^{\frac{1}{K}}.$$

Since $|f(z) - f(z_0)| \geq d_S\big(f(z), f(z_0)\big)$ and $|z - z_0| \leq 2\,d_S(z, z_0)$, where d_S is the spherical distance, for $C \geq \frac{32}{r}$,

$$d_S\big(f(z), f(z_0)\big) \leq C\,[d_S(z, z_0)]^{\frac{1}{K}}, \quad z_0 \in U_{\frac{r}{2}}, z \in U_r, f \in \mathscr{F}.$$

If $z \notin U_r$, then $d_s(z, z_0) > \arctan r - \arctan \frac{r}{2}$ for $z_0 \in U_{\frac{r}{2}}$. Therefore, the above inequality also holds if C is large. Consequently, $\mathscr{F}$ is uniformly continuous with Hölder exponent $\frac{1}{K}$ in $U_{\frac{r}{2}}$.

The second statement is a particular case of the first. It remains to show that the exponent $\frac{1}{K}$ is optimal, what can be seen with $f(z)=z|z|^{\frac{1}{K}-1}$. □

As a consequence, we have the following conclusion.

(3.24) *For families of K-quasiconformal mappings in a region Ω, equicontinuity, uniform Hölder continuity, and normality are equivalent.*

It was proved in (1.10) that the limit of a sequence of uniformly convergent quasiconformal mappings in compact subsets of a region is a quasiconformal mapping if it is a homeomorphism. Now, we prove a stronger result.

(3.25) *If a sequence $\{f_n\}$ of K-quasiconformal mappings in a region $\Omega\subset\mathbb{C}$ is convergent, the limit is a constant or a 2-valued function or a K-quasiconformal mapping. In the second case, the limit function assumes 1 of the 2 values at a single point, and the convergence is uniform in each compact subset of the complement of that point in Ω; in the third case, the sequence converges uniformly in each compact subset of Ω.*

Proof The following three cases can be distinguished:[18]
1st case) f assumes a single value in Ω.

2nd case) f assumes exactly two different values in Ω, in particular at points $a_1, a_2\in\Omega$. Thus, by (3.22.1), $\{f_n\}$ is equicontinuous in $\Omega\backslash\{a_1, a_2\}$, and if $z\in\Omega\backslash\{a_1\}$ is such that $f(z)=f(a_1)$, $\{f_n\}$ is equicontinuous in $\Omega\backslash\{z, a_2\}$. Therefore, $\{f_n\}$ is equicontinuous in $\Omega\backslash\{a_2\}$ and it converges in this set and uniformly in each of its compact subsets. So, f is continuous in $\Omega\backslash\{a_2\}$. As f only assumes two values in Ω, one of them is assumed at just 1 point.

3rd case) f assumes at least three distinct values at points $a_1, a_2, a_3\in\Omega$; thus, there exists $d>0$ such that $d_S\big(f_n(a_j), f_n(a_k)\big)>d$ for distinct $j, k\in\{1, 2, 3\}$, with d_S the spherical distance. By (3.22.2), $\{f_n\}$ converges uniformly in compact subsets of Ω and, therefore, f is continuous on Ω. It remains to prove that f is a K-quasiconformal mapping, for what it suffices to prove that f is injective, since, as a result of the invariance of domain theorem, f is a homeomorphism[19] and, by (1.10)), f is a K-quasiconformal mapping.

If $a, b\in\Omega$ and $f(a)=f(b)$, with $\varepsilon\in]0, d_S(a, b)[$ such that $\mathrm{cl}\, U_\varepsilon\subset\Omega$, where $U_R(x)$ denotes the open ball with center x and radius $R>0$ in the spherical distance, the curve $f(\partial U_\varepsilon)$ separates $f_n(a)$ from $f_n(b)$ for all $n\in\mathbb{N}$, and the minimum of $d_S\big(f_n(z), f_n(a)\big)$ for $z\in\partial U_\varepsilon$ is $d_S\big(f(a), f(b)\big)=0$. Thus, there are points arbitrarily close to but different from a where f has the value $f(a)$. By the equicontinuity

[18] These are also the possible cases if $f_n\to f$ in a set $D\subset\Omega$ dense in Ω.

[19] See Appendix D.

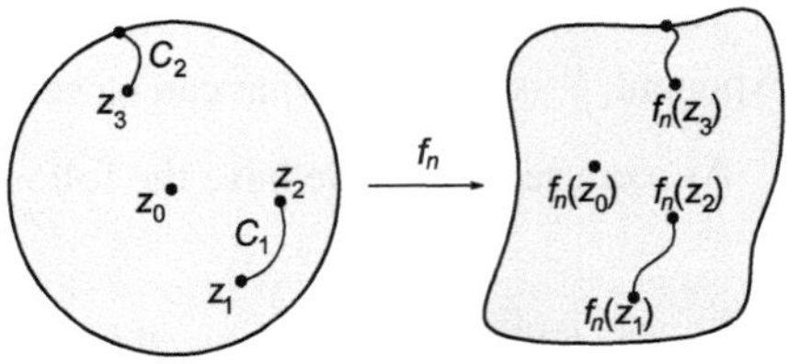

Fig. 3.4 Auxiliar figure for the proof of (3.25)

of $\{f_n\}$, there is $U_r(z_0)$ such that $d_S\big(f_n(z), f_n(z_0)\big) < \frac{\pi}{4}$ for $n \in \mathbb{N}$, $z \in U_r(z_0)$. If f is neither constant nor injective in $U_r(z_0)$, there are $z_1, z_2, z_3 \in U_r(z_0)$ such that $f(z_1) \neq f(z_2) = f(z_3)$. Connect the points z_1 to z_2 and z_3 to $z_4 \in \partial U_r(z_0)$ by subarcs of disjoint Jordan curves, resp., C_1, C_2 included in $U_r(z_0)$ except for the end z_4 of C_2 (Fig. 3.4). $S = U_r(z_0)\backslash(C_1 \cup C_2)$ is a ring whose image by f_n is a subset of $U_{\frac{\pi}{4}}\big(f_n(z_0)\big)$ and separates $f_n(z_1)$, $f_n(z_2)$ from $f_n(z_3)$, $\partial U_{\frac{\pi}{4}}\big(f_n(z_0)\big)$. Since $\lim_{n\to+\infty} d_S\big(f_n(z_2), f_n(z_3)\big) = 0$, $\lim_{n\to+\infty} d_S\big(f_n(z_1), f_n(z_2)\big) > 0$ and the diameters in the spherical distance of the connected components of $\partial f_n(S)$ are $\geq d_S\big(f(z_1), f(z_2)\big)$ for $n \in \mathbb{N}$. By (1.18), $\lim M\big(f_n(S)\big) = 0$, contradicting $M\big(f_n(S)\big) \geq \frac{1}{K} M(S) > 0$. Hence, there is $r > 0$ such that f is constant or injective in $U_r(z_0) \subset \Omega$. Denote E_1 the subset of Ω of points with neighborhoods where f has the constant value $f(a)$ and $E_2 = \Omega\backslash E_1$. E_1, E_2 are open sets, and since Ω is connected, $E_2 = \emptyset$. Thus, f is constant in Ω, contradicting the assumption. Therefore, f is injective in Ω. □

For a sequence $\{E_n\}$ of subsets of $\mathbb{C}$, $N\big(\{E_n\}\big) = \cup_{k=1}^{\infty} \operatorname{int} \cap_{n=k}^{\infty} E_n$ is called the sequence **nucleous**.

$z \in N\big(\{E_n\}\big)$ if and only if there are $N \in \mathbb{N}$ and a neighborhood $U \subset \mathbb{C}$ of z such that $U \subset E_n$ for $n > N$.

(3.26) *If $\{f_n\}$ is a sequence of K-quasiconformal mappings converging to a function f in a region $\Omega \subset \mathbb{C}$ with $\#\partial\Omega \geq 2$, then, alternatively:*

1. *$f(\Omega) = \{w_0\}$ with $w_0 \in \big[\mathbb{C}\backslash N\big(\{f_n(\Omega)\}\big)\big] \cap \big[\mathbb{C}\backslash N\big(\mathbb{C}\backslash\{f_n(\Omega)\}\big)\big]$.*
2. *$f(\Omega) = \{w_1, w_2\}$ with $\# f^{-1}\big(\{w_1\}\big) = 1$ and $N\big(\{f_n(\Omega)\}\big) = \mathbb{C}\backslash\{w_2\}$.*
3. *f is a quasiconformal mapping, $f(\Omega)$ is a connected component of the nucleous of $\{f_n(\Omega)\}$, and $\#\Omega \cap f^{-1}\big(N\big(\{f_n(\Omega)\}\big)\big) \leq 1$.*

Proof If $w_0 \in N\big(f_n(\Omega)\big)$, $f(z_0) = w_0$, and V is a connected neighborhood of w_0 such that there is $n_0 \in \mathbb{N}$ with $V \subset f_n(\Omega)$ for $n \geq n_0$, the functions f_n^{-1}, for $n \geq n_0$, omit 2 values. Thus, by (3.22), $\{f_n^{-1}\}_{n \geq n_0}$ is equicontinuous in V and, as $\lim f_n(z_0) = w_0$, $\lim f_n^{-1}(w_0) = f_n^{-1}(f_n(z_0)) = z_0$. So, if $z_0, z_1 \in \Omega$ and $f(z_0) = f(z_1) = w_0$, then $\lim f_n^{-1}(w_0) = z_0 = z_1$. Therefore, f cannot assume values in $N\big(\{f_n(\Omega)\}\big)$ in more than 1 point of Ω.

The alternatives are the three cases of (3.25):

(1) By the preceding paragraph, for each $n \in \mathbb{N}$, $w_0 \in \mathbb{C} \setminus \operatorname{int} N(\{f_n(\Omega)\})$. Since each neighborhood of w_0 intersects $\mathbb{C} \setminus f_n(\Omega)$ for infinite values of $n \in \mathbb{N}$, we have $w_0 \in [\mathbb{C} \setminus N(\{f_n(\Omega)\})] \cap [\mathbb{C} \setminus N(\mathbb{C} \setminus \{f_n(\Omega)\})]$.

(2) By the first paragraph, $w_2 \notin N(\{f_n(\Omega)\})$ and $w_1 \in N(\{f_n(\Omega)\})$.

(3) If $z_0 \in \Omega$ and U is an open set such that $z_0 \in U$ and $\operatorname{cl} U_\varepsilon \subset \Omega$, by (3.25), $\{w_n\}$ converges to f uniformly in U. Let d_S be the spherical distance and $V_R(z)$ the disk with center z and radius $R > 0$ in that distance. With $r = d_S(f(z_0, \partial f(U))$, there is $n_0 \in \mathbb{N}$ such that $V_{\frac{r}{2}}(z_0) \cap \partial f_n(U) = \emptyset$ for $n \geq n_0$ and $f_n(z_0) \in V_{\frac{r}{2}}(z_0)$. Thus, for $n \geq n_0$, $f_n(U) \subset f_n(\Omega)$ and $f(z_0) \in N(\{f(\Omega_n)\})$. Therefore, $f(\Omega) \subset N(\{f(\Omega_n)\})$. Since $f(\Omega)$ is connected, it is a subset of a connected component of $N(\{f(\Omega_n)\})$. If $f(\Omega)$ were not such a connected component of $N(\{f(\Omega_n)\})$, this set would contain a point $a \in \partial f(\Omega)$. By the definition of nucleous of a sequence, there would exist $n_0 \in \mathbb{N}$ and a connected neighborhood of a, $V \subset f_n(\Omega)$, for $n \geq n_0$, and a sequence $\{a_m\} \subset \Omega$ such that $\{f(a_m)\} \subset V \cap f(\Omega)$ and $\lim f(a_m) = a$. For $n \geq n_0$, the values of functions f_n^{-1} at points of V would omit two points of $\partial\Omega$. By (3.22), $\{f_n^{-1}\}_{n \geq n_0}$ would be an equicontinuous family. Hence, it would be a normal family, and would have a subsequence, still denoted $\{f_n^{-1}\}$, uniformly convergent on each compact subset of V to a continuous function F. By (3.25), F would be constant or a K-quasiconformal mapping, and, by equicontinuity, $F(f(a_m)) = \lim f_n^{-1}(f_n(a_m)) = a_m$ for $m \in \mathbb{N}$. Therefore, F would be a K-quasiconformal mapping in V and $F(V)$ would be a region included in a connected component of $N(\{f_n^{-1}(V)\}) \subset \Omega$. Consequently, $F(a) \in \Omega$ and $f(F(a)) = a$, contradicting $a \in \partial f(\Omega)$. Hence, $f(\Omega)$ is a full connected component of $N(\{f(\Omega_n)\})$. □

If the ranges of all functions f_n coincide, 2 in the preceding result cannot occur and we have the following property.

(3.27) *If $\{f_n\}$ is a convergent sequence of K-quasiconformal mappings in a region $\Omega \subset \mathbb{C}$ with $\#\partial\Omega \geq 2$ and $f_n(\Omega)$ is the same region of Ω' for all $n \in \mathbb{N}$, then $f = \lim f_n$ is a constant function with value in $\partial\Omega'$, or it is a K-quasiconformal mapping.*

It is easy to clarify the possibilities when the boundary of the region where a sequence of K-quasiconformal mappings is defined is $\emptyset$ or 1 point.

If $\{f_n\}$ is a convergent sequence of a K-quasiconformal mappings in $\mathbb{C}$, then $f_n(\Omega) = N(\{f_n(\Omega)\}) = \mathbb{C}$, and trivially $f(\Omega) \subset N(\{f_n(\Omega)\})$.

If $\{f_n\}$ is a convergent sequence of K-quasiconformal mappings in a region Ω and $\partial\Omega = \{a\}$, by (3.1), each f_n can be extended to a K-quasiconformal mapping in $\mathbb{C}$. If $\{f_n\}$ converges in Ω to a K-quasiconformal mapping or to a function assuming only two values, from what was seen for (3.25) and the observation in the first footnote of its proof, $\{f_n\}$ converges in the whole of $\mathbb{C}$. In both cases, $N(\{f_n(\Omega)\}) = \mathbb{C} \setminus \{a\}$, and we have the alternatives 2 and 3 in (3.26) for $\#\partial\Omega \geq 2$.

3.7 Circular Dilatation

The **circular dilatation** of a homeomorphism $f:\Omega\to\Omega'$ preserving orientation, with $\Omega,\,\Omega'\subset\mathbb{C}_\infty$ open sets, is (Fig. 3.5)

$$H_f(z)=\overline{\lim_{r\to0+}}\frac{\max_{\partial B_r(z)}|f-f(z)|}{\min_{\partial B_r(z)}|f-f(z)|},\quad \text{if } z\in\Omega\backslash\{\infty\},\ f(z)\neq\infty,$$

$$H_f(\infty)=H_{\widetilde{f}}(0),\ \text{if } \widetilde{f}(z)=f\left(\tfrac{1}{z}\right),\qquad H_f(z)=H_{\frac{1}{f}}(z),\ \text{if } f(z)=\infty.$$

If $z\in\mathbb{C}$, $f(z)\neq\infty$, and f is C^1 in $\{z\}$ with $f'(z)\neq0$, then $H_f(z)=K_f(z)$, where $K_f(z)$ is the dilatation of f at z, and, by the invariance of H_f with Möbius transformations, the equality holds if $z=\infty$ or $f(z)=\infty$.

If $\Omega\subset\mathbb{C}$ is a region and $f:\Omega\to\mathbb{C}$ is K-quasiconformal mapping, as, by (1.29), f is regular a.e. in Ω, $H_f\leq K$ a.e. in Ω. However, it may be $H_f(z)>K$ for large K and even with an arbitrarily large difference, at points z where f is not regular, as follows from the following result obtained by O. Lehto, K. Virtanen, and J. Väisälä in 1959.

(3.28) *Let* $\lambda(K)=\frac{1}{[M_G^{-1}(K/4)]^2}-1$ *as in the proof of* (3.3).

1. *For* $K\geq1$, $\varepsilon>0$ *there is a K-quasiconformal mapping* f *from* $\mathbb{C}$ *onto* $\mathbb{C}$ *and* $z\in\mathbb{C}$ *such that* $H_f(z)\geq\lambda(K)-\varepsilon$.
2. $\lim_{K\to+\infty}\lambda(K)\,e^{-\pi K}=\frac{1}{16}$.
3. *For every* $\Delta>0$ *and* $K>1$ *large, there are a K-quasiconformal mapping from* $\mathbb{C}$ *onto* $\mathbb{C}$ *and* $z\in\mathbb{C}$ *such that* $H_f(z)>K+\Delta$.

Proof

(1) The module of the quadrilateral in the complex upper half plane with vertices $-\lambda(K)$, 0, 1, ∞ is $4\,M_G\left(\sqrt{1+\lambda(K)}\right)$. By (3.3), there is a K-quasiconformal mapping f from the open upper complex half plane $\mathbb{H}$ onto $\mathbb{H}$ with $f(0)=0$,

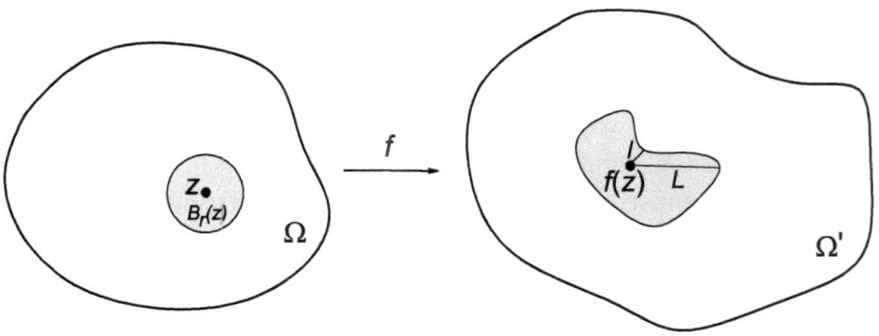

Fig. 3.5 Image of circle with center z and small radius by function f for visualization of approximate circular dilatation $H_f(z)\approx\frac{L}{l}$

$f(1)=1$, $f(-1)=-\lambda(K)$, $f(\infty)=\infty$. Extend f to $\mathbb{C}$ by reflection relative to the real axis. If $B_n=B_n\backslash B_{1/n}$, $B'_n=f(B_n)$ for $n\in\mathbb{N}\backslash\{1\}$, and f_n is a conformal homeomorphism from B'_n onto $A_n=B_{b_n}\backslash B_{a_n}$ with $f_n(1)=1$, the K-quasiconformal mapping $f_n\circ f: B_n \to A_n$ can be extended to $\mathbb{C}\backslash\{0\}$ by successive reflections. By (3.1), $z=0$ and $z=\infty$ are removable singularities, and $f_n\circ f$ has an extension to a K-quasiconformal mapping in $\mathbb{C}$ with $(f_n\circ f)(0)=0$, $(f_n\circ f)(1)=1$, and $(f_n\circ f)(\infty)=\infty$. By (3.22) the sequence $\{f_n\circ f\}$ is a normal family and, by (3.25), it has a subsequence converging uniformly in compact sets to a K-quasiconformal mapping in $\mathbb{C}$. In each compact subset of $\mathbb{C}\backslash\{0\}$ after some order, the functions f_n are conformal homeomorphisms and converge to a conformal homeomorphism. So, there is a conformal homeomorphism h in $\mathbb{C}$ such that $h=\lim f_n$ in $\mathbb{C}\backslash\{0\}$. Hence, $h-1_{\mathbb{C}_\infty}$ is a Möbius transformation[20] and, as $h(0)=0$, $h(1)=1$, $h(\infty)=\infty$, it vanishes at 0, 1, ∞; so, it is equal to 0 and $h=1_{\mathbb{C}_\infty}$. Therefore, there is $N\in\mathbb{N}$ such that

$$\left|(f_N\circ f)(-1)+\lambda(K)\right|=\left|(f_N\circ f)(f^{-1}(-\lambda(K)))+\lambda(K)\right|=\left|(f_N(-\lambda(K)))+\lambda(K)\right|<\varepsilon.$$

Since $(f_N\circ f)(1)=1$, it is $\frac{\max|f_N\circ f|}{\min|f_N\circ f|}>\lambda(K)-\varepsilon$ on the circle $|z|=1$, and the inequality also holds in the circles $|z|=\frac{1}{N^{2k}}$ obtained by the reflections. The inequality in *1* holds for the K-quasiconformal mapping $f_N\circ f$ from $\mathbb{C}$ onto $\mathbb{C}$.

(2) $M_G\left(\frac{1}{\sqrt{\lambda(K)}}\right)=\frac{K}{4}$. By (1.25.4), $\lim\limits_{K\to+\infty}\lambda(K)=+\infty$, and, by (1.25.8)

$$\lim_{K\to+\infty}\left[\frac{K}{4}-\frac{1}{2\pi}\log\left(4\sqrt{\lambda(K)}\right)\right]=0 \Leftrightarrow \lim_{K\to+\infty}\log\frac{e^{K/4}}{4^{1/2\pi}\lambda(K)^{1/4\pi}}=0,$$

which is equivalent to $\lim\limits_{K\to+\infty}\frac{e^{K/4}}{4^{1/2\pi}\lambda(K)^{1/4\pi}}=1$, i.e., $\lim\limits_{K\to+\infty}\frac{e^{\pi K}}{16\lambda(K)}=1$.

(3) It is an immediate consequence of *1* and *2*. □

For the homeomorphism f of $\mathbb{C}$ onto B_1 such that $f(z)=e^{-\frac{1}{|z|}+i\operatorname{Arg} z}$, it is $H_f(0)=1$ and $K_f(0)=\infty$. So, $K_f(z)$ may fail to be finite even if $H(z)$ is. However, in 1957, A. Mori obtained the following upper bound of the circular dilatation of K-quasiconformal mappings in terms of K.

(3.29) *For every K-quasiconformal mapping f, $H_f\le e^{\frac{K}{2}}$.*

Proof Let $f:\Omega\to\Omega'$ be a K-quasiconformal mapping, with Ω, $\Omega'\subset\mathbb{C}$ regions. For $z_0\in\Omega$ and $r>0$ such that $\operatorname{cl} B_r(z_0)\subset\Omega$ denote

$$m_1(r)=\max_{\partial B_r(z_0)}|f-f(z_0)|, \qquad m_2(r)=\min_{\partial B_r(z_0)}|f-f(z_0)|,$$

and, resp., z_1, z_2 points where these values are assumed. Consider $r>0$ such that $B_{m_1(r)}(f(z_0))\subset\Omega'$ and the annulus $A'=B_{m_1(r)}(f(z_0))\setminus\operatorname{cl} B_{m_2(r)}(f(z_0))$. The ring $A=f^{-1}(A')$ separates the points z_0, z_2 from z_1, ∞. As $|z_1-z_0|=|z_2-z_0|$,

[20] See the beginning of Section 5 of Chapter 12 of *CADOVA*.

by Teichmüller ring module theorem and (1.25.5), $M(A) \le 2\,M_G\left(\frac{1}{\sqrt{2}}\right) = \frac{1}{2}$. Since $M(A') = \frac{1}{2\pi}\log\left(\frac{m_1(r)}{m_2(r)}\right)$ and $M(A') \le M(A)$, it is $\frac{m_1(r)}{m_2(r)} \le e^{\frac{K}{2}}$. Thus, it also is $H_f(z_0) \le e^{\frac{K}{2}}$. □

(3.30) *A homeomorphism between regions $\Omega, \Omega' \subset \mathbb{C}_\infty$ preserving orientation is a K-quasiconformal mapping if and only if it has finite circular dilatation H_f in all Ω and $H_f \le K$* a.e. *in Ω.*

Proof Necessity follows from what was obtained earlier in this section.

To prove sufficiency, assume that f is an orientation preserving homeomorphism with finite circular dilatation H_f at all points of Ω satisfying $H_f \le K$ a.e. in Ω.

We start by proving that f is absolutely continuous on lines.[21] Consider a rectangle $R = [a, b] \times [c, d] \subset \Omega$ and, for $y \in]c, d[$, $R_y =]a, b[\times]c, y[$. Since the area $A(y)$ of $f(R_y)$ is an increasing function of y, the derivative A' exists[22] a.e. in $]c, d[$. As $H_f \le K$ a.e. in R, by Fubini theorem, also $H_f \le K$ a.e. in $]a, b[\times\{y\}$ a.e. for $y \in]c, d[$. Therefore, it suffices to prove absolute continuity of f in $]a, b[\times\{y\}$ so that $A'(y)$ exists and $H(x+iy) \le K$ a.e. for $x \in]a, b[$. Let $y_0 \in]c, d[$ be such that $I =]a, b[\times\{y_0\}$ satisfies these conditions and $F \subset I$ is compact.

H is assumed to be upper bounded by N in F. Denote $M(z, r)$ and $m(z, r)$, resp., the maximum and minimum of $s \mapsto |f(s) - f(z)|$ in $\partial B_r(z)$, and for $n \in \mathbb{N}$ such that $\frac{1}{n}$ is less than the distance from F to $-R$, consider

$$F_n = \left\{z \in F : r \in \left]0, \tfrac{1}{n}\right] \Longrightarrow \tfrac{M(z,r)}{m(z,r)} \le N+1\right\}.$$

It is $F = \cup_{n\in\mathbb{N}} F_n$ and $f(F) = \cup_{n\in\mathbb{N}} f(F_n)$. Thus, $\lim \mu\big(f(F_n)\big) = \mu\big(f(F)\big)$, with μ the Lebesgue measure in $\mathbb{R}$. Since F_n is compact, it has a finite cover by open disks $B_{\frac{1}{n}}(z_j) \subset \mathbb{C}$, $z_j \in F_n$, $j \in [\mathbf{m_n}] \stackrel{def}{=} \{1, 2, \ldots, m_n\}$, and $\left\{I \cap B_{\frac{1}{n}}(z_j)\right\}_{j\in[\mathbf{m_n}]}$ is an open cover of $I \cap F_n$ by straight line segments without ends, which still is a cover after eliminating in each subset of intersecting line segments all but 2 suitably chosen. So, without loss of generality, each disk of the cover $\left\{B_{\frac{1}{n}}(z_j)\right\}_{j\in[\mathbf{m_n}]}$ of F_n only intersects another disk of the cover. Since each line segment $I \cap D_n(z_j)$ has length $\frac{2}{n}$, with $D_n = \cup_{j\in[\mathbf{m_n}]} B_{\frac{1}{n}}(z_j)$, it is $\mu\big(I \cap D_n\big) \ge \frac{m_n}{n}$. If U is an open subset of I containing F, for $\frac{1}{n}$ less than the distance of F to $I\backslash U$, then $I \cap D_n \subset U$. Since U can be chosen so that $\mu(U\backslash F)$ is arbitrarily small, $\overline{\lim}\,\mu(I \cap D_n) \le \mu(F)$. As $F_n \subset I \cap D_n$ and, by the Lebesgue dominated convergence theorem, $\lim \mu(F_n) = \mu(F)$, it is $\lim \mu(I \cap D_n) = \mu(F)$. The area of $\cup_{j\in[\mathbf{m_n}]} f(D_n(z_j))$ is greater or equal to $\pi \sum_{j\in[\mathbf{m_n}]} \frac{1}{2}[m(z_j, \frac{1}{n})]^2$ and, since $D_n(z_j) \subset \left]a, b\right[\times \left]y_0 - \frac{1}{n}, y_0 - \frac{1}{n}\right[$,

[21] See last Section of Chap. 1.

[22] See Section 3 of Appendix A.

$$2\left[A\left(y_0+\tfrac{1}{n}\right)-A\left(y_0-\tfrac{1}{n}\right)\right] \geq \pi \sum_{j\in[\mathbf{m_n}]} [m(z_j,\tfrac{1}{n})]^2 \geq \frac{\pi}{(N+1)^2} \sum_{j\in[\mathbf{m_n}]} \left[M(z_j,\tfrac{1}{n})\right]^2$$

$$\geq \frac{\pi}{m_n}\left[\sum_{j\in[\mathbf{m_n}]} \frac{1}{N+1} M(z_j,\tfrac{1}{n})\right]^2 = \frac{\pi}{m_n(N+1)^2}\left[\sum_{j\in[\mathbf{m_n}]} M(z_j,\tfrac{1}{n})\right]^2,$$

where the second inequality is because $\frac{M(z_j,1/n)}{m(z_j,1/n)} \leq N+1$ for $j \in [\mathbf{m_n}]$ and the third is an application of the Cauchy-Schwarz inequality in the space $\mathbb{R}^{m_n}$ with the canonical inner product. Since $\mu\big(I \cap D_n\big) \geq \frac{m_n}{n}$,

$$\mu(I\cap D_n)\,\frac{A(y_0+1/n)-A(y_0-1/n)}{2/n} \geq \frac{\pi}{4(N+1)^2}\left[\sum_{j\in[\mathbf{m_n}]} M(z_j,\tfrac{1}{n})\right]^2.$$

As $\lim \mu(I \cap D_n) = \mu(F)$, the left limit of the preceding inequality when $n \to +\infty$ is $\mu(F)A'(y_0)$. To get the right limit of the inequality, as for each $n_0 \in \mathbb{N}$, it is $f(F_{n_0}) \subset \cup_{j\in[\mathbf{m_n}]} f\big(D_n(z_j)\big)$ for all $n \geq n_0$, and $f\big(D_n(z_j)\big)$ is included in a disk with radius $M\big(z_j,\frac{1}{n}\big)$, there is a cover of $f(F_{n_0})$ by disks with sum of diameters $2\sum_{j\in[\mathbf{m_n}]} M(z_j,\frac{1}{n})$. Since f is uniformly continuous in R, the largest $M\big(z_j,\frac{1}{n}\big)$ approaches 0 when $n \to +\infty$. Hence

$$\big[\mu(f(F))\big]^2 = \lim_{n_0\to+\infty} \big[\mu(f(F_{n_0}))\big]^2 \leq 4 \varliminf_{n\to+\infty} \left[\sum_{j\in[\mathbf{m_n}]} M\big(z_j,\tfrac{1}{n}\big)\right]^2 \leq \tfrac{16}{\pi}(N+1)^2\mu(F)A'(y_0),$$

and, therefore, $\big[\mu\big(f(F)\big)\big]^2 \leq \frac{16}{\pi}(N+1)^2\mu(F)A'(y_0)$ for every compact set $F \subset I$ in which H_f is upper bounded by N.

We extend the inequality at the end of the preceding paragraph from compact subsets F of I to Borel subsets of I. Since $H_f(z)$ is finite for $z \in I$, so is $\Phi(z) = \sup\big\{\frac{M(z,r)}{m(z,r)} : r > 0 \text{ and } z+r,\, z-r \in I\big\}$ for $z \in I$. If E is a compact subinterval of I and $I_N = \{z : \Phi(z) \leq N\}$, the set $E_N = I_N \cap E$ is closed. $H_f(z) \leq N$ for $z \in E_N$ and, by the last inequality in the preceding paragraph, $\mu\big(f(E_N)\big)$ is finite. $E = \cup_{N\in\mathbb{N}} E_N$ and, since I is a countable union of line segments which includes the end points, $f(I)$ is also a countable union of sets with finite measure μ. If B is a Borel subset of I where H_f is upper bounded by N, $f(B)$ is also a Borel set, and there is a sequence of closed sets $\{G'_k\} \subset f(B)$ such that $\lim \mu(G'_k) = \mu(B')$. The preimages of these sets $G_k = f^{-1}(G'_k)$ are also closed. Therefore, the last inequality in the preceding paragraph holds with F replaced by G_k for $k \in \mathbb{N}$. Thus, the inequality also holds with F replaced by B.

Finally, the last inequality of two paragraphs above is extended to Borel subsets of I even if H_f is not upper bounded in this set. Let B_0 be a Borel subset of I with $\mu(B_0) = 0$. In the set $I_N \cap B_0$, the function H_f is upper bounded by N. As $\mu(I \cap B_0) = 0$, also $\mu\big(f(I \cap B_0)\big) = 0$. Since $f(B_0) = \cup_{N\in\mathbb{N}} f(I_N \cap B_0)$, also $\mu\big(f(B_0)\big) = 0$. As $H_f \leq K$ in $I \setminus B_0$ for some $B_0 \subset I$ with $\mu(B_0) = 0$, if B is a Borel subset of I, then $\mu\big(f(B)\big) = \mu\big((B \setminus B_0)'\big)$ and $H_f \leq K$ in $B \setminus B_0$. By the last inequality of three paragraphs above, $\left[\mu\left(f(B)\right)\right]^2 \leq \frac{16}{\pi}(K+1)^2 A'(y_0)\mu(B)$. Therefore, the homeomorphism f is absolutely continuous in I. This concludes the proof that f is absolutely continuous in $]a, b[\times\{y\}$ a.e. for $y \in]c, d[$.

Similarly, we get f absolutely continuous in $]c, d[\times\{x\}$ a.e. for $x \in]a, b[$.

The orientation preserving homeomorphism f is absolutely continuous on lines in Ω. So, to guarantee that f is a K-quasiconformal mapping, by (1.28), it remains to prove that $|f_{\bar{z}}| \leq k|f_z|$ a.e. in Ω, with $k = \frac{K-1}{K+1}$. Since f is absolutely continuous on lines in Ω, it has finite partial derivatives a.e. and, by the Gehring-Lehto theorem (E.3), f is differentiable a.e. as a function of the corresponding region of $\mathbb{R}^2$ to $\mathbb{R}^2$. Let $z \in \Omega$ be a point where f is differentiable. If f is regular in z, then $|f_{\bar{z}}| < |f_z|$. It is $H_f(z) = \frac{|f_z| + |f_{\bar{z}}|}{|f_z| - |f_{\bar{z}}|}$, and $\frac{|f_z| + |f_{\bar{z}}|}{|f_z| - |f_{\bar{z}}|} \leq K$, which is equivalent to $|f_{\bar{z}}| \leq k|f_z|$ with $k = \frac{K-1}{K+1}$. If f is not regular in z, then $|f_{\bar{z}}| = |f_z|$, and $\lim_{r\to 0} \frac{m(z,r)}{r} = 0$. As $H_f(z) \leq K$, it is also $\lim_{r\to 0} \frac{M(z,r)}{r} = 0$. So, $|f_z| + |f_{\bar{z}}| = 0$, $|f_z| = |f_{\bar{z}}| = 0$, and $|f_{\bar{z}}| \leq k|f_z|$. □

As circular dilatation only involves the notion of distance, it can be used to define quasiconformal mapping in any metric space.

3.8 Quasiconformity of Monotone Functions

In 2007, L. Kovalev proved that δ-monotone functions in a Hilbert space of dimension > 1 are quasiconformal mappings, and in 2008, K. Astala, T. Iwaniec, and G.J. Martin simplified the proof for functions in $\mathbb{R}^n$, which is included here for functions in the plane because monotonic functions are simple to describe and useful in elliptic partial differential equations and calculus of variations, and to see them as examples of quasiconformal mappings is likely to increase the readers understanding of quasiconformity.[23]

If H is an Euclidean space and $\delta \in [0, 1]$, a **function δ-monotone** in $\Omega \subset H$ is $\mathbf{f}: \Omega \to H$ such that

$$\langle \mathbf{f}(\mathbf{z}) - \mathbf{f}(\mathbf{w}), \mathbf{z} - \mathbf{w} \rangle \geq \delta \, \|\mathbf{f}(\mathbf{z}) - \mathbf{f}(\mathbf{w})\| \, \|\mathbf{z} - \mathbf{w}\| , \quad \mathbf{z}, \mathbf{w} \in \Omega ;$$

0-monotone functions are called **monotone functions**.

It is geometrically clear that δ-monotone functions[24] are those for which there is a $\delta > 0$ such that cosines of angles of the increases of function values with the

[23] This section follows parts of K. Astala, T. Iwaniec, and G.J. Martin. Monotone maps of $\mathbb{R}^n$ are quasiconformal, *Meth. Appl. Analysis* **85** (2008), 31–38. Kovalev, L.V., Quasiconformal geometry of monotone mappings, *J. Lond. Math. Soc.*, **75** (2007), 391–408.

[24] They were defined for linear operators and $\delta > 0$ in 1957 by Pavel Sobolevsky (1930–2018). The study of monotone functions was introduced by G. Minty (1929–1986) in 1960 for electric circuits and Felix Browder (1927–2016) in 1972 for partial differential equations.

corresponding increases of the independent variable are $\geq \delta$, i.e., the absolute values of such angles are $\leq \arccos\delta$, which for $\delta \in [0, 1]$ are acute or right angles. In $H = \mathbb{R}$, the δ-monotone functions in $\Omega \subset H$ are simply the nondecreasing functions (they may fail to be injective or continuous), but if $\dim H > 1$ and Ω is open, when they are not constant, they are injective, continuous, and open, as proved below.

Since the canonical inner product in $\mathbb{R}^2$ $\langle (x_1, y_1), (x_2, y_2)\rangle = x_1x_2 + y_1y_2$, corresponds in $\mathbb{C}$ with $z_j = x_j + iy_j$ with $x_j, y_j \in \mathbb{R}$ for $j = 1, 2$, to $\mathcal{R}e(z_1\overline{z_2})$, a δ-**monotone function** in $\Omega \subset \mathbb{C}$ is $f : \Omega \to \mathbb{C}$ such that

$$\mathcal{R}e\big([f(z) - f(w)](\overline{z} - \overline{w})\big) \geq \delta\, |f(z) - f(w)|\, |z - w|\,, \quad z, w \in \Omega\,.$$

We want to prove that a δ-monotone function, with $\delta \in]0, 1]$, in a region is a homeomorphism onto its range preserving orientation with finite circular dilatation at all points of its domain bounded a.e., and afterward to apply results of the preceding section. A large part of this proof is the same in $\mathbb{R}^n$ with $n \geq 2$; so, we begin by describing it in this setting but will restrict to $n = 2$ at the end, when it would be necessary for larger n to use stronger results than those included in this book.

If $\mathbf{v}, \mathbf{z} \in \mathbb{R}^n$, with $\mathbf{v} \neq \mathbf{z}$, $n \geq 2$, the **cone with vertices v, symmetry axis z − v, and width** δ is $C^\delta_{\mathbf{v}}(\mathbf{z}) = \big\{\mathbf{w} \in \mathbb{R}^n \setminus \{\mathbf{v}\} : \big\| \frac{\mathbf{w}-\mathbf{v}}{\|\mathbf{w}-\mathbf{v}\|} - \frac{\mathbf{z}-\mathbf{v}}{\|\mathbf{z}-\mathbf{v}\|} \big\| < \frac{\delta}{2}\big\}$, which has angular width θ such that $2\sin\frac{\theta}{4} = \frac{\delta}{2}$, i.e., $\theta = 4\arcsin\frac{\delta}{4}$.

If $\Omega \subset \mathbb{R}^n$, $n \geq 2$, and $\mathbf{f} : \Omega \to \mathbb{R}^n$ is δ-monotone,

$$\begin{aligned}\Big\langle \mathbf{f}(\mathbf{w}) - \mathbf{f}(\mathbf{v}), \tfrac{\mathbf{z}-\mathbf{v}}{\|\mathbf{w}-\mathbf{v}\|}\Big\rangle &= \Big\langle \mathbf{f}(\mathbf{w}) - \mathbf{f}(\mathbf{v}), \tfrac{\mathbf{w}-\mathbf{v}}{\|\mathbf{w}-\mathbf{v}\|}\Big\rangle - \Big\langle \mathbf{f}(\mathbf{w}) - \mathbf{f}(\mathbf{v}), \tfrac{\mathbf{w}-\mathbf{v}}{\|\mathbf{w}-\mathbf{v}\|} - \tfrac{\mathbf{z}-\mathbf{v}}{\|\mathbf{z}-\mathbf{v}\|}\Big\rangle \\ &\geq \big(\delta - \tfrac{\delta}{2}\big)\|\mathbf{f}(\mathbf{w}) - \mathbf{f}(\mathbf{v})\| = \tfrac{\delta}{2}\,\|\mathbf{f}(\mathbf{w}) - \mathbf{f}(\mathbf{v})\|\,, \quad \mathbf{v}, \mathbf{z} \in \Omega\,, \mathbf{v} \neq \mathbf{z}\,, \mathbf{w} \in C^\delta_{\mathbf{v}}(\mathbf{z}) \cap \Omega\,,\end{aligned}$$

Hence, if $\mathbf{x} \in C^\delta_{\mathbf{v}}(\mathbf{z}) \cap C^\delta_{\mathbf{z}}(\mathbf{v}) \cap \Omega$, then

$$\begin{aligned}\|\mathbf{f}(\mathbf{x}) - \mathbf{f}(\mathbf{v})\| + \|\mathbf{f}(\mathbf{x}) - \mathbf{f}(\mathbf{z})\| &\leq \tfrac{2}{\delta}\Big\langle [\mathbf{f}(\mathbf{x}) - \mathbf{f}(\mathbf{v})] - [\mathbf{f}(\mathbf{x}) - \mathbf{f}(\mathbf{z})], \tfrac{\mathbf{z}-\mathbf{v}}{\|\mathbf{x}-\mathbf{v}\|}\Big\rangle \\ &= \tfrac{2}{\delta}\Big\langle \mathbf{f}(\mathbf{z}) - \mathbf{f}(\mathbf{v}), \tfrac{\mathbf{z}-\mathbf{v}}{\|\mathbf{x}-\mathbf{v}\|}\Big\rangle \leq \tfrac{2}{\delta}\,\|\mathbf{f}(\mathbf{z}) - \mathbf{f}(\mathbf{v})\|\,,\end{aligned}$$

the last inequality by the Cauchy-Schwarz inequality. The intersection of the cones $C^\delta_{\mathbf{v}}(\mathbf{z}) \cap C^\delta_{\mathbf{z}}(\mathbf{v})$ includes $B_{r_{\mathbf{z},\mathbf{v}}}\big(\frac{1}{2}(\mathbf{z}+\mathbf{v})\big)$, with (Fig. 3.6)

$$r_{\mathbf{z},\mathbf{v}} = \sin\big(\tfrac{\theta}{2}\big)\tfrac{1}{2}\|\mathbf{z}-\mathbf{v}\| = \sin\big(\tfrac{\theta}{2}\big)\cos\big(\tfrac{\theta}{2}\big)\|\mathbf{z}-\mathbf{v}\| = \tfrac{\delta}{4}\sqrt{1 - \tfrac{\delta^2}{16}}\,\|\mathbf{z}-\mathbf{v}\| > \tfrac{\sqrt{15}}{16}\,\delta\,\|\mathbf{z}-\mathbf{v}\|\,.$$

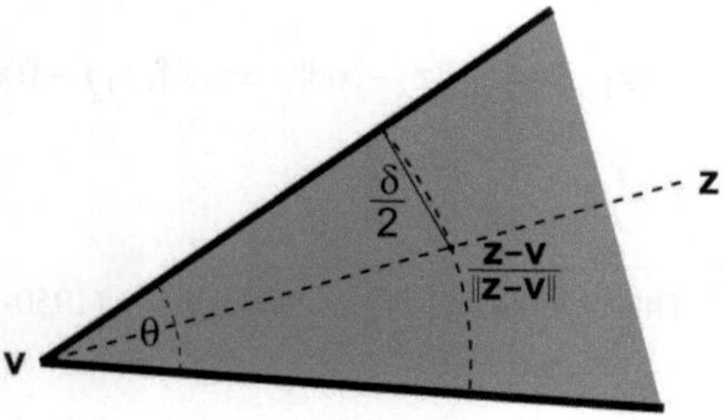

Fig. 3.6 Cone $C^\delta_{\mathbf{v}}(\mathbf{z})$ with vertices at $\mathbf{v}$, symmetry axis $\mathbf{z}-\mathbf{v}$, and width δ, corresponding to angular width $\theta = 4\arcsin\frac{\delta}{4}$

(3.31) *Let $\delta\in\,]0,1]$ and $\Omega\subset\mathbb{R}^n$, $n\geq 2$, be a region including* cl $B_{3/\delta}$. *There is $H>0$ depending only on δ and Ω such that if for $\mathbf{u}\in\partial B_1\subset\mathbb{R}^n$ $\mathcal{F}_{\mathbf{u}}$ is the set of δ-monotone functions in Ω and $\mathbf{h}:\Omega\to\mathbb{R}^n$ is such that $\mathbf{h}(0)=0$ and $\|\mathbf{h}(\mathbf{u})\|=1$, then $\|\mathbf{h}(\mathbf{z})\|\leq H$ for all $\mathbf{h}\in\mathcal{F}_{\mathbf{u}}$, $\mathbf{z},\mathbf{u}\in\partial B_1\subset\mathbb{R}^n$.*

Proof Let $X=\{\mathbf{z}\in\Omega:\ \sup_{\mathbf{h}\in\mathcal{F}_{\mathbf{u}}}\|\mathbf{h}(\mathbf{z})\|\in\mathbb{R}\}$, which includes $\{0,1\}$. Since if $\mathbf{v},\mathbf{z}\in\Omega$, $\mathbf{v}\neq\mathbf{z}$ and, by the last paragraph preceding the statement, if $\mathbf{x}\in C^{\delta}_{\mathbf{v}}(\mathbf{z})\cap C^{\delta}_{\mathbf{z}}(\mathbf{v})\cap\Omega$, then $\|\mathbf{f}(\mathbf{x})-\mathbf{f}(\mathbf{v})\|+\|\mathbf{f}(\mathbf{x})-\mathbf{f}(\mathbf{z})\|\leq\frac{2}{\delta}\|\mathbf{f}(\mathbf{z})-\mathbf{f}(\mathbf{v})\|$, and the set X is the intersection of a convex set with Ω. By (VII.5), if it were $X\neq\Omega$, there would be $\mathbf{z_0}\in\Omega\setminus X$ and $\mathbf{z_1},\mathbf{z_2}\in X$ such that the angle of the vectors $\mathbf{z}_1-\mathbf{z}_0$ and $\mathbf{z}_2-\mathbf{z}_0$ would be arbitrarily close to π. As it would be $\|\mathbf{h}(\mathbf{z}_1)\|$, $\|\mathbf{h}(\mathbf{z}_2)\|\in\mathbb{R}$ and there would exist a sequence $\{\mathbf{h}_m\}\subset\mathcal{F}_{\mathbf{u}}$ such that $\|\mathbf{h}(\mathbf{z}_0)\|\to+\infty$, the sequence of angles of the vectors $\mathbf{h}_m(\mathbf{z}_1)-\mathbf{h}_m(\mathbf{z}_0)$ and $\mathbf{h}_m(\mathbf{z}_2)-\mathbf{h}_m(\mathbf{z}_0)$ would converge to 0 and for every $\varepsilon>0$ there would exist $M\in\mathbb{N}$ such that for some of the $j\in\{1,2\}$ the angle of the vectors $\mathbf{h}_M(\mathbf{z}_j)-\mathbf{h}_M(\mathbf{z}_j)$ and $\mathbf{z}_j-\mathbf{z}_0$ would be $>\frac{\pi}{2}-\varepsilon$, contradicting the δ-monotonicity of $\mathbf{h}_M$ in Ω. So, $X=\Omega$, i.e., at each point $\mathbf{z}\in\Omega$ $\{\|\mathbf{h}(\mathbf{z})\|:\mathbf{h}\in\mathcal{F}_{\mathbf{u}}\}$ is upper bounded and, therefore, it has a supremum $S_{\mathbf{z}}$. We want to prove there is an upper bound independent of $\mathbf{z}$. Since cl $B_{3/\delta}\subset\Omega$, it is $\pm\mathbf{w}\in\left(\frac{3}{\delta},0,\ldots,0\right)\in\Omega$. Denoting $s=\sup_{\mathbf{h}\in\mathcal{F}_{\mathbf{u}}}\{\|\mathbf{h}(\mathbf{w})\|+\|\mathbf{h}(-\mathbf{w})\|\}\in\mathbb{R}$, by the last paragraph before the statement, $B_{r_{\mathbf{w},-\mathbf{w}}}\subset C^{\delta}_{\mathbf{w}}(-\mathbf{w})\cap C^{\delta}_{-\mathbf{w}}(\mathbf{w})$ and

$$\|\mathbf{h}(\mathbf{z})-\mathbf{h}(\mathbf{w})\|+\|\mathbf{h}(\mathbf{z})-\mathbf{h}(-\mathbf{w})\|\leq\tfrac{2}{\delta}\|\mathbf{h}(\mathbf{w})-\mathbf{h}(-\mathbf{w})\|\leq\tfrac{2}{\delta}s\,,\quad\mathbf{z}\in B_{r_{\mathbf{w},-\mathbf{w}}}\cap\Omega\,.$$

By the triangular inequality, $\|\mathbf{h}(\mathbf{z})\|\leq\frac{1}{2}\left(\frac{2}{\delta}s+2s\right)=\left(\frac{1}{\delta}+1\right)s$, $\mathbf{z}\in B_{r_{\mathbf{w},-\mathbf{w}}}\cap\Omega$. Hence, $\|\mathbf{h}(\mathbf{z})\|\leq\left(\frac{1}{\delta}+1\right)s$ for $\mathbf{z}\in\Omega$, $\mathbf{h}\in\mathcal{F}_{\mathbf{u}}$. We want to prove that this upper bound does not depend on $\mathbf{u}$, for what it suffices to verify that δ-monotonicity is invariant under linear mappings with matrix representation in the canonical basis of $\mathbb{R}^n$ by orthogonal matrices O, what results from having, for all $\mathbf{z},\mathbf{w}\in\mathbb{R}^n$ such that $O\mathbf{z}$, $O\mathbf{w}\in\Omega$,

$$\begin{aligned}\langle O^t\mathbf{h}(O\mathbf{z})-O^t\mathbf{h}(O\mathbf{w})\,,\mathbf{z}-\mathbf{w}\rangle&=\langle\mathbf{h}(O\mathbf{z})-\mathbf{h}(O\mathbf{w},O\mathbf{z}-O\mathbf{w}\rangle\\ &\geq\delta\|\mathbf{h}(O\mathbf{z})-\mathbf{h}(O\mathbf{w}\|\,\|O\mathbf{z}-O\mathbf{w}\|=\delta\|O^t\mathbf{h}(O\mathbf{z})-O^t\mathbf{h}(O\mathbf{w})\|\,\|\mathbf{z}-\mathbf{w}\|.\end{aligned}$$

□

A **Weakly quasisymmetric function**[25] in $\Omega\subset\mathbb{R}^n$ is a function $\mathbf{f}:\Omega\to\mathbb{R}^n$ for which there is $S>0$ such that

$$\|\mathbf{z}_1-\mathbf{w}\|\leq\|\mathbf{z}_2-\mathbf{w}\|\implies\|\mathbf{f}(\mathbf{z}_1)-\mathbf{f}(\mathbf{w})\|\leq S\,\|\mathbf{f}(\mathbf{z}_2)-\mathbf{f}(\mathbf{w})\|\,,\qquad\mathbf{z}_1,\mathbf{z}_2,\mathbf{w}\in\Omega\,.$$

[25] This definition is due to Pekka Tukia (1950-) and J. Väisälä in 1980.

The following result was obtained in 1980 by P. Tukia and J. Väisälä.[26]

(3.32) *Every function $\mathbf{f}:\Omega\to\mathbb{R}^n$, with $\emptyset\neq\Omega\subset\mathbb{R}^n$, weakly quasisymmetric and nonconstant is injective and continuous.*

Proof Let $\mathbf{w}\in\Omega$, $A=\mathbf{f}^{-1}(\{\mathbf{f}(\mathbf{w})\})\setminus\{\mathbf{w}\}$ and $B=\Omega\setminus\mathbf{f}^{-1}(\{\mathbf{f}(\mathbf{w})\})$. If there is $\mathbf{z}_2\in A$, by the definition of weakly quasisymmetric function, $\mathbf{f}$ has the value $\mathbf{f}(\mathbf{w})$ in all $\operatorname{cl} B_{\|\mathbf{z}_2-\mathbf{w}\|}(\mathbf{z}_2)\cap\Omega$ and A is open. If there is $\mathbf{z}_1\in B$ and a ball centered at $\mathbf{z}_1$ included in $B=\Omega\setminus\mathbf{f}^{-1}(\{\mathbf{f}(\mathbf{w})\})$, arbitrarily close to $\mathbf{z}_1$ there are points $\mathbf{x}\in A\cup\{\mathbf{w}\}$, and, by the weakly quasisymmetric function definition, $\mathbf{f}$ is not $\mathbf{f}(\mathbf{w})$ in $\Omega\setminus\operatorname{cl} B_{\|\mathbf{z}_1-\mathbf{x}\|}(\mathbf{z}_1)$, contradicting $\mathbf{w}$ to belong to this set for $\|\mathbf{z}_1-\mathbf{x}\|$ small because $\mathbf{z}_1\neq\mathbf{w}$; therefore, there is a ball centered at $\mathbf{z}_1$ included in B and B is open. So, if $A\neq\emptyset$ or $B\neq\emptyset$, it is an open set, and since $\Omega\setminus\{\mathbf{w}\}=A\cup B$ is connected (because Ω is connected and open), it is $A=\emptyset$ or $B=\emptyset$. As $\mathbf{f}$ is not constant, $B\neq\emptyset$, and $A=\emptyset$, i.e., $\mathbf{f}^{-1}(\{\mathbf{f}(\mathbf{w})\})=\mathbf{w}$. Since $\mathbf{w}$ is an arbitrary point of Ω, $\mathbf{f}$ is injective.

If $\mathbf{f}$ were not continuous at $\mathbf{z}_0\in\Omega$, there would exist $\varepsilon>0$ and a sequence $\{\mathbf{z}_m\}\subset\Omega$ such that $\mathbf{z}_m\to\mathbf{z}_0$ and $\|\mathbf{f}(\mathbf{z}_m)-\mathbf{f}(\mathbf{z}_0)\|\geq\varepsilon$, satisfying, without loss of generality, $\|\mathbf{z}_{m+1}-\mathbf{z}_m\|\leq\frac{1}{2}\|\mathbf{z}_m-\mathbf{z}_0\|$ (if necessary, replacing the sequence by a subsequence), and, therefore, $\|\mathbf{z}_{m+1}-\mathbf{z}_0\|\leq\|\mathbf{z}_{m+1}-\mathbf{z}_m\|$. By the triangle inequality and the definition of weakly quasisymmetric function

$$\|\mathbf{f}(\mathbf{z}_m)\|-\|\mathbf{f}(\mathbf{z}_0)\|\leq\|\mathbf{f}(\mathbf{z}_m)-\mathbf{f}(\mathbf{z}_0)\|\leq S\|\mathbf{f}(\mathbf{z}_1)-\mathbf{f}(\mathbf{z}_0)\|\,.$$

Thus, the sequence $\{\mathbf{f}(\mathbf{z}_m)\}$ is bounded and, therefore, its terms belong to a compact subset of $\mathbb{R}^n$, implying the existence of a subsequence converging to some $\mathbf{a}\in\mathbb{R}^n$. From the definition of weakly quasisymmetric function

$$0<\varepsilon\leq\|\mathbf{f}(\mathbf{z}_{m+1})-\mathbf{f}(\mathbf{z}_0)\|\leq S\|\mathbf{f}(\mathbf{z}_{m+1})-\mathbf{f}(\mathbf{z}_m)\|\,.$$

Hence, $0<\varepsilon\leq\lim S\|\mathbf{f}(\mathbf{z}_{m+1})-\mathbf{f}(\mathbf{z}_m)\|=S\|\mathbf{a}-\mathbf{a}\|=0$, what is contradictory. Consequently, $\mathbf{f}$ is continuous. □

(3.33) *If $\mathbf{h}:\Omega\to\mathbb{R}^n$ are δ-monotone injective and continuous nonconstant functions defined in a region $\Omega\subset\mathbb{R}^n$, with $n\geq 2$ and $\delta\in\,]0,1]$, then there is a function $H:\,]0,1]\to\,]0,+\infty[$ such that if $\mathbf{z}\in\Omega$ and $B_R(\mathbf{z})\subset\Omega$ with*

(continued)

[26] In Appendix D, it is shown with the invariance of domain theorem on $\mathbb{R}^n$ that every invertible function $\mathbf{f}$ in a region $\Omega\subset\mathbb{R}^n$ is a homeomorphism of Ω onto $\mathbf{f}(\Omega)$. Thus, the present result of P. Tukia and J. Väisälä states that if $\mathbf{f}$ is a nonconstant weakly quasisymmetric function, it is a homeomorphism of Ω onto $\mathbf{f}(\Omega)$. In Appendix D, for simplicity, the invariance of domain theorem is proved only in $\mathbb{R}^2$, and it is proved that a nonconstant weakly quasisymmetric function $\mathbf{f}$ in a region $\emptyset\neq\Omega\subset\mathbb{R}^2$ is a homeomorphism of Ω onto $\mathbf{f}(\Omega)$.

(3.33) (continued)
$R>0$; so, $\mathbf{h}$ is a weakly quasisymmetric function on $B_{\delta R/6}(\mathbf{z})$ with **circular dilatation**

$$H_{\mathbf{h}}(\mathbf{z}) \overset{def}{=} \overline{\lim_{r\to 0+}} \frac{\max_{\partial B_r(\mathbf{z})} \|\mathbf{h}-\mathbf{h}(\mathbf{z})\|}{\min_{\partial B_r(\mathbf{z})} \|\mathbf{h}-\mathbf{h}(\mathbf{z})\|} \leq H(\delta)\,, \quad r\in]0, \tfrac{\delta}{3}R[\,.$$

Proof If $\mathbf{h}$ were not injective, $X=\{\mathbf{z}\in\Omega\setminus\{\mathbf{v}\}:\mathbf{h}(\mathbf{z})=\mathbf{h}(\mathbf{v})\}\neq\emptyset$ for some $\mathbf{v}\in\Omega$. As in the first paragraph of the proof of (3.31), X would be the intersection of a convex set with Ω and, therefore, $X=\Omega$; so, $\mathbf{h}$ would be constant, contradicting the hypothesis. Thus, $\mathbf{h}$ is injective.

For $\mathbf{z}\in\Omega$, there is $R>0$ such that cl $B_R(\mathbf{z})\subset\Omega$. If $0<r<R$, as $\mathbf{h}$ is injective, $\|\mathbf{h}-\mathbf{h}(\mathbf{z})\|>0$ and it has an infimum $m_r\geq 0$ in $\partial B_r(\mathbf{z})$. Let $\{\mathbf{w}_k\}\subset\partial B_1$ be a sequence such that $\|\mathbf{h}(\mathbf{z}+r\mathbf{w}_k)-\mathbf{h}(\mathbf{z})\|\to m_r$. Since ∂B_1 is compact, there is a subsequence of $\{\mathbf{w}_k\}$ converging to some $\widetilde{\mathbf{w}}_r\in\partial B_1$. Thus, there is a subsequence $\{\mathbf{w}_k\}\subset\partial B_1$ such that

$$\mathbf{w}_k\to\widetilde{\mathbf{w}}_r\,, \quad \|\mathbf{h}(\mathbf{z}+r\mathbf{w}_k)-\mathbf{h}(\mathbf{z})\|\to m_r\,, \quad \|\mathbf{h}(\mathbf{z}+r\widetilde{\mathbf{w}}_k)-\mathbf{h}(\mathbf{z})\|\geq m_r\,.$$

Since $\mathbf{h}$ is injective, the function $\mathbf{g}_r(\mathbf{x})=\frac{\mathbf{h}(\mathbf{z}+r\mathbf{x})-\mathbf{h}(\mathbf{z})}{\|\mathbf{h}(\mathbf{z}+r\widetilde{\mathbf{w}}_r)-\mathbf{h}(\mathbf{z})\|}$ is defined in all $\mathbf{x}\in B_{R/r}$ and satisfies $\mathbf{g}_r(0)=0$ and $\|\mathbf{g}_r(\widetilde{\mathbf{w}}_r)\|=1$. For $r\in]0,\frac{\delta}{3}R[$ and $\varepsilon>0$ small, $\mathbf{g}_r$ is defined in $B_{3/\delta+\varepsilon}$ and, since δ-monotonicity of functions is invariant with translations in the domain or in the range and scalings of the functions values, $\mathbf{g}_r$ is δ-monotone in $B_{3/\delta+\varepsilon}$. Therefore, with a constant H depending only on δ and ε whose existence is established in (3.31), we have $\|\mathbf{g}_r\|\leq H$ in ∂B_1. Thus,

$$\frac{\sup_{\partial B_r(\mathbf{z})}\|\mathbf{h}-\mathbf{h}(\mathbf{z})\|}{\inf_{\partial B_r(\mathbf{z})}\|\mathbf{h}-\mathbf{h}(\mathbf{z})\|}\leq H\,, \quad r\in]0,\tfrac{\delta}{3}R[\,.$$

Since $\varepsilon>0$ is arbitrary, if it is small enough, the constant H can be taken to be a function of only δ, $H=H(\delta)$. Therefore,

$$\|\mathbf{h}(\mathbf{w}_1)-\mathbf{h}(\mathbf{z})\|\leq H\|\mathbf{h}(\mathbf{w}_2)-\mathbf{h}(\mathbf{z})\|\,, \quad \text{if } \|\mathbf{w}_1-\mathbf{z}\|=\|\mathbf{w}_2-\mathbf{z}\|<\tfrac{\delta}{3}R\,.$$

Given three noncollinear points $\mathbf{z}_1,\mathbf{z}_2,\mathbf{w}\in\mathbb{R}^n$, there is a unique circle containing them; if $\mathbf{z}_0$ and d are the circle, resp., center and radius, then

$$\|\mathbf{z}_1-\mathbf{z}_0\|=\|\mathbf{z}_2-\mathbf{z}_0\|=\|\mathbf{w}-\mathbf{z}_0\|=d\,,$$
$$\|\mathbf{z}_1-\mathbf{w}\|\leq\|\mathbf{z}_1-\mathbf{z}_0\|+\|\mathbf{w}-\mathbf{z}_0\|\leq 2d\,, \quad \|\mathbf{z}_2-\mathbf{w}\|\leq\|\mathbf{z}_2-\mathbf{z}_0\|+\|\mathbf{w}-\mathbf{z}_0\|\leq 2d\,.$$

If $\|\mathbf{z}_1-\mathbf{w}\|\leq\|\mathbf{z}_2-\mathbf{w}\|$, the straight line orthogonal to the line segment from $\mathbf{z}_1$ to $\mathbf{z}_2$ and passing through its middle point intersects the line segment from $\mathbf{z}_2$ to $\mathbf{w}$ at a point $\mathbf{z}$ equidistant of $\mathbf{z}_1$ and $\mathbf{z}_2$, and $\|\mathbf{z}-\mathbf{z}_0\|\leq d$ (Fig. 3.7). Since the triangle with vertices $\mathbf{z}_1,\mathbf{z}_2,\mathbf{w}$ is inscribed in the circle with radius d and center $\mathbf{z}_0$, it is

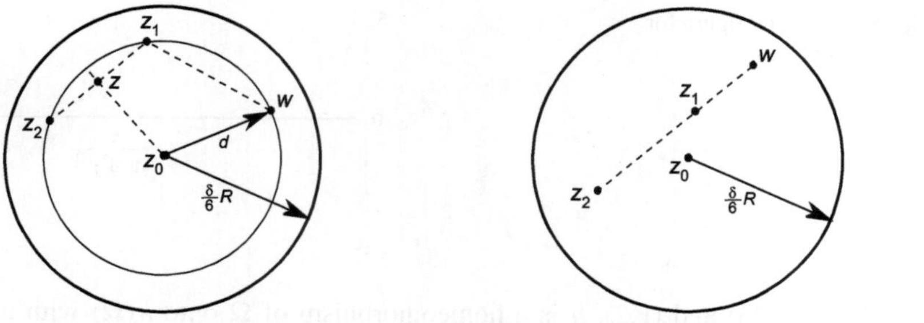

Fig. 3.7 Auxiliar figure for the proof of (3.33), at left for noncollinear $\mathbf{z}_1, \mathbf{z}_2, \mathbf{w}$ and at right collinear

$\|\mathbf{z}_1-\mathbf{z}\| = \|\mathbf{z}_2-\mathbf{z}\| \le 2d$. Therefore, if $R>0$ is such that $B_R(\mathbf{z}_0) \subset \Omega$ and $d < \frac{\delta}{6}R$, for $\mathbf{z}_1, \mathbf{z}_2, \mathbf{w} \in B_{\frac{\delta}{6}R}(\mathbf{z}_0) \subset B_R(\mathbf{z}_0) \subset \Omega$ not lying in a same straight line and such that $\|\mathbf{z}_1-\mathbf{w}\| \le \|\mathbf{z}_1-\mathbf{w}\|$, then

$$\begin{aligned}\|\mathbf{h}(\mathbf{z}_1)-\mathbf{h}(\mathbf{w})\| &\le \|\mathbf{h}(\mathbf{z}_1)-\mathbf{h}(\mathbf{z})\|+\|\mathbf{h}(\mathbf{z})-\mathbf{h}(\mathbf{w})\| \\ &\le H\big(\|\mathbf{h}(\mathbf{z}_2)-\mathbf{h}(\mathbf{z})\|+\|\mathbf{h}(\mathbf{z})-\mathbf{h}(\mathbf{w})\|\big) \le \tfrac{4}{\delta}H\,\|\mathbf{h}(\mathbf{z}_2)-\mathbf{h}(\mathbf{w})\|,\end{aligned}$$

where the second inequality follows by the end of the preceding paragraph and $H \ge 1$ and the third from the inequality obtained for δ-monotone functions in $C^\delta_{\mathbf{z}_2} \cap C^\delta_{\mathbf{w}}$. For $\mathbf{z}_1, \mathbf{z}_2, \mathbf{w} \in B_{\frac{\delta}{6}R}(\mathbf{z}_0) \subset B_R(\mathbf{z}_0) \subset \Omega$ collinear points such that $\|\mathbf{z}_1-\mathbf{w}\| \le \|\mathbf{z}_2-\mathbf{w}\|$ (Fig.3.7), by the inequality obtained for δ-monotone functions in $C^\delta_{\mathbf{z}_2} \cap C^\delta_{\mathbf{w}}$,

$$\|\mathbf{h}(\mathbf{z}_1)-\mathbf{h}(\mathbf{w})\| \le \|\mathbf{h}(\mathbf{z}_1)-\mathbf{h}(\mathbf{w})\|+\|\mathbf{h}(\mathbf{z}_1)-\mathbf{h}(\mathbf{z}_2)\| \le \tfrac{2}{\delta}\,\|\mathbf{h}(\mathbf{z}_2)-\mathbf{h}(\mathbf{w})\|.$$

Therefore, $\mathbf{h}$ is a weakly quasisymmetric function in $B_{\delta R/6}(\mathbf{z}_0)$. By (3.32), $\mathbf{h}$ is continuous and, by the Weierstrass theorem of extrema of continuous functions on compact sets, in the upper bound obtained at the end of the preceding paragraph for the quotient of the supremum and the infimum in $\partial B_r(\mathbf{z})$, these are assumed by $\mathbf{h}$ and, therefore, they are the, resp., maximum and minimum of $\mathbf{h}$ in $\partial B_r(\mathbf{z})$. □

As mentioned at the beginning of this section, in 2007 L. Kovalev proved that δ-monotone functions in a Hilbert space H of dimension > 1 are quasiconformal mappings, and in 2008, K. Astala, T. Iwaniec, and J. Martin simplified the proof for $H = \mathbb{R}^n$ and obtained the optimal dilatation as a function of δ , which for complex functions is the following result.

(3.34) *A function $h : \Omega \to \mathbb{C}$ δ-monotone preserving orientation and nonconstant defined in a region $\Omega \subset \mathbb{C}$, with $0<\delta\le 1$, is a K-quasiconformal mapping with optimal $K = \frac{1+\sqrt{1-\delta^2}}{1+\sqrt{1-\delta^2}}$, since there are regions and δ-monotone functions for which this value of K cannot be reduced.*

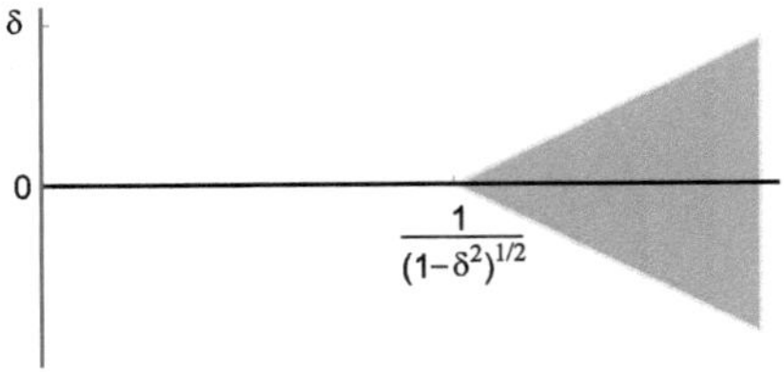

Fig. 3.8 Auxiliar figure for the proof of (3.8)

Proof By (3.33) and (F.2), h is a homeomorphism of Ω onto $h(\Omega)$ with circular dilatation $H_{\mathbf{h}}$ in Ω upper bounded by $H(\delta)$, where $H:]0,1] \to]0,+\infty[$ is a function as in (3.33). As by assumption h preserves orientation, by (3.30), it is a K-quasiconformal mapping with $K \geq H(\delta)$.

Since h is a quasiconformal mapping, by (1.28), it is differentiable a.e. as a function of two real variables and $h_z, h_{\overline{z}} \in L^2_{loc}(\Omega)$. At each point z_0 of differentiability, this derivative is a linear transformation given by $T_{z_0}(z) = \lim\limits_{t\to 0} \frac{1}{t}[h(z_0+tz)-h(z_0)]$, with $t \in \mathbb{R}\setminus\{0\}$. Since h is δ-monotone

$$\lim_{t\to 0} \frac{1}{t^2}\, \mathcal{R}e\big([h(z_0+tz)-h(z_0)]\, t\overline{z}\big) \geq \lim_{t\to 0} \frac{1}{t^2}\, \delta\, |h(z_0+tz)-h(z_0)|\, |t|\, |z|,$$

and $\mathcal{R}e\big(T_{z_0}(z)\,\overline{z}\big) \geq \delta\, |T_{z_0}(z)|\, z|$. Thus, the linear transformation T_{z_0} is a δ-monotone function. The dilatation of h is at each point of differentiability $K_h = \frac{|h_z|+|h_{\overline{z}}|}{|h_z|-|h_{\overline{z}}|}$. To get an upper bound, it suffices to have an upper bound for the linear transformations from $\mathbb{C}$ to $\mathbb{C}$ which are δ-monotone functions.

If $\Omega \subset \mathbb{C}$ and T is a linear transformation, it is of the form $T(z) = az + b\overline{z}$ for some $a, b \in \mathbb{C}$. As δ-monotonicity is invariant under the addition of constants in the domain, to identify conditions under which T is δ-monotone, it suffices to consider the condition in the definition of δ-monotone function in $\mathbb{C}$ with $w = 0$, and since T is a linear transformation, it suffices to check it for $z \in \mathbb{C}$ with $|z| = 1$, in which case this condition is $\mathcal{R}e\big(T(z)\,\overline{z}\big) \geq \delta |T(z)|\, |z|$, i.e., $\mathcal{R}e\big((az+b\,\overline{z})\overline{z}\big) \geq \delta |az+b\overline{z}|$ for $z \in \partial B_1$. If $b \neq 0$ and $\lambda = \frac{\overline{z}}{z}$, this condition is equivalent to $\mathcal{R}e\big(\frac{a}{b}+\lambda\big) \geq \delta \big|\frac{a}{b}+\lambda\big|$ for $\lambda \in \partial B_1$, what geometrically corresponds to the disk with center $\frac{a}{b}$ and radius 1 to be included in the cone $C(\delta) = \big\{x+iy : \delta |y| \leq \big(\sqrt{1-\delta^2}\,\big)x\,,\ \ x, y \in \mathbb{R}\big\}$ with vertices 0. The set of possible disk centers is the cone parallel to $C(\delta)$ with vertices $\frac{1}{\sqrt{1-\delta^2}}$ (Fig. 3.8). Hence, the linear transformation T is δ-monotone if and only if $|b| + \delta\, |\mathcal{I}m\, a| \leq \sqrt{1-\delta^2}\, \mathcal{R}e\, a$. If the linear transformation T is δ-monotone, then $|b| \leq \sqrt{1-\delta^2}\, |a|$, and the dilatation of T is at each point $K_T = \frac{|a|+|b|}{|a|-|b|} \leq \frac{1+\sqrt{1-\delta^2}}{1-\sqrt{1-\delta^2}}$. If $a = 1$ and $b = \sqrt{1-\delta^2}$, the inequality is an equality. It follows that the linear transformations from $\mathbb{C}$ to $\mathbb{C}$ which are δ-monotone also are K-quasiconformal with $K = \frac{1+\sqrt{1-\delta^2}}{1+\sqrt{1-\delta^2}}$ optimal. □

Notice that δ-monotone orientation preserving functions with $\delta \in\,]0,1]$ are a very restricted class of quasiconformal mappings. A very simple example, even with a conformal homeomorphism, is $h(z) = \frac{1}{z}$ defined in the open right complex half plane, which maps the vertical line $R = \{z \in \mathbb{C} : \mathcal{R}e\, z = 1\}$ to an arc of circle C with center and radius $\frac{1}{2}$ except for the point 0, and with $z=1+i$ and $w=1$, the vectors $z-w=1$ and $h(z)-h(w)=\frac{1}{1+i}-1=-\frac{1}{2}(1+i)$ make the obtuse angle $\frac{3\pi}{4}$, so h is not δ-monotone for any $\delta\in\,]0,1]$.

By (3.33) and (F.2), a non-weakly quasisymmetric nonconstant function f in a region $\Omega\subset\mathbb{C}$ is a homeomorphism of Ω onto $f(\Omega)$, and, by (3.30), if it preserves orientation and has bounded circular dilatation H_f, then it is a quasiconformal mapping, but a quasiconformal mapping in Ω may fail to be weakly quasisymmetric. The function $h=h^{-1}$ in the preceding paragraph is a very simple example. $h(C)=R$ and with $w=\frac{1}{2}$, $z_1=1$, $z_2=h(1+iy)$ with $y\in\mathbb{R}$, for $|h(z_2)-h(\frac{1}{2})|\le S|h(z_1)-h(\frac{1}{2})|$ to hold, what, as $h(\frac{1}{2})=2$, is equivalent to $|1+iy-2|\le S|1-2|$, i.e., to $S\ge|iy-1|=1+y^2$, and, since y can be any real number, there is no $S\in\mathbb{R}$ such that h satisfies the definition of weakly quasisymmetric function. Therefore, h is a conformal homeomorphism and it is not a quasisymmetric function. However, it can be proved that for a restricted class of regions $\emptyset\ne\Omega=\mathbb{C}$ which includes $\Omega=\mathbb{C}$, $f:\Omega\to\mathbb{C}$ is a quasiconformal mapping if and only if it is weakly quasisymmetric.[27]

3.9 Application to Leau-Fatou Conjugacy

The measurable Riemann mapping theorem in Chap. 2 and the Mori theorem earlier in this chapter can be applied to obtain the Leau-Fatou conjugacy in the neighborhood of a parabolic fixed point of the dynamics of iterations of holomorphic functions.[28] to the affine function $z\mapsto 1+z$ differently than in *CADOVA*.[29] It is a first example of application of quasiconformal mappings exhibiting the flexibility they allow and an introductory anticipation of quasiconformal surgery, the central subject of Chap. 6.

A function f holomorphic in an open set $U\subset\mathbb{C}$ is assumed to have a fixed point (without loss of generality) at 0 with multiplier 1, $f''(0)\ne 0$ (the cases where the multiplier is 1, $f''(0)=0$ and $f^{(k)}(0)\ne 0$ for some $k\in\mathbb{N}$, or the multiplier is a root of unity of order $p\in\mathbb{N}\setminus\{1\}$ can be treated from the considered case with changes of variables) .

The function $z\mapsto\frac{1}{2}f''(0)\,z$ is a conjugacy of f with a function given by a series of the form $z+z^2+\sum_{j=3}^{\infty}a_jz^j$, and $z\mapsto-\frac{1}{z}$ is a conjugacy of this function with a function given by a series of the form $g(z)=1+z+\sum_{k=2}^{\infty}\frac{b_k}{z^k}$ in a neighborhood

---

[27] For example, see Väisälä, J., The free quasiworld. Freely quasiconformal and related maps in Banach spaces, 55–118, in *Quasiconformal Geometry and Dynamics Banach Center Publications*, vol 48, Institute of Mathematics, Polish Academy of Sciences, Warszawa, 1999, and the references cited in this article.

[28] For these concepts and basic aspects of Complex Dynamics, see Chapter 13 of *CADOVA*.

[29] This section follows the exposition in the book by L.Carleson and T.Gamelin listed in the final bibliography. Carleson, Lennart (1928-). Gamelin, Theodore William (1939-). See Section 9 of Chapter 13 of *CADOVA*.

of ∞. Let $M > 0$ be such that $V = \{z \in \mathbb{C} : \mathcal{R}e\, z > M\}$ is invariant under g, i.e., $g(V) \subset V$, S be the band in $\mathbb{C}$ bounded by the straight line $\mathcal{R}e\, z = M$ and its image by the function g, and Σ the vertical band $\Sigma =]0, 1[\times\mathbb{R}\subset\mathbb{C}$. We consider a diffeomorphism $h : \Sigma \to S$ which on the straight line $\{0\}\times\mathbb{R}\subset\mathbb{C}$ converges to $h(i\eta) = A+i\eta$, on the line $\{1\}\times\mathbb{R}\subset\mathbb{C}$ converges to $h(1+i\eta) = g(A+i\eta)$ and is a conformal homeomorphism when restricted to a disk subset of Σ. h is extended to a diffeomorphism in the open complex right half plane by defining it in successive vertical bands of width 1 by $h(1+\zeta) = g\big(h(\zeta)\big)$.

Since g is holomorphic, the Beltrami coefficient κ of h is invariant under the translation of 1, i.e., $\kappa(\zeta+1) = \kappa(\zeta)$. We extend κ to $\mathbb{C}$ so as to be periodic with period 1. The function κ thus obtained is differentiable in $\mathbb{C}$, $|\kappa| < 1$ everywhere and $\kappa = 0$ on the subset of $\mathbb{C}$ where h is holomorphic, which includes a countable union of disks. By the measurable Riemann mapping theorem, there is a quasiconformal mapping $\psi : \mathbb{C} \to \mathbb{C}$ which is the only solution of the Beltrami equation with coefficient κ, up to left compositions by conformal homeomorphisms with fixed points 0, 1, ∞, and, therefore, there is a unique solution of the Beltrami equation with fixed points 0, 1, ∞.

The field of ellipses corresponding to κ is invariant under the translation by 1, $G(\zeta) = \zeta + 1$. By the chain rule for derivatives, the Beltrami coefficient of $\psi \circ G \circ \psi^{-1}$ is 0 in $\mathbb{C}$. Thus, this function is holomorphic and, as it has derivative $\neq 0$ in $\mathbb{C}$, it is a conformal homeomorphism and, hence, a Möbius transformation. As ∞ is a fixed point of the function and 0 has image 1 and vice versa, this Möbius transformation is an affine function, $cz+1$, for some $c \in \mathbb{C}$. So, $\psi(\zeta+1) = c\,\psi(\zeta)+1$. $\varphi = \psi \circ h^{-1}$ is holomorphic in V and

$$\varphi \circ g = \psi \circ h^{-1} \circ g = \psi \circ G \circ h^{-1} = c(\psi \circ h^{-1})+1 = c\,\varphi+1\,.$$

The function h^{-1} in V can be extended to a quasiconformal mapping in $\mathbb{C}_\infty$. Thus, φ is the restriction to V of a homeomorphism of $\mathbb{C}_\infty$ onto $\mathbb{C}_\infty$. If it were $|a| < 1$, it would be

$$\varphi\big(g^n(z)\big) = a^n\varphi(x) + \sum_{j=1}^{n-1} a^j = a^n\varphi(x) + \tfrac{a^n-1}{a-1}\,,$$

which would have limit $\frac{1}{1-a}$ when $n \to +\infty$, contradicting $\varphi\big(g^n(z)\big) \to \infty$. Thus, $|a| \geq 1$ and, consequently, $\varphi\big(g^n(z)\big)| = O(|a|^n)$ when $n \to +\infty$.

By Mori theorem (3.21), quasiconformal mappings are Hölder continuous, and, consequently, $\frac{1}{\varphi(1/z)}$ also is. Therefore,

$$\big|\varphi\big(g^n(z)\big)\big| \leq C\big|g^n(z)\big|^\alpha = O(n^\alpha)\,, \quad n \to +\infty\,,$$

for some $\alpha \in]0, 1[$, what is incompatible with the asymptotic estimate at the end of the preceding paragraph if $a > 1$. Consequently, $a = 1$ and φ is a conjugacy of f with the function $z \mapsto 1+z$.

Chapter 4
Holomorphic Motions

4.1 Introduction

For the study of structural stability and bifurcation of the dynamics defined by iteration of complex rational functions, the subject of the following chapter, R. Mañé, P. Sad, and D. Sullivan[1] considered in 1983 families of injections in a nonclosed set $\Omega \subset \mathbb{C}_\infty$ depending holomorphically on a parameter τ in a Riemann surface T with a base point $\tau_0 \in T$ where the injection is the identity in Ω, and the extension of these families to the closure of Ω, and they proved the existence of a unique extension of the family function to $\operatorname{cl}\Omega$ with the resulting family also depending holomorphically on the parameter $\tau \in T$, a result they called "λ-lemma."[2] They also proved that, due to the holomorphic dependence on τ and the functions injectivity, these were automatically quasiconformal mappings depending continuously jointly on the parameter and the variable in Ω, in spite of this not being assumed *a priori*.

In 1986, D. Sullivan and W. Thurston[3] imagining the injections moving as a function of $\tau \in T$ from the identity at the base point gave to a family satisfying

---

[1] Mañé, R., Sad, P., Sullivan, D.P., On the dynamics of rational maps, *Ann. Sci. École Norm. Sup.* **16** (1983), 193–217.

[2] This inexpressive name comes from the dependence on a parameter λ, by analogy with the equally inexpressive name given to a basic result of Dynamical Systems describing topologically the saddle point property in a neighborhood of a non-neutral fixed point established in 1967 by Jacob Palis (1940-) in his doctoral thesis and referred by this name in the influential article in the same year of his adviser, S. Smale, Differentiable dynamical systems, *Null. Amer. Math. Society* **73** (1967), 747–817, establishing basic notions and a program for the study of topological and geometric properties of Dynamical Systems, and giving the example of the Smale horseshoe. Stephan Smale (1930-) received the *Fields Medal* in 1966, the *Oswald Veblen Prize in Geometry* of the *AMS* in 1966, and the *John von Neumann Lecture Prize* of the *SIAM* in in 1989.

[3] Sullivan, D., Thurston, W.P., Extending holomorphic motions, *Acta Mathematica* **157** (1986), 243–257.

L. T. Magalhães, *Quasiconformal Mappings in the Plane and Complex Dynamics*,
https://doi.org/10.1007/978-3-031-80115-0_4

those conditions the name holomorphic motion of Ω with parameter in T and base point τ_0. Then, they proved there is a universal $r \in]0, 1]$ such that any holomorphic motion of $\Omega \subset \mathbb{C}_\infty$ with at least four points, and parameter varying in B_1 with base point 0 has an extension to $\mathbb{C}_\infty$ for the parameter varying in B_r, but without guarantee of uniqueness and leaving open the identification of the value of r and whether it could be $r = 1$.

Also in 1986, L. Bers and H. Royden[4] proved the extension of holomorphic motions is valid with $r = \frac{1}{3}$ with uniqueness and obtained a simple upper bound of the Beltrami coefficient associated with each quasiconformal mapping of a holomorphic motion of a region in $\mathbb{C}_\infty$ with parameter in B_1 in terms of the value of the parameter, what C. McMullen and D. Sullivan called "harmonic λ-lemma." It was proved using Teichmüller spaces of subsets of $\mathbb{C}_\infty$ covered by $\mathbb{H}$, in particular with the Bers embedding proved in 1960 by L. Bers stating that Teichmüller spaces of such sets can be embedded in the space of quadratic differentials in $-\mathbb{H}$ by Schwarzian derivatives.

In 1991, Z. Slodkowski[5] proved with several variables complex analysis that the result of extension of holomorphic motions is valid with $r = 1$, without guarantying uniqueness. Later on appeared other proofs, some with methods of one complex variable, e.g., in 1995 also by Z. Slodkowski, in 2001 by K. Astala and G.J. Martin, in 2004 by E.M. Chirka, and in 2015 by F. Gardiner, Y. Jiang, and Z. Wang.[6]

It is useful to extend holomorphic motions preserving symmetries. General examples with evident importance are holomorphic motions of subsets of more general Riemann surfaces, what can be done through the universal covering with covering space conformal to B_1 and the corresponding holomorphic motion in B_1, but this requires the holomorphic motions to be equivalent under the action of the Riemann surface fundamental group. The existence of extension of a holomorphic motion with parameter in B_1 from a subset of $\mathbb{C}_\infty$ to this whole set preserving symmetries was conjectured in 1990 by C. McMullen and was proved in 1994 by C.J. Earle, I. Kra, and S.L. Krushka,l[7] and with a broader hypothesis in 1995 by Z. Slodkowski.

---

[4] Bers, L., Royden, H.L., Holomorphic families of injections, *Acta Math.* **157** (1986), 259–286. Bers, L., Holomorphic families of isomorphisms of Möbius groups, *J. Math. Kyoto Univ.* **26** (1986), 73–76.

[5] Slodkowski, Z., Holomorphic motions and polynomial hulls, *Proc. Amer. Math. Soc.* **111** (1991), 347–355.

[6] Slodkowski, Z. Extensions of holomorphic motions. *Ann. Scuola Norm. Sup. Pisa Cl. Sci.* 1(4) **22** (1995), 185–210. Astala, K., Martin, G.J., Holomorphic motions, *Papers on Analysis; Report. Univ. Jyvaskyla* **83** (2001), 27–40. Chirka, E.M., On the extension of holomorphic motions, *Doklady Mathematics,* **70** (2004), 516–519. Gardiner, F.P., Jiang, Y., Wang, Z., Holomorphic motions and related topics, in *Geometry of Riemann Surfaces,* eds. Gardiner, F.P., González-Diez, G., Kourouniotis, C., London Math. Soc. Lecture Note Ser., **368**, Cambridge Univ. Press, Cambridge, 2010, 156–193.

[7] Earle, C.J., Kra, I., Krushkal, S.L., Holomorphic motions and Teichmüller spaces, *Trans. Amer. Math. Soc.,* **343** (1994), 927–948.

4.2 Holomorphic Motions and Quasiconformal Mappings

A **holomorphic motion** of a set $\emptyset \neq \Omega \subset \mathbb{C}_\infty$ with parameter in a Riemann surface T and base point $\tau_0 \in T$ is a family $\{\phi_\tau\}_{\tau\in T}$ of injections from Ω into $\mathbb{C}_\infty$ such that $\tau \mapsto \phi_\tau(z)$ is holomorphic in T for each $z\in\Omega$ and $\phi_{\tau_0} = 1_\Omega$.

We recall[8] that given an hyperbolic Riemann surface S, normalizing the Poincaré metric of the disk $B_1 \subset \mathbb{C}$ with center 0 and radius 1 in the Euclidean metric so that the Gauss curvature is $K=-1$, with length element $ds = \frac{2}{1-|z|^2}|dz|$, with universal coverings and the uniformization theorem, we get one Poincaré metric in $\widetilde{S}$, called the **Poincaré metric** or **hyperbolic metric** of the hyperbolic Riemann surface S. The **Poincaré length** of a path γ in S is $\int_\gamma ds$, with ds the length element of the Poincaré metric of S. The infimum of the Poincaré lengths of all piecewise regular (or rectifiable) paths with end points $z, w \in S$ is called the **Poincaré distance** of z and w , denoted $\boldsymbol{d_S(z, w)}$.

Although the definition of holomorphic motion does not require continuity of $(\tau, z) \mapsto \phi_\tau(z)$, if Ω has at least six points, it is automatic from the holomorphy of $\tau \mapsto \phi_\tau(z)$ and the injectivity of each ϕ_τ, which also imply that ϕ_τ is a quasiconformal mapping for each $\tau \in T$, as established in 1983 by R. Mañé, P. Sad, and D. Sullivan and proved as follows.

(4.1) *If $\{\phi_\tau\}_{\tau\in T}$ is a holomorphic motion of $\emptyset \neq \Omega \subset \mathbb{C}_\infty$, $\#\Omega \geq 6$ with parameter in a hyperbolic Riemann surface T, then $(\tau, z) \mapsto \phi_\tau(z)$ is continuous in T, $\times \operatorname{cl}\Omega$ and ϕ_τ is a quasiconformal mapping for $\tau \in T$.*

Proof Let $a_1, a_2, a_3, b_1, b_2, b_3$ be distinct points of Ω and τ_0 be the holomorphic motion base point. Without loss of generality, $\phi_\tau(a_1) = 0$, $\phi_\tau(a_2) = 1$, $\phi_\tau(a_3) = \infty$ for all $\tau \in T$, as this can be achieved with compositions with Möbius transformations. Consider the functions $g_k(\tau) = \phi_\tau(b_k)$, $k = 1, 2, 3$, and $g_4(\tau) = \frac{g_1(\tau)-g_2(\tau)}{g_1(\tau)-g_3(\tau)}$, which do not assume the values $0, 1, \infty$. Let $S = \mathbb{C}_\infty \backslash \{0, 1, \infty\}$. By the Pick theorem for Riemann surfaces,[9] the Poincaré distance in S of images of pairs of points in T by these functions is smaller or equal to the Poincaré distance in T of the pair of points; thus, $d_S\big(g_k(\tau_1), g_k(\tau_2)\big) \leq d_T(\tau_1, \tau_2)$, for $\tau_1, \tau_2 \in T$, $k \in \{1, 2, 3, 4\}$. Hence, $d_S\big(\phi_{\tau_1}(b_1), \phi_{\tau_2}(b_1)\big) = d_S\big(g_1(\tau_1), g_1(\tau_2)\big)$ is uniformly small for $d_T(\tau_1, \tau_2)$ small, and $\{\phi_\tau\}_{\tau\in T}$ is a family of equicontinuous functions in $\operatorname{cl}\Omega$ and $(\tau, z) \mapsto \phi_\tau(z)$ is continuous in $T \times \operatorname{cl}\Omega$. For $R \in]0, 1[$, denote $B_{R(\tau_0)}$ the open ball with center τ_0 and radius R in the Poincaré distance in T. For $\tau \in \operatorname{cl} B_R(\tau_0)$, it is $d_S\big(g_4(\tau), g_4(\tau_0)\big) \leq d_T(\tau, \tau_0) \leq R$. Since b_1, b_2, b_3 are arbitrary distinct points of $\Omega\backslash\{a_1, a_2, a_3\}$, the function g_4 is bounded in $\operatorname{cl} B_R$ and, by the circular dilatation characterization of quasiconformal mappings (3.30), ϕ_τ is quasiconformal for all $\tau \in T$. □

[8] See Section "Riemann Surfaces Geometry" in Chapter 12 of *CADOVA*.

[9] See Section 5 of Chapter 12 of *CADOVA*. Pick, Georg (1859–1942).

Conversely, every quasiconformal mapping f in a region $\Omega \subset \mathbb{C}_\infty$ preserving orientation can be connected by a holomorphic motion $\{\phi_\tau\}_{\tau \in B_1}$ from Ω to 1_Ω, e.g., if κ is the Beltrami coefficient of f, with ϕ_τ the solution of the Beltrami equation with coefficient $\tau\kappa$.

If f is a quasiconformal mapping in a region $\Omega \subset \mathbb{C}_\infty$, it satisfies the Beltrami equation in the sense of distributions with coefficient $\kappa \in L^\infty(\Omega)$, $\|\kappa\|_{L^\infty} < 1$, and f is K-quasiconformal if $\frac{1+\|\kappa\|_{L^\infty}}{1-\|\kappa\|_{L^\infty}} \leq K$.

In 1986, L.Bers and H.Royden obtained the following simple upper bound of the dilatation of each quasiconformal mapping ϕ_τ of a holomorphic motion of Ω with parameter in the open ball $B_1 \subset \mathbb{C}$ with center 1 and radius 0 in the usual Euclidean distance of $\mathbb{C}$.

(4.2) *If $\{\phi_\tau\}_{\tau \in B_1}$ is a holomorphic motion of a region $\Omega \subset \mathbb{C}_\infty$ with parameter in $B_1 \subset \mathbb{C}$ and base point 0, then for each $\tau \in B_1$ the Beltrami coefficient κ_τ of ϕ_τ is a holomorphic function, $\kappa_\tau \in L^\infty(\Omega)$, $\|\kappa_\tau\|_{L^\infty(\Omega)} \leq 1$, and ϕ_τ is a K-quasiconformal mapping in Ω with $K \leq \frac{1+|\tau|}{1-|\tau|}$.*

Proof Since the Beltrami coefficient of each quasiconformal mapping ϕ_τ is $\kappa_\tau(z) = \frac{\partial\phi_\tau/\partial\bar{z}}{\partial\phi_\tau/\partial z}$,

$$\kappa_\tau(z) = \frac{\partial\phi_\tau/\partial\bar{z}}{\partial\phi_\tau/\partial z} = \lim_{h\in\mathbb{R},\, h\to 0} \frac{1+i\,\sigma_\tau(z,h)}{1-i\,\sigma_\tau(z,h)}, \qquad \text{with } \sigma_\tau(z,h) = \frac{\phi_\tau(z+ih)-\phi_\tau(z)}{\phi_\tau(z+h)-\phi_\tau(z)}.$$

For $z \in \mathbb{C}\backslash\{0, 1\}$ and $h \in \mathbb{R}$, the function $\tau \mapsto \sigma_\tau(z, h)$ is holomorphic in B_1 and does not assume the values $0, 1, \infty$, and σ_0 is the constant function with value i. By the Schwarz-Pick lemma,[10]

$$d_S\big(\sigma_\tau(z,h), i\big) \leq d_{B_1}(\tau, 0) = 2\,\text{arctanh}\,|\tau| = \ln\frac{1+|\tau|}{1-|\tau|}.$$

Thus, there is $r>0$ such that $\sigma_\tau(z, h)+i \neq 0$ in $\text{cl}\, B_r$, and $\tau \mapsto \frac{1+i\,\sigma_\tau(z,h)}{1-i\,\sigma_\tau(z,h)}$ is uniformly bounded for $\tau \in \text{cl}\, B_r$. By the bounded convergence theorem,[11] for each $g \in L^1(\Omega)$ with compact support the sequence of functions,

$$(\Psi_g)_n : B_1 \to \mathbb{C}, \qquad (\Psi_g)_n(\tau) = \int_{\mathbb{R}^2} g(x+iy)\frac{1+i\,\sigma_\tau(x+iy,1/n)}{1-i\,\sigma_\tau(x+iy,1/n)}\,dxdy,$$

converges uniformly in $\text{cl}\, B_r$ to a function

$$\Psi_g : B_1 \to \mathbb{C}, \qquad \Psi_g(\tau) = \int_{\mathbb{R}^2} g(x+iy)\,\kappa_\tau(x+iy)\,dxdy.$$

Thus, for every $g \in L^1(\Omega)$ with compact support, Ψ_g is holomorphic in $\text{cl}\, B_r$. Since the dual space of[11] $L^1(\Omega)$ is $L^\infty(\Omega)$, the function $\tau \mapsto \kappa_\tau$ from B_1 to $L^\infty(\Omega)$ is

[10] See Section 5 of Chapter 11 of *CADOVA*.

[11] See Appendix A.

holomorphic. As $\kappa_0 = 0$, by the Schwarz lemma,[12] $\|\kappa_\tau\|_{L^\infty} \leq |\tau|$, and the dilatation of ϕ_τ satisfies

$$K_\tau = \frac{1+\|\kappa_\tau\|_{L^\infty}}{1-\|\kappa_\tau\|_{L^\infty}} \leq \frac{1+|\tau|}{1-|\tau|}\,.$$

□

4.3 Holomorphic Motions of Julia Sets

An additional motivation for holomorphic motions is the application to deformations of **filled Julia sets** $K(f_c)$ of quadratic complex polynomials[13] $f_c : \mathbb{C}_\infty \to \mathbb{C}_\infty$ with $f_c(z) = z^2 + c$ and $c \in \mathbb{C}$, which are the sets of points with bounded positive orbits. The **Julia set** of f_c is the boundary $J(f_c)$ of the filled Julia set. The **Mandelbrot set** $\mathcal{M}$ is the set of $c \in \mathbb{C}$ such that $0 \in K(f_c)$ (Fig. 4.1[14]).

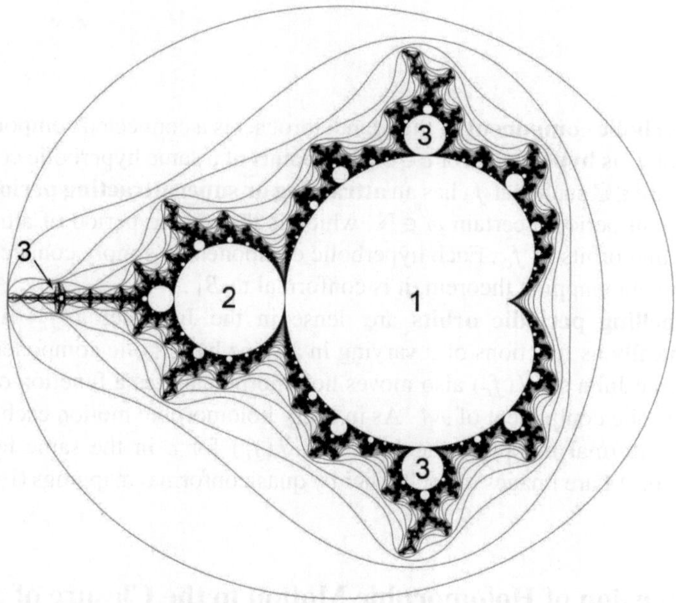

Fig. 4.1 Mandelbrot set $\mathcal{M}$. It includes bounded white components which are hyperbolic of values $c \in \operatorname{int} \mathcal{M}$ such that $f_c(z) = z^2 + c$ has attracting or superattracting periodic orbits with fixed minimum period of the corresponding attracting periodic orbits, with those of minimum period 1, 2, 3 labeled accordingly. The thin line outer curves do not belong to $\mathcal{M}$; they are points whose positive orbits take the same number of iterations to leave the disk $B_2 \subset \mathbb{C}$

[12] See Section 3 of Chapter 10 of *CADOVA*.

[13] For basic aspects of the dynamics of iteration of complex rational functions, see Chapter 13 of *CADOVA*. Here, we follow part of the introduction of Astala, K., Martin, G.J., Holomorphic motions, Papers on analysis, *Rep. Univ. Jyväskylä Dep. Math. Stat.*, **83** (2001), 27–40.

[14] Author and date of figure: Geek3 (January 2009), under License *Creative Commons CC Attribution Share Alike 3.0 Unported,* to which the numbers 1, 2, 3 have been added. https://commons.wikimedia.org/wiki/File:Mandelbrot_Leminiscates_2.png

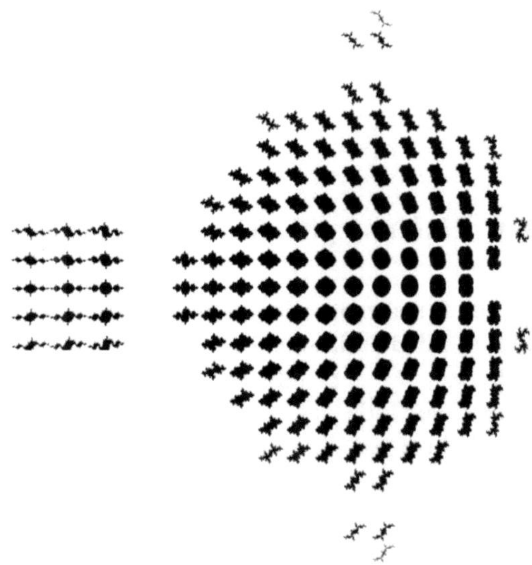

Fig. 4.2 Filled Julia sets in grids of points in components of the Mandelbrot set $\mathcal{M}$ of attracting or superattracting periodic orbits of periods 1, 2, 3 (the bigger)

An **hyperbolic component** of the Mandelbrot set is a connected component U of $\mathcal{M}$ such that f_c is **hyperbolic** for $c \in U$. The points of a same hyperbolic component of $\mathcal{M}$ are the $c \in \mathbb{C}$ such that f_c has an **attracting or superattracting periodic orbit** with minimum period a certain $m \in \mathbb{N}$, which is the unique period of attracting or superattracting orbits in f_c. Each hyperbolic component is simply connected; thus, by the Riemann mapping theorem, it is conformal to B_1.

The **repelling periodic orbits** are dense in the Julia set $J(f_c)$ and move holomorphically as functions of c varying in a same hyperbolic component of $\mathcal{M}$. Therefore, the Julia set $J(f_c)$ also moves holomorphically as a function of c in the same hyperbolic component of $\mathcal{M}$. As in every holomorphic motion each injection is a quasiconformal mapping, the Julia sets $J(f_c)$ for c in the same hyperbolic component of $\mathcal{M}$ are images of each other by quasiconformal mappings (Fig. 4.2[15]).

4.4 Extension of Holomorphic Motion to the Closure of a Domain

The following result was obtained in 1983 by R. Mañé, P. Sad, and D. Sullivan, who named it "λ-lemma," and, independently, by M. Lyubich.[16]

[15] Author of the base figure: George (February 2008), under License *Public Domain*. https://commons.wikimedia.org/wiki/File:Map_Julia_set_625.png

[16] Lyubich, Mikhail (1959-).

(4.3) λ-lemma: *Every holomorphic motion* $\{\phi_\tau\}_{\tau\in B_1}$ *of a nonclosed set* $\emptyset \neq \Omega \subset \mathbb{C}_\infty$ *with parameter in* $B_1 \subset \mathbb{C}$ *and base point* 0 *has a unique extension to a holomorphic motion of* cl Ω *and* ϕ_τ *is a K-quasiconformal mapping in* Ω *with* $K \le \frac{1+|\tau|}{1-|\tau|}$*, for all* $\tau \in T$.

Proof Since $\emptyset \neq \Omega \neq \mathrm{cl}\,\Omega$, Ω is infinite. Proceeding as in the proof of (4.1) with the same notation, for $M > m > 0$ and $0 < R < 1$ if $d_S\big(g_2(0), 0\big) < M$, then $d_S\big(g_2(\tau), 0\big)$ for $|\tau| \le R$ is bounded by a constant depending on M. The function $g_5(\tau) = \frac{g_2(\tau)-g_1(\tau)}{g_2(\tau)}$ does not assume the values 0, 1, ∞, and if $0 < m \le d_S\big(g_2(0), 0\big) \le M$ and $d_S\big(g_5(\tau), 0\big)$ is small, and so is $d_S\big(g_2(\tau), g_1(\tau)\big)$. Therefore, each ϕ_τ is uniformly continuous in $\Omega \cap \big(\mathrm{cl}\, B_M \setminus B_m\big)$. Permuting 0, 1, ∞, the resulting sets cover $\mathbb{C}_\infty$. So, each ϕ_τ has a unique extension to a function $\phi_\tau : \mathrm{cl}\,\Omega \to \mathbb{C}_\infty$, which has a continuous inverse for all $\tau \in T$. For $z \in \mathrm{cl}\,\Omega$, the function $\tau \mapsto \phi_\tau(z)$ is holomorphic in T, as it is a limit of a sequence of uniformly convergent holomorphic functions in cl B_r, $0 < r \le R$. By (4.1), $(\tau, z) \mapsto \phi_\tau(z)$ is continuous, and by (4.2), ϕ_τ is a quasiconformal mapping with dilatation $K \le \frac{1+|\tau|}{1-|\tau|}$. □

4.5 Extension of Holomorphic Motion to $\mathbb{C}_\infty$

As mentioned in the introduction to this chapter, in 1986, D. Sullivan and W. Thurston proved that a holomorphic motion of a set $\Omega \subset \mathbb{C}_\infty$ with $\#\Omega \ge 4$, base point 0, and parameter in B_1 can be extended to all $\mathbb{C}_\infty$ with the parameter in a B_r for some universal $r \in \,]0, 1]$. Then, L. Bers and H. Royden proved that $r \ge \frac{1}{3}$, and for $r = \frac{1}{3}$, the extension is unique and the Beltrami coefficient of each element of the holomorphic motion is a harmonic function (leading C. McMullen and D. Sullivan in 1998 to call this result "harmonic λ-lemma"). And in 1991, Z. Slodkowski proved with several variables complex analysis that $r = 1$ if the requirement of uniqueness is dropped. The proof of this result is presented below with one variable complex analysis, as done by E.M.Chirka in 2004, relying on the presentation of J.Hubbard[17] in 2006, with modifications for self-sufficiency of the presentation.

(4.4) Slodkowski theorem: *Every holomorphic motion* $\{\phi_\tau\}_{\tau\in B_1}$ *of* $\emptyset \neq \Omega \subset \mathbb{C}_\infty$ *with parameter in* B_1 *and base point* 0 *has extension to a holomorphic motion of* $\mathbb{C}_\infty$ *with parameter in* B_1 *and base point* 0.

[17] Hubbard, J., *Teichmüller Theory and Applications to Geometry, Topology and Dynamics, Volume 1 Teichmüller Theory*, Matrix Editions, Ithaca, New York, 2006.

Proof Without loss of generality, take $\#\Omega \geq 3$, $\Omega \supset \{0, 1, \infty\}$, $\phi_\tau(0) = 0$, $\phi_\tau(1) = 1$, and $\phi_\tau(\infty) = \infty$, as it can be achieved adding one or two points to Ω, extending the holomorphic motion and applying Möbius transformations.

The set of complex numbers with real and imaginary parts in $\mathbb{Q}$ is countable and dense in $\mathbb{C}$. Hence, there is a countable dense subset of Ω and, therefore, there is an expansive sequence $\{\Omega_n\}$ of finite subsets of Ω containing $\{0, 1, \infty\}$ such that $\operatorname{cl} \bigcup_{n \in \mathbb{N}} \Omega_n = \Omega$. Similarly, there is a countable subset X of $\mathbb{C} \setminus \operatorname{cl} \Omega$ with closure containing this set.

In 1986, D.Sullivan and W.Thurston, as well as L.Bers and H.Royden, published in articles in consecutive pages of the same issue of *Acta Mathematica* a proof of existence of extensions of holomorphic motions of finite sets with at least three points and parameter in an open disk to $\mathbb{C}_\infty$ by choosing successive points outside that set (what D. Sullivan and W. Thurston called **holomorphic axiom of choice**) and proved that this implies existence of extension of holomorphic motions of any set with at least three points.

More explicitly: *If* $\{\phi_\tau\}_{\tau \in B_r}$ *is a holomorphic motion of a finite set* $\Phi \subset \mathbb{C}_\infty$ *with* $\{0, 1, \infty\}$ *keeping these points fixed, with parameter in* B_r *for some* $r > 0$ *and base point* 0, *and* $\{r_n\} \subset]0, 1[$ *is an increasing sequence with* $\lim r_n = r$, *each holomorphic motion of* Φ *obtained by restriction of the parameter to* B_{r_n} *has extension to a holomorphic motion of* $\mathbb{C}_\infty$ *with parameter in* B_{r_n} *and base point* 0, *then for* $x \in X$, $\rho \in]0, 1[$ *and* $n \in \mathbb{N}$ *large so that* $r_n > \rho$, *the elements of the sequence of functions* $\tau \mapsto (\phi_n)_\tau(x)$ *are holomorphic functions in* B_ρ *omitting the values* $0, 1, \infty$.

By the Montel-Carathéodory theorem, this sequence is a normal family,[18] and, therefore, it has a uniformly convergent subsequence on compact subsets to a function $\tau \mapsto \widehat{\phi_\tau}(z)$ holomorphic in B_ρ for each $z \in \Omega \cup X$ such that $\widehat{\phi_\tau}(z) = \phi_\tau(z)$ for $\tau \in B_\rho$, $z \in \Omega$. Since $\rho \in]0, 1[$ is arbitrary, there is a holomorphic function in B_1 for each $z \in \Omega$, also denoted $\tau \mapsto \widehat{\phi_\tau}(z)$, and $\widehat{\phi_\tau}(z) = \phi_\tau(z)$ for $\tau \in B_1$, $z \in \Omega$. If $z, s \in \Omega \cup X$ and $z \neq s$, the function $\tau \mapsto (\phi_\tau)(z) - (\phi_\tau)(s)$ does not vanish. Hence, by the Hurwitz theorem,[19] $\widehat{\phi_\tau}(z) - \widehat{\phi_\tau}(s) \neq 0$ and, therefore, $\widehat{\phi_\tau}$ is injective for all $\tau \in B_1$. Consequently, $\{\widehat{\phi_\tau}\}_{\tau \in B_1}$ is a holomorphic motion of $\Omega \cup X$ with parameter in B_1 extending the initial holomorphic motion $\{\widehat{\phi_\tau}\}_{\tau \in B_1}$ of Ω. As $(\operatorname{cl} \Omega) \cup X = \mathbb{C}_\infty$, by the λ-lemma proved in the preceding section, there is an extension of the initial holomorphic motion of Ω to a holomorphic motion of $\mathbb{C}_\infty$, both with parameter in B_1 and base point 0.

To finish the proof, it suffices to prove the statement in italics two paragraphs above.[20] Without loss of generality, as r can be scaled, take $r > 1$. We intend to obtain a holomorphic motion in $\mathbb{C}_\infty$ with parameter in $B_{r'}$ for any $r' \in]0, r[$. It is convenient to consider the holomorphic motion with base point ∞ instead of 0,

[18] See Sections 4 of Chapter 10 and 6 of Chapter 11 of *CADOVA*. Montel, Paul Antoine (1876–1975).

[19] See Section 5 of Chapter 6 of *CADOVA*. Hurwitz, Adolf (1859–1919).

[20] In the rest of this proof, the argument of several complex variables used by Z. Slodkowski is replaced by that of one variable introduced by E.M.Chirka in 2004.

changing the parameter by $\widehat{\phi_\tau}(z) = \phi_{r'/\tau}(z)$ and giving the holomorphic motion $\{\widehat{\phi_\tau}\}_{\tau\in\mathbb{C}_\infty\setminus B_r}$ of Φ with base point ∞. For each fixed $z\in\Phi$, the function $\tau\mapsto\widehat{\phi_\tau}$ defined in $\mathbb{C}_\infty\setminus B_r$ can be extended to a continuous function in $\mathbb{C}_\infty$ choosing a $r''\in\,]r', r[$ and defining

$$\widetilde{\phi_\tau}(z) = \begin{cases} \widehat{\phi_\tau}(z)\,, & \tau\in\mathbb{C}_\infty\setminus B_{r'/r''} \\ \widehat{\phi}_{(r'/r'')^2\tau/|\tau|^2}(z)\,, & \tau\in B_{r'/r''}\,. \end{cases}$$

In general, for a fixed $z\in\Phi$, this function is not C^∞ in $\tau\in\partial B_{r'/r''}$, and to get from it a holomorphic motion of $\mathbb{C}_\infty$ extending the holomorphic motion $\{\widehat{\phi_\tau}\}_{\tau\in\mathbb{C}_\infty\setminus B_1}$ of Φ, it is convenient to consider an approximation $\psi_\tau(z)$ C^∞ as a function of $\tau\in\mathbb{C}_\infty$ for each fixed $z\in\Phi$ coinciding with $\widehat{\phi_\tau}(z)$ for $\tau\in(\mathbb{C}_\infty\setminus B_1)\cup B_{r'/r}$ and such that each ψ_τ is also injective in Φ for $\tau\in B_1\setminus B_{r'/r}$, what is possible because Φ is finite and $\frac{r'}{r}<\frac{r'}{r''}<1$. More explicitly, with $h:[0,+\infty[\,\to\,]0,+\infty[$ and $h(x)=\left(\frac{r'}{r''x}\right)^2$ for $0\le x<\frac{r'}{r''}$ and equal to 1 for $x\ge\frac{r'}{r''}$, we have $\widetilde{\phi_\tau}(z)=\widehat{\phi}_{h(|\tau|)\tau}(z)$ for $z\in\Phi$, and the C^∞ approximation $\psi_\tau(z)$ can be defined by $\psi_\tau(p_j)=\widehat{\phi}_{\widetilde{h}(|\tau|)\tau}(z)$, with $\widetilde{h}$ a C^∞ nonincreasing function different from h only in a subset of $\left]\frac{r'}{r''},1\right[$ and with $\max\left|\widetilde{h}-h\right|$ so small that $\max_{j\in 1,\ldots,N}|\psi_\tau(p_j)-\widehat{\phi_\tau}(p_j)|$ is less than half the minimum distance of pairs of points in Φ.

Since for $\tau\in\mathbb{C}_\infty$ the function ψ_τ is injective in Ψ, for all distinct $z, s\in\Phi$, it is $\psi_\tau(z)\neq\psi_\tau(s)$. By the Weierstrass theorem of extrema of continuous functions, for each fixed distinct pair of points $z, s\in\Phi$, the continuous function $\tau\mapsto\psi_\tau(z)-\psi_\tau(s)$ has a minimum in the compact set $\operatorname{cl} B_1$ and, as this function does not vanish, the minimum is a $m_{z,s}>0$. Thus, $\delta=\min\limits_{z,s\in\Phi} m_{z,s}>0$. Let $\sigma:[0,+\infty[\,\to[0,1]$ be a nonincreasing function C^∞ which is 1 in $\left[0,\frac{\delta}{4}\right]$ and 0 in $\left[\frac{\delta}{2},+\infty\right[$, and, for each $\tau\in\mathbb{C}_\infty$, let $F:\mathbb{C}\times\mathbb{C}\to\mathbb{C}$ be such that

$$F(\tau,z)=\sum_{s\in\Phi\setminus\{\infty\}}^{n}\sigma\big(|z-\psi_\tau(s)|\big)\,\frac{\partial\psi_\tau}{\partial\overline{\tau}}(s)\,.$$

This is a C^∞ function with compact support included in $(\operatorname{cl} B_1)\times\mathbb{C}$ as a function of four real variables (the real and imaginary parts of τ and z), and for each $\tau, z\in\mathbb{C}$ only one term of the sum defining $F(\tau,z)$ can be $\neq 0$. Since for each $z\in\Phi$ the function $\tau\mapsto\psi_\tau(z)$ is holomorphic, it is $\frac{\partial\psi_\tau}{\partial\overline{\tau}}(z)=0$ and, therefore, $\frac{\partial\psi_\tau}{\partial\overline{\tau}}(z)=0=F\left(\tau,\frac{\partial\psi_\tau}{\partial\overline{\tau}}(z)\right)$.

By the result following this proof, which we use now, the differential equation $\frac{\partial f}{\partial\overline{\tau}}(\tau)=F\big(\tau,f(\tau)\big)$ with $F:\mathbb{C}^2\to\mathbb{C}$ as a C^∞ function of four real variables with support included in $(\operatorname{cl} B_1)\times K$ and $K\subset\mathbb{C}$ a compact set has one unique solution $f:\mathbb{C}_\infty\to\mathbb{C}$ such that $f(a)=z$, for $(a,z)\in(\mathbb{C}_\infty\setminus\operatorname{cl} B_1)\times\mathbb{C}$. We denote $\widehat{\psi}(\tau,z)$ the unique solution of the ordinary differential equation with F the function of the preceding paragraph and $\widehat{\psi}(\infty,z)=z$, for $z\in\mathbb{C}$, extended by continuity for $z\in\mathbb{C}_\infty$. Defining $\eta_\tau(z)=\widehat{\psi}\left(\frac{r'}{\tau},z\right)$, we have $\eta_0(z)=\widehat{\psi}(\infty,z)=z$ for $z\in\Phi$, and the function $\tau \mapsto \eta_\tau(z)$ satisfies $\frac{\partial \widehat{\psi}}{\partial \overline{\tau}}(\tau, z) = F\big(\tau, \widehat{\psi}(\tau, z)\big)$ for $\tau \in \mathbb{C}_\infty$. Hence, by the uniqueness of solution of the problem considered, $\psi_\tau(z) = \widehat{\psi}(\tau, z)$. Therefore, for $\tau \in B_{r'}$, $z \in \Phi$ is $\eta_\tau(z) = \phi_\tau(z)$, and, also by uniqueness of solution of the problem for the partial differential equation, for $\tau \in \mathbb{C}_\infty \backslash \mathrm{cl}\, B_1$, the function $\widehat{\psi}_\tau$ is injective in $\mathbb{C}_\infty$. Thus, for $\tau \in B_{r'}$, the function η_τ is injective in $\mathbb{C}_\infty$, and for $z \in \mathbb{C}_\infty$, $\tau \in \mathbb{C}_\infty \backslash \mathrm{cl}\, B_1$ the function $\tau \mapsto \widehat{\psi}(\tau, z)$ satisfies $\frac{\partial \widehat{\psi}}{\partial \overline{\tau}}(\tau, z) = F\big(\tau, \widehat{\psi}(\tau, z)\big) = 0$. Therefore, this function is holomorphic in $\mathbb{C}_\infty \backslash \mathrm{cl}\, B_1$ for every $z \in \mathbb{C}_\infty$ and $\tau \mapsto \eta_\tau(z)$ is holomorphic in $B_{r'}$ for every $z \in \mathbb{C}_\infty$. Since $r' \in]0, 1[$ can be chosen arbitrarily close to 1, $\{\eta_\tau\}_{\tau \in B_1}$ is a holomorphic motion of $\mathbb{C}_\infty$ extending the initial holomorphic motion of Φ. □

(4.5) *The partial differential equation* $\frac{\partial f}{\partial \overline{\tau}}(\tau) = F\big(\tau, f(\tau)\big)$, *with* $F : \mathbb{C}^2 \to \mathbb{C}$ C^1 *as a function of four real variables with support a subset of* $(\mathrm{cl}\, B_1) \times K$ *and* $K \subset \mathbb{C}$ *a compact set, has a unique continuous solution* $f : \mathbb{C}_\infty \to \mathbb{C}$ *such that* $f(a) = z$ *for* $(a, z) \in (\mathbb{C}_\infty \backslash \mathrm{cl}\, B_1) \times \mathbb{C}$.

Proof Assume, without loss of generality, $a = \infty$, as this can be achieved with a Möbius transformation.

By the Green theorem for complex functions[21] $g = u + iv$, with u, v and x, y, resp., the real and imaginary parts of, resp., g and z,

$$\int_{\partial D} g(z)\, dz = -\iint_D \big[\big(\tfrac{\partial v}{\partial x} + \tfrac{\partial u}{\partial y}\big) + i\big(\tfrac{\partial v}{\partial y} - \tfrac{\partial u}{\partial x}\big)\big] dx\, dy = i \iint_D \big(\tfrac{\partial g}{\partial x} + i\tfrac{\partial g}{\partial y}\big) dx\, dy = 2i \iint_D \tfrac{\partial g}{\partial \overline{z}}(z)\, dx\, dy,$$

under the hypothesis that $D \subset \mathbb{C}$ is a regular domain[22] and $\frac{\partial g}{\partial \overline{z}}(z)$ is continuous in $\mathrm{cl}\, D$. Thus, if $h : \mathbb{C} \to \mathbb{C}$ is such that $\frac{\partial h}{\partial \overline{z}}(z)$ is continuous with support included in B_R for some $R > 0$, since $z \mapsto \frac{1}{\tau - z}$ is holomorphic in $\mathbb{C} \setminus \{\tau\}$ and, therefore, $\frac{\partial}{\partial \overline{z}}\big(\frac{h(z)}{\tau - z}\big) = \frac{\partial h}{\partial \overline{z}}(z)\, \frac{1}{\tau - z}$ for $z \neq \tau$, of the above formula with $g(z) = \frac{h(z)}{\tau - z}$ and $D = B_R \backslash B_\varepsilon(\tau)$ for $\varepsilon > 0$ small so that $\mathrm{cl}\, B_\varepsilon(\tau) \subset B_R$,

$$\begin{aligned}\big(\tfrac{\partial h}{\partial \overline{z}} * \tfrac{1}{\pi z}\big)(\tau) &= \tfrac{1}{\pi} \lim_{\varepsilon \to 0} \iint_{\mathbb{R}^2 \backslash B_\varepsilon} \tfrac{(\partial h/\partial \overline{z})(z)}{\tau - z}\, dx\, dy = \tfrac{1}{2\pi i} \lim_{\varepsilon \to 0} \int_{\partial B_\varepsilon(\tau)} \tfrac{h(z)}{\tau - z}\, dz \\ &= \tfrac{1}{2\pi i} \lim_{\varepsilon \to 0} \int_0^{2\pi} \tfrac{h(\tau + \varepsilon e^{i\theta})}{\varepsilon e^{i\theta}}\, i\varepsilon\, e^{i\theta} d\theta = h(\tau).\end{aligned}$$

If f is a continuous solution of the differential equation, $\frac{\partial f}{\partial \overline{z}} = F\big(z, f(z)\big)$ with compact support included in B_R for some $R > 0$, and, by the formula at the end of the preceding paragraph,[23]

[21] If necessary, see the beginning of Section 4 of Chapter 4 of *CADOVA*.

[22] A **regular domain** $D \subset \mathbb{C}$ is an open set that is the interior of its closure with boundary a regular Jordan curve.

[23] The complex function $\frac{1}{\pi z}$ is called **Cauchy kernel**.

$$\tfrac{\partial}{\partial\overline{\tau}}\big[\big(F\big(z, f(z)\big)*\tfrac{1}{\pi z}\big)(\tau)\big] = F\big(\tau, f(\tau)\big),$$

and

$$\tfrac{\partial}{\partial\overline{\tau}}\big[f(\tau)-\big(F\big(z, f(z)\big)*\tfrac{1}{\pi z}\big)(\tau)\big] = \tfrac{\partial f}{\partial\overline{\tau}}(\tau)-F\big(\tau, f(\tau)\big) = 0.$$

Thus, $f(\tau)-\big(F\big(z, f(z)\big)*\frac{1}{\pi z}\big)(\tau)$ is holomorphic in $\mathbb{C}_\infty$, and, consequently, it is constant and, with $s=f(\infty)$,

$$f(\tau) = s+\big(F(z, f(z)*\tfrac{1}{\pi z}\big)(\tau) = s+\tfrac{1}{\pi}\iint_{\mathbb{R}^2}\tfrac{F(x+iy\,,\,f(x+iy))}{\tau-(x+iy)}\,dxdy\,.$$

Conversely, if f is a continuous function in $\mathbb{C}_\infty$ satisfying this equality,

$$\tfrac{\partial f}{\partial\overline{\tau}}(\tau)=\tfrac{\partial}{\partial\overline{\tau}}\big[\big(F\big(z, f(z)\big)*\tfrac{1}{\pi z}\big)(\tau)\big], \qquad \tfrac{\partial f}{\partial\overline{\tau}}(\tau)=F\big(\tau, f(\tau)\big),$$

and f is a solution of the partial differential equation with $f(\infty)=s$. Therefore, for $f:\mathbb{C}_\infty\to\mathbb{C}$ continuous, the integral equation in the last expression highlighted above is equivalent to the initial value problem for the partial differential equation $\frac{\partial f}{\partial\overline{\tau}}(\tau)=F\big(\tau, f(\tau)\big)$, $f(\infty)=s$.

To prove the integral equation has a unique solution, we apply an implicit function theorem to

$$\Phi: V\times G_s\to G_0\,, \qquad [\Phi(v, g)](\tau)=g(\tau)-s-\big[v\big(z, g(z)\big)*\tfrac{1}{\pi z}\big](\tau)\,,$$

where $s\in\mathbb{C}$, V is the Banach space of the C^1 functions from $\mathbb{C}^2$ to $\mathbb{C}$ with support included in (cl B_1)$\times K$ and $K\subset\mathbb{C}$ is a compact set with the supremum norm of the modulus of the function and of its derivative, and G_s is the complete metric space of continuous functions from $\mathbb{C}_\infty$ to $\mathbb{C}$ such that $g(\infty)=s$ with the distance of two elements given by the supremum of the magnitude of their difference.

It is now proved that if $f:\mathbb{C}\to\mathbb{C}$ is continuous with support contained in cl' B_R and $0<\alpha<1$, then $g(z)=f(z)*\frac{1}{\pi z}$ is Hölder continuous with exponent α, and there exists $C_{R,\alpha}>0$ such that $\|g\|_\alpha\le C_{R,\alpha}\|f\|_0$ with $C_\alpha\to+\infty$ when $\alpha\to 1$, where $\|g\|_\alpha\overset{def}{=}\sup\big\{\frac{|g(z)-g(\tau)|}{|\tau-z|^\alpha} : z, \tau\in\mathbb{C}\,, z\neq\tau\big\}$. It is

$$|g(\tau_1)-g(\tau_2)|\le\tfrac{1}{\pi}\iint_{B_R}\big|f(x+iy)\big[\tfrac{1}{\tau_1-(x+iy)}-\tfrac{1}{\tau_2-(x+iy)}\big]\big|\,dxdy$$

$$=\tfrac{|\tau_2-\tau_1|}{\pi}\iint_{B_R}\big|f(x+iy)\big[\tfrac{1}{[\tau_1-(x+iy)][\tau_2-(x+iy)]}\big]\big|\,dxdy\,.$$

By the Hölder inequality[24] with $p, q\in[1, +\infty[$ such that $\frac{1}{p}+\frac{1}{q}=1$, the integral in the last term is

[24] See the end of Appendix A.

$$\leq \Big[\iint_{B_R} |f(x+iy)|^p dxdy\Big]^{\frac{1}{p}} \Big[\iint_{B_R} \Big|\frac{1}{[\tau_1-(x+iy)][\tau_2-(x+iy)]}\Big|^q dxdy\Big]^{\frac{1}{q}}$$

$$\leq \|f\|_0 \Big[\iint_{B_R} 1\Big]^{\frac{1}{p}} \Big[\iint_{\mathbb{C}} \Big(\frac{1}{|\tau_2-\tau_1|^2 |x'+iy'|\,|x'+iy'-1|}\Big)^q |\tau_2-\tau_1|^2\, dx'dy'\Big]^{\frac{1}{q}},$$

in the last inequality extending the region of integration of the second integral to $\mathbb{C}$ and changing variables $(x, y)=(\tau_2-\tau_1)(x', y')+\tau_1$. Consequently

$$|g(\tau_1)-g(\tau_2)| \leq \frac{1}{\pi}\Big[\iint_{B_R} 1\Big]^{\frac{1}{p}} \Big[\iint_{\mathbb{C}} \Big(\frac{1}{|x'+iy'|\,|x'+iy'-1|}\Big)^q dx'dy'\Big]^{\frac{1}{q}} |\tau_2 - \tau_1|^{\frac{2}{q}-1} \|f\|_0$$

$$= C_{R,\frac{2}{q}-1} |\tau_2 - \tau_1|^{\frac{2}{q}-1} \|f\|_0,$$

where the second integral converges if and only if $q>1$ and $\frac{2}{q}-1>0$ if and only if $q<2$, conditions satisfied if and only if $\frac{2}{q}-1 \in]0, 1[$. Therefore, for all $\alpha \in]0, 1[$, the function $g=f(z)*\frac{1}{\pi z}$ is Hölder continuous with exponent α and $\|g\|_\alpha \leq C_{R,\alpha}\|g\|_0$, where $C_{R,\alpha}\to +\infty$ when $\alpha \to 1$.

If $(v, g) \in \Phi^{-1}(\{0\})$, then $\frac{\partial g}{\partial \overline{\tau}}(\tau)=0$ for $\tau \in \mathbb{C}_\infty \setminus \mathrm{cl}\, B_1$ and, therefore, g is holomorphic in this set. Thus, it is a constant equal to s in $\mathbb{C}_\infty$ or takes a maximum at some point $\tau \in \mathrm{cl}\, B_1$ with $g(\tau) \in K$. If $C \subset V$ is compact and $g(\tau)=s+\iint_{\mathbb{R}^2} \frac{v((x+iy),\, g(x+iy))}{\tau-(x+iy)}\, dxdy$, the function g is bounded and, by the preceding paragraph, for each $\alpha>1$, we have $\|g\|_\alpha \in [0, +\infty[$, and $\{g : (v, g) \in \Phi^{-1}(\{0\}),\ v \in C\}$ is compact. Thus, with the projection $p : \Phi^{-1}(\{0\}) \to V$ such that $p(v, g)=v$, also $p^{-1}(C)$ is compact. Therefore, preimages of compact subsets of G_0 are compact subsets of the domain of Φ, i.e., p is a proper function.

Φ is C^1 and $\frac{\partial \Phi}{\partial g}(v, g)$ is the linear transformation from G_0 to G_0,

$$\Big(\Big[\frac{\partial \Phi}{\partial g}(v, g)\Big]h\Big)(\tau) = h(\tau) - \Big[D_2 v\big(z, g(z)\big)h(z) * \frac{1}{\pi z}\Big](\tau)$$

If $\big[\frac{\partial \Phi}{\partial g}(v, g)\big]h=0$, then $h = D_2 v\big(z, g(z)\big)h(z) * \frac{1}{\pi z}$, what is equivalent to the initial value problem for the partial differential equation $\frac{\partial h}{\partial \overline{\tau}}(\tau)=\big[D_2 v\big(\tau, g(\tau)\big)\big]h(\tau)$, $h(\infty)=0$. Thus, $\big|\frac{\partial h}{\partial \overline{\tau}}(\tau)\big| \leq 2 \sup \|Dv\|\ \|h\|$. With $\widetilde{h}(z) = -\frac{\partial}{\partial \overline{\tau}} \log h(z)$ if $h(z) \neq 0$ and $\widetilde{h}(z)=0$ if $h(z) = 0$, we have $|\widetilde{h}(z)| \leq 2 \sup \|Dv\|$, and the support of $\widetilde{h}$ is included in B_1. Hence, $k(z)=\widetilde{h}(z) * \frac{1}{\pi z}$ is $\frac{\partial k}{\partial \overline{\tau}} = -\frac{\partial}{\partial \overline{\tau}} \log h$. Therefore, $e^k h$ is continuous and

$$\frac{\partial}{\partial \overline{z}}\big[e^{k(z)}h(z)\big] = e^k\big[h\big(-\frac{\partial}{\partial \overline{\tau}} \log h\big)+\frac{\partial h}{\partial \overline{z}}\big](z) = e^k\big[h\big(-\frac{1}{h}\frac{\partial k}{\partial \overline{\tau}} h\big)+\frac{\partial h}{\partial \overline{z}}\big](z) = 0.$$

Consequently, $e^{k(z)}h(z)$ is holomorphic at points z where $h(z) \neq 0$ and, since $h(\infty)=0$, by the maximum principle, $h(z)=0$ for $z \in \mathbb{C}_\infty$. The continuous linear transformation $\frac{\partial \Phi}{\partial g}(v, g) : G_0 \to G_0$ is injective, and, by the result following this proof, it is an isomorphism. The equation $\Phi(v, g) = 0$ in a neighborhood of any solution (v_0, g_0) defines implicitly g as a function of v. Since $p^{-1}\big(\{0\}\big)$ is a constant function with value s and the range of p is a linear space V, p is a homeomorphism of $\Phi^{-1}(\{0\})$ onto V. Therefore, there is a unique continuous solution $f:\mathbb{C}_\infty\to\mathbb{C}$ of the problem $\frac{\partial f}{\partial\overline{\tau}}=F\big(\tau, f(\tau)\big),\ f(\infty)=s$, given by $(F,f)=p^{-1}(F)$. □

A continuous linear transformation $K: G_0\to G_0$ is **compact** if the images of bounded sequences in G_0 have convergent subsequences.

It is easy to verify that the continuous linear transformation, defined by

$$K: G_0\to G_0, \qquad [K(h)](\tau)=\big[D_2 v\big(z, g(z)\big)h(z)*\tfrac{1}{\pi z}\big](\tau).$$

is compact, and the following result applies directly.

(4.6) *If E is a Banach space, $K:E\to E$ is a compact continuous linear transformation, and 1_E+K is injective, then it is an isomorphism of E onto E.*

Proof If $y\in \mathrm{cl}\,(1_E+K)(E)$, there is a sequence $\{x_n\}\subset E$ with $(1_E+K)(x_n)\to y$. If $\{\|x_n\|\}$ were unbounded, also denoting $\{x_n\}$ a subsequence obtained suppressing terms 0 and with limit $+\infty$, and defining $z_n=\frac{x_n}{\|x_n\|}$,

$$\|z_n\|=1, \qquad \lim\,(1_E+K)(z_n)=\lim\frac{1}{\|x_n\|}(1_E+K)(x_n)=\lim\frac{1}{\|x_n\|}y=0.$$

By the compactness of K, there would exist a subsequence, still denoted $\{z_n\}$, such that $\{K(z_n)\}$ would be convergent to some a and for this subsequence

$$\lim z_n=-\lim K(z_n)=-a,\quad \|a\|=1,\quad (1_E+K)(-a)=\lim\,(1_E+K)(z_n)=-a+a=0,$$

in contradiction with the injectivity of 1_E+K, which vanishes at 0, and $\|-a\|=1$. Hence, the initial sequence $\{x_n\}$ is bounded and, by the compactness of K, it has a subsequence, still denoted $\{x_n\}$, such that $\{K(x_n)\}$ converges to some $b\in E$, and $\lim z_n=y-\lim K(z_n)=y-b$. Thus,

$$(1_E+K)(y-b)=\lim\,(1_E+K)(x_n)=y-b+b=y, \qquad y\in(1_E+K)(E).$$

Therefore, $\mathrm{cl}\,(1_E+K)(E)=(1_E+K)(E)$, i.e., the range of 1_E+K is a closed subspace of E.

We wish to prove that 1_E+K is a surjection onto E. Suppose not and consider $E_n=(1_E+K)^n(E)$. The restriction of K to E_n is compact; so, $E_1\subsetneqq E$. By the preceding paragraph and the injectivity of 1_E+K, the subspaces E_n are closed and $\mathrm{cl}\,E_{n+1}\subsetneqq E_n$ for $n\in\mathbb{N}\cup\{0\}$. Also, $K(E)\subset E$ and if $K(E_n)\subset E_n$ for some n,

$$\begin{aligned} K(E_{n+1}) &= K(1_E+K)^{n+1}(E)=K(1_E+K)(1_E+K)^n(E)\\ &= K(1_E+K)(E_n)\subset K(E_n)=E_{n+1}. \end{aligned}$$

This proves that $K(E_n)\subset E_n$ for $n\in\mathbb{N}\cup\{0\}$. By the Riesz[25] lemma, to be proved next, there is $x_n\in E_n$ at a distance $>\frac{1}{2}$ of E_{n+1} with $\|x_n\|=1$ for $n\in\mathbb{N}\cup\{0\}$. Hence, for $m>n$

$$K(x_m)-K(x_n)=K(x_m)-x_n+(1_E+K)(x_n)\in E_m-\{x_n\}+E_{n+1}\subset E_{n+1}-\{x_n\}.$$

Therefore, $\|K(x_m)-K(x_n)\|>\frac{1}{2}$ and, $\{K(x_n)\}$ does not have any convergence subsequence, while $\{x_n\}$ is bounded, contradicting the compactness of K. Consequently, 1_E+K is a surjection onto E.

We conclude that 1_E+K is a isomorphism of E onto E. □

(4.7) Riesz lemma: *If F is a closed proper subspace of a Banach space E, for every $\varepsilon\in\,]0,1[$ there is $x\in E$ with norm 1 at a distance $>\varepsilon$ of F.*

Proof If $z\in E\setminus F$ is at a distance d from F, there is $y\in F$ such that $\|z-y\|<\frac{d}{\varepsilon}$, and, with $x=\frac{x-y}{\|z-y\|}$, it is $\|x-y\|=\|\frac{x-y(1+\|z-y\|)}{\|z-y\|}\|>\frac{d}{d/\varepsilon}=\varepsilon$. Thus, x has norm 1 and its distance to F is a $>\varepsilon$. □

The Slodkowski theorem gives existence of extension to $\mathbb{C}_\infty$ of any holomorphic motion $\{\pi_\tau\}_{\tau\in T}$ of $\emptyset\neq\Omega\subset\mathbb{C}_\infty$ with parameter on any simply connected Riemann surface[26] T instead of B_1 and any basis point in T, since, by the uniformization theorem,[27] such a Riemann surface is conformal to B_1 or to $\mathbb{C}_\infty$ or $\mathbb{C}$. In the first case, it follows directly from Slodkowski theorem. In the other two cases, as observed at the beginning of the proof for $T=B_1$, it can be assumed without loss of generality $\Omega\supset\{0,1,\infty\}$, $\phi_\tau(0)=0$, $\phi_\tau(1)=1$, $\phi_\tau(\infty)=\infty$, and then for each $z\in\Omega$, the function $\tau\mapsto\phi_\tau(z)$ is holomorphic from T to $\mathbb{C}\setminus\{0,1\}$. Thus, by the Picard little theorem,[28] this function is constant, both if $T=\mathbb{C}$ or $T=\mathbb{C}_\infty$, and the holomorphic motion of Ω can be trivially extended to $\mathbb{C}_\infty$.

Sometimes, it is useful to extend holomorphic motions preserving symmetries. For example, it is useful to consider holomorphic motions of subsets of Riemann surfaces, what can be done through the universal covering with covering space conformal to B_1 and the corresponding holomorphic motion in B_1, but it is necessary that the holomorphic motions be equivalent under the action of the Riemann surface fundamental group.

A simple case of symmetry preservation for a holomorphic motion $\{\phi_\tau\}_{\tau\in B_1}$ of $\emptyset\neq\Omega\subset\mathbb{C}_\infty$ with parameter in B_1 and base point 0 is with the symmetry $\overline{\phi_{\overline{\tau}}(\overline{z})}=\phi_\tau(z)$ for $\tau\in B_1$, $z\in\Omega$ and extension to holomorphic motion of $\mathbb{C}$ with the same symmetry. In this case, it can be seen that it is possible to preserve the

[25] Riesz, Frigyes (1880–1956).

[26] If necessary, see Chapters 11 and 12 of *CADOVA*.

[27] See Chapter 12 of *CADOVA*.

[28] See in the last Section of Chapter 11 of *CADOVA*. Picard, Émile (1856–1941).

symmetry throughout the proof of the given Slodkowski theorem, whereby the holomorphic motion of $\mathbb{C}_\infty$ whose existence is obtained satisfies the symmetry automatically. However, this is not necessarily so for other symmetries, and the proof has to be modified by extending the holomorphic motion to more points and considering full invariant orbits under the symmetry instead of isolated points.

If Γ is a discrete subgroup of $\mathrm{Aut}\,\mathbb{C}_\infty$ and $\rho : B_1\times\Gamma\to\mathrm{Aut}\,\mathbb{C}_\infty$, with $\gamma_\tau=\rho(\tau,\gamma)$ such that $\gamma_0=\gamma$, for each pair of $\gamma\in\Gamma$, $z\in\Omega$, the function $\tau\mapsto\gamma_\tau(z)$ is holomorphic in B_1, and for every fixed $\tau\in B_1$, the function $\gamma\mapsto\gamma_\tau$ is an isomorphism onto a discrete subgroup of Γ, we say that a holomorphic motion $\{\phi_\tau\}_{\tau\in B_1}$ of $\emptyset\neq\Omega\subset\mathbb{C}_\infty$ invariant under Γ with parameter in B_1 and base point 0 is ρ-**equivariant** if $\phi_\tau\big(\gamma(z)\big)=\gamma_\tau\big(\phi_\tau(z)\big)$ for $z\in\Omega$.

As under the conditions of this definition $\gamma_\tau\phi_\tau(z)=\phi_\tau\big(\gamma(z)\big)=\phi_\tau(z)$, *if* z *is a fixed point of* γ, *then* $\phi_\tau(z)$ *is a fixed point of* γ_τ.

(4.8) Slodkowski equivariant theorem:
If Γ *is a discrete subgroup of* $\mathrm{Aut}\,\mathbb{C}_\infty$ *and* ρ *is as in the* ρ*-equivariant definition of holomorphic motion, every holomorphic motion* $\{\phi_\tau\}_{\tau\in B_1}$ *of* $\emptyset\neq\Omega\subset\mathbb{C}_\infty$ *with parameter in* B_1 *and base point* 0 ρ*-equivariant with the fixed points of all* $\gamma\in\Gamma$ *belonging to* Ω *has a* ρ*-equivariant extension to a holomorphic motion of* $\mathbb{C}_\infty$ *with parameter in* B_1 *and base point* 0.

Proof Consider a countable set $X=\{x_1,x_2,\dots\}\subset\mathbb{C}_\infty\backslash\Omega$ such that the orbits of each x_j are disjoint and their union is dense in $\mathbb{C}_\infty\backslash\Omega$. For an argument by induction, consider the holomorphic motion $\{(\Phi_0)_\tau\}_{\tau\in B_1}$ with $(\Phi_0)_\tau=\phi_\tau$ and suppose $\{(\Phi_j)_\tau\}_{\tau\in B_1}$ of $\Omega\cup X_j$ with $X_j=\{x_1,\dots,x_j\}$ is a holomorphic motion ρ-equivariant. By the Slodkowski theorem, the holomorphic motion $\{(\Phi_j)_\tau\}_{\tau\in B_1}$ can be extended to a holomorphic motion of $\mathbb{C}_\infty$, and we can define $(\Phi_{j+1})_\tau\big(\gamma(x_{j+1})\big)=\gamma_\tau\big((\Phi_j)_\tau(x_{j+1})\big)$ for $\tau\in B_1$. The function $\tau\mapsto(\Phi_{j+1})_\tau\big(\gamma(x_{j+1})\big)$ is holomorphic in B_1, and for each $\tau\in B_1$, $\gamma(x_{j+1})\mapsto(\Phi_{j+1})_\tau\big(\gamma(x_{j+1})\big)$ is injective except if there are $\gamma,\delta\in\Gamma$ with $\gamma\neq\delta$ such that $\gamma_\tau\big((\Phi_j)_\tau(x_{j+1})\big)$ is equal, alternatively, to

(1) $\delta_\tau\big((\Phi_j)_\tau(x_k)\big)$, with $k\le j$, (2) $\delta_\tau\big((\Phi_j)_\tau(z)\big)$, with $z\in\Omega$, (3) $\delta_\tau\big((\Phi_j)_\tau(x_{j+1})\big)$;

in each one of these cases, $\delta_\tau^{-1}\gamma_\tau\big((\Phi_j)_\tau(x_{j+1})\big)$ would be equal to, resp.,

(1) $(\Phi_j)_\tau(x_k)$, with $k\le j$, (2) $(\Phi_j)_\tau(z)$, with $z\in\Omega$, (3) $(\Phi_j)_\tau(x_{j+1})$.

(1) and (2) contradict the hypothesis that the image of $\tau\mapsto(\Phi_j)_\tau(x_{j+1})$ is disjoint from $\{(\Phi_j)_\tau(z) : \tau\in B_1,\ z\in\Omega\cup(\Gamma X_j)\}$, and (3) is equivalent to $(\Phi_j)_\tau(x_{j+1})$ to be fixed point of $\delta_\tau^{-1}\gamma_\tau$, in which case x_{j+1} would be a fixed point of $\delta^{-1}\gamma$ in contradiction with all fixed points of elements of Γ to be in Ω. Therefore, $\gamma(x_{j+1})\mapsto(\Phi_{j+1})_\tau\big(\gamma(x_{j+1})\big)$ is injective. Hence, $\{(\Phi_{j+1})_\tau\}_{\tau\in B_1}$ is a holomorphic

motion ρ-equivariant of $\Omega \cup X_{j+1}$. Thus, by induction, there is a ρ-equivariant holomorphic motion of $\Omega \cup X$, and as the closure of this set is $\mathbb{C}_\infty$, by the λ-lemma, this holomorphic motion has a unique extension to a holomorphic motion of $\mathbb{C}_\infty$, which, by continuity, is ρ-equivariant. □

4.6 Unique Extension of Holomorphic Motion to $\mathbb{C}_\infty$

If $\{f_\tau\}_{\tau\in T}$, with T a Riemann surface, is a holomorphic family of rational functions, it is said that a **holomorphic motion** $\{\phi_\tau\}_{\tau\in T}$ **of** Ω **respects the dynamics of** $\{f_\tau\}_{\tau\in T}$ **in** $\Omega \subset \mathbb{C}_\infty$ if it is a conjugacy with the different functions f_τ, i.e., if $\phi_\tau\big(f_\tau(z)\big) = f_{\tau_0}\big(\phi_\tau(z)\big)$ with $\tau, \tau_0 \in T$ and $z \in \Omega$.

In the two following chapters on Complex Dynamics, we need to have conjugacies of rational functions with holomorphic motions, what requires uniqueness of extensions of these holomorphic motions to $\mathbb{C}_\infty$ to guarantee they respect the dynamics of the rational functions mapping orbits to orbits. So, in this section, we prove the "λ-harmonic lemma" established by L. Bers and H. Royden in 1986, for what we need elements of Teichmüller theory.

Teichmüller Space of Hyperbolic Region in $\mathbb{C}_\infty$

To prove the harmonic λ-lemma of L. Bers and H. Royden it suffices to consider the Teichmüller space of regions $\Omega \subset \mathbb{C}_\infty$ with $\infty \in \partial\Omega$ and $\#\mathbb{C}_\infty \setminus \Omega \geq 3$. *For every such region* Ω, *there is a universal covering* $(\mathbb{H}, \pi)$ *with covering space the open complex upper half plane* $\mathbb{H}$ *and element of length of the Poincaré metric of* Ω, $\rho_\Omega(z)|dz|$ *for* $z \in \Omega$, *with* $\rho_\Omega\big(\pi(\zeta)\big)|\pi'(\zeta)| = \frac{2}{\zeta - \bar{\zeta}}$ *for* $\zeta \in \mathbb{H}$. We denote the **covering group** $(\mathbb{H}, \pi)$ by $\boldsymbol{G_\pi}$. It is the subgroup of the group of Aut $\mathbb{H}$ (isomorphic to $PSL(2,\mathbb{R}) = SL(2,\mathbb{R})/\{\pm I_2\}$, with $SL(2,\mathbb{R})$ the special linear group of real matrices 2×2 with determinant 1) consisting of the covering deck mappings,[29] which are homeomorphisms $\psi : \mathbb{H} \to \mathbb{H}$ with $\pi \circ \psi = \pi$.

The **Teichmüller space**[30] **of** Ω is the set **Teich(Ω)** of the equivalence classes $[F]$ of quasiconformal mappings in Ω, with the equivalence of F_1, F_2 if there is a conformal homeomorphism h of $F_1(\Omega)$ to $F_2(\Omega)$ such that the extension of $F_2^{-1} \circ h \circ F_1$ to the boundary is **homotopic to the identity modulo the boundary**, meaning homotopic to the identity with the boundary points fixed. We call **Teichmüller distance** in Teich(Ω) to $\boldsymbol{D_T}$: Teich$(\Omega)^2 \to [0, +\infty[$ with

$$D_T\big([F_1],[F_2]\big) = \tfrac{1}{2}\inf\big\{\log K_{f_1\circ f_2^{-1}} : f_1 \in [F_1],\, f_2 \in [F_2]\big\},$$

---

[29] For basic aspects of coverings, see Sections 2 and 5 of Chapter 12 of *CADOVA*.

[30] The idea of Teichmüller spaces was of O. Teichmüller in 1939, but only for compact Riemann surfaces, for which he obtained important results. L. Ahlfors made precise the bases of the theory in 1954, when he introduced the name Teichmüller space and also made the definition of quasiconformal mapping more flexible, arousing the interest of many mathematicians and leading to remarkable progresses since then.

where K_f is the dilatation of the quasiconformal mapping f in Ω, which in terms of the Beltrami coefficient κ_f of f is $K_f = \frac{1+\|\kappa_f\|_{L^\infty}}{1-\|\kappa_f\|_{L^\infty}}$. This type of Teichmüller space is a particular case of Teichmüller spaces of Riemann surfaces to be considered in Chap. 9.

(4.9) *If* $\Omega \subset \mathbb{C}_\infty$ *has universal covering* $(\mathbb{H}, \pi)$, $[F_1], [F_2] \in \mathrm{Teich}(\Omega)$,

1. $D_T([F_1],[F_2]) = \frac{1}{2}\inf\{\log K_{f_1\circ f_2^{-1}} : f_1 \in [F_1]\}$, *for* $f_2 \in [F_2]$.
2. *There is* $f_1 \in [F_1]$ *such that* $K_{f_1} = \inf\{K_f : f \in [F_1]\}$.
3. $(\mathrm{Teich}(\Omega), D_T)$ *is a complete connected metric space.*
4. $D_T([F_1],[F_2]) \overset{def}{=} \frac{1}{2}\inf\{\log \frac{1+\|(\kappa_{f_1}-\kappa_{f_2})/(1-\overline{\kappa_{f_1}}\kappa_{f_2})\|_{L^\infty}}{1-\|(\kappa_{f_1}-\kappa_{f_2})/(1-\overline{\kappa_{f_1}}\kappa_{f_2})\|_{L^\infty}} : f_1 \in [F_1], f_2 \in [F_2]\}$, *where* κ_f *is the Beltrami coefficient of* f.
5. $D_B([F_1],[F_2]) \overset{def}{=} \frac{1}{2}\inf\{\|\frac{\kappa_{f_1}-\kappa_{f_2}}{1-\overline{\kappa_{f_1}}\kappa_{f_2}}\|_{L^\infty} : f_1 \in [F_1], f_2 \in [F_2]\}$ *is a distance in* $\mathrm{Teich}(\Omega)$ *with the same topology of* D_T.

Proof

(1) Denote the right hand side of the equality $D_{T,f_2}([F_1])$. It is $D_T([F_1],[F_2]) \le D_{T,f_2}([F_1])$. For $\varepsilon > 0$, there are $f_{1,\varepsilon} \in [F_1]$, $f_{2,\varepsilon} \in [F_2]$ such that $\log K_{f_{1,\varepsilon}\circ f_{2,\varepsilon}^{-1}} < 2\, D_T([F_1],[F_2]) + \varepsilon$ and, as

$$\log K_{f_{1,\varepsilon}\circ f_{1,\varepsilon}^{-1}} = \log K_{f_{1,\varepsilon}\circ f_{1,\varepsilon}^{-1}\circ f_2 \circ f_2^{-1}} \ge 2\, D_{T,f_2}([F_1]),$$

$D_T([F_1],[F_2]) + \varepsilon \ge D_{T,f_2}([F_1])$; thus, $D_T([F_1],[F_2]) \ge D_{T,f_2}([F_1])$. By the preceding inequalities, $D_T([F_1],[F_2]) \le D_{T,f_2}([F_1])$.

(2) If $f \in [F]$ is homotopic to the identity modulo the boundary, so that the corresponding function in $\mathbb{H}$ resulting by extension to the boundary and the universal covering $\widetilde{f}$ coincide with the identity in $\partial\mathbb{H} = \mathbb{R} \subset \mathbb{C}_\infty$. By (3.22), any sequence of quasiconformal mappings in $\mathbb{H}$ coinciding with the identity in $\partial\mathbb{H}$ is a normal family. Thus, it has a convergent subsequence to some quasiconformal mapping. Therefore, if $\{f_n\} \subset [F]$ is such that $K_{f_n} \to \inf\{K_f : f \in [F]\}$, there is a subsequence of $\{f_n\}$ converging to a quasiconformal mapping f_0 which is the identity in the boundary, since all functions f_n are. If $\widetilde{f_0}$ is the function in $\mathbb{H}$ corresponding to f_0 through the universal covering, for every $t \in [0,1]$, the values of $t\widetilde{f_0}(\zeta) + (1-t)1_{\mathrm{cl}\,\mathbb{H}}(\zeta)$ in the boundary remain fixed. Thus, $\pi(t\widetilde{f_0}(\zeta) + (1-t)1_{\mathrm{cl}\,\mathbb{H}}(\zeta))$ is a homotopy of f_0 in 1_Ω.

(3) If f is a conformal homeomorphism, so is f^{-1} and $K_{f^{-1}} = K_f$. Thus, if f_1, f_2 are quasiconformal mappings with $K_{f_1\circ f_2^{-1}} = K_{f_2\circ f_1^{-1}}$ and

$$D_T([F_1],[F_2]) = D_T([F_2],[F_1]), \qquad [F_1],[F_2] \in \mathrm{Teich}(\Omega).$$

As $f_1 \circ f_1^{-1} = 1_\Omega$ and $K_{1_\Omega} = 1$, it is $D_T([F_1],[F_1]) = 0$. Conversely, if $D_T([F_1],[F_2]) = 0$, by 2, there are $f_1 \in [F_1]$ and $f_2 \in [F_2]$ with

$K_{f_1\circ f_2^{-1}}=1$, whereby $h=f_1\circ f_2^{-1}$ is a conformal homeomorphism and, as $f_1^{-1}\circ h\circ f_2=1_\Omega$, $[F_1]=[f_1]=[f_2]=[F_2]$. Since for $f_1\in[F_1]$, $f_2\in[F_2]$, $f_3\in[F_3]$,

$$\log K_{f_1\circ f_3^{-1}}=\log K_{f_1\circ f_2^{-1}\circ f_2\circ f_3^{-1}}\leq\log\big(K_{f_1\circ f_2^{-1}}K_{f_2\circ f_3^{-1}}\big)=\log K_{f_1\circ f_2^{-1}}+\log K_{f_2\circ f_3^{-1}}\,.$$

Therefore,

$$D_T\big([F_1],[F_3]\big)\leq D_{T,f_2}([F_1])+D_{T,f_2}([F_3])=D_T\big([F_1],[F_2]\big)+D_T\big([F_2],[F_3]\big)\,.$$

Thus, D_T satisfies the conditions of the definition of distances.

Now, we prove the metric space $\big(\mathrm{Teich}(\Omega),\,D_T\big)$ is complete. If $\{[f_n]\}$ is a Cauchy sequence of this space, by 1, for every $\varepsilon>0$ there is $j\in\mathbb{N}$ such that $\inf\{\log K_{f\circ f_j^{-1}}\,:\,f\in[f_{j+p}],\,p\in\mathbb{N}\}<\delta$. Each term of the sequence f_{j+p} with $p\in\mathbb{N}$ is replaced by an element of the equivalence class $[f_{j+p}]$ such that $\log K_{f_{j+p}\circ f_j^{-1}}<\delta$, with $\delta=\frac{1}{2}$, and afterward, the first j–1 terms of the sequence are suppressed and the terms are renumbered so that f_{j+p} of the preceding sequence is f_p of the new sequence. This process is successively repeated with $\delta=\frac{1}{2^{n+1}}$ for the sequence starting at the term $n+1$ of the preceding sequence, for every $n\in\mathbb{N}$. For the resulting sequence $\{f_n\}$, $\{[f_n]\}$ is a subsequence of the initial Cauchy sequence and $\log K_{f_{n+1}\circ f_n^{-1}}<\frac{1}{2^n}$ for $n\in\mathbb{N}$. So, for $n,\,p\in\mathbb{N}$,

$$\begin{aligned}\log K_{f_{n+p}\circ f_n^{-1}}&=\log K_{f_{n+p}\circ f_{n+p-1}^{-1}\circ f_{n+p-1}\circ\cdots\circ f_{n+2}\circ f_{n+2}^{-1}\circ f_n\circ f_n^{-1}}\\&\leq\log\prod_{j=1}^{p}K_{f_{n+j}\circ f_{n+j-1}^{-1}}=\sum_{j=1}^{p}\log K_{f_{n+j}\circ f_{n+j-1}^{-1}}<\sum_{j=1}^{p}\frac{1}{2^{n+j-1}}<\frac{1}{2^{n-1}}\,.\end{aligned}$$

Since for $p\in\mathbb{N}$ it is $K_{f_{1+p}\circ f_1-1}<1$ and $K_{f_{1+p}\circ f_1-1}<e$,

$$K_{f_{1+p}}=K_{f_{1+p}\circ f_1^{-1}\circ f_1}\leq K_{f_{1+p}\circ f_1-1}K_{f_1}\leq e\,K_{f_1}\,,\qquad p\in\mathbb{N}\,.$$

The terms of $\{f_n\}$ are (eK_{f_1})-quasiconformal mappings. As they have fixed points 0, 1, ∞, by (3.22), $\{f_n\}$ is a normal family. It has a uniformly convergent subsequence in the spherical metric to some (eK_{f_1})-quasiconformal mapping f and $[f]=\lim\,[f_n]$ in the Teichmüller metric.

If $[F]\in\mathrm{Teich}(S)$ and the Beltrami coefficient of F is κ_F, for every $t\in[0,\,1]$, the function $t\kappa_F$ is the Beltrami coefficient of some quasiconformal mapping F_t in Ω and for $t_1,\,t_2\in[0,\,1]$,

$$2D_T\big([F_{t_1}],[F_{t_2}]\big)\leq\log\frac{1+\|(t_1\kappa_F-t_2\kappa_F)/(1-t_1t_2\overline{\kappa_F}\kappa_F)\|_{L^\infty}}{1-\|(t_1\kappa_F-t_2\kappa_F)/(1-t_1t_2\overline{\kappa_F}\kappa_F)\|_{L^\infty}}\leq\log\Big(1+\Big\|\frac{t_1-t_2}{1-\|\kappa_F\|_{L^\infty}^2}\Big\|\Big).$$

Thus, $\lim\limits_{t_2\to t_1}D_T\big([F_{t_1}],[F_{t_2}]\big)=0$, and $t\mapsto[F_t]$ is continuous from $[0,1]$ to $\mathrm{Teich}(\Omega)$. Therefore, it is a path in $\mathrm{Teich}(\Omega)$ and $\mathrm{Teich}(\Omega)$ is path connected. Consequently, $\mathrm{Teich}(\Omega)$ is connected.

(4) This is a consequence of the formula of the dilatation of a quasiconformal mapping f in Ω, $K_f = \frac{1+\|\kappa_f\|_{L^\infty}}{1-\|\kappa_f\|_{L^\infty}}$, with κ_f the corresponding Beltrami coefficient, and of the formula for the Beltrami coefficient of $f_1 \circ f_2^{-1}$ in the end of Sect. 1.2 of Chap. 1, $\|\kappa_{f_1 \circ f_2^{-1}}\|_{L^\infty} = \left\| \frac{\kappa_{f_1} - \kappa_{f_2}}{1-\overline{\kappa_{f_1}}\kappa_{f_2}} \right\|_{L^\infty}$.

5) Just notice that $D_B([F_1],[F_2]) = \tanh D_T([F_1],[F_2])$ and that tanh is strictly increasing in $[0, +\infty[$. □

By 2 in the preceding result, *in the formula defining the Teichmüller distance and also in the formulas in 1, 3, and 5 of this result, infimum can be replaced by minimum.*

Schwarzian Derivative

Given a Möbius transformation $T(z) = \frac{az+b}{cz+d}$ with $ad-bc \neq 0$,

$$T'(z) = \frac{ad-bc}{(cz+d)^2}, \qquad T''(z) = \frac{(ad-bc)(-2c)}{(cz+d)^3}, \qquad T'''(z) = \frac{(ad-bc)6c^2}{(cz+d)^3},$$

$$\Big[\frac{T'''}{T'} - \frac{3}{2}\Big(\frac{T''}{T'}\Big)^2\Big](z) = (ad-bc)\Big[\frac{-3c[-2c(cz-d)^2]}{(cz-d)^4} - \frac{3}{2}\Big(\frac{-2ccz-d)^2}{(cz-d)^3}\Big)^2\Big] = 0.$$

Conversely, if T is a holomorphic function that satisfies $\frac{T'''}{T'} - \frac{3}{2}\big(\frac{T''}{T'}\big)^2 = 0$, since $\big(\frac{T''}{T'}\big)' = \frac{T'''}{T'} - \big(\frac{T''}{T'}\big)^2$, then $\big(\frac{T''}{T'}\big)'\big(\frac{T''}{T'}\big)^{-2} = \frac{1}{2}$ and $\big(\frac{T''}{T'}\big)^{-1}(z) = -\frac{1}{2}z + A$. Thus, $\frac{T''}{T'}(z) = \frac{1}{-z/2+A}$ and, therefore, $\log T'(z) = -2\log\big(-\frac{z}{2}+A\big)+B$, $T'(z) + \frac{e^B}{(-z/2+A)^2}$, $T(z) = \frac{2e^B}{-z/2+A} + C + \frac{-(C/2)z+2e^B+AC}{-z+2A}$, with A, B, C complex constants. Since $-(C/2)(2A) - (2e^B + AC)(-1) + e^B \neq 0$, T is a Möbius transformation. Therefore, a holomorphic function is a Möbius transformation if and only if $\frac{T'''}{T'} - \frac{3}{2}\big(\frac{T''}{T'}\big)^2 = 0$. So, it is natural to define the **Schwarzian derivative**[31] of a holomorphic function g by

$$\mathscr{S}g \stackrel{def}{=} \Big(\frac{g''}{g'}\Big)' - \frac{1}{2}\Big(\frac{g''}{g'}\Big)^2 = \frac{g'''}{g'} - \frac{3}{2}\Big(\frac{g''}{g'}\Big)^2,$$

and we have: *A holomorphic function is a Möbius transformation if and only if it has Schwarzian derivative* 0.

If $\Omega \subset \mathbb{C}_\infty$ is a simply connected region, the **Nehari norm**[32] in the complex linear space $H(\Omega)$ of the holomorphic functions in Ω is

$$\|f\|_{\mathcal{N}_\Omega} = \frac{1}{\eta_\Omega^2} \sup_\Omega |f|,$$

where $\eta_\Omega(z)\,|dz|$ is the element of length of the Poincaré metric[33] of Ω.

[31] It was introduced in 1781 by Joseph-Louis Lagrange (1736–1813) in the publication *Sur la Construction des Cartes Géographiques*, and it also appeared in an article by Ernst Kummer (1810–1893) in 1836. The name after H. Schwarz was given by Arthur Cayley (1821–1895) in 1880 due to 1869 and 1873 articles by H.Schwarz where it is explicitly considered, but H.Schwarz informed A.Cayley it had been introduced by J.-L.Lagrange in 1781.

[32] Nehari, Zeev (1915–1978).

[33] See Section 5 of Chapter 11 of *CADOVA*.

(4.10) *Let $\Omega \subset \mathbb{C}_\infty$ be a simply connected region. The Schwarzian derivative has the following properties:*

1. $\mathscr{S} f = 0$ *for* $f \in H(\Omega)$ *if and only if* f *is a Möbius transformation.*
2. *If* $f(z) \neq 0$, *then* $\mathscr{S}\big(\frac{1}{f}\big)(z) = (\mathscr{S} f)(z)$.
3. ***Chain rule for Schwarzian derivatives:***

$$\mathscr{S}(f \circ g) = \big((\mathscr{S} f) \circ g\big)(g')^2 + \mathscr{S} g\,.$$

4. *Schwarzian derivatives are invariant under left compositions by Möbius transformations.*
5. *If* $\varphi \in H(\Omega)$, *there is a meromorphic function* f *such that* $\mathscr{S} f = \varphi$, *unique modulo left composition with Möbius transformations, i.e., any solution is* $T \circ f$ *with* T *Möbius transformation.*
6. *If* $\Omega \subsetneqq \mathbb{C}$ *and* $f, g \in H(\Omega)$, *then* h^{-1} *is a conformal mapping in* Ω,

$$\|\mathscr{S} f - \mathscr{S} g\|_{\mathcal{N}_\Omega} = \|\mathscr{S}(f \circ h) - \mathscr{S}(g \circ h)\,\|_{\mathcal{N}_{h^{-1}(\Omega)}}\,,$$

$$\|\mathscr{S} f - \mathscr{S} h^{-1}\|_{\mathcal{N}_\Omega} = \|\mathscr{S}(f \circ h)\,\|_{\mathcal{N}_{h^{-1}(\Omega)}}\,,$$

$$\|\mathscr{S} h^{-1}\|_{\mathcal{N}_\Omega} = \|\mathscr{S} h\,\|_{\mathcal{N}_{h^{-1}(\Omega)}}\,.$$

7. *Nehari norms of Schwarzian derivatives are invariant under compositions with Möbius transformations* T *(at left or right):*

$$\|\mathscr{S} f\|_{\mathcal{N}_\Omega} = \|T \circ \mathscr{S} f\|_{\mathcal{N}_\Omega} = \|\mathscr{S}(f \circ T)\|_{\mathcal{N}_{T^{-1}(\Omega)}}\,.$$

8. *If* $f : \mathbb{C}_\infty \to \mathbb{C}_\infty$ *is a quasiconformal mapping with Beltrami coefficient* κ_f, *a conformal homeomorphism in* Ω $\#\partial\Omega \geq 2$

$$\|\mathscr{S}(f_{|\Omega})\|_{\mathcal{N}_\Omega} \leq \sigma_0(\Omega)\,\|\kappa_f\|_{L^\infty},$$

$$\sigma_0(\Omega) = \sup\big\{\|\mathscr{S} g\|_{\mathcal{N}_\Omega} : g \in H(\Omega) \text{ is injective}\big\}.$$

Proof

(1) This was proved in the motivation of Schwarzian derivative.

(2) It is $\big(\frac{1}{f}\big)' = -\frac{f'}{f^2}$, $\big(\frac{1}{f}\big)'' = -\frac{f''f - 2(f')^2}{f^3}$, $\big(\frac{1}{f}\big)''' = -\frac{f'''f - 2f''f'}{f^3} + \frac{4f'f''f - 6(f')^3}{f^4}$

$$\begin{aligned}\mathscr{S}\big(\tfrac{1}{f}\big) &= \tfrac{f'''}{f'} - \tfrac{6f''}{f} + \tfrac{6(f')^2}{f^2} - \tfrac{3}{2}\big(\tfrac{f''}{f'} - 2\tfrac{f'}{f}\big)^2 \\ &= \tfrac{f'''}{f'} - \tfrac{6f''}{f} + \tfrac{6(f')^2}{f^2} - \tfrac{3}{2}\big(\tfrac{(f'')^2}{(f')^2} - 4\tfrac{f''}{f} + 4\tfrac{(f')^2}{f^2}\big) = \tfrac{f'''}{f'} - \tfrac{3}{2}\big(\tfrac{f''}{f'}\big)^2 = \mathscr{S} f\,.\end{aligned}$$

(3) Since

$$(f\circ g)'=(f'\circ g)g', \qquad (f\circ g)''=(f''\circ g)(g')^2+(f'\circ g)g'',$$

$$(f\circ g)'''=(f'''\circ g)(g')^3+3(f''\circ g)g'g''+(f'\circ g)g''',$$

$$\mathscr{S}(f\circ g)=\frac{(f\circ g)'''}{(f\circ g)'}-\frac{3}{2}\left(\frac{(f\circ g)''}{(f\circ g)'}\right)^2$$

$$=\frac{(f'''\circ g)(g')^2}{f'\circ g}+3\frac{(f''\circ g)g''}{f'\circ g}+\frac{g'''}{g'}-\frac{3}{2}\frac{(f''\circ g)^2(g')^2}{(f'\circ g)^2}-3\frac{(f''\circ g)g''}{f'\circ g}-\frac{3}{2}\left(\frac{g''}{g'}\right)^2$$

$$=\frac{(f'''\circ g)(g')^2}{f'\circ g}-\frac{3}{2}\frac{(f''\circ g)^2(g')^2}{(f'\circ g)^2}+\frac{g'''}{g'}-\frac{3}{2}\left(\frac{g''}{g'}\right)^2=\mathscr{S}(f\circ g)=(\mathscr{S}f)\circ g\,(g')^2+\mathscr{S}g\,.$$

(4) If T is a Möbius transformation, $\mathscr{S}T=0$ and, by 2, $\mathscr{S}(T\circ f)=\mathscr{S}f$.

(5) With $g=\frac{f''}{f'}$, the equation $\mathscr{S}(f)=\varphi$ is equivalent to $g'-\frac{1}{2}g^2=\varphi$. With the substitution $g=-2\frac{u'}{u}$, $w''+\frac{1}{2}\varphi w=0$. The initial value problem for this linear differential equation with $w(z_0)=w_0$ and $w'(z_0)=w_0'$ has unique holomorphic solution in a neighborhood of any $z_0\in\Omega$, $w(z)=\sum_{n=0}^{\infty}c_n(z-z_0)^n$. If $\varphi(z)=2\sum_{n=0}^{\infty}a_n(z-z_0)^n$ in a disk with center z_0, w is a solution if and only if $n(n-1)c_n+\sum_{k=0}^{n-2}a_{n-2-k}c_k=0$ for $n\in\mathbb{N}$. With $c_0=0$, $c_1=1$, $0<r<1$, and $M>1$ such that $|a_n|<\frac{M}{r^n}$ for $n\in\mathbb{N}\cup\{0\}$, and $Mr^2<1$

$$n(n-1)|c_n|\le M\sum_{k=0}^{n-2}r^{-n+2+k}|c_k|<r^{-n}\sum_{k=0}^{n-2}c_kr^k\,.$$

If $r^{-k}\ge|c_k|$ for $k=0,\ldots,n-2$, then $r^{-n}\ge n|c_n|>|c_n|$. By induction, this inequality holds for $n\in\mathbb{N}$. Therefore, $w(z)=\sum_{n=0}^{\infty}c_n(z-z_0)^n$ is holomorphic in the disk $B_r(z_0)$. As the region Ω is simply connected, by analytic continuation, we get a holomorphic function w_1 solution of the differential equation. With $c_0=1$ and $c_1=0$, we similarly get a holomorphic function w_2 solution of the differential equation such that w_1, w_2 are a pair of linearly independent solutions. The function $f=\frac{w_1}{w_2}$ is meromorphic and locally injective and

$$\mathscr{S}f=\left(\frac{f''}{f'}\right)'-\frac{1}{2}\left(\frac{f''}{f'}\right)^2=\left(-2\frac{w_2'}{w_2}\right)'-\frac{1}{2}\left(\frac{w_2'}{w_2}\right)^2=-2\frac{w_2''}{w_2}=\varphi\,.$$

By 4, if $\mathscr{S}f=\varphi$ and T is a Möbius transformation, $\mathscr{S}(T\circ f)=\mathscr{S}f=\varphi$.

If also $\mathscr{S}g=\varphi$, as $\mathscr{S}f=\varphi$ and f is locally injective so that we can define f^{-1} locally, by the first equality in 6, $\mathscr{S}(f\circ h)-\mathscr{S}(g\circ h)=0$ for every holomorphic function h and with $h=f^{-1}$ locally, we get

$$\mathscr{S}(g\circ f^{-1})=\mathscr{S}(f\circ f^{-1})=\mathscr{S}(1_U)=0$$

locally. Thus, $T = g \circ f^{-1}$ is locally a Möbius transformation, $g = T \circ f$ locally, and, therefore, also globally in Ω because as g and f are holomorphic, the local Möbius transformation T is the same in open sets with nonempty intersection.

(6) By 3, $\mathscr{S}(f \circ h) - \mathscr{S}(g \circ h) = \big[(\mathscr{S}f)\circ h - (\mathscr{S}g)\circ h\big](h')^2$ and, as $\eta_{h^{-1}(\Omega)} = (\eta_\Omega \circ h)\,|h'|$,

$$|\mathscr{S}(f \circ h) - \mathscr{S}(g \circ h)|\, \eta_{h^{-1}(\Omega)}^{-2} = \big(|\mathscr{S}f - \mathscr{S}g|\, \eta_\Omega^{-2}\big) \circ h\,.$$

Thus, $\|\mathscr{S}(f \circ h) - \mathscr{S}(g \circ h)\|_{\mathcal{N}_{h^{-1}(\Omega)}} = \|\mathscr{S}f - \mathscr{S}g\|_{\mathcal{N}_\Omega}$. The second equality follows from the first with $g = h^{-1}$, as $\mathscr{S}(1_{h^{-1}(\Omega)}) = 0$, and the third equality follows from the second with $f = 1_\Omega$.

(7) By 4, $\|\mathscr{S}f\|_{\mathcal{N}_\Omega} = \|T \circ \mathscr{S}f\|_{\mathcal{N}_\Omega}$. By 6, with $g = 1_\Omega$ and $h = T$,

$$\|\mathscr{S}f\|_{\mathcal{N}_\Omega} = \|\mathscr{S}(f \circ T)\|_{\mathcal{N}_{T^{-1}(\Omega)}}\,.$$

(8) If T is a Möbius transformation, f can be replaced by $f \circ T$ without changing the Beltrami coefficient or the norm and $\sigma_0(\Omega) = \sigma_0\big(T^{-1}(\Omega)\big)$. Thus, without loss of generality, $\infty \in \Omega$ and the corresponding Beltrami coefficient κ_f is nonzero a.e. and it has compact support. For $w \in B_1 \subset \mathbb{C}$, let $F_{\kappa_f,w}$ be the only quasiconformal mapping with Beltrami coefficient $\frac{w\kappa_f}{\|\kappa_f\|_{L^\infty}}$ such that $\lim_{z\to\infty} [F_{\kappa_f,w}(z) - z] = 0$. By the result on existence, uniqueness, and analytic dependence on parameters of solutions of the Beltrami equation (2.1), for each $z \in \Omega$, the derivative $F'_{\kappa_f,w}(z)$ depends analytically on $w \in B_1$. For each $z \in \Omega$, we consider the function $w \mapsto \psi_z(w), = [\mathscr{S}(F_{\kappa_f,w})](z)\,[\eta_\Omega(z)]^{-2}$, where $\eta_\Omega(z)\,|dz|$ is the element of length of the Poincaré metric of Ω. As the Schwarzian derivative of a function is a rational function of the function derivatives of orders 1, 2, and 3, ψ_z is holomorphic in B_1, and $|\psi_z(w)| \le \sigma_0(\Omega)$ with $\psi_z(0) = 0$ because $F_{\kappa_f,0}(z) = z$. By the Schwarz lemma,[34] $\psi_z(w) \le \sigma_0(\Omega)|w|$. With $w = \|\kappa_f\|_{L^\infty}$, we get $\|\mathscr{S}(f_{|\Omega})\|_N \le \sigma_0(\Omega)\|\kappa_f\|_{L^\infty}$. □

The formula in (4.10.2) can be used to define Schwarzian derivative of meromorphic functions at points where they have a simple pole. So, if f is meromorphic and locally injective, its Schwarzian derivative exists and is a holomorphic function. The formula (4.10.3) can be used to define the Schwarzian derivative of meromorphic functions locally injective at ∞, considering the function $\varphi(z) = f\big(\frac{1}{z}\big)$, as by (4.10.3), $z^4(\mathscr{S}\varphi)(z) = (\mathscr{S}f)\big(\frac{1}{z}\big)$, and we can define $(\mathscr{S}f)(\infty) = \lim_{z\to 0} z^4(\mathscr{S}\varphi)(z)$, which is holomorphic at ∞.

A *K*-**quasidisk** is a region with boundary a *K*-**quasicircle** image of a circle by a *K*-quasiconformal mapping.

---

[34] See Section 3 of Chapter 10 of *CADOVA*.

(4.11) *Let C be a K-quasicircle boundary of two regions A_1, A_2.*

1. *If ψ is a K-quasiconformal reflection in C and $z_0 \in C$*

$$\overline{\lim_{z\to z_0}} \left|\frac{\psi(z)-z_0}{z-z_0}\right| \le e^{\frac{K}{2}}.$$

2. *There exists a quasiconformal reflection ψ in C such that for every meromorphic locally injective function f in A_2,*

$$|\psi(z)-z|^2|\psi'(z)||(\mathscr{S}f)(\psi(z))| \le e^{\frac{K}{2}} \|\mathscr{S}f\|_{\mathscr{N}_{A_2}}, \; z\in A_1\backslash\{\infty, \psi^{-1}(\{\infty\})\}.$$

Proof

(1) If $\infty \in C$, by (3.29), the circular dilatation of the K-quasiconformal mapping ψ is upper bounded by $e^{K/2}$, $\overline{\lim}_{z\to z_0} \left|\frac{\psi(z)-z_0}{z-z_0}\right| \le e^{K/2}$.

Since for a Möbius transformation T

$$T(z)-T(w) = [\,T'(z)\,T'(w)\,]^{\frac{1}{2}}\,(z-w),$$

with $\varphi = T\circ\psi\circ T^{-1}$,

$$\varphi\big(T(z)\big)-T(z_0) = T\big(\psi(z)\big)-T(z_0) = \big[\,T'\big(\psi(z)\big)\,T'(z_0)\,\big]^{\frac{1}{2}}[\,\psi(z)-z_0\,].$$

Thus,

$$\frac{\varphi(T(z))-T(z_0)}{T(z)-T(z_0)} = \left(\frac{T'(\psi(z))}{T'(z)}\right)^{\frac{1}{2}} \frac{\psi(z)-z_0}{z-z_0},$$

and for $z_0 \in C$ such that $T(z_0) \neq \infty$, with $\zeta = T(z)$ and $\zeta_0 = T(z_0)$,

$$\overline{\lim_{\zeta\to\zeta_0}} \left|\frac{\varphi(\zeta)-\zeta_0}{\zeta-\zeta_0}\right| = \overline{\lim_{z\to z_0}} \left|\frac{\varphi(T(z))-T(z_0)}{T(z)-T(z_0)}\right| = \overline{\lim_{z\to z_0}} \left|\frac{\psi(z)-z_0}{z-z_0}\right|.$$

(2) If ψ is C^1, then $T'\big(\psi(z)\big)\psi'(z) = T'(z)\,\varphi'\big(T(\psi(z))\big)$ and, since from the proof of 1, $\varphi\big(T(z)\big)-T(z) = \big[T'\big(\psi(z)\big)\,T'(z)\big]^{1/2}[\psi(z)-z]$.

$$[\psi(z)-z]^2\psi'(z)\big[T'\big(\psi(z)\big)\big]^2 = \big[\varphi\big(T(z)\big)-T(z)\big]^2\varphi'\big(T(z)\big).$$

With $\widetilde{f} = f\circ T^{-1}$, we have $(\mathscr{S}\psi)(z) = (\mathscr{S}f)\big(\varphi(T(z))\big)\,T'\big(\psi(z)\big)^2$. Hence,

$$\big|\big[\varphi\big(T(z)\big)-T(z)\big]^2\varphi'\big(T(z)\big)\,(\mathscr{S}\widetilde{f})\big(\varphi(T(z))\big)\big|$$
$$= \big|[\psi(z)-z]^2\psi'(z)\,[(\mathscr{S}f)(\psi(z))]^2\big| \le e^{\frac{K}{2}}\,\|\mathscr{S}f\|_{\mathscr{N}_{A_2}}.$$

□

In 1963, L. Ahlfors proved the following result on extension of meromorphic functions with bounded Nehari norm to injective quasiconformal mappings.

(4.12) *A meromorphic function in a K-quasidisk A with* $\|\mathscr{S}f\|_{\mathcal{N}_A} < e^{-\frac{K}{2}}$ *is injective and has extension to a quasiconformal mapping in* $\mathbb{C}$ *with Beltrami coefficient* κ *such that* $\|\kappa\|_{L^\infty} \le e^{\frac{K}{2}} \|\mathscr{S}f\|_{\mathcal{N}_A}$.

Proof Assume, without loss of generality, that f is locally injective, $\infty \notin \operatorname{cl} A$, $f = \frac{w_1}{w_2}$ with w_1, w_2 solutions of the differential equation $w'' + \frac{1}{2}(\mathscr{S}f)w$ in A such that $w_1 w_2' - w_1' w_2 = 1$, since, from the proof of (4.10.5), $\mathscr{S}f = \mathscr{S}\big(\frac{w_1}{w_2}\big)$, and $f = T \circ \frac{w_1}{w_2}$ with T a Möbius transformation.

We begin supposing that f, w_1, w_2 are holomorphic functions in ∂A. Let $A_1 = \mathbb{C} \setminus \operatorname{cl} A$, ψ be a quasiconformal reflection satisfying the inequality in (4.11.2),

$$g = \frac{w_1 \circ \psi + (1_{A_1} - \psi)\, w_1' \circ \psi}{w_2 \circ \psi + (1_{A_1} - \psi)\, w_2' \circ \psi} \text{ in } A_1\,, \qquad g(\infty) = \frac{(w_1' \circ \psi)(\infty)}{(w_2' \circ \psi)(\infty)}\,.$$

It is $\lim_{z \to z_0} g(z) = f(z_0)$ for $z_0 \in \partial A$ and g is C^1 as a function of variable in $\mathbb{R}^2$ at points $z \in A_1 \setminus \{\infty\}$ where $g(z) \ne \infty$, and

$$\frac{\partial g}{\partial z}(z) = -\frac{1 + (1/2)(\mathscr{S}f) \circ \psi)(z)\,[z - \psi(z)]^2 (\partial\psi/\partial z)(z)}{[\,(w_2 \circ \psi)(z) + [z - \psi(z)]\,(w_2' \circ \psi)(z)\,]^2}$$

$$\frac{\partial g}{\partial \bar z}(z) = -\frac{(1/2)((\mathscr{S}f) \circ \psi)(z)\,[z - \psi(z)]^2 (\partial\psi/\partial \bar z)(z)}{[\,(w_2 \circ \psi)(z) + [z - \psi(z)]\,(w_2' \circ \psi)(z)\,]^2}\,.$$

By the inequality in (4.11.2), the Beltrami coefficient κ_g of g satisfies

$$|\kappa_g| \le \frac{e^{k/2}\|\mathscr{S}f\|_{\mathcal{N}_A}}{2 - e^{K/2}\|\mathscr{S}f\|_{\mathcal{N}_A}} \le e^{k/2}\|\mathscr{S}f\|_{\mathcal{N}_A} < 1\,.$$

As $\|\mathscr{S}f\|_{\mathcal{N}_A} < e^{-K/2}$, $\frac{\partial g}{\partial z}(z) \ne 0$ and the Jacobian $\big|\frac{\partial g}{\partial z}(z)\big|^2 \big(1 - |\kappa_g(z)|^2\big) > 0$. Thus, g is locally injective in some neighborhood of each $z \in A_1 \setminus \{\infty\}$ such that $g(z) \ne \infty$. Hence, $\frac{1}{g}$ is injective in some neighborhood of $z \in A_1 \setminus \{\infty\}$ such that $g(z) \ne 0$. Considering the function $z \mapsto g\big(\frac{1}{z}\big)$, we get that also g is locally injective at ∞. Thus, g is locally injective in A_1. The function F equal to f in $A \cup \partial A$ and to g in A_1 is locally injective and, as $F(z) \ne \infty$ in ∂A, $F^{-1}(\{\infty\})$ is finite. By the upper bound (4.11-1), $\lim_{z \to z_0} \frac{g(z) - f(z_0)}{z - z_0} = f'(z_0)$ for $z_0 \in \partial A$. The inequality obtained in the preceding paragraph applied to $z \mapsto \log(z - z_0)$ gives $(z - z_0)^2 \frac{\partial \psi}{\partial z}(z) \to 0$, $(z - z_0)^2 \frac{\partial \psi}{\partial \bar z}(z) \to 0$. Thus, F is C^1 in ∂A as function of real variables and, therefore, also in $\mathbb{C} \setminus F^{-1}(\{\infty\})$. In ∂A, the Jacobian of F is $|f'|^2$. Hence, F is locally injective into $\mathbb{C}$. By the theorem of existence, uniqueness, and analytic and continuous dependence on parameters of solutions of Beltrami equation (2.1), there is a quasiconformal mapping h in $\mathbb{C}$ with a Beltrami coefficient equal to that of F a.e. in $\mathbb{C}$. The function $\varphi = F \circ h^{-1}$ satisfies $\frac{\partial \varphi}{\partial \bar z}(z) = 0$ a.e. in $\mathbb{C}$. Since the quasiconformal mapping h^{-1} has partial derivatives in L^2 as a function of real variables and the partial derivatives of F are continuous in $\mathbb{C}\setminus F^{-1}(\{\infty\})$, φ has partial derivatives in L^2 as a function of real variables and is holomorphic in $\mathbb{C}\setminus h\big(\mathbb{C}\setminus F^{-1}(\{\infty\})\big)$. Thus, φ is meromorphic in $\mathbb{C}_\infty$. Hence, it is a rational function and, as it is locally injective, it is a Möbius transformation. Therefore, $F=\varphi\circ h$ is a quasiconformal mapping in $\mathbb{C}$, ending the proof for f holomorphic in ∂A.

If f is not holomorphic in ∂A, consider an expansive sequence of quasidisks included in A with union A and boundaries without poles of f. For each term of the sequence, apply what was proved in the preceding paragraph to obtain the validity of the result in A. □

Space of Quadratic Differentials

If g is a holomorphic function, by (4.10.1,3), $(\mathscr{S}g)\circ\gamma)[\gamma']^2=\mathscr{S}g$. Let $\boldsymbol{B(-\mathbb{H}, G_\pi)}$ be the complex linear space of the holomorphic functions g in $-\mathbb{H}$ such that

$$(g\circ\gamma)\,[\gamma']^2=g\,,\quad \gamma\in G_\pi\,,$$

where $\mathbb{H}$ is the open complex upper half plane and G_π is the covering group $(\mathbb{H},\pi)$ of Ω, with the Nehari norm which, with the element of length of the Poincaré metric[35] of $-\mathbb{H}$, $\frac{|dz|}{|\mathcal{I}m\, z|}$, is

$$\|g\|_{\mathcal{N}_{-\mathbb{H}}}\overset{def}{=}\sup_{z\in-\mathbb{H}}\,(\mathcal{I}m\,z)^2\,|g(z)|\in\mathbb{R}\,.$$

$B(-\mathbb{H}, G_\pi)$ with this norm is the complete normed space (i.e.,Banach space) of the **quadratic differentials in Ω**.

Since Möbius transformations have zero Schwarzian derivatives, the Nehari norm of the Schwarzian derivative of a holomorphic function is a possible quantification of how much the function is away from Möbius transformations.

The next property was obtained independently in 1932 by W. Kraus[36] and in 1949 by Z.Nehari.

(4.13) *If $g:-\mathbb{H}\to\mathbb{C}_\infty$ is injective and holomorphic, $\|\mathscr{S}g\|_{N_{-\mathbb{H}}}\le\frac{3}{2}$.*

Proof For a Möbius transformation $T_w(z)=\frac{z-\overline{w}}{z-w}$ with $w\in-\mathbb{H}$, it is $T(-\mathbb{H})=\mathbb{C}\setminus\mathrm{cl}\,B_1$ and $T(w)=\infty$. Since, by (4.10.7), Nehari norms of Schwarzian derivatives are invariant under right compositions with Möbius transformations, we can assume $g(w)=\infty$ and $g'(w)=\frac{1}{T_w'(\infty)}$. Thus, $F=g\circ T_w^{-1}$ is injective in

[35] About Poincaré metrics, see Section 5 of Chapters 11 and 12 of *CADOVA*.

[36] Kraus, Wilhelm. The rest of this subsection follows parts of the E. Faria and W. de Melo book listed in the final bibliography and the book Lehto, O., *Univalent Functions and Teichmüller spaces*, also in the final bibliography.

$\mathrm{cl}\, B_1$ and $F'(\infty) = z$. Hence, $F(z) = z + \sum_{j=1}^{\infty} \frac{b_j}{z^j}$, $F\left(\frac{1}{z}\right) = \frac{1}{z} + \sum_{j=1}^{\infty} b_j z^j$, and $z \mapsto F\left(\frac{1}{z}\right)$ is injective and meromorphic with a simple pole at 0 with residue 1. By the Gronwall area theorem,[37] $|b_j| \le \frac{1}{\sqrt{j}}$. In particular, $|b_1| \le 1$, and consequently, $(\mathscr{S}F)(z) = -\frac{6b_1}{z^4} + \cdots$, where the dots stand for terms in integer powers of $\frac{1}{z}$ of order ≥ 5. Therefore,

$$(\mathscr{S}g)(z) = (-6b_1 + \cdots)\, \frac{1}{[T(z)]^4}\, \frac{4(\mathcal{I}m\, z)^2}{(z-w)^4} = (-6b_1 + \cdots)\, \frac{1}{[T(z)]^4}\, \frac{4(\mathcal{I}m\, z)^2}{(z-\overline{w})^4} = -\frac{6b_1}{4(\mathcal{I}m\, z)^2},$$

and $\|\mathscr{S}g\|_{N_{-\mathbb{H}}} = \sup_{z \in -\mathbb{H}} (\mathcal{I}m\, z)^2 |\mathscr{S}g(z)| \le \frac{3}{2}$. □

The following result is of L. Ahlfors and G.G. Weill[38] in 1962.

(4.14) Ahlfors-Weill section theorem:
If $g : -\mathbb{H} \to \mathbb{C}_\infty$ is holomorphic with $\|g\|_{N_{-\mathbb{H}}} < \frac{1}{2}$, then it coincides in $-\mathbb{H}$ with the Schwarzian derivative of the solution f with fixed points $0, 1, \infty$ of the Beltrami equation with Beltrami coefficient $\kappa(z) = -2(\mathcal{I}m\, z)^2 g(\overline{z})$ for $z \in \mathbb{H}$ and 0 for $z \notin \mathbb{H}$. If g is invariant under a group isomorphic to a discrete subgroup of $PSL(2, \mathbb{R})$, then so is the Beltrami coefficient.

Proof We start by assuming that g is holomorphic in $-\mathrm{cl}\, \mathbb{H} \cup \{\infty\}$ and $|z^4 g(z)|$ is bounded in a neighborhood of ∞.

If y_1, y_2 are linearly independent solutions of the linear ordinary differential equation $y'' + \left(\frac{1}{2}g\right)y = 0$ with Wronskian $y_1 y_2' - y_1' y_2 = 1$, they cannot be simultaneously 0, and $F : \mathbb{C}_\infty \to \mathbb{C}_\infty$ is such that $F = \frac{y_2}{y_1}$ in $-\mathrm{cl}\, \mathbb{H}$ and $F(z) = \frac{y_2(\overline{z}) + (z-\overline{z})\, y_2'(\overline{z})}{y_1(\overline{z}) + (z-\overline{z})\, y_1'(\overline{z})}$ for $z \in \mathbb{H}$. The function F is continuous in the real axis, and it is holomorphic in $-\mathbb{H}$; thus, $\frac{\partial F}{\partial \overline{z}} = 0$ in $-\mathbb{H}$, where also

$$\frac{y_2''}{y_1'} = \frac{-2y_1'(y_2'y_1 - y_2y_1')/y_1^3}{(y_2'y_1 - y_2y_1')/y_1^2} = -2\,\frac{y_1'}{y_1}, \qquad \mathscr{S}\left(\frac{y_2}{y_1}\right) = \left(-2\,\frac{y_1'}{y_1}\right)' - \frac{1}{2}\left(-2\,\frac{y_1'}{y_1}\right)^2 = -2\,\frac{y_1''}{y_1} = g.$$

As g is holomorphic in the real axis, so are y_1 and y_2. Since

$$\begin{aligned}\frac{\partial F/\partial \overline{z}}{\partial F/\partial z}(z) &= \frac{(z-\overline{z})y_2''(\overline{z})[y_1(\overline{z}) + (z-\overline{z})y_1'(\overline{z})] - (z-\overline{z})y_1''(\overline{z})[y_2(\overline{z}) + (z-\overline{z})y_2'(\overline{z})]}{y_2'(\overline{z})[y_1(\overline{z}) + (z-\overline{z})y_1'(\overline{z})] - y_1'(\overline{z})[y_2(\overline{z}) + (z-\overline{z})y_2'(\overline{z})}\\ &= \frac{(z-\overline{z})[-g(\overline{z})/2]y_2(\overline{z})[y_1(\overline{z}) + (z-\overline{z})y_1'(\overline{z})] - (z-\overline{z})[-g(\overline{z})/2]y_1(\overline{z})[y_2(\overline{z}) + (z-\overline{z})y_2'(\overline{z})]}{1}\\ &= 2(\mathcal{I}m\, z)^2[g(y_2y_1' - y_1y_2')](\overline{z}) = -2(\mathcal{I}m\, z)^2 g(\overline{z}) = \kappa(\overline{z}), \quad z \in \mathbb{H},\end{aligned}$$

F satisfies the Beltrami equation with coefficient κ a.e. in $\mathbb{C}_\infty$.

[37] See the final section of Chapter 10 of *CADOVA*. Gronwall, Thomas (1877–1932).

[38] Weill, Georges Gustav (1926-).

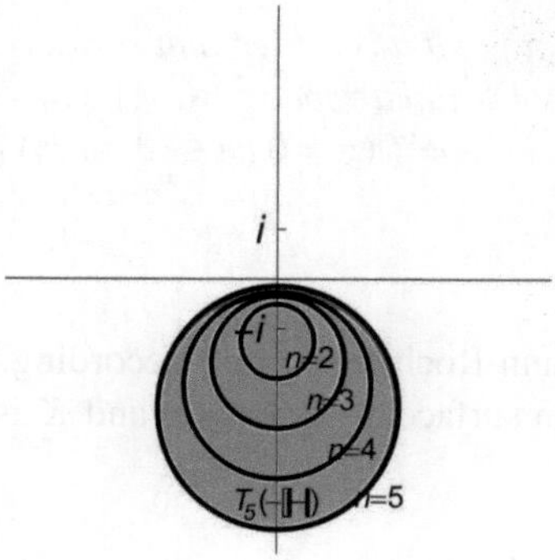

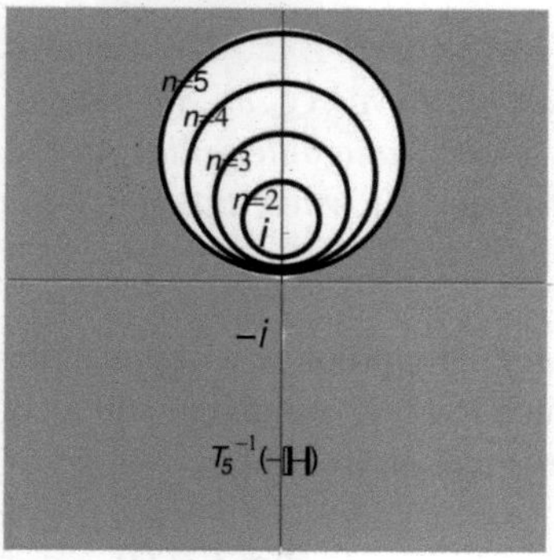

Fig. 4.3 At left, $T_n(-\mathbb{H})$, and at right, $T_n^{-1}(-\mathbb{H})$, $T_n(z)=\frac{nz-i}{iz+i}$, $n\in\mathbb{N}\setminus\{1\}$

Since

$$|\kappa(z)| = 2\left|(\mathcal{I}m\, z)^2 g(\overline{z})\right| \leq 2\,\|g\|_{\mathcal{N}_{-\mathbb{H}}} < 1\,, \quad \text{a.e. in } \mathbb{C}_\infty\,,$$

it is $\|\kappa\|_{L^\infty} < 1$. By (2.8) on the existence and uniqueness of solution of the Beltrami equation on which the proof of the measurable Riemann mapping theorem in Chap. 2 stands, F is a quasiconformal mapping with Beltrami coefficient κ, and there is a Möbius transformation T such that $f = T\circ F$ and $\mathscr{S}f = \mathscr{S}F = g$.

Take now the general case of g holomorphic in $-\mathbb{H}$, with $\{T_n\}_{n\in\mathbb{N}\setminus\{1\}}$ the sequence of Möbius transformations $T_n(z) = \frac{nz-i}{iz+i}$, which map the half plane $-\mathbb{H}$ to a disk with closure in this half plane (Fig. 4.3) and converges to $1_{-\mathbb{H}}$. Each $g_n(z) = g\big(T_n(z)\big)\big(T_n'(z)\big)^2$ is holomorphic in $T_n^{-1}(-\mathbb{H}) \supset -\mathrm{cl}\,\mathbb{H}$ (Fig. 4.3). As

$$\left|\frac{\mathcal{I}m\, T_n(z)}{\mathcal{I}m\, z}\right| = \frac{|(n^2+1)\mathcal{I}m\, z - n(1+|z|^2)|}{|iz+n|^2|\mathcal{I}m\, z|} = \frac{(n^2+1)|\mathcal{I}m\, z| + n(1+|z|^2)|}{|iz+n|^2|\mathcal{I}m\, z|} \geq \frac{n^2+1}{|iz+n|^2}\,, \quad z\in -\mathbb{H}\,,$$

and $|T_n'(z)| = \frac{n^2-1}{|iz+n|^2}$, it is $|T_n'(z)| < \left|\frac{\mathcal{I}m\, T_n(z)}{\mathcal{I}m\, z}\right|$ for $z\in -\mathbb{H}$,

$$|(\mathcal{I}m\, z)^2 g_n(z)| \leq |\mathcal{I}m\, T_n(z)|^2\, \left|g\big(T_n(z)\big)\right|, \quad z\in -\mathbb{H}\,.$$

Thus, $\|g_n\|_{\mathcal{N}_{-\mathbb{H}}} \leq \|g\|_{\mathcal{N}_{-\mathbb{H}}} < \frac{1}{2}$ for $n\in\mathbb{N}$. So, g_n satisfies the conditions assumed in the two preceding paragraphs, and there is a quasiconformal mapping f_n in $\mathbb{C}_\infty$ such that $\mathscr{S}f_n = g_n$ in $-\mathbb{H}$, with the corresponding Beltrami coefficient κ_n such that $\|\kappa_n\|_{L^\infty} \leq \|\kappa\|_{L^\infty} < 1$ for $n\in\mathbb{N}$. As $g_n \to g$ uniformly in compact subsets of $-\mathbb{H}$, we have $\kappa_n \to \kappa$ a.e. in $\mathbb{H}$. Therefore, $\{f_n\}$ converges uniformly in compact subsets of $-\mathbb{H}$ to a quasiconformal mapping f in $\mathbb{C}_\infty$ with Beltrami coefficient κ.

If g is invariant under a group isomorphic to a discrete subgroup of $PSL(2,\mathbb{R})$, then so is the Beltrami coefficient κ. □

A Riemann surface S is said to be of **finite type** (g, n) if it has finite genus g, is included in a compact Riemann surface and its complement in it has a finite number n of points, called **punctures**.

(4.15) *If $\Omega \subset \mathbb{C}_\infty$ has universal covering $(\mathbb{H}, \pi)$ and it is a Riemann surface of finite type (g, n), the dimension of the complex linear space $B(-\mathbb{H}, G_\pi)$ of quadratic differentials in Ω is $3(g-1)+n$ except if $(g=0, n \in \{0, 1, 2\})$ or $(g=1)$, when it is, resp., 0 or $1+n$.*

Proof The proof is an application of the Riemann-Roch theorem[39] according to which if D, K are divisors in a compact Riemann surface S of genus g and K is a canonical divisor, then

$$\dim L_S(D) - \dim L_S(K-D) = \deg D + 1 - g ,$$

where $L_S(D)$ is the set of meromorphic nonzero functions from S to $\mathbb{C}_\infty$ with principal divisor $\operatorname{div}(f) \geq -D$ enlarged with the function identically 0, and

$$\operatorname{div}(f) = \sum_{P \in \mathscr{P}} \operatorname{ord}_P(f)\, P ,$$

with $\operatorname{ord}_P(f)$ the order of the zero or the symmetric of the order of the pole of f in P.

The linear space of holomorphic quadratic differentials in S is $L_S(-2K)$, and the equality of the Riemann-Roch theorem in the preceding paragraph gives for the dimension of the complex linear space $B(-\mathbb{H}, G_\pi)$, with a canonical divisor K in Ω, which has degree $2g-2$,

$$\begin{aligned}\dim B(-\mathbb{H}, G_\pi) &= \dim L_\Omega(-2K) = \deg(2K) + 1 - g + \dim L_\Omega K \\ &= 2(2g-2) + 1 - g + \dim L_\Omega K .\end{aligned}$$

If there are no punctures, for $g \geq 2$, it is $\deg K > 0$, implying $\dim L_\Omega K = 0$, and $\dim B(-\mathbb{H}, G_\pi) = 3g - 3$. For $g = 1$, it is $\deg K = 0$ and

$$\dim B(-\mathbb{H}, G_\pi) = \dim L_\Omega(-2K) = \dim_\Omega K = 1 .$$

For $g = 0$, it is $\deg K = -2$, and $\dim B(-\mathbb{H}, G_\pi) = \dim(-2K) = 0$. If there are n punctures, $\deg K$ is the number of punctures added by n, and we have the values of $\dim B(-\mathbb{H}, G_\pi)$ in the statement. □

Bers Embedding of Teichmüller Space in Space of Schwarzian Derivatives

An **embedding** or **injection** of a topological space X in another Y is a homeomorphism h of X onto $h(X) \subset Y$. If it exists, it means there is a subset of Y topologically identical to X.

L.Bers established in 1960 that if $\Omega \subset \mathbb{C}_\infty$ has a universal covering $(\mathbb{H}, \pi)$, there is an embedding of $\operatorname{Teich}(\Omega)$ in $B(-\mathbb{H}, G_\pi)$ with an open bounded range.

[39] See Section 6 of Chapter 12 of *CADOVA*. Roch, Gustav (1839–1886).

(4.16) Bers embedding: *If* $\Omega \subset \mathbb{C}_\infty$ *has universal covering* $(\mathbb{H}, \pi)$*, then there is a function* $h : \mathrm{Teich}(\Omega) \to B(-\mathbb{H}, G_\pi)$ *homeomorphism of* $\big(\mathrm{Teich}(\Omega), D_T\big)$ *onto* $h\big(\mathrm{Teich}(\Omega)\big) \subset B_{3/2}(0) \subset B(-\mathbb{H}, G_\pi)$ *with the Nehari norm and* $h(0) = 0$. $h\big(\mathrm{Teich}(\Omega)\big)$ *is not a complete metric subspace of the Banach space* $\big(B(-\mathbb{H}, G_\pi), \|\ \|_{\mathcal{N}_{-\mathbb{H}}}\big)$*, its elements are Schwarzian derivatives of quasiconformal mappings, and the largest open ball* $B_r(0) \subset B(-\mathbb{H}, G_\pi)$ *it contains is for* $r = \frac{1}{2}$.

Proof If $[F] \in \mathrm{Teich}(\Omega)$ and $\kappa_F \in L^\infty(\mathbb{H})$ with $\|\kappa_F\|_{L^\infty} < 1$ is the Beltrami coefficient of F, let $\widetilde{\kappa}_F \in L^\infty(\mathbb{C}_\infty)$ be equal to κ_F in $\mathbb{H}$ and to 0 in $\mathbb{C}_\infty \backslash \mathbb{H}$, and $f : \mathbb{C}_\infty \to \mathbb{C}_\infty$ be the unique solution of the Beltrami equation with coefficient $\widetilde{\kappa}_F$ and fixed points 0, 1, ∞. Since κ_F is invariant under the group G_π, also $\widetilde{\kappa}_F$ is. The restriction of f to $-\mathbb{H}$ is an injective conformal homeomorphism. As $\mathscr{S}\big(f_{|-\mathbb{H}} \circ \gamma\big) = \mathscr{S}\big(\gamma \circ f_{|-\mathbb{H}}\big)$, by the Schwarzian derivative rule of composition of functions, $\mathscr{S}\big(f_{|-\mathbb{H}}\big) \circ \gamma\, (\gamma')^2 = \mathscr{S}\big(f_{|-\mathbb{H}}\big)$. With $g_{\kappa_F} = \mathscr{S}\big(f_{|-\mathbb{H}}\big)$, $(g_{\kappa_F} \circ \gamma)(\gamma')^2 = g_{\kappa_F}$ for $\gamma \in G_\pi$, and $g_{\kappa_F} \in B(-\mathbb{H}, G_\pi)$. If $F = 0$, then $\kappa_F = 0$, $f = 0$, and $g_{\kappa_F} = 0$.

If $\widehat{\kappa} \in L^\infty(\mathbb{H})$ with $\|\widehat{\kappa}\|_{L^\infty} < 1$ is the Beltrami coefficient of $\widehat{f} \in [F]$, the restrictions to $\partial\mathbb{H}$ of the solutions $\widehat{f_+}$, f_+ of the Beltrami equation in $\mathbb{H}$ with Beltrami coefficient, resp., $\widehat{\kappa}$, κ_F and fixed points 0, 1, ∞ satisfy $\widehat{f_+}_{|\partial\mathbb{H}} = f_{+|\partial\mathbb{H}}$, and we define the quasiconformal mapping $\widehat{F} : \mathbb{C}_\infty \to \mathbb{C}_\infty$ equal to $(\widehat{f_+})^{-1} \circ f_+$ in $\mathrm{cl}\,\mathbb{H}$ and to $1_{-\mathbb{H}}$ in $-\mathbb{H}$, and the function $W : \mathbb{C}_\infty \to \mathbb{C}_\infty$ with $W = f \circ \widehat{F} \circ (f_+)^{-1}$, which results from sewing the two functions at a Jordan curve.[40] Since $W_{|f(-\mathbb{H})} = \widehat{f} \circ (f^{-1})_{|f(-\mathbb{H})}$, $W_{|f(\mathbb{H})} = \widehat{f} \circ (\widehat{f_+})^{-1} \circ f_+ \circ (f^{-1})_{|f(\mathbb{H})}$, and both $\widehat{f} \circ \big((\widehat{f_+})^{-1}\big)_{|f(\mathbb{H})}$ and $f_+ \circ (f^{-1})_{|f(\mathbb{H})}$ are conformal homeomorphisms, as f_+ and $f_{|\mathbb{H}}$ have the same Beltrami coefficient, and analogously for $\widehat{f}$ and $(\widehat{f_+})_{|\mathbb{H}}$, W is a conformal homeomorphism, and since it has fixed points 0, 1, ∞, $W = 1_{\mathbb{C}_\infty}$. Hence, $\widehat{f}_{|-\mathbb{H}} = f_{|-\mathbb{H}}$, $g_{\widehat{\kappa}} = \mathscr{S}\big(\widehat{f}_{|-\mathbb{H}}\big)$, and g_{κ_F} is independent of the representative of the equivalence class $[F]$ considered.

$h : \mathrm{Teich}(\Omega) \to B(-\mathbb{H}, G_\pi)$ such that $h\big([F]\big) = g_{\kappa_F}$ is injective, $h(0) = 0$ and, by (4.13), $h\big([F]\big) \in B_{3/2} \subset B(-\mathbb{H}, G_\pi)$. To prove that h is a homeomorphism of $\mathrm{Teich}(\Omega)$ onto $h\big(\mathrm{Teich}(\Omega)\big)$, it now suffices to prove that it is a continuous open function, because in this case the preimages of open subsets of $B(-\mathbb{H}, G_\pi)$ by h^{-1} are open in $\mathrm{Teich}(\Omega)$, and therefore, h^{-1} is continuous.

By the Ahlfors-Weill section theorem (4.14), the open ball $B_{1/2}$ of $B(-\mathbb{H}, G_\pi)$ is included in $h\big(\mathrm{Teich}(\Omega)\big)$. Thus, $B_{1/2} \subset \mathrm{int}\, h\big(\mathrm{Teich}(\Omega)\big)$, but we need to prove that all $g \in h\big(\mathrm{Teich}(\Omega)\big)$ belong to $\mathrm{int}\, h\big(\mathrm{Teich}(\Omega)\big)$ and h is continuous or alternatively that h and h^{-1} are continuous.

[40] See Section "Sewing at a Jordan curve" in Chap. 3.

If $[F_1], [F_2] \in \mathrm{Teich}(\Omega)$ and f_1, f_2 are the functions defined from, resp., $[F_1], [F_2]$ as f was defined from $[F]$ at the beginning, then $h([F_j]) = \mathscr{S}(f_{j|-\mathbb{H}})$ for $j \in \{1, 2\}$ and

$$\begin{aligned}\|h([F_1]) - h([F_2])\|_{\mathcal{N}_{-\mathbb{H}}} &= \|\mathscr{S}(f_{1|-\mathbb{H}}) - \mathscr{S}(f_{2|-\mathbb{H}})\|_{\mathcal{N}_{-\mathbb{H}}} \\ &= \|\mathscr{S}(f_{1|-\mathbb{H}} \circ (f_{2|-\mathbb{H}})^{-1})\|_{\mathcal{N}_{f_{2|-\mathbb{H}}(-\mathbb{H})}} \le \sigma_0(-\mathbb{H})\, \|\kappa_{f_{1|-\mathbb{H}} \circ (f_{2|-\mathbb{H}})^{-1}}\|_{L^\infty} \\ &= \sigma_0(-\mathbb{H}) \left\| \frac{\kappa_{f_{1|-\mathbb{H}}} - \kappa_{f_{2|-\mathbb{H}}}}{1 - \overline{\kappa_{f_{1|-\mathbb{H}}}}\, \kappa_{f_{2|-\mathbb{H}}}} \right\|_{L^\infty} \le \sigma_0(-\mathbb{H})\, D_B([F_1], [F_2]),\end{aligned}$$

where the second equality results from (4.10.6), the first from (4.10.8), the following equality from the formula obtained at the end of the second Section of Chap. 1 for the Beltrami coefficient of the composition $f_1 \circ f_2^{-1}$, and the last inequality from the definition of D_B in (4.9). As the distance D_B in $\mathrm{Teich}(\Omega)$ is equivalent to the Teichmüller distance D_T, the function $h : \mathrm{Teich}(\Omega) \to B(-\mathbb{H}, G_\pi)$ is continuous.

To prove continuity of h^{-1}, it is convenient to upper bound $D_B([F_1], [F_2])$ by $\|h([F_1]) - h([F_2])\|_{\mathcal{N}_{-\mathbb{H}}}$ multiplied by a constant. Choose $f_2 \in [F_2]$ with the smallest possible dilatation K_2 and a quasiconformal mapping f_1 such that

$$\|\mathscr{S}(f_{1|-\mathbb{H}}) - \mathscr{S}(f_{2|-\mathbb{H}})\|_{\mathcal{N}_{-\mathbb{H}}} < \varepsilon(K_2)\,,$$

where $\varepsilon(K_2)$ is a constant whose existence is established in (4.12). As seen in the preceding paragraph,

$$\|\mathscr{S}(f_{1|-\mathbb{H}} \circ (f_{2|-\mathbb{H}})^{-1})\|_{\mathcal{N}_{f_{2|-\mathbb{H}}(-\mathbb{H})}} = \|\mathscr{S}(f_{1|-\mathbb{H}}) - \mathscr{S}(f_{2|-\mathbb{H}})\|_{\mathcal{N}_{-\mathbb{H}}} < \varepsilon(K_2)\,.$$

By (4.12), $w = f_{1|-\mathbb{H}} \circ (f_{2|-\mathbb{H}})^{-1}$ has a quasiconformal extension to $\mathbb{C}$, also denoted w, with a Beltrami coefficient κ such that

$$\|\kappa\|_{L^\infty} \le \frac{\|\mathscr{S}(f_{1|-\mathbb{H}}) - \mathscr{S}(f_{2|-\mathbb{H}})\|_{\mathcal{N}_{-\mathbb{H}}}}{\varepsilon(K_2)}\,.$$

The function $f_3 = w \circ f_2$ is a quasiconformal extension of $f_{2|\mathbb{H}}$ to $-\mathbb{H}$ and, therefore, $f_3 \in [F_1]$. Let κ_j denote the Beltrami coefficient of f_j for $j \in \{1, 2, 3\}$. Since $w = f_3 \circ f_2^{-1}$,

$$\|\kappa\|_{L^\infty} = \left| \frac{\kappa_3 - \kappa_2}{1 - \overline{\kappa_3}\, \kappa_2} \right| \ge D_B([F_1], [F_2])\,.$$

By the inequalities obtained in the two preceding paragraphs,

$$\|h([F_1]) - h([F_2])\|_{\mathcal{N}_{-\mathbb{H}}} = \|\mathscr{S}(f_{1|-\mathbb{H}}) - \mathscr{S}(f_{2|-\mathbb{H}})\|_{\mathcal{N}_{-\mathbb{H}}} \ge \varepsilon(K_2)\, D_B([F_1], [F_2])\,.$$

Since the two distances D_B and D_T in $\mathrm{Teich}(\Omega)$ are equivalent, it follows that $h^{-1} : h(\mathrm{Teich}(\Omega)) \to \mathrm{Teich}(\Omega)$ is continuous.

So, h is a homeomorphism of $\mathrm{Teich}(\Omega)$ to $h(\mathrm{Teich}(\Omega)) \subset B(-\mathbb{H}, G_\pi)$.

The Schwarzian derivatives of $f_n : \mathbb{H} \to \mathbb{C}$ with $f_n(z) = z^{\frac{1}{n}}$ are

$$(\mathscr{S} f_n)(z) = \left(1 - \tfrac{1}{n^2}\right) \tfrac{1}{2z^2}$$

and, therefore, $\|\mathscr{S} f_n\|_{\mathcal{N}_{-\mathbb{H}}} = \frac{1}{2}\left(1 - \frac{1}{n^2}\right) < \frac{1}{2}$. The functions f_n have extensions to quasiconformal mappings in $\mathbb{C}$, which are still denoted f_n, and $\mathscr{S} f_n \in B(-\mathbb{H}, G_{1_\mathbb{H}})$. $f(z) = \log z$ does not have quasiconformal extension to $\mathbb{C}$ and $\|\mathscr{S} f_n - \mathscr{S} f\|_{\mathcal{N}_{-\mathbb{H}}} = \frac{2}{n^2} \to 0$. Thus, $\lim \mathscr{S} f_n = \mathscr{S} f \notin B(-\mathbb{H}, G_{1_\mathbb{H}})$. Therefore, the metric space $B(-\mathbb{H}, G_{1_\mathbb{H}})$ is not complete, and $B(-\mathbb{H}, G_\pi)$ is not a complete metric space. □

With this result and the preceding subsection, the Teichmüller space of a region $\Omega \subset \mathbb{C}_\infty$ with universal covering $(\mathbb{H}, \pi)$ is a complex manifold of finite dimension $3g - 3 + n$ except if $(g = 0, n \in \{0, 1, 2\})$ or $(g = 1, n = 0)$, when the dimension is, resp., 0 or 1, where $g \in \mathbb{N} \cup \{0\}$ is the genus of the Riemann surface Ω with $n \in \mathbb{N} \cup \{0\}$ punctures.

Harmonic λ-Lemma
If f is a solution of the Beltrami equation $\frac{\partial f}{\partial \bar{z}} = \kappa_f(z) \overline{\frac{\partial f}{\partial z}}$, with a holomorphic change of variables $z = \gamma(w)$, it is $\frac{\partial \gamma}{\partial \bar{w}} = 0$, $\frac{\partial \bar{\gamma}}{\partial w} = \left(\overline{\frac{\partial \gamma}{\partial \bar{w}}}\right) = 0$,

$$\frac{\partial (f \circ \gamma)}{\partial \bar{w}} = \left(\frac{\partial f}{\partial \bar{z}} \circ \gamma\right) \frac{\partial \bar{\gamma}}{\partial \bar{w}}, \qquad \frac{\partial (f \circ \gamma)}{\partial w} = \left(\frac{\partial f}{\partial z} \circ \gamma\right) \frac{\partial \gamma}{\partial w}, \qquad \kappa_{f \circ \gamma} = (\kappa_f \circ \gamma) \frac{\overline{\gamma'}}{\gamma'}.$$

On the other hand, if $\rho_\Omega(z)\,|dz|$ is the element of length of the Poincaré metric of Ω, with the same change of variables, $\rho_\Omega = (\rho_{\gamma(\Omega)} \circ \gamma)\,|\gamma'|$. Therefore, if $\psi(z) = \rho^2_{\gamma(\Omega)}(z)\,\overline{\kappa_f}(z)$, we have $\psi \circ \gamma = \frac{\rho^2_\Omega \circ \gamma}{|\gamma'|^2}\,(\overline{\kappa_f} \circ \gamma)$, and $(\psi \circ \gamma)|\gamma'|^2 = \psi \circ \gamma$. If $\Omega \subset \mathbb{C}_\infty$ has universal covering $(\mathbb{H}, \pi)$. G_π is covering group, and γ is a conformal mapping invariant under this group, then ψ is a quadratic differential belonging to $B(-\mathbb{H}, G_\pi)$, what motivates the following definition.

A **harmonic Beltrami coefficient** in a region $\Omega \subset \mathbb{C}_\infty$ with universal covering $(\mathbb{H}, \pi)$ is a function $\kappa = \rho^{-2}_{\gamma(\Omega)} \overline{\psi}$ where $\rho_\Omega(z)\,|dz|$ is the element of length of the Poincaré metric of Ω and ψ is a quadratic differential belonging to $B(-\mathbb{H}, G_\pi)$.

The usefulness of harmonic Beltrami coefficients in this context stems from the following uniqueness property.

(4.17) *If $\Omega \subset \mathbb{C}_\infty$ is a region with universal covering $(\mathbb{H}, \pi)$, the quasiconformal mappings with harmonic Beltrami coefficients belonging to a same element of* Teich(Ω) *have the same Beltrami coefficient in $L^\infty(\Omega)$.*

Proof If $f_2 \in [f_1] \in$ Teich(Ω) have harmonic Beltrami coefficients, resp., κ_1, κ_2, then $\mathscr{S} f_1 = \mathscr{S} f_2 \in B(-\mathbb{H}, G_\pi)$ is a holomorphic function in $-\mathbb{H}$ and, by (4.10.5), there is a Möbius transformation T such that $f_2 = T \circ f_1$. Therefore, $\kappa_1 = \kappa_2$ a.e. in Ω. □

The following result is used in the proof of the harmonic λ-lemma.

(4.18) *If $\Omega \subset \mathbb{C}_\infty$ is a region with universal covering $(\mathbb{H}, \pi)$ and for some $a>0$ $(t, z) \mapsto \psi_t(z)$ from $[-a, a] \times (\Omega \cup \partial\Omega)$ to $\Omega \cup \partial\Omega$ is such that $\psi_0 = 1_{\Omega\cup\partial\Omega}$ and for $t \in [-a, a]$ is continuous, $(\psi_t)_{|\partial\Omega} = 1_{\partial\Omega}$ with ψ_t a homeomorphism of $\Omega \cup \partial\Omega$ onto $\Omega \cup \partial\Omega$ and a K-quasiconformal mapping in Ω, then $(\psi_t)_{|\Omega} \in [1_\Omega] \in \mathrm{Teich}(\Omega)$.*

Proof Denote the covering group by G_π. For $t \in [-a, a]$, let $W_t : \mathbb{H} \to \mathbb{H}$ be a homeomorphism such that $\pi \circ W_t = w_t \circ \pi$ and $t \mapsto W_t(i)$ is continuous with $W_0(i) = i$. $W_0 = 1_{\mathbb{H}}$ and for $t \in [-a, a]$ W_t is a K-quasiconformal mapping with continuous extension to a continuous function from $S = \mathbb{H} \cup \mathbb{R}_\infty$ to S with the spherical distance, also denoted W_t. The function $(t, z) \mapsto W_t(z)$ is continuous for $[-a, a] \times \mathbb{H}$ and for $t \in [-a, a]$ W_t is a K-quasiconformal mapping and $[-a, a] \times \{i\}$ is a compact set; thus, $t \mapsto W_t(z)$ is continuous for $z \in \mathbb{R}_\infty$.

If the closure of the set Λ of the attracting fixed points of G_π is $\mathbb{R}_\infty$, for every $g \in G_\pi$, $g_t \overset{def}{=} W_t \circ g \circ W_t^{-1} \in G_\pi$. Thus, $t \mapsto g_t(z)$ is continuous with z fixed and $g_t = g_0$. Since $g_0 = g$, W_t commutes with g. Therefore, W_t fixes the attracting points of all $g \in G_\pi$ and, as the closure of Λ is $\mathbb{R}_\infty$, then $W_t(x) = x$ for $t \in [-a, a]$, $x \in \mathbb{R}$.

If $\mathrm{cl}\,\Lambda \neq \mathbb{R}_\infty$, $\mathbb{R}_\infty \setminus \Lambda$ is open and dense in $\mathbb{R}_\infty$. If $x_0 \in \mathbb{R} \setminus \Lambda$, there is $\varepsilon > 0$ such that π is injective in $B_\varepsilon(x_0) \cap \mathbb{H}$. So, for $0 < \epsilon' < \varepsilon$ in $B_{\varepsilon'}(x_0) \cap \mathbb{H}$, the function π is the quotient of two bounded holomorphic functions.

By the theorem of F. and M. Riesz,[41] there is $S \subset \mathbb{R}_\infty \setminus \Lambda$ with positive measure such that for all $x \in S$ the function $z \mapsto \pi(z)$ has sectorial limit $\pi(x)$ not constant in any subset of S with positive measure. If $g : [0, 1] \to \mathbb{H}$ is continuous, $g(1) = x \in S$ and there is $\varepsilon \in]0, 1]$ such that $g([1-\varepsilon, 1])$ is included in the **Stolz angle**[42] with vertex z_0 and width $m > 0$, $\{z \in \mathbb{C} : |\mathcal{I}m\, z| \geq m |\mathcal{R}e\, z - z_0|\}$ for $t \in [1-\varepsilon, 1]$.

Since f is a quasiconformal mapping, $f\big(g([1-\varepsilon, 1])\big)$ is included in the Stolz angle with vertex $Z_0 = \lim_{t \to 1} f\big(g(t)\big)$ and width $M > 0$. If z_0, $f(z_0) \in \mathbb{R}$, f has an extension to a quasiconformal mapping from $\mathbb{C}_\infty$ to $\mathbb{C}_\infty$ by $f(\overline{z}) = \overline{f(z)}$. With Γ_j the ray in $\mathbb{H}$ with equation $\mathcal{I}m\, z = (-1)^j\, m\, (\mathcal{R}e\, z - z_0)$ for $j \in \{0, 1\}$. The Jordan curves $\Gamma_0 \cup [z_0, +\infty[\cup \{\infty\}$ and $\Gamma_1 \cup]-\infty, z_0] \cup \{\infty\}$ are quasicircles just like their images by f, and $f([z_0, +\infty[)$, $f(]-\infty, z_0])$ are rays included in the real axis. For $x > 0$, $c = f(z_0 + x + imx)$, $b = f(z_0)$, and $a = \mathcal{R}e\, c$. Since the images of the mentioned Jordan curves are quasicircles, there is $M > 0$ such that for small $x > 0$, $|\mathcal{R}e\, c - f(z_0)|\, \frac{1}{M}\, |\mathcal{I}m\, c|$. Hence, the curve $f([z_0, +\infty[)$ has an end point at z_0, and it is a subset of $\{z \in \Omega : |\mathcal{I}m\, z| \geq M |\mathcal{R}e\, z - Z_0|\}$. The analogous result for $f(]-\infty, z_0])$ is obtained similarly. If z_0 or $f(z_0)$ is ∞, we begin by applying

[41] See Section 6 of Chapter 10 of *CADOVA*. Riesz, Marcel (1986–1969).

[42] See Section 3 of Chapter 5 of *CADOVA*. Stolz, Otto (1842–1905).

a change of variables by a Möbius transformation before applying the preceding paragraph and finish reversing the change of variables.

Since $\pi \circ W_t = w_t \circ \pi$, it is $W_t(x) \in S$ and $\pi\big(W_t(x)\big) = w_t\big(\pi(x)\big)$. As the points of $\partial\Omega$ are fixed points of w_t and $\pi(x) \in \partial\Omega$, $\pi\big(W_t(x)\big) = \pi(x)$ to $x \in S$. Since for fixed x , $t \mapsto W_t(x)$ is continuous, $W_0(x) = x$. If $W_t \neq 1_\Omega$ for $t \in [-a, a]$, then $\{W_t(x) : t \in [-a, a]\} \subset \mathbb{R}_\infty$ contains an interval I_0 with positive length. Since $\pi\big(W_t(x)\big) = w_t\big(\pi(x)\big)$, the function π is not constant in $I_0 \cap S$, what is impossible since this set does not have measure 0 . Therefore, $W_t = 1_S$ and, as S is dense in $\mathbb{R}_\infty$, $W_t = 1_{\mathbb{R}_\infty}$ and $W_t(x) = x$ for $t \in [-a, a]$, $x \in \mathbb{R}$. □

As mentioned at the beginning of this section, L.Bers and H.Royden established in 1986 the following harmonic λ-lemma

(4.19) Harmonic λ-lemma: *A holomorphic motion $\{\phi_\tau\}_{\tau \in B_1}$ with base point 0 of a region $\Omega \subset \mathbb{C}_\infty$ with universal covering $(\mathbb{H}, \pi)$ has a unique extension to a holomorphic motion of $\mathbb{C}_\infty$ with base point 0 if the parameter is restricted to $B_{1/3}$. For $\tau \in B_{1/3}$, the Beltrami coefficient is $\kappa_\tau = \rho_U^{-2}\overline{\psi_\tau}$ in each connected component U of $\mathbb{C}_\infty \setminus \mathrm{cl}\,\Omega$, with $\rho_U(z)|dz|$ the element of length of the Poincaré metric of U, ψ_τ and $\tau \mapsto \overline{\psi_\tau}(z)$ for $z \in U$ are holomorphic in, resp., U and $B_{1/3}$, i.e., κ_τ is a harmonic Beltrami coefficient.*

Proof Assume, without loss of generality, that $0, 1, \infty$ are fixed points of the functions ϕ_τ . By the Slodkowski theorem (4.4) and (4.2), the holomorphic motion $\{\phi_\tau\}_{\tau \in B_1}$ of Ω has extension to a holomorphic motion of $\mathbb{C}_\infty$, denoted the same way. For $r \in]0, 1[$ fixed and each τ with $|\tau| \leq r$, the function ϕ_τ is a K-quasiconformal mapping in $\mathbb{C}_\infty$ with $K \leq \frac{1+r}{1-r}$, and the Beltrami coefficient of ϕ_τ is a holomorphic function $\kappa_\tau \in L^\infty(\Omega)$ with $\|\kappa_\tau\|_{L^\infty(\Omega)} \leq 1$.

Suppose that both $\{\widetilde{\phi}_t\}_{t \in B_{1/3}}$ and $\{\widehat{\phi}_t\}_{t \in B_{1/3}}$ are extensions of the given holomorphic motion of Ω to holomorphic motions of $\mathbb{C}_\infty$, but restricting the parameter to $B_{1/3}$, $\{\phi_t\}_{t \in B_{1/3}}$, and the Beltrami coefficients of each element of these holomorphic motions are harmonic in $\mathbb{C}_\infty \setminus \Omega$. If V is a connected component of $\mathbb{C}_\infty \setminus \mathrm{cl}(\Omega)$, then $\widetilde{\phi}_t(V) = \widehat{\phi}_t(V_t)$ for a connected component V_t of $\mathbb{C}_\infty \setminus \widetilde{\phi}_t(\Omega) = \widehat{\phi}_t(\Omega)$. The set $\{t \in B_{1/3} : V_t = V\}$ is nonempty, open, and closed, and since $B_{1/3}$ is connected, it is $B_{1/3}$ itself. For every $t \in B_{1/3}$, the function $g_t : V \to V$ such that $g_t = (\widehat{\phi}_t)^{-1} \circ \widetilde{\phi}_t$ is a quasiconformal mapping with the points of ∂V fixed and the mapping $t \mapsto g_t(z)$ with $z \in V$ fixed is continuous. With $a = \frac{1}{3} - \varepsilon$ and $\varepsilon > 0$ small, for all $\theta \in \mathbb{R}$, the function defined in $(t, z) \in [-a, a] \times (V \cup \partial V)$ by $g_{te^{i\theta}}(z)$ satisfies the hypothesis of (4.18). Therefore, for all $t \in B_{1/3}$, the quasiconformal mappings $\widetilde{\phi}_t$ and $\widehat{\phi}_t$ belong to the same element of $\mathrm{Teich}(V)$. Since, by hypothesis, these functions have harmonic Beltrami coefficients, by (4.17), they have the same Beltrami coefficient in $L^\infty(V)$, and $(\widehat{\phi}_t)^{-1} \circ \widetilde{\phi}_t \in H(V)$. As the points of ∂V are fixed points of this function, for $t \in B_{1/3}$, $\widetilde{\phi}_t = \widehat{\phi}_t$ in V and, therefore, also in $\mathbb{C}_\infty \setminus \mathrm{cl}\,\Omega$, guaranteeing uniqueness of the extension to a holomorphic motion of $\mathbb{C}_\infty$ with parameter in $B_{1/3}$.

It remains to prove there is an extension of the holomorphic motion $\{\phi_t\}_{t\in B_{1/3}}$ of Ω to a holomorphic motion $\{\widetilde{\phi}_t\}_{t\in B_{1/3}}$ of $\mathbb{C}_\infty$ with Beltrami coefficient of each $\widetilde{\phi}_t$ harmonic in $\mathbb{C}_\infty\setminus\Omega$. As the Slodkowski theorem was proved beginning by extending holomorphic motions of each term of an expanding sequence of finite sets $X_n=\{0, 1, \infty, x_1, \dots, x_n\}$ with $n+3$ distinct points and $\cup_{n\in\mathbb{N}}X_n$ is dense in Ω for a holomorphic motions of $\mathbb{C}_\infty$, it is possible to get harmonic Beltrami coefficients for each one of these extensions.

Consider $M_n=\{(z_1, \dots, z_n)\in(\mathbb{C}\setminus\{0, 1\})^n : z_j\neq z_k \quad \text{for} \quad j\neq k\}$ and the function p defined in quasiconformal mappings from $\mathbb{C}_\infty\setminus X_n$ to $\mathbb{C}_\infty$ by $p(F)=\big(F(x_1), \dots, F(x_n)\big)$. Let $F_2\in[F_1]\in\text{Teich}(\mathbb{C}_\infty\setminus X_n)$ and h be a Möbius transformation from $F_1(\mathbb{C}_\infty\setminus X_n)$ onto $F_2(\mathbb{C}_\infty\setminus X_n)$ such that $F_2^{-1}\circ h\circ F_1$ is homotopic to the identity in $\mathbb{C}_\infty\setminus X_n$. Since h is a Möbius transformation with fixed points $0, 1, \infty$, it is the identity. Thus, $F_2^{-1}\circ F_1$ is homotopic to the identity modulo the boundary in $\mathbb{C}_\infty\setminus X_n$, and its extension by continuity, denoted in the same way, satisfies $F_2^{-1}\circ F_1(x_j)=x_j$ for all $x_j\in X_n$. Therefore, the range of p is included in M_n and F_2, F_1 coincide in X_j. So, we can consider the function $P:\text{Teich}(\mathbb{C}_\infty\setminus X_n)\to M_n$ defined by $P([F])=p(F)$. Each point of M_n is $p(F)$ for some conformal homeomorphism defined in $\mathbb{C}_\infty\setminus X_n$. Therefore, P is a surjection.

Each $\mathbb{C}_\infty\setminus X_n$ is a hyperbolic Riemann surface with universal covering $(\mathbb{H}, \pi_n)$. By the Bers embedding (4.16), there is a homeomorphism $h_n:\text{Teich}(\mathbb{C}_\infty\setminus X_n)\to B(-\mathbb{H}, G_{\pi_n})$ onto $h_n\big(\text{Teich}(\mathbb{C}_\infty\setminus X_n)\big)$, which is included in $B_{3/2}$ of $B(-\mathbb{H}, G_{\pi_n})$ with the Nehari norm, $\text{Teich}(\mathbb{C}_\infty\setminus X_n)\supset B_{1/2}$ and, by (4.15), $B(-\mathbb{H}, G_{\pi_n})$ has dimension $(3)-3+(n+3)=n$. If $[F]\in\text{Teich}(\mathbb{C}_\infty\setminus X_j)$, F is a quasiconformal mapping in $\mathbb{C}_\infty$ with fixed points $0, 1, \infty$, κ_F is its Beltrami coefficient, and $\{\varphi_1, \dots, \varphi_j\}$ is a basis of $B(-\mathbb{H}, G_{\pi_n})$ with $\|\varphi_j\|_{\mathcal{N}_{-\mathbb{H}}}<\frac{1}{2}$ for $j\in\{1, \dots, n\}$, by the Ahlfors-Weill section theorem (4.14), there are Beltrami coefficients $\kappa_1, \dots, \kappa_n\in L^\infty\big(\mathbb{C}_\infty\setminus X_n\big)$ such that if F_j is the solution of the Beltrami equation with coefficient κ_j, then it has fixed points $0, 1, \infty$, and $h_n([F_j])=\varphi_j$ for $j\in\{1, \dots, n\}$. There exists $\varepsilon>0$ such that if $(t_1, \dots, t_n)\in\mathbb{C}^n$ and $\sum_{j=1}^n|t_j|^2<\varepsilon^2$, $\eta=\sum_{j=1}^n t_j\kappa_j$ is a Beltrami coefficient in $\mathbb{C}_\infty\setminus X_n$, and we denote f^η the solution of the Beltrami equation with that coefficient and fixed points $0, 1, \infty$. If ν is the Beltrami coefficient of $f^\eta\circ F$, then

$$\nu=\frac{\sum_{j=1}^n t_j\nu_j+\mu}{1-\overline{\mu}\sum_{j=1}^n t_j\nu_j}, \qquad \text{with } \nu_j=(\mu_j\circ F)\,\frac{\overline{\partial F/\partial z}}{\partial F/\partial z}, \quad j\in\{1, \dots, n\}.$$

$(t_1, \dots, t_n)\mapsto[f^\nu]\in\text{Teich}\big(\mathbb{C}_\infty\setminus X_n\big)$ from $B_\varepsilon(0)\subset\mathbb{C}^n$ to a neighborhood of κ is differentiable, and $P:\text{Teich}(\mathbb{C}_\infty\setminus X_n)\to M_n$ defined in the preceding paragraph is differentiable in a neighborhood of $[F]\in\text{Teich}(\mathbb{C}_\infty\setminus X_n)$.

The functions $\widetilde{g}$:$\text{Teich}(\mathbb{C}_\infty\setminus X_n)\to\text{Teich}(\mathbb{C}_\infty\setminus X_n)$ such that $\widetilde{g}([F])=[F\circ g^{-1}]$ with g a quasiconformal mapping from $\mathbb{C}_\infty\setminus X_n$ to $\mathbb{C}_\infty\setminus X_n$ keeping the elements of X_n fixed, with the composition of functions, is a group $\widetilde{G}$. If $[F]$ is a fixed point of one of these functions $\widetilde{g}$, then $g^{-1}=F^{-1}\circ(F\circ g^{-1})$ is homotopic to the identity in $\mathbb{C}_\infty\setminus X_n$, and so is g. Thus, $\widetilde{g}$ is the identity in $\text{Teich}(\mathbb{C}_\infty\setminus X_n)$ and of the elements of the group $\widetilde{G}$ only the identity has fixed points. If $[F], [G]\in\text{Teich}(\mathbb{C}_\infty\setminus X_n)$ are such that $P([F]) = P([G])$, then $g = F^{-1} \circ G$ keeps the elements of X_n fixed, and, therefore, it belongs to $\widetilde{G}$. This is so if and only if $[F], [G]$ belong to the same orbit of the group $\widetilde{G}$. Hence, $\text{Teich}(\mathbb{C}_\infty \setminus X_n)/\widetilde{G}$ is isomorphic to M_n and $\big(\text{Teich}(\mathbb{C}_\infty \setminus X_n), P\big)$ is a covering of $B(-\mathbb{H}, G_{\pi_n})$.

By the three preceding paragraphs, with $P : \text{Teich}(\mathbb{C}_\infty \setminus X_n) \to M_n$, $P([F]) = p(F)$, and $p(F) = \big(F(x_1), \ldots, F(x_n)\big)$, $\big(\text{Teich}(\mathbb{C}_\infty \setminus X_n), P\big)$ is a covering of $B(-\mathbb{H}, G_{\pi_n})$. P is differentiable, $p(0) = 0$ and $P(0) = 0$. Hence, there is a differentiable function $\widehat{p} : B_1 \to \text{Teich}(\mathbb{C}_\infty \setminus X_n)$ with $\widehat{p}(0) = 0$ such that $P \circ \widehat{p} = p$. With the Bers embedding h_n from $\text{Teich}(\mathbb{C}_\infty \setminus X_n)$ to $B(-\mathbb{H}, G_{\pi_n})$ considered two paragraphs above, $h_n \circ \widehat{p} : B_1 \to B(-\mathbb{H}, G_{\pi_n})$ is differentiable and has values in $B_{3/2} \subset B(-\mathbb{H}, G_{\pi_n})$. Since $h_n(0) = 0$ and $\widehat{p}(0) = 0$, it is $(h_n \circ \widehat{p})(0) = 0$. This function is from a ball of $\mathbb{C}$ to a ball of the finite dimensional Banach space $\big(B(-\mathbb{H}, G_{\pi_n}), \| \, \|_{\mathcal{N}_{-\mathbb{H}}}\big)$. As the Schwarz lemma can be generalized for differentiable functions between complex Banach spaces[43] giving $\big\| (h_n \circ \widehat{p})(z) \big\|_{\mathcal{N}_{-\mathbb{H}}} \le \|z\|$, it is $(h_n \circ \widehat{p})\big(B_{1/3}\big) \subset B_{1/2}$ with this last ball in the Nehari norm in $B(-\mathbb{H}, G_{\pi_n})$ which, by the Bers embedding (4.16) and the Ahlfors-Weill section theorem (4.14), consists of quadratic differentials corresponding by the Bers embeddings h_n to harmonic Beltrami coefficients. So, there is an extension of the holomorphic motion $\{\phi_\tau\}_{\tau \in B_1}$ of $X_n = \{x_1, \ldots, x_n\}$ to a holomorphic motion of $\mathbb{C}_\infty$ with parameter in $B_{1/3}$, $\{\phi_\tau\}_{\tau \in B_{1/3}}$, such that the Beltrami coefficient of each ϕ_τ is harmonic.

As in the proof of Slodkowski theorem, we consider an expansive sequence of finite sets X_n with $\cup_{n \in \mathbb{N}} X_n$ dense in Ω, and we get a sequence of extensions of holomorphic motions with convergent subsequence in compact subsets to a holomorphic motion of $\mathbb{C}_\infty$ with parameter in $B_{1/3}$. By the λ-lemma, there is an extension of the holomorphic motion to $\text{cl}\,\Omega$, and it is assumed that Ω is closed. If U is a connected component of $\mathbb{C}_\infty \setminus \text{cl}\,\Omega$ and $\rho_n(z)|dz|$, $\rho_U(z)|dz|$ are the elements of length of the Poincaré metrics of, resp., $\mathbb{C}_\infty \setminus X_n$ and U, then

$$\rho_{n|U} \le \rho_{n+1|U} \le \rho_U, \qquad \rho_\infty(z) \stackrel{def}{=} \lim \rho_n(z) \le \rho_U(z), \quad z \in U.$$

Since the Gaussian curvature of the Poincaré metric is a constant taken to be -1, each ρ_n satisfies $\Delta \log \rho = \rho^2$.

Since

$$\Delta\big(\log \rho_\infty - \log \rho_{n|U}\big) = \rho_\infty^2 - \rho_{n+1|U}^2 \ge 0,$$

in each compact set $K \subset U$ with regular boundary $\frac{\rho_\infty}{\rho_n}$ is bounded by the maximum value[44] $M_n \ge 1$, it assumes in ∂K, and $\lim M_n = 1$,

$$\rho_\infty(z) - \rho_{n|U}(z) = \rho_{n|U}(z)\left(\tfrac{\rho_\infty(z)}{\rho_{n|U}}(z) - 1\right) \le \rho_U(M_n - 1), \quad z \in U.$$

---

[43] See Appendix H.

[44] See Section 6 of Chapter 9 of *CADOVA*.

Thus, the convergence of the sequence $\{\rho_n\}$ is uniform in compact subsets of Ω and, therefore, the second-order partial derivatives of ρ_n also converge, ρ_∞ satisfies the same differential equation, and the metric with element of length $\rho_\infty(z)|dz|$ has Gaussian curvature -1. To prove that $\rho_\infty = \rho_\Omega$, it suffices to obtain that for every rectifiable curve C in Ω with final point $b \in \partial\Omega$ it is $\int_C \rho_\infty(z)|dz| = +\infty$. If $b \in \partial\Omega \setminus \{0, 1, \infty\}$, there is a sequence $\{b_n\}$ such that $b_n \in \Phi_n$ and $\lim b_n = b$. If $\sigma_b(z)|dz|$ is the element of length of the Poincaré metric of $\mathbb{C}_\infty \setminus \{0, 1, \infty, b\}$, then $\int_C \sigma_b(z)|dz| = +\infty$ and $\int_C \sigma_{b_n}(z)|dz| \to +\infty$, since $x \mapsto \sigma_x(z)$ is continuous. For $b \in \{0, 1, \infty\}$, with a Möbius transformation, the role of these points in the normalization of the holomorphic motion can be changed to other three points of $\partial\Omega$, and we can proceed as in the previous case and also obtain $\lim \int_C \sigma_b(z)|dz| = +\infty$. As $\sigma_{b_n}(z) \le \rho_m(z)$ for all $m > n$, $m, n \in \mathbb{N}$, it is $\sigma_{b_n}(z) \le \rho_\infty(z)$ for $n \in \mathbb{N}$. Therefore, $\int_C \rho_\infty(z)|dz| = +\infty$, and $\rho_\infty = \rho_\Omega$.

Let $\kappa_{n,\tau}$ be the Beltrami coefficient of $\phi_{n,\tau}$, where $\{\phi_{n,\tau}\}_{\tau \in B_{1/3}}$ is the holomorphic motion of $\mathbb{C}_\infty$ extending that of $\Omega \setminus X_n$ which, as the elements of this holomorphic motion have harmonic Beltrami coefficients, i.e., $\kappa_{n,\tau}(z) = \rho_n^{-2}(z)\, \overline{\psi_{n,\tau}(z)}$ with $\psi_{n,\tau}$ holomorphic, and $\rho_n \le \frac{2}{d(z, X_n)}$, where $d(z, X_n)$ is the Poincaré distance of z to X_n, since $\|\kappa_{n,\tau}\|_{L^\infty} < 1$, such that

$$\left|\psi_{n,\tau}(z)\right| \le |\kappa_{n,\tau}|\left(\frac{2}{d(z, X_n)}\right)^2 < \frac{4}{d^2(z, X_n)} \,.$$

As $\{\psi_{n,\tau}\}_{n \in \mathbb{N}, \tau \in B_{1/3}}$ is uniformly bounded in compact subsets of $B_{1/3} \times \Omega$, there is a uniformly convergent subsequence on compact subsets of $B_{1/3} \times \Omega$ to a holomorphic function $\psi_\tau : \Omega \to \mathbb{C}_\infty$ for each $\tau \in B_{1/3}$ such that $\tau \mapsto \overline{\psi_\tau}$ is holomorphic. Therefore, the sequence of the corresponding Beltrami coefficients $\{\kappa_{n,\tau}(z)\}_{n \in \mathbb{N},\, \tau \in B_{1/3}}$ converges uniformly for (τ, z) in compact subsets of $B_{1/3} \times \Omega$ to $\kappa_\tau(z) = \rho_V^{-2}(z)\, \overline{\psi_\tau \tau(z)}$ when $n \to +\infty$. This is the Beltrami coefficient of each element of the extension to holomorphic motion of $\mathbb{C}_\infty$ with parameter in $B_{1/3}$, and it is harmonic. □

Chapter 5
Stability and Bifurcation in Complex Dynamics

5.1 Introduction

In 1983, R. Mañé, P. Sad, and D. Sullivan and, independently, M. Lyubich introduced notions of structural stability and bifurcation for families of holomorphic or rational functions[1] in Julia sets and obtained some of their properties, including relationships with hyperbolicity and properties of genericity. In 1998, C. McMullen and D. Sullivan considered various notions of stability and bifurcation and completed the study initiated by R. Mañé, P. Sad, and D. Sullivan 15 years before.

In 1993, C. McMullen and D. Sullivan presented the no invariant line fields conjecture[2] and proved the remarkable result that its validity (of an intrinsic property of each function) implies the validity of the generic hyperbolicity conjecture (of perturbations of functions), obtaining a property of each function that implies its structural stability in Julia sets without having to consider perturbations.

In 1985, R. Mañé gave a remarkable proof that complex rational functions with cycles of Herman rings are not structurally stable by proving that they are not infinitesimally stable, a notion that he then introduced in Complex Dynamics, but which had been introduced for diffeomorphisms on differential manifolds in 1971 by J. Robbin, inspired by an article by J. Moser of 1969, as an intermediate property to prove that hyperbolicity and transversality of diffeomorphisms imply structural stability under C^1 perturbations, and was generalized and used to prove equivalence

[1] R. Mañé, P. Sad, and D. Sullivan article cited at the beginning of the preceding chapter and Lyubich, M.Yu., Some typical properties of the dynamics of rational maps, *Russ. Math. Surv.* **38** (1983) 154–155. See Chapter 13 of CADOVA for a self-contained introduction to the general Complex Dynamics Theory.

[2] It first appeared in the 1993 *Preprint* by C. McMullen and D. Sullivan *Quasiconformal homeomorphisms and dynamics III: the Teichmüller space of a holomorphic dynamical system,* published in 1998 in *Adv.Math.,* **135** (1998) 351–395, and meanwhile in C. McMullen, *Complex Dynamics and Renormalization*, Princeton University Press, New Jersey, 1994.

L. T. Magalhães, *Quasiconformal Mappings in the Plane and Complex Dynamics*,
https://doi.org/10.1007/978-3-031-80115-0_5

of hyperbolicity and transversality to infinitesimal stability for diffeomorphisms by R. Mañé in 1975, who extended it in 1977 to endomorphisms in differential manifolds, proving in 1987 that infinitesimal stability and structural stability are equivalent for C^1 diffeomorphisms. In 1988 and 1989, H. Ikeda obtained for endomorphisms properties of infinitesimal stability similar to those of diffeomorphisms.

5.2 Quasiconformal, Topological, Postcritical and Julia Stability

$f, g \in H(\mathbb{C}_\infty, \mathbb{C}_\infty)$ are **topologically equivalent** functions if there is a homeomorphism h of $\mathbb{C}_\infty$ onto $\mathbb{C}_\infty$, called a **conjugacy** of f with g, such that $h \circ f = g \circ h$; f, g are, resp., **quasiconformally** or **conformally equivalent** if, in addition, h is, resp., a quasiconformal or a conformal homeomorphism.

A **relation of critical orbits** of $f \in H(\mathbb{C}_\infty, \mathbb{C}_\infty)$ is a set

$$\left\{(j, k, m, n) \in \{1, \ldots, N\}^2 \times (\mathbb{N} \cup \{0\})^2 : f^m(c_j) = f^n(c_k)\right\},$$

where $c_1, \ldots, c_N$ are the critical points of f (arbitrarily ordered).

$f, g \in H(\mathbb{C}_\infty, \mathbb{C}_\infty)$ are **postcritically equivalent** functions if they have the same number of critical points and there is a ordering of those of g such that f and g have the same relation of critical orbits, i.e., there is a bijection from the positive orbits of f to those of g .

A **holomorphic family of rational functions** is a set $\mathcal{F} = \{f_\tau\}_{\tau \in T}$, with T a Riemann surface and $(\tau, z) \mapsto f_\tau(z)$ a holomorphic function from $T \times \mathbb{C}_\infty$ to $\mathbb{C}_\infty$.

Holomorphic families of rational functions $\mathcal{F} = \{f_\tau\}_{\tau \in T}$ include the family $\mathcal{R}_d$ of all complex rational functions of degree $d \geq 2$ and the family $\mathcal{P}_d$ of all complex polynomials of degree $d \geq 2$. For $\mathcal{R}_d$, the Riemann surface T has dimension $2d+1$ and is considered with the topology induced by the C^0 topology in $H(\mathbb{C}_\infty)$, and the three dimensional group $G_{\mathcal{R}_d}$ of normalized Möbius transformations $z \mapsto \frac{Az+B}{Cz+D}$, with $AD - BC = 1$, acts by conjugacy $f \mapsto T \circ f \circ T^{-1}$, with $T \in G_{\mathcal{R}_d}$, in $\mathcal{R}_d$. So, the space of functions not equivalent under this conjugacy has dimension $2d-2$, which is the number of critical points of each one of these functions, counting multiplicities. Analogously, if $\mathcal{P}_d$ is the Riemann surface T, its dimension of $d+1$ and the two dimensional group $G_\mathcal{P}$ of affine transformations $z \mapsto Az + B$ acts on $\mathcal{P}$ by the conjugacy $f \mapsto T \circ f \circ T^{-1}$, with $T \in G_\mathcal{R}$. Therefore, the space of polynomials not equivalent under this conjugacy has dimension d–1 , the number of critical points of the polynomials, counting multiplicity.

If $\mathcal{F} = \{f_\tau\}_{\tau \in T}$ is a holomorphic family of rational functions, f_{τ_0} with $\tau_0 \in T$ is said to be a **structurally stable function in** $\mathcal{F}$ if τ_0 has a neighborhood $T_0 \subset T$ such that f_τ is topologically equivalent to f_{τ_0} for $\tau \in T_0$. If f_τ is quasiconformally, conformally, or postcritically equivalent to f_{τ_0}, we say that f_{τ_0} is, resp., **quasiconformally**, **conformally**, or **postcritically stable** in $\mathcal{F}$. The sets of parameters $\tau \in T$ such that f_τ is conformally, quasiconformally, structurally, postcritically stable in $\mathcal{F}$ are denoted, resp.,

$$\boldsymbol{T}^{\mathbf{conf}}(\mathcal{F}) \subset \boldsymbol{T}^{\mathbf{qconf}}(\mathcal{F}) \subset \boldsymbol{T}^{\mathbf{top}}(\mathcal{F}) \subset \boldsymbol{T}^{\mathbf{posc}}(\mathcal{F}) .$$

A function f_τ, with $\tau \in T$, not structurally stable is called a **bifurcation point of** $\mathcal{F}$.

If $\mathcal{F} = \{f_\tau\}_{\tau \in T}$ is a holomorphic family of rational functions and $\{\phi_\tau\}_{\tau \in T}$ is a holomorphic motion of a set $\Omega \subset \mathbb{C}_\infty$ with parameter in T with base point $\tau_0 \in T$, it is said that the **holomorphic motion respects the dynamics in** $\mathcal{F}$ if it is a conjugacy of f_{τ_0} with f_τ in Ω for all $\tau \in T$, i.e., $(\phi_\tau \circ f_{\tau_0})(z) = (f_\tau \circ \phi_\tau)(z)$ whenever $z, f_{\tau_0}(z) \in \Omega$.

In 1983, R.Mañé, P.Sad, and D.Sullivan considered structural stability properties in Julia sets of holomorphic families of rational functions, and to study them, they introduced several other concepts such as persistently neutral periodic points.

If $\mathcal{F} = \{f_\tau\}_{\tau \in T}$ is a holomorphic family of rational functions, f_{τ_0} with $\tau_0 \in T$ is said to be **stable on Julia sets** or simply **Julia-stable** in $\mathcal{F}$ if there is a holomorphic motion of the Julia set $J(f_{\tau_0})$ with parameter in a connected neighborhood $T_0 \subset T$ of τ_0 and base point τ_0 that respects the dynamics in $\mathcal{F}$. The set of parameters $\tau \in T$ such that f_τ is Julia-stable is denoted $\boldsymbol{T^{J\text{stable}}(\mathcal{F})}$.

If $\mathcal{F} = \{f_\tau\}_{\tau \in T}$ is a holomorphic family of rational functions, $z_0 \in \mathbb{C}_\infty$ is said to be a **persistently neutral periodic point of** f_{τ_0} **in** $\mathcal{F}$ if there is a neighborhood $T_0 \subset T$ of τ_0 and a function $Z \in H(T_0, \mathbb{C}_\infty)$ such that $Z(\tau_0) = z_0$, $Z(\tau)$ is a periodic point with the same period m as the periodic point z_0 of f_{τ_0} and $\left|(f_\tau^m)'(Z(\tau))\right| = 1$ for all $\tau \in T_0$. The set of parameters $\tau_0 \in T$ with some neighborhood $T_0 \subset T$ such that for $\tau \in T_0$ the periodic neutral points of f_τ that may exist are persistently neutral is denoted $\boldsymbol{T^{\text{pneutral}}(\mathcal{F})}$.

The following result gives several conditions equivalent to Julia-stability. In essence, it was obtained by R. Mañé, P. Sad, and D. Sullivan in 1983, with proofs simplified later on by C.McMullen.

(5.1) Characterizations of Julia-stability *If $\mathcal{F} = \{f_\tau\}_{\tau \in T}$ is a holomorphic family of rational functions of degree ≥ 2 and $\tau_0 \in T$, then there is a neighborhood $T_0 \subset T$ of τ_0 such that the following are equivalent:*

1. $\tau_0 \in T^{J\text{stable}}(\mathcal{F})$.
2. $\tau_0 \in T^{\text{pneutral}}(\mathcal{F})$.
3. *The number of attracting or superattracting periodic orbits of f_τ is the same, for $\tau \in T_0$.*
4. *The maximum of attracting or superattracting periodic orbits of f_τ is bounded (and constant) for $\tau \in T_0$.*
5. *The function $\mathscr{J}(\tau) = J(f_\tau)$ is continuous from T_0 to $\mathscr{P}(\mathbb{C}_\infty)$ with the Hausdorff distance.*

If, in addition, f_τ has the same number of critical points $c_j(\tau)$, $j = 1, \ldots, N$, these can be ordered so that the functions $\tau \mapsto c_j(\tau)$ are holomorphic from T to $\mathbb{C}_\infty$, and the following conditions are equivalent to those above:

6. *The postcritical orbits of f_τ are normal families for $\tau \in T_0$.*
7. $c_j(\tau) \in J(f_\tau)$ *for all* $\tau \in T_0$ *if and only if* $c_j(\tau_0) \in J(f_{\tau_0})$.

If $\tau_0 \in T^{J\text{stable}}$, there is a holomorphic motion $\{J(f_\tau)\}_{\tau \in T_0}$ of $J(f_\tau)$, with $T_0 \subset T$ a connected neighborhood of τ_0.

Proof The last statement follows directly from the definition of $T^{J\text{stable}}(\mathcal{F})$.

(1 ⇒ 2) Let z_0 be a neutral nonparabolic periodic point of f_τ with period m. If $z_0 \notin J(f_\tau)$, then $|(f_\tau^m)(z)| \neq 1$ and, by the implicit function theorem, for v in some neighborhood $V \subset T$ of τ there is a unique function Z such that $Z(v)$ is the only periodic point with period m of f_v for $v \in V$ and this function is holomorphic in V with $|(f_v^m)'(Z(v))| \neq 1$. By the continuity of Z and the motion of $J(f_\tau)$, for v in a small neighborhood $V \subset T$ of τ, $Z(v) \notin J(f_v)$ and, therefore, $Z(v)$ is not a repelling periodic point of f_v and $|(f_v^m)'(Z(v))| \leq 1$ to $v \in V$. Hence, by the maximum modulus principle, $|(f_v^m)'|$ is constant in V and z_0 is a persistently neutral periodic point. If $z_0 \in J(f_\tau)$, the implicit function theorem can also be applied, and $Z(v) = \phi_v(z) \in J(f_v)$. Therefore, $|(f_v^m)'(Z(v))| \geq 1$ for v in some neighborhood $V \subset T$ of τ and, analogously, $|(f_v^m)'|$ is constant in V and z_0 is a persistently neutral periodic point. Consequently, $\tau_0 \in T^{\text{pneutral}}(\mathcal{F})$.

(2 ⇒ 3) If $V \subset T$ is a neighborhood of τ_0 such that $V \subset T^{\text{pneutral}}(\mathcal{F})$, as a repelling periodic point cannot change to attracting or superattracting without passing through neutral and vice versa, there is a neighborhood $T_0 \subset T$ of τ_0 where the number of attracting or superattracting periodic points of f_τ for $\tau \in T_0$ is constant.

(3 ⇒ 4) By the implicit function theorem, an attracting or superattracting periodic point remains attracting or superattracting with the same period for τ in some neighborhood of τ_0. Thus, the maximum of the minimal periods of these periodic points of f_τ is the same as the maximum of the minimal periods of those of f_{τ_0} for $\tau \in T_0$.

(4 ⇒ 1) If $T_0 \subset T$ is a neighborhood of τ_0 such that the maximum of the minimal periods of attracting or superattracting periodic orbits of f_τ is $n \in \mathbb{N}$ for $\tau \in T_0$, the repelling periodic points of f_{τ_0} with periods $> N$ persist as repelling periodic points of f_τ for $\tau \in T_0$ and cannot change to attracting or neutral periodic points. Thus, the periodic points with period $> N$ move with the parameter τ holomorphically without collisions. As repelling periodic points with periods $> N$ are dense in $J(f_\tau)$, there is a holomorphic motion of $J(f_{\tau_0})$ with parameter in a connected neighborhood $T_0 \subset T$ of τ_0 and base point τ_0 that respects the dynamics in $\mathcal{F}$. By the λ-lemma (4.3), $\tau_0 \in T^{J\text{stable}}(\mathcal{F})$.

The proof of equivalence of the conditions 1 to 4 is complete.

(1 ⇒ 5) If $\tau_0 \in T^{J\text{stable}}(\mathcal{F})$, there is a holomorphic motion of $J(f_{\tau_0})$ with parameter in a connected neighborhood $T_0 \subset T$ of τ_0 and base point τ_0 that respects the dynamics in $\mathcal{F}$. By the λ-lemma (4.3), $\mathcal{J}(\tau) = J(f_\tau)$ is continuous from T_0 to $\mathcal{P}(\mathbb{C}_\infty)$ with the Hausdorff distance.[3]

(5 ⇒ 3) If $T_0 \subset T$ is a connected neighborhood of τ_0 and $\mathcal{J} : T_0 \to \mathcal{P}(\mathbb{C}_\infty)$ is continuous in the Hausdorff metric, by the Siegel theorem,[4] there is a dense set in the circle $S^1 = \partial B_1 \subset \mathbb{C}$ such that periodic points with multiplier in this set are centers

[3] See Section 5 of Appendix A of *CADOVA*.

[4] See Section 11 of Chapter 13 of *CADOVA*.

of Siegel disks. Since the center of a Siegel disk is at a distance >0 from the Julia set, with a small variation of the parameter, it cannot belong to this set and, therefore, cannot become a repelling periodic point and remains a neutral periodic point. So, the number of attracting or superattracting periodic orbits remains constant in T_0.

Therefore, *1* $\Rightarrow$ *5* $\Rightarrow$ *3*, and, as *1* $\Leftrightarrow$ *3*, *5* $\Leftrightarrow$ *1* and *5* $\Leftrightarrow$ *3*. Thus, with the equivalences proved above, 1 to 5 are equivalent.

(6 $\Rightarrow$ 4) If the positive postcritical orbits $\{f_\tau^n(c_j(\tau))\}_{n\in\mathbb{N}}$, $j=1,\dots,N$, of f_τ are normal families for τ in a neighborhood T_0 of τ_0, taking successive subsequences uniformly convergent in compact subsets of T_0 for each $j=1,\dots,N$, leads to a subsequence $\{n_k\}$ of the identity sequence such that $\{f_\tau^{n_k}(c_j(\tau))\}_{k\in\mathbb{N}}$ is uniformly convergent on compact subsets of T_0 to a function $g_j \in H(T_0, \mathbb{C}_\infty)$, for all $j=1,\dots,N$. If f_τ has an attracting or superattracting periodic orbit of period m for some $\tau \in T_0$, as the immediate attracting basin of that periodic orbit contains a critical point,[5] the orbit contains a $g_j(\tau)$ for some $j=1,\dots,N$, and $f^m(g_j(\tau)) = g_j(\tau)$. The attracting or superattracting periodic orbit of period m deforms with small perturbations of the parameter τ to an attracting or superattracting periodic orbit of period m and, with a neighborhood $T_0 \subset T$ of τ_0 smaller if necessary, this holds for $\tau \in T_0$. Therefore, a periodic orbit of f_τ attracting the critical point $c_j(\tau)$ has period $\leq m$. Since the number of critical points is finite, the minimal periods of the attracting or superattracting periodic orbits of f_τ, $\tau \in T_0$, are upper bounded.

(7$\Rightarrow$4) If there are neighborhoods $T_0 \subset T$ of τ_0 and $j \in \{1,\dots,N\}$ such that $c_j(\tau) \in J(f_\tau)$ if and only if $c_j(\tau_0) \in J(f_{\tau_0})$, with a neighborhood $T_0 \subset T$ of τ_0 smaller if necessary, it is possible to choose three additional points $z_k(\tau) \in J(f_\tau)$, $k=1,2,3$, depending holomorphically on $\tau \in T_0$, e.g., by choosing three distinct points in the set of repelling periodic points of $J(f_{\tau_0})$ for $\tau = \tau_0$ and the corresponding points obtained with the implicit function theorem for τ in a neighborhood of τ_0. If, in addition, $\tau_1 \in T_0$ is such that the attracting basin of some attracting or superattracting periodic orbit of f_{τ_1} contains the critical point $c_j(\tau_1)$, then $c_j(\tau_1) \notin J(f_{\tau_1})$ and, by 6, $c_j(\tau) \notin J(f_\tau)$ for $\tau \in T_0$. Therefore, z_k, $k=1,2,3$, are omitted by the sequence $\{f_\tau^n(c_j(\tau))\}_{n\in\mathbb{N}}$ for $\tau \in T_0$. By the Montel-Carathéodory theorem,[6] this sequence is a normal family for each $\tau \in T_0$. Pursuing with the argument of the preceding paragraph, we get that the minimal periods of the attracting or superattracting periodic orbits of f_τ, for $\tau \in T_0$, are upper bounded.

(1$\Rightarrow$7) If $\tau_0 \in T^{J\text{stable}}(\mathcal{F})$, there is a holomorphic motion $\{\phi_\tau\}_{\tau\in T_0}$ of $J(f_{\tau_0})$ with parameter in a connected neighborhood $T_0 \subset T$ of τ_0, and for $\tau \in T_0$, there is a holomorphic motion $\{J(f_\tau)\}_{\tau\in T_0}$ of $J(f_\tau)$ with parameter in a connected neighborhood $T_0 \subset T$ of τ_0. Since $J(f_\tau)$ is a totally invariant perfect set, $z \in J(f_\tau)$ is a critical point of f_τ with multiplicity m if and only if $f_{\tau|J(f_\tau)}$ is locally a function $(m+1)$-to-1. Since $\{\phi_\tau\}_{\tau\in T_0}$ respects the dynamics in $\mathcal{F}$, $\phi_{\tau|J(f_\tau)}$ is a conjugacy of f_{τ_0} in $J(f_{\tau_0})$ with f_τ in $J(f_\tau)$, preserves the critical points in these

---

5 See Section 6 of Chapter 13 of *CADOVA*.

6 See Section 6 of Chapter 11 of *CADOVA*.

Julia sets, and the resp. multiplicities and positive orbits. Thus, $c_j(\tau_1) \in J_{\tau_1}$ for some $\tau_1 \in T_0$ implies $c_j(\tau) \in J(f_\tau)$ for all $\tau \in T_0$, and $\psi\big(c_j(\tau_0)\big) = c_j(\tau)$ for $\tau \in T_0$, $j = 1, \ldots, N$; therefore, 7 holds.

(1⇒6) Pursuing the argument of the preceding paragraph with a choice of points $z_1, z_2, z_3 \in J(f_{\tau_0})$ not postcritical points of f_{τ_0}, each $\phi_\tau(z_j)$ is not a postcritical point of f_τ for $\tau \in T_0$, $j = 1, \ldots, N$. By the Montel-Carathéodory theorem,[6] the postcritical orbits of f_τ are normal families for $\tau \in T_0$, and 6 holds.

As *1* ⇔ *4*, with the implications proved in the last four preceding paragraphs, 1, 4, 6, 7 are equivalent, and, with the equivalences obtained before, the proof of equivalence of the conditions 1 to 7 ends. □

The following result, of R. Mañé, P. Sad, and D. Sullivan in 1983, is used in the final part of the next chapter to get optimal upper bounds of certain sums of the numbers of cycles of attracting or parabolic basins, neutral or Cremer points,[7] Siegel disks, or Herman rings.

(5.2) *Let $\mathcal{F} = \{f_\tau\}_{\tau \in W}$ be a holomorphic family of rational functions, $W_0 \subset W$ a simply connected open set, and $\varphi \in H(W_0, \mathbb{C}_\infty)$ such that $\varphi(\tau)$ does not belong to any postcritical orbit of f_τ for $\tau \in W_0$. If $\varphi(\tau)$ is not a periodic point of f_τ for $\tau \in W_0$ or is a periodic point with the same period $m \in \mathbb{N}$ for $\tau \in W_0$, then $W_0 \subset T^{\mathrm{top}}(\mathcal{F})$.*

Proof If there is $m \in \mathbb{N}$ such that for every $\tau \in T_0$ $\varphi(\tau)$ is a periodic point with period m, i.e., $f_\tau^m\big(\varphi(\tau)\big) = \varphi(\tau)$ and, therefore, $\varphi(\tau)$ is a zero of the rational function $f_\tau^m - 1_{\mathbb{C}_\infty}$, which has degree d^m with d the degree of the rational function f. As $\varphi(\tau)$ does not belong to any postcritical orbit, $(f_\tau^m)'\big(\varphi(\tau)\big) \neq 0$. Therefore, there are d^m possible values of $\varphi(\tau)$ and, by the inverse function theorem, there is an open simply connected subset of T_0, also denoted T_0, and a unique set of d^m functions $\varphi_{m,j} \in H(T_0, \mathbb{C}_\infty)$, $j = 1, \ldots, d^m$, such that $f_\tau^m\big(\varphi(\tau)\big) = \varphi(\tau)$,

$$f_\tau^m\big(\varphi_{m,j}(\tau)\big) = \varphi(\tau), \quad \varphi_{m,j}(\tau) \neq \varphi_{m,k}(\tau), \quad m \in \mathbb{N},\ j \neq k,\ j,\ k = 1, \ldots, d^m,\ \tau \in T_0.$$

For $\tau_0 \in T_0$ and $j \in \{1, \ldots, d^m\}$ such that $f_\tau^{m-1}\big(\varphi_{m,j}(\tau_0)\big) \neq \varphi_{m,j}(\tau_0)$, which exist as the order of the polynomial, $f_\tau^{m-1} - 1_{\mathbb{C}_\infty}$ is $d^{m-1} < d^m$, and the zeros of $f_\tau^m - 1_{\mathbb{C}_\infty}$ are distinct, $f_\tau^{m-1}\big(\varphi_{m,j}(\tau)\big) \neq \varphi_{m,j}(\tau)$ for $\tau \in T_0$, and $\varphi_{m,j}(\tau)$ is not a periodic point of f_τ for all $\tau \in T_0$. Therefore, the proof when there is a $m \in \mathbb{N}$ such that for all $\tau \in T_0$ the point $\varphi(\tau)$ is periodic with period m can be reduced to the proof when $\varphi(\tau)$ is not a periodic point of f_τ for any $\tau \in T_0$.

If $\varphi(\tau)$ is not a periodic point of f_τ for any $\tau \in T_0$, since for $\tau_0 \in T_0$ it is $f_{\tau_0}^n(f_{\tau_0}^{-n}(\varphi(\tau_0))) = \varphi(\tau_0)$, for each $n \in \mathbb{N}$ there is only one $j = , \ldots, d^n$ such that $f_{\tau_0}^{-n}(\varphi(\tau_0)) = \varphi_{n,j}(\tau_0)$. Denote

[7] Cremer, Hubert (1897–1983).

$$\Lambda = \cup_{n=0}^{\infty} f_{\tau_0}^{-n}\big(\varphi(\tau_0)\big) = \big\{\varphi_{n,j}(\tau_0)\colon j=1,\ldots,d^n, n\in\mathbb{N}\cup\{0\}\big\},$$

and define $h: T_0 \times \Lambda \to \mathbb{C}_\infty$ such that $h\big(\tau, \varphi_{n,j}(\tau_0)\big) = \varphi_{m,j}(\tau)$. Since $\varphi(\tau_0)$ is not a periodic point of f_{τ_0}, $\varphi_{m,j}(\tau_0) \neq \varphi_{m,k}(\tau_0)$ for $m \in \mathbb{N}$ and $j \neq k$ with $j, k = 1, \ldots, d^m$, for each $z \in \Lambda$, m and j such that $z = \varphi_{m,j}(\tau_0)$ are unique, h is holomorphic and $h_\tau : \Lambda \to \mathbb{C}_\infty$ such that $h_\tau(z) = h(\tau, z)$ is injective. It is

$$(f_\tau \circ h_\tau)\big(\varphi_{n,j}(\tau_0)\big) = f_\tau\big(\varphi_{n,j}(\tau)\big),$$

$$(h_\tau \circ f_{\tau_0})\big(\varphi_{n,j}(\tau_0)\big) = h_\tau\big(f_{\tau_0}\big(f_{\tau_0}^{-n}(\varphi(\tau_0))\big)\big) = h_\tau\big(f_{\tau_0}^{-(n-1)}(\varphi(\tau_0)\big) = \varphi_{n-1,k}(\tau)\,,$$

for some $k \in \{1, \ldots, d^n\}$, and, by the uniqueness of the functions $\varphi_{n,j}$ and the injectivity of h_τ, $k = j$. As

$$f_\tau^{n-1}\big(f_\tau\big(\varphi_{n,j}(\tau)\big)\big) = f_\tau^{n}\big(\varphi_{n,j}(\tau)\big) = \varphi(\tau)\,,$$

the uniqueness of the function $\varphi_{n,j}$, gives $f_\tau\big(\varphi_{n,j}(\tau)\big) = \varphi_{n-1,j}(\tau)$. $f_\tau \circ h_\tau = h_\tau \circ f_{\tau_0}$ in Λ, and h_τ is a conjugacy of the restriction of f_{τ_0} to Λ with the restriction of f_τ to $h_\tau(\Lambda)$. By the λ-lemma (4.3), each element of the family $\{h_\tau\}_{\tau \in T_0}$ has a quasiconformal extension $h_\tau : \mathrm{cl}\,\Lambda \to \mathbb{C}_\infty$ which is a homeomorphism of $\mathrm{cl}\,\Lambda$ onto $h_\tau\big(\mathrm{cl}\,\Lambda\big)$ such that $f_\tau \circ h_\tau = h_\tau \circ f_{\tau_0}$ in $\mathrm{cl}\,\Lambda$. Therefore, the extension h_τ is a conjugacy of the restriction of f_{τ_0} to $\mathrm{cl}\,\Lambda$ with the restriction of f_τ to $h_\tau\big(\mathrm{cl}\,\Lambda\big)$.

Since $\mathrm{cl}\,\Lambda$ contains all points with positive orbits containing $\varphi(\tau_0)$, $J(f_{\tau_0}) \subset \mathrm{cl}\,\Lambda$. If $z \in J(f_{\tau_0})$ is the limit of a sequence $\{z_n\}$ of repelling periodic points of f_{τ_0} different from z and $\{h_\tau(z_n)\}$ is a sequence of periodic points of f_τ different from $h_\tau(z)$ and converging to this point. Therefore, $h_\tau(z) \in J(f_\tau)$, and $h_\tau\big(J(f_{\tau_0})\big) \subset J(f_\tau)$. Switching the roles of τ and τ_0 gives the corresponding inclusion with h_τ^{-1} instead of h_τ, i.e., $h_\tau^{-1}\big(J(f_\tau)\big) \subset J(f_{\tau_0})$ and, therefore, $h_\tau\big(J(f_\tau)\big) \subset J(f_{\tau_0})$. Combining the inclusions at the end of the two preceding periods, $h_\tau\big(J(f_{\tau_0})\big) = J(f_\tau)$. Hence, h_τ is a conjugacy of the restriction of f_{τ_0} to $J(f_{\tau_0})$ with the restriction of f_τ to $J(f_\tau)$ and, by the continuity of h_τ, the function $\mathscr{J}(\tau) = J(f_\tau)$ from T_0 to $\mathscr{P}(\mathbb{C}_\infty)$ is continuous in a neighborhood $T_1 \subset T$ of each $\tau_1 \in T_0$ with the Hausdorff distance. Hence, the elements of the family $\mathcal{F} = \{f_\tau\}_{\tau \in T_0}$ are structurally stable in $\mathcal{F}$. Consequently, $T_0 \subset T^{\mathrm{top}}(\mathcal{F})$. □

(5.3) *If $\mathcal{F} = \{f_\tau\}_{\tau \in T}$ is a holomorphic family of rational functions and $\tau_0 \in T^{\mathrm{posc}}(\mathcal{F})$, then there is a neighborhood $U_0 \subset T$ of τ_0 and a holomorphic motion of $\mathbb{C}_\infty$ with parameter in U_0 and base point τ_0 that respects the dynamics in $\mathcal{F}$.*

Proof Let V_0 be a neighborhood of τ_0 in $T^{\mathrm{posc}}(\mathcal{F})$. Denoting the different critical points of f_τ by $c_1(\tau), \ldots, c_N(\tau)$, $\{c_1(\tau), \ldots, c_N(\tau)\}_{\tau \in V_0}$ is a holomorphic motion of the set of critical points with parameter in V_0 with base point τ_0. Since f_{τ_0}

is postcritically stable, the function $f_{\tau_0}^m(c_j(\tau_0)) \mapsto f_{\tau_0}^n(c_j(\tau_0))$ is well defined and depends holomorphically in $\tau \in V_0$ for (j, j, m, n) in the corresponding relation of critical orbits. This holomorphic motion has a unique extension compatible with the dynamics defined by the functions f_τ to a holomorphic motion of the set of orbits containing critical points. By the λ-lemma (4.3), this holomorphic motion has extension to the closure $\widetilde{P}(\tau)$ of the union of the orbits of critical points of f_τ for $\tau \in V_0$.

If $\#\widetilde{P}(\tau_0) \leq 2$, for some $n \in \mathbb{N}$, there is a conjugacy of f_τ with $z \mapsto z^n$ for $\tau \in V_0$, giving the result.

If $\#\widetilde{P}(\tau_0) \geq 3$, $J(\tau_0) \subset \widetilde{P}(\tau_0)$ and the Fatou set is the union of open connected subsets of $\mathbb{C}_\infty$, which are hyperbolic Riemann surfaces. By the harmonic λ-lemma (4.19), the holomorphic motion of $\widetilde{P}(\tau_0)$ has a unique extension to a holomorphic motion $\{\phi_\tau\}_{\tau \in V_0}$ of $\mathbb{C}_\infty$ such that the Beltrami coefficient $\mu_\tau(z)$ of $\phi_\tau(z)$ is a harmonic function of $\tau \in \mathbb{C}_\infty \backslash \widetilde{P}(\tau_0)$ and the corresponding holomorphic motion is defined in a neighborhood $U_0 \subset V_0$ of τ_0 with $\frac{1}{3}$ the size of V_0.

For each $\tau \in U_0$, the function $f_\tau : \mathbb{C}_\infty \backslash \widetilde{P}(\tau) \to \mathbb{C}_\infty \backslash \widetilde{P}(\tau)$ is a covering. $\psi_\tau = f_\tau^{-1} \circ \phi_\tau \circ f_{\tau_0}$ in $\mathbb{C}_\infty \backslash \widetilde{P}(\tau)$, where f_τ^{-1} is a continuous branch of inversion of f_τ such that $\psi_\tau = 1_{\mathbb{C}_\infty \backslash \widetilde{P}(\tau)}$, defines another extension of the initial holomorphic motion to a holomorphic motion of $\mathbb{C}_\infty$, leaving it unchanged in $\widetilde{P}(\tau)$ as it respects the dynamics in this set. Since for $\tau \in U_0$ f_τ is a conformal mapping in $\mathbb{C}_\infty \backslash \widetilde{P}(\tau)$ and f_{τ_0} is a local isometry in the Poincaré metric of $\mathbb{C}_\infty \backslash \widetilde{P}(\tau_0)$, by the harmonic λ-lemma (4.19), this holomorphic motion of $\mathbb{C}_\infty$ is an extension of that of $\widetilde{P}(\tau)$. By the uniqueness of such an extension, also guaranteed by the harmonic λ-lemma, $\phi_\tau = \psi_\tau$ for $\tau \in U_0$; thus, the holomorphic motion $\{\phi_\tau\}_{\tau \in U_0}$ respects the dynamics in $\mathcal{F}$. □

(5.4) *If $\mathcal{F}$ is a holomorphic family of rational functions of degree ≥ 2, then*

$$T^{\text{qconf}}(\mathcal{F}) = T^{\text{top}}(\mathcal{F}) = T^{\text{posc}}(\mathcal{F}) \subset T^{J\,\text{stable}}(\mathcal{F}) = T^{\text{pneutral}}(\mathcal{F}).$$

Proof By the preceding result, if $\tau_0 \in T^{\text{posc}}(\mathcal{F})$, there are neighborhoods $U_0 \subset T^{\text{posc}}(\mathcal{F})$ of τ_0 and a holomorphic motion $\{\phi_\tau\}_{\tau \in U_0}$ of $\mathbb{C}_\infty$ with base point τ_0 that respects the dynamics in $\mathcal{F}$. By the λ-lemma (4.3), ϕ_τ is a quasiconformal mapping in $\mathbb{C}_\infty$ for $\tau \in U_0$. Therefore, $T^{\text{posc}}(\mathcal{F}) \subset T^{\text{qconf}}(\mathcal{F})$. Since $T^{\text{qconf}}(\mathcal{F}) \subset T^{\text{top}}(\mathcal{F}) \subset T^{\text{posc}}(\mathcal{F})$, these sets are equal. By the definitions, $T^{\text{top}}(\mathcal{F}) \subset T^{J\,\text{stable}}(\mathcal{F})$, and, the equivalence of characterizations 1 and 2 of Julia-stability in (5.1), $T^{J\,\text{stable}}(\mathcal{F}) = T^{\text{pneutral}}(\mathcal{F})$. □

(5.5) Genericity of stability: *If $\mathcal{F} = \{f_\tau\}_{\tau \in T}$ is a holomorphic family of rational functions of degree $d \geq 2$, then*

$$T^{\text{qconf}}(\mathcal{F}) = T^{\text{top}}(\mathcal{F}) = T^{\text{posc}}(\mathcal{F}) \subset T^{J\,\text{stable}}(\mathcal{F}) = T^{\text{pneutral}}(\mathcal{F})$$

are open dense subsets of T.

Proof Directly from the definition, $T^{\text{posc}}(\mathcal{F}) \subset T$ is an open set.

If $N(\tau)$ is the number of attracting or superattracting periodic orbits of f_τ, there is $M > 0$ (depending on d) such that $N(\tau) \leq M$ for $\tau \in T$. As these orbits persist under small perturbations of τ in T, for $\tau_0 \in T$, it is $N(\tau_0) \leq \overline{\lim}\, N(\tau_n)$ for every sequence $\{\tau_n\} \subset T$ such that $\lim \tau_n = \tau_0$. Thus, τ_0 is the limit point of the set of points where $\tau \mapsto N(\tau)$ has a local maximum, which, by the equivalence of characterizations 1 and 3 of Julia-stability (5.1), are points of $T^{J\,\text{stable}}(\mathcal{F})$; therefore, $T^{J\,\text{stable}}(\mathcal{F})$ is a dense open subset of T.

By the two preceding paragraphs and (5.4), it suffices to prove that $T^{\text{posc}}(\mathcal{F})$ is dense in $T^{J\,\text{stable}}(\mathcal{F})$.

Let T_0 be the set of $\tau_0 \in T^{J\,\text{stable}}(\mathcal{F})$ such that every superattracting periodic orbit of f_{τ_0} persists under small perturbations of the parameter τ_0 in T and the number of critical points of f_τ assumes a local maximum at τ_0. The negation of each one of these conditions corresponds to a level set of a holomorphic function, i.e., to a proper complex submanifold of $T^{J\,\text{stable}}(\mathcal{F})$. Therefore, T_0 is an open dense subset of $T^{J\,\text{stable}}(\mathcal{F})$. Let U be a simply connect open subset of T_0 and $c_1(\tau), \ldots, c_N(\tau)$ be the ordered critical points of f_τ, with the functions $c_j : U \to \mathbb{C}$ holomorphic.

As the intersection of a number finite of open dense sets is an open and dense set, it suffices to prove that for $j, k \in \mathbb{N} \cup \{0\}$ there is a dense open subset of U such that the elements (j, k, m, n) of the relation of critical orbits of f_τ are locally constant for τ in that set.

Since $U \subset T^{J\,\text{stable}}(\mathcal{F})$, every critical point $c_j(\tau_0)$ of f_{τ_0} belonging to $J(f_{\tau_0})$ for $\tau_0 \in U$ is such that $c_j(\tau) \in J(f_\tau)$ for $\tau \in U$, and the elements (j, j, m, n) of the relation of critical orbits f_τ corresponding to critical points in $J(f_\tau)$ are invariant for $\tau \in U$.

If c_j, c_k are different critical points in Fatou sets, for $m, n \in \mathbb{N} \cup \{0\}$ with $m \neq n$, (j, k, m, n) belongs to the relation of critical orbits of f_τ, i.e., $f_\tau^m(c_j(\tau)) = f_\tau^n(c_k(\tau))$, and we denote $U_{jk} \subset U$ the set of $\tau \in U$ such that $f_\tau^m(c_j(\tau)) \neq f_\tau^n(c_k(\tau))$ for all $m, n \in \mathbb{N} \cup \{0\}$ with $m \neq n$. If $U_{jk} = \emptyset$, by the Baire category theorem,[8] some (j, k, m, n) belongs to the relation of critical orbits of f_τ for τ in some open subset of U, hence to $\tau \in U$, and the elements of the relation of critical orbits of f_τ corresponding to critical points in $J(f_\tau)$ are invariant for $\tau \in U$.

We now consider the remaining case, when $\tau_0 \in U$ and (k, k, m, n) belongs to the relation of critical orbits of f_{τ_0}, but not of f_τ for τ outside a proper complex submanifold of U.

If $j = k$, as $J(f_\tau)$ moves continuously with $\tau \in U$ and periodic orbits do not change type (attracting, superattracting, neutral, repelling), there is a bijection between Fatou components of f_τ, for $\tau \in U$, which preserves the types of these components in the Sullivan classification theorem.[9] $c_j(\tau)$ belongs to an attracting basin of an attracting or superattracting periodic orbit or a Siegel disk, since $f_\tau^m(c_j(\tau)) = f_\tau^n(c_k(\tau))$ cannot hold in a Herman ring or a parabolic basin.

---

[8] See Section 3 of Appendix A of *CADOVA*.

[9] See Section 12 of Chapter 13 of *CADOVA*.

Therefore, $f_\tau^m(c_j(\tau_0))$ belongs to the corresponding attracting, superattracting, or neutral periodic orbit. In the two first cases, the orbit of the critical point tends toward this orbit without intersecting it, since τ is close to τ_0, but outside the manifold where $f_\tau^m(c_j(\tau)) = f_\tau^n(c_k(\tau))$. In the case of a Siegel disk, $f_\tau^m(c_j(\tau))$ belongs to the Siegel disk, but it is not its center if $f_\tau^m(c_j(\tau)) \neq f_\tau^n(c_k(\tau))$. Thus, the corresponding postcritical orbit is infinite and $\tau_0 \in \operatorname{cl}\operatorname{int} U_{jj}$.

If $j \neq k$, as the elements (j, k, m, n), with $j = k$, of the relation of critical orbits of f_τ remain constant in a dense open subset of U, this set can be redefined to have this property.

If the orbit of $c_j(\tau)$ under f_τ is finite for some $\tau \in U$, for all $\tau \in U$, the orbit of $c_k(\tau)$ under f_τ is finite, and the two orbits end up in the same periodic orbit for $\tau = \tau_0$. Since each periodic orbit moves holomorphically with $\tau \in U$, we have $f_\tau^m(c_j(\tau)) = f_\tau^n(c_k(\tau))$ to $\tau \in U$.

If the orbits of $c_j(\tau)$ and $c_k(\tau)$ under f_τ are infinite for $\tau \in U$, increasing m and n if necessary, $p = f_{\tau_0}^m(c_j(\tau_0)) = f_{\tau_0}^n(c_k(\tau_0))$ belongs to a periodic Fatou component of f_{τ_0}, which, if it is an attracting basin of an attracting or superattracting periodic orbit or a parabolic basin, can be assumed to be close to a point of the corresponding orbit.

In the cases of an attracting or parabolic basin of a periodic orbit, there is an open ball B containing p such that for τ close to τ_0, the sets $C_\tau = \{f_\tau^n(B) : n \in \mathbb{N} \cup \{0\}\}$ are disjoint and the restrictions of the functions f_τ^n to B are injective for $n \in \mathbb{N}$, by the Koenigs, Böttcher, and Leau-Fatou conjugacies.[10] For τ close to τ_0, but outside the submanifolds where $f_\tau^m(c_j(\tau)) = f_\tau^n(c_k(\tau))$, the positive orbits of $c_j(\tau)$ and $c_k(\tau)$ are disjoint and $\tau_0 \in \operatorname{cl}\operatorname{int} U_{jk}$.

In the case of an attracting basin of a superattracting periodic orbit, the argument of the preceding paragraph does not apply as it is not possible to guarantee injectivity, and it is shown that for τ close to τ_0, $c_j(\tau)$ and $c_k(\tau)$ belong to invariant Jordan curves under distinct f_τ. We choose the local Böttcher coordinate $(\tau, z) \mapsto Z_\tau(z)$ in the neighborhood of (τ_0, p), for local conjugacy of f_τ with the integer power function $Z \mapsto Z^\ell$, where $f_\tau^{(\ell)}(p) \neq 0$ and $f_\tau^{(n)}(p) = 0$ for $n = 1, \ldots, \ell - 1$; thus, $Z_\tau(f_\tau^k(z)) = Z_\tau^\ell(z)$. Since there is a neighborhood $V \subset U$ of τ_0 such that $f_\tau^m(c_j(\tau)) \neq f_\tau^n(c_k(\tau))$ for $\tau \in V \setminus \{\tau_0\}$ and none of the terms of this inequality vanishes, because the corresponding postcritical orbit is infinite, with a neighborhood $V \subset U$ of τ_0 smaller and switching j with k if necessary,

$$\left|Z_\tau\left(f_\tau^m(c_j(\tau))\right)\right|^\ell < \left|Z_\tau\left(f_\tau^m(c_k(\tau))\right)\right| < \left|Z_\tau\left(f_\tau^m(c_j(\tau))\right)\right|^{\frac{1}{\ell}}, \quad \tau \in V.$$

Removing from V the real analytic proper submanifold of V, we get an open set $U_{jk} \subset V$ with $\tau_0 \in \operatorname{cl} U_{jk}$ such that

$$\left|Z_\tau\left(f_\tau^m(c_j(\tau))\right)\right| = \left|Z_\tau\left(f_\tau^m(c_k(\tau))\right)\right|, \quad \tau \in U_{jk},$$

since two points where $\log\log|Z|^{-1}$ differ but by less than $\log \ell$ cannot have intersecting global orbits.

[10] Koenigs, Gabriel (1858–1931). Bottcher, Lucjan (1872–1937).

We now consider a rotation domain with period k where for each $\tau \in U$ the positive orbit of $c_j(\tau)$ determines a union $C_j(\tau)$ of k Jordan curves which is invariant under f_τ. Each of these Jordan curves is a real analytic proper submanifold of U, and there is a holomorphic motion $\{\phi_\tau(C_j(\tau_0))\}_{\tau \in U}$ that respects the dynamics in $\mathcal{F}$. For each $\tau \in T$, as there is a holomorphic conjugacy of f_τ with a rotation, ϕ_τ is holomorphic. If $f_\tau q^n(c_k(\tau)) \in C_j(\tau)$ for τ in some connected neighborhood $V \subset U$ of τ_0, then $g(\tau) = \phi_\tau^{-1}(f_\tau^n(c_k(\tau)))$ is a holomorphic function of V in $C_j(\tau)$ and, by the uniqueness theorem of analytic functions, this function is constant in V. Thus, $f_\tau^m(c_j(\tau)) \neq f_\tau^n(c_k(\tau))$ for $\tau \in V \setminus \{\tau_0\}$ and, therefore, also for $\tau \in U \setminus \{\tau_0\}$. Otherwise, the positive orbits of $c_j(\tau)$ and $c_k(\tau)$ are disjoint for τ outside a real analytic proper submanifold of U where $f_\tau^n(c_k(\tau)) \in C_j(\tau)$. □

A good geometric description of $T^{J\text{stable}}(\mathcal{F}) \setminus T^{\text{top}}(\mathcal{F})$ is as follows.

(5.6) *If $\mathcal{F} = \{f_\tau\}_{\tau \in T}$ is a holomorphic family of rational functions of degree ≥ 2, then $T^{J\text{stable}}(\mathcal{F})$ is a union of a dense open subset of $T^{\text{top}}(\mathcal{F})$ with a countable set of complex proper submanifolds and a set of real analytic submanifolds of $T^{\text{top}}(\mathcal{F})$, and $T^{J\text{stable}}(\mathcal{F}) \setminus T^{\text{top}}(\mathcal{F})$ has Lebesgue measure 0, with real analytic submanifolds if and only if for some $\tau \in T^{J\text{stable}}(\mathcal{F})$ f_τ has a rotation domain or a persistent superattracting periodic orbit.*

5.3 Structural Stability and Hyperbolicity

The following results relate hyperbolicity and stability.

If $\mathcal{F} = \{f_\tau\}_{\tau \in T}$ is a holomorphic family of rational functions of degree ≥ 2, $\boldsymbol{T^{\text{hyperb}}}(\mathcal{F})$ denotes the set of parameters $\tau \in T$ with f_τ hyperbolic.

(5.7) *If $\mathcal{F} = \{f_\tau\}_{\tau \in T}$ is a holomorphic family of rational functions of degree ≥ 2, then $T^{\text{hyperb}}(\mathcal{F}) \subset T^{J\text{stable}}(\mathcal{F})$ and $T^{\text{hyperb}}(\mathcal{F})$ is open and closed relatively to $T^{J\text{stable}}(\mathcal{F})$, and one of the following holds:*

(i) $T^{\text{hyperb}}(\mathcal{F}) = \emptyset$.
(ii) $T^{\text{hyperb}}(\mathcal{F}) = T^{J\text{stable}}(\mathcal{F})$.
(iii) $\{T^{\text{hyperb}}(\mathcal{F}), T^{J\text{stable}}(\mathcal{F}) \setminus T^{\text{hyperb}}(\mathcal{F})\}$ *is a separation of the disconnected set* $T^{J\text{stable}}(\mathcal{F})$.

Proof f_τ is hyperbolic if and only if the critical points of f_τ are in the Fatou set, and the Fatou components are eventually periodic attracting or superattracting.[11] As the

[11] See Section 12 of Chapter 13 of *CADOVA*.

set of $\tau \in T$ such that all critical points of f_τ tend to attracting or superattracting periodic orbits is open and, by conditions 1 and 7 of (5.1), it is included in $T^{J\text{stable}}(\mathcal{F})$, then $T^{\text{hyperb}}(\mathcal{F}) \subset T^{J\text{stable}}(\mathcal{F})$ and $T^{\text{hyperb}}(\mathcal{F})$ is open relatively to $T^{J\text{stable}}(\mathcal{F})$. If $\tau_0 \in T^{J\text{stable}}(\mathcal{F})$, critical points or periodic orbits of $J(f_{\tau_0})$ persist under small variations of $\tau \in T$, and $T^{J\text{stable}}(\mathcal{F}) \setminus T^{\text{hyperb}}(\mathcal{F})$ is open relative to $T^{J\text{stable}}(\mathcal{F})$. Thus, $T^{\text{hyperb}}(\mathcal{F})$ is closed relative to $T^{J\text{stable}}(\mathcal{F})$. The alternative follows from the definition of disconnected set.[12] □

The first part of this result guarantees that the hyperbolicity of each function of a holomorphic family of rational functions of degree ≥ 2, which is a property of each function, implies the structural stability of the family functions, which is a property of perturbations in the family.

By the last two results, the generic hyperbolicity conjecture in the space of rational functions of degree $d \geq 2$ is equivalent to the J-stable hyperbolicity conjecture of rational functions of degree d.

$f \in H(\mathbb{C}_\infty, \mathbb{C}_\infty)$ is said to have an **invariant line field on** $J(f)$ if there is a set with positive Lebesgue measure $A \subset J(f)$ totally invariant under f and a set of lines in the plane $a + \mathbb{C} f'(a)$ a.e. for $a \in A$ and $\{f'(a) : a \in A\}$ is a measurable set.

A necessary and sufficient condition for a quasiconformal conjugacy of complex rational functions to be holomorphic follows.

(5.8) *Let $f, g \in H(\mathbb{C}_\infty, \mathbb{C}_\infty)$ and $h : \mathbb{C}_\infty \to \mathbb{C}_\infty$ be a quasiconformal mapping conjugacy of f in $J(f)$ with g in $J(g)$. $h \in H(\mathbb{C}_\infty, \mathbb{C}_\infty)$ if and only if the restriction of h to $\mathbb{C}_\infty \backslash J(f)$ is holomorphic and f does not have any invariant line field in $J(f)$.*

Proof The necessity of the condition for holomorphy in $\mathbb{C}_\infty$ is immediate. So, it remains to prove sufficiency.

If $J(f)$ has Lebesgue measure 0, there is nothing to prove. So, it is assumed that this is not the case. Considering $M = \mathbb{C}_\infty$ as a real differential manifold, the inverse of the derivative of h^{-1} at each $h(z)$ where it exists, denoted $D_{h(z)} h^{-1} : T_{h(z)} M \to T_z M$, maps circles in the tangent space $T_{h(z)} M$ with center 0 to ellipses in the tangent space $T_z M$ with center 0, all with the same eccentricity and major axis direction, a.e. for $z \in M$. Denoting the circle in $T_{h(z)} M$ with radius 1 and center 0 by $S^1_{h(z)}$ and its image by $D_{h(z)} h^{-1}$ in $T_z M$ by $E_z = D_{h(z)} h^{-1}(S^1_{h(z)})$, as $g \circ h = h \circ f$ in $J(f)$, it is $f'(z) E_z = E_{f(z)}$ a.e. in $J(f)$. If Σ is the set of $z \in J(f)$ such that E_z is not a circle, by the preceding period, $f^{-1}(\Sigma) = \Sigma$. For $z \in \Sigma$, let L_z denote the linear subspace of $T_z M$ of the vectors with the direction of the major axis of E_z. By the condition above, $f'(z) L_z = L_{f(z)}$ for $z \in \Sigma$. Since f does not have any invariant line field in $J(f)$, the Lebesgue measure of Σ is 0 and $E(z)$ is a circle a.e. for $z \in J(f)$, and also a.e. for $z \in \mathbb{C}_\infty$ since f is holomorphic in $\mathbb{C}_\infty \backslash J(f)$. Therefore, $h \in H(\mathbb{C}_\infty, \mathbb{C}_\infty)$. □

[12] See Section 5 of Appendix A of *CADOVA*.

As mentioned in the introduction to this chapter, in 1993, C. McMullen and D. Sullivan stated the no invariant line fields conjecture: A rational function f of degree $\geq$, 2 does not have invariant line fields in $J(f)$ except if f belongs to the set $\mathcal{T}$ defined below and proved that the validity of this conjecture implies that of the generic hyperbolicity conjecture. $\mathcal{T}$ is the set of rational functions f doubly covered by a holomorphic function in the surface of a complex m-torus T_m ($m \in \mathbb{N} \cup \{0\}$) in itself, which are functions $f \in H(\mathbb{C}_\infty, \mathbb{C}_\infty)$ such that $f \circ \wp = \wp \circ F$, with F the function defined in the complex torus surface T_m by $F(z) = nz$ for some $n \in \mathbb{N}$ and $\wp \in H(\mathbb{C}/\Lambda, \mathbb{C}_\infty)$ is such that $\wp(-z) = \wp(z)$ ($\wp$ is unique up to conformal homeomorphisms of $\mathbb{C}_\infty$ onto $\mathbb{C}_\infty$, given by the Weierstrass $\wp$-function). For each $(F, \wp)$, there is a unique function $f \in H(\mathbb{C}_\infty, \mathbb{C}_\infty)$ with this property, $J(f) = \mathbb{C}_\infty$, and f has an invariant line field in[13] $\mathbb{C}_\infty$.

5.4 Instability of Rational Functions with a Herman Ring

If f is a complex rational function with a fixed Herman ring U, i.e., $f(U) = U$, with an invariant Jordan curve bounding a convex region, it is easy to prove that f is not structurally stable.[14] In this case, there is a conformal homeomorphism of an annulus $A(r, R)$ onto U such that the image of each circle with radius $r \in]r, R[$ is a Jordan curve invariant under f, and for every Jordan curve C in u, $f(C) \cap C \neq \emptyset$. If the region bounded by a Jordan curve C invariant under f is convex and f were structurally stable, then every complex rational function g close to f would have a fixed Herman ring $U_g \supset C$. Without loss of generality, $0 \in U_g$, and for $\varepsilon \in]0, 1[$ and $g_\varepsilon = (1-\varepsilon) f$, it is $g_\varepsilon(C) = (1-\varepsilon)\, f(C) = (1-\varepsilon)\, C$. Since the region bounded by C is connected and contains 0, $g_\varepsilon(C) \cap C = \emptyset$, contradicting the end of the first paragraph. Therefore, rational functions with a fixed Herman ring containing an invariant Jordan curve bounding a convex region are not structurally stable.

The hypothesis of existence of a Jordan curve bounding a convex set in the preceding paragraph can be suppressed by considering a quasiconformal conjugacy $h : \mathbb{C}_\infty \to \mathbb{C}_\infty$ of f with a rational function $h \circ f \circ h^{-1}$ with a large ring module $h(U)$ conformal to an annulus $A(r, R)$ with small $r < 1$ and large $R > 1$. Without loss of generality, 0 and ∞ belong to different connected components of $\mathbb{C}_\infty \backslash h(U)$, and 1 is a fixed point of a conformal homeomorphism φ from $A(r, R)$ onto $h(U)$. φ is close to the identity or to its reciprocal $\frac{1}{z}$ in $A\left(\frac{1}{2}, 2\right)$ if r is small and R is large. The region bounded by the Jordan curve $\varphi\left(\partial B_1\right)$ is convex and the argument in the preceding paragraph can be applied, giving that $h \circ f \circ h^{-1}$ is not structurally stable. By (5.4), the quasiconformally stable rational functions coincide with the structurally stable

[13] See examples in Sections 5 and 7 of Chapter 13 of *CADOVA*.

[14] This section follows parts of the article Mañé, R., On the instability of Herman rings, *Invent. Math.* **81** (1985), 459–471.

ones. Thus, complex rational functions with a fixed Herman ring are not structurally stable.

This proof does not work for cycles of Herman rings with period $m > 1$ since iterations of Jordan curves by g_ε may fail to be simple as for a fixed Herman ring. If the reader is only interested in fixed Herman rings, the rest of this chapter can be skipped.

As mentioned in the introduction to this chapter, in 1985, R. Mañé gave a proof, remarkable from several points of view, that complex rational functions with cycles of Herman rings are not structurally stable, by proving that they are not infinitesimally stable.

We call **variation** of a complex rational function $f:\mathbb{C}_\infty\to\mathbb{C}_\infty$ any holomorphic family of rational functions $\mathcal{G}=\{g_\lambda\}_{\lambda\in B_1}$ with $g_0=f$, and **infinitesimal variation associated with the variation** $\mathcal{G}$ **of** f to $\mathcal{I}_{\mathcal{G}}(z)=\frac{\partial[g_\lambda(z)]}{\partial\lambda}_{|\lambda=0}$ for $z\in\mathbb{C}_\infty$, and say that f is **infinitesimally stable** if for every infinitesimal variation $\mathcal{I}_{\mathcal{G}}$ associated with a variation $\mathcal{G}$ of f there is a continuous vector field $V_{\mathcal{G}}:\mathbb{C}_\infty\to\mathbb{C}_\infty$ such that

$$f'(z)\,V_{\mathcal{G}}(z)-V_{\mathcal{G}}\big(f(z)\big)=\mathcal{I}_{\mathcal{G}}(z)\,.$$

If $\mathcal{F}=\{f_\tau\}_{\tau\in T}$ is a holomorphic family of rational functions, the set of $\tau\in T$ such that f_τ is infinitesimally stable is denoted $\boldsymbol{T^{inf}}(\mathcal{F})$.

(5.9) *If $\mathcal{F}=\{f_\tau\}_{\tau\in T}$ is a holomorphic family of rational functions of degree ≥ 2, then $T^{top}(\mathcal{F})\subset T^{inf}(\mathcal{F})$.*

Proof If $f\in T^{top}(\mathcal{F})$ and $\mathcal{G}=\{g_\lambda\}_{\lambda\in B_1}$ is a variation of f, there is a holomorphic motion $\{\varphi_\lambda\}_{\lambda\in B_1}$ of $\mathbb{C}_\infty$ such that $\varphi_0=1_{\mathbb{C}_\infty}$ and $\varphi_\lambda\big(f(z)\big)=g_\lambda\big(\varphi_\lambda(z)\big)$ for $\lambda\in B_1$, $z\in\mathbb{C}_\infty$. With $V_{\mathcal{G}}(z)=-\frac{\partial[\varphi_\lambda(z)]}{\partial\lambda}_{|\lambda=0}$,

$$-V_{\mathcal{G}}\big(f(z)\big)=\frac{\partial[g_\lambda(z)]}{\partial\lambda}_{|\lambda=0}-f'(z)\,V_{\mathcal{G}}(z)\,.$$

Therefore, $f'(z)\,V_{\mathcal{G}}(z)-V_{\mathcal{G}}\big(f(z)\big)=\mathcal{I}_{\mathcal{G}}(z)$ and, as by (4.1) $V_{\mathcal{G}}$ is a continuous vector field in $\mathbb{C}_\infty$, f is infinitesimally stable. □

The general idea of the R.Mañé proof is to transform the rational function f with a Herman ring cycle by a quasiconformal conjugacy giving a function with one of the rings of the Herman ring cycle where the function is close to a rotation and contains an invariant Jordan curve C close to the circle ∂B_1, and to use this Jordan curve to prove that the new function is not infinitesimally stable, and consequently also not structurally stable, what, by (5.4), implies that f is not structurally stable.

(5.10) *If* $f \in H(\mathbb{C}_\infty, \mathbb{C}_\infty)$ *has a Herman ring cycle with period* m, $U_j = f^j(U)$ *for* $j \in \{0, \ldots, m\}$, $C = \cup_{j=0}^m U_j$ *for* $z \in \mathbb{C}_\infty$, *and* $j \in \{0, \ldots, m\}$, *with* $o_j(z) = 1$ *or* $= -1$ *according to* $f_{|U_j} : U_j \to U_{j+1}$ *preserves or reverses orientation of closed invariant curves in* U_j, $\nu_k(z) = \prod_{j=k}^{m-1} o_j(z)$, *then there is* $z \in \mathbb{C}_\infty \setminus C$ *with* $\sum_{k=0}^{m-1} \nu_k(z) \neq 0$.

Proof Let $z, w \in \mathbb{C}_\infty \setminus C$ be points of connected components other than $\mathbb{C}_\infty \setminus U_k$ and in the same connected component of the other $\mathbb{C}_\infty \setminus U_j$ with $j \neq k$. If for $j \in \{0, \ldots, m\}$, $\gamma_j \subset U_j$ are invariant Jordan curves under f with $\mathrm{Ind}_{\gamma_j}(z) = -1$, then $\mathrm{Ind}_{\gamma_j}(w) = -1$ for $j \neq k$ and $\mathrm{Ind}_{\gamma_k}(w) = 1$. Thus, for $k \in \{1, \ldots, m-1\}$,

$$o_k(z) = -\,o_k w\,, \quad o_{k-1}(z) = -o_k w\,,$$
$$o_j(z) = -o_j(w)\,, \qquad j \in \{0, \ldots, m-1\},\ j \neq k,\ j \neq k-1\,;$$

and $\nu_k(z) = -\nu_k(w)$, $\nu_j(z) = \nu_j(w)$, $j \neq k$, $j \neq k-1$. If $\sum_{k=0}^{m-1} \nu_k(z) = 0$

$$\sum_{k=0}^{m-1} \nu_k(w) = \nu_0(w) + \nu_1(w) + \sum_{k=2}^{m-1} \nu_k(w) = \nu_0(w) - \nu_1(w) + \sum_{k=2}^{m-1} \nu_k(w)$$
$$= -2\nu_1(w) + \sum_{k=0}^{m-1} \nu_k(w) = -2\nu_1(w) \neq 0\,.$$

□

(5.11) *If* $\mathcal{F} = \{f_\tau\}_{\tau \in T}$ *is a holomorphic family of rational functions of degree* ≥ 2, *then the set of* $\tau \in T$ *such that* f_τ *does not have a Herman ring cycle is open and dense in* T. *If* f *is a rational function with a Herman ring, it is not structurally stable.*

Proof Let $\tau \in T$ be such that f_τ has a Herman ring cycle with period m, $U_j = f^j(U)$ for $j \in \{0, \ldots, m\}$, $C = \cup_{j=0}^m U_j$. By (5.10), there is $z \in \mathbb{C}_\infty \setminus C$ such that $\sum_{k=0}^{m-1} \nu_k(z) \neq 0$, with ν_k as in (5.10). Without loss of generality, $z = \infty$. Let $\varphi : A(r, R) \to U$ be a conformal homeomorphism mapping circles with positive orientation relative to 0 to invariant Jordan curves with positive orientation relative to $\varphi(0)$. Without loss of generality, $1 \in U$ and $\varphi(1) = 1$.

Let $\{Q_n\}$ be a sequence of quasiconformal mappings of $A\big(\frac{r}{n}, Rn\big)$ onto $A(r, R)$ keeping 1 fixed and commuting with a rotation by an angle θ around 0. Consider the sequence $\{\kappa_n\} \subset L^\infty(\mathbb{C}_\infty)$ with $\|\kappa_n\|_{L^\infty} < 1$ such that the field of ellipses associated with the Beltrami equation with coefficient κ_n restricted to U is mapped by $(\varphi\, Q_n^{-1})'$ to a field of circles and is extended to $\cup_{j=0}^\infty f^{-j}(U)$ by the dynamics defined by iterations of f, i.e., the field of ellipses associated with the equation of Beltrami coefficient κ_n restricted to each $f^j(U)$ is transformed by f^j in the restriction of these fields to U, with $\kappa_n = 0$ in $\mathbb{C}_\infty \backslash \cup_{j=0}^{\infty} f^{-j}(U)$. Since Q_n commutes with the rotation of angle θ around 0, κ_n is invariant under f, i.e., the field of ellipses associated with the Beltrami equation with coefficient κ_n is invariant under f'. By the measurable Riemann mapping theorem (2.9), there is a unique quasiconformal mapping h_n solution of the Beltrami equation with coefficient κ_n and fixed points $0, 1, \infty$. $g_n = h_n \circ f \circ (h_n)^{-1}$ satisfies the Beltrami equation with coefficient 0, and $g_n \in H(\mathbb{C}_\infty, \mathbb{C}_\infty)$. By (5.9), to complete the proof, it suffices to show g_n is not infinitesimally stable for large $n \in \mathbb{N}$.

$(g_n)^j\big(h_n(U)\big)$ for $j \in \{0, \ldots, m\}$ is a Herman ring cycle of g_n with period m and rotation number[15] θ. Since ∞ is a fixed point of g_n, the Herman rings $(g_n)^j\big(h_n(U)\big)$ are subsets of $\mathbb{C}$. Let $\mathcal{P}_n = \{p_{n,\lambda}\}_{\lambda \in B_1}$ be a variation of g_n and $\mathcal{I}_{\mathcal{P}_n}(z) = \frac{\partial[p_{n,\lambda}(z)]}{\partial\lambda}_{|\lambda=0}$ the associated infinitesimal variation. The condition for g_n to be infinitesimally stable is the existence of a continuous vector field $V_{\mathcal{P}_n} : \mathbb{C}_\infty \to \mathbb{C}_\infty$ such that $(g_n)'(z)\, V_{\mathcal{P}_n}(z) - V_{\mathcal{P}_n}\big(g_n(z)\big) = \mathcal{I}_{\mathcal{P}_n}(z)$. This equation is considered in the subset C_n of $\cup_{j=0}^{m-1} (g_n)^j\big(h_n(U)\big)$ where g_n has an inverse and the vector field in C_n is defined by $\widetilde{\mathcal{I}}_{\mathcal{P}_n} = \mathcal{I}_{\mathcal{P}_n} \circ [(g_n)_{|C_n}]^{-1}$. Hence, the equation in C_n can be written

$$(g_n)'\big([(g_n)_{|C_n}]^{-1}(z)\big)\, V_{\mathcal{P}_n}\big([(g_n)_{|C_n}]^{-1}(z)\big) - V_{\mathcal{P}_n}(z) = \widetilde{\mathcal{I}}_{\mathcal{P}_n}(z)\,.$$

With the chain rule and omitting the explicit restriction to C_n to simplify the notation, for $z \in C_n$,

$$\big[(g_n)^m\big]'\,\big([g_n]^{-m}(z)\big)\, V_{\mathcal{P}_n}\big([g_n]^{-m}(z)\big) - V_{\mathcal{P}_n}(z) = \sum_{j=0}^{m-1} [(g_n)^j]'\big([g_n]^{-m}(z)\big)\, \widetilde{\mathcal{I}}_{\mathcal{P}_n}(z)\,.$$

The function $\psi_n = h_n \circ \varphi \circ Q_n$ is a bijection of $A\big(\frac{r}{n}, nR\big)$ onto $h_n(U)$, and it is holomorphic as it satisfies the Beltrami equation with coefficient 0. Since, by the chain rule,

$$\begin{aligned}
&\big[(\psi_n)^{-1} \circ (g_n)^m \circ \psi_n\big]'\big([(\psi_n)^{-1} \circ (g_n)^{-m} \circ \psi_n](z)\big) \\
&\quad = \big[(\psi_n)^{-1}\big]'\big([(g_n)^m \circ \psi_n \circ (\psi_n)^{-1} \circ (g_n)^{-m} \circ \psi_n](z)\big) \\
&\quad\quad \big[(g_n)^m\big]'\big([\psi_n \circ (\psi_n)^{-1} \circ (g_n)^{-m} \circ \psi_n](z)\big)\, \big[\psi_n\big]'\big([(\psi_n)^{-1} \circ (g_n)^{-m} \circ \psi_n](z)\big) \\
&\quad = \big[(\psi_n)^{-1}\big]'\big(\psi_n(z)\big)\, \big[(g_n)^m\big]'\big([(g_n)^{-m} \circ \psi_n](z)\big)\, \frac{1}{\big[(\psi_n)^{-1}\big]'\big([(g_n)^{-m} \circ \psi_n](z)\big)}\,,
\end{aligned}$$

multiplying by $[(\psi_n)^{-1}]'\big(\psi_n(z)\big)$ the equality at the end of the second paragraph above, evaluated at $\psi_n(z)$ instead of z, with the functions defined in $A\big(\frac{r}{n}, nR\big)$ by $\widetilde{V}_{\mathcal{P}_n} = \big([(\psi_n)^{-1}]'\, V_{\mathcal{P}_n}\big) \circ \psi_n$ and $\Phi_n^{(j)} = (g_n)^{-j} \psi_n$ (note that $\Phi_n^{(j)}$ is a conformal

---

[15] See Section 8 of Chapter 13 of *CADOVA*.

homeomorphism of $A\big(\frac{r}{n}, nR\big)$ onto $(g_n)^{-j}\big(h_n(U)\big)$, and for $z \in A\big(\frac{r}{n}, nR\big)$, it is $(\psi_n)^{-1}\circ(g_n)^m\circ\psi_n(z)=e^{i\theta}z$,

$$e^{i\theta}\,\widetilde{V}_{\mathcal{P}_n}(e^{-i\theta}z) - \widetilde{V}_{\mathcal{P}_n}(z) = \sum_{j=0}^{m-1}[(\Phi_n^{(j)})^{-1}]'(\Phi_n^{(j)}(z))\,\widetilde{\mathcal{I}}_{\mathcal{P}_n}\big(\Phi_n^{(j)}(z)\big), \quad z\in A\big(\tfrac{r}{n}, nR\big).$$

Since for a continuous $H:\partial B_1\to\mathbb{C}$, with path $\gamma:[0,2\pi]\to B_1$ such that $\gamma(t)=e^{it}$ and a simple change of variables of integration gives

$$\int_\gamma \frac{e^{i\theta}}{z^2}H(e^{-i\theta}z)\,dz=\int_\gamma \frac{1}{z^2}H(z)\,dz,$$

applying this formula with $H=\widetilde{V}_{\mathcal{P}_n}$,

$$\sum_{j=0}^{m-1}\int_\gamma \frac{1}{z^2}[(\Phi_n^{(j)})^{-1}]'(\Phi_n^{(j)}(z))\,\widetilde{\mathcal{I}}_{\mathcal{P}_n}\big(\Phi_n^{(j)}(z)\big)\,dz = 0\,.$$

Therefore, for f to be infinitesimally stable, this equality must hold. Now, it suffices to prove that this is impossible.

For every $j\in\{0,\ldots,m-1\}$ and $n\in\mathbb{N}$, there are constants $a_n^{(j)}$, $b_n^{(j)}\in\mathbb{C}$ and functions $F_n^{(j)}(z)$ equal to z or $\frac{1}{z}$ with $\varphi_n^{(j)} = a_n^{(j)}F_n^{(j)}(\Phi_n^{(j)}+b_n^{(j)})$ satisfy the hypothesis of the result following this proof. Therefore, $\varphi_n^{(j)}\to 1_{A(r/n,nR)}$ when $n\to+\infty$ uniformly in compact subsets of $\mathbb{C}\backslash\{0\}$. As $\Phi_n^{(j)} = \frac{1}{a_n^{(j)}}F_n^{(j)}\varphi_n^{(j)}+b_n^{(j)}$, and in a neighborhood V of ∂B_1,

$$\Big[(\Phi_n^{(j)})^{-1}\Big]'\big(\Phi_n^{(j)}(z)\big) = c_n^{(j)}a_n^{(j)}G_n^{(j)}(z)\,z^{\rho_n^{(j)}},$$

with $\big(c_n^{(j)},\rho_n^{(j)}\big)$, resp., $(1,0)$ or $(-1,2)$ according to $F_n^{(j)}(z)$ be z or $\frac{1}{z}$, and $G_n^{(j)}\to 1$ uniformly in V when $n\to+\infty$, and

$$\int_\gamma \frac{1}{z^2}[(\Phi_n^{(j)})^{-1}]'(\Phi_n^{(j)}(z))\,\widetilde{\mathcal{I}}_{\mathcal{P}_n}\big(\Phi_n^{(j)}(z)\big)\,dz$$
$$=\int_\gamma \frac{1}{z^2}\,c_n^{(j)}a_n^{(j)}G_n^{(j)}(z)\,z^{\rho_n^{(j)}}\,\widetilde{\mathcal{I}}_{\mathcal{P}_n}\Big(\frac{1}{a_n^{(j)}}F_n^{(j)}(z)\varphi_n^{(j)}(z)+b_n^{(j)}\Big)dz\,.$$

With the variation $\mathcal{V}_n=\{v_{n,\lambda}\}_{\lambda\in B_1}$ of g_n with $v_{n,\lambda}(z)=(1+\lambda)g_n(z)$, whose associated infinitesimal variation is $\mathcal{I}_{\mathcal{V}_n} = \frac{\partial v_{n,\lambda}(z)}{\lambda}_{|\lambda=0} = g_n(z)$ for $z\in\mathbb{C}_\infty$, for which $\widetilde{\mathcal{I}}_{\mathcal{V}_n}=\mathcal{I}_{\mathcal{V}_n}\circ[(g_n)_{|C_n}]^{-1}=1_{C_n}$, the term in the right-hand side of the preceding equality is

$$\int_\gamma \frac{1}{z^2}\,c_n^{(j)}G_n^{(j)}(z)\,z^{\rho_n^{(j)}}\big(F_n^{(j)}(z)\varphi_n^{(j)}(z)+a_n^{(j)}b_n^{(j)}\big)\,dz\,.$$

Since $\varphi_n^{(j)}\to 1_{A(r/n,nR)}$ when $n\to+\infty$ uniformly in compact subsets of $\mathbb{C}\setminus\{0\}$, $G_n^{(j)}\to 1$ uniformly in a neighborhood V of ∂B_1 when $n\to+\infty$, and for big n, the preceding integral can be approximated by

$$c_n^{(j)} \int_\gamma \frac{1}{z^2}\, z^{\rho_n^{(j)}}\, F_n^{(j)}(z)\, dz + c_n^{(j)} a_n^{(j)} b_n^{(j)} \int_\gamma \frac{1}{z^2}\, z^{\rho_n^{(j)}}\, dz = c_n^{(j)} \int_\gamma \frac{1}{z}\, dz = \pm i 2\pi\,,$$

where the last equality is because $z^{\rho_n^{(j)}} F_n^{(j)}(z) = z$, $\rho_n^{(j)} \in \{0, 2\}$, and $c_n^{(j)} = \pm 1$, positive or negative according to $F_n^{(j)}(z)$ be z or $\frac{1}{z}$.

Since $F_n^{(j)}(z)$ is z or $\frac{1}{z}$ according to $(g_n)^j : (g_n)^{-j}\big(h_n(U)\big) \to h_n(U)$ to preserve or invert the orientation of the invariant Jordan curves in the Herman rings, in the sense of (5.10), then $F_n^{(j)}(z)$ is z or $\frac{1}{z}$ according to $\nu_j(\infty)$ defined in that statement is 1 or -1. Consequently, the sum in the left-hand side of the last equality of two paragraphs above tends to $i2\pi \sum_{j=0}^{m-1} \nu_j(\infty) \neq 0$ when $n \to +\infty$, invalidating the possibility of being 0 for n large. This ends the proof that f is not infinitesimally stable and, therefore, not structurally stable. □

Next, we prove the auxiliary result used in the final paragraphs of the preceding proof.

(5.12) *If* $\{r_n\}, \{R_n\} \subset [0, +\infty[$ *are sequences strictly, resp., decreasing to* 0, *increasing to* $+\infty$ *with* $r_1 < R_1$, *and* $\{\varphi_n\} \subset H\big(A(r_n, R_n), \mathbb{C}_\infty\big)$ *is a sequence of injective functions with* $\varphi_n(1) = 1$ *for* $n \in \mathbb{N}$, 0 *belongs to the bounded connected component of* $\mathbb{C} \backslash \varphi_n\big(A(r_n, R_n)\big)$ *for* $n \in \mathbb{N}$, $\mathrm{Ind}_{\gamma_n}(0) = 1$ *with* $\gamma_n : [0, 2\pi] \to \mathbb{C}$ *such that* $\gamma_n(\theta) = \varphi_n(e^{i\theta})$, *then* $\{\varphi_n\}$ *converges to the identity uniformly in compact subsets of* $\mathbb{C} \backslash \{0\}$.

Proof Let $r, R \in]0, +\infty[$ be such that $r < R$ and $\mathcal{S}_{0,\infty}$ be a family of injections $\varphi \in H\big(A(r, R), \mathbb{C}_\infty\big)$ separating $0, \infty$, in the sense that these points belong to different connected components of $\varphi\big(A(r, R)\big)$.

For every sequence $\{\psi_n\} \in \mathcal{S}_{0,\infty}$, if $r_1, R_1 \in]0, +\infty[$ are such that $r_1 < r < R < R_1$, then $\psi_n\big(A(r, R)\big)$ does not contain $A(r_1, R_1)$, as this would imply that $(\psi_n)^{-1}_{|A(r,R)}$ would be a univalent function from $A(r_1, R_1)$ to $A(r, R)$ and the bounded connected component of $\mathbb{C} \backslash (\psi_n)^{-1}\big(A(r_1, R_1)\big)$ would contain B_r, what is impossible. So, for $n \in \mathbb{N}$, there is $\alpha_n \in \mathbb{C}$ with $|\alpha_n| \in \{r_1, R_1\}$ and $\alpha_n \notin \psi_n\big(A(r, R)\big)$. The functions $\frac{\psi_n}{\alpha_n} : A(r, R) \to \mathbb{C}$, $n \in \mathbb{N}$, do not assume 1 or 0. By the Montel-Carathéodory theorem,[16] $\left\{\frac{\psi_n}{\alpha_n}\right\}$ is a normal family, and it converges uniformly in compact subsets of $A(r, R)$ to some $\psi \in H\big(A(r, R), \mathbb{C}\big)$ or to ∞. In the first case, as $\{\alpha_n\} \subset \partial B_{r_1} \cup \partial B_{R_1}$ and this set is compact, by the Bolzano-Weierstrass theorem, there is a subsequence $\{\alpha_{n_j}\}$ convergent to a point α of this set and $\{\psi_{n_j}\}$ converges to $\alpha\psi$. In the second case, as $|\alpha_n| \geq r_1$, $\{\psi_{n_j}\}$ converges to ∞ in compact subsets of $A(r, R)$. It follows that $\mathcal{S}_{0,\infty}$ is a normal family.

If $\{\varphi_n\}$ would not converge uniformly to the identity in compact subsets of $\mathbb{C} \backslash \{0\}$, there would exist a strictly increasing sequence $\{m_j\} \subset \mathbb{N}$ such that

[16] See Section 6 of Chapter 11 of *CADOVA*.

$\{\varphi_{n|A(r_{m_j},R_{m_j})}\}_{n\geq m_j,\, n\in\mathbb{N}}$ would not converge uniformly in compact subsets of $A(r_{m_j}, R_{m_j})$ to the identity. As $\{\varphi_n\}\in\mathcal{S}_{0,\infty}$, by the preceding paragraph, it is a normal family and, by the Cantor diagonal argument, it has a subsequence $\{\varphi_{n_j}\}$ converging uniformly in compact subsets of $A(r_{m_k}, R_{m_k})$ to $\varphi\in H\big(\mathbb{C}\backslash\{0\},\mathbb{C}\big)$. As $\varphi_n(1)=1$ for $n\in\mathbb{N}$, also $\varphi(1)=1$, and as the functions φ_n are injective, by the Hurwitz injection theorem,[17] φ is injective or constant. As $\operatorname{Ind}_{\gamma_n}(0)=1$ for $\gamma_n:[0,2\pi]\to\mathbb{C}$ with $\gamma_n(\theta)=\varphi_n(e^{i\theta})$, also $\operatorname{Ind}_{\gamma}(0)=1$ for $\gamma:[0,2\pi]\to\mathbb{C}$ with $\gamma(\theta)=\varphi(e^{i\theta})$, and φ is not constant; thus, φ is injective. Hence, φ cannot have an essential singularity at 0 nor at ∞. Therefore, φ is an injective rational function, and it can only be of one of the forms $z-a$ or $\frac{1}{z-b}$ or $\frac{z-a}{z-b}$, with $a, b\in\mathbb{C}$. As $\varphi(1)=1$, it cannot be of the last form as it would be identically 1 contradicting injectivity, and for each one of the other forms, it would be, resp., $a=0$, $b=0$. This only leaves the possibilities $\varphi(z)=z$ or $\varphi(z)=\frac{1}{z}$. As $\operatorname{Ind}_{\gamma}(0)=1$, it has to be the first case, and φ is the identity in $\mathbb{C}\backslash\{0\}$. □

[17] See Section 5 of Chapter 6 of *CADOVA*.

Chapter 6
Quasiconformal Surgery and Complex Dynamics

6.1 Introduction

In Autumn 1981, D. Sullivan and also A. Douady and J. Hubbard introduced quasiconformal surgery[1] to study properties of the dynamics of iterations of holomorphic functions with perturbations mediated by quasiconformal mappings, allowing a flexibility impossible with conformal homeomorphisms. The immediate illustration of this idea was the proof that each Mandelbrot set hyperbolic component can be parameterized by the multiplier of the corresponding attracting periodic orbit.[2]

Other applications of quasiconformal surgery followed. The most important was the proof by D. Sullivan in 1985 of the P. Fatou Conjecture, unanswered for about 60 years, that the Fatou components of rational functions of degree ≥ 2 are nonwandering,[3] what had been proven in general false for integer functions in 1976 by I. Baker.[4] Another important application, also in 1985, was the proof by R. Mañé of structural instability of rational functions with Herman rings presented in the last section of the preceding chapter.

These applications were preceded by the anticipation for two decades of the idea of quasiconformal surgery by L. Bers, who in 1960 observed,[5] when dealing with Teichmüller Spaces, the usefulness of extensions of Quasiconformal mappings in $\mathbb{H}$ to the whole complex plane in such a way that they are conformal in $-\mathbb{H}$, as seen in Chap. 4 and strengthened in Chap. 9.

[1] D. Sullivan in a seminar at the *IHES – Institut des Hautes Études Scientifiques*, France, and A. Douady and J. Hubbard in their seminar at *Université de Paris Sud - Orsay*.

[2] Explained in A.Douady, Systèmes dynamiques holomorphes, *Astérisque*, **105–106** (1983), 39–63.

[3] Sullivan, D., Quasiconformal Homeomorphisms and Dynamics, I. Solution of the Fatou-Julia problem on wandering domains, *Annals of Mathematics, Second Series,* **122**, 401–418.

[4] Baker, Irvine (1932–2001).

[5] Bers, L., Simultaneous uniformization. *Bull. Amer. Math. Soc.* **66** (1960), 94–97.

L. T. Magalhães, *Quasiconformal Mappings in the Plane and Complex Dynamics*,
https://doi.org/10.1007/978-3-031-80115-0_6

In 1987, M. Shishikura presented in an article[6] the results of his 1985 master thesis on the application of quasiconformal surgery to:

(1) transform all neutral periodic orbits of a rational function to attracting;
(2) decompose a rational function with Herman rings in functions with Siegel disks;
(3) transform a Siegel disk cycle to a Herman ring cycle;
(4) obtain optimal upper bounds on the number of stable periodic orbits of a rational function of degree d, namely $2(d-1)$, and on the number of Herman rings, namely $d-2$.

As a consequence of (3) he obtained that for each $p \in \mathbb{N}$ there are rational functions of degree 3 with Herman ring of order p, giving an alternative proof of the existence of Herman rings to that of M. Herman in 1979 based on the 1961 Arnold theorem on circle homeomorphisms.[7]

Besides the applications of quasiconformal surgery to the dynamics of complex rational functions mentioned above, this chapter includes:

- modifying the multiplier of an attracting periodic orbit;
- modifying superattracting to attracting periodic orbits;
- modifying the module of a rotation ring;
- relating polynomial type rational functions with polynomials;
- sewing of a *continuum* in a Julia set;
- characterizing components of Julia sets.

An introductory example of quasiconformal surgery was given at end of Chap. 3: the proof of existence of Leau-Fatou conjugacy in a neighborhood of a parabolic fixed point of a holomorphic function, using the Mori theorem and the measurable Riemann mapping theorem.

6.2 Hyperbolic Components of Mandelbrot set

As mentioned in the introduction to this chapter, quasiconformal surgery was initiated almost simultaneously by D. Sullivan and by A. Douady together with J. Hubbard, illustrated by the proof of the following result on parametrization of the hyperbolic components of the **Mandelbrot set** (Fig. 6.1[8]), which is the set of $c \in \mathbb{C}$ such that the critical point 0 of $P_c(z) = z^2 + c$ belongs to the **filled Julia set** $K(P_c)$ of P_c, i.e., of points of $\mathbb{C}$ with bounded orbits.[9]

---

[6] Shishikura, M., On the quasiconformal surgery of rational functions, *Annales scientifiques de l'É.N.S. 4e series*, **20** (1987), 1–29.

[7] See Section 11 of chapter 13 of *CADOVA*.

[8] Author and date of base figure: Claude Heiland-Allen (17.02.2016), under License *Creative Commons CC BY-SA 4.0.* https://es.m.wikipedia.org/wiki/Archivo:Wakes_near_the_period_1_continent_in_the_Mandelbrot_set.png; some outer rays were added by symmetry with the real axis and replaced symbolic dynamics annotations of outer radii $T_{p/q}$ by the fractions p/q.

[9] See Section 15 of Chapter 13 of *CADOVA*. We follow here the book by L. Carleson and T.W. Gamelin listed in the final bibliography.

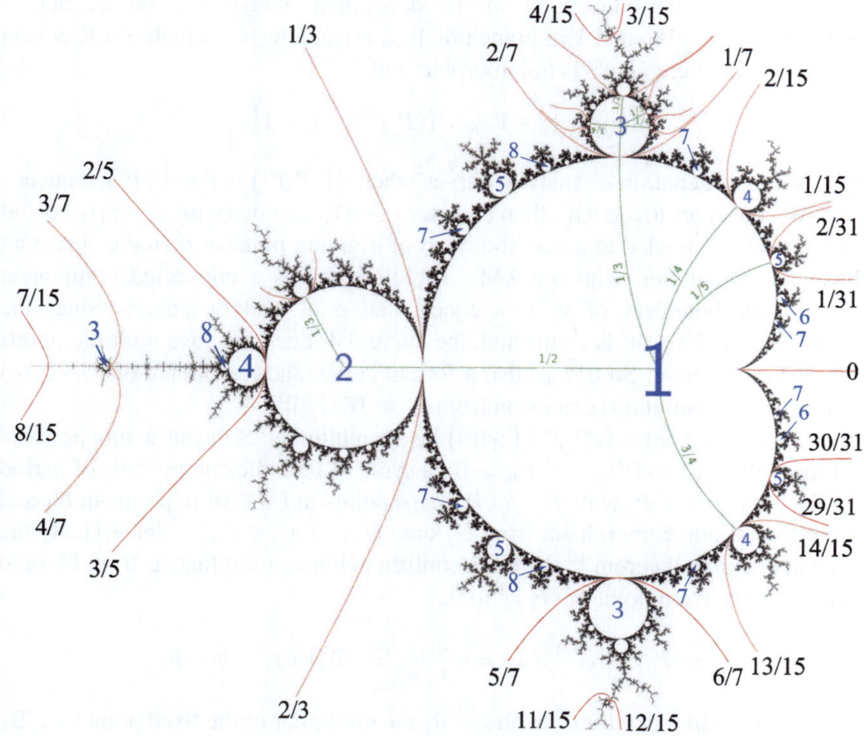

Fig. 6.1 Boundary of the Mandelbrot set $\mathcal{M}$ with numbers in some hyperbolic components of $\mathcal{M}$ indicating the minimal period 1, 2, 3, 4, 5, 6, 7, 8 of the corresponding attracting periodic orbits (there are smaller hyperbolic components corresponding to attracting periodic orbits with those minimal periods except 1 or 2), and with some outer radii that land at points $c \in \partial\mathcal{M}$ for which P_c has a parabolic fixed point, and some interior radii of the main cardioid

> **(6.1)** *In each hyperbolic component Ω of the Mandelbrot set $\mathcal{M}$ the function $\Lambda : \Omega \to B_1$ assigning to each $\omega \in \Omega$ the multiplier of the attracting periodic orbit of P_ω is a conformal homeomorphism with a continuous extension to* $\operatorname{cl}\Omega$, *and a homeomorphism onto* $\operatorname{cl} B_1$.

Proof A hyperbolic component W of $\mathcal{M}$ corresponds to attracting periodic orbits with period m. For $c \in W$ let $Q_{c,m}(z) = (P_c)^m - 1_{\mathbb{C}}$ and $V_{c,m}$ be the set of zeros of $Q_{c,m}$, which is a 1-dimensional complex analytic manifold in $\mathbb{C}^2$ with elements (z, c), z a periodic point of P_c with period dividing m. As $Q_{c,m}$ is a monic polynomial in both z and c and if one of $|z|$, $|c|$ with $(z, c) \in V_{c,m}$ tends to ∞, also the other tends to ∞, $(V_{c,m}, P)$ with $P(z, c) = c$ is a covering of W with 2^m sheets.

If $z_1(c)$ is a periodic point of P_c depending analytically on c, denote $\widetilde{W}=\{(z_1(c), c): c \in W\}$ and V_0 a branch of $V_{c,m}$ containing $\widetilde{W}$, which is a Riemann surface over $\mathbb{C}$ where $(P_c)^m$ is holomorphic. Let

$$\widetilde{V}_{c,m} = \{z \in V_{c,m} : |((P_c)^m)'(z)| < 1\}.$$

$\partial \widetilde{V}_{c,m}$ is a concatenation of analytic curves where $|((P_c)^m)'(z)| = 1$. If a sequence $\{c_n\} \subset W$ converges to $a \in \partial W$, then $(z_1(c), c) \to \partial V_{c,m}$, otherwise $c \mapsto z_1(c)$ could be analytically extended to a neighborhood of a giving point of periodic attracting orbits in contradiction with $a \in \partial \mathcal{M}$. Therefore, $\widetilde{W}$ is a connected component of $V_{c,m}$. The boundary of W is a concatenation of analytic curves. Since the complement of $W \cup \partial W$ is connected, the curve ∂W does not have multiple points and is a Jordan curve. So $\partial \widetilde{W}$ is also a Jordan curve, and the branch of $c \mapsto z_1(c)$ considered has continuous extension from W to $W \cup \partial W$.

For $c \in W$, let $\Lambda(c) = ((P_c)^m)'(z_1(c))$ be the multiplier of the attracting periodic orbit through $z_1(c)$ and $U_0, \ldots, U_m = U_0$ a cycle of periodic components of period m of the Fatou set of P_c with $0 \in U_0$. P_c maps points in $U_0 \setminus \{0\}$ to points in U_1 and is a conformal homeomorphism from U_j onto U_{j+1} for $j \in \{1, \ldots, m-1\}$. By the Riemann mapping theorem,[10] there is a conformal homeomorphism φ from U_0 onto B_1 mapping the fixed point of $(P_c)^m$ to 0.

$$(\varphi \circ (P_c)^m \circ \varphi^{-1})(w) = w \tfrac{w+\lambda}{1+\bar{\lambda} c} \stackrel{def}{=} \mathfrak{B}_\lambda(w), \quad w \in B_1,$$

is called a **Blaschke product**[11] with $\lambda \in B_1$ the multiplier of the fixed point 0 of $\mathfrak{B}_\lambda$ and $\Lambda(c) = \lambda$.

For the inverse relation, for each $\varepsilon \in]0, 1[$, take $\lambda \in B_{1-\varepsilon}$ and combine the dynamics of P_c outside the U_j with that of $\mathfrak{B}_\lambda$ in the U_j in order to obtain the value of the parameter $c(\lambda)$ such that $\Lambda(c(\lambda)) = \lambda$ and $\lambda \mapsto c(\lambda)$ is continuous. Let $r \in]1-\varepsilon, 1[$ be so close to 1 that for every $\lambda \in B_{1-\varepsilon}$, the set $(\mathfrak{B}_\lambda)^{-1}(\partial B_r)$ is a Jordan curve in the annulus $B_1 P \setminus \mathrm{cl}\, B_r$ whose image by $\mathfrak{B}_\lambda$ goes twice around ∂B_r. $\varphi^{-1}(B_r) \subset A_0 \stackrel{def}{=} U_0 \cap (P_c)^{-m}(\varphi^{-1}(B_r))$ and $(A_0, (P_c)^m)$ is a covering of $\varphi^{-1}(B_r)$. With $\varphi_\lambda : \partial A_0 \to (\mathfrak{B}_\lambda)^{-1}(\partial B_r)$ such that $\mathfrak{B}_\lambda \circ \varphi_\lambda = \varphi \circ (P_c)^m$, choosing a branch so that $\lambda \mapsto \varphi_\lambda$ is continuous, φ_λ extends to a diffeomorphism of A_0 onto $(\mathfrak{B}_\lambda)^{-1}(\partial B_r)$ such that $\varphi_\lambda = \varphi$ in $\varphi^{-1}(B_r)$ and $\lambda \mapsto \varphi_\lambda$ is C^1. Consider $A_j \subset U_j$ for $j \in \{1, \ldots, m-1\}$, so that P_c is a conformal homeomorphism of A_j onto A_{j+1} with $A_m = A_0$, and define $g_\lambda = P_c$ outside A_0 and $g_\lambda{}^m = \varphi^{-1} \circ \mathfrak{B}_\lambda \circ \varphi_\lambda$ in A_0; then, $g_\lambda = P_c^{-m+1} \circ (g_\lambda)^m$ in A_0. Consider a field of ellipses invariant under g_λ which is of circles outside of $\cup_{nj=0}^{m-1} A_j$. The positive orbits of each point under g_λ intersect the ring where this function is not a conformal homeomorphism, whereby the corresponding Beltrami coefficients κ_λ satisfy $\|\kappa_\lambda\|_{L^\infty} < 1$ and the function $\lambda \mapsto \kappa_\lambda$ is continuous. By the measurable Riemann mapping theorem (2.9), there is a solution of the Beltrami equation with Beltrami coefficient κ_λ such that

[10] See Section 5 of Chapter 10 of *CADOVA*.

[11] See Section 7 of Chapter 13 of *CADOVA*. Blaschke, Wilhelm (1885–1962).

$\psi_\lambda - 1_{\mathbb{C}_\infty} \to 0$ at ∞. $F_\lambda \stackrel{def}{=} \psi_\lambda \circ g_\lambda \circ (\psi_\lambda)^{-1}$ is holomorphic and in some neighborhood of ∞, $z \mapsto F_\lambda(z) - z^2$ is bounded; thus, $F_\lambda(z) = z^2 + c(\lambda) = F_{c(\lambda)}(z)$. As g_λ has an attracting periodic orbit with period m, also F_λ has such an orbit with equal multiplier, and $\Lambda\big(c(\lambda)\big) = \lambda$. As the solution ψ_λ of the Beltrami equation is continuous in λ, $\lambda \mapsto c(\lambda)$ is continuous.

Therefore, the function $\Lambda : \mathrm{cl}\, W \to \mathrm{cl}\, B_1$ defined as indicated in the statement is surjective and $\Lambda_{|W}$ is holomorphic with range B_1. □

It is possible to know how many hyperbolic components of $\mathcal{M}$ have an attracting periodic orbit with a given minimal period $q \in \mathbb{N}$.

(6.2) *The numbers n_q of hyperbolic components of the Mandelbrot set $\mathcal{M}$ with attracting periodic orbit of minimal period $p \in \mathbb{N}$ are the terms of the sequence recursively defined by $n_1 = 1$ and $n_p = 2^{p-1} - \sum_{\{q:\, q|p,\, q<p\}} n_q$, where $q|p$ means q divides p.*

Proof By the preceding result, each hyperbolic component C of $\mathcal{M}$ has a zero c_0 of the polynomial $F_p(c) = (P_c)^p(0)$. As the degree of this polynomial is 2^{p-1} it has 2^{p-1} zeros, counting multiplicities, and $\Lambda(c) = (P_c^q)'(0) = 2^q \prod_{j=0}^{q} \alpha_j(c)$, where q is the minimal period of the attracting periodic orbit $\{\alpha_j(c)\}_{j=0}^{q}$ in C. With $\varphi(c, z) = f_c^p(z) - z$, $\varphi(c, \alpha_0(c)) = 0$ and $\frac{\partial \varphi}{\partial z}(c_0, 0) \neq 0$, and since the zero $\alpha_0(c)$ of $z \mapsto \varphi(c_0, z)$ is simple, the zero c_0 of F_p is also simple. Therefore, all zeros of F_p are simple, F_p has 2^{p-1} zeros and $\sum_{q|p} n_q = 2^{p-1}$, $n_1 = 2^0 = 1$. □

For instance, the number of hyperbolic components of the Mandelbrot set with attracting periodic orbit of minimal period 1 to 8 is, resp., 1, 1, 3, 6, 15, 27, 62, 127.

The Generic Hyperbolicity Conjecture of Fatou in 1919 is for $P_c(z) = z^2 - c$ equivalent to the nonexistence of nonhyperbolic connected components of $\mathcal{M}$. It remains open more than a century later.

In 1985 A. Douady and J. Hubbard stated the *MLC Conjecture: $\partial\mathcal{M}$ is locally connected.* If true, it would imply the validity of the Generic Hyperbolicity Conjecture, since as $\mathcal{M}$ is bounded and simply connected, by the Riemann mapping theorem, there is a conformal homeomorphism of $\mathcal{M}$ onto B_1 which, if $\partial\mathcal{M}$ were locally connected, could be extended by continuity to $\mathrm{cl}\,\mathcal{M}$ by the result obtained in 1918 by M. Torhorst[12] of extension by continuity of conformal homeomorphisms of simply connected regions onto the disk B_1 to the boundary of the region if and only if this boundary is locally connected, but the *MLC* Conjecture remains open after almost four decades.

[12] Torhorst, Marie (1888–1989).

6.3 Sullivan Nonwandering Domains Theorem

In 1985, D. Sullivan applied quasiconformal surgery to prove the P. Fatou Conjecture, about 60 years before, that Fatou components of rational functions of degree ≥ 2 were nonwandering, a result called Sullivan nonwandering domains theorem. This result has a special significance when we take into account that in 1976 I. Baker had proved the existence of entire functions with wandering Fatou components, i.e., Fatou components U of f such that $f^n(U)$ for $n \in \mathbb{N} \cup \{0\}$ are all distinct. The Sullivan nonwandering domains theorem complements the general information about the dynamics of Fatou components of rational functions, restricting it substantially.

The resulting drastic restriction can be apprehended by observing that if a **conformal structure** is defined in a Fatou component U of f, i.e., a field of ellipses (considered the circles of the structure) varying regularly from point to point,[13] determined by the solutions of the Beltrami equation with coefficient $\kappa \in L^{\infty}(U)$ with $\|\kappa\|_{L^{\infty}} < 1$, and this conformal structure is preserved by f, it is specified in each $f^n(U)$ by successive applications of f, and if U is a non-fixed periodic or eventually periodic Fatou component, as there are different $n_1, n_2 \in \mathbb{N} \cup \{0\}$ such that $f^{n_1}(U) = f^{n_2}(U)$, the conformal structure $f^{n_2}(U)$ obtained from $f^{n_1}(U)$ by iteration of f must match the initial one, what imposes very strong restrictions to conformal structures.

D. Sullivan observed that if there would exist a nonwandering Fatou component, its successive iterations by a rational function would generate an infinite dimensional space of rational functions of the same degree, contradicting that this space is determined by a finite number of complex parameters.

(6.3) Sullivan nonwandering domains theorem:
The Fatou components of a rational function f of degree ≥ 2 in $\mathbb{C}_{\infty}$ are nonwandering: each Fatou component is fixed, periodic, or preperiodic.

Proof For a proof by absurdity, it is assumed that U is a wandering Fatou component of f, and, without loss of generality, $\cup_{n=0}^{\infty} f^n(U)$ is taken without critical points of f, what can be achieved by replacing U with $f^m(U)$ with a large m, since the number of critical points of f is finite. Also without loss of generality, ∞ is taken to belong to U, what can be achieved with a Möbius transformation.

All other Fatou components are subsets of the bounded set $\mathbb{C}_{\infty} \backslash U \subset \mathbb{C}$. Thus, $\{f^n(U)\}_{n \in \mathbb{N}}$ is a normal family. As the area of $f^n(U)$ approaches 0, the limit of

---

[13] Some authors prefer the name "quasiconformal structure" reserving "conformal structure" for fields of circles, but this terminology, although suggestive, is problematic, because every manifold with a "quasiconformal structure" has a "conformal structure." We follow here the John Milnor option in the book listed in the final bibliography.

each convergent subsequence of $\{f_n\}$ in compact subsets of U is a constant function. If γ is a simple closed path in U, the diameter of the compact set $f^n(\gamma^*)$ tends to 0, and also does the diameter of the union K_n of $f^n(\gamma^*)$ with the bounded components of $\mathbb{C}\backslash\gamma^*$. If the image by f of some point of $K_n\backslash\gamma_n^*$ would belong to the complement of K_{n+1}, as this would be an open set with boundary included in γ_{n+1}^*, it would contain the complement of K_{n+1}, what is impossible for K_n and K_{n+1} small. Thus, for large n, K_n is a subset of the Fatou set of f and, therefore, of $f^n(U)$. As γ_n is homotopic to a point in K_n, it is also homotopic to a point in $f^n(U)$. Since $f^n(U)$ does not contain critical points of f, (U_n, f) is a covering of U_{n+1}, and, therefore, γ_n is also homotopic to a point in K_n and also in U. If there would exist a wandering Fatou component U of f, for some $m\in\mathbb{N}$, $f^n\big(f^m(U)\big)$ would be simply connected[14] for $n\in\mathbb{N}$.

It is possible to define infinitely many conformal structures in the disk B_1. Consider the group G of diffeomorphisms with three points fixed in ∂B_1, e.g., ± 1 and i, and the function $E:\mathbb{R}/\mathbb{Z}\to\partial B_1$ such that $E(t)=e^{i2\pi t}$. Every $g\in G$ maps each $E(t)\in\partial B_1$ to $E\big(t+v(t)\big)$, with v belonging to the convex set of C^∞ functions $v:\mathbb{R}/\mathbb{Z}\to\mathbb{R}$ which are 0 in 3 fixed points such that $[t+v(t)]'>0$ for all $T\in\mathbb{R}/\mathbb{Z}$. Of the elements of G, only the identity can be extended to a conformal homeomorphism of cl B_1, because for every $g\in G$ there are four points of ∂B_1 that become four different real numbers with the cross ratio[15] not preserved, what is impossible for conformal homeomorphisms, but g has extension to a diffeomorphism $\widehat{g}$ of cl B_1 onto cl B_1 defined by $\widehat{g}\big(r\,E(t)\big)=r\,E\big(t+\eta(r)\,v(t)\big)$ with $\eta:[0,1]\to[0,1]$ a nondecreasing $\mathbb{C}^\infty$ function equal to 0 in $\big[0,\frac{1}{3}\big]$ and to 1 in $\big[\frac{2}{3},1\big]$.

A conformal mapping φ from U onto the disk B_1 defines a conformal structure in B_1 according to $\widetilde{g}\circ\varphi$, and it can be naturally extended to the **grand orbit** of points of U under f, i.e. to the union of orbits that intersect orbits of points of U (there may be isolated points of this grand orbit whose positive orbits have critical points of f and at these points the conformal structure is not defined, but it does not matter because they are sets with measure 0); at points outside the grand orbit of U, we use the usual conformal structure. The conformal structure obtained in this way has an associated Beltrami coefficient κ defined a.e. in $\mathbb{C}_\infty$ satisfying $\|\kappa\|_{L^\infty}<1$. By the measurable Riemann mapping theorem (2.9), the solutions of the corresponding Beltrami equation are a family of quasiconformal mappings h_g that can be normalized to have fixed points $0, 1, \infty$, and we consider a family of functions f_g so that the following diagram is commutative

[14] In D. Sullivan original proof the possibility of Fatou components not simply connected was not eliminated at the outset, what complicated the proof. The proof given here follows the presentation in the J.W. Milnor book listed in the final bibliography.

[15] A **cross ratio** of 4 points $z_1, z_2, z_3, z_4\in\mathbb{C}_\infty$ is $z_1:z_2:z_3:z_4=\frac{(z_3-z_1)(z_4-z_2)}{(z_2-z_1)(z_4-z_3)}$; it is invariant under Möbius transformations, and it is real if and only if the four points belong to a same line or circle.

$$\begin{array}{ccccccc} B_1 & \xleftarrow{\varphi} & U & \subset & \mathbb{C}_\infty & \xrightarrow{f} & \mathbb{C}_\infty \\ \downarrow{\scriptstyle \widehat{g}} & & \downarrow{\scriptstyle h_{g|U}} & & {\scriptstyle h_g}\downarrow & & \downarrow{\scriptstyle h_g} \\ B_1 & \xleftarrow{\varphi_g} & U_g & \subset & \mathbb{C}_\infty & \xrightarrow{f_g} & \mathbb{C}_\infty \end{array}$$

where the horizontal functions are holomorphic and the vertical functions are quasiconformal. Each f_g is a rational function of the same degree d of f and the function $g \mapsto f_g$ is C^1.

The rational functions of degree d in $\mathbb{C}_\infty$ form a complex manifold R_d with finite dimension N_d . On the other hand, G is an infinite dimensional complex manifold. To work in finite dimensions, consider a submanifold M_G of G of dimension $N_d + 1$ and restrict the function $g \mapsto f_g$ to M_G . Let $g_0 \in M_G$ with the characteristic of the derivative of this function $r_0 \leq N_d$ maximal. There is a neighborhood V of g_0 in M_G with image a complex submanifold M_1 of M_G of dimension r_0 . The preimage in V of a regular value of the function in M_1 is a complex manifold M_2 of dimension $N{-}1{-}r \geq 1$ whose image by the considered function is just the regular value chosen in M_1. Every nonconstant regular path in M_2 $g : [0, 1] \to M_2$ is mapped to a single point in M_1 . As $f_{g(t)} = f_{g(0)}$ for $t \in [0, 1]$, each $H_t = h_{g(t)} \circ (h_{g(0)})^{-1}$ commutes with $f_{g(0)}$. Since the periodic points of $f_{g(0)}$ remain fixed under deformations commuting with $f_{g(0)}$, the restriction of H_t to the Julia set $J_{f_{g(0)}}$ is the identity, leading to a contradiction as it is shown next.

Firstly, let ∂U be a Jordan curve. For any four points of ∂U, consider the **cross ratio with respect to** $\operatorname{cl} U$ defined by the usual cross ratio on B_1 after choosing a conformal homeomorphism of U onto B_1 and extending it by continuity to ∂B_1 . For $t \in [0, 1]$ such that $g(t) \neq g(0)$, choose four points in ∂B_1 with cross ratio not invariant under $[g(t)] \circ [g(0)]^{-1}$. H_t cannot have such four points fixed, in contradiction with this function being a conformal homeomorphism as it is a composition of conformal homeomorphisms.

If ∂U is not a Jordan curve, the Carathéodory compactification[16] $\widetilde{U}$ of U, which is the union of U with the set of its prime ends with the appropriate topology, allows extending the Riemann mapping of U onto B_1 to a homeomorphism of $\widetilde{U}$ onto $\operatorname{cl} B_1$. The cross ratio of any four points of $\partial\widetilde{U}$ relative to $\widetilde{U}$ is well defined. Each prime end is determined by a fundamental chain $\{N_j\}$ in $\widetilde{U}$ of cross section neighborhoods and if the corresponding images $h(N_j)$ by a homeomorphism of the pair $\big(\operatorname{cl} U, \partial U\big)$ are disjoint from N_j , each $(\operatorname{cl} N_j) \cup h(\operatorname{cl} N_j)$ is a region bounded by a Jordan curve. If the homeomorphism h preserves orientation and restricted to ∂U is the identity, then the image of each prime end by h is that same prime end. The same contradiction as in the previous paragraph is obtained when ∂U is not a Jordan curve with an analogous argument. □

This result completes the **Sullivan classification theorem** of periodic Fatou components in the section "Dynamics of Fatou components" of Chapter 13 of *CADOVA*.

[16] See Section 7 of Chapter 10 of *CADOVA*.

By the Sullivan nonwandering domains theorem (6.3), each bounded Fatou component of $P_c(z) = z^2 + c$ is fixed, periodic or preperiodic, and there is, at most, a cycle of bounded Fatou components. As polynomial functions do not have Herman rings, by the Sullivan classification theorem,[17] there are the following possibilities:

1. There is a cycle of attracting Fatou components: alternatively, there is an attracting fixed point and a single bounded Fatou component, or there is a cycle with more than one attracting Fatou component, and in this case, there are infinitely many of such components.
2. There is a cycle of parabolic Fatou components: alternatively, there is a parabolic fixed point with multiplier 1 and a single bounded Fatou component, or there is a cycle with more than one parabolic Fatou component, and in this case, there are infinitely many of such components.
3. There is a cycle of Siegel disks;alternatively, there is a single Siegel disk, or there is a cycle with more than one Siegel disk; in both cases, each Siegel disk has two preimages, each one with two preimages and so on, and there are no other cycles of Fatou components, and each Fatou component is eventually a cycle of Siegel disks.
4. There are no bounded Fatou components (e.g., $f_{-2}(z) = z^2 - 2$ has $J(f_{-2}) = [-2, 2]$ and the Fatou set is unbounded).

In cases 1 and 2, P_c is subhyperbolic and $J(P_c)$ is locally connected, because it is connected; only in cases 3 or 4 $J(P_c)$ may fail to be locally connected.

(6.4) Example: With the Sullivan classification theorem complemented with the Sullivan nonwandering domains theorem, it is possible to better specify the possible Julia sets of Blaschke products $\mathfrak{B}$ of degree ≥ 2, completing what was seen in *CADOVA*:

The Julia set $J(\mathfrak{B})$ of a Blaschke products $\mathfrak{B}$ of degree ≥ 2 is, alternatively: the circle S^1 with radius 1 and center 0 or a totally disconnected subset of S^1; thus, a Cantor set.[18] *If $\mathfrak{B}$ has an attracting or parabolic fixed point with only 1 petal in S^1, then $J(\mathfrak{B})$ is totally disconnected. If there is an attracting fixed point z_0 in B_1, then $\frac{1}{\bar{z}_0} \in \mathbb{C}_\infty$ is also an attracting fixed point, $J(\mathfrak{B})$ separates the, resp., attracting basins and $J(\mathfrak{B}) = S^1$. There remains the possibility of a parabolic fixed point in S^1 with 2 attracting petals, for which it also is $J(\mathfrak{B}) = S^1$. The periodic points of $\mathfrak{B}$ not fixed points are repelling.*

These cases already occur for Blaschke products of degree 2.

---

[17] See Section 12 of Chapter 13 of *CADOVA*.

[18] A **Cantor set** is a nonempty compact set with no isolated points and **totally disconnected** (i.e., whose connected components are the points of the set. See Sections 7, 12, and 15 of Chapter 13 of *CADOVA*.

6.4 Modifying the Multiplier of Attracting Periodic Orbit

Quasiconformal surgery can be used to change the multiplier of an attracting periodic orbit from one value λ_0 to another λ, both in $B_1\setminus\{0\}$. The linearization of the dynamics in the attracting basin of each orbit by the Koenigs conjugacy[19] requires a conjugacy of the linear transformation $z\mapsto\lambda_0 z$ with $z\mapsto\lambda z$, what is impossible with a conformal conjugacy but is possible with a **quasiconformal conjugacy**. We begin by establishing this.[20]

(6.5) *For* $\lambda\in\mathbb{C}$, T_λ *denotes the linear transformation* $T_\lambda:\mathbb{C}\to\mathbb{C}$ *such that* $T_\lambda(z)=\lambda z$. *For* $\nu,\lambda\in B_1\setminus\{0\}$, *there is a quasiconformal mapping* $\varphi_{\lambda,\nu}$ *from* B_1 *onto* B_1 *such that* $T_\lambda=\varphi_{\lambda,\nu}\circ T_\nu\circ(\varphi_{\lambda,\nu})^{-1}$. *A possibility is* $\varphi_{\lambda,\nu}(z)=e^{T_\nu\log(z)}$ *for* $z\in B_1\setminus\{0\}$ *and* $\varphi_{\lambda,\nu}(0)=0$.

Proof Denoting $\mathbb{H}$ the open upper complex half plane and $p:\mathbb{H}\to B_1\setminus\{0\}$ such that $p(z)=e^{iz}$, $(\mathbb{H},p)$ is a universal covering of $B_1\setminus\{0\}$. With $\mathcal{I}m\,\theta\in[0,2\pi[$ such that $\theta=\log\nu$, applying T_ν in $B_1\setminus\{0\}$ corresponds to the translation $w\mapsto w-i\theta$ in $\mathbb{H}$ with fixed point 1. For $\sigma\in\mathbb{H}$, consider the linear transformation L_σ that is a conjugacy of the translations in $\mathbb{H}$ $w\mapsto w-i\theta$ with $w\mapsto w-i\sigma$, i.e., $L_\sigma(z)=az+b\bar{z}$ such that $L_\nu(z)-i\sigma=L_\nu(z-i\vartheta)$ and $L_\nu(1)=1$, where $a=\frac{\sigma+\bar{\theta}}{\theta+\bar{\theta}}$ and $b=\frac{\theta-\sigma}{\theta+\bar{\theta}}$. L_ν is a solution of the Beltrami equation with coefficient $\frac{b}{a}=\frac{\theta-\sigma}{\theta+\bar{\theta}}$. The function $L_\nu\circ\log$, with log the branch of the logarithm with imaginary part in $[0,2\pi[$, is the solution of the Beltrami equation with coefficient

$$\kappa_\nu(z)=\frac{\partial(L_\nu\circ\log)/\partial\bar{z}}{\partial(L_\nu\circ\log)/\partial z}(z)=\frac{b/\bar{z}}{a/z}=\frac{\theta-\sigma}{\theta+\bar{\theta}}\,\frac{z}{\bar{z}},$$

which has constant modulo $|\kappa_\nu|=\left|\frac{\theta-\sigma}{\theta+\bar{\theta}}\right|<1$ and argument varying 2π along circles with center 0, i.e., with field of ellipses with the same ratio of major to minor axis, but orientation of the axes going around a complete circle along circles with center 0. A solution $\varphi_{\lambda,\nu}$ of the Beltrami equation is a conjugacy of T_λ with T_ν by the equation in the statement. It can be seen that a solution of this equation is $\varphi_{\lambda,\nu}(z)=e^{T_\nu\log(z)}$ for $z\in B_1\setminus\{0\}$ and $\varphi_{\lambda,\nu}(0)=0$. □

$\mathbf{Rat}_d$ denotes the set of rational functions $f:\mathbb{C}_\infty\to\mathbb{C}_\infty$ of degree d.

---

[19] See Section 6 of Chapter 13 of *CADOVA*.

[20] In this and the next two sections, we use parts of the book by B. Branner and N. Fagella listed in the final bibliography.

(6.6) *If $f_{\lambda_0} \in \mathrm{Rat}_d$, $d \geq 2$, with an attracting periodic orbit with period $m \in \mathbb{N}$ and multiplier $\lambda_0 \in B_1 \setminus \{0\}$ and $\lambda \in B_1 \setminus \{0\}$, there is a quasiconformal conjugacy of f_{λ_0} with a function $f_\lambda \in \mathrm{Rat}_d$ with the corresponding orbit of period m with multiplier λ. If f_{λ_0} is a polynomial of degree d, the conjugacy can be such that f_λ is a polynomial of degree d with superattracting fixed point at ∞, and if f_{λ_0} is monic, so f_λ can be.*

Proof Let $a_0, \ldots, a_{m-1}$ be the points of an attracting periodic orbit of f_{λ_0} with period m and multiplier λ_0 such that $a_{j+1} = f_{\lambda_0}(a_j)$ and $a_0 = f(a_{m-1})$, and let A be its attracting basin, which, without loss of generality, does not contain ∞ (otherwise, as $A \subsetneqq \mathbb{C}_\infty$, this can be achieved with a Möbius transformation). Let U_0 be a neighborhood of a_0 included in the immediate attracting basin of the fixed point a_0 of $f^m_{\lambda_0}$ such that $\psi_0 \colon U_0 \to B_1$ is the Koenigs conjugacy of $f^m_{\lambda_0}$ with $z \mapsto \lambda_0 z$, which can be extended as a holomorphic function to the entire attracting basin $A_{f^m_{\lambda_0}}(a_0)$ of the fixed point a_0 of $f^m_{\lambda_0}$ with $\psi_0\big(A_{f^m_{\lambda_0}}(a_0)\big) = \mathbb{C}$. Let ν_0 be such that $e^{\nu_0} = \lambda_0$ with $\mathcal{I}m\, \nu_0 \in [0, 2\pi[$, and let $\mathcal{L} \colon U_0 \to \mathbb{C}$ be a branch of the logarithm with $\mathcal{L}(\lambda_0) = \nu_0$.

By the preceding result, changing from the canonical conformal structure (corresponding to Beltrami coefficient $\kappa_0 = 0$) to a conformal structure in $\mathbb{C} \setminus \{0\}$ corresponding to the Beltrami coefficient $\kappa_\lambda(z) = \frac{\mathcal{L}(\lambda) - \nu_0}{\mathcal{L}(\lambda) + \overline{\nu_0}} \frac{z}{\bar{z}}$, which is invariant under the linear transformation $z \mapsto \lambda_0 z$, and considering the solution φ_λ of the Beltrami equation with this coefficient and fixed points 0 and 1, which is a quasiconformal conjugacy of $z \mapsto \lambda_0 z$ with $z \mapsto \lambda z$, i.e., $\lambda\, \varphi_\lambda(z) = \varphi_\lambda\big(\lambda_0 z\big)$, in the attracting basin of the fixed point a_0 of $f^m_{\lambda_0}$, the Beltrami coefficient $\widetilde{\kappa}_\lambda = \kappa_\lambda \circ \psi_0$ can be extended to the attracting basin A of the considered periodic orbit by applying f_{λ_0}. In $\mathbb{C}_\infty \setminus A$ define $\widetilde{\kappa}_\lambda = 0$. The function $f_\lambda = \widetilde{\varphi}_\lambda \circ f_{\lambda_0} \circ (\widetilde{\varphi}_\lambda)^{-1} \colon \mathbb{C}_\infty \to \mathbb{C}_\infty$, with $\widetilde{\varphi}_\lambda$ solution of the Beltrami equation with Beltrami coefficient $\widetilde{\kappa}_\lambda$ is holomorphic, $f_\lambda \in \mathrm{Rat}_d$, and it is conjugated with f_{λ_0} by a quasiconformal mapping. $\{\widetilde{\kappa}_\lambda(a_0), \ldots, \widetilde{\kappa}_\lambda(a_{m-1})\}$ is an attracting periodic orbit of f_λ with period m. By the Weyl lemma (1.26), $\psi_\lambda \overset{def}{=} \varphi_\lambda \circ \psi_0 \circ \big(\widetilde{\varphi}_\lambda\big)^{-1}$ is a holomorphic function of $\widetilde{\varphi}_\lambda(U_0)$ onto B_1, since $\frac{\partial \psi_\lambda}{\partial \bar{z}}(z) = 0$, and as it is a conjugacy of $(f_\lambda)^m$ with $z \mapsto \lambda z$, the multiplier of the periodic orbit $\{\widetilde{\kappa}_\lambda(a_0), \ldots, \widetilde{\kappa}_\lambda(a_{m-1})\}$ is λ.

If f_{λ_0} is a polynomial of degree d, as the conformal structure associated with the Beltrami coefficient $\widetilde{\kappa}_\lambda$ coincides with the canonical conformal structure in a neighborhood of ∞, with the solution $\widetilde{\varphi}_\lambda$ of the Beltrami equation with coefficient $\widetilde{\kappa}_\lambda$, which also has a fixed point ∞, ∞ is a superattracting fixed point of f_λ, and this function is also a polynomial of degree d. If f_{λ_0} is a monic polynomial and the solution $\widetilde{\varphi}_\lambda$ of the Beltrami equation with coefficient $\widetilde{\kappa}_\lambda$ satisfies additionally the condition $\lim\limits_{z \to \infty} \widetilde{\varphi}_\lambda / z = 0$, then

$$\frac{f_\lambda \circ \widetilde{\varphi}_\lambda(z)}{[\widetilde{\varphi}_\lambda(z)]^d} = \frac{\widetilde{\varphi}_\lambda \circ f_{\lambda_0}(z)}{f_{\lambda_0}(z)} \frac{f_{\lambda_0}(z)}{z^d} \frac{z^d}{[\widetilde{\varphi}_\lambda(z)]^d} \to 1, \quad z \to \infty.$$

Thus, the polynomial f_λ of degree d is also monic. □

6.5 Modifying Superattracting to Attracting Orbits

The examples of quasiconformal surgery in the last three sections are related to the dynamics of holomorphic functions through the construction of quasiconformal mappings obtained with Beltrami equations. This section considers a first case of quasiconformal surgery involving cutting and sewing.[21]

The method of modifying the multiplier of an attracting periodic orbit of the preceding section cannot be applied in this case, as it cannot modify a multiplier 0 to one $\neq 0$. To modify a multiplier $\lambda_0 \in B_1$ to another $\lambda \in B_1$ (including the possibility of one of them 0) a different method is needed to modify the dynamics in an arbitrarily small neighborhood of the periodic orbit. First, consider Blaschke products.

(6.7) *Consider the family of Blaschke products with attracting fixed point 0 with multiplier $\lambda \in B_1$, $\mathfrak{B}_\lambda(z) = z\,\frac{z+\lambda}{1+\bar\lambda z}$. For every $\lambda_0, \lambda \in B_1$, there is a quasiconformal conjugacy of $\mathfrak{B}_\lambda$ in B_1 with a quasiconformal mapping in B_1 that coincides with $\mathfrak{B}_{\lambda_0}$ in the complement of an arbitrarily small neighborhood of 0 in B_1 and with $\mathfrak{B}_\lambda$ in a smaller neighborhood of 0.*

Proof Consider the family of Blaschke products $\mathfrak{B}_\lambda(z) = z\,\frac{z+\lambda}{1+\bar\lambda z}$ with $\lambda \in B_1$, which are rational functions of degree 2 with fixed points $0, \infty$ and $\frac{1-\lambda}{1-\bar\lambda} \in \partial B_1$, the two first attracting with multipliers, resp., λ and $\bar\lambda$ and the last one repelling. The function $\frac{1}{z}$ is a conjugacy of $\mathfrak{B}_\lambda$ with $\mathfrak{B}_{\bar\lambda}$. B_1 and $\mathbb{C}\backslash \mathrm{cl}\, B_1$ are the attracting basins of the fixed points, resp., 0 and ∞, and the circle ∂B_1 is the Julia set of $\mathfrak{B}_\lambda$. Since $\frac{\mathfrak{B}_\lambda(z)}{z}$ maps B_1 to itself, and, with $B_r = B_r$, $\mathrm{cl}\, B_r \subset \mathfrak{B}_\lambda^{-1}(B_r)$. If $\lambda \in B_1\backslash\{0\}$, for small $r > 0$ $\mathfrak{B}_\lambda^{-1}(B_r)$ has two connected components and if $r \in]|\lambda|, 1[$ has only one connected component. Thus, $\mathfrak{B}_\lambda^{-1}(B_r)\backslash B_r$ is a ring, and in this case $-\lambda \in B_r$. Therefore, for $\lambda \in B_1$ and $r > |\lambda|$ (Fig. 6.2),

$$\mathfrak{B}_\lambda(B_r) \subset B_r\,, \qquad \mathfrak{B}_\lambda\big(\mathfrak{B}_\lambda^{-1}(B_r)\big) = B_r\,, \qquad \mathfrak{B}_\lambda\big(B_1\backslash \mathfrak{B}_\lambda^{-1}(B_r)\big) = B_1\backslash B_r\,.$$

For $\lambda_0 \in B_1$, $r \in]|\lambda_0|, 1[$ and $\lambda < r$ define $g_\lambda : B_1 \to B_1$ with g_λ equal to

$$\mathfrak{B}_{\lambda_0} \text{ in } B_1\backslash \mathfrak{B}_{\lambda_0}^{-1}(B_r)\,, \qquad \mathfrak{B}_\lambda \text{ in } \mathrm{cl}\, B_r\,, \qquad h_\lambda \text{ in } \mathrm{cl}\, \mathfrak{B}_{\lambda_0}^{-1}(B_r)\backslash B_r\,,$$

where h_λ is a quasiconformal mapping in $\mathrm{cl}\, \mathfrak{B}_{\lambda_0}^{-1}(B_r)\backslash B_r$ equal to $\mathfrak{B}_{\lambda_0}$ in $\partial\big(\mathfrak{B}_{\lambda_0}^{-1}(B_r)\big)$ and to $\mathfrak{B}_\lambda$ in ∂B_r, whose existence is guaranteed by (3.5) (Fig. 6.3). These functions are holomorphic of degree 2 in the considered boundaries, which are analytic curves, so h_λ can be such that $\big(\mathfrak{B}_\lambda^{-1}(B_r)\backslash B_r, h_\lambda\big)$ is a covering with two sheets of $\mathfrak{B}_{\lambda_0}^{-1}(B_r)\backslash B_r$, and for each $z \in \mathfrak{B}_{\lambda_0}^{-1}(B_r)\backslash B_r$, $h_\lambda(z)$ depends continuously on λ.

21 Some authors call the first type of quasiconformal surgery "soft surgery" and the second type "cut and paste surgery."

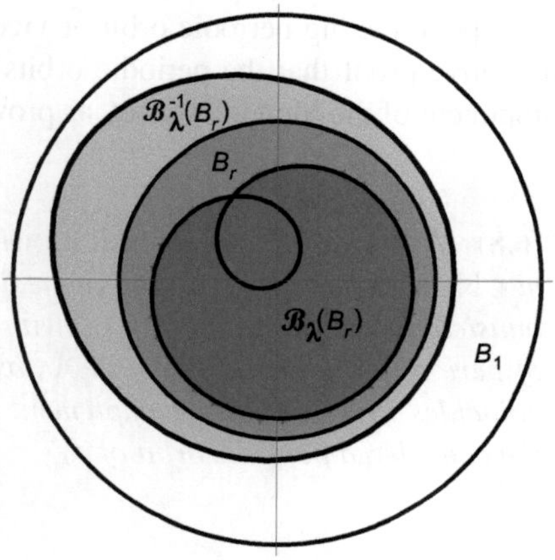

Fig. 6.2 Blashcke product $\mathfrak{B}_\lambda(z) = z\,\frac{z+\lambda}{1+\bar\lambda z}$ with $\lambda \in B_1$ and $r > |\lambda|$

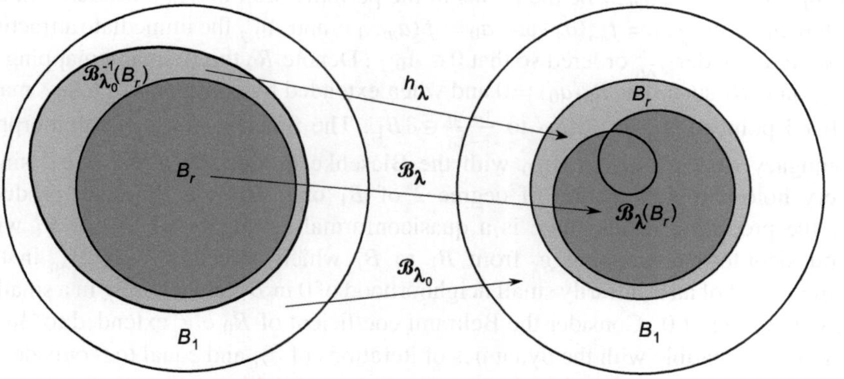

Fig. 6.3 Pasting of two Blashcke products $\mathfrak{B}_\lambda(z) = z\,\frac{z+\lambda}{1+\bar\lambda z}$, with $\lambda_0 \in B_1$, $r \in]|\lambda_0|, 1[$ and $|\lambda| < r$, by quasiconformal mappings

Construct a Beltrami coefficient κ_λ corresponding to the canonical conformal structure in B_r, i.e., with $\kappa_\lambda = 0$ in B_r equal to the Beltrami coefficient of h_λ in ring $\operatorname{cl} \mathfrak{B}_{\lambda_0}^{-1}(B_r) \setminus B_r$ and defined in the remainder of B_1 so as to be compatible with the iteration dynamics of g_λ. For $\lambda \in B_r$, $\|\kappa_\lambda\|_{L^\infty} < 1$ in B_1 and for each $z \in B_1$ the function $\lambda \mapsto \kappa_\lambda(z)$ is continuous at B_r. By the measurable Riemann mapping theorem (2.9), for $\lambda \in B_1$ the Beltrami equation with coefficient κ_λ has a solution $\varphi_\lambda : \operatorname{cl} B_1 \to \operatorname{cl} B_1$ with fixed point 0, unique up to composition at the left by rotations, and it can be chosen so that $\varphi_\lambda\big(\frac{1-\lambda_0}{1-\bar\lambda_0}\big) = \frac{1-\lambda}{1-\bar\lambda}$. $G_\lambda = \varphi_\lambda \circ g_\lambda \circ \varphi_\lambda^{-1}$ is a holomorphic function of B_1 onto B_1. Thus, it is a Blashcke product, and, as $\frac{1-\lambda}{1-\bar\lambda}$ is a fixed point of G_λ, it is $G_\lambda = \mathfrak{B}_\lambda$. The multiplier of the fixed point 0 of G_λ is λ. □

The following result not only establishes the possibility of transforming periodic orbits with any multiplier in B_1 to periodic orbits with the same period with any other multiplier in B_1, including the transformation of an attracting periodic orbit to a superattracting periodic orbit or vice versa, but also that it can be used for an alternative proof that the periodic orbits multiplier parameterizes each hyperbolic component of the Mandelbrot set, as proved in Sect. 6.2.

(6.8) *If $\lambda_0, \lambda \in B_1$ and $f_{\lambda_0} \in \mathrm{Rat}_d$ with $d \geq 2$ has a periodic orbit of period $m \in \mathbb{N}$, with multiplier λ_0 and immediate attracting basin A_0, then there is a quasiconformal conjugacy of f_{λ_0} with some $f_\lambda \in \mathrm{Rat}_d$ having a periodic orbit of period m and multiplier λ in A_0 by a quasiconformal mapping in A_0 that coincides with f_{λ_0} in the complement of an arbitrarily small neighborhood of the considered periodic orbit of f_{λ_0}.*

Proof Let $a_0, \ldots, a_{m-1}$ be the points of the periodic orbit of f_{λ_0} considered in the statement with $a_{j+1} = f_{\lambda_0}(a_j)$ and $a_0 = f(a_{m-1})$, and $A_{0,j}$ the immediate attracting basin of a_j under $f_{\lambda_0}^m$ ordered so that $0 \in A_{0,j}$. Denote R_0 the Riemann mapping of $A_{0,0}$ onto B_1 such that $R_0(a_0) = 0$ and when extended by continuity to $\partial A_{0,0}$ maps a fixed point of $f_{\lambda_0}^m$ in $\partial A_{0,0}$ to $\frac{1-\lambda_0}{1-\overline{\lambda_0}} \in \partial B_1$. The function R_0 is a holomorphic conjugacy of $f_{\lambda_0} : A_{0,0} \to A_{0,0}$ with the Blaschke product $\mathfrak{B}_{\lambda_0} : B_1 \to B_1$, since every holomorphic function of degree 2 of B_1 onto B_1 is a Blashcke product. By the preceding result, there is a quasiconformal conjugacy of $\mathfrak{B}_\lambda$ to B_1 with a quasiconformal mapping g_λ from B_1 to B_1 which coincides with $\mathfrak{B}_{\lambda_0}$ in the complement of an arbitrarily small neighborhood of 0 in B_1 and with $\mathfrak{B}_\lambda$ in a smaller neighborhood of 0. Consider the Beltrami coefficient of $R_0 \circ g_\lambda$ extended to A_0 so as to be compatible with the dynamics of iterations of f_{λ_0} and equal to 0 outside A. By (2.8), the solution $\widetilde{\varphi}_\lambda$ of the Beltrami equation with this coefficient, fixed points 0 and ∞ and such that $\lim_{z \to \infty} \widetilde{\varphi}_\lambda(z)/z = 0$, and $\widetilde{\varphi}_\lambda \circ f_\lambda \circ \widetilde{\varphi}_\lambda^{-1}$ is a polynomial of degree d with a periodic orbit $\{\widetilde{\varphi}_\lambda(a_0), \ldots, \widetilde{\varphi}_\lambda(a_{m-1})\}$ of period m and multiplier λ. □

6.6 Modifying a Rotation Ring Module

A **rotation ring** of a complex rational function f is a ring with finite module that is a periodic component of $\mathbb{C}_\infty \backslash \mathrm{cl}\, PC(f)$, where $PC(f)$ is the set of postcritical points of f, equal to the closure of the union of positive orbits of critical points of f. Rotation rings are subsets of Siegel disks or Herman rings.[22]

As seen in Chap. 1, there is a conformal homeomorphism between rings if and only if they have the same module. Therefore, it is not possible to modify the module of a rotation ring of a rational function f with conformal homeomorphisms. With quasiconformal surgery, it can be modified, giving a quasiconformal conjugacy of

[22] See Section 12 of Chapter 13 of *CADOVA*.

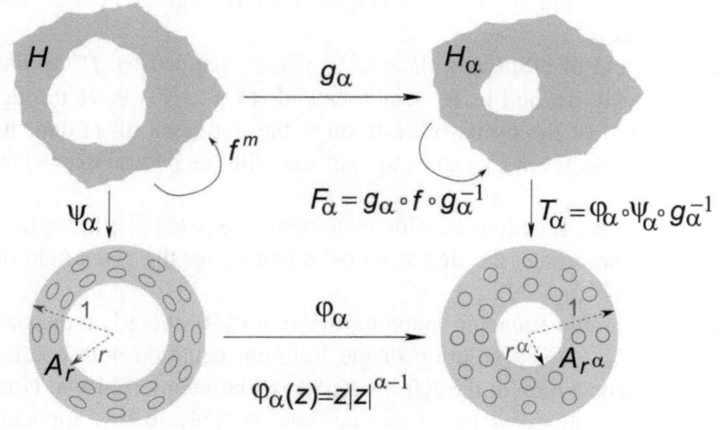

Fig. 6.4 Modifying the module of a Herman ring (6.9); ψ_α, T_α are conformal homeomorphisms and g_α, φ_α are quasiconformal mappings

the dynamics of f in the initial and final conformal structures, as observed in 2003 by N. Fagella and L. Geyer[23] (Fig. 6.4).

(6.9) *If H is a rotation ring of $f \in \mathrm{Rat}_d$ with $d \geq 2$ and module $M(H)$, then there is a family $\{F_\tau\}_{\tau \in \mathbb{H}} \subset \mathrm{Rat}_d$, with $\mathbb{H}$ the open upper complex half plane, such that:*

1. *F_τ has a rotation ring with module $-\frac{1}{2\pi}\,\mathcal{R}e\,\tau$;*
2. *there is a quasiconformal conjugacy of f with F_τ, conformal in $\mathbb{C}_\infty$ for $\tau_0 = -\,2\pi M(H)$, and outside the set of points with positive orbits intersecting H, for $\tau \in \mathbb{H}$.*

Proof For $r = e^{-2\pi M(f)}$ denote $A_r = B_1 \setminus \mathrm{cl}\, B_r$, and, for $\alpha > 0$, $\varphi_\alpha : A_r \to A_{r^\alpha}$, $\varphi_\alpha(z) = z|z|^{\alpha-1}$, which is a conjugacy of the rotation by an angle θ around 0 in A_r with R_θ in A_{r^α}. We have

$$\frac{\partial \varphi_\alpha}{\partial \bar z}(z) = \frac{\alpha-1}{2}\, z^{\frac{\alpha+1}{2}}\, \bar z^{\frac{\alpha-1}{2}-1}\,, \qquad \frac{\partial \varphi_\alpha}{\partial z}(z) = \frac{\alpha+1}{2}\, z^{\frac{\alpha+1}{2}-1}\, \bar z^{\frac{\alpha-1}{2}}\,.$$

Therefore, φ_α is a quasiconformal mapping with Beltrami coefficient $\kappa_\alpha = \frac{\alpha-1}{\alpha+1}\frac{z}{\bar z}$, $\|\kappa_\alpha\|_{L^\infty} = \left|\frac{\alpha-1}{\alpha+1}\right| < 1$. A field of ellipses in A_r mapped by φ_α to a field of circles in A_{r^α} is invariant under R_θ.

There is a homeomorphism ψ_α of H onto A_r and a field of ellipses in H mapped by ψ_α to the field of ellipses defined in A_r. As there is $m \in \mathbb{N}$ such that $f^m(H) = H$,

[23] Geyer, Lukas.

ψ_α is a conformal conjugacy of f^m in H with R_θ in A_r, and the field of ellipses in H is invariant under f^m.

There is a field of ellipses in $\mathcal{H} = \cup_{n=1}^m f^n(K)$ mapped by f^m to the field of ellipses previously defined in H, which extends to $f^{-1}(\mathcal{H}) \supset \mathcal{H}$ through f and then to the union of the orbits of H through the iterations of f, and, finally, by complementing this set in $\mathbb{C}_\infty$ so as to coincide with the natural extension there as a field of circles.

As the successive functions considered to relate the fields of ellipses in A_r and in $f^{-1}(\mathcal{H})$ are holomorphic, the Beltrami coefficient $\widetilde{\kappa}_\alpha$ for the given field of ellipses in $\mathbb{C}_\infty$ satisfies $\|\widetilde{\kappa}_\alpha\|_{L^\infty} < 1$.

By the measurable Riemann mapping theorem (2.9), there is a quasiconformal mapping $g_\alpha : \mathbb{C}_\infty \to \mathbb{C}_\infty$ solution of the Beltrami equation with coefficient $\widetilde{\kappa}_\alpha$, unique up to composition at the left by conformal homeomorphisms. Normalizing the solution for uniqueness (what can be done suitably to this application), the function $\alpha \mapsto g_\alpha$ is analytic.

$F_\alpha \overset{def}{=} g_\alpha \circ f \circ g_\alpha^{-1} : \mathbb{C}_\infty \to \mathbb{C}_\infty$ satisfies $\partial F_\alpha / \partial \overline{z} = 0$, since f is holomorphic, and, by the Weyl lemma (1.26), it is holomorphic in $\mathbb{C}_\infty$; so, it is a rational function. The function g_α is a quasiconformal conjugacy of f with F_α, and this conjugacy is holomorphic outside the set of points with positive orbits intersecting the orbit of the rotation ring H. $H_\alpha \overset{def}{=} g_\alpha(H)$ is a rotation ring for F_α and $T_\alpha \overset{def}{=} \varphi_\alpha \circ \psi \circ g_\alpha^{-1} : H_\alpha \to A_{r^\alpha}$, which is a quasiconformal mapping, preserving the canonical conformal structure associated with Beltrami coefficient 0 and has derivative $\partial / \partial \overline{z}$ equal to 0. Thus, by the Weyl lemma (1.26), it is a conformal homeomorphism. Therefore, the module of the rotation ring H_α satisfies

$$M(H_\alpha) = \tfrac{1}{2\pi} \log \tfrac{1}{r^\alpha} = \alpha M(H) .$$

The modules of the rotation rings H_α and H differ if $\alpha \neq 1$, and, therefore, there is no conformal conjugacy of F_α with f except if $\alpha = 1$. □

This result is now exemplified with cubic rational functions.

(6.10) Example: Consider the family of cubic rational functions

$$f_{t,a}(z) \overset{def}{=} e^{i2\pi t} z \tfrac{1-az}{1-a/z}, \quad t \in \mathbb{R}, a \in \left]0, \tfrac{1}{3}\right[.$$

The restriction of $f_{t,a}$ to the circle S^1 with center 0 and radius 1 is an orientation preserving analytic diffeomorphism. For $t \in \mathbb{Z}$, the function $f_{t,a}$ has fixed point 1 with rotation number[24] 0. For each $\theta \in [0, 1[\setminus \mathbb{Q}$ and $a \in \left]0, \tfrac{1}{3}\right[$, there is a unique $t \overset{def}{=} T_\theta(a) \in]0, 1[$ such that the restriction of $f_{t,a}$ to S^1 has rotation number. By the Denjoy-Wolff theorem,[25] there is a homeomorphism $h : S^1 \to S^1$ that is a conjugacy of f with the rotation R_θ of angle θ around 0, i.e., $(f_{t,a} \circ h)(z) = h\big(e^{i2\pi\theta} z\big)$. By

---

[24] See Section 8 of Chapter 13 of *CADOVA*.

[25] See Section 7 of Chapter 13 of *CADOVA*. Denjoy, Arnaud (1884–1974). Wolff, Julius (1888–1945).

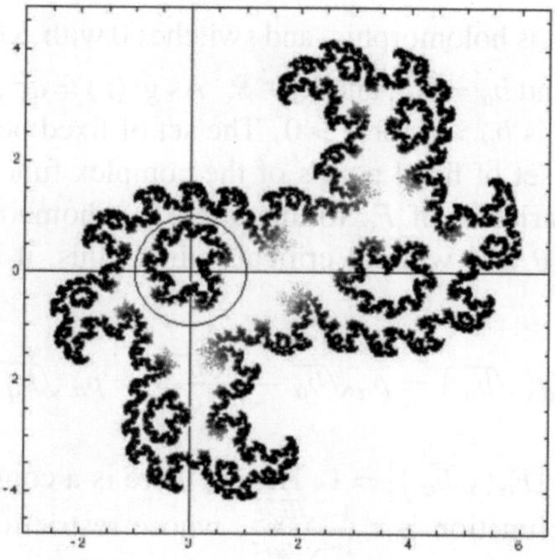

Fig. 6.5 Julia set J of $f(z)=e^{i2\pi/6}z\,\frac{1-az}{1-a/z}$, with $a=\frac{\sqrt{43}}{25}$, and invariant circle S^1 with center 0 and radius 1 (computed with *Mathematica*). There is a Herman ring $H_a\supset S^1$. Despite the approximate symmetry relative to 1.5 , there are a ring A_1 around the component of J around $3.5-i\,\frac{1}{4}$ whose points go in one iteration to H_a and are captured in it in the following iterations, and Rings A_2 , A_2' around the two components of J immediately smaller whose points go in two iterations to H_a with a first iteration in A_1, and similarly around the four components of J immediately smaller in three iterations, and so on

the Arnold theorem for homeomorphisms[24] in S^1, θ is a **Diophantine number**, i.e., there are k, $M>0$ such that

$$\forall\ \tfrac{q}{m} \text{ with } q\in\mathbb{Z}\,,\ m\in\mathbb{N} \text{ in minimal terms is } \left|\theta-\tfrac{q}{m}\right|>\tfrac{M}{m^k}\,.$$

For every $R\in\,]0,1[$, the function h has an extension to a conformal Homeomorphism in the annulus $A_R\overset{def}{=}B_R\backslash\mathrm{cl}\,B_{1/R}$, still denoted h, such that $(f_{t,a}\circ h)(z)=h\big(e^{i2\pi\theta}z\big)$ for $z\in A_R$. Hence, $h(A_R)$ is invariant under $f_{t,a}$, and, therefore, it is a subset of the Fatou set of $f_{t,a}$. Since each connected component of $\mathbb{C}_\infty\backslash f_{t,a}(A_R)$ contains one of the superattracting fixed points 0 and ∞ of $f_{t,a}$, $h(A_R)$ is not a Siegel disk. Therefore, it is a subset of a Herman ring H_a (Fig. 6.5).

By (6.9) and the, resp., proof, for $\alpha>0$, there is a unique quasiconformal mapping $g_\alpha:\mathbb{C}_\infty\to\mathbb{C}_\infty$ normalized to have fixed points 0 and ∞ and $g_\alpha\big(\frac{1}{a}\big)=1$ such that $F_\alpha\overset{def}{=}g_\alpha\circ f_{t,a}\circ g_\alpha^{-1}:\mathbb{C}_\infty\to\mathbb{C}_\infty$ is a cubic rational function with superattracting fixed points 0 and ∞, and with a Herman ring $H_\alpha=g_\alpha(H)$ with module $M(H_\alpha)=\alpha M(H)$. Since $f\big(\frac{1}{a}\big)=0$, $F_\alpha(1)=0$ and as f has a pole at a, F_α has pole at $b_\alpha=g_\alpha(a)$. Hence, there exists $\rho_\alpha\in\mathbb{C}\backslash\{0\}$ analytically dependent on α such that $F_\alpha(z)=\rho_\alpha z\,\frac{1-z}{1-b_\alpha/z}$.

The rational function $f_{t,a}$ commutes with the complex function $\tau(z)=\frac{1}{z}$. With $\tau_r=\tau\big(\frac{z}{r}\big)$ and the notation of the proof of (6.9), for the homeomorphism ψ of H onto A_r, $\psi\circ\tau\circ\psi^{-1}=\tau_r$ and $\varphi_\alpha\circ\tau_r=\tau_{r^\alpha}\circ\varphi_\alpha$. The fields of ellipses defined in the proof of (6.9) in, resp., A_r and H are invariant under, resp., τ_r and τ.

Therefore, $\overline{g_\alpha \circ \tau \circ g_\alpha^{-1}}$ is holomorphic, and switches 0 with ∞ and 1 with b_α. Thus, $g_\alpha \circ \tau \circ g_\alpha^{-1}(z) = \frac{b_\alpha}{\bar{z}}$ and $b_\alpha = \overline{b_\alpha}$, and $b_\alpha \in \mathbb{R}$. As $g_1(z) = az$, it is $b_1 = a^2 > 0$ and, as $b_\alpha \neq 0$ for $\alpha > 0$, it is $b_\alpha > 0$ for $\alpha > 0$. The set of fixed points of τ is the circle S^1, and $g_\alpha(S^1)$ is the set of fixed points of the complex function $\frac{b_\alpha}{\bar{z}}$, which is the circle $\partial B_{\sqrt{b_\alpha}}$. The restriction of F_α to this circle is a homeomorphism onto itself with rotation number θ and without critical points; thus, it is a diffeomorphism. Since

$$F_\alpha\big(\sqrt{b_\alpha}\big) = \rho_\alpha \sqrt{b_\alpha}\, \frac{1-\sqrt{b_\alpha}}{1-b_\alpha/\sqrt{b_\alpha}} = \rho_\alpha \sqrt{b_\alpha}\,,$$

it is $|\rho| = \big|\rho_\alpha \sqrt{b_\alpha}\,\big| = \big|F_\alpha\big(\sqrt{b_\alpha}\big)\big| = 1$. Hence, there is a conjugacy of the rational function F_α with the function $\rho_\alpha z\, \frac{1-\sqrt{b_\alpha}\, z}{1-\sqrt{b_\alpha}/z}$, whose restriction to the circle S^1 is a diffeomorphism onto S^1 with rotation number θ. Therefore, $\sqrt{b_\alpha} \in \left]0, \frac{1}{3}\right[$ and $\rho_\alpha = e^{i2\pi T_\theta(\sqrt{b_\alpha})}$.

As the Herman ring H_α of F_α separates the zero of F_α at 1 from its pole at b_α, and $M(H_\alpha) = \alpha M(H) \to +\infty$ when $\alpha \to +\infty$, $\lim_{\alpha \to +\infty} b_\alpha = 0$.

The function F_α depends analytically on α. Thus, also b_α, $\sqrt{b_\alpha}$ and $\rho_\alpha = \frac{F_\alpha(\sqrt{b_\alpha})}{\sqrt{b_\alpha}}$ do. Therefore, the restriction of T_θ to the curve described by $\alpha \mapsto \sqrt{b_\alpha}$ is analytic. If $(t_0, a_0) \in]0, 1[\times]0, \frac{1}{3}[$ is such that f_{t_0, a_0} has a Herman ring containing S^1 with rotation number θ, then $t_0 = T_\theta(a_0)$, and T_θ is analytic in a neighborhood of a_0. For every $a \in]0, a_0]$, there is $\alpha > 0$ such that $b_\alpha = a$, and the function $\sqrt{b_\alpha}\, z$ is a conjugacy of $f_{T_\theta(a), a}$ with F_α. Since F_α has a Herman ring including the circle with center 0 and radius $\sqrt{b_\alpha}$, $f_{T_\theta(a), a}$ has a Herman ring including S^1.

The conclusion is: *For each* $\theta \in]0, 1[\setminus\mathbb{Q}$ *and* $a_\theta \in \left[0, \frac{1}{3}\right]$ *the circle* S^1 *is invariant under* $f_{t,a} \stackrel{def}{=} e^{i2\pi t} z\, \frac{1-az}{1-a/z}$, *with* $t \in \mathbb{R}$, $a \in \left]0, \frac{1}{3}\right[$, *and there are a unique* $T_\theta(a) \in]0, 1[$ *such that the restriction of* $f_{t,a}$ *to* S^1 *has rotation number* θ *and* $a_\theta \in \left[0, \frac{1}{3}\right]$ *such that* T_θ *is a real valued analytic function of a real variable in* $]0, a_\theta[$, *and for* $(t, a) \in]0, 1[\times\left]0, \frac{1}{3}\right[$ *the function* $f_{t,a}$ *has a Herman ring* H_a *with rotation number* θ *if and only if* $a \in]0, a_\theta[$ *and* $t = T_\theta(a)$. *For* $\alpha > 0$ *there is a quasiconformal conjugacy of* $f_{t,a}$ *with* $F_\alpha(z) = \rho_\alpha z\, \frac{1-z}{1-b_\alpha/z}$, *where* $\rho_\alpha = e^{i2\pi T_\theta(\sqrt{b_\alpha})}$ *and* $b_\alpha \in \left]0, \frac{1}{9}\right[$ *is such that* $\lim_{\alpha \to +\infty} b_\alpha = 0$, *which has a Herman ring* H_α *with module with* $\lim_{\alpha \to +\infty} M(H_\alpha) = +\infty$.

6.7 Making a Herman Ring from a Siegel Disk

One of the first applications of quasiconformal surgery of "cut and paste" type was the construction of Herman rings from Siegel disks[26] in 1987 by M. Shishikura, giving an alternative construction of examples of rational functions with Herman

[26] See Sections 7 and 12 of Chapter 13 of *CADOVA*.

rings, 8 years after M. Herman gave the first construction with the 1961 Arnold theorem on circle homeomorphisms.[27] The construction of M. Shishikura is more elaborate than the presented in this section, because it is for Herman ring cycles from Siegel disk cycles (as in the following Sections 6.12 and 6.13), while this section is for a single Herman ring.

(6.11) *If f is a rational function with a Siegel disk D, there is a rational function g with a Herman ring A such that there are conjugacies of $f_{|D}$ and $f_{|A}$ with the same rotation number around 0, in, resp., an open disk and an open annulus. Each Brjuno number is the rotation number of a Herman ring of some rational function.*

Proof Without loss of generality, take the Siegel disk center of f at 0.

Begin by cutting the Siegel disk along a Jordan curve $C \subset D$ invariant under f separating $\mathbb{C}_\infty$ in two regions, E_0, E_∞, containing, resp., the Siegel disk center and ∞. Next, replace the dynamics in E_0 by that of a function conjugated by a diffeomorphism ψ of E_0 onto E_∞ with a rational function $\widetilde{f}$ reversing the orientation of the dynamics of f in E_∞ and mapping ∞ to 0, so that the pasting along C is such that in the intersection of a neighborhood of C with E_0, and, therefore, in the complement of a compact subset $K \neq \emptyset$ of E_0, the dynamics matches that of f. The resulting dynamics in $\mathbb{C}_\infty$ is of a function g, and the *continuum* of invariant Jordan curves under g in $(\mathbb{C}_\infty \cap D)\setminus K$ extends to a proper subset of E_0, because the dynamics in E_0 is obtained from that of $f_{|E_\infty}$ and $E_\infty \cap D$ is a proper subset of E_∞ (Fig. 6.6).

Denote the multiplier of the fixed point 0 of f by λ. The function $\widetilde{f}(z) = \overline{f(\overline{z})}$ has a Siegel disk $\widetilde{D}$ with center 0 and multiplier $\overline{\lambda}$. There are conformal homeomorphisms $\varphi, \widetilde{\varphi}$ of, resp., D and $\widetilde{D}$ onto B_1 which are conjugacies of f and $\widetilde{f}$ with rotations in B_1 around 0, resp., R_λ and $R_{\overline{\lambda}}$ given by multiplying each point by, resp., λ and $\overline{\lambda}$. For $0 < r, \widetilde{r} < 1$ let $C \subset D$ and $\widetilde{C} \subset \widetilde{D}$ be curves corresponding through the, resp., conjugacies to circles with center 0 and radius, resp., r and $\widetilde{r}$. Let ψ be a diffeomorphism between annular neighborhoods of C and $\widetilde{C}$ inverting orientation and defined through the conjugacies by $\Psi : B_1 \to B_1$ with $\Psi(z) = \frac{r\widetilde{r}}{z}$. This function Ψ is a conjugacy of the rotation with multiplier λ with that with multiplier $\overline{\lambda}$ and maps the circle $\varphi(C)$ with radius r onto the circle $\widetilde{\varphi}(\widetilde{C})$ with radius $\widetilde{r}$, reversing orientation. Hence, $\psi(C) = \widetilde{C}$ and $\psi \circ f = \widetilde{f} \circ \psi$ in a neighborhood of C. The pasting function in the curve C is $h = \widetilde{\varphi}^{-1} \circ \Psi \circ \varphi$ (Fig. 6.7).

Extend ψ to a diffeomorphism of E_0 onto E_∞, so that ψ is holomorphic in a neighborhood of 0 and $\psi(0) = \infty$. The result of the pasting is the function g equal to f in E_0 and to $\psi^{-1} \circ \widetilde{f} \circ \psi$ in E_∞. In a neighborhood of C, it is $f = \psi^{-1} \circ \widetilde{f} \circ \psi = g$.

[27] See Section 11 of Chapter 13 of *CADOVA*, which also includes the definition of Brjuno number. Brjuno, Alexander (1940–).

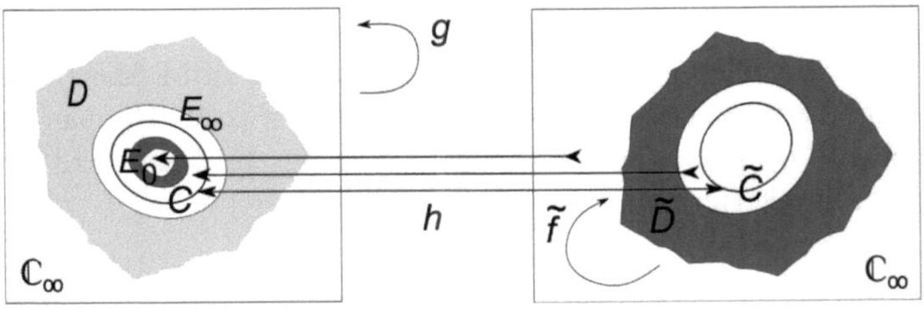

Fig. 6.6 Result of "cut and paste" to get a Herman ring from a Siegel disk D; the function $g:\mathbb{C}_\infty \to \mathbb{C}_\infty$ is in E_0 the conjugacy of $\widetilde{f}$ by diffeomorphism ψ of E_0 onto E_∞ constructed in the proof of (6.11) and coinciding with f in the complement of a compact subset $K \neq \emptyset$ of E_0

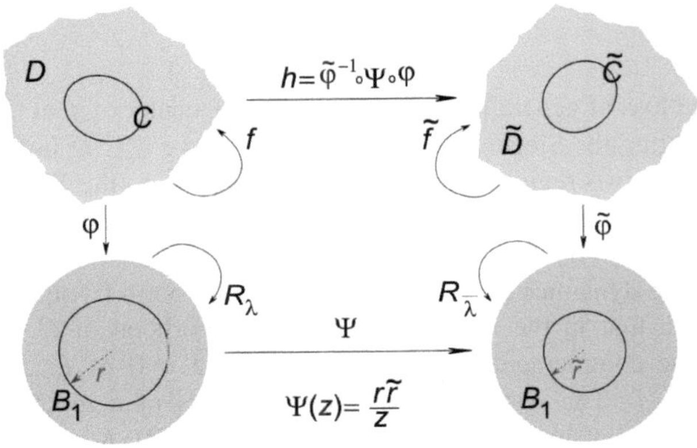

Fig. 6.7 Pasting function h to get a Herman ring from a Siegel disk

Thus, g is holomorphic in the complement of a compact subset $U \neq \emptyset$ properly contained in the bounded region the complement of C (Fig. 6.6).

As we want a rational function with Herman ring obtained from the Siegel disk of f, it is convenient to obtain a conjugacy F of g with a rational function, what can be done with a quasiconformal mapping F solution of an appropriate Beltrami equation. Consider E_∞ with a field of circles and E_0 with a field of ellipses mapped by ψ to a field of circles. As f and $\widetilde{f}$ are holomorphic; this field is invariant under g. For F, solution of the Beltrami equation with coefficient corresponding to $f_0 = F \circ g \circ F^{-1}$ is a rational function. If A is the ring intersection of D with the unbounded region in the complement of C and $A' = F(A)$, as $g(A) = A$, then $f_0(A') = A'$ and A' is contained in a Fatou component U_0 of f_0, and this is a rotation ring of f_0. By the definition of Fatou set, $\{f^n\}$ and $\{\widetilde{f}^n\}$ are not normal families outside of, resp., D and $\widetilde{D}$. Therefore, $\{f_0^n\}$ is not a normal family in any connected component of the complement of A', and U_0 is not a Siegel disk of f_0. Thus, U_0 is a Herman ring of the rational function f_0 (of degree $2d$ if d is the degree of f).

As there is a conjugacy of $f_{|C}$ with $f_{0|F(C)}$, if θ is the rotation number of f in C, it also is that of f_0 in $F(C)$, and there is a conjugacy of $f_{0|U_0}$ with the rotation $z \mapsto e^{i2\pi\theta}z$ restricted to an annulus centered at 0. By Brjuno theorem[27], if ω is a Brjuno number, there is a rational function with Siegel disk with rotation number ω, and there is a rational function with Herman ring with rotation number ω. □

6.8 General Principle of Quasiconformal Surgery

In 1987, M. Shishikura stated the following lemma in an article where he presented the results obtained in his master thesis.

(6.12) Fundamental lemma of quasiconformal surgery:
If $f:\mathbb{C}_\infty \to \mathbb{C}_\infty$ *is a quasiregular mapping and, for disjoint open sets* $U_1,\ldots,U_p \subset \mathbb{C}_\infty$, $f(U_j)=U_{j+1}$, $f(U_p)\subset U_1$, $\psi : U \stackrel{def}{=} \cup_{j=1}^{p} U_j \to \widetilde{U} \subset \mathbb{C}_\infty$ *is a quasiconformal mapping,* $H:\widetilde{U}\to\widetilde{U}$ *is quasiregular with* H^p *holomorphic and* $f_{|U} = \psi^{-1}\circ H \circ \psi$ *with* $\frac{\partial f}{\partial \bar{z}} = 0$ a.e. *in* $\mathbb{C}_\infty \backslash f^{-N}(U)$, $N \in \mathbb{N}\cup\{0\}$, *then* f *is a* ***quasirational function****, i.e., such that it has a quasiconformal conjugacy with a rational function. If* f *is* K_1*-quasiregular and* ψ *is* K_2*-quasiconformal, the conjugacy of* f *with the rational function is* $K_1K_2^N$*-quasiconformal* (Fig. 6.8).

A consequence is the following easier to apply lemma.

(6.13) *If* $f:\mathbb{C}_\infty \to \mathbb{C}_\infty$ *is a quasiregular mapping such that for an open set* $U\subset\mathbb{C}_\infty$ *and a* $N\in\mathbb{N}$ f *is a quasiconformal mapping in* U*, each orbit of a point of* U *has less than* N *points in* U *and* $\frac{\partial f}{\partial \bar{z}} = 0$ a.e. *in* $\mathbb{C}_\infty\backslash U$*, then* f *is a quasirational function.*

However, in 1983, D. Sullivan[28] had sketched the proof of the Sullivan straightening theorem, which implies the two preceding statements. It is a **general principle of quasiconformal surgery** and gives a necessary and sufficient condition for a quasiregular mapping to be quasirational.[29] The proof uses the following auxiliary result.[30]

---

[28] Sullivan, D., Conformal dynamical systems, *Geometric dynamics* (Proc, of Rio de Janeiro conference, 1981), 725–752, Lecture Notes in Math., Springer-Verlag, Berlin, 1983.

[29] The proof given here follows that of P. Tukia in 1986. The result can be extended to nonrational quasiregular mappings, but that is not done here.

[30] What is called Hausdorff distance is a pseudodistance, as it can be 0 for different sets (see Section 4 of Appendix A of *CADOVA*). Young William Henry (1863–1942).

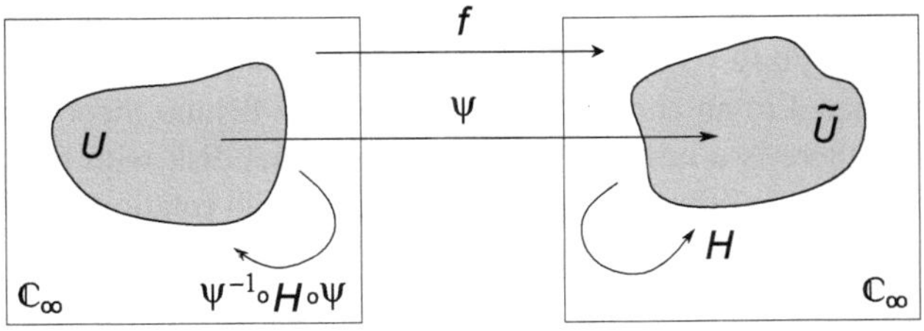

Fig. 6.8 Fundamental lemma of quasiconformal surgery ($N=0$, $p=1$): If f is a quasiregular mapping, $f_{|U}=\psi^{-1}\circ H\circ\psi$ with ψ quasiconformal, H a holomorphic function and $\frac{\partial f}{\partial \bar z}=0$ a.e. in $\mathbb{C}_\infty\setminus f^{-1}(U)$, then f is a quasirational function

(6.14) *If $S\subset B_1$ is bounded in the Poincaré metric of B_1 and $\#S\geq 2$, there is 1 unique disk with minimal radius in this metric containing S, whose center c_S is called* ***circumcenter*** *of S. The function $S\mapsto c_S$ is continuous with the Hausdorff distance.*

Proof Let $r_S>0$ be half the diameter of S in the Poincaré metric d_{B_1} of B_1 and $\{D_{r_n}(z_n)\}$ be a sequence of open disks in the metric space (B_1, d_{B_1}) with $D_{r_n}(z_n)\supset S$ and $\lim r_n = r_S$. If $\varepsilon>0$, there are $r>R_S$ and $r'\in\,]0, r_S[$ such that in (B_1, d_{B_1}) no geodesic in the annulus centered at 0 $D_r(0)\setminus D_{r'}(0)$ has length $>\frac{\varepsilon}{2}$. Since Möbius transformations preserving B_1 are isometries in (B_1, d_{B_1}), this is also the case for annuli (B_1, d_{B_1}) centered at any $z\in S$. If $m, n\in\mathbb{N}$ are large enough so that $r_m, r_n<r$, γ^* is the geodesic segment in (B_1, d_{B_1}) with end points z_m, z_n, since there is $z\in S$ such that $d_{B_1}(z, y)>r'$ for $y\in S$, by the convexity of d_{B_1}, less than half of γ^* in (B_1, d_{B_1}) is in $D_r(z)\setminus D_{r'}(z)$, and $d_S(z_m, z_n)\leq\varepsilon$. Therefore, $\{x_n\}$ is a Cauchy sequence in the complete metric space (B_1, d_{B_1}) and $\{x_n\}$ converges to the unique circumcenter c_S of S.

To prove the last sentence in the statement consider an arbitrary sequence $\{S_n\}\subset B_1$ of sets uniformly bounded in (B_1, d_{B_1}) converging to S in the Hausdorff distance. Let c_{S_n} and r_{S_n} be, resp., the circumcenter and the corresponding minimal radius of disks (B_1, d_{B_1}) with center c_{S_n} containing S_n, and $r\overset{def}{=}\underline{\lim}\, r_n$. Since the sequence $\{c_{S_n}\}$ is contained in a compact set of (B_1, d_{B_1}), it has a subsequence converging to a $c\in B_1$. Therefore, $S\subset \mathrm{cl}\, D(c, r)$ and $r_S\leq r$. If it were $r_S<r$, there would exist $r''\in\,]r_S, r[$ such that for $n\in\mathbb{N}$ big $r_n>r''$ and $S_n\subset D(c, r'')$, contradicting the definition of r_n; thus, $r=r_S$, and by the uniqueness of circumcenter, $c=c_S$, proving the continuity of the mapping $S\mapsto c_S$ with the Hausdorff distance. □

(6.15) Sullivan straighthening theorem:
A rational quasiregular mapping $f:\mathbb{C}_\infty\to\mathbb{C}_\infty$ is quasirational if and only if for some $K\geq 1$ the f^n, $n\in\mathbb{N}$ are K-quasiregular mappings.

Proof The necessity of the condition follows directly from the conjugacy of f with a holomorphic function F by a K-quasiconformal mapping as φ implies that $f^n = \varphi^{-1} \circ F \circ \varphi$, $n \in \mathbb{N}$, is a K^2-quasiconformal mapping. It remains to prove sufficiency.

$\mathcal{D}(f)$ is the set of points $z \in \mathbb{C}_\infty$ where f as a function of real variables is not differentiable or has a noninvertible derivative, which is a set of measure 0 since f is a quasiregular mapping. For $z \in \mathbb{C}_\infty \setminus \mathcal{D}(f)$, the Beltrami coefficient of f is denoted $\kappa_f(z)$ and $\lambda(z) = \frac{\partial f/\partial z}{\overline{\partial f/\partial z}}(z)$. Let $\mathcal{B}$ denote the set of Beltrami coefficients κ in $\mathbb{C}_\infty$ such that $\|\kappa\|_{L^\infty} < 1$, and for each $z \in \mathbb{C}_\infty$, let $\mathcal{B}_z$ denote the subset of B_1 of the values of every $\kappa \in \mathcal{B}$ at z. Since

$$\kappa_{g \circ f} = \frac{\kappa_f + (\kappa_g \circ f)\,[(\partial \overline{f}/\partial \overline{z})/(\partial f/\partial z)]}{1 + (\kappa_g \circ f)\,[(\partial \overline{f}/\partial z)/(\partial f/\partial z)]} = \frac{\kappa_f + (\kappa_g \circ f)\,(1/\lambda)}{1 + (\kappa_g \circ f)\,(\overline{\lambda\, \kappa_f})},$$

consider $\widetilde{f} : \mathcal{B} \to \mathcal{B}$ such that

$$\big[\widetilde{f}(\kappa)\big](z) = \frac{1}{\lambda(z)}\, \frac{\lambda(z)\,\kappa_f(z) + \kappa(f(z))}{1 + \kappa(f(z))\, \overline{\lambda(z)\,\kappa_f(z)}}, \qquad \text{a.e. in } \mathbb{C}_\infty .$$

The set $\mathcal{N}(f) \overset{def}{=} \cup_{n \in \mathbb{Z}} f^n\big(\mathcal{D}(f)\big)$ has measure 0 and is positively invariant under f. Denote $\widetilde{\mathcal{B}} \overset{def}{=} \big\{ \widetilde{f}^n(\kappa_0) : n \in \mathbb{Z} \big\} \subset \mathcal{B}$, which is a bounded set, and $\widetilde{\mathcal{B}}_z \overset{def}{=} \widetilde{\mathcal{B}} \cap \mathcal{B}_z$, so that $\widetilde{f}\big(\widetilde{\mathcal{B}}_{f(z)}\big) = \widetilde{\mathcal{B}}_z$.

As for every $z \in \mathbb{C}_\infty \setminus \mathcal{N}(f)$, $\widetilde{\mathcal{B}}_z \subset B_1$ is bounded in the Poincaré metric of B_1, by the preceding result, there is a unique circumcenter of $\widetilde{\mathcal{B}}_z$. Since $\widetilde{f}_{|\mathcal{B}_{f(z)}}$ is an isometry in that metric, as $\widetilde{f}\big(\widetilde{\mathcal{B}}_{f(z)}\big) = \widetilde{\mathcal{B}}_z$ and by the uniqueness of circumcenter, $\big[\widetilde{f}(\kappa)\big]\big(f(z)\big) = \kappa(z)$ to $z \in \mathbb{C}_\infty \setminus \mathcal{N}(f)$. For $z \in \mathcal{N}(f)$, define $\kappa(z) = \kappa_0(z)$; κ is invariant under f and $\|\kappa\|_{L^\infty} < 1$.

For $n \in \mathbb{N} \cup \{0\}$, denote $\widetilde{B}(n) \overset{def}{=} \big\{ \widetilde{f}^k(\kappa_0) : k \in \mathbb{Z}, |k| \le n \big\}$, $\widetilde{B}_z(n) \overset{def}{=} \widetilde{B}(n) \cap \widetilde{B}(z)$ (a finite set), and κ_z^n the circumcenter of $\widetilde{B}_z(n)$ which, by the preceding result, exists and is unique. Since $\mathcal{N}(f)$ is measurable and for $k \in \mathbb{Z}$ $\widetilde{f}^k$ is measurable, $z \mapsto \kappa_z^n$ defined in $\mathbb{C}_\infty \setminus \mathcal{N}(f)$ is measurable. By the preceding result, $\kappa_z^n \to \kappa_z$ when $n \to +\infty$; thus, κ is measurable. As $\mathcal{N}(f)$ has measure 0, $\widetilde{f}(\kappa) = \kappa$ a.e..

By the measurable Riemann mapping theorem (2.9), the Beltrami equation with coefficient κ has solutions, and they are quasiconformal mappings differing from one of them φ by composition at the left by holomorphic functions. Since κ is invariant under f, $f \circ \varphi$ is also a solution of the Beltrami equation, and there is a holomorphic function T such that $\varphi \circ f = T \circ \varphi$, i.e., $\varphi \circ f \circ \varphi^{-1} = T$. A quasiconformal mapping is a conjugacy of f with the holomorphic function T in $\mathbb{C}_\infty$, which is a rational function. Therefore, f is a quasirational mapping. □

6.9 Sewing a Siegel Disk at an Invariant Jordan Curve

One of the first uses of quasiconformal surgery was in 1984 by E.Ghys[31] to obtain a holomorphic function with a Siegel disk containing 0 with boundary the image by a

[31] Ghys, Étienne (1954–).

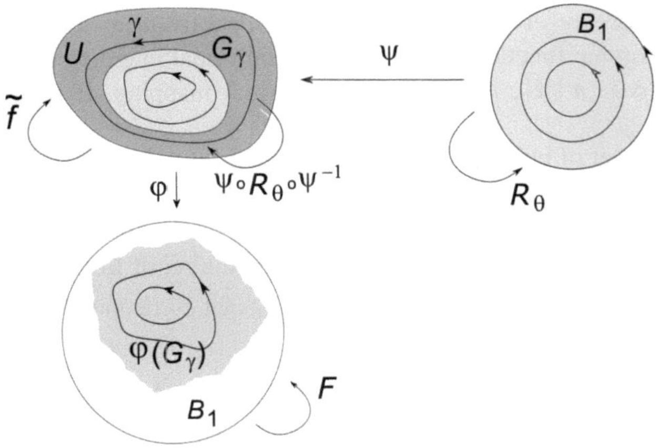

Fig. 6.9 Sewing a Siegel disk D (6.16) at an invariant Jordan curve γ with $D=\varphi(G_\gamma)$ (and $V=U$), $\widetilde{f}=f$ in $U\subset G_\gamma$ and $\widetilde{f}=\psi\circ R_\theta\circ\psi^{-1}$

quasiconformal mapping of the circle S^1 (Fig. 6.9), when trying to prove that there can be Siegel disks with boundary a Jordan curve without critical points, under the hypothesis of existence of a class of such functions from S^1 to S^1, what was proved by M. Herman, but only in 1986.

(6.16) *If U, V are neighborhoods of a Jordan curve γ^* in $\mathbb{C}_\infty$, $f\in H(U,V)$ and $f_{|\gamma^*}$ is a homeomorphism preserving orientation with rotation number $\theta\in[0,1[\backslash\mathbb{Q}$ relative to points in the region G_γ encircled in the positive orientation by γ, and there is a quasiconformal mapping conjugacy of $f_{|\gamma^*}$ with a rigid rotation R_θ in $S^1=\partial B_1$ with rotation number θ, then there is a quasiconformal mapping $\varphi: G_\gamma\cup U\to B_1$ and a holomorphic function $F: B_1\to\mathbb{C}_\infty$ with a Siegel disk with rotation number θ containing $\varphi(G_\gamma)$ and $F=\varphi\circ f\circ\varphi^{-1}$ in $B_1\backslash\varphi(G_\gamma)$.*

Proof There is a quasisymmetric function $\psi: S^1\to\gamma^*$ conjugacy of R_θ with $f_{|\gamma^*}$, By (3.3) and the Riemann mapping theorem, it has a quasiconformal extension to $\mathrm{cl}\,B_1$, also denoted ψ. The function $\widetilde{f}: G_\gamma\cup U\to G_\gamma\cup V$ coinciding with f in $U\setminus G_\gamma$ and with $\psi\circ R_\theta\circ\psi^{-1}$ in G_γ, by the sewing in a Jordan curve theorem (3.17), is a quasiregular mapping. An adaptation of the Shishikura fundamental lemma of quasiconformal surgery (6.12) gives the result. □

If in this result f is a rational function defined in $\mathbb{C}_\infty$, so is F.

One application is to obtain a Siegel disk from a Herman ring, which was one of the results of M. Shishikura in 1987.

(6.17) *If f is a complex rational function of degree d with Herman ring A with rotation number $\theta \in]0,1[$, γ^* is any of the Jordan invariant curves in A, and G_γ is one of the connected components separated in $\mathbb{C}_\infty$ by γ^*, then there is a quasiconformal mapping φ and a rational function F of degree d equal to $\varphi \circ f \circ \varphi^{-1}$ with a Siegel disk with rotation number θ or $1-\theta$ which contains $\varphi(G_\gamma)$.*

6.10 Polynomial-Like Functions

The Newton method for computing zeros of functions applied to complex cubic polynomials P corresponds to the iteration of the cubic rational function

$$f(z) = z - \frac{P(z)}{P'(z)}.$$

To study the convergence of this method, the polynomials can be normalized to the one parameter family $P_\lambda(z) = (z-1)[(z+\frac{1}{2})^2 - \lambda^2]$, considering the iteration of the corresponding cubic rational function $f_\lambda(z) = \frac{2z^3+1/4-\lambda^2}{3z^2-3/4-\lambda^2}$. Then,

$$P'_\lambda(z) = \left(z+\tfrac{1}{2}\right)^2 - \lambda^2 + 2(z-1)\left(z+\tfrac{1}{2}\right), \quad P''_\lambda(z) = 4\left(z+\tfrac{1}{2}\right) + 2(z-1) = 6z,$$

$$f'_\lambda = \frac{P_\lambda P''_\lambda}{[P'_\lambda]^2}, \qquad f''_\lambda = \frac{(P'_\lambda)^3 P''_\lambda + P P' P''' - 2P(P'')^2}{[P'_\lambda]^3}.$$

Orbits of points in the Julia set $J(f_\lambda)$ of f_λ do not converge to any of the zeros of P_λ. In 1983 J. Curry, L. Garrett and D. Sullivan posed the question of the existence of orbits outside the Julia set $J(f_\lambda)$ of f_λ not converging to one of the zeros of P_λ. Such an orbit must converge to an attracting periodic orbit and as such orbits of rational functions have a critical point[32] and $f'_\lambda = \frac{P_\lambda P''_\lambda}{[P'_\lambda]^2}$, the critical points are the zeros of P_λ, which are fixed points of f_λ, and the zero of $P''_\lambda(z)$, which is 0 since $P''_\lambda(z) = 6z$, and this is the only critical point of f_λ where P_λ does not vanish. So, the question corresponds to asking whether the orbit of 0 is periodic, and, if so, it is known that the set of initial points such that the Newton method does not converge to a zero of the polynomial is the union of the Julia set $J(f_\lambda)$ with the attracting basin of the orbit of 0. With exhaustive numerical search in a grid of values for λ, they observed that there is a set of values of λ for which the orbit of 0 is periodic and this set is a union of sets "similar" to the Mandelbrot set $\mathcal{M}$ of the family $Q_c(z) = z^2 + c$ (Fig. 6.10[33]).

---

[32] See Section 6 of Chapter 13 of *CADOVA*.

[33] Base figure author and date: Gertbuschmann (01.04.2011), under License *Creative Commons Public Domain* https://commons.wikimedia.org/wiki/File:Juliasetsdkpictnewman.jpg; the colors of the original figure were replaced by shades of gray that distinguish regions of parameter values for which the Newton method converges to different zeros or diverges (in this case in black).

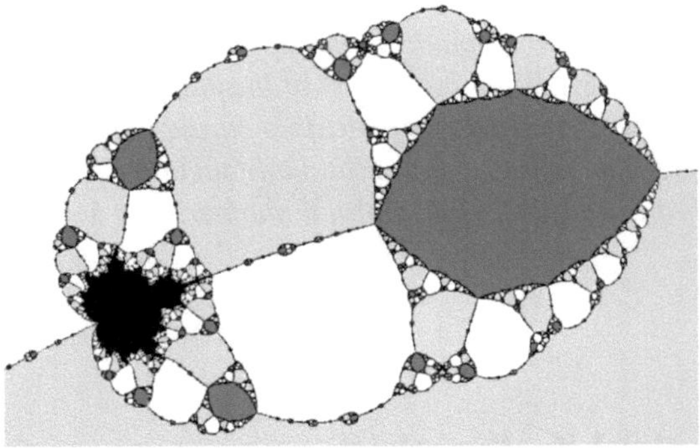

Fig. 6.10 Part of the set of the parameter $\lambda \in \mathbb{C}$ analogous to the Mandelbrot set but for f_λ whose iterations are those of the Newton method applied to the polynomial $P_\lambda(z) = (z-1)[(z+\frac{1}{2})^2 - \lambda^2]$, with a zone (in black in the figure) "similar" to the Mandelbrot set of $Q_c(z) = z^2 + c$ for which the Newton method does not converge to any zero of P_λ. The regions in white are values of λ for which the Newton method converges to the zero of P_λ at 1; each of the regions in the two shades of gray is of values of λ for which Newton method converges to each one of the other zeros of P_λ, $-\frac{1}{2} \pm \lambda$, de P_λ

The heuristics for this possibility is that when the orbit of 0 is periodic with period k, since

$$f_\lambda(0) = \frac{1/4-\lambda^2}{-3/4-\lambda^2}, \quad f_\lambda'(0) = 0, \quad f_\lambda''(0) = \frac{P_\lambda(0)P_\lambda'''(0)}{[P_\lambda'(0)]^2} = \frac{6 f_\lambda(0)}{P_\lambda'(0)} = \frac{1/4-\lambda^2}{(-3/4-\lambda^2)^2},$$

the periodic orbit is superattracting, and there is a conjugacy of f_λ^k in a neighborhood of 0 with z^2, for small neighborhoods U of 0, $f_\lambda^k(U) \subset U$, and there is a larger neighborhood V such that $f_\lambda^k(V) \supset V$, what also happens for small perturbations of λ and in this case the dynamics in V is "similar" to that of Q_c with c depending on λ.

Figures 6.11, 6.12, 6.13, 6.14, and 6.15 illustrate the "similarity" of Fatou components of f_λ attracting basins of the periodic orbit of 0 (for λ such that it exists) to filled Julia sets of Q_c, with c chosen to illustrate the "similarity." Each Figs. 6.11, 6.12, 6.13, 6.14, and 6.15 has 3 figures for the same value of λ, the 1st and 2nd (an enlargement) of Julia and Fatou sets of f_λ and the 3rd the Julia set of the quadratic polynomial $Q_c(z) = z^2 + c$ with c chosen to show the "similarity." The value of λ for each figure corresponds to a point in one of the hyperbolic connected components of the Mandelbrot set $\mathcal{M}$ of Q_c, more specifically, in increasing order of figure number, to the hyperbolic component of functions f_λ (see in Fig. 6.1):

1. with attracting fixed point (the main cardioid of $\mathcal{M}$),
2. with attracting periodic orbit of minimal period 2 (the disk immediately to the left of the main cardioid of $\mathcal{M}$),
3. with attracting periodic orbit of minimal period 3 (the hyperbolic component immediately smaller than the previous ones adjacent to the main cardioid at the top),

4. with attracting periodic orbit of minimal period 4 (the hyperbolic component immediately smaller than the previous ones adjacent to the main cardioid at the top),
5. with attracting periodic orbit of minimal period 4 (but the hyperbolic component immediately to the left of the circle of 2).

In 1985, A. Douady and J. Hubbard[34] analyzed theoretically this question, for which they introduced the notion of polynomial-like function and obtained several general results for them.

If U_1, U_2 are simply connected bounded regions of $\mathbb{C}$ with analytic boundaries and $\mathrm{cl}\, U_1 \subset U_2$, a **polynomial-like function** of degree $d \in \mathbb{N}$ from U_1 to U_2 is $f \in H\big(\overline{U_1}, U_2\big)$ such that $f(U_1) = U_2$, $f(\partial U_1) = \partial U_2$, and (U_1, f) is a covering of U_2.

(6.18) Example: Consider the complex function $f(z) = \cos z - 2$ and the set

$$U_1 = \{z \in \mathbb{C} : |\mathcal{R}e\, z| < 2\,,\, |\mathcal{I}m\, z| < 3\}\,.$$

f is a polynomial-like function of degree 2 from U_1 to $U_2 = f(U_1)$ (Fig. 6.16).

Analogously to polynomial functions,[35] the **filled Julia set** of a polynomial-like function f from U_1 to U_2 is $K(f) = \bigcap_{n=0}^{\infty} f^{-n}(U_1)$, and the **Julia set** of f is $J(f) = \partial K(f)$.

Many properties of the dynamics of polynomial functions extend to polynomial-like functions with analogous proofs, in particular if they use arguments based on properties of holomorphic functions and not in the rigidity of polynomial functions, such as:

1. *The attracting basin of any attracting periodic orbit has at least one critical point.*
2. *The filled Julia set is connected if and only if it contains all critical points; if not, it is a Cantor set.*
3. *The orbit of a critical point in the filled Julia set converges to an attracting periodic orbit if and only if the function is hyperbolic.*

Polynomial-like functions f, g are said to be **hybridly conjugated**[36] if there is a quasiconformal mapping ψ from an open set $U \supset K(f)$ to an open set $V \supset K(g)$ Holomorphic in $K(f)$ and a conjugacy of f with g; ψ is said to be a **hybrid conjugacy**.

In the 1985 article where A. Douady and J. Hubbard introduced polynomial-like functions, it was proved that they are hybridly conjugated to polynomial functions,[37] allowing to enlarge the properties of the dynamics of polynomial functions that can be extended.

---

[34] Douady, A., Hubbard, J.H., On the dynamics of polynomial-like mappings. *Ann. Sci. École Norm. Sup. 4éme série* **18** (1985), 287–343.

[35] See Section 15 of Chapter 13 of *CADOVA*.

[36] This definition is a little different from the original one in A. Douady and J. Hubbard.

[37] They called this result “straightening theorem.”

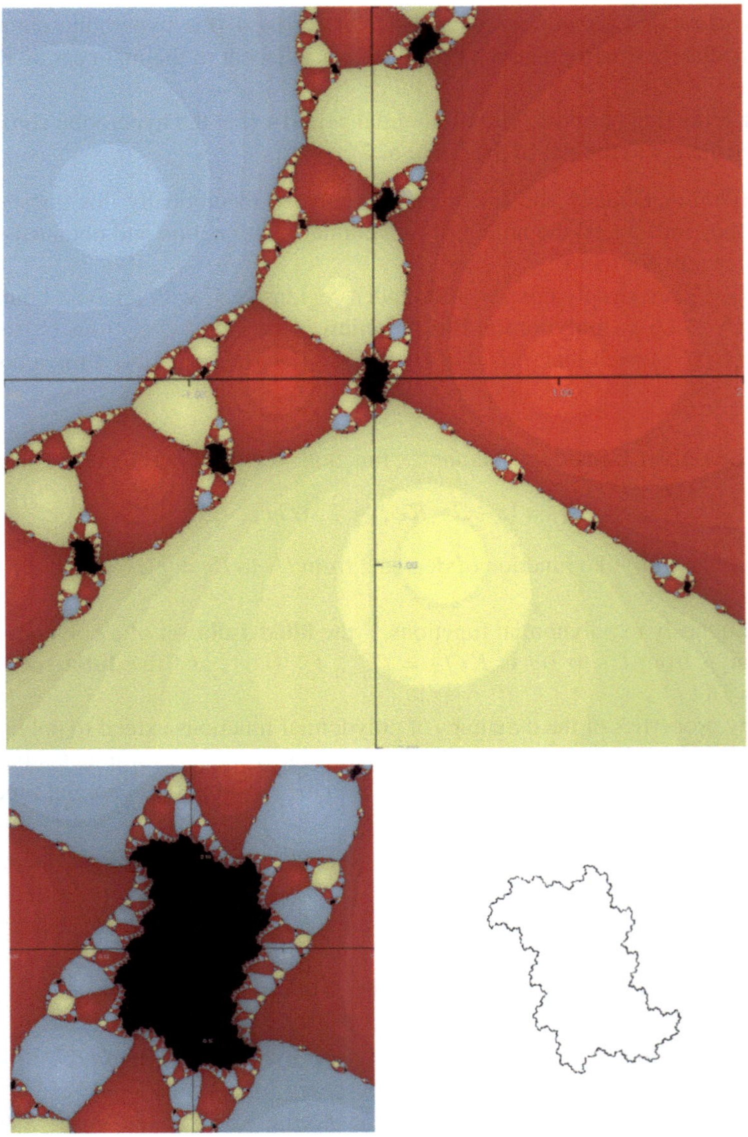

Fig. 6.11 The first and second figures (computed with a program made available by Mark McClure in the Internet) are Julia and Fatou sets of f_λ whose iterations are those of Newton method applied to the polynomial $P_\lambda(z) = (z-1)[(z+\frac{1}{2})^2 - \lambda^2]$ with $\lambda = -0.883 + i\,0.933$, which is a parameter for which there are Fatou components of initial values for which the Newton method converges to a fixed point of f_λ which is not a zero of P_λ. The Julia set is the boundary of the regions with different shades in the figure whose union is the Fatou set; the four different shades correspond to initial conditions of the Newton method that iterated converge to each one of the three zeros of P_λ or (the darkest one) to the fixed point of f_λ that is not a zero of P_λ. The third figure (computed with *Mathematica*) is of the Julia set of $Q_c(z) = z^2 + c$ with c chosen to show the "similarity"

Fig. 6.12 Figures analogous to Fig. 6.11, but for $\lambda = -0.903 + 0.915$, which is a parameter for which there are Fatou components of initial values of the Newton method for which there is no convergence to a zero of P_λ, but to a periodic orbit of period 2 of f_λ. The regions in the darkest shade correspond to initial conditions that iterated converge to such a periodic orbit

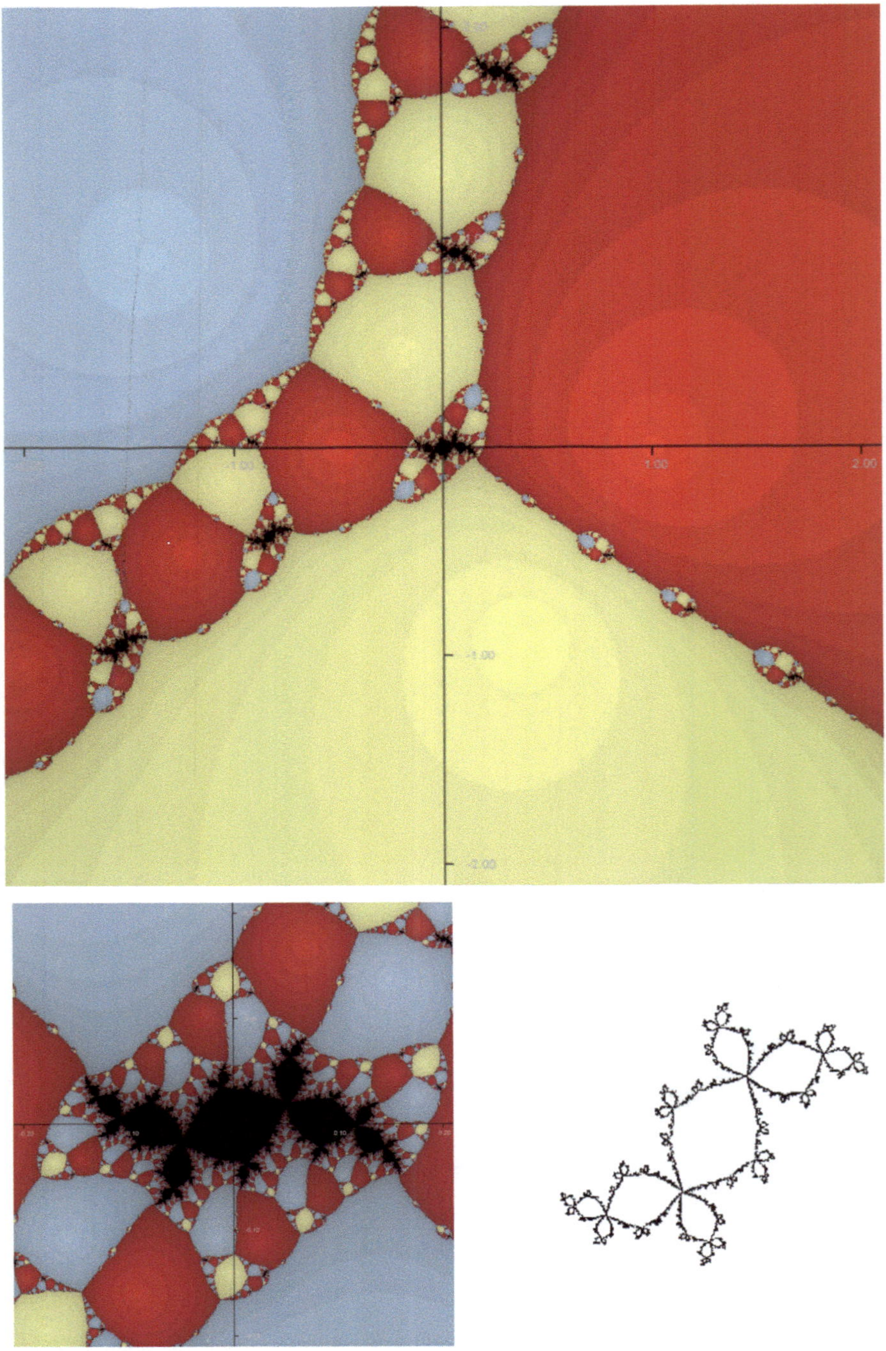

Fig. 6.13 Figures analogous to Fig. 6.11, but for $\lambda = -0.884 + 0.908$, which is a parameter for which there are Fatou components of initial values of the Newton method for which there is no convergence to a zero of P_λ, but to a periodic orbit of period 3 of f_λ. The regions in the darkest shade correspond to initial conditions that iterated converge to such a periodic orbit

Fig. 6.14 Figures analogous to Fig. 6.11, but for $\lambda = -0.887+0.937$, which is a parameter for which there are Fatou components of initial values of the Newton method for which there is no convergence to a zero of P_λ, but to a periodic orbit of period 4 of f_λ. The regions in the darkest shade correspond to initial conditions that iterated converge to such a periodic orbit

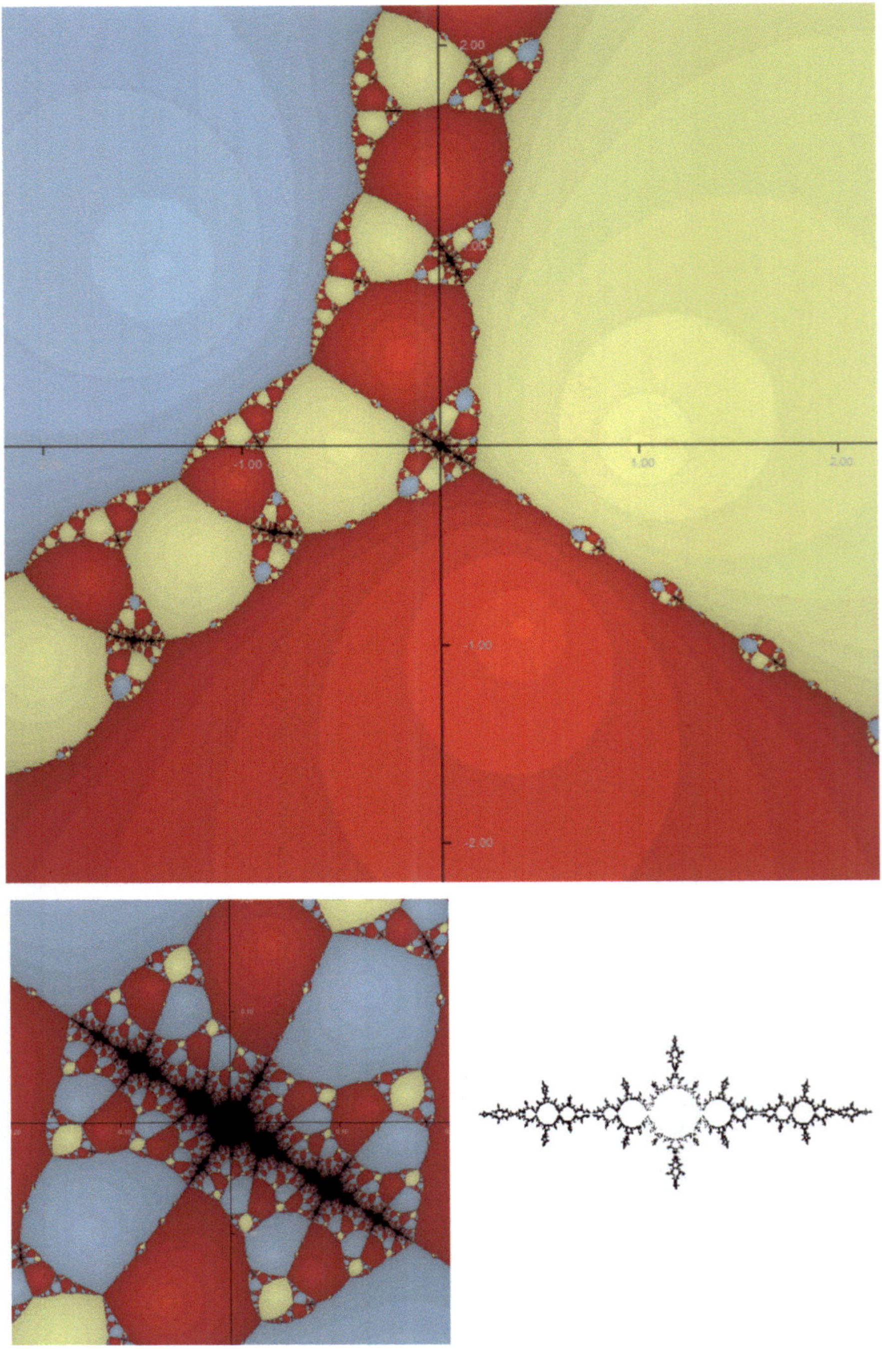

Fig. 6.15 Figures analogous to Fig. 6.11, but for $\lambda = 0.9051 + 0.9142$, which is a parameter for which there are Fatou components of initial values of the Newton method for which there is no convergence to a zero of P_λ, but to a periodic orbit of period 4 of f_λ. The regions in the darkest shade correspond to initial conditions that iterated converge to such a periodic orbit

Fig. 6.16 $U_1 = \{z \in \mathbb{C} : |\mathcal{R}e\, z| < 2\,, |\mathcal{I}m\, z| < 3\}$ and $U_2 = f(U_1) \subset U_2$ of the Example (6.18) of polynomial-like function $f(z) = \cos z - 2$ de U_1 in U_2

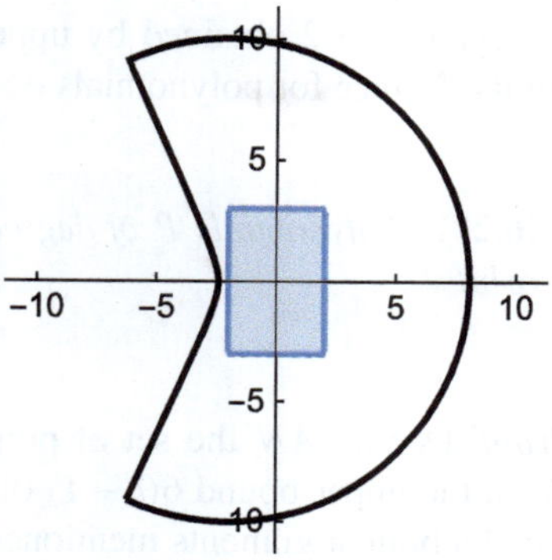

(6.19) *Every polynomial-like function of degree $d \in \mathbb{N}$ is hybridly conjugated with a polynomial of the same degree.*

Proof Let f be a polynomial-like function of degree d from U_1 to U_2, both simply connected bounded regions of $\mathbb{C}$, and $\rho > 1$. By the Riemann mapping theorem, there is a conformal homeomorphism Φ from $\mathbb{C}_\infty \backslash U_2$ onto $\mathbb{C}_\infty \backslash \mathrm{cl}\, B_{\rho^d}$ with $\Phi(\infty) = \infty$. Since the composition of f with an analytic simple path as a function of one real variable describing ∂U_1 is a path describing d turns around ∂U_2, Φ can be defined in $z \in \partial U_1$ such that $\Phi^d(z) = \Phi\big(f(z)\big)$ and then extended to $U_2 \backslash \mathrm{cl}\, U_1$ so as to be holomorphic in U_2.

Consider $g : \mathbb{C} \to \mathbb{C}$ such that $g = f$ in U_1 and $g = \Phi^{-1} \circ \Phi^d$ outside U_1. Since g is a composition of a quasiconformal mapping with a holomorphic function, its extension by continuity to $\mathbb{C}_\infty$ is quasiregular and $\mathbb{C}_\infty \backslash U_1$ is invariant under g. By the Shishikura fundamental lemma of quasiconformal surgery (6.12), g is a quasirational function. Therefore, g is conjugated by a quasiconformal mapping with a rational function, and since g is polynomial-like in U_1, it is conjugated with a polynomial by a conjugacy holomorphic in U_2 (a hybrid conjugacy) and the polynomial has degree d. □

This result, together with the observation in the 2nd paragraph of this section, justifies the relationship of the dynamics of f_λ for values of λ where the Newton method does not converge to a zero of the polynomial P_λ with the dynamics of quadratic polynomials, since in this case and for λ such that 0 is a point of a periodic orbit of period k in a neighborhood of 0 f_λ^k is a polynomial-like function of degree 2 and, therefore, hybridly conjugated to a polynomial of degree 2.

With the flexibility allowed by perturbations with polynomial-like functions when compared with polynomial functions, it is possible to transform neutral periodic orbits to attracting and to easily obtain the optimal upper bound $d-1$ of the number of attracting or neutral periodic orbits of polynomials of degree d, which had been obtained in 1982 by A. Douady with a more complicated argument. Thus, the upper bound $5d-1$ resulting from the upper bound $6(d-1)$ for rational functions of degree $d \geq 2$ obtained by upper bounding algebraically the number of critical points,[38] since for polynomials ∞ is an attracting fixed point with multiplicity d.

(6.20) *Polynomials P of degree d have* $\leq d-1$ *attracting or neutral periodic orbits.*

Proof Denote AN the set of points of attracting or neutral periodic orbits of P. From the upper bound $6(d-1)$ of the number of these orbits obtained purely with the algebraic arguments mentioned above, AN is a finite set. Let m be the number of neutral periodic orbits and $z_1, \ldots, z_m$ points in each one of them. Consider a polynomial Q whose zeros are the points of AN such that $\sum_{j=1}^{m} \mathcal{R}e \, \frac{Q'(z_j)}{P'(z_j)} < 0$, where $z_1, \ldots, z_m$, what is possible with a polynomial Q of large degree because $P'(z_j) \neq 0$ for $j \in \{1, \ldots, m\}$. For $\varepsilon > 0$, define $f_\varepsilon = P + \varepsilon Q$. As $Q = 0$ in AN, for small $\varepsilon > 0$, the attracting or neutral periodic orbits of f_ε and P are the same. When $\varepsilon \to 0$,

$$\sum_{j=1}^{m} \log |f_\lambda'(z_j)|$$
$$= \sum_{j=1}^{m} \log |P'(z_j)| + \sum_{j=1}^{m} \log \left|1 + \varepsilon \frac{Q'(z_j)}{P'(z_j)}\right| = 0 + \varepsilon \sum_{j=1}^{m} \mathcal{R}e \, \frac{Q'(z_j)}{P'(z_j)} + O(\varepsilon^2);$$

so, the neutral periodic orbits for $\varepsilon = 0$ turn attracting for small $\varepsilon > 0$.

Without loss of generality, the polynomial P is taken to be monic. For $r > 0$, big and $\varepsilon > 0$ small f_ε is a proper function of $f_\varepsilon^{-1}(B_{rd})$ onto B_{rd} and defines a relation d-to-1. Hence, the restriction of f_λ to $f_\varepsilon^{-1}(B_{rd})$ is a polynomial-like function of degree d from $f_\varepsilon^{-1}(B_{rd})$ to B_{rd}. By the preceding result, there is a hybrid conjugacy of f_ε with a polynomial S of degree d. So, there is a bijection of the periodic attracting orbits of f_λ and those of S. As $d-1$ upper bounds the number of attracting periodic orbits (not counting ∞) of any degree d polynomial[39], it also upper bounds that number of f_λ, and the number of attracting or neutral periodic orbits of P. □

(6.21) Example: If P is a quadratic polynomial with a Siegel disk cycle of minimal period $m \in \mathbb{N}$, $P^j(U)$, $j \in \{0, 1, \ldots, m-1\}$, $P^m(U) = U$, for every j, there is a Fatou component $V_j \neq U_j$ with $V_j = f(U_{j-1})$. By the preceding result, there are no other cycles of sets with attracting or neutral orbits besides $\{\infty\}$. Thus, by the

[38] See Section 10 of Chapter 13 of *CADOVA*. The proof of the following result is presented in the book by L. Carleson and T.W. Gamelin listed in the final bibliography. In 1987, M. Shishikura published the optimal upper bound $\leq 2(d-1)$ of the number of attracting or neutral periodic orbits of rational functions of degree ≥ 2 he had obtained with quasiconformal surgery in his 1985 master thesis, as shown in the next section. This upper bound implies for polynomials the upper bound $\leq d-1$, coinciding with that obtained by A. Douady in 1982.

Sullivan classification theorem,[39] successive iterations of each Fatou component bounded by P converge to the Siegel disk cycle.

6.11 Upper Bound of Number of Attracting or Neutral Periodic Orbits

In this section, the optimal upper bound $\leq 2(d-1)$ of the number of attracting or neutral periodic orbits of rational functions of degree ≥ 2 is proved with quasiconformal surgery as published by M. Shishikura in[40] 1987 and obtained in his master thesis of 1985.

Given a rational function f, denote

$$n_A(f), n_N(f), n_P(f), n_I(f), n_C(f), n_{AB}(f), n_{PB}(f),\ n_{SD}(f),\ n_{HR}(f), n_{CP}(f), n_{CPER}(f)$$

the number of cycles, resp., attracting, neutral, parabolic, irrationally neutral, of Cremer points,[41] of attracting basins of periodic orbits, of parabolic basins of periodic orbits, of Siegel disks, of Herman rings, of critical points in $\mathbb{C}_\infty$, and of critical points with nonperiodic eventually periodic repelling orbits. Adopt a similar notation for the corresponding numbers of orbits or cycles contained in a $D \subsetneqq \mathbb{C}_\infty$ just replacing (f) by (f, D).

As in the proof of the upper bound $n_A(P, \mathbb{C}) + n_N(P, \mathbb{C}) \leq d-1$ for polynomials P of degree d at the end of the preceding section, the idea of the proof is to perturb the function transforming neutral cycles to attracting, but obtaining such perturbations of rational functions is much more difficult. P. Fatou conjectured this upper bound in 1920 and had the idea of the perturbation and applied it, but only managed to get half of the neutral cycles. So, as he had proved

$$n_A(f)+n_N(f) \leq n_{AB}(f)+n_{PB}(f) \leq n_{PC}\big(f, \mathbb{C}_\infty \setminus J(f)\big) \leq 2(d-1),$$

P. Fatou obtained in 1920 the upper bound $n_A(f)+\frac{1}{2}\, n_N(f) \leq 2(d-1)$.

As each cycle of attracting or neutral periodic orbits of a rational function f of degree d has a finite number of attracting or neutral periodic orbits, it suffices to obtain successive perturbations of f, as long as there are neutral periodic orbits, transforming each one of them, by arbitrary order, in an attracting periodic orbit

---

[39] See Section 12 of Chapter 13 of *CADOVA*.

[40] We follow mainly the remarkable article Shishikura, M., On the quasiconformal surgery of rational functions, *Annales scientifiques de l'É.N.S. 4e series,* **20** (1987), 1–29, and parts of the book by B. Branner and N. Fagella in the final bibliography. This result reduces to $\frac{1}{2}$ and $\frac{1}{3}$ the upper bounds of the number of periodic orbits, resp., neutral with multiplier $\neq 1$ and attracting or neutral got without resorting to quasiconformal surgery in Section 10 of Chapter 13 of *CADOVA*.

[41] An **irrational neutral point** is a point p of a neutral periodic orbit of period $m \in \mathbb{N}$ with multiplier $\lambda = e^{i2\pi\omega}$, $\omega \in [0, 1[\setminus \mathbb{Q}$; it is called a **Siegel point** if f^m is linearizable in a neighborhood of p, i.e., it can be conjugated with the rotation $z \mapsto \lambda z$, and otherwise it is called a **Cremer point** (see Section 11 of Chapter 13 of *CADOVA*).

without modifying f in the vicinity of repelling periodic orbits, and to apply quasiconformal surgery to ensure there is a quasiconformal conjugacy of the function obtained at each step with a rational function of degree d. As in each step the number of attracting periodic orbits increases and it is upper bounded by $2(d-1)$, the process ends in a finite number of steps.

For each periodic orbit $O(z_0) = \{z_0, \ldots, z_{p-1}\}$, with $f(z_j) = z_{j+1}$ for $j \in \{0, \ldots, p-2\}$ and $f(z_{p-1}) = z_0$, associated with an attracting or neutral cycle, we consider a small perturbation changing f to a quasirational function g_δ so that $O(z_0)$ be an attracting periodic orbit of g_δ. Without loss of generality, ∞ does not belong to the set of attracting or neutral periodic points, what can be achieved by changing variables with a Möbius transformation, ensuring that the attracting or neutral periodic points belong to a compact subset of $\mathbb{C}$, so that f remains unchanged in the complement of a neighborhood V_ϵ of ∞ in $\mathbb{C}_\infty$. To do this, let A_δ be a transition ring from the perturbation of f in a small neighborhood of $O(z_0)$ to $g_\delta = f\circ(1_{\mathbb{C}_\infty}+\delta P)$ with $\delta > 0$ small and P an appropriate polynomial in $O(z_0)$ for the neighborhood V_δ where f remains unchanged (Fig. 6.17). Thus, $O(z_0)$ is a periodic orbit of g_δ and $\left|(g_\delta^p)'\right| = \left|(f^p)'(1+\delta P')^p\right|$. If $O(z_0)$ is a neutral periodic orbit of f, choosing P so that P' is -1 in $O(z_0)$, in this set, we have $\left|(g_\delta^p)'\right| = (1-\delta)^p < 1$ for $\delta < 1$, and $O(z_0)$ is an attracting periodic orbit of g_δ. If $O(z_0)$ is an attracting periodic orbit of f, it is $|f^p| \le M < 1$ in $O(z_0)$ and $\left|(g_\delta^p)'\right| = M|1+\delta P'|^p$, which for $\delta > 0$ small is < 1, and $O(z_0)$ is the attracting periodic orbit of g_δ.

(6.22) *If f is a rational function of degree $d \ge 2$ and $\{z_0, \ldots, z_N\}$ is the union of the attracting with the neutral periodic orbits of f, then for $\varepsilon > 0$ there are rational functions f_ε of degree d and $z_0^\varepsilon, \ldots, z_N^\varepsilon \in \mathbb{C}_\infty$ such that:*

(i) $f_\varepsilon \to f$ and $z_j^\varepsilon \to z_j$ when $\varepsilon \to 0$, for $j \in \{0, \ldots, N\}$; (ii) if $f(z_j) = z_k$, then $f_\varepsilon(z_j^\varepsilon) = z_k^\varepsilon$ and each z_j^ε is a point in the attracting periodic orbit of f_ε with the same period as the periodic orbit of f containing z_j. Thus,

$$n_A(f_\varepsilon) \ge n_A(f) + n_N(f), \qquad n_A(f) + n_N(f) \le 2(d-1).$$

Proof If P is a polynomial of degree k, for $\varepsilon \in \mathbb{C}$ with small $|\varepsilon|$, $H_\varepsilon : \mathbb{C}_\infty \to \mathbb{C}_\infty$, $H_\varepsilon(z) = z + \varepsilon P(z)\,\rho\left(|\varepsilon|^{1/k}|z|\right)$ for $z \in \mathbb{C}$ and $H_\varepsilon(\infty) = \infty$, where ρ is a C^∞ function $\rho : [0, +\infty[\to [0, 1]$ with support in $[0, 2]$ and 1 in $[0, 1]$, satisfies the Beltrami equation with C^∞ coefficient κ_{H_ε} such that $\|\kappa_{H_\varepsilon}\|_{L^\infty} < 1$ and $\lim_{\varepsilon \to 0} \|\kappa_{H_\varepsilon}\|_{L^\infty} = 0$ (so, it is Quasiconformal) and $H_\varepsilon(z) \to 1_{\mathbb{C}_\infty}$ uniformly (in the spherical metric) in $\mathbb{C}_\infty$ when $\varepsilon \to 0$.

Let $\delta_0 > 0$ and for $\delta \in [0, \delta_0[$ let $E_\delta \subset \mathbb{C}$ be uniformly bounded open sets with $f(\infty) \in E_0 \subset E_\delta$ and E_δ invariant under $f \circ (1_{\mathbb{C}} + \delta P)$, and let $g_\delta = f \circ H_\delta$, $V_\delta = \mathbb{C}_\infty \backslash \mathrm{cl}\, B_{\delta^{-1/k}}$. For $\delta > 0$ small, $E_\delta \cap \mathrm{cl}\, V_\delta = \emptyset$ and $g_\delta(\mathrm{cl}\, V_\delta) \subset E_\delta$. Since for $\delta > 0$ small g_δ is a quasiregular mapping and $\frac{\partial g_\delta}{\partial \bar{z}} = 0$ in $E_\delta \cup \left(\mathbb{C}_\infty \backslash g_\delta^{-1}(E_\delta)\right) \subset \mathbb{C}_\infty \backslash \mathrm{cl}\, V_\delta$, by the Shishikura fundamental lemma of quasiconformal surgery (6.12) (with $f = g_\delta$, $\psi = 1_{E_\delta}$, $N = 1$ in that lemma), g_δ is

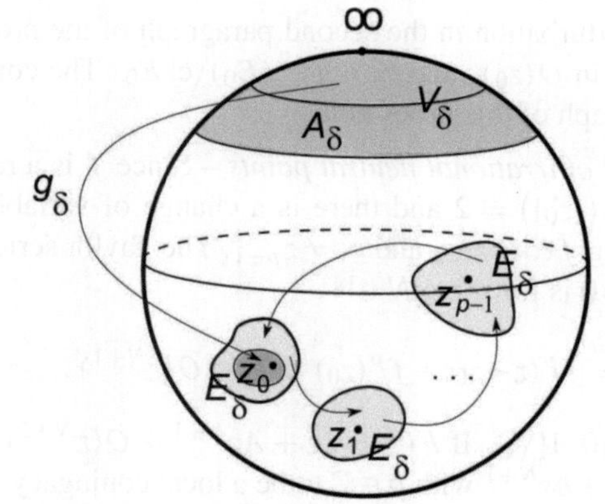

Fig. 6.17 Perturbation scheme of a rational function f to a quasirational function g_δ to ensure that an attracting or neutral periodic orbit of period p, $O(z_0)=\{z_0, \ldots, z_{p-1}\}$, is an attracting periodic orbit of g_δ which in a neighborhood of $O(z_0)$ is $g_\delta = f \circ (1_{\mathbb{C}_\infty}+\delta P)$ without changing f in a neighborhood V_δ of ∞ containing repelling periodic orbits with a transition ring A_δ

a quasirational function. Thus, it is conjugated by a quasiconformal mapping φ_δ with a rational function f_δ, i.e., $f_\delta = \varphi_\delta \circ g_\delta \circ \varphi_\delta^{-1}$. The function φ_δ satisfies the Beltrami equation with a coefficient κ_{φ_δ} such that $|\kappa_{\varphi_\delta}| \le \frac{K_\delta - 1}{K_\delta + 1}$ with H_δ a K_δ-quasiconformal mapping. As $H_\delta \to 1_{\mathbb{C}_\infty}$ uniformly when $\delta \to 0$, we have $K_\delta \to 1$ and $\|\kappa_{\varphi_\delta}\|_{L^\infty} \to 0$ when $\delta \to 0$. By the continuous dependence on parameters of the solutions of the Beltrami equation converging to the identity at ∞, $\varphi_\delta \to 1_{\mathbb{C}}$ and, therefore, $f_\delta \to f$ uniformly, when $\delta \to 0$. Consequently: *If $\delta_0 > 0$ and for $\delta \in [0, \delta_0[$, $E_\delta \subset \mathbb{C}$ are uniformly bounded open sets with $f(\infty) \in E_0 \subset E_\delta$ and E_δ invariant under $f \circ (1_{\mathbb{C}} + \delta P)$, then for small $\delta > 0$ there are mappings g_δ quasirational and φ_δ quasiconformal in $\mathbb{C}_\infty$ such that $f_\delta = \varphi_\delta \circ g_\delta \circ \varphi_\delta^{-1}$ are rational functions and $\varphi_\delta \to 1_{\mathbb{C}}$, $f_\delta \to f$ uniformly, when $\delta \to 0$.*

These perturbations are constructed below separately for $O(z_0)$ *(1)* attracting, *(2)* parabolic, *(3)* of irrational neutral points:

(1) Attracting periodic orbit – With $E_\delta = E_0$ a union of small disks with centers at each z_j, $j \in \{0, \ldots, p-1\}$, invariant under f with $\infty \in f^{-1}(E_0) \setminus \mathrm{cl}\, E_0$ and P a polynomial with $P(z_j)=0$, $j \in \{0, \ldots, p-1\}$, and the conditions at the end of the 2nd paragraph of this proof hold.

(2) Parabolic periodic orbit – There is an attracting petal[42] $\mathcal{P}$ of f at z_0. With $E_0 = \cup_{j=1}^{p-1} f^j(\mathcal{P})$, $f^p(E_0) \subset E_0 \cup O(z_0)$. The elements of $O(z_0)$ are attracting periodic points of g_δ and there is a neighborhood $N_\delta(z_0)$ of z_0 such that $g_\delta\big(N_\delta(z_0)\big) \subsetneqq N_\delta(z_0)$. Let $E_{\delta,0} = \mathcal{P} \cup N_\delta(z_0)$. For small $\delta > 0$, $g_\delta^p(\mathrm{cl}\, E_{\delta,0}) \subset E_{\delta,0}$; define $E_{\delta,j} = g_\delta(E_{\delta,j-1})$ for $j \in \{1, \ldots, p-1\}$, $E_\delta = \cup_{j=0}^{p-1} E_{\delta,j}$, and choose $N_\delta(z_0)$ to be $E_\delta \subset \mathrm{cl}\, B_{\delta^{-1/k}}$, where k is the degree of the polynomial P to be

---

[42] See Section 9 of Chapter 13 of *CADOVA*.

considered in the perturbation in the second paragraph of the proof, chosen so that P is 0 and P' is -1 in $O(z_0)$, and $\infty \in f^{-1}(E_0)\backslash \mathrm{cl}\, E_0$. The conditions at the end of the second paragraph of this proof hold.

(3) Periodic orbit of irrational neutral points – Since f is a rational function of degree $d \geq 2, \#f^{-1}(\{z_0\}) = 2$ and there is a change of variables with a Möbius transformation giving $f(\infty)=z_0$ and $\infty \neq z_{p-1}$. The Taylor series of $f(z+z_0)$ in a neighborhood of $z=0$ is for some $N \in \mathbb{N}$,

$$F(z) \stackrel{def}{=} f^p(z+z_0) - f^p(z_0) = \lambda z + O(z^{N+1}), \quad z \to 0,$$

with $\lambda = e^{i2\pi\omega}$, $\omega \in [0, 1[\backslash\mathbb{Q}$. If $F(z) = \lambda z + Az^{N+1} + O(z^{N+2})$, with $A \neq 0$ when $z \to 0$, for $T(s) = s + bs^{N+1}$ with $b \in \mathbb{C}$ to be a local conjugacy of F with

$$(T^{-1} \circ F \circ T)(s) = \sum_{j=1}^{N+1} c_j s^j + O(s^{N+2}),$$

it must be $(F \circ T)(s) = T\left(\sum_{j=1}^{N+1} c_j s^j + O(s^{N+2})\right)$, when $s \to 0$, i.e.,

$$\begin{aligned}\lambda(s+bs^{N+1}) &+ A(s+bs^{N+1})^{N+1} + O\left((s+bs^{N+1})^{N+2}\right) \\ &= \sum_{j=1}^{N+1} c_j s^j + O(s^{N+2}) + b\left(\sum_{j=1}^{N+1} c_j s^j + O(s^{N+2})\right)^{N+1}, \quad s \to 0,\end{aligned}$$

or

$$\lambda s + \lambda b s^{N+1} + A s^{N+1} + O(s^{N+2}) = \sum_{j=1}^{N+1} c_j s^j + b c_1^{N+1} s^{N+1} + O(s^{N+2}), \quad s \to 0.$$

Thus, $c_1 = \lambda$, $c_j = 0$ for $j \in \{2, \ldots, N\}$ and $\lambda b + A = c_{N+1} + b\lambda^{N+1}$. With $c_{N+1} = 0$ and $b = \frac{A}{\lambda^{N+1} - \lambda}$ (note that $\lambda^{N+1} - \lambda \neq 0$ since ω is irrational) if for some $N \in \mathbb{N}$ it is $F(z) = \lambda z + O(z^{N+1})$ when $z \to 0$, then there is a local conjugacy T of F with $\lambda z + O(z^{N+2})$ when $z \to 0$. Applying this successively for increasing values of N gives that for every $N \in \mathbb{N}$, there is a polynomial local conjugacy Q of F with $\lambda z + O(z^{N+2})$, when $z \to 0$. If P is a polynomial of degree k equal to 0 with P' equal to -1 in the orbit of z_0 and $N = k+1$,

$$(Q^{-1} \circ g_\delta^p \circ Q)(z) = [(Q^{-1} \circ F \circ Q)(z) - \lambda z] + \lambda z(1-\delta)^p + O(\delta z^2), \quad z \to 0.$$

Therefore,

$$(Q^{-1} \circ g_\delta^p \circ Q)(z) = \lambda z\,[(1-\delta)^p + O(\delta z) + O(z^{k+2}), \quad z \to 0.$$

Consequently, with $E'_\delta = Q\left(B_{\delta^{1/(k+1)}}\right)$ and $E_\delta = \cup_{j=0}^{p-1} g_\delta^j(E'_\delta)$, for $\delta > 0$ small, $g_\delta^p(\mathrm{cl}\, E'_\delta) \subset E'_\delta$; thus, $g_\delta^p(E_\delta) \subset E_\delta$. For all $a > 0$, the distance of z_0 to the attracting basin of the orbit of z_0 tends to 0 in order smaller than δ^a when $\delta \to 0$.

Hence, there does not exist $E_0 \neq \emptyset$ as in the second paragraph of this proof, and in this case, the argument in that paragraph cannot be applied. However, the conclusion of that paragraph also holds because $g_\delta(\mathrm{cl}\, V_\delta) \subset E_\delta$ and $E_\delta \cap V_\delta = \emptyset$ for $\delta > 0$ small. The statement is proved proceeding successively as indicated in the paragraph immediately preceding it with the corresponding perturbations constructed above. At the end, the number of attracting or neutral periodic orbits of f is equal to the number of attracting periodic orbits of f_δ, which is bounded by $2(d-1)$, since f_δ can be conjugated with a rational function of degree d. □

The following result helps to get the upper bounds in the next section.

(6.23) *If f is a rational function of degree $d \geq 2$ with Julia set $J(f)$, $P_{HR,f}$ is the union of preimages of the Herman rings of f, $D_f = \mathbb{C}_\infty \setminus (J(f) \cup P_{f,HR})$ and for $\delta > 0$ small f_δ is the final perturbation of f constructed in the preceding result, then*

$$n_I(f) + n_{CP}(f, D_f) \leq n_{CP}(f_\delta, D_{f_\delta}) \leq n_{CP}(f) \leq 2(d-1).$$

Proof Let H_δ be as in the proof of the preceding result, φ_δ, f_δ as at the end of that proof, and $\widetilde{E}_\delta$ the union of the sets E_δ considered in each step of the proof, which is included in the Fatou set of f_δ, i.e., $\widetilde{E}_\delta \subset \mathbb{C}_\infty \setminus J(f_\delta)$. Let $O(z_0) = \{z_0, \ldots, z_{p-1}\}$ be an attracting or neutral periodic orbit of f of period p. For each of the three cases considered for the orbit $O(z_0)$ in the proof of the preceding result, we have:

(1) Attracting periodic orbit – Let $c_1, \ldots, c_m$ be the critical points of f whose positive orbits eventually enter the attracting basin cycle of the attracting periodic orbit of one of the points z_j under f. For each critical point c_k, there is $N \in \mathbb{N}$ such that $f_\delta^N(\varphi_\delta(c_k)) \in \widetilde{E}_\delta$ and the positive orbit of c_k under f_δ eventually enters the attracting basin cycle of the attracting periodic orbit of $\varphi_\delta(z_j)$. Therefore, the number of these critical points of f in D_f does not change with the perturbation f_δ for small $\delta > 0$.

(2) Parabolic periodic orbit – Analogous to *(1)* replacing attracting basin by parabolic basin.

(3) Periodic orbit of irrational neutral points – If the z_j are Siegel points and $c_1, \ldots, c_m$ are the critical points of f whose positive orbits eventually enter the Siegel disk S_j containing z_j, since by the Riemann mapping theorem there are conformal homeomorphisms $\psi_j : S_j \to B_1$ with $\psi_j(z_j) = 0$. For $r \in]0, 1[$, $A_j = \psi_j^{-1}(\mathrm{cl}\, B_r)$ and $B_p = B_0$ are $f(B_j) = B_{j+1}$ for $j \in \{0, \ldots, p-1\}$. Choose r such that $S_j \setminus \mathrm{int}\, B_j$ does not contain postcritical points. There is $N \in \mathbb{N}$ such that $f^N(c_k) \in \bigcup_{j=0}^{p-1} B_j$ and for $\delta > 0$ small the orbit of $\varphi_\delta(c_k)$ under f_δ eventually enters the attracting basin cycle of the periodic orbit of $\varphi_\delta(z_j)$.

For small $\delta > 0$, the set $\varphi_\delta(B_0)$ is simply connected, and the restriction of f_δ^p to $\varphi_\delta(B_0)$ is an injective function of this set into itself with $\varphi_\delta(z_j)$ an attracting fixed point. If the positive orbits of the critical points of f_δ would not intersect

$\varphi_\delta(B_0)\backslash f_\delta(\varphi_\delta(B_0))$, we could obtain recursively for $n \in \mathbb{N}$ holomorphic functions g_n such that $f_\delta^n \circ g_n = 1_{\varphi_\delta(B_0)}$ and $g_n(z_j) = z_j$. By the Montel-Carathéodory theorem,[43] $\{g_n\}$ would be a normal family as it would omit at least three points (in particular, of the Julia set $J(f_\delta)$), contradicting $\lim_{n\to+\infty} g_n'(z_j) = \lim_{n\to+\infty} \frac{1}{[f'(z_0)]^n} = \infty$. Consequently, there are critical points c of f_δ and $N \in \mathbb{N}$ with $f_\delta^N(c) \in \varphi_\delta(B_0) \setminus f_\delta(\varphi_\delta(B_0))$, and $\varphi_\delta(B_0)$ and c belong to the same connected component of $\cup_{n=0}^{\infty} f_\delta^{-n}(\varphi_\delta(B_0))$. For $\delta > 0$ small, there is a critical point of f_δ different from $\varphi_\delta(c_k)$ for $k \in \{1, \ldots, m\}$ in the attracting basin cycle of the periodic orbit of $\varphi_\delta(z_j)$.

If the z_j are Cremer points, the attracting basin cycle of the periodic orbit of $\varphi_\delta(z_j)$ contains at least one critical point of f_δ.

We conclude that for small $\delta > 0$, the number of critical points of f_δ in D_{f_δ} is larger than that of f in $D_{\tilde{f}}$ at least by a number equal to the number of irrational cycles of f. □

6.12 Upper Bound of Number of Herman Ring Cycles

In this section, the optimal upper bound $d-2$ of the number of Herman ring cycles of rational functions of degree $d \geq 2$ is obtained with quasiconformal surgery, as published in 1987 by M. Shishikura and established in 1985 in his master thesis. The lengthy proof is preceded by a coarse upper bound that can be obtained simply by comparing dimensions[44] analogously to the proof of the Sullivan nonwandering domains theorem (6.3). This excludes infinitely many Herman rings that is necessary for the proof of the optimal upper bound.

(6.24) *A rational function of degree d has* $\leq 4d+2$ *Herman ring cycles.*

Proof Let f be a rational function of degree d with Herman ring A, $E \subset A$ an invariant compact set under f conformal to the closed annulus $C = \mathrm{cl}\, B_R \backslash B_1$ with $R > 1$, and $G : C \to C$ such that $G(z) = \frac{z^2}{|z|^2}$. For $t \in]0, 1[$ the field of ellipses corresponding to the Beltrami equation with coefficient tG is invariant under rotations around 0, and the ellipses have a major axis parallel at every point to the direction of the rotation. This Beltrami equation has a solution $|z|^{2t/(1-t)}z$.

Let $A_1, \ldots, A_N$ be Herman rings, one of each of N Herman ring cycles of f. For $j \in \{1, \ldots, N\}$, consider E_j, C_j, R_j, G_j, t_j defined as in the preceding paragraph. For each $t_j \in]0, 1[$, the corresponding Beltrami coefficients $t_j G_j$ are extended in a compatible way with the preimages of A_j by f, so that the corresponding field

[43] See Section 6 of Chapter 11 of *CADOVA*.

[44] The following proof is from the L. Carleson and T.W. Gamelin book listed in the final bibliography.

of ellipses is invariant under f, and set the Beltrami coefficient in the complement of the union of these preimages equal to 0. Without loss of generality, $\infty \notin \cup_{j=1}^{N} E_j$, and the Beltrami coefficient $\kappa_{\mathbf{t}}$, with $\mathbf{t} = (t_1, \ldots, t_N)$ as above, has compact support in $\mathbb{C}$. Let $\varphi_{\mathbf{t}}$ be the solution of the Beltrami equation with coefficient $\kappa_{\mathbf{t}}$ such that $\varphi_{\mathbf{t}}(z) - z \to 0$ when $z \to \infty$. The function $F_{\mathbf{t}} = \varphi_{\mathbf{t}} \circ f \circ \varphi_{\mathbf{t}}^{-1}$ is rational of degree d and depends continuously on $\mathbf{t}$. Each $\varphi_{\mathbf{t}}(A_j)$ is a Herman ring of $F_{\mathbf{t}}$ with module a strictly increasing function of each t_j with the others fixed. Hence, there is an open set $\emptyset \neq U \subset]0, 1[^N$ such that the rational functions $F_{\mathbf{t}}$, for $\mathbf{t} \in U$, are all different. The rational functions of degree d are parameterized by $2d+1$ complex parameters, coefficients of the polynomials in the numerator and denominator with this one monic. Since $\mathbb{C}^{2d+1}$ is a linear space isomorphic to $\mathbb{R}^{2(2d+1)}$, we get a continuous injective function of the open set $U \subset]0, 1[^N$ in $\mathbb{R}^{4d+2}$, and it must be $N \leq 4d+2$. □

The optimal upper bounds are obtained by constructing an auxiliar semidynamical system[45] with conjugacies by quasiconformal mappings (in general different at each instant of a cycle) so that each connected component of the set obtained by separating each Herman ring in two by an invariant Jordan curve for the dynamics defined by the auxiliary function is contained in a Siegel disk of the auxiliary semidynamical system, and this system does not have Herman rings[46] (each ring modified with quasiconformal surgery in two Siegel disks, one in each of the regions bounded by the invariant Jordan curve). To construct such an auxiliary semidynamical system, it is convenient to introduce some notation.[47]

The set of the Herman rings of a complex rational function f is denoted[48] $\boldsymbol{\mathcal{A}_0}$. To each Herman ring A associate, an analytic Jordan curve $\boldsymbol{\gamma_A}$ such that $f(\gamma_A) = \gamma_{f(A)}$. If A is an element of a Herman ring cycle of period p, i.e., $f^p(A) = A$, then $f^p(\gamma_A) = \gamma_A$. $\boldsymbol{\mathcal{A}}$ denotes the set of connected components of $A \backslash \gamma_A$ for all Herman rings A of f and, for each $n \in \mathbb{N} \cup \{0\}$, $\boldsymbol{\Gamma_n}$ denotes the set of connected components of $f^{-n}(\gamma_A)$ for all Herman rings A of f, which is a union of analytic Jordan curves

---

45 A **dynamical system** is an evolution in time (discrete or continuous) deterministic for both future and past and depending continuously on initial conditions; if there is determinism for the future but not necessarily for the past, the evolution is said to be a **semidynamical system**.

46 If f is a rational function, ψ such that $\psi(j, f^j) = (j+1, f^{j+1})$ for $(j, z) \in \mathbb{N} \cup \{0\}$ satisfies $\psi(0, f) = (0, f)$ and $\psi(j+k, f^{j+k}) = \psi(k, \psi(j, f^j))$; so, ψ is a semidynamical system in the set of functions $\{f^n\}_{\mathbb{N} \cup \{0\}}$. Thus, it has properties of semidynamical systems that can be useful to consider the evolution around cycles of f where the value of the function varies from point to point and, therefore, from instant to instant. Dynamical or semidynamical systems defined from a given dynamical or semidynamical system based on this idea are called **skewproduct flows** or **semiflows** or **cocycles** and are useful in several situations, e.g., to describe as a dynamical or semidynamical system the evolution defined by solutions of an ordinary differential equation with time varying coefficients, or the evolution in a neighborhood of an orbit of a dynamical or semidynamical system, e.g.,of a periodic orbit.

47 To facilitate references to the new notations, they appear in bold where they are defined.

48 We follow part of the M. Shishikura article mentioned in the preceding section.

each one oriented according to the application of f. $\mathcal{X}$ is the union of $\mathcal{A}$ with the attracting or neutral periodic orbits of f.

For each $n \in \mathbb{N} \cup \{0\}$, every $X \in \mathcal{X}$ is contained in a connected component of $\mathbb{C}_\infty \backslash \cup_{\gamma \in \Gamma_n} \gamma$. Since $\mathcal{X}$ is finite, there is a $\boldsymbol{N} \in \mathbb{N} \cup \{0\}$ such that if $X_1, X_2 \in \mathcal{X}$ are in a same connected component of $\mathbb{C}_\infty \backslash \cup_{\gamma \in \Gamma_n} \gamma$, then $f(X_1)$, $f(X_2)$ also are in a same connected component of that set. $\mathcal{D}$ denotes the set of connected components of $\mathbb{C}_\infty \backslash \cup_{\gamma \in \Gamma_N} \gamma$ containing some element of $\mathcal{X}$, and $\mathcal{D}^c$ denotes the set of the remaining connected components of $\mathbb{C}_\infty \backslash \cup_{\gamma \in \Gamma_N} \gamma$. As each $X \in \mathcal{X}$ belongs to a cycle of f, there are $L \in \mathbb{N}$ and, for each $j \in \{1, \dots, L\}$, $m_j \in \mathbb{N}$ such that $\mathcal{D}$ can be partitioned in a set of disjoint cycles $\{\boldsymbol{D_{j,k}} : j \in \{1, \dots, L\}, k \in \{0, \dots, m_j - 1\}\}$, where for every $X \in \mathcal{X}$ with $X \subset D_{j,k}$, it is $f(X) \subset D_{j,k+1}$, identifying $D_{j,m_j} = D_{j,0}$. *It is always* $f(D_{j,k}) \supset D_{j,k+1}$.

For an oriented Jordan curve $\gamma \subset \mathbb{C}_\infty$, $\mathbf{Int}\,\boldsymbol{\gamma}$ denotes the region with boundary γ surrounded by this curve with positive orientation.

Let $\gamma \in \Gamma_N$, $\gamma \subset \partial D_{j,k}$ with $D_{j,k} \subset \operatorname{Int} \gamma$. If $N \in \mathbb{N}$, then $f(\gamma) \cap f(D_{j,k}) = \emptyset$ and $f(z) \in \operatorname{Int} f(\gamma)$. So, $f(D_{j,k}) \cap \operatorname{Int} f(\gamma) \neq \emptyset$, $D_{j,k+1} \subset f(D_{j,k}) \subset \operatorname{Int} f(\gamma)$. If $N = 0$, then $D_{j,k}$ contains some $A \in \mathcal{A}$ with $\gamma \subset \partial A$. Therefore, $f(A) \cap \operatorname{Int} f(\gamma) \neq \emptyset$ and, as $f(A) \subset D_{j,k+1}$ and $D_{j,k+1} \cap f(\gamma) = \emptyset$, it is $D_{j,k+1} \subset \operatorname{Int} f(\gamma)$. *In any case, if* $\gamma \in \Gamma_N$ *and* $\gamma \subset \partial D_{j,k}$ *with* $D_{j,k} \subset \operatorname{Int} \gamma$, *then* $D_{j,k+1} \subset \operatorname{Int} \gamma$.

For $m \in \mathbb{N}$, denote $\boldsymbol{m}\mathbb{C}_\infty = \{0, \dots, m-1\} \times \mathbb{C}_\infty$, and for $k \in \{0, \dots, m-1\}$, denote $\boldsymbol{P_k}$ the function defined in $\mathbb{C}_\infty$ by $P_k(z) = (k, z)$. If $S \subset \mathbb{C}_\infty$ and $g(\{k\} \times S) \subset \{k+1\} \times S$ for $k \in \{0, \dots, m-1\}$, identifying $\{m\} \times \mathbb{C}_\infty$ with $\{0\} \times \mathbb{C}_\infty$, $g : \{0, \dots, m-1\} \times S \to m\mathbb{C}_\infty$ is said to be a **cyclic function**. For each $j \in \{1, \dots, L\}$, denote

$$\widetilde{\boldsymbol{D}}_{\boldsymbol{j,k}} = P_k(D_{j,k}), \qquad \widetilde{\boldsymbol{D}}_{\boldsymbol{j}} = \cup_{k=0}^{m-1} \widetilde{D}_{j,k},$$

$$\widetilde{\boldsymbol{\Gamma}}_{\boldsymbol{n}} = \{P_k(\gamma) : \gamma \in \Gamma_n, \gamma \subset \operatorname{cl} D_{j,k}\}, \qquad \widetilde{\boldsymbol{\mathcal{A}}}_j = \{P_k(A) : A \in \mathcal{A}, A \subset D_{j,k}\}.$$

The following properties are used to prove the optimal upper bound of the number of Herman cycles. To simplify notation, j is omitted.

(6.25) *There are rational functions* $f_0, f_1, \dots, f_{m-1}$ *and quasiconformal mappings* $\varphi_0, \dots, \varphi_{m-1}$ *in* $\mathbb{C}$ *which, with* $\widetilde{f}_\ell : \widetilde{D} \to m\mathbb{C}_\infty$ *such that* $\widetilde{f}_\ell(k, z) = (k+1, f(z))$ *and* $F, \varphi : m\mathbb{C}_\infty \to m\mathbb{C}_\infty$ *such that* $F(\ell, z) = (\ell+1, f_\ell(z))$, $\varphi(\ell, z) = (\ell, \varphi_\ell(z))$ *satisfy:*

1. $\varphi \circ \widetilde{f} \overset{def}{=} F \circ \varphi$ *in* $\widetilde{D}$;
2. $\kappa_{\varphi_\ell} = 0$ *for* $\ell \in \{0, \dots, m-1\}$ a.e. *in* $K(\widetilde{f}) = \cup_{n=0}^{\infty} \widetilde{f}^{-n}(\widetilde{D})$, *with* κ_{φ_ℓ} *the Beltrami coefficient of* φ_ℓ;
3. $F(m\mathbb{C}_\infty \backslash \varphi(\widetilde{D})) \subset m\mathbb{C}_\infty \backslash \varphi(\widetilde{D})$;
4. *For* $A \in \mathcal{A}$, $\varphi(A)$ *is included in a Siegel disk of* F;
5. $\exists_{M \in \mathbb{N}} : F^M(m\mathbb{C}_\infty \backslash \varphi(\widetilde{D}))$ *is included in the attracting basins of Siegel disks of* F, *and* $m\mathbb{C}_\infty \backslash \varphi(\widetilde{D}) \subset \mathbb{C}_\infty \backslash J(F)$.

Proof The Shishikura fundamental lemma of quasiconformal surgery (6.12) can be extended to cyclic functions by replacing $\mathbb{C}_\infty$ with $m\mathbb{C}_\infty$ for $m \in \mathbb{N}$ and the quasiconformal conjugacy $\varphi : \mathbb{C}_\infty \to \mathbb{C}_\infty$ with a rational function can be extended by $\varphi : m\mathbb{C}_\infty \to m\mathbb{C}_\infty$ such that $\varphi(\ell, z) = (\ell, \varphi(z))$ with each function $z \mapsto \varphi(\ell, z)$ for $\ell \in \{0, \ldots, m-1\}$ fixed a quasiconformal conjugacy with a rational function. It is this extension that is applied in this proof.

Extend by continuity each $\widetilde{f}$ to $\partial\widetilde{D}$, i.e., to $P_k(\gamma)$ with $\gamma \in \Gamma_N$. Denote $\widetilde{\boldsymbol{\Gamma}}'_N = \widetilde{\Gamma}_N \cup \{P_{k+1}(f(\gamma)) : \gamma \in \Gamma_N, \gamma \subset \partial D_k\}$, each element with the orientation induced by the functions P_k. By the property two paragraphs before the statement, for $\gamma \in \Gamma_N$ and $\gamma \subset \partial D_k$ with $D_k \subset \operatorname{Int}\gamma$, it is $D_{k+1} \subset \operatorname{Int}\gamma$. If necessary inverting orientations, the functions $\widetilde{f}$ respect the orientations of the Jordan curves $\gamma \in \widetilde{\Gamma}'_N$ and it is $D_k \cap (\mathbb{C}_\infty \backslash \operatorname{Int}\gamma) = \emptyset$.

For each $\gamma \in \widetilde{\Gamma}'_N$, denote $\boldsymbol{E_\gamma} = \operatorname{cl}(\mathbb{C}_\infty \backslash \operatorname{Int}\gamma)$ and $\mathscr{E} = \{E_\gamma : \gamma \in \widetilde{\Gamma}_N\}$. So, $m\mathbb{C}_\infty$ is the disjoint union of D and the elements of $\mathscr{E}$. If $E \in \mathscr{E}$, then $\partial E \in \widetilde{\Gamma}_N$ and $\widetilde{f}(\partial E) \subset m\mathbb{C}_\infty \backslash \widetilde{D}$. Therefore, there is only one $E' \in \mathscr{E}$ such that $\widetilde{f}(\partial E) \subset E'$. Define $R : \mathscr{E} \to \mathscr{E}$ for $R(E) =$e$'$. Next, define a function g mapping E onto E' and such that $g = \widetilde{f}$ in ∂E.

Denote $\mathscr{E}_1 = \{E_\gamma : \gamma \in \widetilde{\Gamma}_0\}$. If $E_\gamma \in \mathscr{E}_1$, then $R(E_\gamma) \in \mathscr{E}_1$ and $\widetilde{f}(\partial E_\gamma) = \partial R(E_\gamma)$. For every $\gamma \in \widetilde{\Gamma}_0$, there is $p \in \mathbb{N} \cup \{0\}$ such that $\widetilde{f}^p(\gamma) = \gamma$. Hence, there are $E^{(s)} \in \mathscr{E}_1$ with $E^{(s)} = R(E^{(s-1)})$ distinct for $s \in \{1, \ldots, p\}$, $E^{(p)} = E^{(0)}$, and $\mathscr{E}_1$ is the union of these cycles. For each cycle, by the definition of Herman ring, there is an analytic diffeomorphism as a function of real variables $\psi_0 : \gamma_0 \to S^1$ preserving orientation, with S^1 the circle with radius 1 and center 0 oriented negatively relative to 0 such that $\psi_0 \circ \widetilde{f}^p \circ \psi_0^{-1}(z) = \lambda z$ with $|\lambda| = 1$.

Denote $\boldsymbol{\psi_s} : \partial E^{(s)} \to S^1$ such that $\boldsymbol{\psi_s} = \psi_0 \circ \widetilde{f}^{p-s}$ for $s \in \{1, \ldots, p\}$, with $\psi_p = \psi_0$. Then, $(\psi_s \circ \widetilde{f}^s)(z) = \lambda\, \psi_0(z)$ for $z \in \gamma_0$, $s \in \{1, \ldots, p\}$. Each $\boldsymbol{\psi_s}$ has an extension to a quasiconformal mapping of $E^{(s)}$ to $\operatorname{cl} B_1$, also denoted ψ_s. Let $\boldsymbol{\Psi} : \cup_{s=0}^{p-1} E^{(s)} \to \operatorname{cl} B_1$ be equal to ψ_s in each $E^{(s)}$.

Define $\boldsymbol{g}$ in $\cup_{s=0}^{p-1} E^{(s)}$ successively for $s = 0, \ldots, p-1$: in $E^{(0)}$ by $\boldsymbol{g} = \psi_1^{-1} \circ (\lambda\, \psi_0)$ with values in $E^{(1)}$, and in $E^{(s)}$ by $\boldsymbol{g} = \psi_{s+1}^{-1} \circ \psi_s$ with values in $E^{(s+1)}$. Let $\mathscr{E}_2 = \{E^{(0)} \in \mathscr{E} \setminus \mathscr{E}_1 : \exists_{s \in \{0, \ldots, p-1\}} E^{(s)} \in \mathscr{E}\}$. If $E^{(0)} \in \mathscr{E}_2$, with $E^{(0)}, \ldots, E^{(p-1)}$ distinct and $E^{(p)} = E^{(0)}$, $E^{(0)}, \ldots, E^{(p-1)}$ would belong to $\mathscr{E}_2$, and if $\widetilde{f}(\partial E^{(s)}) = \partial E^{(s+1)}$ for $s \in \{0, \ldots, p-1\}$, it would be $\widetilde{f}^p(\partial E^{(0)}) = \partial E^{(0)}$, and $\partial E^{(0)} \in \widetilde{\Gamma}_0$, in contradiction with $\partial E^{(0)} \notin \mathscr{E}_2$. Thus, there is $s_0 \in \{0, \ldots, p-1\}$ such that $\widetilde{f}(\partial E^{(s_0)}) \neq \partial E^{(s_0+1)}$. In the interior of $E^{(s)}$ we consider an open disk $E'^{(s)}$ such that $\widetilde{f}(\partial E^{(s_0)}) \subset E'^{(s_0+1)}$ and $E'^{(p)} = E'^{(0)}$. Define the function $\boldsymbol{g}$ in each $E^{(s)}$ so as to be a holomorphic quasiregular mapping in $E'^{(s)}$, coinciding with $\widetilde{f}$ in $\partial E^{(s)}$ and satisfying $\boldsymbol{g}(E^{(s)}) \subset E^{(s+1)}$ and $\boldsymbol{g}(E'^{(s)}) \subset E'^{(s+1)}$. Since $g(E(k_0)) \subset E'^{(k_0+1)}$, it is $g^p(E(s)) \subset E'^{(s)}$. Define $\boldsymbol{\Psi}$ in $E'^{(s)}$ equal to the identity. Define $\boldsymbol{g}$ in $E \in \mathscr{E} \setminus (\mathscr{E}_1 \cup \mathscr{E}_2)$ to be a quasiregular mapping such that $\boldsymbol{g} = \widetilde{f}$ in ∂E and $\boldsymbol{g}(E) \subset R(E)$. Since the set $\mathscr{E}$ is finite, there is $E^{(s)} \in \mathscr{E}$ for $s \in \{0, \ldots, r\}$ such that $E^{(r)} = R^r(E) \in \mathscr{E}_1 \cup \mathscr{E}_2$. Define $\boldsymbol{g} = \widetilde{f}$ in $\widetilde{D}$. The function g defined in this way in $m\mathbb{C}_\infty$ is a continuous quasiregular mapping.

Define $\mathscr{E}^{(1)}$ as the union of the elements of $\mathscr{E}_1$, $\mathscr{E}^{(2)}$ as the union of the sets E' such that $E \in \mathscr{E}_2$ and $\boldsymbol{E^*} = \mathscr{E}^{(1)} \cup \mathscr{E}^{(2)}$.

The set $m\mathbb{C}_\infty \backslash \widetilde{D}$ is invariant for g, and there is $M \in \mathbb{N}$ such that $g^M(m\mathbb{C}_\infty \backslash \widetilde{D}) \subset E^*$, so that in $m\mathbb{C}_\infty \backslash g^{-M}(E^*)$ is $\frac{\partial g}{\partial \bar{z}} = \frac{\partial \widetilde{f}}{\partial \bar{z}} = 0$.

The conditions for applying the Shishikura fundamental lemma of quasiconformal surgery (6.12) extended to cyclic functions hold. So, there are quasiconformal mappings $\varphi_0, \ldots, \varphi_{m-1}$ in $\mathbb{C}$ such that the corresponding function φ defined in $m\mathbb{C}_\infty$ as in the statement we are proving is such that $F = \varphi \circ g \circ \varphi^{-1}$ is holomorphic and the corresponding rational functions $f_0, \ldots, f_{m-1}$ are such that $F(\ell, z) = \big(\ell + 1, f_\ell(z)\big)$, and it is easy to verify that the conditions 1 to 3 in the statement hold.

If $A \in \widetilde{\mathcal{A}}$, $\gamma \in \widetilde{\Gamma}_0$, and $\gamma \subset \partial A$, then $S = \varphi(E_\gamma \cup A)$ is a connected open set. For some $q \in \mathbb{N}$, $F^q(S) = S$ and $(F^q)_{|S}$ are conjugated with a rotation around 0 in a given disk centered 0 by multiplication with $e^{2\pi\omega}$, where $\omega \in \mathbb{R} \setminus \mathbb{Q}$. Hence, S is a Siegel disk of F and 4 holds.

By the Schwarz lemma[49], for each $E \in \mathscr{E}_2$ there is a point in $\varphi(E')$ with attracting periodic orbit and attracting basin containing $\varphi(E)$, and 5 holds. □

(6.26) *If f is a rational function of degree $d \in \mathbb{N}$, then*

$$n_{AB}(f) + n_{PB}(f) + n_{SD}(f) + 2n_{HR}(f) + n_C(f) \le 2(d-1).$$

Proof If f does not have Herman rings, the inequality follows from (6.23), as $n_{AB}(f) + n_{PB}(f) \le n_{CP}\big(f, \mathbb{C}_\infty \backslash J(f)\big)$, $n_I(f) = n_{SD}(f) + n_C(f)$.

Now consider there are Herman rings. Denote $\boldsymbol{D}$, $\boldsymbol{D^c}$, $\boldsymbol{D}$ the unions of the elements of, resp., $\mathscr{D}$, $\mathscr{D}^c$, $\big\{D_k : k \in \{0, \ldots, m-1\}\big\}$, and let φ and F be as in the preceding result. F does not have Herman rings. By 5 in the preceding result, the critical points of F in $\mathbb{C}_\infty \backslash \varphi\big(\widetilde{D}\big)$ belong to the Fatou set of F. Each neutral irrational cycle of F is in $\varphi\big(\widetilde{D}\big)$ or is the cycle of centers of a cycle of Siegel disks containing some $\varphi(A)$ with $A \in \widetilde{\mathcal{A}}$. Denoting $\boldsymbol{n_{cy}}\big(\widetilde{\mathcal{A}}\big)$ the number of cycles of $\widetilde{\mathcal{A}}$, by (6.23),

$$n_{CP}\Big(F, \varphi\big(\widetilde{D} \cap (\mathbb{C}_\infty \backslash J(F))\big)\Big) + n_I\big(F, \varphi(\widetilde{D})\big) + n_{cy}\big(\widetilde{\mathcal{A}}\big) \le n_{CP}\big(F, \varphi(\widetilde{D})\big).$$

As we want the upper bound for f and not for F, it is necessary to relate the Julia sets of both functions.

Let $z \in D_k \cap (\mathbb{C}_\infty \backslash J(f))$ and $\widetilde{z} = (\varphi \circ P_k)(z)$. We consider a metric in $m\mathbb{C}_\infty$ and denote by $\boldsymbol{d_0}$ the distance of the sets $J(F)$ and $m\mathbb{C}_\infty \backslash \varphi\big(\widetilde{D}\big)$. If there are points of the positive orbit of $\widetilde{z}$ under F at a smaller distance of $m\mathbb{C}_\infty \backslash \varphi\big(\widetilde{D}\big)$, then $\widetilde{z}$ belongs to the Fatou set of F. Otherwise, as the functions f^n are equicontinuous in a neighborhood of z since this point belongs to the Fatou set of f, there is a neighborhood U of z

[49] See Section 2 of Chapter 10 of *CADOVA*.

such that $f^n(U) \subset D_{k+n}$ for $n \in \mathbb{N} \cup \{0\}$, considering $k+n$ modulo m. By 1 in the preceding result, $F^n \circ \varphi \circ P_k = \varphi \circ P_{n+k} \circ f^n$ in U, and $\widetilde{z}$ belongs to the Fatou set of F. Therefore,

$$\cup_{k=0}^{m-1} (\varphi \circ P_k)\big(D_k \cap (\mathbb{C}_\infty \setminus J(f))\big) \subset \mathbb{C}_\infty \setminus J(F).$$

By the definitions of N and D_k before the statement of the preceding result, for all $n \in \mathbb{N} \cup \{0\}$ if z in the Fatou set of f, it belongs to the union of the attracting or neutral periodic orbits of f; so, $f^n(z) \in D_{n+k}$ and $F^n(\widetilde{z}) = (\varphi \circ P_{n+k})(z)$, and $\widetilde{z}$ belong to a periodic orbit of F. It is easy to verify that $\widetilde{z}$ is an attracting, parabolic, Siegel or Cremer fixed point if and only if z is.

By the two preceding paragraphs,

$$n_{CP}\big(f, \mathbb{C}_\infty \setminus (J(f) \cap D)\big) \leq n_{CP}\big(F, \mathbb{C}_\infty \setminus (J(F) \cap \varphi(\widetilde{D}))\big),$$
$$n_I(f, D) \leq n_I\big(F, \varphi(\widetilde{D})\big).$$

As the number of cycles of $\mathcal{A}$ contained in D is equal to $n_{\text{cy}}(\widetilde{\mathcal{A}})$, by the preceding inequality,

$$n_{CP}\big(f, \mathbb{C}_\infty \setminus (J(f) \cap D)\big) + n_I(f, D) + n_{\text{cy}}(\widetilde{\mathcal{A}}) \leq n_{CP}(f, D).$$

Adding both terms of this inequality for all $j \in \{1, \dots, L\}$ and recalling each Herman ring were separated in two connected components of $\mathcal{A}$,

$$n_{CP}\big(f, \mathbb{C}_\infty \setminus (J(f) \cap \mathscr{D})\big) + n_I(f) + 2n_{HR}(f) \leq n_{CP}(f, \mathscr{D}).$$

Since $n_{PC}\big(f, \mathbb{C}_\infty \setminus (J(f) \cap \mathscr{D}^c)\big) \leq n_{PC}(f, \mathscr{D}^c)$ and $n_{CP}(f) \leq 2(d-1)$, the inequality in the statement holds. □

This result gives the upper bound of the number of Herman ring cycles $n_{HR}(f) \leq d-1$, but the following stronger upper bound holds.

(6.27) *If f is a rational function of degree $d \in \mathbb{N}$, then $n_{HR}(f) \leq d-2$.*

Proof If there is at least one critical or irrational periodic point in the Fatou set of f, the result follows from (6.26). Thus, it suffices to prove that if there are not critical or irrational periodic points, it is impossible to have equality in the inequality in (6.26).

A **minimal Herman ring** A_0 of f is a Herman ring A such that Int γ_A does not intersect any other Herman ring of f, or if there is not such A, a Herman ring that satisfies this condition by inverting the orientation of the curves γ_A.

If p is the order of a Herman ring cycle containing A_0, then $A_p = A_0$. If A_0 is a minimal Herman ring of order p and every Herman ring $f^j(A_0)$ with $j \in \{0, \dots, p-1\}$ is minimal, for some $j \in \{1, \dots, p\}$ it is

$$f\left(\operatorname{Int}\gamma_{f^{j-1}(A_0)}\right)\not\subseteq f\left(\operatorname{Int}\gamma_{f^{j}(A_0)}\right),$$

and the connected component C of $\mathbb{C}_\infty\backslash f^{-1}\big(f^j(A_0)\big)$ with $f^{j-1}(A_0)\cap\operatorname{Int}f^{j-1}(A_0)$ is not simply connected. If, on the other hand, some $f^j(A_0)$ is not a minimal Herman ring, can be chosen a $j\in\{1,\ldots,p\}$ so that $f^{j-1}(A_0)$ is a minimal Herman ring, and, as $\operatorname{Int}\gamma_{f^{j-1}(A_0)}$ contains an element of $\mathcal{A}_0$ with image by f a subset of the complement of $\operatorname{Int}\gamma_{f^j(A_0)}$, the connected component C of $\mathbb{C}_\infty\backslash f^{-1}\big(f^j(A_0)\big)$ containing $f^{j-1}(A_0)\cap\operatorname{Int}f^{j-1}(A_0)$ it is not simply connected. Thus, if A_0 is a minimal Herman ring of order p, there is $k\in\{1,\ldots,p\}$ such that the connected component C_k of $\mathbb{C}_\infty\backslash f^{-1}\big(f^k(A_0)\big)$ containing $f^{k-1}(A_0)\cap\operatorname{Int}f^{k-1}(A_0)$ is not simply connected and $f^k(A_0)$ is a minimal Herman ring.

With C_k as in the preceding paragraph, $(C_k, f_{|C_k})$ is a covering of $f(C_k)$. If C_k had 1 or 0 critical points of f, it would be a simply connected region. Therefore, C_k contains at least two critical points.

Let D_k be the element of $\mathcal{D}$ containing $f^{\ell-1}(A_0)\cap\operatorname{Int}f^{\ell-1}(A_0)$. Since $f^\ell(A_0)$ is a minimal Herman ring, the only element of $\mathcal{A}$ such that $f(C_\ell)=\operatorname{Int}\gamma_{f^\ell(A_0)}$ is $f^\ell(A_0)\cap\operatorname{Int}f^\ell(A_0)$. So, the only element of $\mathcal{A}\cap C_\ell$ is $f^{\ell-1}(A_0)\cap\operatorname{Int}f^{\ell-1}(A_0)$. Also, the only element of $\mathcal{D}$ that C_ℓ contains is D_k. Hence, $D_k\subset C_\ell\subset\operatorname{cl}(D_k\cup\mathcal{D}^c)$, and, by the preceding paragraph, $n_{CP}(f, D_k)+n_{CP}(f,\mathcal{D}^c)\geq 2$; so, $n_{HR}\big(f, D\big)=1$. Therefore, $n_{CP}(f, D_k)\geq 2$ or $n_{CP}(f,\mathcal{D}^c)\geq 1$, and in either case, there is no equality in the inequality in (6.26). □

In particular, rational functions of degree 2 do not have Herman rings.

6.13 Making a Herman Ring Cycle from a Siegel Disk Cycle

As mentioned in Sect. 6.7, in 1987, M. Shishikura gave a method of quasiconformal surgery to get a rational function with a Herman ring cycle from a rational function with a Siegel disk cycle that is presented now (Figure 6.18).

(6.28) *If f_0 is a complex rational function with a Siegel disk cycle of order $p\in\mathbb{N}$ and rotation number θ, $S_0,\ldots,S_{p-1}$ with $f_0(S_j)=S_{j+1}$ for $j\in\{0,\ldots,p-1\}$ (identifying $S_p=S_0$), and $f_1,\ldots,f_p$ rational functions such that $\widetilde{f}=f_p\circ\cdots\circ f_1$ has a Siegel disk with rotation number $-\theta$, $\widetilde{S}_0$ with $\widetilde{f}_0(\widetilde{S}_0)=\widetilde{S}_0$, then there is a rational function F with a cycle of Herman rings with rotation number θ, $A_0,\ldots,A_{p-1}$, $F(A_j)=A_{j+1}$ for $j\in\{0,\ldots,p-1\}$, identifying $A_p=A_0$, and $A_j\in\mathbb{C}_\infty\backslash A_k$ for $j,k\in\{0,\ldots,p-1\}$, $j\neq k$.*

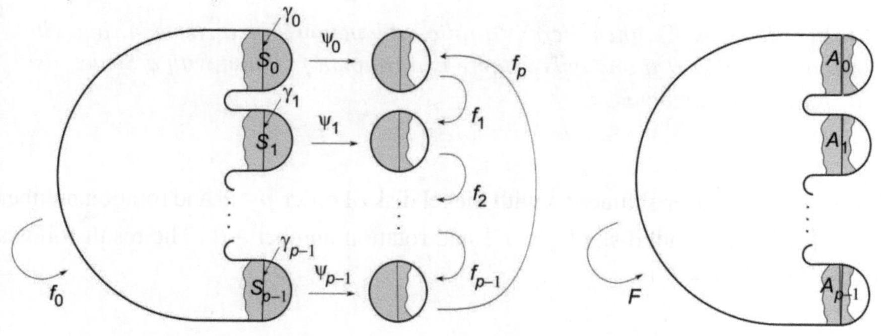

Fig. 6.18 Making a cycle of Herman rings from one of Siegel disks (6.28)

Proof Let $\gamma_0, \widetilde{\gamma}_0$ be Jordan curves analytic as described by functions of real variable in, resp., S_0 , $\widetilde{S}_0$ and invariant under, resp., f_0^p , $\widetilde{f}$. For $j \in \{1, \dots, p-1\}$ denote $\gamma_j = f_0^j(\gamma_0)$, $\widetilde{\gamma}_j = f_j \circ \cdots \circ f_1(\widetilde{\gamma}_0)$, $\widetilde{S}_j = \widetilde{f}^j(\widetilde{S}_0)$. There are conjugacies of f_0 with f_j by diffeomorphisms analytic as functions of real variables $\widetilde{\psi}_j : \gamma_j \to \widetilde{\gamma}_j$ and $\widetilde{\psi}_0 = \widetilde{\psi}_{p-1} \circ \cdots \circ \widetilde{\psi}_1$ is a conjugacy of f_0^p with $\widetilde{f}$. Let $C_j, \widetilde{C}_j$ be the connected components of, resp., $\mathbb{C}_\infty \setminus \gamma_j, \mathbb{C}_\infty \setminus \widetilde{\gamma}_j$ included in, resp., $S_j, \widetilde{S}_j$. Let $\widehat{\psi}_j : \mathbb{C}_\infty \setminus \gamma_j \to \mathbb{C}_\infty \setminus \widetilde{\gamma}_j$ be conformal homeomorphisms $\widehat{\psi}_j(C_j) = \mathbb{C}_\infty \setminus \mathrm{cl}\, \widetilde{C}_j$ and $\widehat{\psi}_j(\mathbb{C}_\infty \setminus \mathrm{cl}\, C_j) = \widetilde{C}_j$, and consider their extensions by continuity to γ_j , which are denoted as before. Modify each $\widehat{\psi}_j$ to obtain a quasiconformal mapping ψ_j equal to $\widehat{\psi}_j$ in a neighborhood of $\mathbb{C}_\infty \setminus (S_j \cap \widehat{\psi}_j(\widetilde{S}_j))$ and to $\widetilde{\psi}_j$ in γ_j .

The function g equal to f_0 in $\mathbb{C}_\infty \setminus \cup_{j=0}^{p-1} C_j$ and to $\psi_{j+1}^{-1} \circ f_j \circ \psi_j$ in C_j is a continuous quasiregular mapping. With

$$E_0 = \cup_{j=0}^{p-1} (S_j \setminus \mathrm{cl}\, C_j) , \qquad \Phi_0 = 1_{E_0} ,$$

$$E_{j+1} = C_j \cap \psi_j^{-1}(\widetilde{S}_j) , \qquad \Phi_{j+1} = \psi_{j|E_j} , \qquad j \in \{0, \dots, p-1\} ,$$

for $R = \cup_{j=0}^{p} E_j$ is $g(E) = E$ and, as for $j \in \{0, \dots, p-1\}$ each ψ_j is a conformal homeomorphism in a neighborhood of $\mathbb{C}_\infty \setminus (E \cup \gamma_j)$, it is $\frac{\partial g}{\partial \overline{z}} = 0$ in $\mathbb{C}$. The conditions of the Shishikura fundamental lemma of quasiconformal surgery (6.12) hold. So, there is a quasiconformal mapping φ such that $F = \varphi \circ g \circ \varphi^{-1}$ is a rational function and $A_j = \varphi\big(S_j \cap \psi_j^{-1}(\widetilde{S}_j)\big)$, $j = 0, \dots, p-1$ is a cycle of Herman rings of order p of F . □

(6.29) Example: For any appropriate $p \in \mathbb{N}$ and θ, there are $c_1, c_2 \in \mathbb{C}$ such that the functions $f_k(z) = z^2 + c_k$ for $k \in \{1, 2\}$ and $f_j = 1_{\mathbb{C}_\infty}$ for $j \in \{2, \dots, p\}$ satisfy the hypothesis of the preceding result. Since the critical points of f_0 and f_1 are 0 and ∞ and $f_2, \dots, f_p$ have no critical points, g and F have <5 critical points. So, there is a rational function F of degree 3 with a cycle of Herman rings of any order $p \in \mathbb{N}$.

The previous result in this section has the following consequence.

(6.30) *If $\theta \in \mathbb{R}\setminus\mathbb{Q}$, then there is a rational function with a Herman ring with rotation number θ if and only if there is a rational function with a Siegel disk with rotation number θ.*

Proof If f is a rational function with Siegel disk of order $p \in \mathbb{N}$ and rotation number θ, $\overline{f^p(\overline{z})}$ has a Siegel disk of order 1 and rotation number $-\theta$. The result follows by (6.28) and (6.25). □

6.14 Optimal Upper Bound of Number of Cycles

(6.31) *The upper bounds in* (6.26), (6.27) *and* (6.22) *are optimal.*

Proof By the Runge theorem,[50] we can construct the polynomial perturbations used to prove (6.22) and (6.23), so that each critical point c_δ of the perturbation f_δ of f corresponding to a critical point c of f whose orbit is a nonperiodic eventually periodic repelling orbit (the number of these orbits was denoted at the beginning of Section 11 of this chapter by $n_{CPER}(f)$ satisfies these properties for small δ. As the upper bound in (6.26) is based on the upper bound of the number of critical points,

$$n_{AB}(f) + n_{PB}(f) + n_{SD}(f) + 2n_{HR}(f) + n_C(f) + n_{CPER}(f) \le 2(d-1)\,.$$

Thus, it suffices to prove for $N_1, \ldots, N_7 \in \mathbb{N}\cup\{0\}$ with $\sum_{j=1}^{7} N_j + N_4 \le 2(d-1)$ and $N_4 \le d-2$, that there is a rational function f of degree d such that the numbers on the left-hand side of the inequality preceding the last one are, resp., equal to those of the last inequality (it is only needed to proved that each of the terms in the first inequality is $\ge$ the resp. terms of the second inequality, because, then, the first inequality implies equality of the, resp., pairs of terms).

Firstly, consider cases without Herman rings, i.e., $N_4 = 0$.

(1) For $p, q \in \mathbb{N}$ relatively prime to $d-1$, λ_1, λ_2 roots of unity of order, resp., p, q with $\lambda_1 \ne \overline{\lambda_2}$ and $\boldsymbol{\varepsilon} = (\varepsilon_1, \varepsilon_2) \in \mathbb{C}^2$ small, let

$$f_{\boldsymbol{\varepsilon}}(z) = z\,\frac{\lambda_1(1+\varepsilon_1) + z^{d-1}}{1 + \lambda_2(1+\varepsilon_2)z^{d-1}}\,.$$

Comparing $f_{\mathbf{0}}^p \circ f_{\mathbf{0}}(z)$ with $f_{\mathbf{0}} \circ f_{\mathbf{0}}^p(z)$, it follows easily that there is $c_0 \in \mathbb{C}\setminus\{0\}$ and $n \in \mathbb{N}$ such that

$$f_{\mathbf{0}}^p(z) = z\,[1 + c_0 z^{np} + O(z^{np+1})]\,, \quad z \to 0\,.$$

[50] See in one of the exercises at the end of Chapter 11 of *CADOVA*. Runge, Carl (1856–1927).

Denote $R(z)=e^{i2\pi/(d-1)}z$. Since $d-1$ and p are relatively prime and R commutes with f_0, n is a multiple of $d-1$. By the parabolic flower theorem,[51] f_0 has np petals at 0 forming n cycles, and, as it is analogous in ∞ and f_0 has $\leq 2(d-1)$ cycles of parabolic basins, $n=d-1$,

$$f_{\boldsymbol{\varepsilon}}^p(z) = z\,[(1+\varepsilon_1)^p + c(\varepsilon_2)z^{(d-1)p} + O(\varepsilon_1 z) + O(z^{(d-1)p+1})], \quad z, \boldsymbol{\varepsilon} \to 0,$$

with $\varepsilon_2 \mapsto c(\varepsilon_2)$ holomorphic and $\varepsilon_2(0) = c_0$. For $\boldsymbol{\varepsilon}$ small the equation $f_{\boldsymbol{\varepsilon}}^p(z) = z$ has in a small neighborhood of 0 $(d-1)p$ zeros of the form

$$\zeta_{j,k}^{\boldsymbol{\varepsilon}} = e^{i2\pi k/(d-1)} f_{\boldsymbol{\varepsilon}}^j(\zeta^{\boldsymbol{\varepsilon}}), \quad j,k \in \{0,\ldots,d-2\}.$$

Since $[(f_{\boldsymbol{\varepsilon}}^p)'(\zeta^{\boldsymbol{\varepsilon}})]^{d-1} = \prod_{j,k}(f_{\boldsymbol{\varepsilon}}^p)'(\zeta_{j,k}^{\boldsymbol{\varepsilon}})$ is a rational function of the coefficients of the function on the left-hand side of the equation $f_{\boldsymbol{\varepsilon}}^p(z) - z = 0$, the function $\alpha_1(\boldsymbol{\varepsilon}) = (f_{\boldsymbol{\varepsilon}}^p)'(\zeta^{\boldsymbol{\varepsilon}})$ is independent of the possible choices of $\zeta^{\boldsymbol{\varepsilon}}$, and it is holomorphic in some neighborhood of $0 \in \mathbb{C}^2$. Consider also the function α_2 similarly defined in a neighborhood of ∞. It is easily verified that

$$\alpha_1(\varepsilon_1,\varepsilon_2) = 1-(d-1)p^2\varepsilon_1 + o(\varepsilon_1), \quad \varepsilon_1,\varepsilon_2 \to 0,$$

and similarly for α_2. Hence, $\frac{\partial(\alpha_1,\alpha_2)}{\partial(\varepsilon_1,\varepsilon_2)}(0) = -m\,\mathrm{diag}(p^2,q^2)$, and, by the inverse function theorem, (α_1,α_2) is, as a function of real variables, a diffeomorphism in a neighborhood of $0 \in \mathbb{C}^2$ with $(\alpha_1,\alpha_2)(0) = (1,1)$. If θ_1, θ_2 are small Diophantine numbers, there is a unique small $\boldsymbol{\varepsilon} \in \mathbb{C}^2$ such that $(\alpha_1,\alpha_2)(\boldsymbol{\varepsilon}) = \left(e^{i2\pi\theta_1}, e^{i2\pi\theta_1}\right)$. For small $\boldsymbol{\varepsilon}$, $f_{\boldsymbol{\varepsilon}}$ has $2(d-1)$ cycles of Siegel disks of order p or q (p can be 1).

(2) Let f be a rational function with $n_{SD}(f) \geq 2$, and $(z_0,\ldots,z_{p-1})$, $(z_0',\ldots,z_{p-1}')$ centers of Siegel disks of f of order p with Diophantine rotation number. Since f is a rational function of degree $d \geq 2$, $f^{-1}(\{z_0\})$ has two distinct points. Without loss of generality, $f(\infty) = z_0$ with $z_{p-1} \neq \infty$, as this can be achieved with Möbius transformations. Let P be a polynomial such that $P(z_j') = 0$ and $P'(z_j') = 1$, P and P' equal to 0 in the set of periodic attracting or neutral points of f that are not one of the points z_j', and P is 0 in the positive orbits of all critical points with a nonperiodic eventually periodic repelling orbit.

Consider polynomial perturbations H_δ of the identity, g_δ a quasiregular mapping and V_δ a neighborhood of ∞ defined as in the proof of (6.22), but with complex δ. There are δ_0, $\Delta > 0$ such that if $|\delta| < \delta_0$, g_δ has a Siegel disk $S_\delta \supset B_\Delta$. If $E_\delta = \cup_{j=0}^{p-1} g_\delta^j(S_\delta)$, since for small δ, it is $g_\delta\left(\mathrm{cl}\, V_\delta\right) \subset B_\Delta \subset E_\delta$, we get as in the proof of (6.22) a rational function f_δ perturbation of f and a quasiconformal conjugacy φ_δ of g_δ with f_δ. The multiplier of $\varphi_\delta(z_j')$ is $(1+\delta)^p(f^p)'(z_j')$. The cycle of Siegel disks with centers at $z_0',\ldots,z_{p-1}'$ can be perturbed so that $n_{SD}(f_\delta) = n_{SD}(f) - 1$ without changing the number of other cycles of attracting or neutral basins or $n_{CPER}(f)$, i.e., so that

$$n_{AB}(f_\delta) + n_{PB}(f_\delta) + n_C(f_\delta) = n_{AB}(f) + n_{PB}(f) + n_C(f) + 1.$$

---

[51] See Section 9 of Chapter 13 of *CADOVA*.

(3) It is proved there is a small $\delta \in \mathbb{C}$ with $n_{CPER}(f_\delta) = n_{CPER}(f)+1$. Suppose there is no critical point c_ς of f_δ with arbitrarily small δ such that the preceding equality holds. Consider the set of holomorphic functions $\{g_{\delta|B_R} : |\delta| < \delta_1\}$ for $\delta_1 \in]0, \delta_0[$ and $R > 0$ such that $V_\delta \subset \mathbb{C}_\infty \setminus B_R$ and $g_\varsigma(\mathbb{C}_\infty \setminus B_R) \subset S_\delta$ for $|\delta| < \delta_1$. If ζ_0 is a repelling periodic point of f not belonging to any positive orbit of critical points, there is $\delta_2 \in]0, \delta_1[$ and a function ζ holomorphic in B_{δ_2} with $\zeta(0) = \zeta_0$ such that $\zeta(\delta)$ is a repelling periodic point of $g_{\delta|B_R}$. Since $g_\delta^n(\mathbb{C}_\infty \setminus B_R) \subset E_\delta \setminus \{\zeta(\delta) : |\delta| < \delta_2\}$ for $n \in \mathbb{N} \cup \{0\}$, there are branches of $\{g_\delta^{-n} \circ \zeta\}$ holomorphic in B_{δ_2}. As in the proof of (5.2) we get that for $\delta_2 > 0$ smaller if necessary, $\{g_\delta\}_{\delta \in B_{\delta_2}}$ is structurally stable, and so is $\{f_\delta\}_{\delta \in B_{\delta_2}}$, in contradiction with the possible variation of the multiplier of z'_j established in the preceding paragraph. Thus, there is δ with $|\delta| < \delta_2$ such that $n_{CPER}(f_\delta) = n_{CPER}(f) + 1$.

(4) Denote $\mathscr{R}_d$ the space of rational functions of degree d. Let $f_0 \in \mathscr{R}_d$ and $(z_0, \ldots, z_{p-1})$ be the centers of a cycle of Siegel disks of order p of f_0. If ζ is a neutral periodic point of period q with multiplier λ, there are neighborhoods $U \subset \mathbb{C}_\infty$ of ζ and $W \subset \mathscr{R}_d$ of f_0 such that

$$\mathscr{H}_\zeta = \{ f \in W : f \text{ has only 1 periodic point } z \in U \text{ of period } q, \text{ with Multiplier } \lambda \}$$

is an analytic manifold in W. If c is a critical point of f_0 whose positive orbit is a nonperiodic eventually periodic repelling orbit of period q, there are neighborhoods $U' \subset \mathbb{C}_\infty$ of c and $W \subset \mathscr{R}_d$ of f_0 so that

$$\mathscr{H}'_c = \{ f \in W : f \text{ has only 1 Critical Point } c' \in U' \text{ and } f^{n+q}(c') = f^n(c') \text{ for } n \in \mathbb{N} \}.$$

is an analytic manifold in W. The intersection X of the two manifolds for all periodic neutral points ζ of f_0 outside $\{z_0, \ldots, z_{p-1}\}$ and all critical points c of f_0 whose positive orbits are nonperiodic but are eventually repelling periodic orbits is also an analytic manifold in W.

If W is small, there is an analytic function $z(f)$ of $f \in W$ (i.e., analytic in the $2d-1$ ordered coefficients of rational functions of degree d with monic polynomials in denominators) such that $f^p(z(f)) = z(f)$ and $z(f_0) = z_0$. Denote $\lambda(f)$ the Multiplier of $z(f)$. $f \mapsto \lambda(f)$ is analytic in the same sense. Since f_0 can be perturbed as above for $z(f)$ to be an attracting periodic point without changing the conditions in the definitions of $\mathscr{H}_\zeta$ and $\mathscr{H}'_c$, the function $f \mapsto \lambda(f)$ is not constant and $\lambda_{|X}$ maps open sets to open sets. Thus, f can be perturbed to f_δ with $n_{SD}(f_\delta) = f_{SD} - 1$ and

$$n_{AB}(f_\delta) + n_{PB}(f_\delta) + n_C(f_\delta) = n_{AB}(f) + n_{PB}(f) + n_C(f) + 1.$$

By *(1)* to *(4)*, the result is proved for cases without Herman rings, i.e., with $N_4 = 0$, and at least one of N_1, N_2, N_3, N_5 is not 0. If these are all 0, it is necessary to check that it is possible to have $n_{CPER}(f) = 2(d-1)$, for which $f(z) = \frac{2}{\zeta}\left(\frac{z-2}{z}\right)^d$, with $\zeta \neq 1$ a root of unity of order d, are examples.

Consider now the possibility of existence of Herman rings, i.e., of $N_4 \geq 1$. Suppose that f_0 is a rational function with a Siegel disk of order 1 with Diophantine rotation number $-\theta_1$, and analogously for f_{N_4} instead of f_0 and θ_{N_4} instead of θ_0. Let θ_2, θ_{N_4-1} be Diophantine numbers such that $\lambda_j = e^{i2\pi\theta_j}$ satisfy $\lambda_j \neq \lambda_{j+1}$ for $j \in \{1, \dots, N_4-1\}$. Each function $f_j(z) = z\,\frac{\lambda_j + z}{1+\lambda_{j+1} z}$, $j \in \{1, \dots, N_4-1\}$, has two Siegel disks of order 1 with rotation numbers θ_j and $-\theta_{j+1}$. With quasiconformal surgery as in the previous section applied to the functions $f_0, \dots, f_{N_4}$, we get a rational function f with N_4 Herman rings of order 1 and rotation numbers, resp., $\theta_1, \dots, \theta_{N_4}$, and

$$n_{HR}(f) \geq N_4\,,\, n_{SD}(f) \geq n_{SD}(f_0) + n_{SD}(f_{N_4}) - 2\,,\, n_{AB}(f) \geq n_{AB}(f_0) + n_{AB}(f_{N_4}),$$

$$n_{PB}(f) \geq n_{PB}(f_0) + n_{PB}(f_{N_4})\,, \qquad n_C(f) \geq n_C(f_0) + n_C(f_{N_4})\,.$$

Counting the critical points,

$$2(\deg(f)-1) = 2\big(\deg(f_0)-1\big) + 2\big(\deg(f_{N_4})-1\big) + 2(N_4-1)\,.$$

If $n_{AB} + n_{PB} + n_{SD} + 2n_{HR} + n_C + n_{CPER} \leq 2(d-1)$ for f_0 and f_{N_4}, the preceding inequalities hold and, from what we got without Herman rings, we conclude that f_0 and f_{N_4} can be chosen, so that $\deg(f) = d$ and $n_{AB}, n_{PB}, n_{SD}, n_{HR}, n_C, n_{CPER}$ for f are any numbers of $\mathbb{N} \cup \{0\}$ such that the inequality in the beginning of the paragraph holds. □

6.15 Sewing a *Continuum* in a Julia Set

The **ideal boundary** of a simply connected region $U \subset \mathbb{C}_\infty$ with $\#\mathbb{C}_\infty \backslash U \geq 2$ is the set $\mathcal{I}(U)$ of its prime ends.[52] A set $C \subset \mathbb{C}_\infty$ is called a **nondegenerate *continuum*** if it is a compact connected set with more than one point. If it is so, each connected component of $\mathbb{C}_\infty \setminus C$ is a simply connected set conformal to B_1. The **ideal boundary of a nondegenerate *continuum*** C is the union $\mathcal{I}(C)$ of the ideal boundaries of the connected components of $\mathbb{C}_\infty \backslash C$.

In the previous sections of this chapter, "cut and paste" quasiconformal surgery was applied with sewing at quasicircles. In 1988, C. McMullen[53] extended the application of the method to sewing at ideal boundaries, which may be not locally connected *continua*, in particular, possible subsets of sets of Julia of rational functions, with the aim of studying the group of Möbius transformations commuting with a given rational function.

In this section[54] we consider maps for which C is a positively invariant connected component of the Julia set of a rational function f. The connected components of

[52] See Section 7 of Chapter 10 of *CADOVA*.

[53] McMullen, C., Automorphisms of rational maps. *Holomorphic functions and moduli,* Vol. I (Berkeley, CA, 1986), 31–60. Math.Sci.Res.Inst.Publ. 10, Springer-Verlag, New York, 1988.

[54] This section and the following one are mostly about results of C. McMullen of 1988, following closely their presentation in B. Branner and N. Fagella book listed in the final bibliography.

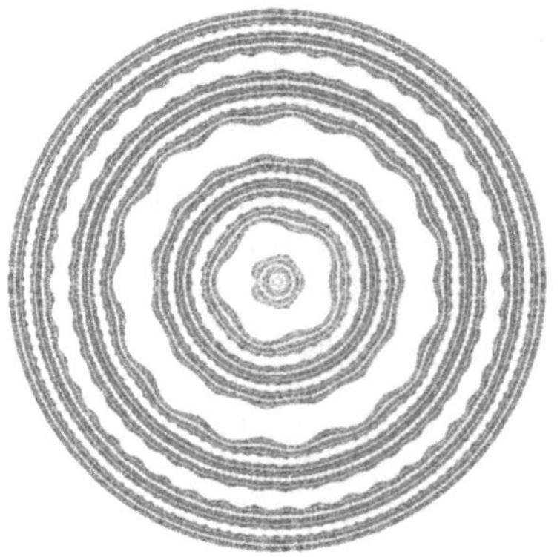

Fig. 6.19 Julia set $J(f_a)$ of $f_a(z) = z^3 + \frac{0.0001}{z^2}$. It is an uncountable Cantor set of quasicircles, one (unique) of them a buried component of $J(f)$ (calculated with a program made available by Mark McClure in the Internet)

$\mathbb{C}_\infty \setminus C$ can be Fatou components of f, but they can also be unions of some of these components with some components of the Julia set of f. For example, C can be the boundary of the immediate attracting basin of ∞ of a cubic rational function with a Herman ring, as in Example (6.10), and the complement of this attracting basin with infinite Fatou components and the corresponding boundaries. C can also be an invariant set of f disjoint of the Fatou components boundaries, such as a quasicircle accumulation of quasicircles which are connected components of boundaries of an infinite family of Fatou components, e.g., for $f_a(z) = z^3 + \frac{a}{z^2}$ with $a \in \mathbb{C} \setminus \{0\}$ small; such a C with $f(C) = C$ is called a **buried component** of the Julia set of f (Fig. 6.19). Another example is $C = S^1$ for $f_a(z) = z^3 \frac{z-a}{1-\bar{a}z}$, $a = 2.05 + i\,0.49$, where the two connected components of $\mathbb{C} \setminus C$ contain infinitely many Fatou components and infinitely many connected components of the Julia set accumulating at C (Fig. 6.20). On the other hand, if f is a polynomial with connected $J(f)$, then $C = J(f)$.

Ideal, $I(U)$, and topological, ∂U, boundaries are related as follows.

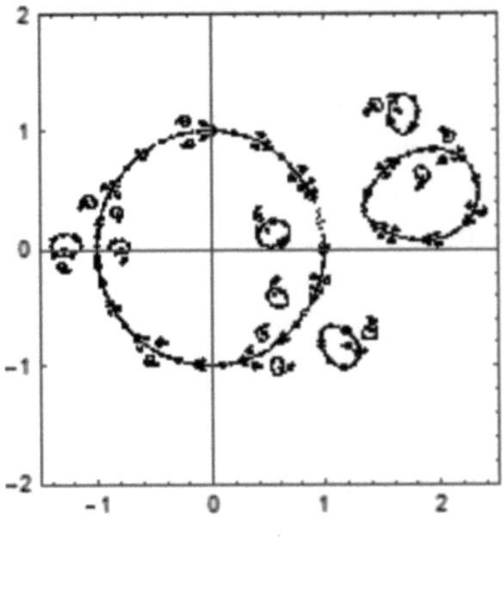

Fig. 6.20 Julia set $J(f_a)$ of $f_a(z) = z^3 \frac{z-a}{1-\bar{a}z}$, $a = 2.05 + i\,0.49$. 0 and ∞ are superattracting fixed points and the positive orbits of the other two fixed points converge to ∞. $J(f_a)$ is symmetric relative to S^1 and has infinitely many connected components (computed with *Mathematica*)

(6.32) *If $U \subset \mathbb{C}_\infty$ is a simply connected region and $\#\mathbb{C}_\infty \backslash U \geq 2$, every homeomorphism h of* cl U *onto* cl U *preserving orientation and the identity on ∂U, is also the identity in $\mathcal{I}(U)$; every quasiconformal mapping $\varphi : U \to U$ inducing the identity in $\mathcal{I}(U)$ has a quasiconformal extension to $\mathbb{C}_\infty$ with $\varphi_{|\mathbb{C}_\infty \backslash U} = 1_{\mathbb{C}_\infty \backslash U}$, $\varphi_{|\partial U} = 1_{\partial U}$.*

Proof If $\{N_j\}$ is a fundamental chain in U, i.e., a sequence of Crosscut neighborhoods of U with corresponding crosscuts with distinct end points belonging to a prime end[55] of U, by uniform continuity, $\{h(N_j)\}$ is an element of another prime end and the end points of the crosscuts corresponding to each C_j and to $h(C_j)$ are the same. If Q is a quadrilateral in U bounded by $C_j \cup C_{j+1}$ for large $j \in \mathbb{N}$, as h preserves orientation, Q and $h(Q)$ belong to a same connected component of $U \backslash C_j$. Therefore, the sequences $\{N_j\}$ and $\{h(N_j)\}$ belong to a same prime end, and $h_{|\mathcal{I}(U)} = 1_{|\mathcal{I}(U)}$.

The Poincaré distances of U of each z to $\varphi(z) \in U$ are upper bounded. As the Poincaré distance tends to ∞ when points approach ∂U in the spherical distance, the extension by continuity of φ to ∂U coincides in this set with the identity. Extend φ to $\mathbb{C}_\infty$ to be equal to the identity in $\mathbb{C}_\infty \backslash U$. The proof ends by applying (3.8). □

If f is a rational function and C is a positively invariant connected component of $J(f)$, the set of connected components of $f^{-1}(C)$ is finite. If U is a connected component of $\mathbb{C}_\infty \setminus C$ and $V = f(U)$, by the Riemann mapping theorem, there are conformal homeomorphisms T_U from U onto B_1 and T_V from V onto B_1, and the holomorphic function $f_{\mathcal{I}} = T_V \circ f \circ T_U^{-1}$ extends to the neighborhood of $S^1 = \partial B_1$ by continuity and the Schwarz reflection principle. This induces an analytic function of real variables $f_{\mathcal{I}} : \mathcal{I}(C) \to \mathcal{I}(C)$ mapping connected components to connected components. If U is a connected component of $\mathbb{C}_\infty \setminus C$ and there are no connected components of $f^{-1}(C)$, $f_{\mathcal{I}}$ is a holomorphic function from $\mathcal{I}(U)$ onto $\mathcal{I}(V)$. If U does not contain any connected component of $f^{-1}(C)$ and U' is the connected component of $f^{-1}(C)$ containing ∂U, then $V' = f(U')$ is included in some connected component V of $\mathbb{C}_\infty \backslash C$ and the image of one of the connected components of the complement of ∂U in its connected neighborhood W by f is the connected component of the complement of $f(\partial U)$ in the connected neighborhood $f(W)$ of $f(\partial U)$. The function $f_{\mathcal{I}}$ in $\mathcal{I}(C)$ can be obtained analogously to the previous case by extending the function defined between the considered components and then extending it with the Schwarz reflection principle, leading to $f_{\mathcal{I}} : \mathcal{I}(C) \to \mathcal{I}(C)$ which is conjugated by a quasiconformal mapping with a function analytic as function of real variables.

If f is a quasirational function, as it is conjugated by a quasiconformal mapping with the rational function of the preceding paragraph and applying such conjugacy,

55 See Section 7 of Chapter 10 of *CADOVA*.

we also obtain a corresponding function $f_{\mathcal{I}}:\mathcal{I}(C)\to\mathcal{I}(C)$ analytic as function of real variables, which is called the **function induced by f in the ideal boundary** of C.

The quasiconformal surgery to be considered in this section consists of replacing f with a function conjugated by a quasiconformal mapping with a Blaschke product $\mathfrak{B}$ from cl B_1 onto cl B_1.

A quasirational function f defined in a simply connected region $U\subset\mathbb{C}_\infty$ with $\#\mathbb{C}_\infty\backslash U\geq 2$ and a Blaschke product $\mathfrak{B}$ from cl B_1 onto cl B_1 are said to be **compatible on the ideal boundary of U through quasiconformal mappings φ_U and φ_V** if $V=f(U)$, $\varphi_U:\mathcal{I}(U)\to S^1$, $\varphi_V:\mathcal{I}(V)\to S^1$ are quasiconformal mappings and the functions they induce on the corresponding ideal boundaries $\widetilde{\varphi}_U$, $\widetilde{\varphi}_V$ are such that $\varphi_V\circ f_{\mathcal{I}}=\mathfrak{B}\circ\varphi_U$.

If $V=U$, φ_U, φ_V in the preceding paragraph may be equal, and then $f_{\mathcal{I}}$ and $\mathfrak{B}$ are conjugated by a quasisymmetric function $\widetilde{\varphi}_U$.

The following results are of C. McMullen in 1988 (Figs. 6.21, 6.22).

(6.33) McMullen Sewing Lemma: *If $f:\mathbb{C}_\infty\to\mathbb{C}_\infty$ is a quasirational function, $C\subset\mathbb{C}_\infty$ is a nondegenerate* continuum, *$f(C)=C$, $U\subset\mathbb{C}_\infty\backslash C$ is a simply connected region, $V=f(U)$, alternatively:*

(i) $\mathcal{I}(U)$ is a fixed connected component of $\mathcal{I}(C)$ under the function $f_{\mathcal{I}}$ induced by f in the ideal boundary of U;
(ii) $\mathcal{I}(U)$ is strictly preperiodic under $f_{\mathcal{I}}$ and $f^n(V)\cap U=\emptyset$ for $n\in\mathbb{N}\cup\{0\}$.

If f and the Blaschke product $\mathfrak{B}$ from cl B_1 *onto* cl B_1 *are compatible on the ideal boundary of U through quasiconformal mappings φ_U, φ_V, then the function g equal to $\varphi_V^{-1}\circ\mathfrak{B}\circ\varphi_U$ in U and to f in $\mathbb{C}_\infty\backslash U$ is a quasirational function. In case (i), $\varphi_U=\varphi_V$ and there is a quasiconformal conjugacy of $g_{|U}$ with $\mathfrak{B}$.*

Proof $f^{-1}\circ g$ in a neighborhood of ∂U, with the branch of f^{-1} such that the points of $\mathbb{C}_\infty\backslash U$ are fixed is a quasiconformal mapping in U and is the identity in $\mathbb{C}_\infty\backslash U$.

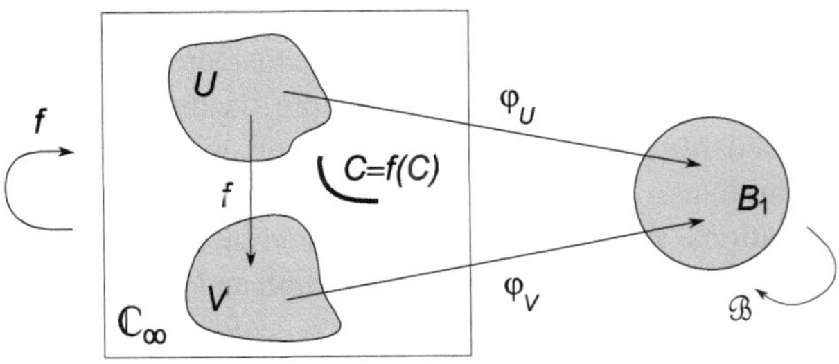

Fig. 6.21 McMullen sewing lemma (6.33). C is a nondegenerate *continuum*, f is a quasirational mapping compatible with Blaschke product $\mathfrak{B}$ on $\mathcal{I}(U)$ through quasiconformal mappings φ_U, φ_V, $g=\varphi_V^{-1}\circ\mathfrak{B}\circ\varphi_U$ in U, $g=f$ in $\mathbb{C}_\infty\backslash U$ is a quasirational mapping

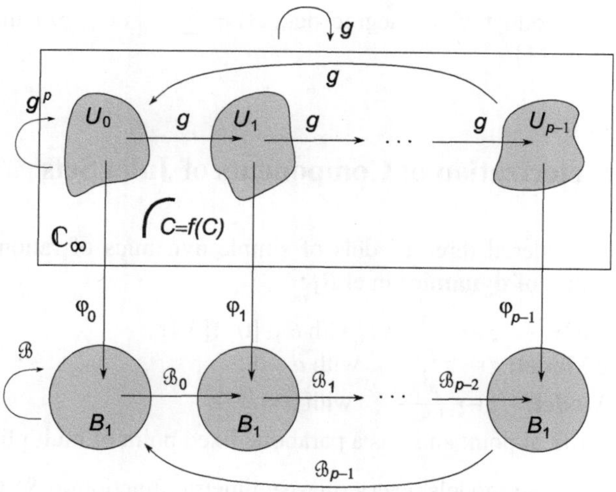

Fig. 6.22 McMullen sewing lemma for cycles (6.34). C is a nondegenerate *continuum*, f is a quasirational mapping compatible with Blaschke product $\mathfrak{B}_j$ on $\mathcal{I}(U_j)$ through quasiconformal mappings φ_j, $g=\varphi_{j+1}^{-1}\circ\mathfrak{B}_j\circ\varphi_j$ in U_j, $j\in\{0,\ldots,p-1\}$, $\mathfrak{B}=\mathfrak{B}_{p-1}\circ\cdots\circ\mathfrak{B}_0$, $g=f$ in $\mathbb{C}_\infty\backslash\cup_{j=0}^{p-1}U_j$ is a quasirational mapping

$f_{\mathcal{I}}=\widetilde{\varphi}_V^{-1}\circ\mathfrak{B}\circ\widetilde{\varphi}_U^{-1}:\mathcal{I}(U)\to\mathcal{I}(U)$ also induces the identity in $\mathcal{I}(U)$. By (6.32), $f^{-1}\circ g$ is locally a quasiconformal mapping. Composing it at the left with f, we get that g is locally a quasiconformal mapping except in small neighborhoods of the branching points, and g is quasiregular. The dilatation K_g of g is upper bounded by $\max\{K_f, K_{\varphi_U}K_{\varphi_V}\}$.

If $f\big(\mathcal{I}(U)\big)_{|f(\mathbb{H})}=\mathcal{I}(U)$, then $g(U)=U$, $g^n(U)=\varphi^{-1}\circ\mathfrak{B}^n\circ\varphi(U)$, $\varphi_U=\varphi_V\overset{def}{=}\varphi$. So $K_{g^n}=K_\varphi^2$ in U and $K_{g^n}\le K_{f^n}K_\varphi^2$. Since f is quasirational, there is $C\ge 1$ such that $K_{f^n}\le C$. Thus, $K_{g^n}\le C K_\varphi^2$ and g are quasirational.

If $f_{\mathcal{I}}\big(\mathcal{I}(U)\big)=\mathcal{I}(V)$, $f^n(V)\cap U=\emptyset$ for $n\in\mathbb{N}$, then $K_{g^n}\le C K_{\varphi_U}K_{\varphi_V}$, since no orbit passes more than once in U. □

(6.34) McMullen Sewing Lemma for Cycles: *With the hypothesis of (6.33), if $U_0,\ldots,U_{p-1}$ are connected components of $\mathbb{C}_\infty\backslash C$ such that $f_{\mathcal{I}}^j\big(\mathcal{I}(U_0)\big)=U_j$ for $j\in\{0,\ldots,p\}$, identifying $U_p=U_0$ and with each $f_{|U_j}$ compatible on ideal boundary to a Blaschke product $\mathfrak{B}_j:B_1\to B_1$ through quasiconformal mappings $\varphi_j:U_j\to B_1$ for $j\in\{0,\ldots,p-1\}$ and with $\varphi_p=\varphi_0$, then the function g equal to $\varphi_{j+1}^{-1}\circ\mathfrak{B}_j\circ\varphi_j$ in U_j for $j\in\{0,\ldots,p-1\}$ and equal to f in $\mathbb{C}_\infty\backslash\cup_{j=0}^{p-1}U_j$ is a quasirational mapping.*

Proof As in the proof of the preceding result, g is a quasiregular mapping. For every $n\in\mathbb{N}\cup\{0\}$, there is a conjugacy by quasiconformal mappings of g^n

with a Blaschke product $\mathfrak{B}$ of degree $\deg(\mathfrak{B}) = \sum_{k=0}^{p-1} \deg(\mathfrak{B}_k)$ in each of the U_j, $j \in \{0, \dots, p-1\}$. □

6.16 Characterization of Components of Julia Sets

C. McMullen considered three models of simple dynamics of rational functions, called **rigid models of dynamics** in cl B_1 :

1. **Elliptic Model:** $z \mapsto e^{i2\pi\theta} z$, with $\theta \in [0, 1[\setminus \mathbb{Q}$;
2. **Hyperbolic Model:** $z \mapsto z^d$, with $d > 1$;
3. **Parabolic Model:** $z \mapsto \frac{z^d + a}{1 + az^d}$, with $a = \frac{d-1}{d+1}$
   (only 0 is a critical point and 1 is a parabolic fixed point of multiplicity 3).

If f is one of these models, every quasisymmetric function in S^1 Conjugacy of $f_{|S^1}$ with itself is a rotation $z \mapsto \alpha z$ with $|\alpha| = 1$. For the elliptic model $\alpha \in S^1$, for the hyperbolic model $\alpha^d = 1$, and for the parabolic model $\alpha = 1$. This is why they are called rigid models.

The following results were proved by C. McMullen in 1988.

(6.35) Dynamics Induced in the Ideal Boundary of a Positively Invariant Connected Component of the Julia Set:
If C is a connected component of the Julia set $J(f)$ positively invariant under a complex rational function f and $\#C \geq 2$, then:

1. *For each connected component A of $\mathbb{C}_\infty \setminus C$, except a finite number of such components, $f_{|A}$ is a quasiconformal mapping.*
2. *Every connected component of $\mathcal{I}(C)$ is preperiodic, and the set of periodic connected components of $\mathcal{I}(C)$ is finite.*
3. *For every connected component A of $\mathcal{I}(C)$, there are $d \in \mathbb{N}$ and quasisymmetric functions ψ_1, ψ_2 of, resp., A and $f_{\mathcal{I}}(A)$ onto S^1, one of which can be arbitrarily chosen, such that $\psi_2 \circ f_{\mathcal{I}} \circ \psi_1^{-1}$ matches $z \mapsto z^d$.*

Proof The set $\mathcal{V}$ of connected components of $\mathbb{C}_\infty \setminus C$ intersecting $f^{-1}(C)$ or containing a critical point is finite.

(1) If U is a connected component of $\mathbb{C}_\infty \setminus C$ not belonging to $\mathcal{V}$, (U, f) is a covering of $f(U)$, which also is a component of $\mathbb{C}_\infty \setminus C$. Since U is conformal to a circle, $f_{|U}$ is a quasiconformal mapping onto $f(U)$.

(2) If U is a component of $\mathbb{C}_\infty \setminus C$ not belonging to $\mathcal{V}$ and $f^n(U) \notin \mathcal{V}$ for all $n \in \mathbb{N} \cup \{0\}$, by the Sullivan nonwandering domains theorem (6.3), $\{f^n(U)\}$ is eventually a cycle of degree p. So, eventually $f^n(U)$ is a Fatou component, and the restriction of f^p to $f^n(U)$ is a quasiconformal mapping and

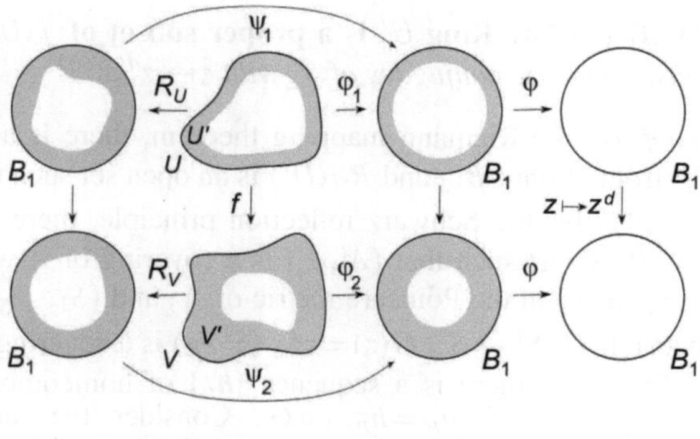

Fig. 6.23 Dynamics induced in the ideal boundary of a positively invariant connected component of the Julia set in (6.35.3)

$f^n(U), \ldots, f^{n+p-1}(U)$ is a Siegel disk cycle. Therefore, $\mathcal{I}(U)$ is preperiodic under $f_{\mathcal{I}}$. If eventually $f^n(U) \in \mathcal{V}$ and some further iteration of f is a $\mathcal{I}(W)$ with $W \notin \mathcal{V}$, by the preceding paragraph, eventually $f^m(W)$ is a Siegel disk or an element of $\mathcal{V}$. As $\mathcal{V}$ and the number of cycles of Siegel disks are finite, $f_{\mathcal{I}}^k(\mathcal{I}(U))$ is eventually a cycle, $\mathcal{I}(U)$ is preperiodic under $f_{\mathcal{I}}$ and the set of periodic connected components of $\mathcal{I}(C)$ is finite.

(3) If $U \notin \mathcal{V}$, then $f_{|U}$ is a quasiconformal function, and there is a quasisymmetric function ψ such that $\psi \circ f_{\mathcal{I}} = 1_{\mathcal{I}(U)}$. If $U \in \mathcal{V}$, then there is a ring $U' \subset U$ such that $\partial U \subset \partial U'$ and (U', f) are a covering of $f(U')$, since as the sets of critical points of f in U and of connected components of $f^{-1}(C)$ contained in U are finite, U' can be chosen avoiding the elements of these sets. There are conformal homeomorphisms φ_1, φ_2 of, resp., U', $f(U')$ onto annuli, resp., $A_1 = B_1 \backslash \mathrm{cl}\, B_{r_1}$, $A_2 = B_1 \backslash \mathrm{cl}\, B_{r_2}$, with $r_1, r_2 \in]0, 1[$. Therefore, $(A_1, \varphi_2 \circ f \circ \varphi_1^{-1})$ is a covering of A_2, and there is a conjugacy φ of $\varphi_2 \circ f \circ \varphi_1^{-1}$ with $z \mapsto z^d$. By the Riemann mapping theorem, there are conformal homeomorphisms R_U, $R_{f(U)}$ of, resp., U, $f(U)$ onto B_1 and $\psi_1 \overset{def}{=} \varphi \circ \varphi_1 \circ R_U^{-1}$, $\psi_2 \overset{def}{=} \varphi \circ \varphi_2 \circ R_{f(U)}^{-1}$ have extension by continuity to S^1. Hence $\psi_2 \circ f_{\mathcal{I}} \circ \psi_1^{-1} : B_1 \to B_1$ coincides with $z \mapsto z^d$ (Fig. 6.23). The arbitrariness of these quasisymmetric functions results from quasisymmetric functions with the same domain to differ by conjugacy with a quasisymmetric function. □

Under the hypothesis of (6.35), if U is a connected component of $\mathbb{C}_\infty$ such that $\mathcal{I}(U)$ is periodic under $f_{\mathcal{I}}$ with period $p \in \mathbb{N} \cup \{0\}$, then $f_{\mathcal{I}}(\mathcal{I}(U)) = \mathcal{I}(U)$ and, as seen in the proof of (6.35.3), there is a ring $U' \subset U$ such that $\partial U \subset \partial U'$ and (U', f) is a covering of $f(U')$. Therefore, to clarify the dynamics induced in the ideal boundary of each connected component U of $\mathbb{C}_\infty$ such that $\mathcal{I}(U)$ is periodic under $f_{\mathcal{I}}$, it suffices to consider the case with $\mathcal{I}(U)$ a fixed component and to separately analyze the cases with ∂U not contained in the boundary of a Fatou component of f, which can be: *(1)* the Ring U' is a proper subset of $f(U')$ or *(2)* ∂U is a buried component of $J(f)$, and the cases of ∂U being contained in the boundary of a Fatou component $V \subset U$ of f according to the possibilities of Fatou components given by the Sullivan classification theorem: *(3)* Attracting or superattracting, *(4)* Parabolic with fixed point outside ∂U, *(5)* Siegel disk or Herman ring, *(6)* Parabolic with fixed point in ∂U.

CASE 1 – The Ring U' is a proper subset of $f(U')$: *There are $d \in \mathbb{N}$ and a quasisymmetric conjugacy of $f_{\mathcal{I}}$ with $z \mapsto z^d$ in S^1.*

Proof By the Riemann mapping theorem, there is a conformal homeomorphism R_U from U onto B_1, and $R_U(U')$ is an open set such that $R_U \circ f \circ R_U^{-1}(R_U(U')) \subset R_U(U')$. By the Schwarz reflection principle, there are rings $A \subsetneqq A' \supset S^1$ and $F \in H(A', A)$ such that (A', F) is a covering of A with $d \in \mathbb{N}$ sheets. Thus, $F_{|S^1}$ is expansive in the Poincaré metric of A, and $(S_1, F_{|S^1})$ is a covering of S^1 with d sheets. If $G: S^1 \to S^1$, $G(z) = z^d$, (S^1, g) is a covering of S^1 with d sheets.

Therefore, there is a sequence $\{h_n\}$ of homeomorphisms of S^1 onto S^1 with $h_0 = 1_{S^1}$ and $F \circ h_n = h_{n-1} \circ G$. Consider the universal covering $(\mathbb{R}, p)$ of S^1 with $p(x) = e^{i2\pi x}$ and functions $\widetilde{F}, \widetilde{G}: \mathbb{R} \to \mathbb{R}$ such that $p \circ \widetilde{F} = F \circ p$ and $p \circ \widetilde{G} = G \circ p$. Since F and G are expansive, there are $K > 0$, $\rho \in]0, 1[$ such that $|(\widetilde{F}^{-n})'|$, $|(\widetilde{G}^{-n})'| \le K\rho^n$ for $n \in \mathbb{N}$. The function T defined in the metric space X of continuous nondecreasing functions $\varphi: \mathbb{R} \to \mathbb{R}$ such that $\varphi(x+1) = \varphi(x) + 1$ with the distance $D(\varphi_1, \varphi_2) = \sup |\varphi_1 - \varphi_2|$ (the supremum is a real number because for $\varphi \in X$ the function $\varphi - 1_{\mathbb{R}}$ is periodic) by $T(\varphi) = (\widetilde{F}^{-1}) \circ \varphi \circ \widetilde{G}$ is a contraction and as (X, D) is a complete metric space, by the Banach fixed point theorem, T has a unique fixed point $\widetilde{H} \in X$, which satisfies $\widetilde{F} \circ \widetilde{H} = \widetilde{H} \circ \widetilde{G}$, and there is a function $H: S^1 \to S^1$ such that $p \circ \widetilde{H} = H \circ p$ and $F \circ H = H \circ G$. Therefore, H is a conjugacy of[56] F with G. Hence, there is a conjugacy h of $f_{\mathcal{I}}$ with G in S^1.

It remains to prove that $h: S^1 \to S^1$ is a quasisymmetric function. Let I_1, I_2 be contiguous arcs of S^1 with the same length. There is a smaller $n \in \mathbb{N} \cup \{0\}$ such that $G^{n+1}(I_1 \cup I_2) \supset S^1$. As $\frac{|G^n(x)|}{|G^n(y)|}$ with x, $y \in I_1 \cup I_2$ is upper bounded by some $M > 0$, the quotients of the lengths of the arcs $G^n(I_1)$ and $G^n(I_2)$ are upper bounded by M. Also, the quotients of the lengths of the arcs $F^n(I_1)$ and $F^n(I_2)$ are upper bounded. By the choice of n, the lengths of the arcs $G^n(I_1)$ and $G^n(I_2)$ are bounded by 2π, and, as h is uniformly continuous, the quotients of the lengths of the arcs $h \circ G^n(I_1)$ and $h \circ G^n(I_2)$ are upper bounded. As $F^n \circ h = h \circ G^n$, the quotients of the arc lengths of $h(I_1)$ and $h(I_2)$ are also upper bounded. So, h is a quasisymmetric function. □

CASE 2 – ∂U is a buried component of $J(f)$: *There are $d \in \mathbb{N}$ and a quasisymmetric function conjugacy of $f_{\mathcal{I}}$ with $z \mapsto z^d$ in S^1.*

Proof Since ∂U is a proper compact subset of $\mathbb{C}_\infty$, for every $n \in \mathbb{N}$, there is a Jordan curve $\gamma_n \subset \mathbb{C} \backslash \partial U$ such that the distance of γ_n to ∂U is $< \frac{1}{n}$. Since ∂U is disjoint of the union of the boundaries of the Fatou components of f, the set of the Fatou components including a γ_n is infinite. For large $n \in \mathbb{N}$, the Fatou components including some γ_n are nonperiodic and the ring V' with boundary $\gamma_n \cup \partial U$ does not contain critical values of f. Denote U' the (unique) connected component of $f^{-1}(V')$ included in U and bounded by $\partial U \cup f^{-k}(\gamma_n)$ for some $k \in \mathbb{N}$. As γ_n is disjoint of the periodic Fatou components, it also is disjoint of their preimages by f; so, alternatively, $V' \subset U'$ or $V' \supset U'$. Since $f(U') = V'$ and both U' and V' are

[56] This argument was given in 1969 by Michael Shub (1943–).

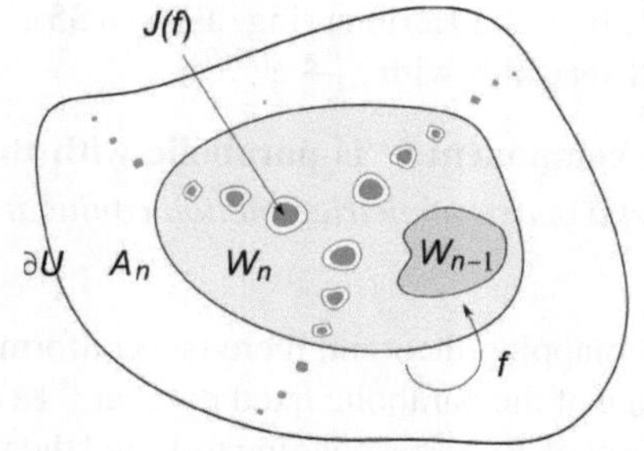

Fig. 6.24 Regions W_n and rings A_n in the proof of Case 3 – The Fatou component V is attracting or superattracting, showing a connected component of ∂U and connected components W_n of $f^{-n}(W_0)$ for W_0 a disk centered at a fixed point. $f(W_0) \subsetneq W_0$ and cl $W_0 \cap \partial U = \emptyset$

rings, their modules satisfy $M(V') \geq M(U')$, and, therefore, $U' \subsetneq V'$. We conclude by applying Case 1. □

CASE 3 – The Fatou component V is attractive or superattractive: *There are $d \in \mathbb{N} \backslash \{1\}$ and a quasisymmetric function conjugacy of $f_{\mathcal{I}}$ with $z \mapsto z^d$ in S^1.*

Proof Let $W_0 \subset V$ be the disk centered at the fixed point in V such that $f(W_0) \subsetneq W_0$ and (cl $W_0) \cap \partial U = \emptyset$, and W_n be the connected component of $f^{-n}(W_0)$ containing W_0. Then, $W_{n-1} \subset W_n$ for $n \in \mathbb{N}$ and $V = \cup_{n=1}^{\infty} W_n$ (W_n may be multiply connected and if V is so, W_n also is so for n large). For each $n \in \mathbb{N}$, there is a connected component of ∂W_n which with ∂U bounds a ring A_n disjoint of W_n (A_n may contain infinitely many connected components of $J(f)$) (Fig. 6.24). The distance of ∂W_n and ∂U tends to 0, as otherwise there would exist a *continuum* in ∂V included in U with closure intersecting ∂U, contradicting that C is a connected component of $J(f)$. For n large so that A_n does not contain critical points of f it must be $A_{n+1} \subsetneq A_n$ and $f(A_{n+1}) = A_n$. We conclude by applying Case 1. □

CASE 4 – The Fatou component V is parabolic with a fixed point outside ∂U: *there are $d \in \mathbb{N} \backslash \{1\}$ and a quasisymmetric function conjugacy of $f_{\mathcal{I}}$ with $z \mapsto z^d$ in S^1.*

Proof As for Case 3, with W_0 an attracting petal with $\partial W_0 \cap \partial V$ a parabolic fixed point, ∂U without points with positive orbits containing the fixed point, as $f(\partial U) = \partial U$ and $W_n \cap \partial U = \emptyset$, $n \in \mathbb{N}$. □

CASE 5 – The Fatou component V is a Siegel disk or a Herman ring: *There is an analytic function as a function of real variables conjugacy of $f_{\mathcal{I}}$ with $z \mapsto e^{i2\pi\theta}$ in S^1, with $\theta \in [0, 1[\backslash \mathbb{Q}$ the rotation number.*

Proof If V is a Siegel disk or a Herman ring, by the Riemann mapping theorem, there is a conformal homeomorphism of V onto, resp., B_1 or $B_1 \backslash$cl B_r with $r \in]0, 1[$ conjugating f with a rotation $z \mapsto e^{i2\pi\theta} z$ in B_1. If V is a Siegel disk, $f_{\mathcal{I}}$ in S^1 is the same function in S^1. If V is a Herman ring, as in (6.35), $f_{\mathcal{I}}$ is conjugated by an analytic function of real variables with $z \mapsto e^{i2\pi\theta} z$. □

CASE 6 – The Fatou component V is parabolic with the fixed point in ∂U: *There are $d \in \mathbb{N} \backslash \{1\}$ and a quasisymmetric function conjugating $f_{\mathcal{I}}$ with $z \mapsto \frac{z^d+a}{1+az^d}$ in S^1, where $a = \frac{d-1}{d+1}$.*

Proof By the Riemann mapping theorem, there is a conformal homeomorphism of U onto B_1 with the image of the parabolic fixed point at 1. Keep the notations V and f also for, resp., the subset of B_1 corresponding to V and the function corresponding to $f_{|U}$ that cannot be defined in all B_1, but it is defined in V and in the intersection of a neighborhood of S^1 with B_1. By the Schwarz reflection principle, f can be extended to a neighborhood of S^1 and $f_{\mathcal{I}}$ is the restriction of this extension to S^1.

Let $\mathcal{P}$ be a petal with $1 \in \partial\mathcal{P}$ and $\partial\mathcal{P}$ a C^∞ curve not intersecting postcritical orbits. Denote by W_n the component of $f^{-n}(\mathcal{P})$ containing $\mathcal{P}$. For $n \in \mathbb{N}$ large, W_n contains all critical points of f in V, and there is only one connected component γ_n of ∂W_n in the region A_n between W_n and ∂B_1, and it intersects S^1 at $f^{-n}(\{1\})$. The curves γ_n and γ_{n+1} bound a region differing from a ring only, because there is a finite number of points of S^1 that also are points of γ_n and γ_{n+1}, whereby the "ring" is tightened in such a way that the radius of B_1 that contains these points does not intersect the interior of the "ring" (Fig. 6.25). The positive orbits of points in the region between γ_{n+1} and S^1 pass once and only once in $A_n \cup \gamma_n$.

It is analogous for the function $P_d(z) = \frac{z^d+a}{1+az^d}$ in B_1, with d the degree of $f_{\mathcal{I}}$ in S^1; denote the corresponding sets by W'_n, γ'_n, A'_n.

For large n, there is a C^1 diffeomorphism φ from γ_n onto γ'_n with fixed point 1 mapping each point of $\gamma_n \cap S^1$ preimage of 1 by iterations of f to the corresponding points of $\gamma_{n+1} \cap S^1$ and satisfies $P_d \circ \varphi = \varphi \circ f$. Extend φ to be a quasiconformal mapping from $\cup_{n=0}^{\infty} A_n$ to $\cup_{n=0}^{\infty} A'_n$ satisfying $P_d \circ \varphi = \varphi \circ f$. The result is a quasiconformal bijection φ from B_1 onto B_1 with a continuous extension to S^1 whose restriction to S^1 is quasisymmetric in S^1 and conjugates $f_{\mathcal{I}}$ with P_d. □

Summarizing what was proved in the six cases:

(6.36) Induced dynamics on a periodic connected component of the ideal boundary of a positively invariant connected component of the Julia set: *If C is a positively invariant connected component of the Julia set of a complex rational function f with $\#C \geq 2$, for each periodic connected component of $\mathcal{I}(C)$ with period $p \in \mathbb{N}$, there is a quasisymmetric function conjugating $f_{\mathcal{I}}^p$ with a rigid model.*

With these results, we can prove the following.

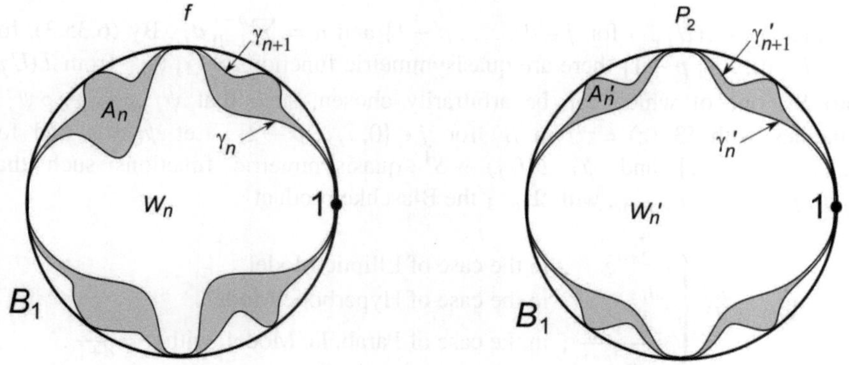

Fig. 6.25 Regions W_n , curves γ_n and rings A_n in the proof of Case 6 – The Fatou component V is parabolic with the fixed point in ∂U mapped to 1 by a conformal homeomorphism from U to B_1 (with $d=2$)

(6.37) *If C is a positively invariant connected component of the Julia set $J(f)$ of a complex rational function f with $\#C \geq 2$, there is a rational function g and a quasiconformal mapping $\varphi:\mathbb{C}_\infty \to \mathbb{C}_\infty$ conjugacy of $g_{|J(g)}$ with $f_{|C}$, $J(g)$ is connected, the periodic Fatou components of g are superattracting, parabolic or Siegel disk cycles, and in each periodic Fatou component of g with period p, there is a conjugacy of g^p with one of the rigid models; if c is a critical point in one of the nonperiodic and eventually periodic Fatou components of g and $n \in \mathbb{N}$ is the minimum number such that $g^n(c)$ belongs to a periodic Fatou component U of g, then:*

1. *If U is an attracting basin of a superattracting periodic orbit, then $g^n(c)$ is a periodic superattracting point;*
2. *If U is a parabolic basin and k is the first in $\mathbb{N}$ such that $g^k(U)$ contains a Critical Point, then $g^{n+k}(c)$ is a critical point of g;*
3. *If U is a Siegel disk, then $g^n(c)$ is its center.*

Proof If $\mathcal{I}(U)$ is periodic under $f_\mathcal{I}$ with period $p \in \mathbb{N}$, by (6.36), there is a quasisymmetric function $\widetilde{\varphi}:\mathcal{I}(U) \to S^1$ such that $\widetilde{\varphi} \circ f_\mathcal{I}^p \circ \widetilde{\varphi}^{-1}$ is one of the rigid models of dynamics at the beginning of this section, denoted M_U. Extend $\widetilde{\varphi}$ to a quasiconformal mapping in cl B_1, also denoted by $\widetilde{\varphi}$. By the Riemann mapping theorem, there is a conformal homeomorphism from U onto B_1. The restrictions of the functions f^p and M_u to S^1 are bijectively related through the function $\varphi = \widetilde{\varphi} \circ R_U$.

If $p = 1$, the function equal to $\varphi^{-1} \circ M_U \circ \varphi$ in U and to f in $\mathbb{C}_\infty \setminus U$, by the McMullen sewing lemma (6.33), is quasirational.

If $p \in \mathbb{N} \setminus \{1\}$, we need to replace f in components U_j such that $\mathcal{I}(U_j) = f_\mathcal{I}^j(\mathcal{I}(U))$, $j \in \{0, \ldots, p-1\}$, by functions compatible on ideal boundaries with $f_\mathcal{I}$ through a function φ and equal to $\varphi^{-1} \circ M_U \circ \varphi$ in U. Denote d_j the degree of

$f_{\mathcal{I}} : \mathcal{I}(U_j) \to \mathcal{I}(U_{j+1})$ for $j \in \{0, \dots, p-1\}$ and $d = \sum_{j=0}^{p-1} d_j$. By (6.35.3), for each $j \in \{0, \dots, p-1\}$ there are quasisymmetric functions $\psi_{j,1}, \psi_{j,2}$ from $\mathcal{I}(U_j)$ onto S^1, one of which can be arbitrarily chosen, such that $\psi_{j+1,2} \circ f_{\mathcal{I}} \circ \psi_{j,1}^{-1}$ coincides with $\mathfrak{B}_j(z) = z^{d_j}$ in S^1 for $j \in \{0, \dots, p-2\}$. Let $\psi_0 = \varphi$ and for $j \in \{1, \dots, p-1\}$ and $\psi_j : \mathcal{I}(U_j) \to S^1$ quasisymmetric functions such that $\psi_j \circ f_{\mathcal{I}} = \mathfrak{B}_{j-1} \circ \psi_{j-1}$, with $\mathfrak{B}_{p-1}$ the Blaschke product

$$\mathfrak{B}_{p-1} \overset{def}{=} \begin{cases} e^{i2\pi\theta} z & \text{, in the case of Elliptic Model,} \\ z^{d_{p-1}} & \text{, in the case of Hyperbolic Model,} \\ \frac{z^{d_{p-1}} + a}{1 + a z^{d_{p-1}}} & \text{, in the case of Parabolic Model, with } a = \frac{d-1}{d+1} \,. \end{cases}$$

By the McMullen sewing lemma for cycles (6.34), the function equal to $\psi_{j+1}^{-1} \circ \mathfrak{B}_j \circ \psi_j$ in U_j for $j \in \{0, \dots, p-1\}$ and to f in $\mathbb{C}_\infty \setminus \cup_{j=0}^{p-1} U_j$, is a quasirational mapping.

It remains the case of a connected component $\mathcal{I}(V)$ of $\mathcal{I}(C)$ nonperiodic under $f_{\mathcal{I}}$. By the Sullivan nonwandering domains theorem, $\mathcal{I}(V)$ is preperiodic under $f_{\mathcal{I}}$. If $f_{\mathcal{I}}^k\big(\mathcal{I}(V)\big) = \mathcal{I}(W)$ is periodic under $f_{\mathcal{I}}$, there are quasisymmetric bijections $\psi_1 : W \to S^1$, $\psi_2 : f_{\mathcal{I}}(W) \to S^1$ such that $\psi_2 \circ f_{\mathcal{I}} \circ \psi_1$ is equal to $z \mapsto z^d$ in S^1 for some $d \in \mathbb{N}$. If $d = 1$, in each $f^j(V)$, $j \in \{0, \dots, k-1\}$, f is not modified. If $d \in \mathbb{N} \setminus \{1\}$, replace f in $f^j(V)$ as follows: successively backward from $j = k-1$ to $j = 0$ consider conformal homeomorphisms R_j, R_{j+1} from, resp., $f^j(V)$, $f^{j+1}(V)$ onto B_1, and define as above quasirational functions ψ_j, ψ_{j+1} and

$$g : f^j(V) \to f^{j+1}(V) \,, \qquad g = \psi_{j+1}^{-1} \circ (\psi_j)^d \,. \qquad \square$$

Consequently, the restriction of a complex rational function f to a positively invariant component of the Julia set $J(f)$ with more than one point is conjugated by a quasiconformal mapping in the Julia set $J(f)$ with a rational function, and if $J(f)$ is connected, the dynamics in each Fatou component is conjugated with one of the rigid models.

Chapter 7
Dimensions of Mandelbrot Set Boundary and of Julia Sets

7.1 Introduction

The boundary of the Mandelbrot set is intricate. Numerical computation in the neighborhood of certain of its points exhibits a density (Fig. 7.1) that led B. Mandelbrot to conjecture in 1985 that its Hausdorff dimension is 2, what was proved by M. Shishikura in an article[1] published in 1998 using holomorphic motions (of Chap. 4), Écalle cylinders (with the measurable Riemann mapping theorem of Chap. 2) and structural stability and bifurcation (of Chap. 5).

It had also been conjectured that the Hausdorff dimension of the Julia set of $P_c(z) = z^2 + c$ is close to 2 for certain $c \in \mathbb{C}$ (Fig. 7.2), and M. Shishikura proved that it is 2 for certain values c in the boundary of the Mandelbrot set, abundant in the sense of being a generic subset, therefore dense, of this boundary.

The proofs of M. Shishikura are based on features of the orbits geometry in neighborhoods of parabolic fixed points not considered in *CADOVA*, because they are not applied there, but useful for analyzing bifurcations of such points, e.g., the fixed point can bifurcate to two fixed points and orbits can pass between them (Fig. 7.3), for which the Écalle cylinders theory is used. This is a good opportunity to present this important topic of general interest, initiated in 1984–1985 by A. Douady and J. Hubbard and continued in 1989 by P. Lavaurs and in 1998 by M. Shishikura, that was not included in *CADOVA*, because it is somewhat intricate and was not needed there. The name Écalle cylinders was given by A. Douady due to its connection with J. Écalle work in 1975 on holomorphic functions tangent to the identity.

[1] M. Shishikura, The Hausdorff dimension of the boundary of the Mandelbrot set and Julia sets, *Annals of Mathematics* **147** (1998), 225–267, with *preprint* disseminated in 1991. This section follows parts of this article.

L. T. Magalhães, *Quasiconformal Mappings in the Plane and Complex Dynamics*,
https://doi.org/10.1007/978-3-031-80115-0_7

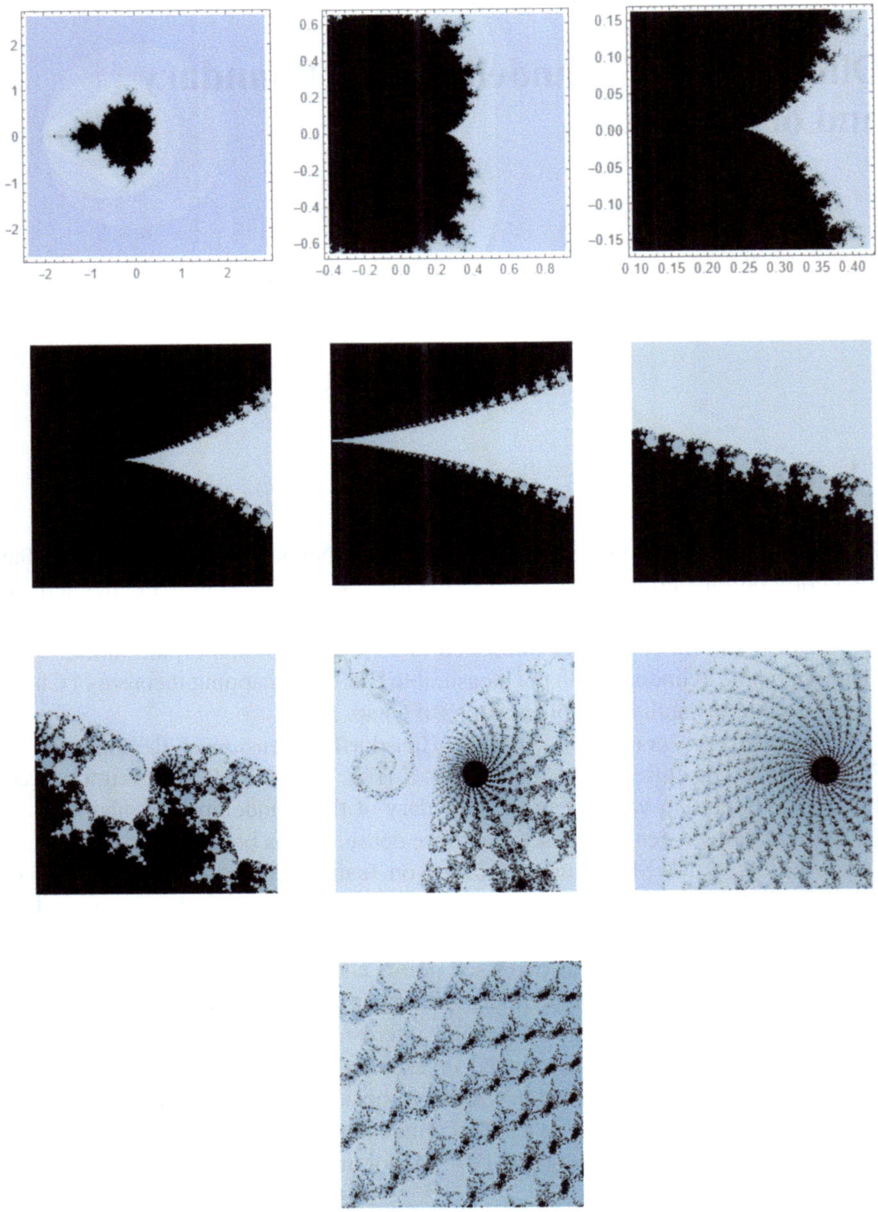

Fig. 7.1 Enlargements of the Mandelbrot set (in black) centered at the center of each figure successively to the quadruple of lengths, illustrating how in certain parts the boundary is intricate and dense, with the density in the last figure increasing as it approaches the point visible in the three preceding figures around which there are spiral structures, suggesting it may have dimension 2 (computed with *Mathematica*)

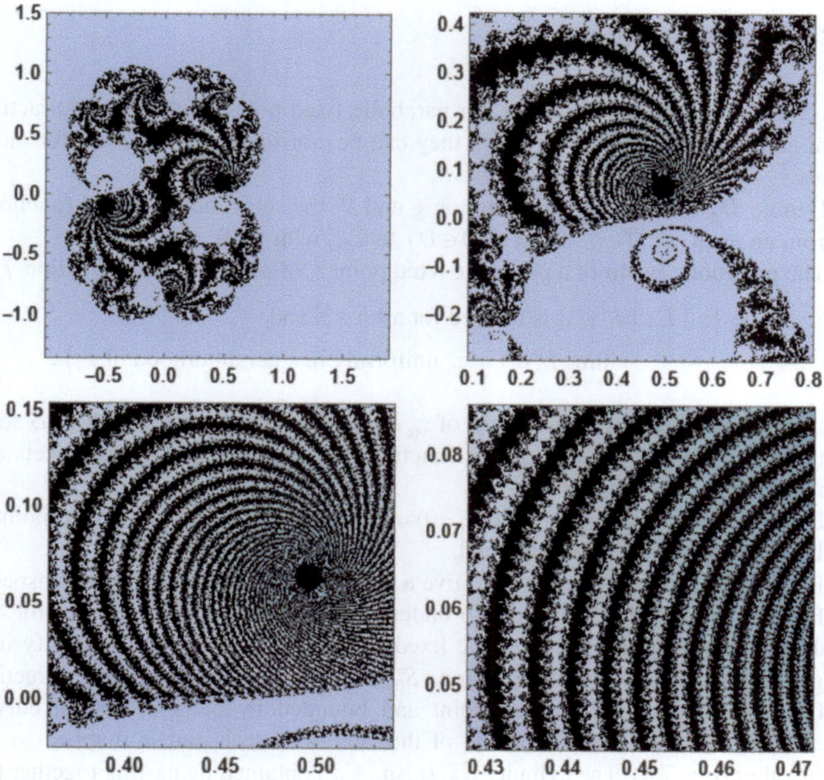

Fig. 7.2 Enlargements of the filled Julia set (in black) of $P_c(z) = z^2 + c$, for $c = 0.25394 + i\,0.00048$ near the boundary of the Mandelbrot set, centered at the center of each successive figure for the quadruple of lengths, illustrating how the boundary is intricate and dense in certain parts, with the density in the last figure increasing as it approaches the point visible in the three preceding figures around which there are spiral structures, suggesting that it may have dimension 2 (computed with *Mathematica*)

Fig. 7.3 Possible bifurcation of parabolic fixed point

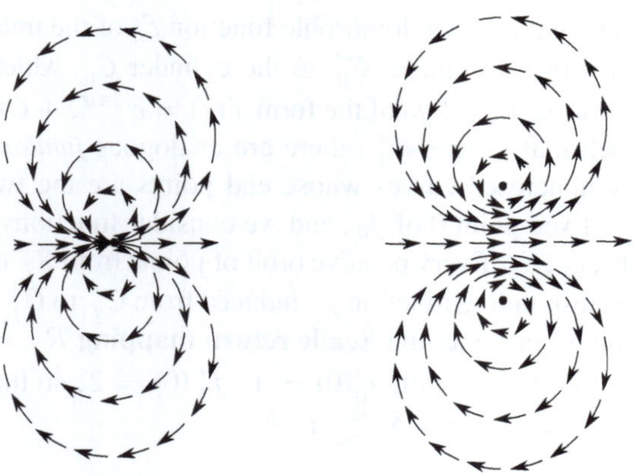

7.2 Écalle Cylinders

We consider Écalle cylinders first for parabolic fixed points with only 1 attracting petal and 1 repelling petal, although they can be considered analogously for more petals.[2]

Denote $\boldsymbol{D_g}$ the domain of a function g and $\mathcal{F}$ the set of holomorphic functions f from an open set $D_f \subset \mathbb{C}_\infty$ with $0 \in D_f$ to $\mathbb{C}_\infty$ with $f(0) = 0$.

The **parabolic basin** of a parabolic fixed point z_0 of a holomorphic function f is

$$\mathcal{B} = \Big\{ z \in \mathbb{C} \colon \{f^n(z)\} \text{ is defined for all } n \in \mathbb{N} \text{ and } \lim_{n \to +\infty} f^n(z) = z_0 \text{ uniformly in a neighborhood of } z_0 \Big\}.$$

We call **immediate parabolic basin** of z_0 each connected component B of $\mathcal{B}$ such that $f(B) = B$ and $f_{|B}$ is a proper function, i.e., preimages of compact sets are compact sets.

Let $f_0 \in \mathcal{F}$ with $f_0'(0) = 1$ and $f_0''(0) \neq 0$, which has a parabolic fixed point 0 with 1 attracting and 1 repelling petal.

Before presenting the theory, we give a simplified general idea of some aspects of Écalle cylinders that may help to understand them and their relevance for the analysis of bifurcation of a parabolic fixed point 0. It is possible to identify in a neighborhood of 0 *fundamental regions* S_0^-, S_0^+ contained in the, resp., attracting and repelling petal of the fixed point and bounded by pairs of Jordan curves intersecting only at 0 such that one of the curves of each pair is mapped by f_0 to the other (Fig. 7.4). The cylinder C_0^- (resp., C_0^+) obtained by pasting together the pair of curves bounding S_0^- (resp., S_0^+) and identifying each point $z \neq 0$ of one of the curves with the point $f(z)$ of the other, is called *attracting (resp., repelling) Écalle cylinder*. Each one of these cylinders is conformal to $\mathbb{C}\backslash\{0\}$ or $\mathbb{C}/\mathbb{Z}$.

A conformal homeomorphism of $\mathbb{C}/\mathbb{Z}$ onto C_0^- (resp., C_0^+) can be lifted to a function $\Phi_0 : \mathcal{B} \to \mathbb{C}$ with $\mathcal{B}$ the parabolic basin of 0 such that $(\mathcal{B}, \Phi_0)$ (resp., $(D_{\varphi_0}, \varphi_0)$) is a universal covering of $\mathbb{C}/\mathbb{Z}$. The restriction of f_0 to the intersection of one of the two boundary curves of S_0^+, excluding 0 with a small neighborhood of 0, induces a holomorphic function $\mathcal{E}_0$ of the union $D_{\mathcal{E}_{f_0}}$ of neighborhoods of the ends of the cylinder C_0^+ to the cylinder C_0^- which is called *Écalle mapping*. For perturbations of f_0 of the form $f(z) = e^{i2\pi\alpha} z + O(z^2)$ when $z \to 0$, with $\alpha \in \mathbb{C}\backslash\{0\}$ such that $|\mathrm{Arg}\,\alpha| < \frac{\pi}{4}$, there are analogous *fundamental regions* S_f^-, S_f^+ bounded by nonclosed curves whose end points are the two fixed points bifurcating from the fixed point 0 of f_0, and we consider functions $\Phi_f, \varphi_f, \mathcal{E}_f$ analogous to, resp., $\Phi_0, \varphi_0, \mathcal{E}_0$. Every positive orbit of points from S_f^- under f eventually enters S_f^+, the corresponding function χ_f induced from C_f^- to C_f^+ is a conformal homeomorphism, and we consider the **Écalle return mapping** $\mathcal{R}_f = \chi_f \circ \mathcal{E}_f : C_f^+ \to C_f^+$.

Let $f_0 \in \mathcal{F}$ with $f_0'(0) = 1$, $f_0''(0) = 2$, so that in a neighborhood of $0 \in \mathbb{C}$ is $f_0(z) = z + z^2 + \sum_{j=3}^{\infty} a_j z^j$.

[2] This chapter mainly follows parts of the article M. Shishikura, The Hausdorff dimension of the boundary of the Mandelbrot set and Julia sets, *Annals of Mathematics* **147** (1998), 225–267, and

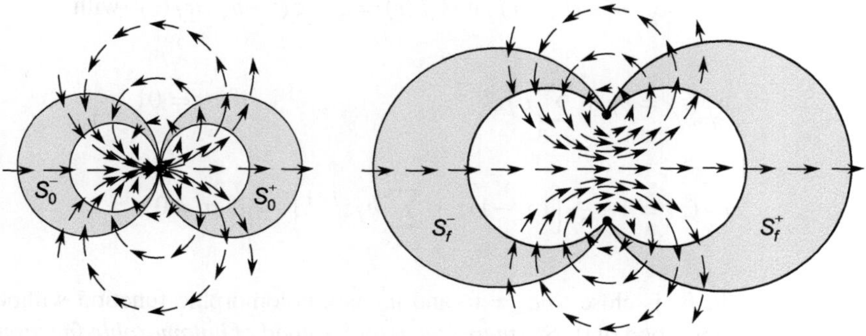

Fig. 7.4 Fundamental regions S_0^-, S_0^+ included in the, resp., attracting and repelling petal of the parabolic fixed point 0 of a holomorphic function $f_0(z) = z + a_2 z^2 + \cdots$, $a_2 \neq 0$, associated Écalle mapping $\mathcal{E}_0$ defined from the union of neighborhoods of the ends of the corresponding repelling Écalle cylinder C_0^+ to the attracting Écalle cylinder C_0^-, and analogous regions S_f^-, S_f^+ for a perturbation $f(z) = e^{i2\pi\alpha} z + O(z^2)$, $z \to 0$, with $\alpha \in \mathbb{C} \setminus \{0\}$ such that $|\operatorname{Arg} \alpha| < \frac{\pi}{4}$, and associated Écalle mapping $\mathcal{E}_f$ and Écalle return mapping $\mathcal{R}_f$ defined from the corresponding repelling Écalle cylinder C_f^+ in itself

By the Leau-Fatou conjugacy for a parabolic fixed point,[3] there is a neighborhood $U \subset \mathbb{C}$ of 0 such that there is a conjugacy $h \in H(U, \mathbb{C})$, $h(0) = 0$, of $f_{0|U}$ with $g(w) = 1 + w$ for $w \in \mathcal{P}$ if $\mathcal{P}$ is the attracting petal of 0 and $w \in \mathcal{P} \cap f_0^{-1}(\mathcal{P})$ if $\mathcal{P}$ is the repelling petal of 0, and h is unique modulo composition with translations. Such a local conjugacy uniquely extends to the attracting basin of $\mathcal{P}$ if this is an attracting petal or to a holomorphic function H such that $f_0 \circ H = H \circ g$ if it is a repelling petal. In this case, if $f_0''(0) = 1$, changing to the Fatou coordinate $w = -\frac{1}{z}$ transforms f_0 to $F_0(w) = 1 + w + O\big(\frac{1}{w}\big)$, when $|w| \to +\infty$, $w \in \mathbb{C} \setminus]-\infty, 0]$ and $h(w) = w + O\big(\log |w|\big)$, $|w| \to +\infty$.

Given an open set $D_f \subset \mathbb{C}_\infty$ and a function $f \in H(D_f, \mathbb{C}_\infty)$, the set of holomorphic functions $g : D_g \to \mathbb{C}_\infty$, with $D_g \subset \mathbb{C}_\infty$ open, such that there are $\varepsilon > 0$ and $\emptyset \neq K \subset D_f$ compact with $\sup_K d_S(g, f) < \varepsilon$, where d_S is the spherical distance in $\mathbb{C}_\infty$, is called a **neighborhood of holomorphic functions** of f. If $f \in \mathcal{F}$, it is a holomorphic function with $D_f \subset \mathbb{C}_\infty$ open and $f(0) = 0$. In a neighborhood of $0 \in \mathbb{C}$, $f(z) = \sum_{j=1}^{\infty} b_j z^j$. If, additionally, f belongs to a neighborhood of holomorphic functions of f_0, there are $\varepsilon > 0$ and $\emptyset \neq K \subset D_f$ compact with $\sup_K d_S(f, f_0) < \varepsilon$. Hence,

$$f(z) = z + z^2 + (b_1 - 1)z + (b_2 - 1)z^2 + \sum_{j=3}^{\infty} b_j z^j$$

uses what was obtained for rational fixed points in Section 9 of Chapter 13 of *CADOVA*, sometimes without making it explicit.

[3] See last Section of Chapter 2 and Section 9 of Chapter 13 of *CADOVA*.

and, therefore, with $\sigma_f = -(b_1-1)$, it is $f(z) = z + z(z-\sigma_f)\,u_f(z)$, with

$$u_f(z) = b_2 + \sum_{j=3}^{\infty} b_j z^{j-2}, \qquad \text{if } \sigma_f = 0,$$

$$u_f(z) = 1 - \tfrac{1}{\sigma_f}\Big[(b_2-1)z + \sum_{j=3}^{\infty} b_j z^{j-1}\Big], \quad \text{if } \sigma_f \neq 0.$$

For $\varepsilon > 0$ small, b_2 is close to $a_2 = 1$, and u_f is a holomorphic function without zeros in a neighborhood of 0. So, *there is a neighborhood of holomorphic functions of* f_0, $\mathcal{N} \subset \mathcal{F}$, *and a neighborhood* $\mathcal{V} \subset \mathbb{C}$ *of* 0 *such that if* $f \in \mathcal{N}$, *then* $\mathrm{cl}\,\mathcal{V} \subset D_f$,

$$f(z) = z + z(z-\sigma_f)\,u_f(z),$$

with $\sigma_f \in \mathcal{V}$, *and* u_f *is a holomorphic function without zeros in a neighborhood of* $\mathrm{cl}\,\mathcal{V}$ *in* $\mathbb{C}$. *So, the fixed points of* f *in* $\mathcal{V}$ *are* 0 *and* σ_f, *with* $\sigma_{f_0} = 0$, $u_{f_0}(z) = \frac{f_0(z)-z}{z^2}$, *and the functions* $f \mapsto \sigma_f$, $f \mapsto u_f(z)$ *are continuous in the topology of* $\mathcal{F}$ *defined by the neighborhoods of holomorphic functions defined as in the preceding paragraph with* $D_{u_f} = \mathcal{V}$ *fixed. Besides, if* $\alpha_f \in \mathbb{C}$ *is such that* $e^{i2\pi\alpha_f} = f'(0) = 1 - \sigma_f u_f(0)$, *then* $\sigma_f = -i\,2\pi\alpha_f[1+o(1)]$ *when* $f \to f_0$.

For $f \in \mathcal{N}$ with $\alpha_f \neq 0$, i.e., $\sigma_f \neq 0$, define

$$z = \tau_f(w) \stackrel{def}{=} \frac{\sigma_f}{1-e^{-i2\pi\alpha_f w}},$$

$$F_f(w) \stackrel{def}{=} w + \frac{1}{i2\pi\alpha_f}\log\Big(1 - \frac{\sigma_f u_f(z)}{1+z\,u_f(z)}\Big), \quad T_f(w) \stackrel{def}{=} w - \frac{1}{\alpha_f},$$

where $F_f(w)$ is defined for w such that $\big|\frac{\sigma_f u_f(z)}{1+z\,u_f(z)}\big| \le \frac{1}{2}$ with the branch of the logarithm with imaginary part of its values in $]-\pi, \pi]$. It is left to the reader to verify that these functions have the following properties:

There are $R_0 > 0$ *and a neighborhood of holomorphic functions of* f_0, $\mathcal{N}' \subset \mathcal{N}$, *such that if* $f \in \mathcal{N}'$ *and* $\alpha_f \in \mathbb{C}\setminus\{0\}$, *then:*

1. $(\mathbb{C}, \tau_f)$ *is a universal covering of* $\mathbb{C}_\infty\setminus\{0, \sigma_f\}$ *with group of deck mappings generated by* T_f, $\tau_f \circ T_f = \tau_f$, *and* $\tau_f(w) \to 0$ *when* $\mathcal{I}m(\alpha_f w) \to +\infty$, *and* $\tau_f(w) \to \sigma_f$ *when* $\mathcal{I}m(\alpha_f w) \to -\infty$.
2. $w \in \mathbb{C}\setminus\cup_{n\in\mathbb{Z}} T_f^n B_{R_0} \;\Rightarrow\; \big(\tau_f(w) \in \mathcal{V},\ \big|\frac{\sigma_f u_f(z)}{1+z\,u_f(z)}\big| \le \frac{1}{2}, |F_f(w) - (1+w)| < \frac{1}{4}$, $|F_f'(w) - 1| < \frac{1}{4}$, $F_f(w) = 1 + w + O\big(\frac{1}{w^2}\big)$ *when* $\mathcal{I}m(\alpha_f w) \to \infty\big)$.
3. $f \circ \tau_f = \tau_f \circ F_f$ and $T_f \circ F_f = F_f \circ T_f$.
4. $\tau_f \to \tau_0$ uniformly in $\big\{w \in \mathbb{C} : |\mathcal{R}e(\alpha_f w)| < \frac{3}{4}\big\} \setminus B_{R_0}$ *and* $F_f \to F_0$ *uniformly in* $\mathbb{C}\setminus\cup_{n\in\mathbb{Z}} T_f^n B_{R_0}$, *when* $f \to f_0$.

(7.1) *If F is holomorphic in $\mathcal{Q}=\mathcal{Q}(b_1,b_2)$ with $b_1,b_2\in\mathbb{C}\cup\{-\infty,+\infty\}$ and $\mathcal{R}e\, b_1<\mathcal{R}e\, b_2-2$, where*

$$\mathcal{Q}(b_1,b_2)\overset{def}{=}\{z\in\mathbb{C}\colon \mathcal{R}e(z-b_1)>-|\mathcal{I}m(z-b_1)|\,,\ \mathcal{R}e(z-b_2)<|\mathcal{I}m(z-b_2)|\,\},$$

without the condition in b_1 (resp., b_2) if $b_1=-\infty$ (resp., $b_2=+\infty$),

$$|F(z)-(1+z)|<\tfrac{1}{4}\,,\quad |F'(z)-1|<\tfrac{1}{4}\,,\qquad z\in\mathcal{Q}\,,$$

then:

1. *F is univalent.*
2. *If $z_0\in\mathcal{Q}(b_1,b_2)$ with $\mathcal{R}e\, b_1<\mathcal{R}e\, z_0<\mathcal{R}e\, b_2-\frac{5}{4}$ and S is the closed subset of $\mathbb{C}$ bounded by the vertical line ℓ passing on z_0 and by the curve $F(\ell)$, then for each $z\in\mathcal{Q}$, there is a unique $n\in\mathbb{Z}$ such that $F^n(z)\in S\backslash F(\ell)$.*
3. *There is a univalent function $\Phi_F:\mathcal{Q}\to\mathbb{C}$ such that $\Phi_F\circ F=\Phi+1$ in the set where the two sides of the equation are defined, and it is unique up to addition of constants.*
4. *If for a fixed $z_0\in\mathcal{Q}$ the function Φ_F satisfies 3 and $\Phi_F(z_0)=0$, then $F\mapsto\Phi_F$ is continuous in the topology with basis the neighborhoods of holomorphic functions defined above.*

Proof

(1) If $b_1,b_2\in\mathbb{C}$, $\mathcal{Q}$ is the region between, on the left (resp., right), the union of two orthogonal rays with slopes ±1 and origin at b_1 (resp., b_2) (Fig. 7.5); if $b_1=-\infty$ and $b_2\in\mathbb{C}$ (resp., $b_1\in\mathbb{C}$ and $b_2=+\infty$) $\mathcal{Q}$ is the region to the left (resp., right) of the rays with origin at b_2 (resp., b_1) previously described; if $b_1=-\infty$ and $b_2=+\infty$, then $\mathcal{Q}=\mathbb{C}$.

It is easy to see that F cannot have the same value at points $z_1\neq z_2$ that can be connected by a line segment included in $\mathcal{Q}$, since

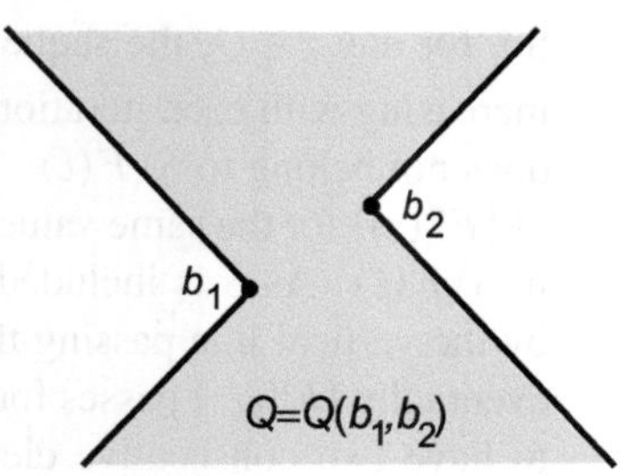

Fig. 7.5 Region $\mathcal{Q}(b_1,b_2)$ considered in (7.1) for $b_1,b_2\in\mathbb{C}$ with $\mathcal{I}m\, b_1<\mathcal{I}m\, b_2$

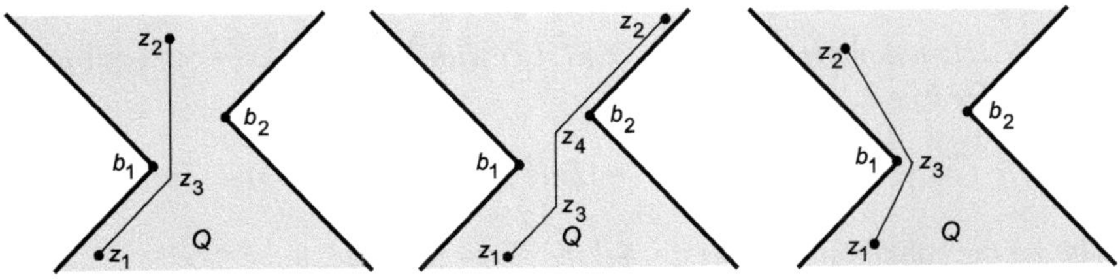

Fig. 7.6 Polygonal lines connecting pairs of points of $Q(b_1, b_2)$, with $b_1, b_2 \in \mathbb{C}$ and $\mathcal{I}m\, b_1 < \mathcal{I}m\, b_2$, in the possible positions in the proof of (7.1.1)

$$|(F(z_1)-z_1)-(F(z_2)-z_2)| = \left| \int_0^1 \left[F'\big(tz_1+(1-t)z_2\big)-1\right](z_1-z_2)\,dt \right|$$
$$\le \int_0^1 \tfrac{1}{4}|z_1-z_2|\,dt = \tfrac{1}{4}|z_1-z_2|\,.$$

Hence, if $F(z_1) = F(z_2)$, then $|z_1-z_2| \le \frac{1}{4}|z_1-z_2|$, and $z_1 = z_2$. Therefore, the restriction of F to any convex subset of Q is univalent.

As Q is not convex, it is necessary to consider the cases where the line segment connecting $z_1 \ne z_2$ in Q is not included in Q. All such cases are of pairs of points that can be connected by a polygonal line of two or three line segments with (the 1 or 2) angles of consecutive line segments obtuse (Fig. 7.6; other possibilities are reduced to these by reflections relative to lines). In an obtuse triangle, the sum of the lengths of the sides forming the obtuse angle is less than $\sqrt{2}$ times the length of the other side. Thus, if $F(z_1) = F(z_2)$ for any $z_1, z_2 \in Q$, using also the estimates in the preceding paragraph, $|z_1-z_2| \le 2\sqrt{2}\,\frac{1}{4}|z_1-z_2| < |z_1-z_2|$. Therefore, $z_1 = z_2$, proving that F is univalent.

(2) Since $|F'(z) - 1| < \frac{1}{4}$ for $z \in Q$, $\frac{3}{4} < \mathcal{R}e\, F'(z) < \frac{5}{4}$, for $n \in \mathbb{Z}$ such that $z, F^{n-2}(z), F^{n-1}(z), F^n(z) \in Q$,

$$\mathcal{R}e\, F^n(z) - \mathcal{R}e\, F^{n-1}(z) > \tfrac{3}{4}\left[\mathcal{R}e\, F^{n-1}(z) - \mathcal{R}e\, F^{n-2}(z)\right].$$

So, for any $z \in Q$, the sequence $\{\mathcal{R}e\, F^n(z)\}$ with $n \in \mathbb{Z}$, $F^n(z) \in Q$ is strictly increasing with each iteration $> \frac{3}{4}$. Hence, if $F^n(z) \in S\backslash F(\ell)$, $F^k(z)$ for $k \ne n$ does not belong to $S\backslash F(\ell)$. Therefore, there cannot be more than one element of $\{F^n(z)\}$ for the same values of $n \in \mathbb{Z}$ considered for the sequence $\{\mathcal{R}e\, F^n(z)\}$ in $S\backslash F(\ell)$. As S is included in a vertical strip of width $\frac{5}{4}$ bounded at the left by the vertical line passing through z_0 (Fig. 7.7), if z is to the left of this strip, eventually $\{F^n(z)\}$ passes for $n > 0$ to the right of the strip, and it is not possible to have two consecutive elements of $\{F^n(z)\}$ one left and the other right of the strip. So, there is $n \in \mathbb{N}$ such that $F^n(z) \in S\backslash F(\ell)$. If z lies to the right of the strip, eventually $\{F^n(z)\}$ moves for $n < 0$ to the left of the strip, and, analogously, there is $-n \in \mathbb{N}$ such that $F^n(z) \in S\backslash F(\ell)$.

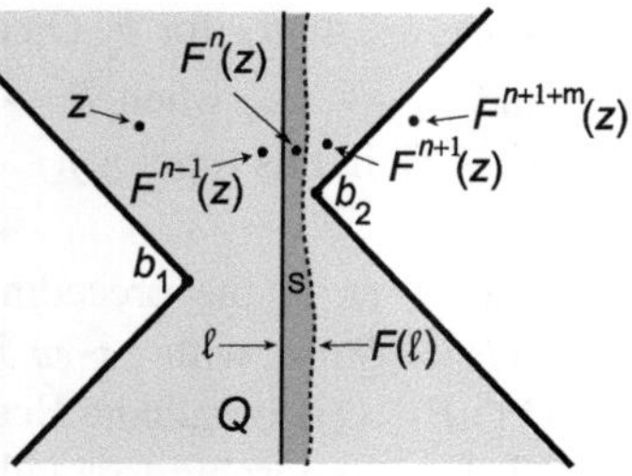

Fig. 7.7 $S \setminus F(\ell)$ in (7.1.2) and vertical strip including it

(3) For $z_0 \in Q$ and $h_1: [0, 1] \times \mathbb{R} \to Q$ such that

$$h_1(x+iy) = (1-x)(z_0+iy) + xF(z_0+iy) \quad x \in [0, 1],\, y \in \mathbb{R},$$

the range of h_1 is the set S of 2, for the fixed value of z_0 and

$$\frac{\partial h_1}{\partial x} = F(z_0+iy) - (z_0+iy), \qquad \frac{\partial h_1}{\partial y} = ixF'(z_0+iy) + i(1-x),$$

$$\left|\frac{\partial h_1}{\partial z} - 1\right| = \frac{1}{2}\left|[F(z_0+iy) - (z_0+iy+1)] + x[F'(z_0+iy) - 1]\right| \le \frac{1}{4}$$

$$\left|\frac{\partial h_1}{\partial \bar{z}}\right| = \frac{1}{2}\left|[F(z_0+iy) - (z_0+iy+1)] + x[F'(z_0+iy) - 1]\right| \le \frac{1}{4}.$$

Hence, $\left|\frac{\partial h_1}{\partial \bar{z}} / \frac{\partial h_1}{\partial z}\right| \le \frac{1}{3}$. Therefore, h_1 is a quasiconformal mapping and $h_1^{-1}(F(z)) = h_1^{-1}(z) + 1$ for $z \in \ell$. Let σ_F be the conformal structure of $[0, 1] \times \mathbb{R}$ such that $h_1 \circ \sigma_F$ is the canonical conformal structure σ_0 of $\mathbb{C}$. With $T(z) = z+1$, extend σ_F to $[-n, -n+1] \times \mathbb{R}$; then, $T^n \circ \sigma_F = \sigma_F$. By the measurable Riemann mapping theorem (2.9), there is a unique quasiconformal mapping $h_2: \mathbb{C} \to \mathbb{C}$ such that $h_2 \circ \sigma_F = \sigma_0$ and $h_2(0) = 0$, $h_2(1) = 1$. As $T \circ \sigma_F = \sigma_F$, it is $h_2 \circ T \circ h_2^{-1} \circ \sigma_0 = \sigma_0$. Thus, $h_2 \circ T \circ h_2^{-1}$ is an affine function. Since T has no fixed points, also $h_2 \circ T \circ h_2^{-1}$ does not, and this function is a translation; as $h_2 \circ T \circ h_2^{-1}(0) = 1$, it is $h_2 \circ T \circ h_2^{-1} = T$. Define $\Phi = h_2 \circ h_1^{-1}$ in S, and extend it to Q by $\Phi(z) = \Phi \circ F^{-1}(z) + 1$ for $z \in Q \setminus S$. The function Φ is a homeomorphism, and it is holomorphic outside the orbit of ℓ. By the Morera theorem,[4] Φ is holomorphic in Q, and, therefore, it satisfies the conditions stated in 3. If $\widetilde{\Phi}$ is another function satisfying these conditions, $\widetilde{\Phi} \circ \Phi^{-1}$ commutes with T in $\Phi(Q)$. Thus, $\widetilde{\Phi} \circ \Phi^{-1} + 1_{\Phi(Q)}$ to $\mathbb{C}$ as a periodic function, and the corresponding extension of $\widetilde{\Phi} \circ \Phi^{-1}$ to $\mathbb{C}$ is holomorphic and commutes with T, as well as the corresponding Φ^{-1}. Hence, $\widetilde{\Phi} \circ \Phi^{-1}$ is an affine function, and since affine functions commuting with translations are translations, there is $c \in \mathbb{C}$ such that $\widetilde{\Phi} \circ \Phi^{-1}(z) = z + c$. Therefore, $\widetilde{\Phi}(w) = \Phi(w) + c$.

(4) Let F_0 be a function satisfying the conditions in the hypothesis for F and h_{10}, h_{20}, Φ_0 the functions and σ_{F_0} a conformal structure for F_0 corresponding

[4] See Section 2 of Chapter 6 of *CADOVA*.

to h_1, h_2, Φ, σ_F for F. On compact subsets of $[0,1]\times\mathbb{R}$, $\frac{\partial h_1}{\partial \bar z}/\frac{\partial h_1}{\partial z} \to \frac{\partial h_{10}}{\partial \bar z}/\frac{\partial h_{10}}{\partial z}$ and $\sigma_F \to \sigma_{F_0}$ when $F \to F_0$ in the topology considered. So, $\Phi \to \Phi_0$ when $F \to F_0$ in this topology. □

A set S as in the preceding statement is called a **fundamental region** and $\mathcal{C} = S/{\sim} = Q/{\approx}$, with $z \sim w$ if $w = F(z)$ and $z \in \ell$, and $z \approx w$ if $w = F(z)$ and $z \in Q \cap F^{-1}(Q)$ is called a **Écalle cylinder** for $F_{|Q}$.

$\mathcal{C}$ is homeomorphic to a cylinder, and it is a Riemann surface with coordinate systems in coordinate neighborhoods of ℓ given by F. With $E(z) = e^{i2\pi z}$ and $\widehat{\Phi}_F \overset{def}{=} E \circ \Phi_F : \mathcal{C} \to \mathbb{C}\setminus\{0\}$, $(\mathcal{C}, \widehat{\Phi}_F)$ is a covering of $\mathbb{C}\setminus\{0\}$ inducing an isomorphism of the fundamental groups of $\mathcal{C}$ and $\mathbb{C}\setminus\{0\}$.

(7.2) *For a region $\Omega \subset \mathbb{C}$, let $\Phi, v \in H(\Omega)$ be functions with Φ univalent, $|v(z) - 1| < \frac{1}{4}$ for $z \in \Omega$ and $\Phi\big(z + v(z)\big) = \Phi(z) + 1$ for $z \in \Omega$ with $z + v(z) \in \Omega$.*

1. *There are $R_1, C_1, C_2 > 0$ such that if $\Omega =]z_0 - R, z_0 + R[\times\mathbb{R}$ with $z_0 \in \Omega$ and $R \geq R_1$, then*

$$\Big|\Phi'(z_0) - \tfrac{1}{v(z_0)}\Big| \leq C_1\big(\tfrac{1}{R^2} + v'(z_0)\big) \leq \tfrac{C_2}{R}.$$

2. *If $\Omega = \{z \in \mathbb{C}\setminus\{0\} : \theta_1 < \arg z < \theta_2\}$, $\theta_2 - \theta_1 < 2\pi$ and $K, v > 0$ are such that $|v'(z)| \leq \frac{K}{|z|^{1+v}}$ for $z \in \Omega$, then for $z_0 \in \Omega$ and $\theta_1', \theta_2' \in]\theta_1, \theta_2[$, $\theta_1' < \theta_2'$ there are $R_2, C_3 > 0$ and $\xi \in \mathbb{C}$ such that*

$$\Big|\,\Phi(z) - \int_{z_0}^{z} \tfrac{1}{v(s)}\,ds - \xi\,\Big| \leq C_3\big(\tfrac{1}{|z|} + \tfrac{K}{|z|^{v}}\big),$$

for $z \in \Omega$ with $\theta_1' < \arg z < \theta_2'$ and $d(z, \mathbb{C}\setminus\Omega) > R_2$, d the canonical Euclidean distance and C_3 only depends on $\theta_1, \theta_2, \theta_1', \theta_2'$.

Proof

(1) Take z_0 as the origin of the coordinates. Since Φ is univalent in $B_2(z)$, $|\Phi'(z)| > 0$ and $f(w) = \frac{\Phi(z+2w) - \Phi(z)}{\Phi'(z)}$ is univalent in B_1, $f(0) = 0$ and $f'(0) = 1$. So, f is a *Schlicht* function and the distortion theorem holds.[5] For large $R > 2$,

$$\frac{|v(z)|}{[1 + |v(z)/2]^2} \leq \Big|\frac{\Phi(z+v(z)) - \Phi(z)}{\Phi'(z)}\Big| \leq \frac{|v(z)|}{[1 - |v(z)/2]^2}, \quad z \in B_{R-2}.$$

[5] See the last Section of Chapter 10 of *CADOVA*.

Hence, there are constants C, C' such that $0 < C < |\Phi'(z)| < C'$ for $z \in B_{R-2}$. For $R > 4$ and $\gamma : [0, 2\pi] \to \mathbb{C}$ such that $\gamma_r(\theta) = z + re^{i\theta}$, by the Cauchy formula,[6]

$$|\Phi''(z)| = \left|\frac{1}{i2\pi}\int_{\gamma_{R/2}} \frac{\Phi'(w)}{(w-z)^2}\,dw\right| \le \frac{1}{2\pi}\int_0^{2\pi} \frac{C'}{(R/2)^2}\,\frac{R}{2}d\theta = \frac{2C'}{R}, \quad z\in B_{R/2}.$$

Since

$$\Phi(z+a) = \Phi(z) + a\Phi'(z) + a^2\int_0^1 (1-t)\Phi''(z+at)\,dt,$$

$|1-\Phi'(z)\,v(z)| \le v^2(z)\frac{R}{4} < \frac{5^2}{4^3R}$ for $z \in B_{R/2-5/4}$. For $R > 5$ and again by the Cauchy formula,

$$|\Phi''(z)\,v(z) + \Phi'(z)\,v'(z)| = \left|[1-\Phi'(z)\,v(z)]'\right| = \left|\frac{1}{i2\pi}\int_{\gamma_{R/4}} \frac{1-\Phi'(z)\,v(z)}{(w-z)^2}\,dw\right|$$
$$\le \frac{1}{2\pi}\int_0^{2\pi} \frac{5^2}{4^3R(R/4)^2}\,\frac{R}{4}d\theta = \frac{5^2}{4^2R^2}, \quad z\in B_{R/4}.$$

Also by the Cauchy formula,

$$|v'(z)| = \left|\frac{1}{i2\pi}\int_{\gamma_{R/2}} \frac{v(w)}{(w-z)^2}\,dw\right| \le \frac{1}{2\pi}\int_0^{2\pi} \frac{5}{4(R/2)^2}\,\frac{R}{2}d\theta = \frac{5}{2R}, \quad z\in B_{R/2},$$
$$|v''(z)| = \left|\frac{1}{i2\pi}\int_{\gamma_{R/2}} \frac{v'(w)}{(w-z)^2}\,dw\right| \le \frac{1}{2\pi}\int_0^{2\pi} \frac{5}{2R(R/2)^2}\,\frac{R}{2}d\theta = \frac{5}{R^2}, \quad z\in B_{R/2}.$$

Hence, $|v'(z)| \le |v'(0)| + \frac{5^2}{4R^2}$ for $z \in B_{5/4}$. Therefore,

$$|\Phi''(z)\,v(z)| \le \frac{5^2}{4^2R^2} + \frac{5C'}{2R}, \qquad |\Phi''(z)| \le \frac{10}{3}\Big(\frac{5}{8R^2} + \frac{C'}{R}\Big), \qquad z\in B_{5/4}.$$

Using again the formula in the third period of this proof, now with $a = v(0)$, $|1-\Phi'(0)\,v(0)| \le v^2(0)\frac{10}{3}\big(\frac{5}{8R^2} + \frac{C'}{R}\big)$, $|\Phi'(0) - \frac{1}{v(0)}| \le \frac{25}{6}\frac{2}{3}\big(\frac{5}{8R^2} + \frac{C'}{R}\big)$, establishing the estimate in the statement.

(2) For $z \in S = \{z \in \mathbb{C}\backslash\{0\} : \theta_1' < \arg z < \theta_2'\}$, it is $d(c, \mathbb{C}\backslash\Omega) > C_4 z$ for some constant $C_4 \in]0, 1[$. By 1, for the region $\{w : |w-z| < C_4|z|\}$,

$$\left|\Phi'(z) - \frac{1}{v(z)}\right| \le C_1\Big(\frac{1}{(C_4|z|)^2} + v'(z)\Big) \le \frac{C_1}{C_4^2}\Big(\frac{1}{|z|^2} + \frac{C_4^2K}{|z|^{1+v}}\Big) < \frac{C_1}{C_4^2}\Big(\frac{1}{|z|^2} + \frac{K}{|z|^{1+v}}\Big).$$

[6] See Section 2 of Chapter 6 of *CADOVA*.

Integrating on a ray S with origin at z, there is $C_3 > 0$ such that

$$\left| \int_z^\infty \left(\Phi'(s) - \tfrac{1}{v(s)}\right) ds \right| \leq C_3\left(\tfrac{1}{|z|} + \tfrac{K}{|z|^\nu}\right).$$

Since

$$\int_z^\infty \left(\Phi'(s) - \tfrac{1}{v(s)}\right) ds = -\Phi(z) + \int_{z_0}^z \tfrac{1}{v(s)}\, ds + \left[\Phi(z_0) + \int_{z_0}^\infty \left(\Phi'(s) - \tfrac{1}{v(s)}\right) ds\right],$$

with the inequality we get the estimate in the statement.[7] □

(7.3) *If $f_0 \in \mathcal{F}$, $f_0' = 1$, $f_0'' = 1$, there are $\Omega_\pm$, φ_0, Φ_0, $\mathcal{E}_{f_0}$ such that:*

1. *$\Omega_+, \Omega_- \subset D_{f_0}$ are simply connected regions bounded by Jordan curves and containing 0, $\Omega_+ \cup \Omega_- \cup \{0\}$ is a neighborhood of 0, $f_0(\mathrm{cl}\,\Omega_+) \subset \Omega_+ \cup \{0\}$ and $f_0(\Omega_- \cup \{0\}) \supset \mathrm{cl}\,\Omega_-$; f_0 is injective into $\Omega_+ \cup \Omega_- \cup \{0\}$; $\Omega_+ \cap \Omega_-$ has 2 connected components; $f_0^n \to 0$ uniformly in Ω_+; if $\mathcal{B}$ is the parabolic basin of 0 for f_0, $z \in \mathcal{B}$ if and only if $n \in \mathbb{N} \cup \{0\}$ and $f_0^n(z) \in \Omega_+$* (Fig. 7.8).
2. *$\varphi_0 : D_{\varphi_0} \to \mathbb{C}_\infty$ is holomorphic, $\varphi_0(w+1) = f_0 \circ \varphi_0(w)$ with the domains of definition of both sides of the equality coincident;*

$$D_{\varphi_0} \supset Q_0 \stackrel{def}{=} \left\{w \in \mathbb{C} : \tfrac{\pi}{3} < \arg(w + \xi_0) < \tfrac{5\pi}{3}\right\}$$

(Fig. 7.9) *and $\{w \in \mathbb{C} : |\mathcal{I}m\, w| > \eta_0\}$ for large $\xi_0, \eta_0 > 0$; $\varphi_0(Q_0) = \Omega_-$; φ_0 is injective into Q_0; if $\widetilde{\varphi}_0 = \frac{1}{\varphi_0}$, there are constants $a, b \in \mathbb{C}$ such that $\widetilde{\varphi}_0(w) = w + a \log w + b + o(1)$ when $w \to \infty$ in Q_0; if f_0 is a rational or entire function, then $D_{\varphi_0} = \mathbb{C}$.*

3. *$\Phi_0 \in H(\mathcal{B})$, $\Phi_{0|\Omega_+}$ is univalent and $\Phi_0 \circ f_0(z) = \Phi_0(z) + 1$, $z \in \mathcal{B}$.*
4. *If $T(z) = z + 1$ and $\widetilde{\mathcal{B}} = \varphi_0^{-1}(\mathcal{B})$, then $T(\widetilde{\mathcal{B}}) = \widetilde{\mathcal{B}}$ and there is $\eta_0 > 0$ such that $\widetilde{\mathcal{B}} \supset \{w \in \mathbb{C} : |\mathcal{I}m| > \eta_0\}$. $\widetilde{\mathcal{E}}_{f_0} : \mathcal{B} \to \mathbb{C}$ such that $\widetilde{\mathcal{E}}_{f_0} = \Phi_0 \circ \varphi_0$ satisfies $\widetilde{\mathcal{E}}_{f_0}(w+1) = \widetilde{\mathcal{E}}_{f_0}(w) + 1$, $w \in \widetilde{\mathcal{B}}$. With $E(z) = e^{i2\pi z}$,*

$$\mathcal{E}_{f_0} = E \circ \widetilde{\mathcal{E}}_{f_0} \circ E^{-1} : E(\widetilde{\mathcal{B}}) \cup \{0, \infty\} \to \mathbb{C}_\infty,$$

$\mathcal{E}_{f_0}(0) = 0$, $\mathcal{E}_{f_0}(\infty) = \infty$, is holomorphic with $\mathcal{E}'_{f_0}(0) \neq 0$, $\mathcal{E}_{f_0}(\infty) \neq 0$; it is called the ***Écalle mapping*** *of f_0 at 0. For constants $c, c' \in \mathbb{C}$, $w \mapsto \varphi_0(w+c)$ and $z \mapsto \Phi_0(z) + c'$ satisfy the conditions of, resp., φ_0, Φ_0 in, resp., 2, 3. Adding an appropriate constant to Φ_0, we get $\mathcal{E}_{f_0}$ with $\mathcal{E}'_{f_0}(0) = 1$, which is called the* ***normalized Écalle mapping****.*

[7] A similar estimate had been obtained 10 years before in Yoccoz, J.-C., Linéarisation des germes de diffeomorphismes holomorphes de $(\mathbb{C}, 0)$, *C.R. Acad. Sci. Paris, Ser. I*, **306** (1988), 55–58.

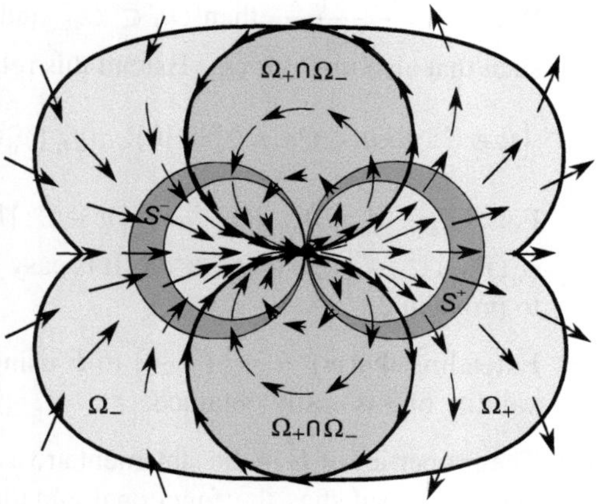

Fig. 7.8 Attracting Ω_- and repelling Ω_+ petals of parabolic flower in (7.3.1)

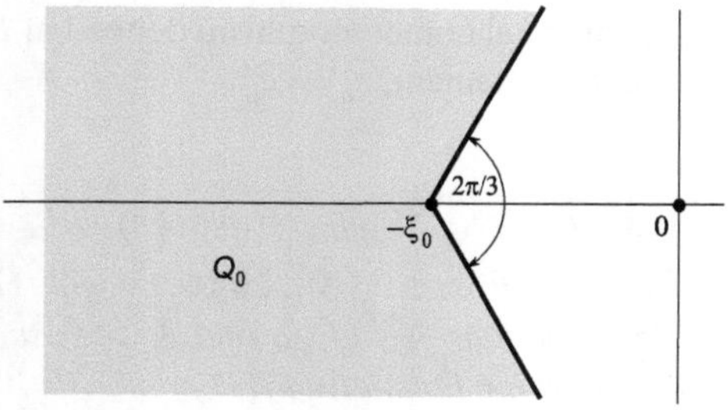

Fig. 7.9 Set Q_0 in (7.3.2)

Proof

(1) Let $Q_0^- \overset{def}{=} Q(\xi_1, -\infty)$ and $Q_0^+ \overset{def}{=} Q(\xi_1, +\infty)$, with $\xi_1 > 0$. For large $\xi_1 > 0$, F_0 obtained from f_0 changing to the Fatou coordinates $w = -\frac{1}{z}$ satisfies the conditions of (7.1). So, there are univalent functions $\Phi_0^- : Q_0^- \to \mathbb{C}$ and $\Phi_0^+ : Q_0^+ \to \mathbb{C}$ such that $\Phi_0^\pm\big(F_0(w)\big) = \Phi_0^\pm(w) + 1$. If

$$F_0(w) = w + 1 + \tfrac{a}{w} + O\big(\tfrac{1}{w^2}\big), \qquad |w| \to +\infty,$$

by (7.2.2),

$$\Phi_0^\pm(w) = w - a\log w + c_0^\pm + o(1), \quad |w| \to +\infty,$$

in a sector of the type considered there with $\Omega = Q_0^\pm$, where $c_0^\pm \in \mathbb{C}$ and the branches of the logarithm for $\Phi_0^\pm$ are chosen to coincide in the upper part of $Q_0^- \cap Q_0^+$ and differ by $i2\pi$ in the lower part. For large $\xi_0 > 0$, Q_0 considered in 2 satisfies $Q_0 \subset \Phi_0^-(Q_0^-)$ and $\Omega_+ = -\frac{1}{T(Q_0^+)}$, $\Omega_- = -\frac{1}{(\Phi_0^-)^{-1}(Q_0)}$, where $T(z) = z + 1$, if necessary for larger $\xi_0 > 0$, the conditions in 1 hold.

Jean-Christophe Yoccoz (1957–2016) received the *Fields Medal* in 1994 for contributions to the Dynamical Systems Theory.

(2) With $\varphi_0 = -\frac{1}{(\Phi_0^-)^{-1}}$, then $Q_0 \subset D_{\varphi_0}$ and $\varphi_0(w+1) = f_0 \circ \varphi_0(w)$ for $w \in Q_0$ such that also $w+1 \in Q_0$. Extend this relation φ_0 to

$$\{w \in \mathbb{C}\colon w-n \in Q_0, n \in \mathbb{N}\cup\{0\}, \quad f_0^j(\varphi_0(w-n)) \in D_{f_0}, \quad j \in \{0,\ldots,n-1\}\}$$

For large $\eta_0 > 0$, $R_{\eta_0} \overset{def}{=} \{w \in \mathbb{C}\colon |\mathcal{R}e\, w| \le \frac{1}{2}, |\mathcal{I}m\, w| > \eta_0\} \subset Q_0$ and $\varphi_0(R_{\eta_0}) \subset \{-\frac{1}{w} : w \in Q_0^+\} \subset \mathcal{B}$. It is easy to check the validity of what remains to prove in 2.

(3) Extending $\Phi_0(w) = \Phi_0^+(-\frac{1}{w})$ to $\mathcal{B}$ using the functional equation for Φ_0^+ the validity of 3 is easily obtained.

(4) The properties of $\widetilde{\mathcal{B}}$ in the statement are a consequence of what was established above. $\mathcal{E}_{f_0}$ satisfies the functional equation in the statement and is univalent in $\{w \in \mathbb{C} : |\mathcal{I}m\, w| > \widetilde{\eta}_0\}$ for large $\widetilde{\eta}_0 > 0$. Hence, $\mathcal{E}_{f_0}$ can be extended to a conformal homeomorphism defined in $E(\widetilde{\mathcal{B}}) \cup \{0, \infty\}$. With the normalization in the statement, $c_0^- = c_0^+$. □

(7.4) *Let $f \in \mathcal{F}$ with $f'(0) \neq 0$, $\alpha_f \in \left]-\frac{1}{2}, \frac{1}{2}\right] \times \mathbb{R}$ for $e^{i2\pi\alpha_f} = f'(0)$ and $\mathcal{F}_1 = \{f \in \mathcal{F}\colon \alpha_f \in \mathbb{C}\setminus\{0\}, |\arg \alpha_f| < \frac{\pi}{4}\}$. There are a neighborhood of holomorphic functions $\mathcal{N}_0$ of f_0 and $\xi_0 > 0$ such that for $f \in \mathcal{N}_0 \cap \mathcal{F}_1$ there are holomorphic functions $\varphi_f\colon D_{\varphi_f} \to \mathbb{C}$, $\mathcal{R}_f\colon D_{\mathcal{R}_f} \to \mathbb{C}_\infty$ and:*

1. *$\varphi_f(w) \in D_f$ if $w, w+1 \in D_{\varphi_f}$, and $\varphi_f(w+1) = f \circ \varphi_f(w)$;*

   $$D_{\varphi_f} \supset Q_f \overset{def}{=} \left\{w \in \mathbb{C} : \tfrac{\pi}{3} < \arg(w+\xi_0) < \tfrac{5\pi}{3}, \left|\arg\left(w + \tfrac{1}{\alpha_f} - \xi_0\right)\right| < \tfrac{2\pi}{3}\right\}$$

   (Fig. 7.10)*; $\varphi_f(w) \to 0$ when $\mathcal{I}m\, w \to +\infty$ in Q_f and $\varphi_f(w) \to \sigma_f$ when $\mathcal{I}m\, w \to -\infty$ in Q_f, where the neighborhood $\mathcal{N}_0$ of f_0 is sufficiently small for f to have just 2 fixed points close to 0 (counting multiplicity), 0 and σ_f, with $\sigma_f = -i2\pi\alpha_f(1+o(1))$ when $f \to f_0$* (see the paragraph following the definition of neighborhood of holomorphic functions).

2. $0, \infty \in D_{\mathcal{R}_f}$, $\mathcal{R}_f(0) = 0$, $\mathcal{R}_f(\infty) = \infty$; $\mathcal{R}_f(D_{\mathcal{R}_f} \cap (\mathbb{C}\setminus\{0\})) \subset \mathbb{C}\setminus\{0\}$; $(\mathcal{R}_f)'(0) = e^{-i2\pi/\alpha_f}$ *and, therefore,* $\alpha_{\mathcal{R}_f} = -\frac{1}{\alpha_f} \pmod{\mathbb{Z}}$.

3. *If $w, w' \in D_{\varphi_f}$, $\mathcal{R}_f(e^{i2\pi w}) = e^{i2\pi w'}$, then $\left|\arg\left(w' + \frac{1}{2\alpha_f} - w\right)\right| < \frac{2\pi}{3}$. So, there is $n \in \mathbb{N}\cup\{0\}$ such that $f^n(\varphi_f(w)) = \varphi_f(w')$. If $U, U' \subset D_{\varphi_f}$ are connected sets such that $\mathcal{R}_f^m(E(U)) \subset E(U')$ for some $m \in \mathbb{N}$, $\varphi_{f|U}$, $E_{|U'}$ are injective and $\left|\arg\left(w' + \frac{1}{2\alpha_f} - w\right)\right| < \frac{2\pi}{3}$ for $w \in U, w' \in U'$, where $E(z) = e^{i2\pi z}$, then there is $n \in \mathbb{N}$ with $n > m$ such that $f^n = \varphi_f \circ (E_{|U'})^{-1} \circ \mathcal{R}_f^m \circ E \circ (\varphi_{f|U})^{-1}$ in $\varphi_f(U)$.*

(continued)

Fig. 7.10 Set Q_f in (7.4.1)

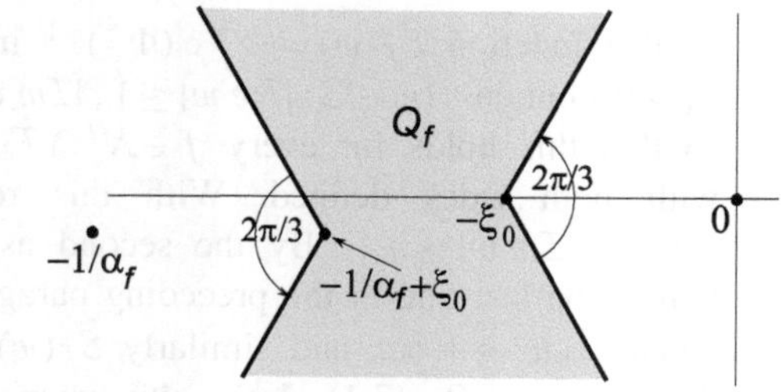

4. With P_1, the natural projection of $\mathbb{C}$ on $\mathbb{C}/\mathbb{Z}$, $P_2:\mathbb{C}/\mathbb{Z}\to\mathbb{C}\backslash\{0\}$ such that $P_2\circ P_1=E$, $\widehat{\mathcal{E}}_{f0}=P_2^{-1}\circ\mathcal{E}_{f0}\circ P_2$ and $\widehat{\mathcal{R}}_f\circ P_2$, if $f\in\mathcal{N}_0\cap\mathcal{F}_1$, then $\varphi_f\to\varphi_{f0}$, $e^{i2\pi/\alpha_f}\mathcal{R}_f\to\mathcal{E}_{f0}$, $\widehat{\mathcal{R}}_f+\frac{1}{\alpha_f}\to\widehat{\mathcal{E}}_{f0}$ when $f\to f_0$ in the topology of neighborhoods of holomorphic functions.

5. With $D(\eta)=\{w\in\mathbb{C}:|w-i\eta|\le\frac{|\eta|}{4}\}$ for $\eta\in\mathbb{R}$, and

$$D'(\eta)=\{w\in\mathbb{C}:\ |w-i\eta|\le\tfrac{|\eta|}{2}\},$$

for $\eta>0$ large, there is a neighborhood $\mathcal{N}_{f0}(\eta)\subset\mathcal{N}_0$ of f_0 such that if $f\in\mathcal{N}_{f0}(\eta)\cap,\mathcal{F}_1$, and φ_f is defined and is injective in $D(\pm\eta)$, $-\frac{1}{\varphi_f}(D(\pm\eta))\subset D'(\pm\eta)$ and $\frac{1}{2}<|(-\frac{1}{\varphi_f})'|<2$ in $D(\pm\eta)$.

Proof

(1) Let $\mathcal{N}'$ be a neighborhood of holomorphic functions of f_0 with the properties preceding (7.1), $f\in\mathcal{N}'\cap\mathcal{F}_1$, $T_f(w)=w-\frac{1}{\alpha_f}$ as defined there, and for $\xi_1>0$,

$$\mathcal{Q}_f^+=\mathcal{Q}\big(\xi_1,-\xi_1+\tfrac{1}{\alpha_f}\big),\qquad \mathcal{Q}_f^-=\mathcal{Q}\big(\xi_1-\tfrac{1}{\alpha_f},-\xi_1\big)=T_f(\mathcal{Q}_f^+).$$

$\mathcal{N}'$ can be chosen so small and ξ_1 so large that $\mathcal{R}e\,\frac{1}{\alpha_f}>2\xi_1+2$ and $\mathcal{Q}_f^\pm\subset D_{F_f}$; then, F_f satisfies the conditions of F in (7.1) in $\mathcal{Q}_f^\pm$. By (7.1), there are univalent functions $\Phi_f^\pm:\mathcal{Q}_f^\pm\to\mathbb{C}$ such that $\Phi_f^\pm(F_f(w))=\Phi_f^\pm(w)+1$. Choose $w_\pm\in\mathcal{Q}_0^\pm$ and normalize $\Phi_f^\pm$ so that $\Phi_f^\pm(w_\pm)=\Phi_0^\pm(w_\pm)$. By property 2 before (7.1), $F_f(w)=w+1+O\big(\frac{1}{w^2}\big)$ when $\mathcal{I}m(\alpha_f w)\to\infty$. By (7.2), there are $c_f^\pm\in\mathbb{C}$ dependent on f such that $\Phi_f^\pm(w)=w+c_f^\pm+o(1)$ when $|w|\to+\infty$ in $\{w\in\mathbb{C}:\theta_1'<\theta_1'<\frac{3\pi}{4}\}\subset\mathcal{Q}_f^\pm$, with $\frac{\pi}{3}<\theta_1'<\theta_1'<\frac{3\pi}{4}$ and $w_0\in\mathbb{C}$. By property 3 before (7.1), $\Phi_f^-\circ T_f:\mathcal{Q}_f^+\to\mathbb{C}$ satisfies the same functional equation satisfied by Φ_f^+, and by the uniqueness in (7.1.3), there is $d_1\in\mathbb{C}$ dependent on f such that $\Phi_f^-\circ T_f=\Phi_f^++d_1$. By the last asymptotic formula in the preceding period, $d_1=c_f^--c_f^+-\frac{1}{\alpha_f}$.

The function $\widetilde{\mathcal{E}}_f(w) = \Phi_f^+ \circ (\Phi_f^-)^{-1}$ in $\Phi_f^-(Q_f^+ \cap Q_f^-)$, which for large $\eta_0 > 0$ contains $\{w \in \mathbb{C} : |\mathcal{R}e\, w| \le 1, |\mathcal{I}m\, w| \ge \eta_0\}$, where η_0 can be chosen so that this holds for every $f \in \mathcal{N}' \cap \mathcal{F}_1$, satisfies $\widetilde{\mathcal{E}}_f(w+1) = \widetilde{\mathcal{E}}_f(w) + 1$ with both sides defined. With this relation $\widetilde{\mathcal{E}}_f$ can be extended to $\{w \in \mathbb{C} : |\mathcal{I}m\, w| > \eta_0\}$. By the second asymptotic formula in the sentence before the last one in the preceding paragraph, $\widetilde{\mathcal{E}}_f(w) = w + c_f^+ - c_f^- + o(1)$ when $\mathcal{I}_m\, w \to +\infty$, and similarly $\widetilde{\mathcal{E}}_f(w) - w$ approaches a constant when $\mathcal{I}_m\, w \to -\infty$. By (7.1), $\Phi_f^\pm \to \Phi_0^\pm$ when $f \to f_0$; thus, $\widetilde{\mathcal{E}}_f \to \widetilde{\mathcal{E}}_{f0}$ uniformly in $[0, 1] \times \{\eta_0\}$. Hence,

$$c_f^+ - c_f^- = \int_{i\eta_0}^{1+i\eta_0} \left[\widetilde{\mathcal{E}}_f(w) - w\right] dw \to \int_{i\eta_0}^{1+i\eta_0} \left[\widetilde{\mathcal{E}}_{f0}(w) - w\right] dw = c_0^+ - c_0^- = 0,$$

with the integrals on the line segment from $i\eta_0$ to $1 + i\eta_0$.

By (7.1.4), the definition of F_f preceding the statement of this result, and by (7.2), for large ξ_0, η_0, the curve bounding Q_f at the right lies in $\Phi_f^-(Q_f^-)$, and that bounding $T_f^{-1}(Q_f)$ at the left in $\Phi_f^+(Q_f^+)$. Since $Q_f^- = T_f(Q_f^+)$ and with the last equalities in the two preceding paragraphs, for larger ξ_0, η_0 if necessary, the curve bounding Q_f at the left lies in Q_f^-. So, $Q_f \subset \Phi_f^-(Q_f^-)$. With $\tau_f(w) \overset{def}{=} \frac{\sigma_f}{1-e^{i2\pi\alpha_f w}}$, $\varphi_f \overset{def}{=} \tau_f \circ (\Phi_f^-)^{-1}$, $Q_f \subset D_{\varphi_f}$ and $\varphi_f(w+1) = f \circ \varphi_f(w)$ if $w, w+1 \in Q_f$. The extension of φ_f to

$$\left\{w \in \mathbb{C} : w - n \in Q_f, n \in \mathbb{N} \cup \{0\}, \quad f^j\left(\varphi_f(w-n)\right) \in D_f, \quad j \in \{0, \ldots, n-1\}\right\}$$

satisfies 1.

(2) The function $\widetilde{\mathcal{R}}_f \overset{def}{=} \Phi_f^- \circ T_f \circ (\Phi_f)^{-1}$ is equal to $\widetilde{\mathcal{R}}_f + c_f^- - c_f^+ - \frac{1}{\alpha_f}$ in the domain of this function and $\widetilde{\mathcal{R}}_f(w+1) = \widetilde{\mathcal{R}}_f(w) + 1$. So, by the asymptotic formula two paragraphs above, $\mathcal{R}_f = E \circ \widetilde{\mathcal{R}}_f \circ E^{-1}$ has extension as a holomorphic function to 0 and ∞. Since $\widetilde{\mathcal{R}}_f(w) = w - \frac{1}{\alpha_f} + o(1)$ when $\mathcal{I}m\, w \to +\infty$, it is $\widetilde{\mathcal{R}}'_f(0) = e^{-i2\pi\alpha_f}$.

(3) Restrict $D_{\widetilde{\mathcal{R}}_f}$ so that $\left|\widetilde{\mathcal{R}}_f(w) - w + \frac{1}{\alpha_f}\right| < \frac{1}{2|\alpha_f|}$ for $w \in D_{\widetilde{\mathcal{R}}_f}$. If $w, w' \in D_{\varphi_f}$ and $\mathcal{R}_f^m\left(E(w)\right) = E(w')$ for some $m \in \mathbb{N}$, there is $n \in \mathbb{N}$ such that $\widetilde{\mathcal{R}}_f^m(w) + n = w'$. By the last inequality, $n > m$. From the definitions of φ_f and $\widetilde{\mathcal{R}}_f$, the relation $\tau_f \circ T_f = \tau_f$, and the functional equation for φ_f, we get $f^n\left(\varphi_f(w)\right) = \varphi_f(w')$. This ends the proof of 3.

(4) We have $\varphi_f \to \varphi_0$ and $\widetilde{\mathcal{E}}_f \to \widetilde{\mathcal{E}}_{f_0}$; so, $\widetilde{\mathcal{R}}_f(w) + \frac{1}{\alpha_f} = \widetilde{\mathcal{E}}_f + c_f^- - c_f^+ \to \widetilde{\mathcal{E}}_{f_0}$, and $e^{i2\pi/\alpha_f}\mathcal{R}_f \to \mathcal{E}_{f_0}$, when $f \to f_0$ in $\mathcal{N}' \cap \mathcal{F}_1$.

(5) It is a consequence of 2 and 4. □

A similar result can be obtained for $f \in \mathcal{F}$ with $\frac{3}{4}\pi < \operatorname{Arg}\alpha_f < \frac{5}{4}\pi$ instead of $f \in \mathcal{F}_1$ with the following modifications: the lower end of the cylinder $\mathbb{C}/\mathbb{Z}$

corresponds to 0; in (7.4.1), the limits of φ_f when $\mathcal{I}m\, w \to \pm\infty$ are switched, and $\frac{1}{\alpha_f}$ is replaced by its symmetric, and in (7.4.3) and (7.4.4); in (7.4.2), replacing $(\mathcal{R}_f)'(0)$ by $(\mathcal{R}_f)'(\infty)$.

Therefore, we have the factorization

$$\widehat{\mathcal{R}}_f = \widehat{\mathcal{T}}_{-1/\alpha_f} \circ \big(\widehat{\mathcal{R}}_f + \tfrac{1}{\alpha_f}\big),$$

where $\widehat{\mathcal{T}}_{-1/\alpha_f}(w) = w - \frac{1}{\alpha_f}$ in $\mathbb{C}/\mathbb{Z}$ and $\widehat{\mathcal{R}}_f + \frac{1}{\alpha_f}$ are a nonlinear function defined in a subset of $\mathbb{C}/\mathbb{Z}$ which approaches $\widehat{\mathcal{E}}_{f_0}$. If $\{f_n\}_{n\in\mathbb{N}} \subset \mathcal{N}_0 \cap \mathcal{F}_1$ with $f_n \to f_0$ and $\{k_n\}_{n\in\mathbb{N}} \subset \mathbb{N}$ with $\frac{1}{\alpha_{f_n}} - k_n \to -c$, then $\widehat{\mathcal{R}}_{f_n} = \widehat{\mathcal{E}}_{f_0} + c$ in $\mathbb{C}/\mathbb{Z}$.

The Écalle mapping $\mathcal{E}_{f_0}$ was defined between different spaces, so it does not make sense to consider it a dynamical system. However, for every $c \in \mathbb{C}$, $w \mapsto \widehat{\mathcal{E}}_{f_0}(w) + c$ in $\mathbb{C}/\mathbb{Z}$ is obtained as limit of the Écalle return mappings of the functions f_n.

Consider now the maximal extension of the Écalle mapping to the set $\boldsymbol{\mathcal{F}_0}$ of functions $f \in \mathcal{F}$ such that $f'(0) = f''(0) = 1$ and f has an immediate parabolic basin with only one critical point.

Note that, except for the negative orbit identically 0, each $w \in D_{\varphi_0}$ has only one negative orbit $\{z_{-j}\}_{j\in\mathbb{N}\cup\{0\}}$ under f_0 such that $\lim z_{-j} = 0$.

(7.5) *For $w \in D_{\varphi_0}\setminus\{0\}$, $\{z_{-j}\}_{j\in\mathbb{N}\cup\{0\}}$, with $z_{-j} = \varphi_0(w-j)$, is a negative orbit of $w = z_0$ under f_0, and it is the only one such orbit with $\lim z_{-j} = 0$; if it does not contain critical points of f_0, then $\varphi_0'(w) \neq 0$.*

Proof By (7.4), $w - j \in D_{\varphi_0}$ for $j \in \mathbb{N}\cup\{0\}$ and $\lim z_{-j} = 0$. If $\{s_{-j}\}_{j\in\mathbb{N}\cup\{0\}}$ is a negative orbit not identically 0 and $\lim s_{-j} = 0$, by (7.3), for $j_0 \in \mathbb{N}$ large and $j \geq j_0$, it is $j_{-j} \in \Omega_+ \cup \Omega_-$. Since $f_0^n \to 0$ uniformly in Ω_+, there is $j_0 \in \mathbb{N}$ such that $z_{-j} \in \Omega_-$ for $j \geq j_0$. $w' = \big(\varphi_{0|Q_0}\big)^{-1}(z_{-j}) + j$ is independent of $j \geq j_0$ and if the negative orbit of w' by f_0 coincides with that of a w'', with $j \in \mathbb{N}$ such that $w' - j$, $w'' - j \in Q_0$, by the injectivity of φ_0 in Q_0, it is $w' = w''$. As for $w \in D_{\varphi_0}$, $\varphi_0(w) = f_0^n \circ \varphi_0(w-n)$ and $\varphi_0' \neq 0$ in Q_0, if the negative orbit of w does not contain critical points of f_0, by the chain rule for derivatives, $\varphi_0'(w) \neq 0$. □

It was seen in *CADOVA*[8] that every parabolic periodic point of a rational function has immediate attracting basins and if its multiplicity is $q+1$, there are at most q attracting petals $V^{(1)}, \ldots, V^{(q)}$, which are disjoint simply connected open sets, and the spaces of orbits $V^{(k)}/\sim$, with $z \sim f(z)$ if z, $f(z) \in V^{(k)}$, are conformal to $\mathbb{C}/\mathbb{Z}$, for each $k \in \{1, \ldots, q\}$. For other holomorphic functions, immediate attracting basins may not exist, but if they do, their number is $\leq q$.

[8] See Section 9 of Chapter 13 of *CADOVA*.

(7.6) *If $f \in \mathcal{F}$, $f'(0) = 1$, $f^{(j)}(0) = 0$ for $j \in \{2, \ldots, q\}$ and $f^{(q+1)} \neq 0$ with $q \in \mathbb{N}$, then 0 has $\leq q$ immediate attracting basins under f and each one has at least 1 critical value of f, except if f is a parabolic Möbius transformation; if an immediate parabolic basin has just 1 critical point or 1 critical value of f, it is simply connected.*

Proof If B is an immediate parabolic basin and B' is a parabolic basin of 0, as if f is a rational function, $B \cap \cup_{j=1}^{q} V^{(j)} \neq \emptyset$ and $\cup_{j=1}^{q} V^{(j)} \subset B'$. As B is a connected component of B', it contains some of the $V^{(j)}$. Therefore, there cannot be more than q immediate parabolic basins.

Define inductively $V_n^{(k)}$ by $V_0^{(k)} = V^{(k)}$ and, if $f\big(V_n^{(k)}\big) \subset V_n^{(k)}$, define $V_{n+1}^{(k)}$ to be the connected component of $f^{-1}\big(V_n^{(k)}\big)$ containing $V_n^{(k)}$, which exists since $f\big(V_{n+1}^{(k)}\big) = V_n^{(k)} \subset V_{n+1}^{(k)}$. Since for every $k \in \{1, \ldots, q\}$ the union $\cup_{n \in \mathbb{N}} V_n^{(k)}$ is a connected component of B', if $B = \cup_{n \in \mathbb{N}} V_n^{(k)}$ is an immediate parabolic basin, then $\big(B, f_{|B}\big)$ is a branched covering of B. If B does not contain any critical point or critical value of f, then $f\big(V_{n+1}^{(k)}\big) \subset V_n^{(k)}$, and $\big(V_n^{(k)}, f\big)$, $\big(V_n^{(k)}, f^n\big)$ are coverings of, resp., $V_n^{(k)}$ and $V^{(k)}$. They induce a deck mapping from B onto $V^{(k)}/\sim$. If $V_n^{(k)}$ are simply connected, B also is. It is conformal to $\mathbb{C}$ and f is a Möbius transformation. Analogously, if B contains just 1 critical point or 1 critical value, it can be proved by induction that the sets $V_n^{(k)}$ are simply connected, and, therefore, also B is. □

(7.7) *If $f_0 \in \mathcal{F}_0$ and B' is the parabolic basin of a fixed point 0 for f_0, then there is 1 unique immediate parabolic basin B of 0 for f_0. Denoting $\widetilde{B} = \varphi_0^{-1}(B)$, $\widetilde{B}' = \varphi_0^{-1}(B')$, for some $\eta_0 > 0$, $\{w \in \mathbb{C} : |\mathcal{I}m\, w| > \eta_0\} \subset \widetilde{B}' \cap \widetilde{B}$ and for $\eta \geq \eta_0$ the connected components $\widetilde{B}^{\pm}$ of $\widetilde{B}$ containing, resp., $\{w \in \mathbb{C} : \pm\, \mathcal{I}m\, w > \eta\}$ satisfy $T\big(\widetilde{B}^{\pm}\big) = \widetilde{B}^{\pm}$, denoted $B^{\pm} = E\big(\widetilde{B}^{\pm}\big)$, where $T(z) = z + 1$ and $E(z) = e^{i2\pi z}$. $\big(B^{-} \cup B^{+}, \mathcal{E}_{f_0}\big)$ is a covering of $\mathbb{C} \backslash \{0\}$ branched at a single $v \in \mathbb{C} \backslash \{0\}$ with infinite leaves, $\widetilde{B}^{\pm}$, $\widetilde{B}^{-} \cup \{\infty\}$, $\widetilde{B}^{+} \cup \{0\}$ are simply connected and $B^{-} \cap B^{+} = \emptyset$.*

Proof The uniqueness of the immediate parabolic basin follows by (7.6), and the existence of $\eta_0 > 0$ such that $\{w \in \mathbb{C} : |\mathcal{I}m\, w| > \eta_0\} \subset \widetilde{B}'$ is a consequence of (7.3.4). As for every $w \in \widetilde{B}'$, there is $n \in \mathbb{N} \cup \{0\}$ such that $T^n(w) \in \widetilde{B}$, also $\{w \in \mathbb{C} : |\mathcal{I}m\, w| > \eta_0\} \subset \widetilde{B}$.

Let Ω_+, Φ_0 be as in (7.3). By (7.3.3), $\Phi_{0|\Omega_+}$ is univalent, and, therefore, for $n \in \mathbb{N} \cup \{0\}$, $\big(f_0^{-n}(\Omega_+) \cup B, \Phi_0\big)$ is a covering of $T^{-n}\Phi_0(\Omega_+)$ branched only at integer translations of the image of the critical value of f_0 by the function Φ_0.

Hence, (B, Φ_0) is a branched covering of $\mathbb{C}$. If $z \in B$, then U, U' are simply connected neighborhoods of z such that $\operatorname{cl} U \subset U'$ and U' does not contain more than one positive orbit of the critical point $z' \in \varphi_0^{-1}(U) \cap \left(\widetilde{B}^- \cup \widetilde{B}^+\right)$ of f_0. If U'_n denotes the connected component of $f_0^{-n}(U')$ containing $\{\varphi_0(z'-n) : n \in \mathbb{N} \cup \{0\}\}$, then there is $m \in \mathbb{N}$ such that for $n \in \mathbb{N}$ with $n \geq m$ the set U'_n does not contain critical points. Therefore, for $k \in \mathbb{N}$, there are branches f_{0,U'_m}^{-k} of the inverse relation of f_0^k defined in U'_m and with values in U'_{m+k}, and the sequence $\{f_{0,U'_m}^{-k}\}_{k\in\mathbb{N}}$ omits 0 and the orbit of the critical point. So, it omits at least three points of $\mathbb{C}_\infty$. By the Montel-Carathéodory theorem,[9] this sequence is a normal family. Since it converges uniformly to 0 near $\varphi_0(z'-m)$, it converges to 0 uniformly in compact sets. Hence, there is $n \in \mathbb{N}$ with $n \geq m$ such that $U_n = f_{0,U'_m}^{m-n}(U_m) \subset \Omega_- = \varphi_0(Q_0)$. Therefore, $z' \in V \overset{def}{=} T^n \circ \left(\varphi_{0|Q_0}\right)^{-1}(U_n)$ and $\varphi_{0|V} = (f_0^n)_{|U_n} \circ (\varphi_{0|Q_0}) \circ T^{-n}$. As there is at most one critical point in U, (V, φ_0) is a branched covering with no more than one branching point. Each connected component of $\varphi_0^{-1}(U)$ is unbranched or branched at a common point of U, and, therefore, $\left(\widetilde{B}^- \cup \widetilde{B}^+, \varphi_0\right)$ is a branched covering of B. With $\widetilde{\mathcal{E}}_{f_0} = \Phi_0 \circ \varphi_0$, $\left(\widetilde{B}^- \cup \widetilde{B}^+, \widetilde{\mathcal{E}}_{f_0}\right)$ and $\left(B^- \cup B^+, \mathcal{E}_{f_0}\right)$ are branched coverings of, resp., $\mathbb{C}$ and $\mathbb{C}\backslash\{0\}$, and the latter branches only at $v = E \circ \Phi_0(c)$, where c is the only critical point of f_0.

If $U \subset B$, $U \neq \Omega_+$ is a connected component of $f_0^{-1}\left(f_0(\Omega_+)\right)$, then $f_0^n(U) \cup U = \emptyset$ for $n \in \mathbb{N}$. By the functional equation for φ_0, for each connected component V of $\varphi_0^{-1}(U)$, $E_{|V}$ is injective. The preimages of each point in the range of

$$E \circ \Phi_{0|U} = E \circ \Phi_{0|\Omega_+} \circ f_{0|U} \quad E \circ \Phi_{0|\Omega_+}$$

are infinite; so, the covering $\left(B^- \cup B^+, \mathcal{E}_{f_0}\right)$ of $\mathbb{C}\backslash\{0\}$ has infinite leaves.

If γ is a closed path in $\widetilde{B}^+$, for some $n \in \mathbb{N}$, then $\widetilde{\gamma}^* = T^{-n}(\gamma^*) \subset Q_0$. Denote by W the region with boundary $\varphi_0(\widetilde{\gamma}^*)$ not containing 0. Since both B and Ω_- are simply connected and do not contain 0, $W \subset B \cap \Omega_-$. Therefore, $\widetilde{\gamma}^*$ is homotopic to a constant path in $\widetilde{B}^+$ and, consequently, so is γ^*. Thus, $\widetilde{B}^+$ and hence also $B^+ \cup \{0\}$ are simply connected. Analogously, $\widetilde{B}^-$ and $B^- \cup \{0\}$ are simply connected.

We now prove that $\widetilde{B}^-$ and $\widetilde{B}^+$ are connected components of different $\widetilde{B}$. Otherwise, since they are simply connected, invariant under T and contain a complex half plane, they would be $\mathbb{C}$, and $\Omega_+ \subset B$, $\Omega_- = \varphi_0(Q_0)$. Therefore, B would be the complement of $\{0\}$ in a neighborhood of 0, and, as it is simply connected, it would be $B = \mathbb{C}_\infty \backslash \{0\}$. Since $f_0 \in H(\mathbb{C}_\infty)$ and it does not have periodic points in B, it would be a parabolic Möbius transformation. So, it would not have critical points, contradicting the hypothesis. Therefore, $\widetilde{B}^-$ and $\widetilde{B}^+$ are disjoint and, consequently, so are B^- and B^+. □

---

[9] See Section 6 of Chapter 11 of *CADOVA*.

(7.8) *If $f_0 \in \mathcal{F}_0$ and $\mathcal{E}_{f_0}$ is the normalized Écalle mapping with $\mathcal{E}'_{f_0}(0)=1$ in* (7.3.4), then $\mathcal{E}''_{f_0}(0) \neq 0$ and $\mathcal{E}_{f_0}$ have a simply connected immediate parabolic basin with only 1 critical point of f_0, $\mathcal{E}_{f_0} \in \mathcal{F}_0$ after linear scaling of coordinates, and if v is the only critical value of $\mathcal{E}_{f_0}$, then $\mathcal{E}^n_{f_0}(v)$ is defined for $n \in \mathbb{N}$ and $\lim_{n\to+\infty} \mathcal{E}^n_{f_0}(v)=0$.

Proof Since $\mathcal{E}_{f_0}$ is not the identity, there are attracting petals $V^{(j)}$, $j \in \{1, \ldots, q\}$ of the fixed point 0 for $\mathcal{E}_{f_0}$, where $q+1$ is the multiplicity of the parabolic fixed point of $\mathcal{E}_{f_0}$. Since $(B^- \cup B^+, \mathcal{E}_{f_0})$ is a branched covering of $\mathbb{C}\backslash\{0\}$, for each $j \in \{1, \ldots, q\}$, as in the proof of (7.6), there is a sequence $\{V^j_n\}_{n \in \mathbb{N}\cup\{0\}}$ with $(V^j_{n+1}, \mathcal{E}_{f_0})$ a branched covering of V^j_n with at most one critical point. The sets V^j_n are simply connected, and for each $j \in \{1, \ldots, q\}$, the sequence $\{\deg g_{0|V^j_n}\}_{n\in\mathbb{N}\cup\{0\}}$ is possibly constant. Therefore, $B_j = \cup_{n=0}^{\infty} V^j_n$ is simply connected and $(B_j, \mathcal{E}_{f_0})$ is a branched covering of B_j with at most one critical point; so, B_j is an immediate parabolic basin of 0 for $\mathcal{E}_{f_0}$. By (7.6), $q=1$, i.e., $\mathcal{E}''_{f_0}(0) \neq 0$. The rest of the proof is left as an exercise. □

(7.9) *If $f_0 \in \mathcal{F}_0$ and $\mathcal{E}_{f_0}$ are the normalized Écalle mapping with $\mathcal{E}'_{f_0}(0)=1$, then at each $w_0 \in \mathbb{C}\setminus\{0\}$ pass 2 nonintersecting negative orbits $\{w_{-j}\}_{j\in\mathbb{N}}$, $\{w'_{-j}\}_{j\in\mathbb{N}}$ $(w'_0 = w_0)$ of $\mathcal{E}_{f_0}$ converging to 0 when $j \to +\infty$, with none of their points a critical point of $\mathcal{E}_{f_0}$ except possibly w_{-1} or w'_{-1}, and if $w_0 \notin D_{w_0}$, also w_{-1}, w'_{-1} are not critical points of $\mathcal{E}_{f_0}$.*

Proof $\mathcal{E}^{-1}_{f_0}(\{w_0\})$ cannot have more than one periodic point, and if it has two points of the positive orbit of v under $\mathcal{E}_{f_0}$, one of them has to be periodic. So,

$$\#\left(\mathcal{E}^{-1}_{f_0}(\{w_0\}) \cap \{\mathcal{E}^n_{f_0}(v)\}_{n\in\mathbb{N}\cup\{0\}} \leq 2\right..$$

By (7.7), with B^-, B^+ as defined there, $\left(B^- \cup B^+, \mathcal{E}_{f_0}\right)$ is a covering of $\mathbb{C}\backslash\{0\}$ branched at a single point $v \in \mathbb{C}\backslash\{0\}$ and with infinite leaves. So, there are two nonperiodic points $w_{-1}, w'_{-1} \in \mathcal{E}^{-1}_{f_0}(\{w_0\})\backslash\{\mathcal{E}^n_{f_0}(v)\}_{n\in\mathbb{N}\cup\{0\}}$.

By (7.7), 0 is a parabolic fixed point of $\mathcal{E}_{f_0}$. So, there is a negative orbit $\{w''_{-j}\}_{j\in\mathbb{N}}$ under $\mathcal{E}_{f_0}$ such that $w''_{-j} \neq 0$ for $j \in \mathbb{N}$, $\lim w''_{-j} = 0$ and

$$\{w''_{-j}\}_{j\in\mathbb{N}} \cap \{\mathcal{E}^n_{f_0}(v)\}_{n\in\mathbb{N}\cup\{0\}} = \emptyset.$$

By (7.8), $\mathcal{E}^n_{f_0}(v) \to 0$. Hence, there is a simply connected open set

$$S \subset (\mathbb{C}\backslash\{0\})\backslash\{\mathcal{E}^n_{f_0}(v)\}_{n\in\mathbb{N}\cup\{0\}}$$

such that $w_{-1}, w'_{-1}, w''_{-1} \in S$.

Since $(B^- \cup B^+, \mathcal{E}_{f0})$ is a covering of $\mathbb{C}\backslash\{0\}$ branched at a single point $v \in \mathbb{C}\backslash\{0\}$, for every $j \in \mathbb{N}$, there is a branch $G_j : S \to \mathbb{C}\backslash\{0\}$ of an inverse of $\mathcal{E}^j_{f0}$ such that $\mathcal{E}^j_{f0} \circ G_j = 1_S$ and $G_j(w''_{-1}) = w''_{-(j+1)}$. As the sequence $\{G_j\}$ omits at least the three points $0, \infty, v$, by the Montel-Carathéodory theorem,[10] it is a normal family. Therefore, since $w''_{-(j+1)} = G_j(w''_{-1}) \to 0$, it is $G_j \to 0$ in S. The sequences defined by $w_{-(j+1)} = G_j(w_{-1})$ and $w'_{-(j+1)} = G_j(w'_{-1})$ for $j \in \mathbb{N}$ are negative orbits of $w_0 = e^{i2\pi \widetilde{w}_0}$ under $\mathcal{E}_{f0}$ with $\lim w_{-j} = \lim w'_{-j} = 0$, $\{w_{-j}\}_{j\in\mathbb{N}} \cap \{w'_{-j}\}_{j\in\mathbb{N}} = \emptyset$ and for $j \geq 2$, w_{-j}, w'_{-j} are not critical points of $\mathcal{E}_{f0}$. If $w_0 \notin D_{\mathcal{E}_{f0}}$, then also w_{-1}, w'_{-1} are not critical points of $\mathcal{E}_{f0}$. □

So far, we have considered the case of f_0 with 1 parabolic fixed point at 0 with multiplier 1. We now consider the possibility of other multipliers, i.e., of f_0 holomorphic in a neighborhood of 0 with $f_0(0) = 0$ and $f'_0(0) = e^{i2\pi p/q}$, $p, q \in \mathbb{N}$ relatively prime.

If f_0^q is not the identity, then $f_0^q(z) = z + a_{\nu q+1} z^{\nu q+1} + O(z^{\nu q+2})$ when $z \to 0$, where $\nu \in \mathbb{N}$ and $a_{\nu q+1} \neq 0$, and with a change of variables,

$$f_0^q(z) = z + z^{q+1} + O\left(z^{2q+1}\right), \quad z \to 0.$$

Changing variables by $w = -\frac{1}{qz^q}$ leads to $F_0(w) = w+1+O\left(\frac{1}{w}\right)$ when $|w| \to +\infty$, where f_0 has q branches. The corresponding Riemann surface $\mathcal{W}$ is obtained considering for $k \in \{1, \ldots, q\}$ q copies $Q_0^{(k\pm)}$ of the $Q_0^{\pm}$ of the proof of (7.3) whose intersection $Q_0^+ \cap Q_0^-$ has two connected components, an upper sector, and a lower sector, and the upper sectors of $Q_0^{(k+)}$ and $Q_0^{(k-)}$ as well as the lower sectors of $Q_0^{((k-1)+)}$ and $Q_0^{(k-)}$ are identified. The function $w \mapsto z = -\frac{1}{qw^q}$ is a conformal homeomorphism of the complement of $\{0\}$ to a neighborhood of 0 in $\mathbb{C}$, and the function F_0 maps points of $Q_0^{(k+)}$ to points of $Q_0^{((k+p)+)}$ and points of $Q_0^{(k-)}$ minus a neighborhood of the boundary into $Q_0^{((k+p)-)}$.

With perturbations $f \in \mathcal{F}_1$ of f_0, the fixed point 0 of f_0 can bifurcate in up to $q+1$ fixed points, where with a change of variables one of them can be fixed at 0 and the orbits of f_0 and f can be as in Fig. 7.11. The change of variables in a neighborhood of 0 is chosen, so that the periodic points $\sigma_f^{(k)}$, $k \in \{1, \ldots, q\}$ satisfy $\sigma_f^{(k)} = e^{i2\pi k/q} \sigma_f^{(0)}$, requiring $\prod_{j=1}^k \left(z - \sigma_f^{(j-1)}\right) = z^q - \sum_{j=1}^q b_f^{(j-1)} z^{j-1}$, and

$$b_f^{(0)} = -i2\pi\alpha_f\left(1 + o(1)\right), \quad b_f^{(j)} = O(\alpha_f), \; j \in \{1, \ldots, q-1\}, \quad f \to f_0.$$

Changing variables by $z' = z/\left(\sum_{j=1}^q \frac{b_j}{b_0} z^{j-1}\right)^{\frac{1}{q}}$, $[f^q(z)]^q = z^q + z^q[z^q - (\sigma_f^{(0)})^q] u_f(z)$, where u_f is a holomorphic function without zeros in a neighborhood of 0. The further change of variables $e^{i2\pi\alpha_f w} = 1/[1 - (\sigma_f^{(0)}/z)^q]$ gives the function

[10] See Section 6 of Chapter 11 of *CADOVA*.

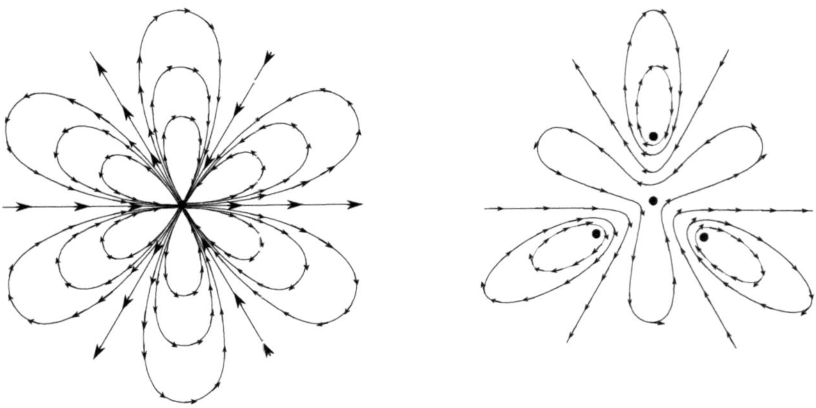

Fig. 7.11 Possible bifurcation of a parabolic fixed point with multiplier $e^{i2\pi p/q}$ with $p, q \in \mathbb{N}$ relatively prime (example with $q = 3$)

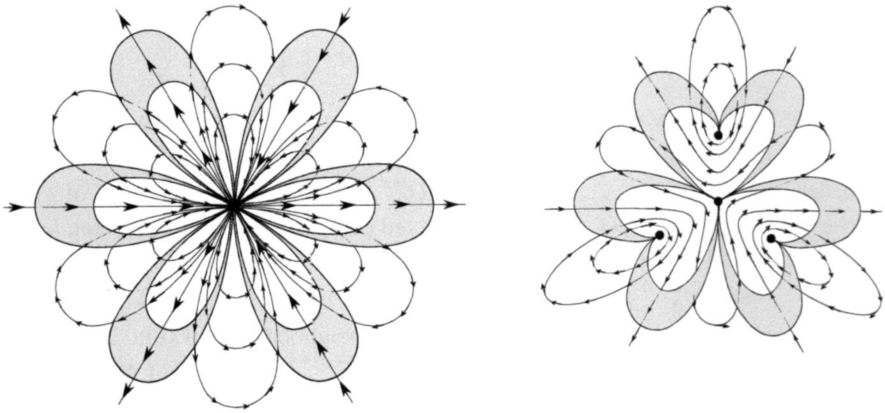

Fig. 7.12 q attracting Étalle cylinders and q repelling Étalle cylinders corresponding to Fig. 7.11

$$F_f(w) = w + \frac{1}{i2\pi\alpha_f} \log\left(1 - \frac{\sigma_f^{(0)} u_f(z)}{1+z^q u_f(z)}\right),$$

with w in a Riemann surface conformal to $B_1 \backslash \{0, \sigma_f^{(1)}, \dots, \sigma_f^{(q)}\}$. Results similar to (7.1)–(7.4) can be obtained for these functions, what is left as an exercise.

This gives q attracting Étalle cylinders $\mathcal{C}_0^{k-}$ and q repelling Étalle cylinders $\mathcal{C}_0^{k+}$, with $k \in \{1, \dots, q\}$ (see Fig. 7.12). Each Écalle mapping maps the upper end of $\mathcal{C}_0^{k+}$ to the upper end of $\mathcal{C}_0^{k-}$ and the lower end of $\mathcal{C}_0^{k-}$ to the lower end of $\mathcal{C}_0^{k+}$.

The modifications of the previous statements for parabolic fixed point at 0 with multiplier 1 are (what is unchanged is not mentioned):

(7.3.1): *Instead of* 2 $\Omega_\pm$ *regions, there are* $2q$ *regions* $\Omega_\pm^{(k)}$, $k\in\{1,\dots,q\}$; $\cup_{k=1}^{q}\big(\Omega_-^{(k)}\cup\Omega_+^{(k)}\cup\{0\}\big)$ *is the neighborhood of* 0 *where* f_0 *is injective; for* $k\in\{1,\dots,q\}$, $f_0\big(\mathrm{cl}\,\Omega_+^{(k)}\big)\subset\Omega_+^{(k+p)}$ *and* $f_0\big(\Omega_-^{(k)}\cup\{0\}\big)\supset\mathrm{cl}\,\Omega_-^{(k+p)}$, $\Omega_+^{(j)}\cap\Omega_-^{(k)}$ *is nonempty and connected if* $j=k$ *or* $j=k-1$, *and it is empty otherwise, and* $f_0^n\to 0$ *uniformly in* $\Omega_+^{(k)}$; *the parabolic basins of* 0 *for* f *are defined by* $\mathcal{B}^{(k)}=\cup_{n=0}^{\infty}f^{-nq}\big(\Omega_+^{(k)}\big)$ *and a point belongs to* $\cup_{k=1}^{q}\mathcal{B}^{(k)}$ *if and only if it has a neighborhood included in* $\cap_{n=1}^{\infty}D_{f^n}$ *where* $f^{nq}\to 0$ *uniformly.*
(7.3.2): φ_0, f_0, Ω_- *are replaced by, resp.,* $\varphi_0^{(k)}$, f_0^q, $\Omega_-^{(k)}$ *and* $\widetilde{\varphi}_0$ *is replaced by* $\widetilde{\varphi}_0^{(k)}=-\frac{1}{q\,[\varphi_0^{(k)}]^q}$, *with* $\widetilde{\varphi}_0^{(k)}(w)=w+O(w^{1-1/q})$ *when* $w\to\infty$, *for* $k\in\{1,\dots,q\}$.
(7.3.3): Φ_0, $\mathcal{B}$, f_0, Ω_+ *are replaced by* $\Phi_0^{(k)}$, $\mathcal{B}^{(k)}$, f_0^q, $\Omega_+^{(k)}$, $k\in\{1,\dots,q\}$.
(7.3.4): $\widetilde{\mathcal{B}}$ *is replaced by the* $2q$ *sets*

$$\widetilde{\mathcal{B}}^{(k+)}=\big(\varphi_0^{(k)}\big)^{-1}\big(\mathcal{B}^{(k)}\big),\qquad \widetilde{\mathcal{B}}^{(k-)}=\big(\varphi_0^{(k)}\big)^{-1}\big(\mathcal{B}^{(k-1)}\big),$$

which are invariant under T *and satisfy* $\widetilde{\mathcal{B}}^{k\pm}\supset\{w\in\mathbb{C}:\pm\mathcal{I}m\,w>\eta_0\}$ *for some* $\eta_0>0$, $\widetilde{\mathcal{E}}_{f0}$ *is replaced by* $\widetilde{\mathcal{E}}_{f0}^{(k\pm)}:\widetilde{\mathcal{B}}^{(k\pm)}\to\mathbb{C}$ *such that* $\widetilde{\mathcal{E}}_{f0}^{(k+)}=\Phi_0^{(k)}\circ\varphi_0^{(k)}$ *and* $\widetilde{\mathcal{E}}_{f0}^{(k-)}=\Phi_0^{(k-1)}\circ\varphi_0^{(k)}$, *which satisfy the same functional equation satisfied by* $\widetilde{\mathcal{E}}_{f0}$ *in* $\widetilde{\mathcal{B}}^{k\pm}$, $\mathcal{E}_{f0}$ *is replaced by* $\mathcal{E}_{f0}^{k+}=E\circ\widetilde{\mathcal{E}}^{k+}\circ E^{-1}$, *which is holomorphic in* $E\big(\widetilde{B}\big)\cup\{0\}$ *with values in* $\mathbb{C}$ *and satisfies* $\big(\mathcal{E}_{f0}^{k+}\big)'(0)\neq 0$, *and by* $\mathcal{E}_{f0}^{k-}=E\circ\widetilde{\mathcal{E}}^{k+}\circ E^{-1}$ *which is holomorphic in* $E\big(\widetilde{B}\big)\cup\{\infty\}$ *with values in* $\mathbb{C}_\infty\setminus\{0\}$ *and satisfies* $\big(\mathcal{E}_{f0}^{k-}\big)'(\infty)\neq 0$, *for* $k\in\{1,\dots,q\}$.
(7.4): α_f *is replaced for* $f\in\mathcal{F}$ with $f'(0)\neq 0$ *by* $\alpha_f\in\left]-\frac{1}{2},\frac{1}{2}\right]\times\mathbb{R}$ *for* $e^{i2\pi(p+\alpha_f)/q}=f'(0)$.
(7.4.1): φ_f *is replaced by* ${\varphi_f}^{(k)}$ *and the functional equation that* φ_f *satisfies is replaced by* ${\varphi_f}^{(k)}(w+1)=f^q\circ{\varphi_f}^{(k)}(w)$, *for* $k\in\{1,\dots,q\}$; *the fixed point* $\sigma_f\neq 0$ *near* 0 *is replaced by periodic points of* f *with period* q *near* 0 *denoted* ${\sigma_f}^{(k)}=[-i2\pi q\,\alpha_f]^{1/q}e^{i2\pi k/q}\big(1+o(1)\big)$ *when* $f\to f_0$, *where* $\big|\mathrm{Arg}\big([-i2\pi q\,\alpha_f]^{1/q}\big)\big|<\frac{\pi}{q}$, *and* $f\big({\sigma_f}^{(k)}\big)={\sigma_f}^{(k+p)}$ for $k\in\{1,\dots,q\}$.
(7.4.2): $\mathcal{R}_f$ *is replaced by* ${\mathcal{R}_f}^{(k+)}$, *which satisfies* $\big({\mathcal{R}_f}^{(k+)}\big)'(0)=e^{-i2\pi k/(p\,\alpha_f)}$ *and* $\alpha_{\mathcal{R}_f^{(k+)}}=-\frac{k}{p\,\alpha_f}\ (\mathrm{mod}\ \mathbb{Z})$, *for* $k\in\{1,\dots,q\}$.
(7.4.3): f^n *is replaced by* f^{nq+r}, $r\in\{1,\dots,n-1\}$, $rp=-1\ (\mathrm{mod}\ q)$.
(7.4.4): $\mathcal{E}_{f0}$ *is replaced by* $\mathcal{E}_{f0}^{(k+)}$, *for* $k\in\{1,\dots,q\}$.
(7.4.5): $-\frac{1}{\varphi_f}$ *is replaced by* $-\frac{1}{q[\varphi_f^{(k)}]^q}$, *for* $k\in\{1,\dots,q\}$.

In the two paragraphs following Fig. 7.10, replace $\mathcal{R}_f$ by $\mathcal{R}_f^{(k-)}$ and $\mathcal{E}_{f0}$ by the $\mathcal{E}_{f0}^{(k-)}$ obtained in the modification of (7.4.4) indicated above, for $k\in\{1,\dots,q\}$, and f^n as in the modification of (7.4.3).

(7.6) *to* (7.8): *The parabolic basin* $\mathcal{B}$ *is replaced by* q *parabolic basins* $\mathcal{B}^{(k)}$, *with* $k \in \{1, \ldots, q\}$, $\mathcal{F}_0$ *is replaced by the set of functions* $f_0 \in \mathcal{F}$ *with an immediate parabolic basin* $B^{(k)}$ *in each* $\mathcal{B}^{(k)}$ *which contains just* 1 *critical point of* f_0^q , *the set* $\widetilde{B}^+$ *is replaced by the connected components* $\widetilde{B}^{k+}$ *of the sets, resp.,* $\widetilde{\mathcal{B}}^{k+}$, *obtained in the modification of* (7.3.4) *indicated above that contain* $\{w \in \mathbb{C} : \mathcal{I}m\, w > \eta_0\}$, *the set* B^+ *is replaced by* $B^{(k+)} = E\big(\widetilde{\mathcal{B}}^{k+}\big)$, *and* $\mathcal{E}_{f_0}^+$ *is replaced by* $\mathcal{E}_{f_0}^{k+}$ *obtained in the modification of* (7.3.4) *indicated above, for* $k \in \{1, \ldots, q\}$.

7.3 Hausdorff and Hyperbolic Dimensions

The **Hausdorff measure**[11] of dimension $d \geq 0$ of $S \subset \mathbb{C}_\infty$ is

$$H^d(S) = \lim_{\delta \to 0} H^d_\delta(S) \in [0, +\infty[\cup \{\infty\}, \text{ with}$$

$$H^d_\delta(S) = \inf\left\{ \sum_{j=1}^{\infty} (\operatorname{diam} U_j)^d : S \subset \{U_j\}_{j \in \mathbb{N}},\ U_j \subset X,\ \operatorname{diam} U_j < \delta \right\},$$

for $\delta > 0$, with the diameters in the spherical metric of $\mathbb{C}_\infty$.

It is left as an exercise to check that *if* $S \subset X$, $0 \leq d < D$, $d, D \in \mathbb{R}$, *then* $H^d(S) \in \mathbb{R} \Longrightarrow H^D(S) = 0$ *and* $H^D(S) > 0 \Longrightarrow H^d(S) = \infty$, and *in* $\mathbb{R}^n$ *the Hausdorff measure of dimension n coincides with the outer Lebesgue measure divided by the Lebesgue measure of the ball* B_1.

The **Hausdorff dimension**[12] of $S \subset \mathbb{C}_\infty$ is

$$\dim_H(S) = \inf\big\{ d \in [0, +\infty[: H^d(S) \big\}.$$

Given a complex rational function $f : \mathbb{C}_\infty \to \mathbb{C}_\infty$, a closed set $\emptyset \neq X \subset \mathbb{C}_\infty$ is said to be an **hyperbolic set for**[13] f if $f(X) \subset X$, and there are $c > 0$ and $k > 1$ such that $\|(f^n)'\| \geq ck^n$ in X for $n \in \mathbb{N} \cup \{0\}$, with $\|\ \ \|$ the operator norm in the spherical metric in $\mathbb{C}_\infty$.

The **hyperbolic dimension**[14] of a rational function $f : \mathbb{C}_\infty \to \mathbb{C}_\infty$ is

$$\dim_{Hyp}(f) = \sup\big\{ \dim_H(X) : X \subset \mathbb{C}_\infty \text{ is a Hyperbolic Set for } f \big\}.$$

[11] This definition applies to any metric space X. It is an **outer measure** as theorized by C. Carathéodory in 1918, i.e., a function μ^* of the set of subsets of X in $[0, +\infty[\cup \{\infty\}$ such that $\mu^*(\emptyset) = 0$ and $A \subset \cup_{j=1}^{\infty} B_j \subset X \Rightarrow \mu^*(A) \leq \sum_{j=1}^{\infty} \mu^*(B_j)$. The corresponding **measure** can be obtained as usual for outer measures, on the set of **measurable** subsets of X, i.e., the sets $S \subset X$ such that $\mu^*(S) = \mu^*(S \cap A) + \mu^*(S \setminus A)$ for all $A \subset X$.

[12] The Hausdorff dimension was introduced in 1918 by F. Hausdorff.

[13] See the notion of hyperbolic rational function in Section 13 of Chapter 13 of *CADOVA*.

[14] The notion of hyperbolic dimension was introduced in 1991 by M. Shishikura in a preprint of the article cited at the beginning of this chapter and has been widely used since then.

To establish fundamental properties of hyperbolic sets and dimensions, the following notions of more general interest are used, in this case for a complex rational function $f:\mathbb{C}_\infty \to \mathbb{C}_\infty$:

A **stable (resp., unstable) set of** $z \in \mathbb{C}_\infty$ under f is

$$W^s(z) = \{x \in \mathbb{C}_\infty : \lim_{n\to+\infty} d_S(f^n(z), f^n(x)) = 0\},$$

(resp., $W^u(z)$ defined analogously, but with $\lim_{n\to-\infty}$), and for $\varepsilon > 0$ denote

$$W^s_\varepsilon(z) = \{x \in \mathbb{C}_\infty : d_S(f^n(z), f^n(x)) \leq \varepsilon\,,\ \text{to } n \in \mathbb{N} \cup \{0\}\},$$

and analogously $W^u_\varepsilon(z)$, but for $-n \in \mathbb{N} \cup \{0\}$.

For $\alpha > 0$, a **α-pseudoorbit under** f is $\{z_j\}_{j=a}^b \subset \mathbb{C}_\infty$ with $a, b \in \mathbb{Z} \cup \{-\infty, +\infty\}$ and $a \leq b$ such that $d_S(f(z_j), z_{j+1}) < \alpha$ for all $j \in \mathbb{Z}$ with $a \leq j < b-1$. For $\beta > 0$, we call **β-shadoworbit of a α-pseudoorbit** a orbit under f of a $z \in \mathbb{C}_\infty$ such that $d_S(f(z), z_j) \leq \beta$ for $j \in \mathbb{Z}$ with $a \leq j \leq b$, where d_S is the spherical distance in $\mathbb{C}_\infty$.

Hyperbolic sets and dimension have the following properties.

(7.10) *If $f:\mathbb{C}_\infty \to \mathbb{C}_\infty$ is a complex rational function and X is a hyperbolic set for f, then:*

1. *X has no Critical Points of f.*
2. *$X \subset J(f)$ and $\dim_H J(f) \geq \dim_{Hyp}(f)$.*
3. *For every $\beta > 0$, there is $\alpha > 0$ such that every α-pseudoorbit under f has a β-shadoworbit.*
4. *X is **stable**, i.e., there is a neighborhood U of f in the space of complex rational functions with the same degree such that for all $g \in U$, there is a hyperbolic set X_g for g homeomorphic to X with homeomorphism ϕ_g from X onto X_g which is a conjugacy of f with g, and $\{\phi_g\}_{g\in U}$ is a holomorphic motion of X.*
5. *The function $g \mapsto \dim_{Hyp}(g)$ defined in the space of complex rational functions with the same degree is **lower semicontinuous**, i.e., for $k \in [0, +\infty[$, $\{g : \dim_{Hyp}(g) > k\}$ is an open set.*
6. *If $\{\phi_\lambda\}_{\lambda\in B_1}$ is a holomorphic motion of X and $v: B_1 \to \mathbb{C}_\infty$ is a holomorphic function such that $v(0) = z_0 \in X$, and the functions $\lambda \mapsto v(\lambda)$ and $\lambda \mapsto \phi_\lambda(z_0)$ do not coincide, then*

$$\dim_H(\lambda \in B_1 : v(\lambda) \in \phi_g\lambda(X)) \geq \lim_{r\to 0} \dim_H(X \cap B_r(z_0)).$$

Proof

(1) It is direct from the definition.

(2) Since the derivatives of f^n grow exponentially with $n \in \mathbb{N}$, the sequence $\{f^n\}$ cannot be a normal family on any open set intersecting X and, therefore, the inequality holds.

(3) Let $\{z_j\}_{j=a}^{b} \subset X$ be a α-pseudoorbit under f, $\varepsilon > 0$ and $\delta \in]0, \varepsilon[$ be such that $W^s_\varepsilon(x) \cap W^u_\varepsilon(y) \cap X \neq \emptyset$ for all $x, y \in X$ such that $d_S(x, y) \leq \delta$. For $M \in \mathbb{N}$ so large that $k^M \varepsilon < \frac{\delta}{2}$ and $\alpha > 0$ such that if $\{x_j\}_{j=0}^{M} \subset X$ is a α-pseudoorbit, then $d_S\big(f^j(x_0), x_j\big)_{|f(\mathbb{H})} < \frac{\delta}{2}$ for $j \in \{0, \dots, M\}$. For $\{y_j\}_{j=0}^{rM} \subset X$ a α-pseudoorbit with $r \in \mathbb{N}$ define recursively[15] y'_{jM} for $j \in \{0, \dots, r\}$ for $y'_0 = y_0$ and

$$y'_{(j+1)M} = W^u_\varepsilon\big(f^M(y'_{jM})\big) \cap W^s_\varepsilon\big(f^M(y_{(j+1)M})\big) \in X,$$

and denote $y = f^{-rM} y'_{rM}$. For $\ell \in \{0, \dots, rM\}$ and $s \in \mathbb{N} \cup \{0\}$ such that $\ell \in \{sM, sM + 1, \dots, (s + 1)M\}$,

$$\begin{aligned} &d_S\big(f^\ell(y), f^{\ell - sM}(y'_{sM})\big) \\ &\quad \leq \sum_{t=s+1}^{r} d_S\big(f^{\ell - tM}(y'_{tM}), f^{\ell-(t-1)M}(y_{(t-1)M})\big) \leq \sum_{t=s+1}^{r} k^{tM-\ell} \varepsilon \leq \tfrac{k\varepsilon}{1-k}, \end{aligned}$$

as $y'_{tM} \in W^u_\varepsilon\big(f^M(y'_{(t-1)M})\big)$. Since $y'_{sM} \in W^s_\varepsilon(y_{sM})$, it is

$$d_S\big(f^{\ell - sM}(y'_{sM}), f^{\ell - sM}(y_{sM})\big) \leq \varepsilon,$$

and $d_S\big(f^{\ell - sM}(y_{sM}), y_\ell\big) < \frac{\delta}{2}$. Thus, with the triangular inequality,

$$d_S\big(f^\ell(y), y_\ell\big) \leq \tfrac{k\varepsilon}{1-k} + \varepsilon + \tfrac{\delta}{2} < \beta,$$

taking $\varepsilon > 0$ small. Every α-pseudoorbit $\{y_j\}_{j=0}^{n} \subset X$ can be extended to $\{y_j\}_{j=0}^{rM} \subset X$ with $rM \geq n$ by $y_j = f^{j-n}(x_n)$ for $j \in \{n+1, \dots, rM\}$. If the orbit of $y \in X$ under f is a β-shadoworbit of this extended α-pseudoorbit, it is also a β-shadoworbit of the original orbit. If $\{y_j\}_{j=a}^{b}$ is a finite α-pseudoorbit, so is $\{y_{j+a}\}_{j=0}^{b-a}$, and if the orbit of $y \in X$ under f is a β-shadoworbit of this α-pseudoorbit, then the orbit of $f^{-a}(y)$ under f is also a β-shadoworbit of the original orbit. Thus, the result is proved for finite α-pseudoorbits. If $\{y_j\}_{j \in \mathbb{Z}} \subset X$ is a α-pseudoorbit and $y^{(m)} \in X$ is such that its orbit under f is a β-shadoworbit of $\{y_j\}_{j=-m}^{m}$, with $y \in X$ a limit point of the sequence $\{y^{(m)}\}$, then the orbit of y under f is a β-shadoworbit of $\{y_j\}_{j \in \mathbb{Z}}$.

(4) There is a neighborhood V of X such that every complex rational function g close to f is expansive in V. So, for every $x \in X$, the positive orbit $\{f^n(x)\}_{n=0}^{\infty}$ under f is a pseudoorbit under g and, by 3, the orbit of some y under g is a shadoworbit of that pseudoorbit. Denote $\phi_g(x) \stackrel{def}{=} g^{-1}_{x_0} \circ g^{-1}_{x_1} \circ \cdots g^{-1}_{x_n}(x_{n+1})$, with $x_j = f^j(x)$ and g^{-1}_x a branch of g^{-1} defined in a neighborhood of $f(z)$ such that $g^{-1}_z\big(f(z)\big)$ is close to z. As g close to f is expansive, the function ϕ_g is well defined, and it is a conjugacy of $f_{|X}$ with $g_{|X_g}$, where $X_g = \phi_g(X)$. As the convergence in the limit above is uniform, $g \mapsto \phi_g(x)$ defined in a neighborhood

[15] Checking the consistency of this definition is left as an exercise.

U of f in the space of complex functions with the same degree is holomorphic. Therefore, $\{\phi_g\}_{g\in U}$ is holomorphic motion of X.

(5) It suffices to prove the function $g \mapsto \dim_H(X_g)$ defined in $B_R \subset U$ is continuous. By the harmonic λ-lemma (4.19), ϕ_g can be extended to a $K(|g|/R)$-quasiconformal mapping, where $K:]0,1[\to]1,+\infty[$ is nonincreasing and $\lim_{t\to 0} K(t)=1$. By the Mori theorem (3.21), ϕ_g and ϕ_g^{-1} are Hölder continuous with exponent $\frac{1}{K(|g|/R)}$. For $g_0 \in B_R$, $\{\phi_g \circ \phi_\tau g_0{}^{-1}\}_{g\in B_R}$ is a holomorphic motion of X_{g_0}. Therefore, for g close to g_0, the function $\phi_g \circ \phi_{g_0}^{-1}$ and its inverse are Hölder continuous with exponent $\alpha'(g_0,g) \in]0,1]$ such that $\lim_{g\to g_0} \alpha'(g_0,g) = 1$. Hence,

$$\alpha'(g_0,g)\dim_H(X_{g_0}) \le \dim_H(X_g) \le \frac{1}{\alpha'(g_0,g)}\dim_H(X_{g_0}).$$

So, the function $g \mapsto \dim_H(g)$ is continuous, and, by the definition of $\dim_{Hyp}$, $g \mapsto \dim_{Hyp}(g)$ is lower semicontinuous.

(6) Without loss of generality, $z_0 = 0$ and $\phi_\lambda(0) = 0$ for all $\lambda \in B_1$, as this can be ensured with a Möbius transformation.

Consider first $v'(0) \neq 0$. There are constants $a>0$ and $\rho \in]0,1[$ such that v is injective into B_ρ and $a|\lambda| \le v(\lambda) < \infty$. Denote

$$b_r = \sup\left\{|\phi_g\lambda(z) : z \in X \cap B_r(0),\, |\lambda| \le \rho\right\}.$$

By the Montel theorem[16] $\lim_{r\to 0} b_r = 0$. Hence, there is $r_0>0$ such that $a\rho > b_r$ for $0<r<r_0$. For such r, $z \in X \cap B_r(0)$ and $\mu \in B_{a\rho/b_r} \subset \mathbb{C}$ consider for $\lambda \in B_{\min\{\rho,\rho/|\mu|\}} \subset \mathbb{C}$ the equation $v(\lambda) - \phi_{\lambda\mu}(z) = 0$. As v and $\lambda \mapsto \phi_{\lambda\mu}(z)$ are holomorphic in, resp., $B_{\min\{\rho,\rho/|\mu|\}}$ and $\partial B_{\min\{\rho,\rho/|\mu|\}}$,

$$|\phi_{\lambda\mu}(z)| \le b_r < a\min\{\rho, \rho/|\mu| \le |v(\lambda)|.$$

Since $\lambda=0$ is the only solution of the equation $v(\lambda)=0$ in $B_{\min\{\rho,\rho/|\mu|\}}$, by the Rouché theorem,[17] the equation $v(\lambda) - \phi_{\lambda\mu}(z) = 0$ has a unique solution $j_\mu^r(v^{-1}(z))$ in $B_{\min\{\rho,\rho/|\mu|\}}$, and for every $z \in X \cap B_r$, the function $\mu \mapsto j_\mu^r(v^{-1}(z))$ is holomorphic in $B_{a\rho/b_r}$; as ϕ_λ is injective, for $\mu \in B_{\min\{\rho,\rho/|\mu|\}}$ the function $j_\mu^r \circ v^{-1}$ is injective into $X \cap B_r$. Denote Y_μ^r the set of solutions of $v(\lambda) - \phi_{\lambda\mu}(z) = 0$ for each $z \in X \cap B_r$. Then, $\{Y_\mu^r\}_{\mu\in B_{a\rho/b_r}}$ is a holomorphic motion with $Y_0^r = v^{-1}(X \cap B_r)$. So, $\dim_H(Y_0^r) = \dim_H(X \cap B_r)$ and $Y_1^r \subset v^{-1}(\phi_\lambda(X) \cap B_1)$. Analogously to what was seen at the beginning of the proof of 5, the functions $j_\mu^r : Y_0^r \to Y_\mu^r$ and $(j_\mu^r)^{-1}$ are Hölder continuous with exponent $\alpha(|\mu| b_r/(a\rho))$. Therefore,

[16] See Section 4 of Chapter 10 of *CADOVA*.

[17] See Section 4 of Chapter 8 of *CADOVA*. Rouché, Eugène (1832–1910).

$$\dim_H\big(v^{-1}(\phi_\lambda(X)\cap B_1)\big) \geq \dim_H(Y_1^r) \geq \alpha\big(\tfrac{b_r}{a\rho}\big)\dim_H\big(X\cap B_r\big),$$

which results in the inequality in the statement.

Consider now $v'(0)=0$. By assumption, v is not identically 0. Without loss of generality, take $\infty\in X$ and $\phi_{\lambda\mu}\lambda(\infty)=\infty$ for all $\lambda\in B_1$, as this can be ensured with a Möbius transformation. Denote m the order of the analytic function v, $G(z)=z^m$, $\widetilde{X}_\lambda = G^{-1}(X_\lambda)$, and $\widetilde{X}=\widetilde{X}_0$. $(\widetilde{X}_\lambda, G)$ is a covering[18] of X_λ branched at 0 and ∞. The lifts $\widetilde{v}: B_R \to \mathbb{C}_\infty$ and $\widetilde{\phi}_\lambda : \widetilde{X}\to\widetilde{X}_\lambda$, which satisfy $v = G\circ\widetilde{v}$, $\widetilde{v}'(0)\neq 0$ and $\phi_\lambda\circ G = G\circ\widetilde{\phi}_\lambda$. By the preceding paragraph, the inequality in the statement holds for $\widetilde{v}$ and $\widetilde{\phi}_\lambda$. Since $v^{-1}(X_\lambda)=\widetilde{v}^{-1}(\widetilde{X}_\lambda)$ and $\dim_H\big(X\cap B_r\big) = \dim\big(\widetilde{X}\cap G^{-1}(B_r)\big)_{|f(\mathbb{H})}$, as G is locally Lipschitz- ian in $\mathbb{C}_\infty\backslash\{0,\infty\}$, the stated inequality for v and ϕ_λ holds. □

Now, we consider the hyperbolic dimension of an element of a 1-parameter holomorphic family $\mathcal{F}$ of complex rational functions of degree ≥ 2 with Hausdorff dimension of the set of parameters where the functions are not Julia-stable in $\mathcal{F}$ with a hyperbolic set containing a postcritical orbit.

(7.11) *If $\mathcal{F}=\{f_\lambda\}_{\lambda\in\Lambda}$ is a holomorphic family of complex rational functions of degree ≥ 2 with $\Lambda\subset\mathbb{C}$ open and $\lambda_0\in\Lambda$ is such that f_{λ_0} is not Julia-stable in $\mathcal{F}$, then*

$$\dim_{Hyp}(f_{\lambda_0}) \leq \dim_H\big\{\lambda\in\Lambda:\ f_\lambda \text{ is not Julia-stable in } \mathcal{F} \text{ and there is a Hyperbolic Set for } f_\lambda \text{ which contains a Postcritical orbit}\big\}.$$

Proof For $\varepsilon>0$, there is a hyperbolic set X for f_{λ_0} such that

$$\dim_H(X) \geq \dim_{Hyp}\big(f_{\lambda_0}\big) - \varepsilon\,.$$

Since X is compact, there is $z_0\in X$ such that $\lim\limits_{r\to 0}\dim_H\big(X\cap B_r(z_0)\big) = \dim_H(X)$. By 4 in the preceding result, there is a neighborhood $\Lambda'\subset\Lambda$ of λ_0 and a holomorphic motion $\{\phi_\lambda\}_{\lambda\in\Lambda'}$ of X such that $\phi_\lambda\circ f_{\lambda_0} = f_\lambda\circ\phi_\lambda$ and $X_\lambda=\phi_\lambda(X)$ is a hyperbolic set for f_λ. Choose Λ' so that

$$\lim_{r\to 0}\dim_H\big(X_\lambda\cap B_r\big(\phi_\lambda(z_0)\big)\big) > \dim_H(X)-\varepsilon\,,\qquad \lambda\in\Lambda'\,,$$

and the critical points of f_λ do not bifurcate at $\lambda\in\Lambda'\backslash\{\lambda_0\}$.

By (5.2), there are $\lambda_1\in\Lambda'\backslash\{\lambda_0\}$, $N\in\mathbb{N}$ and a critical point c of f_{λ_1} such that $f_{\lambda_1}^N(c)=\phi_{\lambda_1}(z_0)$. There is a branch of critical points c_λ of f_λ in a neighborhood $\Lambda''\subset\Lambda'$ of λ_1 with $c_{\lambda_1}=c$, and $\lambda\mapsto c_\lambda$ is meromorphic, where λ_1, c, Λ'' are such that the functions defined in Λ'' by $\lambda\mapsto f_\lambda^N(c_\lambda)$ and $\lambda\mapsto\phi_\lambda(z_0)$ do not coincide. By

---

18 See Section 2 of Chapter 12 of *CADOVA*.

6 in the preceding result, with $v(\lambda) = f_\lambda^N(c_\lambda)$, after an affine change of parameter,

$$\dim_H\{\lambda \in \Lambda'': f_\lambda^N(c_\lambda) \in X_\lambda\} \geq \lim_{r\to 0} \dim_H\big(X_{\lambda_1} \cap B_r(\phi_{\lambda_1}(z_0))\big) > \dim_{Hyp}(f_{\lambda_0}) - 2\varepsilon.$$

If $f_\lambda^N(c_\lambda) \in X_\lambda$, since the functions defined in Λ'' by $\lambda \mapsto f_\lambda^N(c_\lambda)$ and $\lambda \mapsto \phi_\lambda(z)$ do not coincide, f_λ is not Julia-stable in $\mathcal{F}$. The proof ends by noting that $\varepsilon > 0$ is arbitrary. □

7.4 Dimension of the Mandelbrot Set Boundary

Consider complex quadratic polynomials $P_c(z) = z^2 + c$ with $c \in \mathbb{C}$, and denote[19] the **filled Julia set** of P_c

$$K_c = \{z \in \mathbb{C}: P_c^n(z) \nrightarrow \infty \text{ when } n \to +\infty\},$$

the **Julia set** of P_c

$$J_c = \partial K_c = \mathrm{cl}\,\{\text{periodic repelling points } P_c\},$$

the **Mandelbrot set** of P_c

$$\mathcal{M} = \{c \in \mathbb{C}: 0 \in K_c\} = \{c \in \mathbb{C}: K_c \text{ is connected}\}.$$

(7.12) *Except for values of $c \in \mathcal{M}$ such that the orbit of 0 under P_c is periodic (so, superattracting), which belong to the union of the hyperbolic connected components of $\mathcal{M}$, $\partial\mathcal{M}$* ***is the locus of the bifurcations of*** *$\{P_c\}_{c\in\mathbb{C}}$; more precisely,*

$$\partial\mathcal{M} = \{c \in \mathbb{C}: P_c \textit{ is not Julia-Stable in} \{P_c\}_{c\in\mathbb{C}}\},$$

and $\partial\mathcal{M} \supset \{c \in \mathbb{C}: P_c \text{ has a neutral periodic orbit}\}$.
Besides, $\partial\mathcal{M} = \mathrm{cl}\,\{c \in \mathbb{C}: P_c \text{ has a parabolic periodic orbit}\}$.

Proof Since[20] $\mathcal{M} \subset \mathrm{cl}\, B_2$, for $c \in \mathrm{int}\,\mathcal{M}$, the postcritical orbit of P_c is bounded. For $c \in \mathrm{ext}\,\mathcal{M}$, the postcritical orbit of P_c tends to ∞ uniformly for c in compact subsets of $\mathrm{ext}\,\mathcal{M}$. For $c \in \mathbb{C}\setminus\partial\mathcal{M}$, $\{P_c^n(0)\}_{n\in\mathbb{N}}$ is a normal family.[21] The postcritical orbit of P_c for $c \in \partial\mathcal{M}$ is bounded, but for parameter values c' arbitrarily close to c, the postcritical orbit of $P_{c'}$ tends to ∞, and $\{P_c^n(0)\}_{n\in\mathbb{N}}$ is not a normal family. By (5.1.1,6), $\partial\mathcal{M}$ is the set of $c \in \mathbb{C}$ such that P_c is not stable in Julia sets in $\{P_c\}_{c\in\mathbb{C}}$.

[19] See Section 15 of Chapter 13 of *CADOVA*.
[20] See Section 16 of Chapter 13 of *CADOVA*.
[21] See Section 4 of Chapter 10 of *CADOVA*.

By (5.1.1,2), if P_c has a neutral periodic orbit and $c \in \mathbb{C} \setminus \partial\mathcal{M}$, this orbit would be persistently neutral, and, therefore, P_c would have a neutral periodic orbit for all $c \in \mathbb{C}$, contradicting that P_0 has just one periodic orbit: that of the fixed point 0 which is superattracting. Therefore, if P_c has a neutral periodic orbit, then $c \in \partial\mathcal{M}$.

Let $c_0 \in \partial\mathcal{M}$, V be a neighborhood of c_0 and $\{O_1, \ldots, O_{N(c_0)}\}$ be the set of the attracting orbits of P_{c_0} in V. There is at least one of these orbits, and they are finitely many; so, $N(c_0) \in \mathbb{N}$. For any neighborhood U of the set of attracting periodic orbits of P_{c_0}, which is a disjoint union of neighborhoods U'_j of each one of the orbits O_j for $j \in \{1, \ldots, N(c_0)\}$, there is a small neighborhood $V' \subset V$ of c_0 such that each U_j contains a unique attracting periodic orbit of P_{c_0}, and the number $N(c)$ of attracting periodic orbits of P_c in U satisfies $N(c) \geq N(c_0)$ for $c \in V'$. By (5.1.1,3), $N(c)$ is not constant in V', and some attracting point of P_c for some $c \in V'$ ceases to be attracting at some $c' \in V' \setminus \{c_0\}$. Thus, there is a periodic point $z \in P_{c'}$ with multiplier $e^{i2\pi p/q}$, $p, q \in \mathbb{N}$. □

Since, by (7.12), $\partial\mathcal{M}$ is the *locus* of the bifurcations of $\{P_c\}_{c \in \mathbb{C}}$, whose points are the $c \in \mathbb{C}$ for which P_c has a neutral periodic orbit, and the subset of the c for which P_c has a parabolic periodic orbit is dense in $\partial\mathcal{M}$, to prove $\dim_H \partial\mathcal{M} = 2$, it is natural to explore the dynamics in neighborhoods of parabolic fixed points and bifurcations of parabolic periodic orbits.

In addition to what was established in *CADOVA*, we use the Écalle cylinders theory of the previous section. We will get that if P_{c_0} has a periodic parabolic orbit of period k with multiplier $e^{i2\pi p/q}$, $p, q \in \mathbb{N}$ relatively prime, there are $c \in \partial\mathcal{M}$ arbitrarily close to c_0 such that P_c^k has a parabolic orbit with multiplier $e^{i2\pi\alpha_n}$, with $\alpha_n = \frac{p}{q} \pm \frac{1}{nq}$, $n \in \mathbb{N}$, in the vicinity of which there are functions f with $\dim_{Hyp}(f)$ close to 2. So, we use *secondary bifurcations* of parabolic periodic orbits.

Since, by (7.11), the Hausdorff dimension of the set of parameters $c \in \partial\mathcal{M}$ in a neighborhood of a $c_0 \in \partial\mathcal{M}$ such that there is a hyperbolic set of P_c that contains the postcritical orbit is $\geq \dim_{Hyp}(P_{c_0})$, it is convenient to construct hyperbolic sets for which it is possible to get relatively simple lower bounds of their Hausdorff dimension arbitrarily close to 2. For this, we use hyperbolic sets of the following type.

(7.13) *If $f : \mathbb{C}_\infty \to \mathbb{C}_\infty$ is a complex rational function of degree ≥ 2, $U \subset \mathbb{C}_\infty$ is simply connected with $\# \mathbb{C}_\infty \setminus U \geq 2$, $U_1, \ldots, U_N$ are simply connected disjoint open sets with closures included in U, and $n_1, \ldots, n_N \in \mathbb{N}$, $N \in \mathbb{N}$, are such that f^{n_j} are bijections of U_j onto U, then there is a minimal nonempty closed set $X_0 \subset \cup_{j=1}^N \operatorname{cl} U_j$ such that $X_0 = \cup_{j=1}^N \left(f^{n_j}{}_{|U_j}\right)^{-1}(X_0)$. X_0 is a Cantor set, $X = \cup_{j=1}^M f^{j-1}(X_0)$ with $M = \max\{n_1, \ldots, n_N\}$ is a hyperbolic set of f, and*

$$\dim_H X_0 \geq \frac{\log N}{\log(\max_{j \in \{1,\ldots,N\}} \sup_{U_j} |(f^{n_j})'|)}.$$

Proof By the Schwarz-Pick lemma,[22] $\left(f^{n_j}{}_{|U_j}\right)^{-1}: U \to U_j,\ j\in\{1,\dots,N\}$, is a contraction in the Poincaré metric d_U of U in $\cup_{j=1}^{N}\mathrm{cl}\,U_j$. Let $\mathcal{K}$ be the set of closed subsets of $\cup_{j=1}^{N}\mathrm{cl}\,U_j$ excluding $\emptyset$. With the Hausdorff distance defined with the Poincaré metric of U,

$$\delta_H(A,B)=\sup\left\{d_U(a,B),\,d_U(b,A): a\in A\,,\,b\in B\right\},\qquad A,B\subset\mathcal{K}\,,$$

$(\mathcal{K},\delta_H)$ is a complete metric space. For $A\subset\mathcal{K}$, the function $\mathcal{S}:\mathcal{K}\to\mathcal{K}$ such that $\mathcal{S}(A)=\cup_{j=1}^{N}\left(f^{n_j}{}_{|U_j}\right)^{-1}(A)$ satisfies

$$\begin{aligned}\delta_H\big(\mathcal{S}(A)\,,\mathcal{S}(A)\big)&=\delta_H\big(\cup_{j=1}^{N}\left(f^{n_j}{}_{|U_j}\right)^{-1}(A),\cup_{j=1}^{N}\left(f^{n_j}{}_{|U_j}\right)^{-1}(B)\big)\\&\le\max_{j\in\{1,\dots,N\}}\delta_H\big(\left(f^{n_j}{}_{|U_j}\right)^{-1}(A)\,,\left(f^{n_j}{}_{|U_j}\right)^{-1}(B)\big)\le\Big(\max_{j\in\{1,\dots,N\}}\lambda_j\Big)\,\delta_H(A,B)\,,\end{aligned}$$

where $\lambda_j=\sup_{U_j}\left|\left[\left(f^{n_j}\right)^{-1}\right]'\right|<1$; so, $\mathcal{S}$ is a contraction in the distance δ_H, and, therefore, it has one unique fixed point. Hence, there is $X_0\subset\mathcal{K}$ satisfying $X_0=\cup_{j=1}^{N}\left(f^{n_j}{}_{|U_j}\right)^{-1}(X_0)$. As this set is fully disconnected, it is a Cantor set. X is a closed set and

$$f^{M-1}(X_0)=f^{M}(X_0)=f^{M}\big(\cup_{j=1}^{N}\left(f^{n_j}{}_{|U_j}\right)^{-1}(X_0)\big)=\cup_{j=1}^{N}f^{M-n_j}(X_0)\subset X\,;$$

thus, $f(X)\subset X$. f is expansive in X as $f^{n_j}{}_{|U_j}$ is expansive in $X_0\cap U_j$ in the Poincaré metric of U, equivalent to the spherical metric in X_0, and

$$f^{n}=f^{k_1}\circ f^{n_{j_1}}{}_{|U_{j_1}}\circ\cdots\circ f^{n_{j_\ell}}{}_{|U_{j_\ell}}\circ f^{k_1}\text{ in }X\,,\,j_1,\dots,j_\ell\in\{1,\dots,N\}\,,k_1,k_2\in\{0,\dots,M\}$$

are such that

$$f^{k_2}\in X_0\cap\left(f^{n_{j_\ell}}{}_{|U_{j_\ell}}\right)^{-1}\circ\cdots\circ\left(f^{n_{j_1}}{}_{|U_{j_1}}\right)^{-1},\qquad k_1+k_2+\sum_{s=1}^{\ell}n_{j_s}=n\,.$$

The real function of real variable $s\mapsto\sum_{j=1}^{N}\lambda_j^s$ is continuous and strictly decreasing from $[N,0[$ to $[0,+\infty[$. There is 1 unique $s>0$ such that $\sum_{j=1}^{N}\lambda_j^s=1$. If $d>s$, $n\in\mathbb{N}$, $K_n=\{1,\dots,N\}^n$, B is a closed disk containing X_0 and $\varepsilon=2\Big[\max_{j\in\{1,\dots,N\}}\left|\left[\left(f^{n_j}\right)^{-1}\right]'(\mathrm{diam}\,B)\right|\Big]^n$, then

$$H_\varepsilon^d(X_0)\le\sum_{\mathbf{k}\in K_n}H_\varepsilon^d\big(F_{\mathbf{w}}(X_0)\big)\le\sum_{\mathbf{k}\in K_n}H_\varepsilon^d\big(F_{\mathbf{w}}(B)\big)\le\big(2\,\mathrm{diam}\,B\big)^d\Big(\sum_{j=1}^{N}\lambda_j^d\Big)^n.$$

The limit of the last term of these inequalities when $n\to+\infty$ is 0; so, $H_\varepsilon^d(X_0)=0$. Hence, $H^d(X_0)=\lim_{\varepsilon\to0}H_\varepsilon^d(X_0)=0$ and $\dim_H X_0\le s$. So, if $D=\dim_H X_0$, then

[22] See Section 5 of Chapter 11 of *CADOVA*.

$$N \max_{j\in\{1,\dots,N\}} \sup_{U_j} \left|(f^{n_j})'\right|^{-D} \leq \sum_{j=1}^{N} \inf_{U} \left|[(f^{n_j})^{-1}]'\right|^{D} \leq 1\,,$$

resulting in the lower bound of $\dim_H X_0$ in the statement. □

Denote the domain of a function f by $\boldsymbol{D_f}$, the set of holomorphic functions f from some $D_f \subset \mathbb{C}_\infty$ with $0 \in D_f$ to $\mathbb{C}_\infty$ with fixed point 0 by $\boldsymbol{\mathcal{F}}$, the set of functions $f \in \mathcal{F}$ such that $f'(0) = f''(0) = 1$ and f has an immediate parabolic basin with just 1 critical point by $\boldsymbol{\mathcal{F}_0}$, the complex function $E(z) = e^{i2\pi z}$ by $\boldsymbol{E}$, the natural projection of $\mathbb{C}$ on the cylinder $\mathbb{C}/\mathbb{Z}$ by $\boldsymbol{P}$, and the bijection of $\mathbb{C}/\mathbb{Z}$ onto $\mathbb{C}\backslash\{0\}$ such that $E = \widetilde{E} \circ P$ by $\boldsymbol{\widetilde{E}}$ (if $\zeta \in \mathbb{C}/\mathbb{Z}$ approaches the upper (resp., lower) end of $\mathbb{C}/\mathbb{Z}$, $\widetilde{E}(\zeta)$ approaches 0 (resp., ∞)).

To construct a hyperbolic set as in (7.13), we use negative orbits of the fixed point 0 of $f_0 \in \mathcal{F}_0$, the normalized Écalle mapping $\mathcal{E}_{f0}$, the associated functions φ_{f0}, $\varphi_{\mathcal{E}_{f0}}$, and the normalized Écalle mapping $\mathcal{E}_{\mathcal{E}_{f0}}$.

(7.14) *If $f_0 \in \mathcal{F}_0$ is a rational function of degree ≥ 2, then there is a negative orbit $\{z_{-j}\}_{j\in\mathbb{N}\cup\{0\}} \subset D_f$ of $z_0 = 0$ under f_0 such that $z_{-j} \neq 0$ for $j \in \mathbb{N}$ and $z_{-j} \to 0$, there is just 1 $\widetilde{w}_0 \in D_{\varphi_0}$ such that $\varphi_0(\widetilde{w}_0 - j) = z_{-j}$ and $E(\widetilde{w}_0) \notin D_{\mathcal{E}_{f0}}$.*

Proof Let v be the only critical value of the normalized Écalle mapping $\mathcal{E}_{f0}$. As[23] $0 \in J(f_0)$, there is $z_{-1} \in D_{f0}\backslash\{0\}$ outside the immediate parabolic basin of 0. Thus, there is a negative orbit $\{z_{-j}\}_{j\in\mathbb{N}\cup\{0\}} \subset D_f$ of $z_0 = 0$ under f_0 such that $z_{-j} \neq 0$ for $j \in \mathbb{N}$ and $z_{-j} \to 0$. By (7.5), there is only 1 $\widetilde{w}_0 \in D_{\varphi_0}$ such that $\varphi_0(\widetilde{w}_0 - j) = z_{-j}$. Since $z_0 = 0$ does not belong to the parabolic basin of 0 for f_0, it is $E(\widetilde{w}_0) \notin D_{\mathcal{E}_{f0}}$. □

As $\mathcal{E}_{f0} \in \mathcal{F}_0$, there is a single pair of points $\boldsymbol{\widetilde{\zeta}_0}$, $\boldsymbol{\widetilde{\zeta}'_0} \in D_{\varphi_{\mathcal{E}_{f0}}}$ corresponding as in preceding result to the negative orbits, resp., $\{w_{-j}\}$, $\{w'_{-j}\}$, with $w_0 = \mathcal{E}_{f0}(\widetilde{\zeta}_0)$, $w'_0 = \mathcal{E}_{f0}(\widetilde{\zeta}'_0)$, and $\mathcal{E}_{f0}'(\widetilde{\zeta}_0) \neq 0$, $\mathcal{E}_{f0}'(\widetilde{\zeta}'_0) \neq 0$. Denote $\boldsymbol{\widehat{\zeta}_0 = P(\widetilde{\zeta}_0)}$ and $\boldsymbol{\widehat{\zeta}'_0 = P(\widetilde{\zeta}'_0)}$.

(7.15) *If $f_0 \in \mathcal{F}_0$ is a rational function of degree ≥ 2, $h_0 = \mathcal{E}_{\mathcal{E}_{f0}}$, and $b > 0$, then there are neighborhoods $\mathcal{N}_2$ of $\mathcal{E}_{\mathcal{E}_{f0}}$, neighborhoods W, W' in $\mathbb{C}/\mathbb{Z}$ of, resp., $\widehat{\zeta}_0$, $\widehat{\zeta}'_0$ homeomorphic to an open disk and constants $C_0, C_1, C_2 > 0$ such that for $h_1 \in \mathcal{N}_2$, $\beta \in [0, 1[\times]-b, b[$, $W_0 \in \{W, W'\}$, $\widetilde{h} = \widetilde{E}^{-1} \circ h_1 \circ \widetilde{E} - \beta$, there is a sequence of disjoint sets $W_j \subset D_{\widetilde{h}} \subset \mathbb{C}/\mathbb{Z}$ homeomorphic to an open disk with $\widetilde{h}(W_j) = W_{j-1}$ and $\widetilde{h}_{|W_j}$ injective, for $j \in \mathbb{N}$,*

$$C_1 < \left|(\widetilde{h}_{|W_j})'\right| < C_2\,,\ \operatorname{diam} W_j < \tfrac{1}{2}\,,\ \operatorname{dist}(W_j, W_{j+1}) < C_0\,,\ \ j \in \mathbb{N}\cup\{0\},$$

and $\{W_j\}$ approaches one of the cylinder $\mathbb{C}/\mathbb{Z}$ ends as $j \to +\infty$.

[23] See Section 7 of Chapter 13 of *CADOVA*.

Proof Let $h = e^{i2\pi\beta}\mathcal{E}_{\mathcal{E}_{f0}}$. Denote $\widetilde{B}^+, \widetilde{B}^-$ the, resp., upper and lower, domains of $\widetilde{E}_{\mathcal{E}_{f0}}$, and $B^\pm = E(\widetilde{B}^\pm)$, analogously to (7.7). By (7.7), B^+, B^- are disjoint. So, at least one of them, say B^+, does not contain the unique critical value of h. Also by (7.7), the local inverse of h in some neighborhood of 0 can be extended to $H : B^+ \cup \{0\} \to B^+ \cup \{0\}$ such that $h \circ H = 1_{B^+\cup\{0\}}$, $H(0) = 0$ and $|H'(0)| < 1$. By the Schwarz lemma,[24] 0 is an attracting or superattracting fixed point of H, and by the Koenigs and Böttcher conjugacies,[25] there is a conformal homeomorphism L defined in a neighborhood of 0 with $L(0) = 0$ and $L'(0) = 1$ such that $L \circ H = H'(0)L$.

Consequently, there are $y_0 \in \mathbb{R}$, $C_1'', C_2'' > 0$, and a holomorphic function $\widetilde{L} : Y \overset{def}{=} \{\zeta \in \mathbb{C}/\mathbb{Z} : \mathcal{I}m\,\zeta > y_0\} \to \mathbb{C}/\mathbb{Z}$ with $\widetilde{H} \overset{def}{=} \widetilde{E}^{-1} \circ H \circ \widetilde{E}$ defined in Y and satisfying in this set $\mathcal{I}m\big(\widetilde{H} - 1_Y\big) < C_1''$ and $\widetilde{L} \circ \widetilde{H} = \widetilde{L} + a$, where $a = \frac{1}{i2\pi} \log H'(0)$, $\mathcal{I}m\, a > 0$, and $\frac{1}{C_2''} < \big|\widetilde{L}'\big| < C_2''$ (if B^- does not contain the critical value, it is similar with $\mathcal{I}m$ replaced by $-\mathcal{I}m$). Denote $\widehat{\zeta}_0 = P\big(\widetilde{\zeta}_0\big)$ and $\widehat{\zeta}_0' = P\big(\widetilde{\zeta}_0'\big)$. Since at least one of $\widetilde{\zeta}_0, \widetilde{\zeta}_0'$, say $\widetilde{\zeta}_0$, is not a critical value, we can choose $\widehat{\zeta}_1 \in \widetilde{h}^{-1}\big(\widehat{\zeta}_0\big) \cap B^+$ and denote $\widehat{\zeta}_j = H^{j-1}\big(\widehat{\zeta}_1\big)$ for $j \in \mathbb{N}\setminus\{1\}$, so there is $j_0 \in \mathbb{N}$ such that $\zeta_{j_0} \in Y$. There is a connected neighborhood W_0 of $\widetilde{\zeta}_0$ in $\mathbb{C}/\mathbb{Z}$ disjoint of Y such that successive preimages $W_j = \widetilde{h}^{-1}(W_{j-1})$ for $j \in \{1, \dots, j_0\}$ have disjoint closures contained in B^+ without critical points of $\widetilde{h}$, $W_{j_0} \subset Y$, and $\widetilde{h}$ is a bijection from W_j onto W_{j-1} for $j \in \{1, \dots, j_0\}$. These properties hold under perturbations, i.e., for $h = e^{i2\pi\beta_1}h_1$ with h_1 and β_1 in neighborhoods of, resp., $\mathcal{E}_{\mathcal{E}_{f0}}$ and β, with uniform constants in such neighborhood. As the cylinder $\big(\mathbb{R} \times [-b, b]\big)/\mathbb{Z}$ is a compact set, there is a neighborhood $\mathcal{N}_2$ of $\mathcal{E}_{\mathcal{E}_{f0}}$ and connected neighborhoods W, W' of, resp., $\widehat{\zeta}_0, \widehat{\zeta}_0'$ in $\mathbb{C}/\mathbb{Z}$ such that for $h_1 \in \mathcal{N}_2$ and $\beta \in [0, 1[\times]-b, b[$ the properties above hold with uniform constants and $W_0 \in \{W, W'\}$. There are $C_0, C_1, C_2, C_3 > 0$ such that

$$C_1 < \big|\big(\widetilde{h}_{|W_j}\big)'\big| < C_2,\ \operatorname{diam} W_j < C_3,\ \operatorname{dist}(W_j, W_{j+1}) < C_0,\ \ j \in \{0, \dots, j_0\}.$$

Since $\big|\big(\widetilde{h}^{j-j_0}\big)'(\zeta)\big| = \big|\widetilde{L}'(\zeta)\big|\,\big|\widetilde{L}'\big(\widetilde{h}^{j-j_0}(\zeta)\big)\big|^{-1}$, with $W_j = \widetilde{H}^{j-j_0}(W_{j_0})$ for $j \in \mathbb{N}$, $j > j_0$, as 0 is an attracting or superattracting fixed point of H, for $j \in \mathbb{N}$ large $\{W_j\}$ tends to the union of the cylinder $\mathbb{C}/\mathbb{Z}$ ends as $j \to +\infty$, but because $\operatorname{dist}(W_j, W_{j+1}) < C_0$, it tends to one of them, W, W' can be chosen so that $\operatorname{diam} W_j < \frac{1}{2}$ and we have analogous estimates with possibly different constants, uniformly for $j \in \mathbb{N} \cup \{0\}$. □

(7.16) *Under the conditions of the hypothesis and with the sequence* $\{W_j\} \subset \mathbb{C}/\mathbb{Z}$ *of* (7.15), *there is* $\gamma > 0$ *only dependent on* C_0 *such that for* $\eta > 0$ *large, there are neighborhoods* $\mathcal{N}_{f_0}(\eta)$ *of* f_0 *and* $\mathcal{N}_{\mathcal{E}_{f0}}$ *of* $\mathcal{E}_{f_0}$ *and sets*

(continued)

[24] See Section 3 of Chapter 10 of *CADOVA*.

[25] See section 6 of Chapter 13 of *CADOVA*.

$U_1, \ldots, U_N$ homeomorphic to an open disk, with $N \geq \gamma\eta(e^{2\pi\eta})^2$, such that for $f \in \mathcal{N}_{f_0}(\eta) \cap \mathcal{F}_1$ and $g \in \mathcal{N}_{\mathcal{E}_{f_0}}(\eta) \cap \mathcal{F}_1$,

$$\operatorname{cl} U_k \subset \varphi_f\big(B_{|\eta|/4}(i\eta)\big) \subset B_{|\eta|/2}(i\eta)\,,$$

$$V_k = E \circ \big(\varphi_{f|B_{|\eta|/4}(i\eta)}\big)^{-1}(U_k) \subset \varphi_g\big(B_{1/4}(i\widetilde{\eta})\big) \subset B_{1/2}(i\widetilde{\eta})\,,$$

$$W_{j(k)} = P \circ \big(\varphi_{g|B_{1/4}(i\widetilde{\eta})}\big)^{-1}(V_k)\,, \text{ for some } j(k) \in \mathbb{N}\,,$$

where $\widetilde{\eta}$ is $e^{i2\pi\eta}$ or $-e^{i2\pi\eta}$ according to $\{W_j\}$ approaches the upper or the lower end of $\mathbb{C}/\mathbb{Z}$ when $j \to +\infty$. With

$$U_k^* = \rho(U_k)\,,\ \widetilde{V}_k = \big(\varphi_{f|B_{|\eta|/4}(i\eta)}\big)^{-1}(U_k)\,,\quad V_k^* = \rho(V_k)\,,\qquad k \in \{1, \ldots, N\}\,,$$

we have the composition of functions (Fig. 7.13)

$$U_k \xrightarrow{\ \rho\ } U_k^* \xrightarrow{(\varphi^*_{f|B_{|\eta|/4}(i\eta)})^{-1}} \widetilde{V}_k \xrightarrow{\ E^*\ } V_k^* \xrightarrow{P\circ(\varphi^*_{g|B_{1/4}(i\widetilde{\eta})})^{-1}} W_{j(k)}$$

and the estimates for the, resp., derivatives:

$$\tfrac{1}{4}\eta^2 < |(\rho_{|U_k})'| < \tfrac{9}{4}\eta^2\,,\qquad \tfrac{1}{2} < \big|[(\varphi^*_{f|B_{|\eta|/4}(i\eta)})^{-1}]'\big| < 2\,,$$

$$\pi e^{2\pi\eta} < \big|(E^*_{|\widetilde{V}_k})'\big| < 3\pi e^{2\pi\eta}\,,\qquad \tfrac{1}{2} < \big|[P\circ(\varphi^*_{f|B_{|\eta|/4}(i\eta)})^{-1}]'\big| < 2\,.$$

There is a neighborhood $\mathcal{N}_4(\eta)$ of f_0 such that for $f \in \mathcal{N}_4(\eta)$, there are an open set $\mathcal{U}' \subset \mathcal{U}$ and $C, C' > 0$ independent of η with $f^\ell\colon \mathcal{U}' \to \mathcal{U}$ bijective and $C' \leq \eta^{-(1+\frac{1}{\nu})}|(f^\ell)'| \leq C$ in $\mathcal{U}'$.

Proof (7.3) and (7.4) also apply to $g_0 = \mathcal{E}_{\mathcal{E}_{f_0}}$ instead of f_0 and the corresponding functions, resp., g and f. Let $\rho(z) = -\frac{1}{z}$, $\varphi_f^* = \rho \circ \varphi_f$, $\varphi_g^* = \rho \circ \varphi_g$, $E^* = \rho \circ E$, and $\widetilde{\eta} = e^{i2\pi\eta}$ or $\widetilde{\eta} = -e^{i2\pi\eta}$ according to $\{W_j\}$ tends toward the upper or lower end of $\mathbb{C}/\mathbb{Z}$. As the area of $B_{1/4}(i\widetilde{\eta})$ is $\frac{\pi}{16}(\widetilde{\eta})^2$, there is $\gamma' > 0$ such that $B_{1/4}(i\widetilde{\eta})$ contains a number of connected components of preimages of W_j, $j \in \mathbb{N}$, by P greater than $\gamma'(\widetilde{\eta})^2$. By (7.4.5), φ_g maps injectively these components to subsets of $B_{|\widetilde{\eta}|/2}(i\widetilde{\eta})$. The connected components of the preimage of $B_{1/2}(i\widetilde{\eta})$ by E^* are contained in $\mathcal{R} = \big]n, n+1\big[\times \big]\eta - \frac{\log 2}{2\pi}, \eta + \frac{\log(3/2)}{2\pi}\big[$ for some $n \in \mathbb{Z}$. For $\eta > 0$ large, $B_{|\eta|/4}(i\eta)$ contains more than $\frac{1}{3}\eta$ such connected components. φ_f^* maps injectively $B_{|\eta|/4}(i\eta)$ to a subset of $B_{|\eta|/2}(i\eta)$. So, for $\eta > 0$ large and $j \in \mathbb{N}$, with $\gamma = \frac{1}{3}\gamma'$, $B_{|\eta|/4}(i\eta)$ contains more than $\frac{1}{3}\eta\,\gamma'(\eta')^2 = \gamma\,\eta(\widetilde{\eta})^2$ connected components of preimages of W_j by

$$P \circ \big(\varphi^*_{g|B_{1/4}(i\widetilde{\eta})}\big)^{-1} \circ E^* \circ \big(\varphi^*_{f|B_{|\eta|/4}(i\eta)}\big)^{-1} = P \circ \big(\varphi_{g|B_{1/4}(i\widetilde{\eta})}\big)^{-1} \circ E \circ \big(\varphi^*_{f|B_{|\eta|/4}(i\eta)}\big)^{-1}.$$

It is left as an exercise to obtain the estimates for the derivatives, $\left|(E^*_{|\widetilde{V}_k})'\right|$ for $\widetilde{V}_k \subset \mathcal{R}$.

For the last sentence in the statement, consider $\{z_{-j}\}_{j\in\mathbb{N}\cup 0}$, $\widetilde{w}_0$ as in (7.14), $w_0 = e^{i2\pi\widetilde{w}_0}$. There is $\ell \in \mathbb{N}$ such that for $j \geq \ell$, the z_{-j} are not critical points of f_0. Denote E^{-1} the inverse of the restriction of E to a neighborhood of w_0 where E is injective such that $E^{-1}(w_0) = \widetilde{w}_0 - \ell$, and by P^{-1}, the inverse of the restriction of P to a union of neighborhoods of $P(\widetilde{\zeta}_0)$ and of $P(\widetilde{\zeta}_0')$ where P is injective. By (7.5), $\varphi_0'(\widetilde{w}_0 - \ell) \neq 0$, $\varphi_{g_0}'(\widetilde{\zeta}_0) \neq 0$, $\varphi_{g_0}'(\widetilde{\zeta}_0') \neq 0$. If needed, replace W, W' by subsets so that there are neighborhoods $U, \mathcal{N}_3(f_0), \mathcal{N}_3(g_0)$ of, resp., $z_{-\ell}$, f_0, g_0, and constants C_2, C_2' such that for $f \in \mathcal{N}_3(f_0) \cap \mathcal{F}_1$ and $g \in \mathcal{N}_3(g_0) \cap \mathcal{F}_1$, $\varphi_f \circ E^{-1} \circ \varphi_g \circ P^{-1}$ is defined and injective in W and in W', with the image of each one of these sets containing U and $C_2 < \left|\left(\varphi_f \circ E^{-1} \circ \varphi_g \circ P^{-1}\right)'\right| < C_2'$ in $W \cup W'$. We get the validity of the last sentence in the statement with $\nu = \deg_{z_{-\ell}} f_0^\ell$ and $\mathcal{U} = \rho\left(B_{|\eta|/2}(i\eta)\right)$. □

The following result ensures that complex rational functions with Julia sets with Hausdorff dimension close to 2 can be obtained through secondary bifurcation of orbits at parabolic fixed points, in the sense of bifurcating from a sequence of functions with multipliers $e^{i2\pi\alpha_n}$ which bifurcate from f_0, with $\alpha_n = \frac{p}{q} \pm \frac{1}{nq}$.

(7.17) *If $f_0 : \mathbb{C}_\infty \to \mathbb{C}_\infty$ is a rational function of degree $d \geq 2$ with parabolic fixed point ζ with multiplier $e^{i2\pi p/q}$, $p, q \in \mathbb{N}$ are relatively prime and the immediate parabolic basin of ζ has only one critical point of f_0, then for $\varepsilon > 0$, $b > 0$ there are neighborhoods $\mathcal{N}$ of f_0 in the space of complex rational functions of degree d, a neighborhood V of ζ in $\mathbb{C}_\infty$ and $N_1, N_2 \in \mathbb{N}$ such that if $f \in \mathcal{N}$ has a fixed point in V with Multiplier $e^{i2\pi\alpha}$ and $\alpha = \frac{p}{q} \pm \frac{1}{q}\left(\frac{\beta + a_2}{a_1\beta + a_1 a_2 \pm 1}\right)$, $a_1, a_2 \in \mathbb{N}$, $a_1 \geq N_1$, $a_2 \geq N_2$, $\beta \in \mathbb{C}$, $0 \leq \mathcal{R}e\,\beta < 1$, $|\mathcal{I}m\,\beta| \leq b$, and $\dim_{Hyp}(f) > 2 - \varepsilon$.*

Proof It is first proved for multiplier 1, i.e., for $q = 1$ and $p = 0$.

For $f_0 \in \mathcal{F}_0$ a rational function, $b > 0$ and $\eta > 0$ large, for each holomorphic function close of f_0, there is a conjugacy with a function in $\mathcal{F}$ close to f_0 by a translation close to identity. So, just consider functions $f \in \mathcal{F}$ close to f_0. If $\alpha_f = \frac{1}{a_1 - 1/(a_2+\beta)}$ with $a_1, a_2 \in \mathbb{N}$ large and $\beta \in [0, 1[\times[-b, b]$ (different signs in the expression for α_f can be treated analogously using complex conjugates or with the analogous process for the lower end of the cylinder $\mathbb{C}/\mathbb{Z}$ instead of the upper end). If f is close to f_0 and a_1 is large, then $|\mathrm{Arg}\,\alpha_f| < \frac{\pi}{4}$, $\mathcal{R}_f$ (see (7.4)) is defined and $e^{i2\pi/\alpha_f}\mathcal{R}_f$ is close to $g_0 = \mathcal{E}_{\mathcal{E}_{f_0}}$, i.e., $\widehat{R}_f + \frac{1}{\alpha_f}$ is close to $\widehat{g}_0 = \widehat{\mathcal{E}}_{f_0}$, with the notation following the proof of (7.4). For large a_2, $\mathcal{R}_f$ is close to g_0, $\left|\mathrm{Arg}\,\alpha_{\mathcal{R}_f}\right| = \left|\mathrm{Arg}\,\frac{1}{a_2+\beta}\right| < \frac{\pi}{4}$, $\mathcal{R}_g$ is well defined and $h_1 = e^{i2\pi/\alpha_g}\mathcal{R}_g = e^{i2\pi\beta}h$ is close to $h_0 = \mathcal{E}_{g_0}$ (by (7.4.2), $\alpha_g = -\frac{1}{\alpha_f} = \frac{1}{a_2+\beta}$ (mod $\mathbb{Z}$)). Hence, for f close to f_0 and large a_1, a_2,

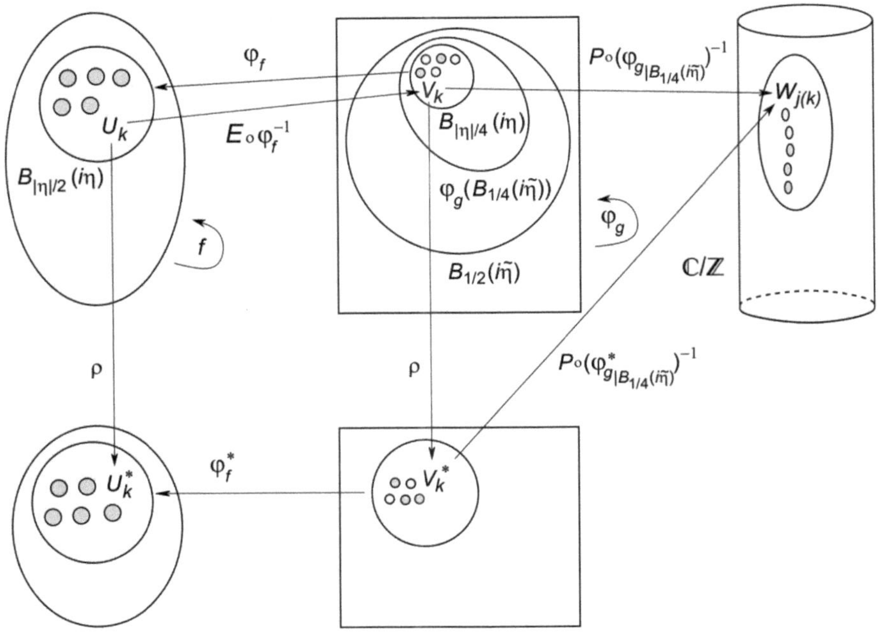

Fig. 7.13 Écalle cylinder in the proof of (7.16)

$$f \in \mathcal{N}_{f_0}(\eta) \cap \mathcal{N}_3(f_0) \cap \mathcal{N}_4(\eta)\,, \quad g \in \mathcal{N}_{g_0}(\eta') \cap \mathcal{N}_3(g_0)\,, \quad h_1 \in \mathcal{N}_2\,.$$

The arguments of (7.15) can be applied to $\widetilde{h} = \widetilde{E}^{-1} \circ h \circ \widetilde{E}$. With $U_1, \dots, U_N$ as in (7.16) and ℓ, U as in the two paragraphs following the proof of that result, the chain of function compositions obtained in (7.16) can be extended to the right as in

$$W_{j(k)} \xrightarrow{\ \widetilde{h}^{j(k)}\ } W_0 \xrightarrow{\ \varphi_{\mathcal{E}_{f_0}} \circ P^{-1}\ } V \xrightarrow{\ \varphi_f \circ E^{-1}\ } U' \supset U \xrightarrow{\ f^\ell\ } f^\ell(U)$$

where $V = \varphi_{\mathcal{E}_{f_0}} \circ P^{-1}(W_0)$ and $U' = \varphi_f \circ E^{-1}(V)$ (Fig. 7.13). The functions of the complete chain of compositions are injective except the last one, f^ℓ, although, by (7.16), $f^\ell_{|\mathcal{U}'}$ is a bijection onto $\mathcal{U}$ and $\mathcal{U}' \subset U$. For each $k \in \{1, \dots, N\}$, denote the composition of all functions in the complete chain of compositions by Ψ_k and $\mathcal{U}_k = \Psi_k^{-1}(\mathcal{U}) \subset U_k$.

Since for $\widetilde{\eta} > 0$ large, $\left|\arg\left(z' + \frac{1}{2\alpha_{\mathcal{E}_f}} - z\right)\right| < \frac{2\pi}{3}$ for $z \in B_{1/4}(i\widetilde{\eta})$, $z' \in P^{-1}(W \cup W')$, by (7.4.3), for $k \in \{1, \dots, N\}$ there are $m_k, n_k \in \mathbb{N}$ and

$$(\mathcal{E}_f)^{m_k} = \varphi_{\mathcal{E}_f} \circ E^{-1} \circ \widetilde{h}^{j(k)} \circ E \circ (\varphi_{\mathcal{E}_f | B_{1/4}(i\widetilde{\eta})})^{-1}\,, \quad \text{in } V_k\,,$$

$$f^{n_k} = \varphi_f \circ E^{-1} \circ (\mathcal{E}_f)^{m_k} \circ E \circ (\varphi_{f | B_{|\eta|/4}(i\eta)})^{-1}\,, \quad \text{in } U_k\,.$$

Therefore, $f, \mathcal{U}, \mathcal{U}_k$, $k \in \{1, \dots, N\}$, satisfy the conditions in (7.13), and with this result, we obtain a hyperbolic set X_f. By (7.16),

$$N \geq \gamma\eta(e^{2\pi\eta})^2, \; |(f^{n_k}_{|\mathcal{U}_k})'| \leq C\eta^{1+\frac{1}{\nu}} e^{2\pi\eta},$$

$$\dim_H X_f \geq \frac{\log(4\pi\gamma\eta^2)}{\log(2\pi\, C\,\eta \log \eta^{1+1/\nu})} = \frac{\log\gamma + \log\eta + 4\pi\eta}{\log C + (1+1/\nu)\log\eta + 2\pi\eta}.$$

When $\eta \to +\infty$, the numerator and the denominator of the fraction in the last term in this inequality are dominated by their, resp., last terms, and, therefore, the limit of the fraction is $\lim\limits_{\eta\to+\infty} \frac{4\pi\eta}{2\pi\eta} = 2$, and $\dim_H X_f \to 2$ when $\eta \to +\infty$, concluding the proof for the case of multiplier 1, since

$$\dim_{Hyp}(f) = \sup\bigl\{\dim_H(X) \colon X \subset \mathbb{C}_\infty \text{ is a Hyperbolic set for } f\bigr\} \leq 2.$$

For multiplier $\neq 1$, modify φ_f as indicated at the end of preceding paragraph in the modification of (7.4), replace $\rho_{|U_k}$ by $-\frac{1}{q z^q}$ and the estimate of $|(\rho_{|U_k})'|$ in (7.16) by multiplying both sides by possibly different positive constants and the estimate $|(f^{n_k}_{|\mathcal{U}_k})'| \leq C\eta^{1+\frac{1}{\nu}} e^{2\pi\eta}$ by $|(f^{n_k}_{|\mathcal{U}_k})'| \leq C\eta^{q+\frac{1}{\nu}} e^{2\pi\eta}$, and $\dim_{Hyp} X_f = 2$ when $\eta \to +\infty$. □

(7.18) *If $U \subset \mathbb{C}$ is a neighborhood of $c \in \partial\mathcal{M}$, then*

$$\dim_H \partial\mathcal{M} \cap U \geq \dim_H\bigl\{c \in \partial\mathcal{M} \cap U \colon 0 \text{ is a nonwandering point of } P_c\bigr\} \geq \dim_{Hyp} P_c.$$

If a periodic orbit of P_c is parabolic, then there is a sequence $\{c_n\} \subset \partial\mathcal{M}$ such that $c_n \to c$ and $\dim_{Hyp} P_{c_n} \to 2$ when $n \to +\infty$.

Proof The first sentence follows directly by (7.11) and by $\partial\mathcal{M}$ to be the *locus* of the bifurcations of $\{P_c\}_{c\in\mathbb{C}}$ (7.12).

If P_{c_0} has a parabolic periodic orbit with minimal period k, $f_0 = P^k_{c_0}$ has a parabolic fixed point, and as P_{c_0} has just 1 critical point in $\mathbb{C}$, f_0 satisfies the hypothesis of (7.17). Thus, there is a perturbation $P_{c'}$ of P_{c_0} in $\{P_c\}_{c\in\mathbb{C}}$ with c' close of c_0 such that $P^k_{c'}$ has a fixed point with multiplier $e^{i2\pi\alpha}$ with α as in (7.17). If, in the formula for α, $\beta \in \mathbb{R}$, such perturbation $P_{c'}$ is not Julia-Stable in $\{P_c\}_{c\in\mathbb{C}}$. So, by (7.17), there is a sequence $\{c_n\} \subset \partial\mathcal{M}$ with the properties in the statement.

□

(7.19) $\dim_H \partial\mathcal{M} = 2$, *and even* $\dim_H(\partial\mathcal{M} \cap U) = 2$ *for all open set $U \subset \mathbb{C}$ intersecting $\partial\mathcal{M}$.*

Proof By (7.12), the set of $c \in \mathbb{C}$ such that P_c has some parabolic periodic orbit is dense in $\partial\mathcal{M}$. Thus, (7.18) implies the result. □

7.5 Generic Dimension of the Julia Set of $z^2 + c$

A property is **generic** or holds **generically** in a metric space X if it holds in a **residual** set in X, i.e., an intersection of countable dense open subsets of X. By the Baire theorem,[26] residual sets in a complete metric space are dense in the space, so the generic properties in the following two results hold in dense subsets.

(7.20) $\dim_H J_c = 2$ *generically for* $c \in \partial\mathcal{M}$.

Proof If $\mathcal{R}_n = \{c \in \partial\mathcal{M} : \dim_{Hyp} P_c > 2 - \frac{1}{n}\}$ for $n \in \mathbb{N}$, then

$$\mathcal{R} = \cap_{n=0}^{\infty} \mathcal{R}_n = \{c \in \partial\mathcal{M} : \dim_{Hyp} P_c = 2\} \subset \{c \in \partial\mathcal{M} : \dim_H J_c = 2\}.$$

By (7.10.5), each $\mathcal{R}_n$ is open relative to $\partial\mathcal{M}$, and, by (7.18), the set of $c \in \mathbb{C}$ such that P_c has a parabolic periodic orbit is dense in $\partial\mathcal{M}$. So, $\mathcal{R}$ is residual in $\partial\mathcal{M}$. □

(7.21) *There is a residual subset* $\mathcal{R}'$ *of* $\mathbb{R}/\mathbb{Z}$ *such that if* P_c *has a periodic point with multiplier* $e^{i2\pi\alpha}$ *and* $\alpha \in \mathcal{R}'$, *then* $\dim_H J_c = 2$.

Proof If W is a hyperbolic component of $\mathcal{M}$, i.e., a component of the set of $c \in \mathcal{M}$ such that P_c has an attracting periodic point, by (6.1), cl W is homeomorphic to cl B_1 with the restriction of a homeomorphism of W onto B_1 the conformal homeomorphism with value in $c \in W$ equal to the multiplier of a non-repelling periodic orbit of P_c. Thus, the set of $c \in \partial\mathcal{M}$ such that P_c has a parabolic periodic orbit is dense in $\partial\mathcal{M}$. As in the proofs of the second statement of (7.18) and (7.20), there is a residual subset $\mathcal{R}'$ of $\mathbb{R}/\mathbb{Z}$ such that if $c \in \partial W$ and P_c has a periodic point with multiplier $e^{i2\pi\alpha}$, $\alpha \in \mathcal{R}'$, and dim $J_c = 2$. The countability of the hyperbolic components of $\mathcal{M}$ ends the proof. □

By (7.17), we also have $\sup_{c \in W_0} \dim_H J_c = \sup_{\mathbb{C} \setminus \mathcal{M}} \dim_H J_c = 2$, where $W_0 = \{c \in \mathbb{C} : P_c \text{ has an attracting fixed point}\}$. Similar results can be obtained for other families of complex rational functions with only one critical point in the

[26] See Appendix A of *CADOVA*.

parabolic basin, e.g., $f_c(z) = z^3 + cz^2$ or $g_c(z) = \frac{z^2+c}{z^2-1}$ with $c \in \mathbb{C}$. Besides, the results obtained have immediate consequences for the continuity of functions $c \mapsto \dim_K J_c$ and $c \mapsto \dim_{Hyp} P_c$. By Julia-stability, the two functions are continuous at $c_0 \in \mathbb{C} \backslash \partial \mathcal{M}$. If $c_0 \in \partial \mathcal{M}$ and $\dim_{Hyp} P_{c_0} = 2$, then $\dim_H J_{c_0} = 2$ and, by (7.10.2) and (7.10.5), also the two functions are continuous on c_0. If $\dim_H J_{c_0} < 2$, then, by (7.10.2), $\dim_{Hyp} P_{c_0} < 2$, and, by the proof of (7.17), the two functions are discontinuous at c_0. If $\dim_H J_{c_0} = 2$ and $\dim_{Hyp} P_{c_0} < 2$, by the proof of (7.17), the function $c \mapsto \dim_{Hyp} P_c$ is discontinuous at c_0, and with the arguments used, it remains unclear whether the continuity of $c \mapsto \dim_H J_c$ in c_0 and whether that is possible.

Chapter 8
Postcritically Finite Rational Functions and Combinatorial Equivalence

8.1 Introduction

Postcritically finite rational functions are important, because many aspects of their dynamics can be derived from information about postcritical orbits, e.g., in Chap. 5, it was seen that an element of a holomorphic family of rational functions of degree ≥ 2 is structurally stable if and only if it is postcritically stable, and it is Julia-stable if and only the postcritical orbits of the functions in the family in a neighborhood of the parameter giving the function are normal families.

Notice that all periodic components of the Fatou set of a complex rational function of degree ≥ 2 are related to postcritical points, since the attracting basins of attracting, superattracting or parabolic periodic orbits have critical points, and the boundaries of Siegel disks or Herman rings, as well as irrational neutral periodic orbits in the Julia set (Cremer points), are contained in the closure of the set of postcritical points, and each connected component of the complement of this set in $\mathbb{C}_\infty$ is hyperbolic.[1] One consequence is that postcritically finite rational functions do not have Siegel disks, Herman rings, or Cremer points, as these are only possible with infinitely many postcritical points.

The usefulness of postcritically finite rational functions is strengthened observing that, excluding the set of Lattès functions[2] (empty for polynomials), the set of postcritically finite rational functions is small. In fact, as attracting basins of periodic orbits contain critical points, if there would exist a non-postcritical attracting periodic orbit, there would exist an infinite postcritical orbit. Thus, postcritically finite rational functions do not have non-postcritical attracting orbits. For example,

[1] See Sections 6, 9, 11, 12, and 13 of Chapter 13 of *CADOVA*.

[2] These are rational functions for which there are holomorphic conjugacies from the surface of a complex torus with affine functions in $\mathbb{C}_\infty$. S. Lattès gave examples in 1918, but Ernst Schröder (1841–1902) had described examples in 1871, almost half a century earlier, and Lucjan Böttcher (1872–1937) cited them in 1904. See an example in Section 5 of Chapter 13 of *CADOVA*.

L. T. Magalhães, *Quasiconformal Mappings in the Plane and Complex Dynamics*,
https://doi.org/10.1007/978-3-031-80115-0_8

for a quadratic complex polynomial $P_c(z) = z^2 + c$ to be postcritically finite, as there must exist $m, n \in \mathbb{N}$ such that $P_c^m(0) = P_c^n(0)$, the parameter c must be an integer number and, as it must belong to the Mandelbrot set, which is included in the disk with radius 2 and center 0, c must belong to a finite set[3] of integer numbers and determine that this set is $\{0, -1, -2\}$, and for every quadratic polynomial P_c with $c \in \mathbb{C}$, there is a conjugacy with one of those 3.

The complex rational functions are, as functions in $\mathbb{C}_\infty$, branched coverings of $\mathbb{C}_\infty$ with functions defined in $\mathbb{C}_\infty$. In 1985, W. Thurston defined combinatorial equivalence of postcritically finite branched coverings if one of them is topologically conjugated with a deformation of the other by isotopy, (i.e., homotopy of homeomorphisms) of postcritically finite branched coverings with the images of fixed postcritical points. This relation is weaker than topological conjugacy, but it gives useful information on the dynamics of the functions associated with the branched coverings, since the global dynamics of one of the functions is determined by that of the other in terms of a topological semiconjugacy if there is a bijection of periodic critical points of both functions respecting their minimal periods.

The notion of orbifold of Riemann surface considered in *CADOVA* in connection with complex subhyperbolic rational functions,[4] in particular with canonical orbifolds associated with postcritically finite rational functions, extends to branched coverings, allowing the use of orbifolds in this context.

In 1982, W. Thurston came up with a topological characterization of postcritically finite rational functions not Lattès functions relating them bijectively with the homotopy classes of branched coverings $\left(\mathbb{C}_\infty, f\right)$ of $\mathbb{C}_\infty$ with a finite set of branching points provided they do not satisfy a condition—called Thurston obstructions—that certain matrices associated with finite sets of disjoint Jordan curves have spectral radius less than 1. This result was presented in 1983 by W. Thurston at a conference, and the proof was published only 10 years later by A. Douady and J. Hubbard[5] based on notes they took from W. Thurston presentations at that conference. The proof in this chapter is of a slightly more general result and was published in 2011 by X. Buff, G. Cui, and L. Tan.[6] Thus, to construct rational functions with given combinatorial properties, one can construct a branched covering with these properties (what is easier) and then check if it avoids Thurston obstructions (what may be difficult).

There is a type of Thurston obstructions relatively simple to check, called Levy cycles,[7] analyzed by S. Levy in his PhD thesis in 1985 where he proved that for a

---

[3] The proof uses the Northcott theorem of Number Theory, established by Douglas Northcott (1916–2005) in 1948.

[4] See Section 14 of Chapter 13 of *CADOVA*.

[5] Douady, A., Hubbard, J., A proof of Thurston's topological characterization of rational functions, *Acta Math.* **171** (1993), 263–297.

[6] Buff, X., Cui, G. Tan L., Teichmüller spaces and holomorphic dynamics, *in Handbook of Teichmüller Theory,* Vol. IV, ed. Athanase Papadopoulos, IRMA Lectures in Mathematics and Theoretical Physics, European Mathematical Society, 2011, 717–756. Papadopoulos, Athanase (1957–).

[7] Levy, Silvio (1959–).

branched covering $(\mathbb{C}_\infty, \Omega)$ of $\mathbb{C}_\infty$ marked by a finite set Ω to be associated with a rational function f of degree ≥ 2 with Ω the set of critical values of f there cannot exist Levy cycles. The converse does not hold, but for topological polynomials, which are functions f from $\mathbb{C}_\infty$ to $\mathbb{C}_\infty$ such that $(\mathbb{C}_\infty, f)$ is a branched covering of $\mathbb{C}_\infty$ and the preimage of ∞ by f is ∞, the absence of Levy cycles is enough for the absence of Thurston obstructions, and for f to be a rational function. A particularly simple condition guaranteeing the absence of Levy cycles for postcritically finite topological polynomials, also given in S. Levy PhD thesis, is that all orbits of critical points be eventually periodic with a periodic orbit containing a critical point.

As the characterization of postcritically finite rational functions not Lattès functions just mentioned is under the condition of nonexistence of Thurston obstructions, this is a sufficient (but not necessary) condition for such characterization. An interesting necessary and sufficient condition was obtained only in 2020 by D. P. Thurston.[8]

C. McMullen observed that for marked branched coverings associated with rational functions of degree > 1 as in the two preceding paragraphs, the spectral radii of the matrices considered for defining the Thurston obstructions are always ≤ 1. So, the condition of absence of Thurston obstructions may fail only marginally, in the sense that there can be spectral radii equal to 1 but not greater. C. McMullen proved that in this case, if the function is not a Lattès function, there are rotation domains (i.e., Siegel disks or Herman rings), using meromorphic quadratic differentials,[9] as they are particularly useful for foliations of sets of Jordan curves, such as those of rotation domains. For this reason, a subsection of this chapter is on meromorphic quadratic differentials with almost all trajectories closed, with the main results obtained by J. Jenkins in 1957, H. Renelt in 1976 and K. Strebel in 1967 and 1976.

The chapter ends with a funny interesting application of the Thurston theorem to harmonic polynomials of complex variable of degree $n > 1$, including that the number of zeros of such polynomials is upper bounded by $3n-2$, as proved with Complex Dynamics by D. Khavinson and G. Świątek in 2003. The optimality of this upper bound was proved in 2008 by L. Geyer. The upper bound $5n-5$ for the analogous problem with harmonic rational functions of complex variable of degree $n > 1$ was obtained in 2006 by D. Khavinson and G. Neumann, also with Complex Dynamics, which, by a contribution in 2003 of S.H. Rhie with a method of small perturbations when studying gravitational lensing,[10] was already known

[8] Thurston, D.P., A positive characterization of rational maps, *Annals of Math.* **192** (2020), 1–46. The necessary and sufficient condition is specified in terms of extreme length and elastic graphs.

[9] Quadratic differentials were introduced in 1938 by Menahem Schiffer (1911–1997) in the context of univalent holomorphic functions and, independently, by O. Teichmüller in the context of Teichmüller spaces, and had important developments by Albert Charles Schaeffer (1907–1957) and Donald Spencer (1912–2001) in 1950, J. Jenkins and D. Spencer in 1951, J. Jenkins in 1957 and K. Strebel in 1967 and 1973–1976.

[10] If the light source is perfectly aligned with a single point mass from the observer, the gravitational lensing effect results in the observation of a point light source as a circle centered

that could not be smaller. The relationship of the number of zeros and the degree n of harmonic polynomials of complex variable is radically different from that of polynomials of one complex variable. The former can have more than n or even infinite zeros, contrasting[11] with polynomials of one complex variable, which, by the fundamental theorem of algebra[12] and polynomial division, if not constant, have

at the observation axis, called *Einstein ring* (Albert Einstein (1879–1955) received the Nobel Prize in Physics in 1921 for his services to Theoretical Physics, and especially for his discovery of the law of the photoelectric effect), as was predicted in 1924 by Orest Chwolson (or Khvolson) (1852–1934) but was observed only in 1989. However, the images of a gravitationally lensed light source depend on the mass distribution between the light source and the observer and the masses relative velocities, and often a small number of bright images corresponding to a single light source are observed instead of a circle. Gravitational lensing can magnify images, allowing to watch stars otherwise unobservable. The first observation of this type was in 1979 by Dennis Walsh (1933–2005), Robert Carswell and Ray Weymann (1934-) of a *quasar* with two images, called a *twin quasar*. The most distant stars watched so far were detected with gravitational lensing with the *Hubble Space Telescope*, Icarus (14 billion light-years from Earth) in 2018 and Earendel (28 billion light-years from Earth) in 2022, the latter was also seen in July 2022 with the *James Webb Space Telescope* in the month it entered service. The first calculations of gravitational lensing with Newtonian Mechanics were by Johann Soldner (1776–1833) in 1804, with independent work of Henry Cavendish (1731–1810) in 1804–1810, and precursor work of I. Newton in 1721 and John Mitchell (1724–1793) in 1783. A. Einstein calculated the effect with Special Relativity in 1911 and with General Relativity in 1915, obtaining for the latter a deflection angle about twice the calculated with Newtonian Mechanics. Observations in a solar eclipse of 1919 at Príncipe Island in São Tomé e Princípe and at Sobral in the State of Ceará in Brazil, by the astronomers Arthur Eddington (1882–1944) and Frank Dyson (1868–1939), are considered the second experimental confirmation of General Relativity, after the explanation of the Mercury perihelion precession that had been observed in 1859 by Urbain Le Verrier (1811–1877), with a magnitude not explained with Newtonian Mechanics, but fully explained with General Relativity, as evaluated by A. Einstein in 1915. Rhie, Sun Hong (1955–2013).

[11] What may raise an even greater appreciation of the fundamental theorem of algebra.

[12] The fundamental theorem of algebra has a long history beginning with attempted proofs with petition of principle or serious gaps in 1746 by Jean le Rond d'Alembert (1717–1783), in 1749 by Leonhard Euler (1707–1783), in 1759 by François de Foncenex (1734–1799), in 1772 by J.-L. Lagrange, in 1795 by P.-S. Laplace, in 1799 by C.F. Gauss, in 1806 by Jean-Robert Argand (1768–1822), for polynomials with real coefficients, but with those of J.R. d'Alembert, J.-R. Argand, and C.F. Gauss with fertile ideas that led to proofs much later: the first two were completed at the turn of the 19th to the 20th century and that of C.F. Gauss only in 1920. C.F. Gauss was the first to consider polynomials with complex coefficients in 1816. He published three other proof attempts, two in 1816 and one in 1849, also incomplete. The first proof to be completed was an intricate algebraic proof by C.F. Gauss in 1816, using the Bolzano theorem on intermediate values of continuous real functions defined in an interval formulated by Bernard Bolzano (1781–1848) in the following year, but without a complete proof until 1878 since it would have to wait for a definition of complex and real numbers, and the two first definitions of real numbers (axiomatic by G. Cantor and completing the rational numbers by Julius Dedekind (1831–1916)) appeared in that year. Many other proofs were presented based on Complex Analysis (see Sections 3 and 4 of Chapter 6 of *CADOVA*) or topology. The author obtained in 2018 a proof with elementary Linear Algebra published in 2021: Magalhães, L.T., Simple proof of existence of a complex eigenvalue of a complex square matrix ... and yet another proof of the fundamental theorem of algebra with linear algebra, *Linear and Multilinear Algebra,* **70** (2022), 5329–5333, likely the most elementary of all.

n zeros counting multiplicities. The proofs of the mentioned upper bounds use a generalization of the argument principle to harmonic functions analogous to that for meromorphic functions, which was obtained in 2000 by T.J. Suffridge and J.W. Thompson, following a generalization to functions without poles obtained in 1996 by P. Duren, W. Hengartner, and R.S. Laugesen.[13]

8.2 Branched Coverings

A **branched covering** of a Riemann surface[14] S is $\mathscr{R} = (\widetilde{S}, f)$ with $\widetilde{S}$ a Riemann surface and $f \in H(\widetilde{S}, S)$ a holomorphic function such that each point of S has a connected neighborhood U with a restriction $f_{|C}$ to each connected component C of $f^{-1}(U)$ a **proper function.**[15] The **local degree** of a branched covering $\mathscr{R}$ at $z_0 \in \widetilde{S}$ is $\delta_{\mathscr{R}}(z_0)$ equal to the smallest $n \in \mathbb{N}$ with nonzero coefficient of $(z-z_0)^n$ in the Taylor series of f at z_0 in local coordinates. A **universal branched covering** of a Riemann surface S is a branched covering (S', P) of S with S' simply connected and local degree of P in each $z \in S'$ equal to $\delta_{\mathscr{R}}(P(z))$.

The complex rational functions considered as defined from $\mathbb{C}_\infty$ to $\mathbb{C}_\infty$ are simple examples of branched coverings.

In 1965, R. Thom[16] proved that every branched covering $(\mathbb{C}_\infty, f)$ of $\mathbb{C}_\infty$ corresponds to a complex rational function through changes of variables due to possibly different homeomorphisms in the domain and range, but without ensuring compatibility with the dynamics defined by the functions of the branched coverings.

(8.1) *For every branched covering $(\mathbb{C}_\infty, f)$ of $\mathbb{C}_\infty$, there are homeomorphisms h_0, h_1 and a complex rational function F with $h_1 \circ f = F \circ h_0$.*

Proof Let $h_1 = 1_{\mathbb{C}_\infty}$ and h_0 be a solution of the Beltrami equation

$$\frac{\partial h_0/\partial \bar{z}}{\partial h_0/\partial z}(z) = \frac{\partial f/\partial \bar{z}}{\partial f/\partial z}(z) = \kappa(z)\,.$$

Since f is locally at each point a composition of a holomorphic function with a diffeomorphism, $\|\kappa\|_{L^\infty} < 1$, h_0, f are quasiconformal mappings differing

---

[13] Suffridge, Ted Joe (1932-). Thompson, John Walker. Duren, Peter (1935–2020). Hengartner, Walter (1936–2003). Laugesen, Richard S. (1967-).

[14] This and the following section use parts of McMullen, C., *Riemann surfaces, dynamics and geometry*, Course Notes, October 7, 2020, made available by Curtis McMullen in the Internet.

[15] A **proper function** in metric spaces is one whose preimages of compact sets are compact sets.

[16] René Thom (1923–2002) was awarded the *Fields Medal* in 1958 for contributions to the foundations of Cobordism Theory, and the *John von Neumann Lecture Prize* of the *SIAM - Society of Industrial and Applied Mathematics* in the USA in 1976.

by composition by a conformal homeomorphism T, i.e., $f = T \circ h_0$, and $F = h_1 \circ f \circ h_0^{-1} = T$. So, F is holomorphic. □

D. Sullivan in 1978 and more explicitly P. Tukia in 1980 and A. Hinkanen[17] in 1996 gave the following necessary and sufficient condition for a branched covering $(\mathbb{C}_\infty, f)$ of $\mathbb{C}_\infty$ to correspond to a rational function via quasiconformal conjugacy.

(8.2) *A branched covering $(\mathbb{C}_\infty, f)$ of $\mathbb{C}_\infty$ is conjugated by a quasiconformal mapping with a complex rational function if and only if the dilatation of f^n is uniformly bounded for $n \in \mathbb{N} \cup \{0\}$.*

Proof If $f = Q \circ F \circ Q^{-1}$ with Q a quasiconformal mapping and F are a complex rational function and then $f^n = Q \circ F^n \circ Q^{-1}$ for $n \in \mathbb{N} \cup \{0\}$, and the dilatation of f is upper bounded by the square of that of Q.

Conversely, for every $z \in \mathbb{C}_\infty$, with $H_z = SL_2(T_z)/SO_2(T_z)$, the hyperbolic plane of the conformal structures in the tangent space $T_z\mathbb{C}_\infty$, every f^n, $n \in \mathbb{N}$, is differentiable a.e. in $\mathbb{C}_\infty$ and the tangent spaces along the grand orbits of each z can be identified with $T_z\mathbb{C}_\infty$ through Df^n, carrying the conformal structure along the grand orbit to a set of conformal structures $E_z \subset H_z$. Since f^n, $n \in \mathbb{N} \cup \{0\}$, is a quasiconformal mapping with uniformly bounded dilatation, E_z is bounded, and with $\mu(z)$ the center of the smallest hyperbolic disk containing E_z, as $\mu(z)$ is invariant under f, with φ solution of the Beltrami equation with coefficient μ, $F = \varphi \circ f \circ \varphi^{-1}$ is a complex rational function. □

A **critical point** of a branched covering $\mathscr{R} = (\widetilde{S}, f)$ of a Riemann surface is an element of $C(\mathscr{R}) = \delta_{\mathscr{R}}^{-1}(\mathbb{N} \backslash \{1\})$. A **postcritical set** of a branched covering $\mathscr{R}$ is $PC(\mathscr{R}) = \mathrm{cl} \bigcup_{n=1}^{\infty} \bigcup_{c \in C(\mathscr{R})} f^n(c)$. The branched covering $\mathscr{R}$ is said to be **postcritically finite** if its postcritical set is finite. Denote $\boldsymbol{C_f}$ the set of critical points of[18] f and $\boldsymbol{PC(f)} = \mathrm{cl} \bigcup_{n=1}^{\infty} \bigcup_{c \in C_f} f^n(c)$ the closure of the postcritical set of f.

(8.3) *If f is a postcritically finite complex rational function, the topologically attracting periodic orbits are superattracting, and if C is the set of periodic critical points of f, then, alternatively:*

1. *$C = \emptyset$, then $J(f) = \mathbb{C}_\infty$ and f is **Ergodic** in $\mathbb{C}_\infty$, i.e., $\mathbb{C}_\infty$ is a disjoint union of sets $X_1, X_2 \neq \emptyset$ with $f^{-1}(X_j) \subset X_j$ for $j \in \{1, 2\}$ and one of the sets X_1, X_2 with measure 0.*
2. *$C \neq \emptyset$, Area $(J(f)) = 0$ and all $z \in \mathbb{C}_\infty \setminus J(f)$ are attracted to C.*

[17] Hinkanen, Aimo (1958-).

[18] See Section 11 of Chapter 13 of *CADOVA*.

Proof By the Sullivan classification theorem, the periodic Fatou components are superattracting.[19] Therefore, all attracting periodic orbits are superattracting.

(1) As there are no periodic critical points, there are no superattracting periodic orbits; so $J(f) = \mathbb{C}_\infty$ and all periodic orbits are repelling. By the Sullivan nonwandering domains theorem (6.3), every orbit is eventually a repelling periodic orbit. Thus, $\overline{\lim_{n\to+\infty}}\, d_S\big(f^n(z), PC(f)\big) > 0$ for z outside any grand orbit of a point in $PC(f)$. If $X_1 \subset \mathbb{C}_\infty$ is invariant under f and has measure > 0, and $z \in X_1$ is a Lebesgue density point[20] outside the grand orbits of points of $PC(f)$, then $\{f^n(z)\}$ is an eventually repelling periodic orbit. So, $\|(f^n)'(z)\| \to +\infty$ in the Poincaré metric of $\mathbb{C}_\infty \backslash PC(f)$, and small balls in the Poincaré metric centered at z are expanded and have images by f^n containing large balls. As $f^n(z)$ is at a distance $\geq \delta$ for n in an infinite set and some $\delta > 0$, also in the spherical metric small balls are expanded and contain large balls. Therefore, with μ the Lebesgue measure, there is a ball B in the spherical metric such that $\mu(B \cap X_1) = \mu(B)$ and, as there is $k \in \mathbb{N}$ such that $f^k(B) = \mathbb{C}_\infty$, with $X_2 = \mathbb{C}_\infty \backslash X_1$ and by the Lebesgue density theorem,[20] $\mu(X_2) = 0$ and f is ergodic.

(2) The orbits of points of C are superattracting periodic orbits, the Fatou set is the union of the, resp., attracting basins and the positive orbit of each $z \in \mathbb{C}_\infty \backslash J(f)$ tends to C. By the Sullivan nonwandering domains theorem (6.3), the orbit of each nonfixed point of $PC(f)$ is eventually periodic; thus, it is a superattracting periodic orbit. All critical points are in the Fatou set, f is hyperbolic and[21] $\mathrm{Area}\big(J(f)\big) = 0$. □

8.3 Combinatorial Equivalence

The definition of combinatorial equivalence of postcritically finite branched coverings $(\mathbb{C}_\infty, f)$ of $\mathbb{C}_\infty$ by W. Thurston in 1985 is extended to **combinatorial equivalence of marked branched coverings** $\big((\mathbb{C}_\infty, f_1), \Omega_1\big)$, $\big((\mathbb{C}_\infty, f_2), \Omega_2\big)$ of $\mathbb{C}_\infty$, where $\Omega_j \subset \mathbb{C}_\infty$ is finite, $\#\Omega_j \geq 3$ and $\Omega_j \subset PC(f_j)$ for $j \in \{1, 2\}$, with the equivalence relation defined by the existence of an ordered pair (φ, ψ) of homeomorphisms $\varphi, \psi : \mathbb{C}_\infty \to \mathbb{C}_\infty$ **isotopic relative to** Ω_1 (i.e., homotopic to $1_{\mathbb{C}_\infty}$ by homeomorphisms with fixed images of Ω_1, such that $\varphi \circ f_1 = f_2 \circ \psi$, $\varphi(\Omega_1) = \psi(\Omega_1) = \Omega_2$, $\varphi_{|\Omega_1} = \psi_{|\Omega_1}$). This extension coincides with the combinatorial equivalence of W. Thurston if $\Omega_j = PC(f_j)$, $j \in \{1, 2\}$.

The notion of combinatorial equivalence of W. Thurston applied to complex rational functions is weaker than topological conjugacy, but it guarantees topologi-

[19] See Section 12 of Chapter 13 of *CADOVA*.

[20] See Section 3 of Appendix B of *CADOVA*.

[21] See Section 13 of Chapter 13 of *CADOVA*.

cal semiconjugacy (i.e., by coordinate changes with continuous surjective functions instead of homeomorphisms) if there is a bijection of the sets of periodic critical points of both functions respecting minimal periods, as shown below.[22] So, the global dynamics of one of the functions is determined by that of the other, but the converse may fail.

(8.4) *If $\mathscr{R}=(\mathbb{C}_\infty, g)$ is a postcritically finite branched covering of $\mathbb{C}_\infty$ combinatorially equivalent to the branched covering associated with a complex rational function f and there is a bijection of the sets of periodic critical points of f and g respecting minimal periods, then there is a continuous family of homeomorphisms $\{\Psi_t\}_{t\in[0,+\infty[}$ with $\Psi = \lim_{t\to+\infty} \Psi_t$ a continuous surjection from $\mathbb{C}_\infty$ onto $\mathbb{C}_\infty$ whose restriction to a neighborhood of the positive orbits of the periodic critical points in $PC(f)$ is a homeomorphism onto a neighborhood of the corresponding points in $PC(\mathscr{R})$ and $\Psi\circ g = f\circ\Psi$.*

Proof As periodic orbits of critical points are superattracting and their union is a finite set, there are homeomorphisms $\Psi_0 : \mathbb{C}_\infty \to \mathbb{C}_\infty$ and neighborhoods U_f, U_g of the positive periodic orbits of critical points of, resp., f, g such that $\Psi_0(U_g) = U_f$ and $\Psi_{0|U_g}$ is a topological conjugacy of the corresponding restrictions of f and g. Therefore, there is a path of homeomorphisms $f_t, t\in[0, 1]$, such that

$$f_0 = \Psi_0\circ g\circ\Psi_0^{-1}, \quad f_1 = f\,, \quad \Psi(U_g) = U_f, \quad PC(f_t) = PC(f)\,,$$
$$f_{t|U_f} = f_{|U_f}, \quad t\in[0, 1]\,,$$

For $t\in[0, 1]$, define Ψ_t such that $f_t\circ\Psi_t = \Psi_0\circ g$, whereby

$$f\circ\Psi_1 = \Psi_0\circ g\,, \qquad \Psi_{1|U_g} = \Psi_{0|U_g}\,, \qquad \psi_1\big(g^{-1}(U_g)\big) = f^{-1}(U_f)\,,$$

and, then, Ψ_t is defined for $t\in]n, n+1]$ and successive $n\in\mathbb{N}$ by the dynamics of f and g, *i.e.*, so that $f^n\circ\Psi_{t+n} = \Psi_t\circ g^n$, $\Psi_{t+n|U_g} = \Psi_{0|U_g}$ $n\in\mathbb{N}$, and $\Psi_{t+n|g^{-n}(U_g)} = \Psi_{n|g^{-n}(U_g)}$ for $t\in[0, 1]$, $n\in\mathbb{N}$. Hence,

$$f^{n+1}\circ\Psi_{n+1} = \Psi_0\circ g^{n+1} = f^n\circ\Psi_n\circ g\,, \quad f\circ\Psi_{n+1} = \Psi_n\circ\Psi_n\circ g\,, \quad \text{in } U_g\,,$$

and $f\circ\Psi_{n+1} = \Psi_n\circ\Psi_n$ for $n\in\mathbb{N}\cup\{0\}$. For each $y\in\mathbb{C}_\infty\backslash U_g$, denote $F_{y,n}$ a local inverse of the restriction of f^n to $\{\Psi_u(y) : u\in[0, 1]\}$, which is a set disjoint of U_f, and choose $C>0$ and $\lambda\in]0, 1[$ such that

$$C\lambda^n \ge \sup\big\{d_S\big(F_{y,n}\Psi_u(y)\,, F_{y,n}\Psi_0(y)\big):\ y\in\mathbb{C}_\infty\backslash U_g\,, u\in[0, 1]\big\}\,,$$

[22] We follow parts of Rees, M., A partial description of parameter space follows rational maps of degree two: Part I, *Acta Math.*, **168** (1992), 11–87. Rees, Mary (1953-).

where d_s is the spherical metric in $\mathbb{C}_\infty$. So, $d_S(\Psi_{n+t}(z), \Psi_n(z)) \le C\lambda^n$ for $n \in \mathbb{N}$, $t \in [0, 1]$. $\{\Psi_n\}$ is uniformly convergent in $\mathbb{C}_\infty$ to a continuous function Ψ and the other claims in the statement hold. □

8.4 Orbifolds

We begin reviewing definitions and results presented in *CADOVA*.[23]

A **Riemann surface orbifold** is $\mathcal{O} = (S, \nu)$ with S a Riemann surface and a **branching function** $\nu : S \to \mathbb{N}$ with $\nu_j = \nu(r_j) \ge 2$ in a countable set $\mathcal{R}_\nu = \{r_j\} \subset S$ of isolated **branching points** and $\nu(z) = 1$ for $z \in S \backslash \mathcal{R}_\nu$. As we only consider here this type of orbifolds (with a Riemann surface base space), for simplicity, the qualifier "Riemann Surface" is omitted in what follows, but it is always implicit.

Riemann surfaces S are orbifolds (S, ν) with $\nu(x) = 1$ for $x \in S$.

A **Riemann surface orbifold metric** is a metric on an orbifold (S, ν) with element of length in local coordinates $\sigma(z)\,|dz|$, with $\sigma > 0$ holomorphic except at each branching point r_j where there is a **branching Index** $\nu_j \in \mathbb{N} \backslash \{1\}$ and for a local covering $z(w) = r_j + w^{\nu_j}$ the metric with element of length $\mu(w)\,|dw|$ with $\mu(w) = \sigma(z(w))\,\left|\frac{dz}{dw}(w)\right|$ is holomorphic and positive in a neighborhood of 0.

The **signature** of an orbifold (S, ν) is the ordered multiple of the values of $\nu(p) \ge 2$ for $p \in \mathcal{R}_\nu$, ordered increasingly.

The **Euler characteristic** of the orbifold (S, ν) is

$$\chi\big((S, \nu)\big) = \chi(S) - \sum_{r \in \mathcal{R}_\nu} \left(1 - \tfrac{1}{\nu(r)}\right),$$

which is a rational number if $\chi(S)$ and $\#\mathcal{R}$ are finite; otherwise, $\chi\big((S, \nu)\big) = -\infty$. The orbifold (S, ν) is said to be **hyperbolic**, **parabolic** (or **Euclidean**), **Elliptic** if $\chi\big((S, \nu)\big)$ is, resp., < 0, $= 0$, > 0.

Orbifolds (S, ν) are usually hyperbolic: they just might not be if $\#\mathcal{R}_\nu \le 4$ and, even then, for a small set of signatures.

To each complex rational function f of degree ≥ 2 such that each postcritical orbit is finite or converges to an attracting or superattracting periodic orbit associate the **canonical orbifold** $\mathcal{O}_f = (S_f, \nu_f)$, where S is the complement in $\mathbb{C}_\infty$ of the union of the attracting or superattracting periodic orbits and the set of branching points is $PC(f)$. The points of $PC(f)$ are isolated in S_f, because each postcritical orbit is finite or converges to an attracting or superattracting periodic orbit and, therefore, does not have accumulation points in S_f (it can have them in $\mathbb{C}_\infty$).

---

[23] See Section 14 of Chapter 13 of *CADOVA*. The name orbifold is used in the more general context of manifolds with local structure of quotient space of a Euclidean space by the action of a finite group. They appeared with H. Poincaré, but the theory was mainly developed in 1956 and 1957 by Ichirō Satake (1927–2014), who called them *V-manifolds*, and after 1976 by W. Thurston.

The **local degree** $\delta_f(z_0)$ of f at $z_0 \in \mathbb{C}_\infty$ is the smallest $n \in \mathbb{N}$ with nonzero coefficient of $(z - z_0)^n$ in the Taylor series of f at z_0, and for each $a \in PC(f)$ denote $\nu(a)$, the least common multiple of the local degrees $\delta_{f^m}(c)$ such that c is a critical point of f and $f^m(c) = a$.

A **covering of an orbifold** (S, ν) is $\big((\widetilde{S}, \widetilde{\nu}), p\big)$ with $(\widetilde{S}, \widetilde{\nu})$ an orbifold such that $(\widetilde{S}, p)$ is a branched covering of the Riemann surface S and $\nu\big(p(z)\big) = n(z)\,\widetilde{\nu}(z)$ for $z \in \widetilde{S}$, with $n(z)$ the local degree of p in z. A **universal covering of an orbifold** (S, ν) is a branched covering $(\widetilde{S}', p)$ of the Riemann surface S with $\widetilde{S}'$ simply connected and local degree of p in each $z \in \widetilde{S}'$ equal to $\nu\big(p(z)\big)$.

Every canonical orbifold $\mathcal{O}_f = (S_f, \nu_f)$ *associated with a postcritically finite complex rational function* f *with degree* ≥ 2 *has a branched universal covering* $(\widetilde{S}'_\nu, p_{\nu_f})$, *unique modulo conformal homeomorphisms, and it has Euler characteristic* $\chi(\mathcal{O}_f) \leq 0$. *Thus,* $\mathcal{O}_f$ *is hyperbolic or Euclidean, and* $\widetilde{S}'_\nu$ *is conformal to, resp.,* B_1 *or* $\mathbb{C}$.

If $\mathcal{O}_f$ *is Euclidean, then* $(\mathcal{O}_f, f)$ *is a covering of the orbifold* $\mathcal{O}_f$, *and if in this case* $\mathcal{R}_{\nu_f} \neq \emptyset$, S_f *is conformal to* $\mathbb{C}_\infty$ *with* $\#\mathcal{R}_{\nu_f} \in \{2, 3, 4\}$ *and one of the signatures* $(2, 3, 6)$, $(2, 4, 4)$, $(3, 3, 3)$, $(2, 2, 2, 2)$.

If $\mathcal{O}_f$ *is hyperbolic and* $\mathcal{R}_{\nu_f} \neq \emptyset$, *then* S_f *is conformal to* B_1 *with* $\#\mathcal{R}_{\nu_f} \in \{1, 2\}$ *and one of the signatures* $(2, 2)$ *or* (n), *with* $n \in \mathbb{N} \setminus \{1\}$.

Another useful property is: *If* $(\widetilde{S}'_{\nu_f}, p_{\nu_f})$, $(\widetilde{S}'_\mu, \mu)$ *are universal coverings of orbifolds, resp.,* $\mathcal{O}_f = (S_f, \nu_f)$ *and* (S_μ, μ), *then* $\big((S_\mu, p_m u), p\big)$ *is a covering of a orbifold* $\mathcal{O}_f$ *if and only if there is a lift* $p_\mu \circ p$ *by the projection* p_{ν_f} *which is a conformal homeomorphism of* $\widetilde{S}'_\nu$ *over* $\widetilde{S}'_\nu$; *if yes and* p *has finite degree* δ, *then* $\chi(S_\mu, \mu) = \delta\, \chi(\mathcal{O}_f)$.

(8.5) Examples:

1. $f(z) = z^n$ with $n \in \mathbb{N}$, $n \geq 2$, only has critical points 0 and ∞, which are fixed points; therefore, $PC(f) = \{0, \infty\}$. The local degree of f^m at 0 is $\delta(f^m, 0) = nm$, $m \in \mathbb{N}$, whose least common multiple is ∞; so $\nu(0) = \infty$. Similarly, $\delta(f^m, \infty) = nm$ and $\nu(\infty) = \infty$. Thus, the signature of $\mathcal{O}_f$ is (∞, ∞), and the orbifold $\mathcal{O}_f$ is Euclidean.

2. $f(z) = z^2 - 1$ only has critical points 0 and ∞. The positive orbit of 0 is the periodic orbit with period 2, $(0, -1)^\infty$, and ∞ is a fixed point; therefore, $PC(f) = \{0, -1, \infty\}$. As $\delta(f^{2k}, 0) = 4k$ for $k \in \mathbb{N}$, whose least common multiple is ∞, $\nu(0) = \infty$ and similarly $\nu(-1) = \infty$. Since ∞ is a fixed point of f, $\delta(f^m, \infty) = 2m$ for $m \in \mathbb{N}$ and $\nu(\infty) = \infty$. Thus, the signature of $\mathcal{O}_f$ is (∞, ∞, ∞), and the orbifold $\mathcal{O}_f$ is hyperbolic.
3. $f(z) = z^2 - 2$ only has critical points 0 and ∞, the positive orbit of 0 is $(0, -2)(2)^\infty$ with fixed point 2, and ∞ is a fixed point of f. Therefore, $PC(f) = \{-1, 2, \infty\}$. Thus, $f^m(0) = -1$ if and only if $m = 1$, $f'(0) = 0$, and $f''(0) = 2$, we have $\delta(f, 0) = 2$ and $\nu(-1) = 2$. As $f^m(0) = 2$ if and only if $m \in \mathbb{N} \setminus \{1\}$,

$$(f^m)'(0) = [f'(2)]^{m-2} f'(-2) f'(0) = 0\,,$$
$$(f^m)''(0) = [f'(2)]^{m-2} f'(-2) f''(0) = -32 \neq 0\,, \quad m \geq 2\,, m \in \mathbb{N}\,,$$

it is $\delta(f^m, 0) = 2$ for $m \geq 2\,, m \in \mathbb{N}$, and $\nu(2) = 2$. As in 2, $\nu(\infty) = \infty$. So, the signature of $\mathcal{O}_f$ is $(2, 2, \infty)$, and the orbifold $\mathcal{O}_f$ is Euclidean.

4. $f(z) = z^2 + i$ has derivative $f'(z) = 2z$. So, it only has critical points 0 and ∞. The positive orbit of 0 is $(0, i)(-1 + i, -i)^\infty$, and ∞ is a fixed point. Thus, $PC(f) = \{i, -1+i, -i, \infty\}$. As $f^m(0) = i$ if and only if $m = 1$, and $\delta(f, 0) = 2$, it is $\nu(i) = 2$. As the orbit of $-1+i$ has period 2 and $f(0) = i$, $f^2(0) = -1 +$, $f^m(0) = -1 + i$ if and only if $m = 2k$ with $k \in \mathbb{N}$. Since $f'(z) = 0$ if and only if $z = 0$, $f''(0) = 2$,

$$(f^{2k})'(0) = [f'(-i) f'(-1+i)]^{k-1} f'(i) f'(0) = 0\,,$$
$$(f^{2k})''(0) = [f'(-i) f'(-1+i)]^{k-1} f'(i) f''(0) \neq 0\,,$$

it is $\delta(f^{2k}, 0) = 2$ for $k \in \mathbb{N}$, and, therefore, $\nu(-1+i) = 2$. Analogously, as $f^m(0) = -i$ if and only if $m = 2k+1$ with $k \in \mathbb{N}$, and

$$(f^{2k+1})'(0) = [f'(-1+i) f'(-i)]^{k-1} f'(-1+i) f'(i) f'(0) = 0\,,$$
$$(f^{2k+1})''(0) = [f'(-1+i) f'(-i)]^{k-1} f'(-1+i) f'(i) f''(0) = 0\,,$$

also $\nu(-i) = 2$. Since ∞ is a fixed point of f, $\delta(f^m, \infty) = 2m$ for $m \in \mathbb{N}$ and $\nu(\infty) = \infty$. Thus, the signature of $\mathcal{O}_f$ is $(2, 2, 2, \infty)$, and the orbifold $\mathcal{O}_f$ is hyperbolic.

5. $f(z) = \left(\frac{z-i}{z+i}\right)^2$ is a Lattès function with $f'(z) = 4i\,\frac{z-i}{(z+i)^3}$ and a single pole at $-i$ of order 2. Hence, f only has critical points $\pm i$, whose positive orbits are $(i, 0, 1)(-1)^\infty$, since $\left(\frac{1-i}{1+i}\right)^2 = -1$, which is a fixed point of f, and $(-i, \infty, 1)(-1)^\infty$. So, $PC(f) = \{0, 1, -1, \infty\}$. As $f^m(i) = 0$ if and only if $m = 1$, $f'(i) = 0$, $f''(z) = -8i\,\frac{1}{(z+i)^3}$ and $f''(i) = 1$, it is $\delta(f, i) = 2$ and $\nu(0) = 2$. Since $f^m(i) = 1$ if and only if $m = 2$,

$$(f^2)'(z) = f'\big(f(z)\big) f'(z)\,,$$
$$(f^2)''(z) = f''\big(f(z)\big) f'(z) + f'\big(f(z)\big) f''(z)\,,$$

and, therefore, $(f^2)'(i) = f'(0) f'(i) = 0$, $(f^2)''(i) = f''(0) \neq 0$, $\delta(f^2, i) = 2$ and $\nu(1) = 2$. As $f^m(i) = -1$ if and only if $m = k+2$ with $k \in \mathbb{N}$,

$$(f^{k+2})'(z) = (f^k)'\big(f^2(z)\big) f'\big(f(z)\big) f'(z)\,,$$
$$(f^{k+2})'(i) = (f^k)'(1) f'(0) f'(i) = 0\,,$$
$$(f^{k+2})''(i) = [f'(-1)]^{k-1} f'(1) f'(0) f''(i) = [-\tfrac{4}{(i-1)^2}]^{k-1} \tfrac{4}{(1+i)^2} 4 \neq 0;$$

thus $\delta(f^{k+2}, i) = 2$, for $k \in \mathbb{N}$, and $\nu(-1) = 2$. Since $f^m(-i) = \infty$ if and only if $m = 1$, with

$$g(w) = \frac{1}{f(1/w)} = \left(\frac{w-i}{w+i}\right)^2, \quad g'(z) = \frac{4i(w+i)}{(w-i)^3}, \quad g'(-i) = 0,$$
$$g''(w) = \frac{-8i}{(w-i)^3}, \quad g''(-i) = -1,$$

it is $\delta(f, \infty) = \delta(g, 0) = 2$. Therefore, the signature of $\mathcal{O}_f$ is $(2, 2, 2, 2)$, and the orbifold $\mathcal{O}_f$ is Euclidean.

Orbifolds $\mathcal{O}_{\mathscr{R}}$ **of branched coverings** $\mathscr{R}$ can be analogously defined to canonical orbifolds $\mathcal{O}_f$ of complex rational functions f.

8.5 Teichmüller Space of $\mathbb{C}_\infty$ Marked by a Subset

The **Teichmüller space of** $\mathbb{C}_\infty$ **marked by** Ω, **Teich**$(\mathbb{C}_\infty, \boldsymbol{\Omega})$, is the set of the equivalence classes of homeomorphisms of $\mathbb{C}_\infty$ onto $\mathbb{C}_\infty$ with two such homeomorphisms φ, ψ equivalent if there is a Möbius transformation M such that $M \circ \varphi_{|\Omega} = \psi_{|\Omega}$ and $M \circ \varphi = \psi \circ h$ with h a homeomorphism isotopic relative to Ω. Similarly to Chap. 4, we consider in Teich$(\mathbb{C}_\infty, \Omega)$ the **Teichmüller distance**

$$D_T\big([F_1], [F_2]\big) = \tfrac{1}{2} \inf\big\{\log K_H : H : \mathbb{C}_\infty \to \mathbb{C}_\infty \text{ is a Quasiconformal Mapping}$$
$$\text{and } F_1^{-1} \circ H \circ F_2 \text{ is isotopic relative to } \Omega\big\},$$

where K_H is the dilatation of the quasiconformal mapping H, which in terms of the Beltrami coefficient κ_H of H is $K_H = \frac{1 + \|\kappa_H\|_{L^\infty}}{1 - \|\kappa_H\|_{L^\infty}}$. The metric space $\big(\text{Teich}(\mathbb{C}_\infty, \Omega), D_T\big)$ is complete.[24]

Let f be a complex rational function of degree ≥ 2 considered as a function from $\mathbb{C}_\infty$ to $\mathbb{C}_\infty$.

The **extended Julia set** of f is the closure $\widetilde{\boldsymbol{J}}(\boldsymbol{f})$ of the grand orbits of the critical points of f. Thus, $J(f) \cup PC(f) \subset \widetilde{J}(f)$.

The space $\mathscr{M}(\boldsymbol{f})$ of the equivalence classes by the conformal relation of the rational functions quasiconformally conjugated to f is called the **moduli space of** f.

Denote $\boldsymbol{QC(f)}$ the group of quasiconformal mappings from $\mathbb{C}_\infty$ onto $\mathbb{C}_\infty$ commuting with f, $\boldsymbol{QC_0(f)}$ its a normal subgroup of the quasiconformal mappings

[24] See basic aspects of the Teichmüller theory in a particular case at the end of Chap. 4. In this section, we mainly use parts of the article of the 1998 C. McMullen and D. Sullivan cited in the introduction of Chap. 5, with some results of C. McMullen published in 1987 and 1988, and of D. Sullivan in a *Preprint* of 1993.

$g : \mathbb{C}_\infty \to \mathbb{C}_\infty$ isotopic to $1_{\mathbb{C}_\infty}$, in the sense that there is a continuous function $h : [0, 1] \times \mathbb{C}_\infty \to \mathbb{C}_\infty$ such that $(t, z) \mapsto \big(t, h(t, z)\big)$ is a homeomorphism onto $[0, 1] \times \mathbb{C}_\infty$ and for each fixed $t \in [0, 1]$ $z \mapsto h_t(z) = h(t, z)$ is a homeomorphism, $h_0 = 1_{\mathbb{C}_\infty}, h_1 = g$.

Denote **Mod(*f*)** $= QC(f)/QC_0(f)$ the **modular group** of f. Its elements are the equivalence classes $[\varphi]$ of quasiconformal mappings $\varphi : \mathbb{C}_\infty \to \mathbb{C}_\infty$ commuting with f with the equivalence relation $\varphi_1 \sim \varphi_2$ if there is $h \in QC_0(f)$ such that $\varphi_2 = \varphi_1 \circ h$.

If h_t is as in the preceding paragraph, its restriction both to the set of periodic points of f and to $PC(f)$ is $1_{\mathbb{C}_\infty}$. So, also the restriction of h_t to $\widetilde{J}(f)$ is $1_{\mathbb{C}_\infty}$ for all $t \in [0, 1]$.

The **Teichmüller space** of f is the set **Teich(*f*)** of the equivalence classes of ordered pairs (g, ψ) such that g is a rational function and ψ is a quasiconformal conjugacy of f with g with the equivalence relation $(g_1, \psi_1) \sim (g_2, \psi_2)$ if there are a Möbius transformation M and $h \in QC_0(f)$ such that $g_1 = M \circ g_2 \circ M^{-1}$ and $M \circ \psi_2 = \psi_1 \circ h$, i.e.,the following diagram is commutative.

$$
\begin{array}{ccccccc}
 & & & \xrightarrow{\qquad\qquad M \qquad\qquad} & & & \\
(\mathbb{C}_\infty, \widetilde{J}(g_2)) & \xleftarrow{\psi_2} & (\mathbb{C}_\infty, \widetilde{J}(f)) & \xrightarrow{h} & (\mathbb{C}_\infty, \widetilde{J}(f)) & \xrightarrow{\psi_1} & (\mathbb{C}_\infty, \widetilde{J}(g_1)) \\
{\scriptstyle g_2}\big\downarrow & & {\scriptstyle f}\big\downarrow & & \big\downarrow{\scriptstyle f} & & \big\downarrow{\scriptstyle g_1} \\
(\mathbb{C}_\infty, \widetilde{J}(g_2)) & \xleftarrow{\psi_2} & (\mathbb{C}_\infty, \widetilde{J}(f)) & \xrightarrow{h} & (\mathbb{C}_\infty, \widetilde{J}(f)) & \xrightarrow{\psi_1} & (\mathbb{C}_\infty, \widetilde{J}(g_1)) \\
 & & & \xrightarrow[\qquad\qquad M \qquad\qquad]{} & & &
\end{array}
$$

The group **Aut(*f*)** *of the Möbius transformations commuting with f is a subgroup of* Mod(f). *If $M \in$* Aut(f), *then $(f, M) \sim (f, 1_{\mathbb{C}_\infty})$ and*

$$\mathscr{M}(f) = \mathrm{Teich}(f)/\mathrm{Mod}(f)\,.$$

(8.6) *If f is a complex rational function, $\varphi \cdot (g, \psi) = (g, \psi \circ \varphi^{-1})$, with φ and (g, ψ) representing elements of, resp.,* Mod(f) *and* Teich(f), *induces an isometric action of* Mod(f) *on* Teich(f) *in the Teichmüller metric.*

Proof It is $\varphi \circ f = f \circ \varphi$ and $\psi \circ f = f \circ \psi$. Thus,

$$(\psi \circ \varphi^{-1}) \circ f = \psi \circ f \circ \varphi^{-1} = \psi \circ (\psi^{-1} \circ g \circ \psi) \circ \varphi^{-1} = g \circ (\psi \circ \varphi^{-1})\,.$$

Therefore, $(g, \psi \circ \varphi^{-1}) \in \mathrm{Teich}(f)$.

If φ_1, φ_2 belongs to a same element of Mod(f) and $(g_1, \psi_1), (g_2, \psi_2)$ to a same element of Teich(f), then $\varphi_j \in QC(f)$ for $j \in \{1, 2\}$ and $\varphi_2 = \varphi_1 \circ h_1$ with

$h_1 \in QC_0(f)$, and $g_1 = M \circ g_2 \circ M^{-1}$, $M \circ \psi_2 = \psi_1 \circ h_2$ with $M \in \mathrm{Aut}(f)$ and $h_2 \in QC_0(f)$. Hence,

$$M \circ \psi_2 \circ \varphi_2^{-1} = \psi_1 \circ h_2 \circ \varphi_2^{-1} = \psi_1 \circ h_2 \circ h_1^{-1} \circ \varphi_1^{-1} = \psi_1 \circ \varphi_1^{-1} \circ \varphi_1 \circ h_2 \circ h_1^{-1} \circ \varphi_1^{-1}.$$

Since $h_2 \circ h_1^{-1} \in QC_0(f)$, which is a normal subgroup of $QC(f)$, it is

$$h = \varphi_1 \circ h_2 \circ h_1^{-1} \circ \varphi_1^{-1} \in QC(f), \qquad M \circ (\psi_2 \circ \varphi_2^{-1}) = (\psi_2 \circ \varphi_2^{-1}) \circ h.$$

Hence, $(g_1, \psi_1 \circ \varphi_1^{-1})$ and $(g_2, \psi_2 \circ \varphi_2^{-1})$ belong to a same element of $\mathrm{Teich}(f)$.

It remains to prove that the action is an isometry in $\mathrm{Teich}(f)$. The distance of $\psi_1 \circ \varphi_1^{-1}$ to $\psi_2 \circ \varphi_2^{-1}$ is $D_T\big([\psi_1 \circ \varphi_1^{-1}], [\psi_2 \circ \varphi_2^{-1}]\big)$. If H is a conformal homeomorphism such that $(\psi_1 \circ \varphi_1^{-1})^{-1} \circ H \circ (\psi_2 \circ \varphi_2^{-1})$, then it is homotopic to $1_{\mathbb{C}_\infty}$, represents an element of $QC_0(f)$ and

$$\varphi_1^{-1} \circ (\psi_1 \circ \varphi_1^{-1})^{-1} \circ H \circ (\psi_2 \circ \varphi_2^{-1}) \circ \varphi_2 = \psi_1^{-1} \circ H \circ \psi_2,$$

where the left-hand side of the equality is also representative of an element of $QC_0(f)$, as this is a normal subgroup of $QC(f)$. Thus, $\psi_1^{-1} \circ H \circ \psi_2$ is also isotopic to 1_∞ relative to Ω. Therefore, $D_T\big([\psi_1 \circ \varphi_1^{-1}], [\psi_2 \circ \varphi_2^{-1}]\big)$ and $D_T\big([\psi_1], [\psi_2]\big)$ are given by infimum of the same set, and, consequently, they are equal. □

Denote **Rat**$_d$ the space of complex rational functions of degree $d \geq 2$. As these functions are quotients of relatively prime polynomials with one of them monic of degree d and the other of degree $\leq d$, Rat_d is a complex differential manifold of dimension equal to the maximum number of coefficients of the two polynomials minus 1, as one of the polynomials can be taken monic, $\dim \mathrm{Rat}_d = 2d - 1$, where we can consider a metric given by the usual Euclidean distance of ordered multiples of the $2d - 1$ coefficients of the two polynomials. The group of Möbius transformations $\mathrm{Aut}(\mathbb{C}_\infty)$ acts on Rat_d by conjugacy, i.e., if $f \in \mathrm{Rat}_d$ and $\varphi \in \mathrm{Aut}(\mathbb{C}_\infty)$, then $\varphi^{-1} \circ f \circ \varphi \in \mathrm{Rat}_d$.

The **deformation space** of $f \in \mathrm{Rat}_d$ is the set **Def**(f) of equivalence classes of ordered pairs (g, ψ) with $g \in \mathrm{Rat}_d$ and $\psi : \mathbb{C}_\infty \to \mathbb{C}_\infty$ a quasiconformal mapping with equivalence relation $(g_1, \psi_1) \sim (g_2, \psi_2)$ if there is a conformal homeomorphism T such that $\psi_2 = T \circ \psi_1$.

If $f \in \mathrm{Rat}_d$ with $d \geq 2$, *then* $\mathrm{Teich}(f) = \mathrm{Def}(f)/QC_0(f)$.

In the 1998 article of C. McMullen and D. Sullivan cited at the beginning of this section, they gave the following parametrization of quasiconformally conjugated rational functions with a complex rational function of degree ≥ 2.

(8.7) *If f is a rational function of degree ≥ 2, then the action of the group* $\mathrm{Mod}(f)$ *on* $\mathrm{Teich}(f)$ *considered in* (8.6) *is properly discontinuous, and the rational functions quasiconformally conjugated with f are parameterized by a holomorphic injection from the orbifold $\mathcal{M}(f)$ to the orbifold* $\mathrm{Rat}_d/\mathrm{Aut}(\mathbb{C}_\infty)$.

Proof By the compactness of quasiconformal mappings with uniformly bounded dilatation, the image of $\mathrm{Mod}(f)$ by the considered action is a closed subgroup of the isometry group of $\mathrm{Teich}(f)$. Since $\mathrm{Teich}(f)$ is a finite dimensional differential manifold, also $\mathrm{Mod}(f)$ is if it is not discrete, and any regular path $\alpha_1 : [0, 1] \to \mathrm{Mod}(f)$, which exists if $\mathrm{Mod}(f)$ is not discrete, could be lifted to $\mathrm{Def}(f)$. So, there would exist a simple path $\beta_1 : [0, 1] \to \mathcal{QC}_0(f)$ with the restriction of the continuous projection associated with the lift to the curve β_1^* described by β_1 onto the curve α_1^* described by α_1, but this is impossible since all points of α_1^* would be isotopic to $1_{\mathbb{C}_\infty}$, which is not homeomorphic to any curve represented by a simple path. Therefore, $\mathrm{Mod}(f)$ is discrete.

Since the action of $\mathrm{Mod}(f)$ on $\mathrm{Teich}(f)$ is properly discontinuous, by the preceding paragraph, $\mathcal{M}(f)$ and $\mathrm{Rat}_d/\mathrm{Aut}(\mathbb{C}_\infty)$ are orbifolds. The existence of injection of the first orbifold to the second follows, recalling that $\mathscr{M}(f) = \mathrm{Teich}(f)/\mathrm{Mod}(f)$ and relating $\mathrm{Teich}(f)$ with Rat_d and $\mathrm{Mod}(f)$ with $\mathrm{Aut}(\mathbb{C}_\infty)$. □

Consequently: *For every $K > 1$ the set of nonisotopic quasiconformal automorphisms of f with dilatation $\leq K$ is finite.*

8.6 Characterization of Postcritically Finite Rational Maps

This section considers the topological characterization of postcritically finite rational maps not Lattès maps formulated for branched coverings by W. Thurston in 1982 and presented at a conference in 1983, whose proof was published in 1993 in an article by A. Douady and J. Hubbard. In 2011, X. Buff, G. Cui, and L. Tan[25] published a proof of a slightly more general result for marked branched coverings, but similar to the proof in the A. Douady and J. Hubbard article.

In summary, it is associated with a branched covering $(\mathbb{C}_\infty, f)$ of $\mathbb{C}_\infty$ the Teichmüller space of $\mathbb{C}_\infty$ marked by a set $\Omega \subset \mathbb{C}_\infty$ containing the postcritical points and it is considered a function—the **Thurston** ***pullback***— associating, through the branched covering, points of $\mathrm{Teich}\big(\mathbb{C}_\infty \backslash f^{-1}(\Omega)\big)$ to complex structures in $\mathbb{C}_\infty \backslash \Omega$ represented by points of $\mathrm{Teich}\big(\mathbb{C}_\infty \backslash \Omega\big)$. Then, we get conditions, expressed in terms of the spectral radius of an associated matrix to be < 1, sufficient for the Thurston *pullback* to be a contraction in the complete metric space $\mathrm{Teich}\big(\mathbb{C}_\infty \backslash \Omega\big)$, whose fixed point corresponds to a complex structure in $\mathbb{C}_\infty \backslash \Omega$ invariant under f, up to isotopies, leading to a holomorphic extension of f to $\mathbb{C}_\infty$; so, to a complex rational function. The opposite conditions are called *Thurston Obstructions.*

Contraction of Quadratic Differentials by Rational Functions *Pullback*
For $X \subset \mathbb{C}_\infty$ with $\#X \geq 3$, denote $\mathcal{Q}(X)$ the space of quadratic differentials in $\mathbb{C}_\infty$ holomorphic in $\mathbb{C}_\infty \backslash X$, i.e., the meromorphic quadratic differentials in $\mathbb{C}_\infty$ with

[25] This section follows mainly parts of the 2011 article of X. Buff, G. Cui and L. Tan cited in the introduction to this chapter, and the article cited in the preceding footnote.

possibly existing poles simple. By (4.15), as a result of the Riemann-Roch theorem,[26] $\mathcal{Q}(X)$ is a complex linear space and $\dim \mathcal{Q}(X) = \#X - 3$. This space is considered with the norm $\|q\| = \int_{\mathbb{C}_\infty} |q| = \int_{\mathbb{R}^2} |q(x+iy)|\, dxdy$. If $\psi + \mathbb{C}_\infty \to \mathbb{C}_\infty$ is an element of an equivalence class $\tau \in \mathrm{Teich}(\mathbb{C}_\infty, \Omega)$, the cotangent space $T_\tau^*\mathrm{Teich}(\mathbb{C}_\infty, \Omega)$ of $\mathrm{Teich}(\mathbb{C}_\infty, \Omega)$ at τ is canonically identified with $\mathcal{Q}(\psi(\Omega))$. Consider in the tangent space $T_\tau\mathrm{Teich}(\mathbb{C}_\infty, \Omega)$ the norm

$$\|\nu\| = \sup\left\{|\nu(q)| : q \in \mathcal{Q}(\psi(\Omega)),\ \|q\| \le 1\right\}, \quad \nu \in T_\tau\mathrm{Teich}(\mathbb{C}_\infty, \Omega).$$

If $f : \mathbb{C}_\infty \to \mathbb{C}_\infty$ is a rational function, define the functions ***pullback* and *pushforward* by** f of a meromorphic quadratic differential in $\mathbb{C}_\infty$, which in local coordinates is $q = b(y)\, dy^2$ with b meromorphic function in $\mathbb{C}_\infty$ with possible poles simple, by, resp., $f^*q = a(x)\, dx^2$ with $a = (b \circ f)(f')^2$, and $f_*q = c(z)\, dz^2$ with $c(z) = \sum_{y \in f^{-1}(z)} \frac{b(y)}{[f'(y)]^2}$. So,

$$\begin{aligned} f\big(\{\text{poles of } f^*q\}\big) &\subset \{\text{poles of } q\},\\ f^{-1}\big(\{\text{poles of } q\}\big) &\subset \{\text{poles of } f^*q \cup C_f\},\\ \{\text{poles of } f_*q\} &\subset f\big(\{\text{poles of } q\} \cup f(C_f)\big). \end{aligned}$$

W. Thurston observed that the *pullback* by $f \in \mathrm{Rat}_d$ contracts meromorphic quadratic differentials q in $\mathbb{C}_\infty$ except if $f^* f_* q = d\, q$, when it preserves the distance of points to 0, the poles of quadratic differentials $\neq 0$ fixed points of f_* are the critical points of f and the signature of the orbifold $\mathcal{O}_f$ is $(2, 2, 2, 2)$; this is the only Euclidean canonical orbifold associated with the rational function of degree ≥ 2 with four branching points.[27]

(8.8) *If $f \in \mathrm{Rat}_d$ with $d \in \mathbb{N}\backslash\{1\}$ and q is an integrable meromorphic quadratic differential in $\mathbb{C}_\infty$, then $\|f_*q\| \le \|q\|$ with equality if and only if $f^* f_* q = d\, q$. If $q = f_*q$ with $q \neq 0$, then the set of poles of q is $PC(f)$ and the signature of the orbifold $\mathcal{O}_f$ is $(2, 2, 2, 2)$.*

Proof If $U \subset \mathbb{C}_\infty \backslash CV(f)$ and $\mathbb{C}_\infty \backslash U$ have measure 0, and $\{g_j\}_{j \in \{1,\dots,d\}}$ are inverses of the branches of f in U, then

$$\int_{\mathbb{C}_\infty} |f_*q| = \int_U |f_*q| = \int_U \left|\sum_{j=1}^d g_j^* q\right| \le \sum_{j=1}^d \int_U |g_j^* q| = \int_{f^{-1}(U)} |q| \le \int_{\mathbb{C}_\infty} |q|\,;$$

[26] See Section 6 of Chapter 12 of *CADOVA*.

[27] See Section 3 of this chapter.

therefore, $\|f_*q\| \leq \|q\|$. In case of equality, $\left|\sum_{j=1}^{d} g_j^* q\right| = \sum_{j=1}^{d} |g_j^* q|$, the terms $g_j^* q$ in the sum have the same argument, and $f^* f_* q$ is the product of q by a real meromorphic function. Hence, by a constant and, therefore, it is equal to d.

If $q = f_* q$, then $\|f_* q\| = \|q\|$ and $f^* q = f^* f_* q = d\, q$. Therefore, the set P of the poles of q satisfies $f(P) \subset P$ and $f^{-1}(P) \subset P \cup C_f$. Hence,

$$\#P + (2d-2) \geq \#P + \#C_f \geq \#f^{-1}(P) \geq d\#P - (2d-2).$$

So, $4(d-1) \geq (d-1)\#P$. Since $d \geq 2$, it is $\#P \leq 4$. If $q \neq 0$, then $\#P \geq 4$, $\#P = 4$ and the inequalities in the above formula are equalities, as the values of the extreme terms in the expression are equal, implying $\#C(f) = 2d-2$, $P \cap C_f = \emptyset$ and that the critical points of f are simple. The equality in the middle of the expression implies $f^{-1}(P) = P \cup C_f$. Therefore, $CV_f \subset P$. As if z is the pole of q, then $f(z)$ is the pole of $f_* q$, and $f(P) \subset P$. Therefore, $PC(f) = \cup_{n=0}^{\infty} f^n\big(CV_f\big) \subset P$. Since $f^{-1}\big(P \backslash PC(f)\big) \subset P \cup C_f$ and $f^{-1}\big(P \setminus PC(f)\big) \cap \big(C_f \cup PC(f)\big) = \emptyset$,

$$f^{-1}\big(P \backslash PC(f)\big) \subset P \backslash PC(f), \quad f^{-n}\big(P \backslash PC(f)\big) \subset P \backslash PC(f), \quad n \in \mathbb{N}.$$

As $\#f^{-1}(\{z\}) = d$ for $z \notin CV(f)$ and $\#\big(P \backslash PC(f)\big) \leq 4$, $P \backslash PC(f) = \emptyset$. Therefore, $\mathcal{O}_f$ has signature (2, 2, 2, 2) and $P = PC(f)$. □

Thurston *Pullback* in Teichmüller Space

Let $(\mathbb{C}_\infty, f)$ be a branched covering of $\mathbb{C}_\infty$, and C_f, CV_f, $PC(f)$ be the sets of, resp., critical points, critical values, closure of the set of postcritical points, and $\Omega \subset \mathbb{C}_\infty$ be a finite set such that $\Omega \supset CV_f$ and $\#\Omega \geq 3$. The Thurston ***Pullback* from *Teich*$\big(\mathbb{C}_\infty \backslash \Omega\big)$** to ***Teich*$\big(\mathbb{C}_\infty \setminus f^{-1}(\Omega)\big)$** is the function $\mathcal{T}_f$ defined as follows: for each $\tau \in \mathrm{Teich}\big(\mathbb{C}_\infty \setminus \Omega\big)$ consider a homeomorphism $\varphi : \mathbb{C}_\infty \to \mathbb{C}_\infty$ such that $\tau = [\varphi]$ is the complex structure it induces in $\mathbb{C}_\infty$, removing singularities to define it in neighborhoods of the critical points, and consider in $\mathbb{C}_\infty$ the complex structure mapped through the covering to the initially complex structure. By the uniformization theorem,[28] there is a conformal homeomorphism ψ in $\mathbb{C}_\infty$ with the complex structure obtained in this way onto $\mathbb{C}_\infty$. The function $\mathcal{T}_f$ is defined so as to map $[\varphi] \in \mathrm{Teich}\big(\mathbb{C}_\infty \setminus \Omega\big)$ to $[\psi] \in \mathrm{Teich}\big(\mathbb{C}_\infty \setminus f^{-1}(\Omega)\big)$, for what it is necessary to guarantee independence of the choice of elements in equivalence classes. The function $\varphi \circ f \circ \psi$ is holomorphic in $\mathbb{C}_\infty$, and, therefore, it is a complex rational function F. Let $\varphi_1, \varphi_2 \in \tau \in \mathrm{Teich}\big(\mathbb{C}_\infty \setminus \Omega\big)$, $M \in \mathrm{Aut}(\mathbb{C}_\infty)$ and $h : \mathbb{C}_\infty \to \mathbb{C}_\infty$ is an isotopy relative to Ω such that $\varphi_1 \circ h \circ \varphi_2^{-1} = M$, and $F_1 = \varphi \circ f \circ \psi$. As $\Omega \supset CV_f$, there is a lift $k : \mathbb{C}_\infty \to \mathbb{C}_\infty$ which is an isotopy relative to $f^{-1}(\Omega)$ such that $h \circ f = f \circ k$. ψ_1, ψ_1, F_2, N are defined, so that the following diagram is commutative.

[28] See Chapter 12 of *CADOVA*.

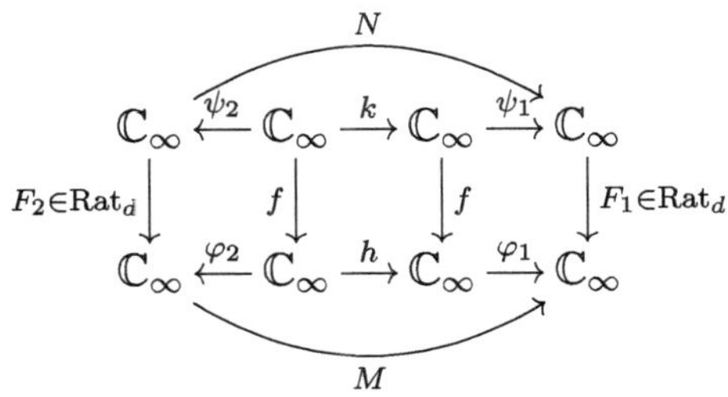

As M, F_1, F_2 are holomorphic functions, so is $N = \psi_1 \circ k \circ \psi_2^{-1}$, and it is a Möbius transformation. Therefore, $[\psi_1] = [\psi_2] \in \mathrm{Teich}\big(\mathbb{C}_\infty \backslash f^{-1}(\Omega)\big)$.

(8.9) *If* $(\mathbb{C}_\infty, f)$ *is a branched covering of* $\mathbb{C}_\infty$ *and* $\Omega \subset \mathbb{C}_\infty$ *is a finite set,* $\Omega \supset CV_f$ *and* $\#\Omega \geq 3$*, then the Thurston* pullback $\mathcal{T}_f$ *from* $\mathrm{Teich}\big(\mathbb{C}_\infty \backslash \Omega\big)$ *to* $\mathrm{Teich}\big(\mathbb{C}_\infty \backslash f^{-1}(\Omega)\big)$ *is a holomorphic function.*

Proof $\mathrm{Teich}(\mathbb{C}_\infty \backslash \Omega)$ and $\mathrm{Teich}(\mathbb{C}_\infty \backslash f^{-1}(\Omega))$ can be canonically identified with quotients of the ball B_1 of the space L^∞ via Beltrami coefficients of the quasi-conformal mappings elements of these spaces. The function corresponding to $\mathcal{T}_f$ through these identifications is a linear transformation in the complex linear space L^∞. Therefore, it is trivially holomorphic. So, also $\mathcal{T}_f$ is holomorphic. □

If $(\mathbb{C}_\infty, f)$ is a branched covering of $\mathbb{C}_\infty$ and $\Omega \subset \mathbb{C}_\infty$ is a finite set with $\#\Omega \geq 3$, $\Omega \supset CV_f$, and $\Sigma \subset f^{-1}(\Omega)$ with $\#\Sigma \geq 3$, there is a holomorphic **embedding** $\sigma : \mathrm{Teich}\big(\mathbb{C}_\infty \backslash f^{-1}(\Omega)\big) \to \mathrm{Teich}\big(\mathbb{C}_\infty \backslash \Sigma\big)$ (i.e., a holomorphic function with derivative at each $\tau \in \mathrm{Teich}\big(\mathbb{C}_\infty \backslash f^{-1}(\Omega)\big)$ surjective from the tangent space $T_\tau \mathrm{Teich}\big(\mathbb{C}_\infty \backslash \Sigma\big)$ onto the tangent space $T_{\sigma(\tau)}\mathrm{Teich}\big(\mathbb{C}_\infty \backslash \Sigma\big))$ that corresponds to forgetting points of $f^{-1}(\Omega) \backslash \Sigma$. The function $\widetilde{\mathcal{T}}_f = \sigma \circ \mathcal{T}_f$ is called the **Thurston *pullback* from Teich$\big(\mathbb{C}_\infty \backslash \Omega\big)$ to Teich$\big(\mathbb{C}_\infty \backslash \Sigma\big)$**. If the branched covering is postcritically finite, then take $\Sigma = \Omega = PC(f)$.

If $\tau \in \mathrm{Teich}\big(\mathbb{C}_\infty \backslash \Omega\big)$, $(\widetilde{\mathcal{T}}_f)^j(\tau) = [\varphi_j]$ with $f_j = \varphi_j \circ f \circ \varphi_{j+1}^{-1}$ a rational function, $g = f_0 \circ \cdots \circ f_{\#\Omega - 1}$ and $G = f^{\#\Omega}$, then $g \circ \varphi_{\#\Omega} = \varphi_0 \circ G$. Thus,

$$(\widetilde{\mathcal{T}}_f)^{\#\Omega}(\tau) = [\varphi_{\#\Omega}] = \sigma\big(\big[G(\varphi)\big]\big) = \sigma\big(\mathcal{T}_G(\tau)\big),$$

and, therefore, $\boldsymbol{(\widetilde{\mathcal{T}}_f)^{\#\Omega} = \widetilde{\mathcal{T}}_{f^{\#\Omega}}}$.

If $(\mathbb{C}_\infty, f)$ is a branched covering of $\mathbb{C}_\infty$, $\Omega \subset \mathbb{C}_\infty$ is a finite set with at least 3 distinct points x_0, x_1, x_2, and $CV_f \subset \Omega \subset f^{-1}(\Omega)$, each $\tau \in \mathrm{Teich}(\mathbb{C}_\infty \backslash \Omega)$ can be represented by a unique homeomorphism $\varphi : \mathbb{C}_\infty \to \mathbb{C}_\infty$ such that the images of x_0, x_1, x_2 are, resp., $0, 1, \infty$ and whose restriction φ_τ to Ω only depends on τ, and each $\mathcal{T}_f(\tau) \in \mathrm{Teich}\big(\mathbb{C}_\infty, f^{-1}(\Omega)\big)$ can be represented by a unique homeomorphism $\psi : \mathbb{C}_\infty \to \mathbb{C}_\infty$ such that the images of x_0, x_1, x_2 are, resp., $0, 1, \infty$, and

$g = \varphi \circ f \circ \psi^{-1}$ is a rational function with restriction ψ_τ to $f^{-1}(\Omega)$ also only depending on τ, just as the rational function $g = g_\tau$ only depends on τ. If d_x denotes the local degree of f in $x \in \mathbb{C}_\infty$, the rational function g_τ can be written $g_\tau = k_\tau \frac{P_\tau}{Q_\tau}$. Thus,

$$P_\tau(z) = \prod_{\substack{x \in f^{-1}(\{x_0\}) \\ x \neq x_2}} [z - \psi_\tau(x)]^{d_x}, \qquad Q_\tau(z) = \prod_{\substack{x \in f^{-1}(\{x_2\}) \\ x \neq x_2}} [z - \psi_\tau(x)]^{d_x},$$

and k_τ is the value of $\frac{P_\tau}{Q_\tau}$ at any point of $\psi_\tau\big(f^{-1}(\{x_1\})\big)$.

Consider the space

$$\mathbb{T}_{f,\Omega} \stackrel{def}{=} \mathrm{Rat}_d \times (\mathbb{C}_\infty)^\Omega \times (\mathbb{C}_\infty)^{f^{-1}(\Omega)},$$

where $(\mathbb{C}_\infty)^\Omega$, $(\mathbb{C}_\infty)^{f^{-1}(\Omega)}$ denote the sets of injections of, resp., Ω, $f^{-1}(\Omega)$ to $\mathbb{C}_\infty$ which map x_0, x_1, x_2 to, resp., 0, 1, ∞, and whose elements are (g, φ, ψ) such that $\varphi \circ f = g \circ \psi$ and the local degree of f in x is the same as g_{τ_n} in $\psi_{\tau_n}(x)$ for all $x \in f^{-1}(\Omega)$. It is a metric space with distance the sum of the distances in the spaces of the corresponding components of the ordered triples of the space, where the distance of two elements of $(\mathbb{C}_\infty)^\Omega$ is the maximum of the spherical distance $d_{\mathbb{C}_\infty}$ in $\mathbb{C}_\infty$ of the, resp., values at each point of $\mathbb{C}_\infty$. For each $c > 0$, denote $\mathbb{T}_{f,\Omega}(c)$ the set of $(g, \varphi, \psi) \in \mathbb{T}_{f,\Omega}$ such that $d_{\mathbb{C}_\infty}(z_1, z_2) \geq c$ for distinct $z_1, z_2 \in \psi\big(f^{-1}(\Omega)\big)$.

As the Thurston *pullback* $\mathcal{T}_f$ is holomorphic, the function defined in $\mathrm{Teich}(\mathbb{C}_\infty \backslash \Omega)$ by $\tau \mapsto (g_\tau, \varphi_\tau, \psi_\tau) \in \mathrm{Rat}_d \times (\mathbb{C}_\infty)^\Omega \times (\mathbb{C}_\infty)^{f^{-1}(\Omega)}$ also is.

If S is a hyperbolic Riemann surface and γ is a geodesic Jordan curve in S in the Poincaré metric of S, denote $\boldsymbol{w_S(\gamma)} = -\log \ell_S(\gamma)$. For γ a Jordan curve in $\mathbb{C}_\infty \backslash \Omega$ such that the two regions separated by γ in $\mathbb{C}_\infty$ contain ≥ 2 points of Ω and $[\varphi] = \tau \in \mathrm{Teich}(\mathbb{C}_\infty \backslash \Omega)$, denote $\boldsymbol{\ell_S(\gamma, \tau)}$ the length in the Poincaré metric of S of the geodesic Jordan curve in this metric in $\mathbb{C}_\infty \backslash \varphi(\Omega)$ homotopic to $\varphi(\gamma)$, and

$$\boldsymbol{w_S(\tau)} = \sup\big\{w_S(\gamma, \tau) : \gamma \text{ is the geodesic Jordan curve in } \mathbb{C}_\infty \backslash \Omega \text{ in the Poincaré Metric of } S\big\}.$$

For $C > 0$ denote $\mathbf{Teich}_C(\mathbb{C}_\infty \backslash \Omega) = \big\{\tau \in \mathrm{Teich}_C(\mathbb{C}_\infty \backslash \Omega) : w_S(\tau) \leq C\big\}$ (C is omitted when the context is clear).

(8.10) *If $(\mathbb{C}_\infty, f)$ is a branched covering of $\mathbb{C}_\infty$, $\Omega \subset \mathbb{C}_\infty$ is a finite set with distinct $x_0, x_1, x_2 \in \Omega$ and $CV_f \subset \Omega \subset f^{-1}(\Omega)$, then:*

1. *For all $c > 0$, $\mathbb{T}_{f,\Omega}(c)$ is a compact subset of $\mathbb{T}_{f,\Omega}$, with this set defined as in three paragraphs above and x_0, x_1, x_2 mapped to, resp., 0, 1, ∞ by injections into the last two components of each ordered triple of $\mathbb{T}_{f,\Omega}$.*

(continued)

2. *For every* $C > 0$, *there is* $c > 0$ *such that*

$$\tau \in \mathrm{Teich}_C(\mathbb{C}_\infty \backslash \Omega) \Longrightarrow (g_\tau, \varphi_\tau, \psi_\tau) \in \mathbb{T}_{f,\Omega}(c),$$

where $(g_\tau, \varphi_\tau, \psi_\tau)$ *represents* τ *in* $\mathbb{T}_{f,\Omega}$.

Proof

(1) For any sequence $\{(g,\varphi_n \psi_n)\} \subset \mathbb{T}_{f,\Omega}(c)$, let $Y_n = \varphi_n(\Omega)$,

$$X_n = \psi_n\big(f^{-1}(\Omega)\big) = f_n^{-1}(Y_n)$$

and order the elements of Ω as $\omega_1, \ldots, \omega_{\#\Omega}$. Since $\mathbb{C}_\infty$ is a compact space, the sequence $\{\varphi_n(\omega_1)\} \subset \mathbb{C}_\infty$ has some subsequence convergent to a point of $\mathbb{C}_\infty$. Proceeding successively to subsequent points of Ω with the previously obtained converging subsequences in the end, we get a subsequence of $\{\varphi_n\}$ convergent to an injection φ of Ω into $\mathbb{C}_\infty$, and proceeding analogously for the subsequence of $\{\psi_n\}$ whose terms are those of the orders retained in the subsequence of $\{\varphi_n\}$, there is a subsequence $\{\varphi_n\}$ convergent to an injection ψ of Ω in $\mathbb{C}_\infty$. Ignoring in the convergent subsequence of $\{\varphi_n\}$ the terms of the initial sequence $\{\psi_n\}$ of the excluded orders, we have a subsequence of $\{(\varphi_n, \psi_n)\}$ convergent to (φ, ψ). The sets $Y = \varphi(\Omega)$ and $X = \psi\big(f^{-1}(\Omega)\big) \subset \mathbb{C}_\infty$ are finite. As the spherical distance of distinct points in X_n is lower bounded by c, ψ is a bijection of $f^{-1}(\Omega)$ onto X and the spherical distance of distinct points in X is also lower bounded by c. The subsequence of $\{g_n\}$ with the terms of the retained orders for $(\varphi_n, \psi_n) \to (\varphi, \psi)$ converges to $g = k\frac{P}{Q}$, where

$$P(z) = \prod_{\substack{x \in f^{-1}(\{x_0\}) \\ x \neq x_2}} [z - \psi(x)]^{d_x}, \qquad Q(z) = \prod_{\substack{x \in f^{-1}(\{x_2\}) \\ x \neq x_2}} [z - \psi(x)]^{d_x},$$

with k the value of $\frac{P}{Q}$ at an arbitrary point of $\psi\big(f^{-1}(\{x_1\})\big)$. In $f^{-1}(\Omega)$, $g \circ \psi = \lim g_n \circ \psi_n(x) = \lim \varphi_n \circ f = \varphi \circ f$. The local degree of g at a point $\psi(x)$ is $\geq d_x$ and for each $y \in Y$, the number of preimages of $\{y\}$ by g counting multiplicities is

$$d = \sum_{x \in (\varphi\circ f)^{-1}(\{y\})} \deg_{\psi(x)} g \geq \sum_{x \in (\varphi\circ f)^{-1}(\{y\})} d_x \geq \sum_{z \in \varphi^{-1}(\{y\})} \sum_{x \in f^{-1}(\{z\})} d_x = d\,\#\varphi^{-1}(\{y\}).$$

Thus, $\#\varphi^{-1}(\{y\}) = 1$ and φ are injective. The local degree of g in $\psi(x)$ is d_x and $(g, \varphi, \psi) \in \mathbb{T}_{f,\Omega}(c)$. So, $\mathbb{T}_{f,\Omega}(c)$ is a compact subset of $\mathbb{T}_f$.

(2) Let $Y_\tau = \varphi_\tau(\Omega)$ and $X_\tau = \psi_\tau\big(f^{-1}(\Omega)\big) = g_\tau^{-1}(Y_\tau)$. The length of any geodesic Jordan curve in the Poincaré metric of $\mathbb{C}_\infty \backslash Y_\tau$ is lower bounded by e^{-C}. As $Y_\tau \supset CV_{g_\tau}$, $(\mathbb{C}_\infty \backslash X_\tau, g_\tau)$ is a covering of $\mathbb{C}_\infty \backslash Y_\tau$. Hence, the length of any

geodesic Jordan curve in the Poincaré metric of $\mathbb{C}_\infty \backslash X_\tau$ is also lower bounded by e^{-C}. Therefore, there is $c>0$ such that $d_{\mathbb{C}_\infty}(\psi_\tau(x), \psi_\tau(y)) \geq c$ uniformly for $\tau \in \text{Teich}_C(\mathbb{C}_\infty \backslash \Omega)$ and $x, y \in f^{-1}(\Omega)$ distinct, concluding the proof. □

(8.11) *If $\mathscr{R} = (\mathbb{C}_\infty, f)$ is a branched covering of $\mathbb{C}_\infty$ of degree $d \geq 2$, $\Omega \subset \mathbb{C}_\infty$ is a finite set with $\#\Omega \geq 4$ such that $PC(f) \subset \Omega$ and $f(\Omega) \subset \Omega$, then $X \subset \Omega$ with $\#X \geq 4$, $(f^{\#\Omega})^{-1}(\Omega) \subset \Omega \cup C_{f^{\#\Omega}}$, and the orbifold signature of $\mathcal{O}_{\mathscr{R}}$ is $(2, 2, 2, 2)$.*

Proof The proof is a simple counting. Define recursively $X_0 = X$ and $X_{j+1} = f^{-1}(X_j) \backslash C_f$. It is $X_{\#\Omega} = (f^{\#\Omega})^{-1}(X) \backslash C_{f^{\#\Omega}} \subset \Omega$ and $\#X_{\#\Omega} \leq \#\Omega$. Since $f^{-1}(X_j) \subset X_{j+1} \cup C_f$ for $j \in \{0, \ldots, \#\Omega - 1\}$, and $\#C_f = 2(d-1)$,

$$\#X_{j+1} + 2(d-1) \geq \#f^{-1}(X_i) \geq d\,\#X_i - 2(d-1), \quad j \in \{0, \ldots, \#\Omega - 1\}.$$

Therefore, $\#X_{j+1} - 4 \geq d(\#X_j - 4)$ for $j \in \{0, \ldots, \#\Omega - 1\}$, which applied recursively gives $d^{\#\Omega - 4} > \#\Omega - 4 \geq d^{\#\Omega}(\#X - 4) \geq 0$. It is $\#X_j = 4$ for $j \in \{0, 1, 2, 3, 4\}$ with equality in the inequalities for $j \in \{0, 1, 2, 3\}$, and
$\#C_f = 2(d-1)$; so, the critical points of f are simple.
$\#X_{j+1} + \#C_f = \#f^{-1}(X_j)$, and $C_f \cap X_{j+1} = \emptyset$ for $j \in \{0, 1, 2, 3\}$.
Since $\#f^{-1}(X_j) = d\,\#X_j - 2(d-1)$, $CV_f \subset X_j$ for $j \in \{0, 1, 2, 3\}$.
Consequently, $X_1 \subset PC(f)$, otherwise a point of $X_1 \backslash PC(f)$ would have $d^3 \geq 8$ preimages in X_4 in contradiction with $\#X_4 = 4$, but also $PC(f) \subset X_1$, since

$$CV_f \subset VC(f^2) \subset VC(f^3) \subset VC(f^4) \subset X_0,$$

implies the number of elements of these sets is between 2 and 4, and, as this number is nondecreasing, it cannot be strictly increasing. Thus, for some $j \in \{0, 1, 2, 3\}$, it is $VC(f^j) = VC(f^k)$ for $j < k \leq 4$, $k \in \{1, 2, 3, 4\}$, and $PC(f) = VC(f^j) = CV(f^3) \subset X_1$. Therefore,

$$PC(f) = X_1, \qquad \#PC(f) = 4, \qquad C_f \cap PC(f) = C_f \cap X_1 = \emptyset,$$

and the critical points are all simple. □

The **adjoint of the derivative** $D_\tau \widetilde{\mathcal{T}}_f$ defined from $T_\tau \text{Teich}(\mathbb{C}_\infty \backslash \Omega)$ to $T_{\widetilde{\mathcal{T}}_f(\tau)} \text{Teich}(\mathbb{C}_\infty \backslash \Sigma)$ corresponds to the *pushforward* $f_* : \mathcal{Q}(\psi(X)) \to \mathcal{Q}(\varphi(Y))$ by canonical identifications of the cotangent spaces $T^*_\tau \text{Teich}(\mathbb{C}_\infty \backslash \Omega)$ and $T^*_{\widetilde{\mathcal{T}}_f(\tau)} \text{Teich}(\mathbb{C}_\infty \backslash \Sigma)$ with the spaces of quadratic differentials, resp., $\mathcal{Q}(\varphi(\Omega))$ and $\mathcal{Q}(\psi(\Sigma))$. Thus,

$$[\varphi] = \tau \in \text{Teich}(\mathbb{C}_\infty \backslash \Omega), \quad [\psi] = \widetilde{\mathcal{T}}_f(\tau) \in \text{Teich}(\mathbb{C}_\infty \backslash \Sigma),$$
$$\|D_\tau \widetilde{\mathcal{T}}_f\| = \|(D_\tau \widetilde{\mathcal{T}}_f)^*\| = \|f_*\|.$$

(8.12) *If $\mathscr{R}=(\mathbb{C}_\infty, f)$ is a branched covering of $\mathbb{C}_\infty$ of degree $d\geq 2$, $\Omega\subset\mathbb{C}_\infty$ is a finite set with $\#\Omega\geq 4$, $PC(f)\subset\Omega$, $f(\Omega)\subset\Omega$ and the signature of the orbifold $\mathcal{O}_{\mathscr{R}}$ is not $(2,2,2,2)$, then the composition of the Thurston* pullback $\widetilde{\mathcal{T}}_f$ *from* $\text{Teich}(\mathbb{C}_\infty\backslash\Omega)$ *to* $\text{Teich}(\mathbb{C}_\infty\backslash\Omega)$ $\#\Omega$ *times with itself is for each $C>0$ a uniform contraction in* $\text{Teich}_C(\mathbb{C}_\infty\backslash\Omega)$, *i.e. there are $\lambda\in]0,1[$ such that $\|D_\tau(\widetilde{\mathcal{T}}_f)^{\#\Omega}\|<\lambda$ for all $\tau\in$* $\text{Teich}_C(\mathbb{C}_\infty\backslash\Omega)$.

Proof We argue by absurdity. Assume there is $\{\tau_n\}\subset\text{Teich}_C(\mathbb{C}_\infty\backslash\Omega)$ such that $\|D_{\tau_n}(\widetilde{\mathcal{T}}_f)^{\#\Omega}\|\to 1$. Fixing $x_0,x_1,x_2\in\Omega\subset f^{-1}(\Omega)$, to each τ_n corresponds a $(g_{\tau_n},\varphi_{\tau_n},\psi_{\tau_n})\in\mathbb{T}_{f^{\#\Omega},\Omega}=\text{Rat}_d\times\mathbb{C}_\infty^{\Omega}\times\mathbb{C}_\infty^{(f^{\#\Omega})^{-1}(\Omega)}$. Let $X_n=\psi_{\tau_n}(\Omega)$ and $Y_n=\varphi_{\tau_n}(\Omega)$. It is[29] $\|D_{\tau_n}(\widetilde{\mathcal{T}}_f)^{\#\Omega}\|=\|(g_{\tau_n})_*\|$, with $(g_{\tau_n})_*:\mathcal{Q}(X_n)\to\mathcal{Q}(Y_n)$. Therefore, there is a sequence of quadratic differentials $\{q_n\}\subset\mathcal{Q}(X_n)$ with $\|q_n\|=1$ such that $\|(g_{\tau_n})_*q_n\|\to 1$. By (8.10), $\{(g_{\tau_n},\varphi_{\tau_n},\psi_{\tau_n})\}$ is included in a compact subset of $\mathbb{T}_{f^{\#\Omega},\Omega}$. Therefore, some subsequence of the considered sequence converges to a point (g,φ,ψ) of this set. By (8.8), $\|g_*\|\leq 1$, with equality if and only if there is $q\in\mathcal{Q}(\psi(\Omega))$ such that $q=d^{-\#\Omega}g^*(g_*q)$. Suppose $\|g_*\|=1$. Let $Y\subset\varphi(\Omega)$ be the set of poles of g_*q. Each point of $g^{-1}(Y)$ is a pole of q or a critical point of g. Therefore, $g^{-1}(Y)\subset\psi(\Omega)\cup C_g$. Hence, $X=\varphi^{-1}(Y)$ satisfies $X\subset\Omega$ and $(f^{\#\Omega})^{-1}(X)\subset\Omega\cup C_{f^{\#\Omega}}$. By (8.11), the signature of the orbifold $\mathcal{O}_{\mathscr{R}}$ is $(2,2,2,2)$, contradicting the hypothesis. Consequently, $\|g_*\|<1$.

The poles of the quadratic differentials of the terms of the subsequence of $\{q_n\}$ corresponding to the convergent subsequence considered in the previous paragraph are simple, and there is $\delta>0$ such that the minimum spherical distance of pairs of poles is $\geq\delta$ for all $n\in\mathbb{N}$. Hence, there is a subsequence of that subsequence converging uniformly on compact subsets of $\mathbb{C}_\infty\backslash X$ to some $q\in\mathcal{Q}(X)$ with $\|q\|=1$, and the corresponding sequence $\{(g_n)_*q_n\}$ converges uniformly on compact subsets of $\mathbb{C}_\infty\backslash Y$ to $g_*q\in\mathcal{Q}(Y)$. As the poles of $(g_n)_*q_n$ belong to Y_n, there is $\epsilon>0$ such that the minimum spherical distance of pairs of poles is $\geq\epsilon$. So, $\|g_*q\|=\lim\|(g_n)_*q_n\|=1=\|q\|$, contradicting $\|q_*\|<1$. Thus, there is $\lambda\in:]0,1[$ with $\|D_\tau(\widetilde{\mathcal{T}}_f)^{\#\Omega}\|<\lambda$ for $\tau\in\text{Teich}_C(\mathbb{C}_\infty\backslash\Omega)$. □

It can be proved that the composition of $\widetilde{\mathcal{T}}_f$ k times with itself is a uniform contraction, since $k\geq 2$ and $d^{k-2}>\#\Omega\backslash PC(f)$. In the W. Thurston original version of the theorem, $\Omega=PC(f)$ and we have a uniform contraction with $k=2$.

Geodesic Jordan Curves in Hyperbolic Riemann Surfaces

The proof of the Thurston theorem of characterization of postcritically finite rational functions presented here uses estimates involving lengths of geodesic Jordan curves in the Poincaré metric of hyperbolic Riemann surfaces which are established in this section.

[29] See the preceding subsection.

If S is a hyperbolic Riemann surface and γ is a geodesic Jordan curve in S in the Poincaré metric of S, denote $\boldsymbol{\ell_S(\gamma)}$ the length of γ in this metric. S is omitted if the context is clear.

The **equator of a ring** A with module $M(A)$ contained in a Riemann surface S is the unique geodesic Jordan curve in the Poincaré metric of A, with length $\frac{M(A)}{\pi}$ in this metric.

(8.13) *If S is a hyperbolic Riemann surface, then:*

1. *If $A \subset S$ is an open ring with equator homotopic to a geodesic Jordan curve $\widetilde{\gamma}$ in the Poincaré metric of S, then $M(A) = \frac{\pi}{\ell_A(\widetilde{\gamma})} \le \frac{\pi}{\ell_S(\widetilde{\gamma})}$, there is a function $\eta :]0, +\infty[\to]0, +\infty[$ strictly decreasing from $+\infty$ to 0, called the* ***collar ring width function****, such that geodesics in the Poincaré metric of A disjoint from γ apart from each other $< \eta(\ell_A(\gamma))$ are not simple curves and $\eta(\ell_A(\gamma))$ is the supremum of the real numbers satisfying this. If $A_\ell(\delta) = \{z \in A_\ell : d_{A_\ell}(z, \gamma) < \delta\}$, with A_ℓ the ring containing a geodesic Jordan curve γ of length ℓ in the Poincaré metric of A_ℓ, then $M(A_\ell(\eta(\ell))) > \frac{\pi}{\ell} - 1$ for $\ell > 0$.*
2. *If $\gamma_1, \gamma_2 \subset S$ are distinct geodesic closed curves in the Poincaré metric of S such that $\ell_S(\gamma_j) < 2\log(1+\sqrt{2})$ for $j \in \{1, 2\}$, then γ_1, γ_2 are disjoint Jordan curves.*
3. *If $\gamma_1, \ldots, \gamma_n$ are disjoint geodesic Jordan curves in the Poincaré metric of S, there are disjoint rings $C_S(\gamma_1), \ldots, C_S(\gamma_n) \subset S$, with γ_j the equator of $C_S(\gamma_j)$ (called the* ***collar of*** *γ_j) isometric to $A_{\ell_j}(\eta(\ell_j))$ for $j \in \{1, \ldots, n\}$* (Fig. 8.1).
4. *If $\Omega \subset S$ is a finite set, $0 < L < 2\log(1+\sqrt{2})$, γ is a geodesic Jordan curve in the Poincaré metric of S, $\{\widetilde{\gamma}_j\}_{j \in J}$, $J \subset \mathbb{N} \cup \{0\}$ (possibly $\emptyset$) is the set of geodesic Jordan curves in $S' = S \backslash \Omega$ in the Poincaré metric of S' homotopic to γ in S such that $\ell_{S'}(\widetilde{\gamma}_j) < L$ for $j \in J$, then:*
   (a) $\ell_S(\gamma) \le \ell_{S'}(\widetilde{\gamma}_j)$ for $j \in J$, $\#J \le \#\Omega + 1$, and J is finite.
   (b) $\frac{1}{\ell_S(\gamma)} - \frac{1}{\pi} - \frac{\#\Omega+1}{L} < \sum_{j \in J} \frac{1}{\ell_{S'}(\widetilde{\gamma}_j)} < \frac{1}{\ell_S(\gamma)} + \frac{\#\Omega+1}{L}$
   (if $J = \emptyset$, then $\frac{1}{\ell_S(\gamma)} - \frac{1}{\pi} - \frac{\#\Omega+1}{L} < 0$).

Proof

(1) If γ is the equator of A, $M(A) = \frac{\pi}{\ell_A(\gamma)}$. Since $A \subset S$ and the Poincaré distances of pairs of points do not increase with application holomorphic functions,[30] $\ell_S(\gamma) \le \ell_A(\gamma)$. Hence, $M(A) = \frac{\pi}{\ell_A(\gamma)} \le \frac{\pi}{\ell_S(\gamma)}$. Throughout homotopies, the length of geodesics cannot decrease. So, if $\widetilde{\gamma} \subset S$ is a

[30] See Section 5 of Chapter 12 of *CADOVA*.

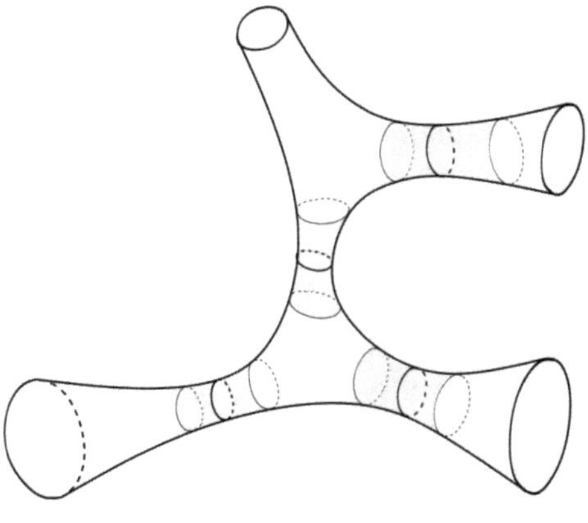

Fig. 8.1 Collars of geodesic Jordan curves in hyperbolic Riemann surface

geodesic in the Poincaré metric of S homotopic to γ, then $\ell_S(\widetilde{\gamma}) \le \ell_S(\gamma)$, and $M(A) \le \frac{\pi}{\ell_S(\gamma)} \le \frac{\pi}{\ell_S(\widetilde{\gamma})}$.

The function $z \mapsto w = \tanh^{-1} z = \frac{1}{2}\log\frac{1+z}{1-z}$ is a conformal homeomorphism of B_1 onto $\mathbb{B} = \mathbb{R} \times \left]-\frac{\pi}{4}, \frac{\pi}{4}\right[$. As the element of length of the Poincaré metric on B_1 is $\frac{2\,|dz|}{1-|z|^2}$ and $z = \tanh w = \frac{e^w - e^{-w}}{e^w + e^{-w}}$ gives $dz = \frac{4}{(e^w+e^{-w})^2}\,dw$, with $w = x + iy$ and $x, y \in \mathbb{R}$, this length element is

$$\frac{2}{1-|(e^w-e^{-w})/(e^w+e^{-w})|^2}\,\frac{4\,|dw|}{|e^w+e^{-w}|^2} = \frac{8\,|dw|}{|e^w+e^{-w}|^2-|e^w-e^{-w}|^2}$$

$$= \frac{8\,|dw|}{(e^w+e^{-w})(e^{\overline{w}}+e^{-\overline{w}})-(e^w-e^{-w})(e^{\overline{w}}-e^{-\overline{w}})} = \frac{8\,|dw|}{2(e^{i2y}+e^{-i2y})} = \frac{2\,|dw|}{\cos(2\,\mathcal{I}m\,w)}.$$

In particular, on the real axis, the Poincaré metric of $\mathbb{B}$ is twice the Euclidean metric. The semicircle included in ∂B_1 in the open upper complex half plane $\mathbb{H}$ (resp., lower complex half plane $-\mathbb{H}$) is mapped to the upper (resp., lower) horizontal line bounding the band $\mathbb{B}$ and the points ± 1 are mapped to $\pm\infty$. The geodesics in the Poincaré metric of $\mathbb{B}$ are the images of the geodesics in the Poincaré metric of B_1, and, therefore, are diameters or arcs of circles in B_1 orthogonal to ∂B_1.

If $A \subset S$ is an open ring with Equator γ of length $\ell_A(\gamma)$, there is a covering $(\mathbb{B}, p)$ of A such that $p(\gamma) = \mathbb{R}$, and $A = \mathbb{B}/(\ell\,\mathbb{Z})$. For a geodesic ζ in the Poincaré metric of A orthogonal to γ, let ζ_1, ζ_2 be lifts of ζ to successive sheets of the covering intersecting the line $\mathcal{I}m\,w = \frac{\pi}{4}$ at points b_1, b_2 with $\mathcal{R}e\,b_2 - \mathcal{R}e\,b_1 = \ell_A(\gamma)$. The geodesic $\alpha \subset \mathbb{B}$ in the Poincaré metric of $\mathbb{B}$ with end points b_1, b_2 is disjoint of $\mathbb{R}$, because distinct geodesics in the Poincaré metric of B_1 cannot intersect in more than 1 point where they intersect transversally and they can only have tangential intersections in ∂B_1. Therefore, distinct geodesics in $\mathbb{B}$ cannot intersect in more than 1 point, and

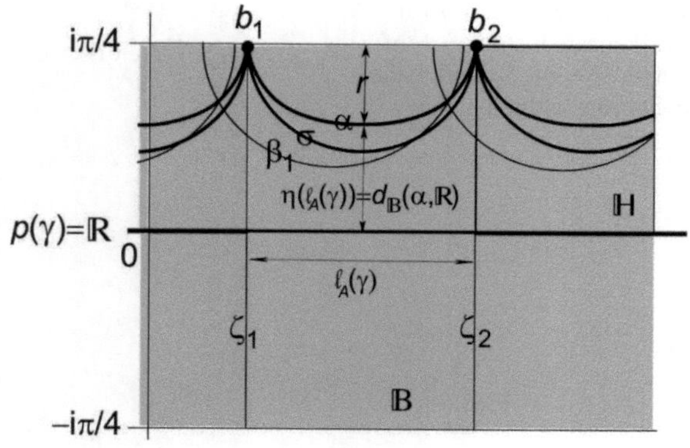

Fig. 8.2 Geodesics in covering space $\mathbb{B} = \mathbb{R}\times\left]-\frac{\pi}{4}, \frac{\pi}{4}\right[$ of ring A with equator γ

transversally if the intersection point does not belong to $\partial\mathbb{B}$. So, $\alpha \subset \mathbb{B}\cap\mathbb{H}$ (Fig. 8.2).

Define $\eta\big(\ell_A(\gamma)\big) = d_{\mathbb{B}}(\alpha, \mathbb{R})$. η is strictly decreasing[31] and a geodesic $\beta_1 \subset \mathbb{B}\cap\mathbb{H}$ in the Poincaré metric of $\mathbb{B}$ intersects its horizontal translation by $\ell_A(\gamma)$ if and only if the Euclidean distance of its end points is $< \ell_A(\gamma) = \frac{\pi}{M(A)}$ (i.e., $d_{\mathbb{B}}(\beta_1, \mathbb{R}) > \eta\big(\ell_A(\gamma)\big)$).In this case, β_1 is not a simple curve (Fig. 8.2).

Let h_0 be the bounded harmonic function in the lower complex half plane $\mathbb{L}$ bounded by the line $\mathcal{I}m\, w = \frac{\pi}{4}$ which is 0 in $[b_1, b_2] \times \left\{\frac{\pi}{4}\right\}$ and 1 at the other points of $\mathbb{R} \times \left\{\frac{\pi}{4}\right\}$, and h a harmonic function in $\mathbb{B}$ which is 0 in $[b_1, b_2] \times \left\{\frac{\pi}{4}\right\}$ and 1 at the other points of $\partial\mathbb{B}$. By the maximum principle for harmonic functions,[32] $h > h_0$ in $\mathbb{B}$ .The geodesic in the poincaré metric of $\mathbb{L}$ with end points b_1, b_2 is the semicircle σ in $\mathbb{L}$ with center $\frac{b_1+b_2}{2}$ which is the level curve $\frac{1}{2}$ of h_0 .The geodesic α in the Poincaré metric of $\mathbb{B}$ with end points b_1, b_2 is the level curve $\frac{1}{2}$ of h, and, therefore, it is included in the half disk bounded by $[b_1, b_2] \times \left\{\frac{\pi}{4}\right\}$ and by the semicircle in $\mathbb{B}$ with center $\frac{b_1+b_2}{2}$ and radius $\frac{b_2-b_1}{2}$. Hence, if r is the Euclidean distance from the horizontal line that bounds $\mathbb{B}$ from above to the furthest away point of α , then $r \le \frac{\ell_A(\gamma)}{2}$ (Fig. 8.2).

There is a conformal homeomorphism T defined from the rectangle $]0, \ell[\times\left]-\frac{\pi}{4}, \frac{\pi}{4}\right[$ onto the ring $A = B_R \backslash \mathrm{cl}\, B_{1/R}$ with $R > 0$ and $M(A) = \frac{1}{2\pi}\log\frac{R}{1/R} = \frac{\log R}{\pi}$ whose only geodesic Jordan curve in the, resp., Poincaré metric is the disk[33] with radius 1 and center 0 and length in this metric $\ell_A(\gamma) = \frac{\pi}{M(A)} = \frac{\pi^2}{\log R}$. T maps the rectangle $]0, \ell[\times\left]-\frac{\pi}{4} + r, \frac{\pi}{4} - r\right[$

[31] It can be proved that $\eta(\ell) = \frac{1}{2}\log\frac{\cosh(\ell/2)+1}{\cosh(\ell/2)-1}$.

[32] See Section 6 of Chapter 9 of *CADOVA*.

[33] See Example (12.30.2) in Chapter 12 of *CADOVA* .

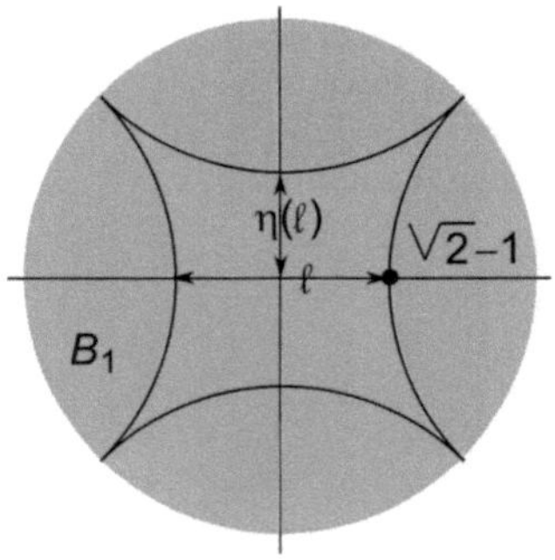

Fig. 8.3 Geodesic in B_1 in the Poincaré metric of B_1 of length $\ell^* = \eta(\ell^*)$, with η the collar ring width function

to an annulus $B_\rho \backslash \mathrm{cl}\, B_{1/\rho}$ with $0 < \rho < R$. The distance of the straight lines $\mathcal{I}m\, z = \frac{\pi}{4}$ and $\mathcal{I}m\, z = \frac{\pi}{4} - R$ in the Poincaré metric of $\mathbb{B}$ is $\int_{\pi/4-r}^{\pi/4} \frac{2}{\cos(2y)}\, dy$, and, due to the Poincaré distance invariance under conformal homeomorphisms, this distance is equal to that of the circles ∂B_R and B_ρ in the Poincaré metric of the annulus $B_R \backslash \mathrm{cl}\, B_{1/R}$, namely[33],

$$\int_\rho^R \pi/\big[2(\log R)\, x \cos\big(\tfrac{\pi \log x}{2\log R}\big)\big]\, dx = \int_{\frac{\pi \log \rho}{4 \log R}}^{\frac{\pi}{4}} \frac{2}{\cos(2t)}\, dt\,,$$

by the change of integration variable $2t = \frac{\pi \log x}{2 \log R}$. By the equality of the two distances in the mentioned Poincaré metrics, $\frac{\pi \log \rho}{4 \log R} = \frac{\pi}{4} - R$. If $\ell = \ell_A$, since $\frac{4r}{\pi} < \frac{\ell_A(\gamma)}{\pi} = \frac{1}{M(A)}$,
$M\big(A_\ell(\eta(\ell))\big) = \frac{\log \rho}{\pi} = \frac{4 \log R}{\pi^2}\left(\frac{\pi}{4} - r\right) = M(A)\left(1 - \frac{4r}{\pi}\right) > M(A) - 1 = \frac{\pi}{\ell} - 1\,.$

(2) Since η is strictly decreasing in $]0, +\infty[$ from $+\infty$ to 0, the equation $\ell = 2\eta(\ell)$ has a unique solution ℓ^*. If $\gamma_1, \gamma_2 \subset S$ are distinct intersecting geodesic closed curves in the Poincaré metric of S, then $\ell_S(\gamma_1) \geq 2\eta\big(\ell_S(\gamma_2)\big)$ and $\ell_S(\gamma_2) \geq 2\eta\big(\ell_S(\gamma_1)\big)$. So, if $\ell_S(\gamma_1), \ell_S(\gamma_1) < \ell^*$, then $\gamma_1 \cap \gamma_2 = \emptyset$. Applying to $\ell_A(\gamma) = 2\eta\big(\ell_A(\gamma)\big)$, which holds in $\mathbb{B}$, the conformal homeomorphism $\tanh z$, we get in B_1 Fig. 8.3. Thus,

$$\ell^* = 2\int_0^{\sqrt{2}-1} \frac{2}{1-x^2}\, dx = 2\int_0^{\sqrt{2}-1} \left[\frac{1}{1+x} + \frac{1}{1-x}\right] dx = 2\left[\log \frac{1+x}{1-x}\right]_0^{\sqrt{2}-1}$$
$$= 2 \log\big(\sqrt{2} + 1\big)\,.$$

(3) For every $j \in \{1, \ldots, n\}$, there is an open ring $A_{\gamma_j} \subset S$ where γ_j is the unique geodesic Jordan curve in the Poincaré metric of A_{γ_j}. So, there is a conformal homeomorphism T of the annulus $B_R \backslash \mathrm{cl}\, B_{1/R}$ with $\ell_{A_{\gamma_j}} = \frac{\log R}{\pi}$ onto A_{γ_j}. Consider the function $F : A_{\gamma_1} \times \cdots \times A_{\gamma_n} \to S$ with

$$F(z_1, \ldots, z_n) = \big(T^{-1}(z_1), \ldots, T^{-1}(z_n)\big)\,.$$

There are collars[34] of the curves γ_j isometric to $A_{\ell_j}(\eta(\ell_j))$ for $j \in \{1, \dots, n\}$ if and only if F is injective.

Suppose $F(w_j) = F(w_k)$ with $w_j \in A_{\ell_j}(\eta(\ell_j))$ and $w_k \in A_{\ell_j}(\eta(\ell_k))$ for some $j, k \in \{1, \dots, n\}$. Let δ_j, δ_k be geodesics in the Poincaré metric of, resp., $A_{\ell_j}(\eta(\ell_j))$ and $A_{\ell_j}(\eta(\ell_j))$ with end points, resp., w_j, w_k and a point of, resp., γ_j, γ_k. Then, $\ell_{A_{\ell_j}(\eta(\ell_s))} < \eta(\ell_{A_{\ell_j}(\eta(\ell_s))})$ for $s \in \{j, k\}$. Consider a universal covering (B_1, p) of $\cup_{j=1}^n T^{-1}(A_{\gamma_j})$. The image by T^{-1} of the geodesic Jordan curve of $A_{\ell_j}(\eta(\ell_j))$ and of $A_{\ell_j}(\eta(\ell_k))$ is the circle of radius 1 in $B_R \backslash \text{cl}\, B_{1/R}$. The lifts $\widetilde{\gamma}_j, \widetilde{\gamma}_k$ by p of, resp., γ_j, γ_k are geodesics in the Poincaré metric of B_1 distant in this metric $< \eta(\ell_j) + \eta(\ell_k)$. Since $\widetilde{\gamma}_j, \widetilde{\gamma}_k$ are different lifts of a same Jordan curve in $B_R \backslash \text{cl}\, B_{1/R}$ or different lifts of disjoint Jordan curves in $B_R \backslash \text{cl}\, B_{1/R}$, it is $\widetilde{\gamma}_j \cap \widetilde{\gamma}_k = \emptyset$. Let $\alpha, \beta_1, \beta_2, \widetilde{\beta}_1, \widetilde{\beta}_2$ be geodesic arcs in the Poincaré metric of B_1 with α orthogonal to both $\widetilde{\gamma}_j, \widetilde{\gamma}_k$, the arcs β_1, β_2 different orthogonals to $\widetilde{\gamma}_j$ at the distance $\frac{\ell_j}{2}$ from α, and $\widetilde{\beta}_1, \widetilde{\beta}_2$ different orthogonals to $\widetilde{\gamma}_k$ at a distance $\frac{\ell_k}{2}$ from α, with $\beta_1, \widetilde{\beta}_1$ in the same side of α in B_1. Given the symmetry on α, a priori there would be four possible configurations: I, II, III, and IV in Fig. 8.4. However, denoting ρ_θ the reflection relative to a geodesic θ, with the automorphisms $g = \rho_{\beta_1} \circ \rho_\alpha$ and $\widetilde{g} = \rho_{\widetilde{\beta}_1} \circ \rho_\alpha$ of B_1, $h = \widetilde{g} \circ g^{-1} = \rho_{\widetilde{\beta}_1} \circ \rho_{\beta_1}$ belongs to the fundamental group of S. For configuration II, h would have a fixed point that is impossible; for configuration III $g(\widetilde{\gamma}_k)$ is a geodesic of B_1 intersecting $\widetilde{\gamma}_k$ transversally, what is impossible since this curve is a simple geodesic, and similarly for the configuration IV. So, only configuration I is possible. Hence, it must be $w_j = w_k$, implying that F is injective, ending the proof of 3.

(4a) Since $S' \subset S$, it is $\ell_S(\widetilde{\gamma}_j) \leq \ell_{S'}(\widetilde{\gamma}_j)$, and $\widetilde{\gamma}_j$ is homotopic to γ in S, which is a geodesic in the Poincaré metric of S, $\ell_S(\gamma) \leq \ell_S(\widetilde{\gamma}_j) \leq \ell_{S'}(\widetilde{\gamma}_j)$. By 2, the curves $\widetilde{\gamma}_j$, $j \in J$, are disjoint, and, as they are homotopic in S and disjoint, they bound a ring in S. As they are not homotopic in S', this ring contains at least one point of Ω; therefore, $\#J \leq \#\Omega + 1$.

(4b) By 3, there are disjoint collars $C_{S'}(\widetilde{\gamma}_j)$. Hence, there is an open ring $A \subset S$ containing $\cup_{j \in J} C_{S'}(\widetilde{\gamma}_j)$ with equator homotopic to γ in S. If $J = \emptyset$, the right-hand side of the inequality in the statement holds trivially. If $J \neq \emptyset$, since $\#J \leq \#Q + 1$,

$$\sum_{j \in J} \frac{1}{\ell_{S'}(\widetilde{\gamma}_j)} - \frac{\#Q+1}{L} \leq \sum_{j \in J} \left(\frac{1}{\ell_{S'}(\widetilde{\gamma}_j)} - 1\right) < \sum_{j \in J} M\big(C_{S'}(\widetilde{\gamma}_j)\big) \leq M(A) \leq \frac{\pi}{\ell_S(\gamma)},$$

where the last inequality is due to 1 and the one before follows from the monotonicity of the ring module in (1.13).

---

[34] The usefulness of the general property of existence of collars of geodesic Jordan curves in hyperbolic Riemann surfaces was noticed by Linda Keen (1940-) in 1974, with proofs by John Peter Matelski in 1976, Peter Buser (1946-) in 1978 and others later on.

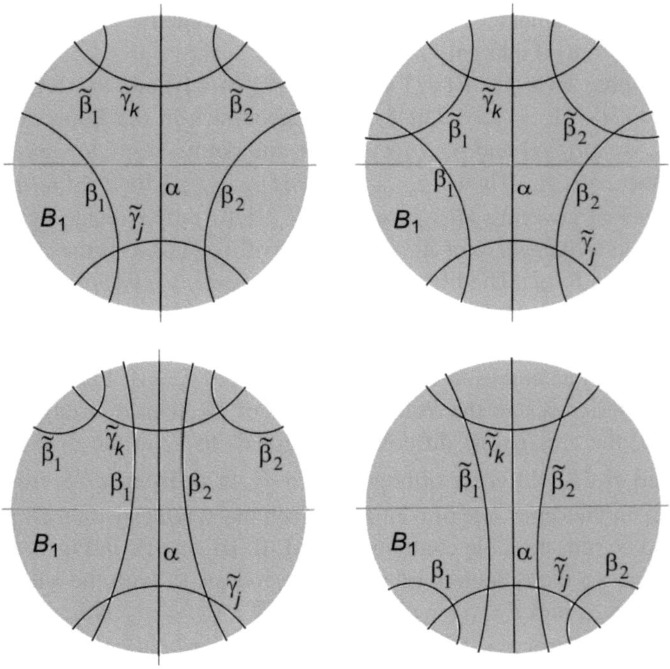

Fig. 8.4 The 4 *a priori* possible configuration for 4 geodesics in the Poincaré metric of B_1 orthogonal to disjoint geodesics $\widetilde{\gamma}_j$ and $\widetilde{\gamma}_k$ and symmetric relatively to a geodesic α orthogonal to both $\widetilde{\gamma}_j, \widetilde{\gamma}_k$

For the inequality on the left-hand side, consider the collar $C_S(\gamma)$ conformal to $A_{\eta(\ell_S(\gamma))(\gamma)}$ whose existence is assured by 3. This collar decomposes as a disjoint union of collars $C_1, \dots, C_k \subset S'$ with any pair of the equators nonhomotopic in S', for $k \in \{1, \dots, \#Q+1\}$, and $\sum_{m=1}^{k} M(C_m) = M\big(C_S(\gamma)\big)$. For $m \in \{1, \dots, k\}$, denote δ_m the geodesic in the Poincaré metric of S' homotopic in S' to the equator of C_m.

$$\begin{aligned} M\big(C_S(\gamma)\big) &= \sum_{m=1}^{k} M(C_m) = \sum_{\{m:\, C_m \leq \frac{\pi}{L}\}} M(C_m) + \sum_{\{m:\, C_m > \frac{\pi}{L}\}} M(C_m) \\ &\leq \tfrac{(\#Q+1)\pi}{L} + \sum_{\{m:\, C_m > \frac{\pi}{L}\}} \tfrac{\pi}{\ell_{S'}(\delta_m)} . \end{aligned}$$

By 3, $C_S(\gamma)$ is conformal to $A_{\ell_S(\gamma)}\big(\eta(\ell_S(\gamma))\big)$, and with 1,

$$M\big(C_S(\gamma)\big) = M\Big(A_{\ell_S(\gamma)}\big(\eta(\ell_S(\gamma))\big)\Big) > \tfrac{\pi}{\ell_S(\gamma)} - 1 .$$

If $\{m : C_m > \frac{\pi}{L}\} \neq \emptyset$, since for m in this set $\ell_{S'}(\delta_m) < L$ and there is $j \in J$ such that $\delta_m = \widetilde{\gamma}_j$, then $\frac{\pi}{\ell_S(\gamma)} - 1 < \frac{(\#Q+1)\pi}{L} + \sum_{j \in J} \frac{\pi}{\ell_{S'}(\widetilde{\gamma}_j)}$ and $J \neq \emptyset$. If $J = \emptyset$, also $\{m : C_m > \frac{\pi}{L}\} = \emptyset$, and $\frac{\pi}{\ell_S(\gamma)} - 1 < \frac{(\#Q+1)\pi}{L}$. □

The following result establishes that the function of $\mathrm{Teich}(\mathbb{C}_\infty \backslash \Omega)$ with the Teichmüller distance for $\mathbb{R}$, $\tau \mapsto w_S(\gamma, \tau) = -\log \ell_S(\gamma, \tau)$, with γ a Jordan curve in $S = \mathbb{C}_\infty \backslash \Omega$, and $\Omega \subset \mathbb{C}_\infty$ with $\#\Omega \geq 3$ and w_S as defined before[35] (8.10), is Lipschitzian with Lipschitz constant[36] 2.

(8.14) *If* $\tau_1, \tau_2 \in \mathrm{Teich}(\mathbb{C}_\infty \backslash \Omega)$, $\Omega \subset \mathbb{C}_\infty$ *is finite,* $\#\Omega \geq 3$, *and* γ *is a Jordan curve in* $S = \mathbb{C}_\infty \backslash \Omega$, *then* $\left|w(\gamma, \tau_1) - w(\gamma, \tau_2)\right| \leq 2\, D_T(\tau_1, \tau_2)$. *If* $\tau_1 \in \mathrm{Teich}_C(\mathbb{C}_\infty \backslash \Omega)$, *then* $\tau_2 \in \mathrm{Teich}_{C+2\, D_T(\tau_1, \tau_2)}(\mathbb{C}_\infty \backslash \Omega)$.

Proof For $D > D_T(\tau_1, \tau_2)$, $[\varphi_1] = \tau_1$, $[\varphi_2] = \tau_2 \in \mathrm{Teich}(\mathbb{C}_\infty \backslash \Omega)$, there is a quasiconformal mapping $h : \varphi_1(\mathbb{C}_\infty \setminus \Omega) \to \varphi_2(\mathbb{C}_\infty \backslash \Omega)$ homotopic to $\varphi_2 \circ \varphi_1^{-1}$ with $\log K_h \leq 2\,D$. Let $S_j = \varphi_j(\mathbb{C}_\infty \backslash \Omega)$, $j \in \{1, 2\}$, γ_1 be a geodesic Jordan curve in S_1 in the Poincaré metric of S_1, γ_2 a geodesic Jordan curve in S_2 in the Poincaré metric of S_2 homotopic to $h(\gamma_1)$, and (A_j, p_j) a covering of S_j with A_j a ring. The corresponding ring modules[37] are $M(A_j) = \frac{1}{2\pi} \log R_j = \frac{\pi}{\ell_{S_j}(\gamma_j)}$, where $R_j > 1$ is such that A_j is conformal to $B_{R_j} \backslash \mathrm{cl}\, B_1$, since the Poincaré length of the geodesic circle in this set around 0, equal to $\ell_{S_j}(\gamma_j)$, is[38] $\frac{2\pi^2}{\log R_j}$, for $j \in \{1, 2\}$. h has lift to a K_h-quasiconformal mapping from A_1 onto A_2. By (1.16, $\frac{1}{K_h} \leq \frac{M(A_1)}{M(A_2)} \leq K_h$, and

$$\left| \log \frac{\ell_{S_1}(\gamma_1)}{\ell_{S_2}(\gamma_2)} \right| = \left| \log \frac{M(A_1)}{M(A_2)} \right| \leq \log K_h .$$

Thus,

$$|w(\gamma, \tau_1) - w(\gamma, \tau_2)| = |\log \ell(\gamma, \tau_1) - \log \ell(\gamma, \tau_2)| \leq \log K_h \leq 2D .$$

As $D > D_T(\tau_1, \tau_2)$ is arbitrary, the stated inequality holds. □

Thurston Obstructions

For $\Omega \subset \mathbb{C}_\infty$, a Jordan curve in $\mathbb{C}_\infty$ is said to be **peripheral** to Ω if both connected components it separates in $\mathbb{C}_\infty$ have ≥ 2 points of Ω.

A **multicurve** of a marked branched covering $\big((\mathbb{C}_\infty, f), \Omega\big)$ of $\mathbb{C}_\infty$, with $\Omega \subset \mathbb{C}_\infty$ finite, $\#\Omega \geq 3$ and $\Omega \supset PC(f)$, is a finite set $\Gamma = \{\gamma_1, \ldots, \gamma_n\}$ of disjoint non-peripheral Jordan curves to $\mathbb{C}_\infty \backslash \Omega$ nonhomotopic (rel Ω) to any of the others

[35] Recall that $\ell_S(\gamma, \tau)$ is the length in the Poincaré metric of S of the geodesic Jordan curve in this metric in $\mathbb{C}_\infty \setminus \varphi(\Omega)$ homotopic to $\varphi(\gamma)$ with $\varphi \in \tau \in \mathrm{Teich}(\mathbb{C}_\infty \backslash \Omega)$.

[36] Lipschitz, Rudolf (1832–1903).

[37] See Section "Quasiconformal Transformations and Ring Module" in Chapter 1.

[38] See Example (12.30.2) in Chapter 12 of *CADOVA*.

nor to a point, where **homotopic (rel Ω)** abbreviates homotopic relative to Ω, meaning with the points of Ω fixed. Γ is said to be a **stable multicurve** if each curve in $f^{-1}(\Gamma)$ is homotopic to one of the curves in Γ (rel Ω) or is peripheral to $\mathbb{C}_\infty \backslash \Omega$. If so, each curve in $f^{-m}(\Omega)$, $m \in \mathbb{N}$, is homotopic (rel Ω) to a curve of Γ or is peripheral to $\mathbb{C}_\infty \backslash \Omega$.

A **transition matrix M_Γ of a stable multicurve** Γ is

$$M_\Gamma = \left[\sum_{\delta=1}^{\Delta_{jk}} \frac{1}{d_{jk\delta}} \right]_{j,k=1}^{n,n},$$

with $\{\gamma_{jk\delta} : \delta \in \{1, \dots, \Delta_{jk}\}\}$ the set of the $\Delta_{jk} \in \mathbb{N}$ connected components of $f^{-1}(\gamma_k)$ homotopic (rel Ω) to γ_j and $d_{jk\delta}$ the degree of the covering $(\gamma_{jk\delta}, f)$ of γ_k. Let $\boldsymbol{\lambda_\Gamma}$ denote the principal eigenvalue of M_Γ which, by the Perron-Frobenius theorem,[39] is $\lambda_\Gamma \geq 0$, and it is the spectral radius of M_Γ. A **Thurston obstruction** is a stable multicurve Γ with principal eigenvalue of the transition matrix $\lambda_\Gamma \geq 1$.

(8.15) *If* $((\mathbb{C}_\infty, f), \Omega)$ *with* $\Omega \subset \mathbb{C}_\infty$ *and* $\#\Omega \geq 4$ *is a marked branched covering of* $\mathbb{C}_\infty$ *without Thurston obstructions,* $\tau_0 \in \mathrm{Teich}(\mathbb{C}_\infty \backslash \Omega)$ *and* $\tau_n = (\widetilde{T}_f)^n(\tau_0)$, *with* $\widetilde{T}_f$ *the Thurston* pullback *from* $\mathrm{Teich}(\mathbb{C}_\infty \backslash \Omega)$ *to* $\mathrm{Teich}(\mathbb{C}_\infty \backslash f^{-1}(\Omega))$, *then there are* $m \in \mathbb{N}$ *functions of* $(f, \#\Omega)$ *and* $C > 0$ *function of* $(f, \#\Omega, D_T(\tau_0, \tau_1))$ *such that*

$$w(\tau_n) > C \implies w(\tau_{n+m}) < w(\tau_n).$$

Proof For $d \in \mathbb{N}$, the number of transition matrices M_Γ of a stable multicurve Γ for f of degree d and $\#\Gamma \leq \#\Omega - 3$ is finite. Let m be the minimum $m \in \mathbb{N}$ such that $\|(M_\Gamma)^m\| < \frac{1}{2}$, with the operator norm for the norm

$$\|(x_1, \dots, x_{\#\Gamma})\|_\infty = \max\left\{|x_j| : j \in \{1, \dots, \#\Gamma\}\right\} \quad \text{in } \mathbb{R}^{\#\Gamma},$$

what is possible by Gelfand formula[40] for spectral radius, $\mathrm{spr}(M_\Gamma) = \lim \|(M_\Gamma)^k\|^{\frac{1}{k}}$ and $\mathrm{spr}(M_\Gamma) = \lambda_\Gamma < 1$, $C > K$ and $w(\tau_n) > C$, where w is as at the end of the preceding subsection. Let

$$A = -\log\left(2\log\left(\sqrt{2}+1\right)\right), \quad J = m\left[\log d + 2D_T(\tau_0, \tau_1)\right], K = (\#\Omega - 3)J + A,$$

[39] Perron, Oskar (1880–1975). Frobenius, Ferdinand (1849–1917).

[40] Gelfand, Israel (1913–2009).

$\varphi_n : \mathbb{C}_\infty \to \mathbb{C}_\infty$ such that $\tau_n = [\varphi_n]$ and $F_{n+1} = \varphi_n \circ f \circ \varphi_{n+1}^{-1}$ is a rational function, $P_n = \varphi_n(\Omega)$, L_n is the set of $w(\gamma, \tau_n)$ such that γ is a non-peripheral Jordan curve to $\mathbb{C}_\infty \backslash \Omega$.

For a the smallest real number for which there is a maximum interval $]a, b[$ of length $> J$ in $[A, +\infty[\backslash L_n$, let Γ be the set of non-peripheral Jordan curves γ in $\mathbb{C}_\infty \backslash \Omega$ with $w(\gamma, \tau_n) \in]a, +\infty[$. Then, $w(\gamma, \tau_n) \geq b$. By (8.13), the Jordan curves γ in $\mathbb{C}_\infty \backslash \Omega$ with $w(\gamma, \tau_n) > A$ are disjoint and not peripheral. Thus, there are $\geq \#\Omega - 3$ elements of L_n greater than A, and some element of L_n is greater than $(\#\Omega - 3)J + A = K$. Therefore, $b < K$ and $\Gamma \neq \emptyset$.

For $\gamma \in \Gamma$, let ζ be a non-peripheral Jordan curve in $\mathbb{C}_\infty \backslash f^{-1}(\gamma)$ and ξ', ξ'' be geodesic Jordan curves in the Poincaré metric homotopic to $\varphi_{n+1}(\zeta)$ in, resp., $\mathbb{C}_\infty \backslash \varphi_{n+1}(\Omega)$, $\mathbb{C}_\infty \backslash \varphi_{n+1}\big(f^{-1}(\Omega)\big)$. Since $\big(\mathbb{C}_\infty \backslash \varphi_{n+1}\big(f^{-1}(\Omega)\big), f\big)$ is a holomorphic covering of $\mathbb{C}_\infty \backslash \varphi_{n+1}(\Omega)$ which is included in $\mathbb{C}_\infty \backslash \varphi_{n+1}\big(f^{-1}(\Omega)\big)$. With $F \stackrel{def}{=} F_{n+1} = \varphi_n \circ f \circ \varphi_{n+1}^{-1}$ and $P \stackrel{def}{=} \varphi_{n+1}(\Omega)\ \varphi\big(f^{-1}(\Omega)\big) = (F)^{-1}(P)$,

$$\ell_{\mathbb{C}_\infty \backslash \varphi_n(f^{-1}(\Omega))}(\xi'') = \deg(f_{|\eta})\, \ell_{\mathbb{C}_\infty \backslash P}\big(F(\xi'')\big) = \deg(f_{|\eta})\, \ell(\gamma, \tau_n),$$

$$\ell_{\mathbb{C}_\infty \backslash \varphi_n(f^{-1}(\Omega))}(\xi'') \geq \ell_{\mathbb{C}_\infty \backslash \varphi(\Omega)}(\xi'') \geq \ell_{\mathbb{C}_\infty \backslash \varphi(\Omega)}(\xi') = \ell(\eta, \tau_{n+1}).$$

Thus,

$$w(\eta, \tau_{n+1}) \geq w(\gamma, \tau_n) - \log \deg(f_{|\eta}) \geq w(\gamma, \tau_n) - \log d \geq b - \log d.$$

By the triangular inequality, (8.14) and (8.12), and $b - a > J$,

$$w(\eta, \tau_n) \geq b - \log d - 2\, D_T(\tau_n, \tau_{n+1}) \geq b - \log d - 2\, D_T(\tau_0, \tau_1) = b - J > a.$$

Therefore, η is homotopic (rel Ω) to an element of the multicurve Γ, i.e., Γ is a stable multicurve of the branched covering $\big((\mathbb{C}_\infty, f), \Omega\big)$.

Let $g = f^m$ and

$$G = \phi_n \circ g \circ \varphi_m^{-1} \in \mathrm{Rat}_{d^m}, \quad P' = \varphi_m(\Omega) \subset P'' = \varphi_m\big(g^{-1}(\Omega)\big) = G^{-1}(P).$$

If β is a geodesic Jordan curve in the Poincaré metric of $\mathbb{C}_\infty \backslash P''$ with length $< d^m e^{-b}$, $G(\beta)$ is a geodesic Jordan curve in the Poincaré metric of $\mathbb{C}_\infty \backslash P$ with length $< e^{-b}$ and, therefore, $w\big(G(\beta)\big) \geq b$. There is a unique Jordan curve $\gamma \in \Gamma$ homotopic (rel P) to $\varphi_n(\gamma)$. The critical values of G belong to P. Lifting the corresponding homotopy by G, we get that every geodesic Jordan curve in the Poincaré metric of $\mathbb{C}_\infty \backslash P''$ with length $< d^m e^{-b}$ is homotopic (rel P'') to a connected component of $G^{-1}\big(\varphi_n(\gamma)\big)$ for a unique element γ of the stable multicurve Γ.

If $L = d^m e^{-K}$ (function of $(m, \#\Omega, d, D_T(\tau_0, \tau_1))$), by the definitions of A, J, K in the first paragraph of the proof and $\#\Omega \geq 4$,

$$L = d^m e^{-(\#\Omega - 3)J + A} \leq e^{-A} = 2 \log\big(\sqrt{2} + 1\big).$$

Let $s = \#\Gamma$, $\Gamma = \{\gamma_1, \ldots, \gamma_s\}$, $v, v' \in \mathbb{R}^s$ with components $v_j = \frac{1}{\ell(\gamma_j, \tau_n)}$ and $v'_j = \frac{1}{\ell(\gamma_j, \tau_{n+m})}$, $j \in \{1, \ldots, n\}$, $q = \#P'' \setminus P' = \#P'' - \#\Omega$. Since P'' has at least two critical points of G,

$$\#P'' = \#G^{-1}(P) < d^m \#P - 1 = d^m \#\Omega - 1\,,$$

and, therefore, $q + 1 \le (d^m - 1)\,\#\Omega$. By (8.13.4(b)),

$$v'_j < \sum_{\beta \in W_j} \frac{1}{\ell_{\mathbb{C}_\infty \setminus P''}(\beta)} + \frac{1}{\pi} + \frac{(d^m - 1)\,\#\Omega}{L}\,, \quad j \in \{1, \ldots, s\}\,,$$

with W_j the set of geodesic Jordan curves in the Poincaré metric of $\mathbb{C}_\infty \setminus P''$ with length $<L$ and homotopic (rel P') to $\varphi_{n+1}(\gamma_j)$.

If $\beta \in W_j$, by two paragraphs above, it is a Jordan curve homotopic to a connected component $\varphi_{n+1}(\eta)$ of $G^{-1}(\varphi_n(\gamma)) = \varphi_{n+1}(g^{-1}(\gamma))$ for a unique element $\gamma \in \Gamma$. Since γ is not peripheral in $\mathbb{C}_\infty \setminus \Omega$, no pair of curves in $g^{-1}(\gamma)$ are homotopic (rel $g^{-1}(\Omega)$), and, therefore, also no pair of curves in $\varphi_{n+1}(g^{-1}(\gamma))$ are homotopic (rel P''). Thus, η is the unique component of $g^{-1}(\gamma)$ such that β is homotopic (rel P'') to $\varphi_{n+1}(\eta)$. So, the relation $\beta \mapsto \eta$ is a function, and, since each curve in $\varphi_{n+1}(g^{-1}(\Gamma))$ is homotopic (rel P'') to a unique geodesic Jordan curve in the Poincaré metric of $\mathbb{C}_\infty \setminus P''$, it is an injection. Since β and $\varphi_{n+1}(\eta)$ are homotopic (rel P''), they also are homotopic (rel P'), and, as β and $\varphi_{n+1}(\gamma_j)$ are homotopic (rel P''), they also are homotopic (rel P').

Since $(\mathbb{C}_\infty \setminus P'', G)$ is a holomorphic covering of $\mathbb{C}_\infty \setminus P$, $G(\beta)$ is a geodesic Jordan curve in the Poincaré metric of $\mathbb{C}_\infty \setminus P$ homotopic (rel P) to $\varphi_n(\gamma)$, and

$$\ell_{\mathbb{C}_\infty \setminus P''}(\beta) = \deg(G_{|\beta})\,\ell_{\mathbb{C}_\infty \setminus P}(G(\beta)) = \deg(G_{|\eta})\,\ell(\gamma, \tau_n)\,.$$

By the inequality at the end of two paragraphs above,

$$v'_j < \sum_{\gamma \in \Gamma} \sum_{\eta \sim_\Omega \gamma} \frac{1}{\deg(G_{|\eta})\,\ell(\gamma, \tau_n)} + \frac{1}{\pi} + \frac{(d^m - 1)\,\#\Omega}{L}\,, \quad j \in \{1, \ldots, s\}\,,$$

where $\eta \sim_\Omega \gamma$ means η homotopic (rel Ω) to γ. As $g = f^m$ and m were chosen in the first paragraph of the proof so that $\|(M_\Gamma)^m\| < \frac{1}{2}$,

$$\|v'\|_\infty \le \tfrac{1}{2}\|v\|_\infty + \tfrac{1}{\pi} + \frac{(d^m - 1)\,\#\Omega}{L}\,.$$

So, $\|v'\|_\infty < \|v\|_\infty$, equivalent to $w(\tau_{n+m}) < w(\tau_n)$, as $w(\tau_n) = \log\|v\|_\infty$ and $w(\tau_{n+m}) = \log\|v'\|_\infty$ if $\|v\|_\infty > 2\big(\frac{1}{\pi} + \frac{(d^m - 1)\,\#\Omega}{L}\big)$. Consequently,

$$C > \max\Big\{K, \log\Big(2\big(\tfrac{1}{\pi} + \tfrac{(d^m - 1)\,\#\Omega}{L}\big)\Big)\Big\}, \qquad w(\tau_n) > C \Longrightarrow w(\tau_{n+m}) < w(\tau_n)\,.$$

□

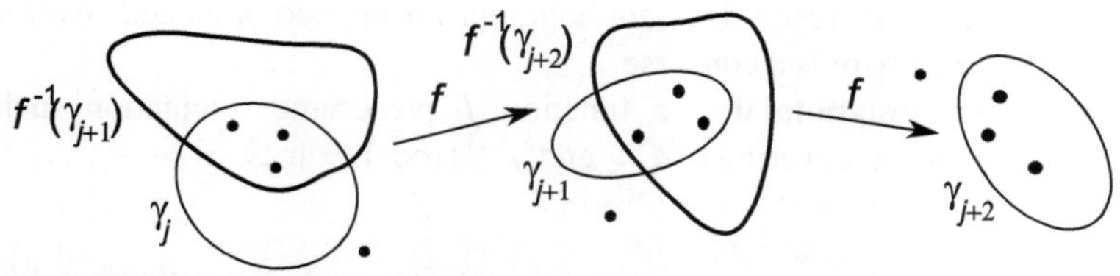

Fig. 8.5 Part of a Levy Cycle. The isolated points belong to $PC(f)$, the curves γ_k belong to the Levy cycle, and the curve labeled $f^{-1}(\gamma_k)$ is the unique connected component of that preimage of f homotopic (rel $PC(f)$) to γ_k

Levy Cycles and Topological Polynomials

Next we define a type of Thurston obstruction useful to analyze whether a branched covering corresponds to a polynomial.[41]

A **Levy cycle** is a subset $\Lambda = \{\gamma_0, \ldots, \gamma_{k-1}\}$ of a stable multicurve Γ of a marked branched covering $\big((\mathbb{C}_\infty, f),\, PC(f)\big)$ of $\mathbb{C}_\infty$ which is a Thurston obstruction such that γ_j is homotopic (rel $PC(f)$) to one and only one component $\widetilde{\gamma}$ of $f^{-1}(\gamma_{j+1})$ and the degree of the covering $\big(\widetilde{\gamma}, f\big)$ of γ_{j+1} is 1, for $j \in \{0, \ldots, k-1\}$, $\gamma_k = \gamma_0$ (Fig. 8.5).

(8.16) *If $f \in \mathrm{Rat}_d$ with $d \geq 2$, the associated marked branched covering $\big((\mathbb{C}_\infty, f),\, PC(f)\big)$ of $\mathbb{C}_\infty$ does not have Levy cycles.*

Proof For a Levy cycle to exist, there must exist a multicurve and, since each region separated in $\mathbb{C}_\infty$ by each one of its elements contains at least two points of $PC(f)$, $\#PC(f) \geq 4$.

If d is the degree of the covering,

$$\# f^{-1}\big(PC(f)\big) = d\, \# PC(f) - 2(d-1)\,.$$

So, if it were $PC(f) = f^{-1}\big(PC(f)\big)$, it would be $(d-1)[\#PC(f) - 2] = 0$, which is impossible because $d \geq 2$ and $\#PC(f) \geq 4$. Therefore, if a Levy cycle Λ exists, then $PC(f) \subsetneqq f^{-1}\big(PC(f)\big)$.

If $\gamma_1, \ldots, \gamma_{k-1}$ were geodesic Jordan curves in the Poincaré metric of $\mathbb{C}_\infty \backslash PC(f)$, each one homotopic to each one of the elements of the Levy cycle Λ and labeled by the corresponding order, the connected component $\widetilde{\gamma}_j$ of $f^{-1}(\gamma_{j+1})$ homotopic to γ_j would have the same length as γ_{j+1} in the Poincaré metric of $\mathbb{C}_\infty \backslash f^{-1}\big(PC(f)\big)$ and smaller length in the Poincaré metric of $\mathbb{C}_\infty \backslash PC(f)$. So, in this Poincaré metric, the length of each γ_j would be smaller than that of γ_{j+1} for $j \in \{0, \ldots, k-1\}$, contradicting $\gamma_k = \gamma_0$. Therefore, there are not Levy cycles. □

---

[41] This subsection follows part of Bielefeld, B., Fisher, Y., Hubbard, J., The classification of critically preperiodic polynomials as dynamical systems, *J. American Mathematical Society* **5** (1992), 721–762. Bielefeld, Benjamin. Fisher, Yuaval (1962-).

The converse of this result does not hold, but for marked branched coverings of the following type, there is a converse.

A **topological polynomial** is a function f preserving orientation such that $(\mathbb{C}_\infty, f)$ is a branched covering of $\mathbb{C}_\infty$ and $f^{-1}(\{\infty\}) = \{\infty\}$.

(8.17) *Every topological polynomial f with a Thurston obstruction has a Levy cycle.*

Proof [42] Let Γ be a Thurston obstruction of a marked branched covering $((\mathbb{C}_\infty, f), \Omega)$ of $\mathbb{C}_\infty$ and $P:\Gamma \to \mathscr{P}(\Gamma)$ be a function mapping each $\gamma \in \Gamma$ to the set of elements of Γ homotopic (rel $PC(f)$) to a connected component of $f^{-1}(\gamma)$.

If the range of P is a set of disjoint subsets of Γ not containing $\emptyset$, then P is a permutation of Γ, and a cycle of this permutation is a Levy cycle. The range of P may not satisfy these conditions, but it is proved below that there is a nonempty subset of Γ for which the corresponding function P satisfies the conditions.

Without loss of generality, Γ is a minimal Thurston obstruction in the sense of not properly containing any Thurston obstruction, implying that every curve $\gamma \in \Gamma$ is homotopic (rel $PC(f)$) to a component of $f^{-1}(\widetilde{\gamma})$ for some $\widetilde{\gamma} \in \Gamma$.

The transition matrix M_Γ of the stable multicurve Γ defines a function T of the set of homotopy classes of non-peripheral Jordan curves γ in $\mathbb{C}_\infty \backslash PC(f)$. Denote $\mathbf{\Gamma}_n$ the set of curves $\gamma \in \Gamma$ such that $T^j([\gamma]) \to 0$, and $\mathbf{\Gamma}_e = \Gamma \backslash \Gamma_n$. If $\gamma \in \Gamma_n$, every non-peripheral connected component of $f^{-1}(\gamma)$ is a curve homotopic (rel $PC(f)$) to an element of Γ_n, and if $\gamma \in \Gamma_e$, some connected component of $f^{-1}(\gamma)$ is a curve homotopic (rel $PC(f)$) to one element of Γ_e.

The range of $P_e : \Gamma_e \to \mathscr{P}(\Gamma_e)$ with $P_e(\gamma) = P(\gamma) \cap \Gamma_e$ does not contain $\emptyset$.

For each Jordan curve η not going through ∞, let $D(\eta)$ be the connected component of $\mathbb{C}_\infty \backslash \{\eta\}$ not containing ∞, which is homeomorphic to an open circle. Denote $\mathbf{\Gamma}_i$ the nonempty set of curves $\gamma \in \Gamma_e$ such that $D(\gamma) \cap \Gamma_e = \emptyset$.

If $\gamma \in \Gamma_i$ and $\widetilde{\gamma}$ are a connected component of $f^{-1}(\gamma)$ homotopic to $\eta \in \Gamma_e$, then $\eta \in \Gamma_i$. If $f^{-1}(D(\eta))$ were not a disjoint union of sets homeomorphic to open disks, $\mathbb{C}_\infty \backslash f^{-1}(D(\eta))$ would have ≥ 2 connected components. If Z were one such component without ∞, it would be $f(Z) = X = \mathbb{C}_\infty \backslash D(\eta)$, which is a connected set. Since f is a topological polynomial, it is an open function (i.e., images of open sets are open sets) and a proper function (i.e., preimages of compact sets are compact sets). So, $f(Z)$ would be open and closed relatively to $\mathbb{C}_\infty$, and since this set is connected, it would be $f(Z) = \mathbb{C}_\infty$, contradicting $\infty \notin f(Z)$. Thus, $f^{-1}(D(\eta))$ is a disjoint union of sets homeomorphic to open disks. There is only one of these sets homeomorphic to an open disk $\widetilde{U}$ with boundary $\widetilde{\gamma}$ not containing ∞. One of the connected components U of $\mathbb{C}_\infty \backslash \{\gamma\}$ contains the same critical values as $\widetilde{U}$ and $\infty \notin U$.

[42] This proof follows an idea of M. Rees.

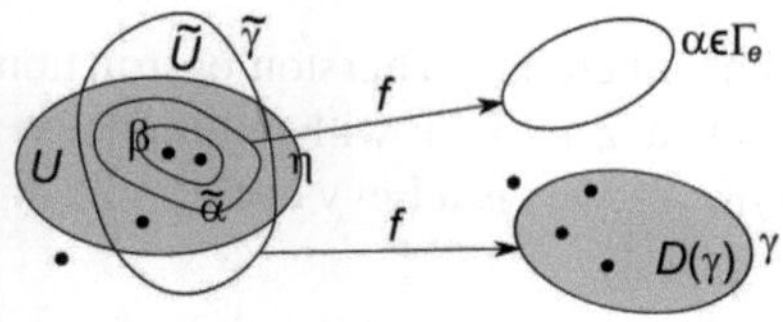

Fig. 8.6 Jordan curve $\gamma \in \Gamma_i$ with a connected component of $f^{-1}(\gamma)$ homotopic to a Jordan curve $\eta \in \Gamma_e$ implies $\eta \in \Gamma_i$

If there would exist $\beta \in \Gamma_e$ in U, it would separate some critical values in U from others, and since Γ is a minimal Thurston obstruction, β would be homotopic (rel $PC(f)$) to a connected component $\widetilde{\alpha} \in f^{-1}(\alpha)$ of some curve $\alpha \in \Gamma_e$ not belonging to $D(\gamma)$ because $\gamma \in \Gamma_i$ (Fig. 8.6). Therefore, $\widetilde{\alpha}$ would be outside $\widetilde{U}$ and could not separate any critical value from others in $\widetilde{U}$, and the same would happen for β homotopic (rel $PC(f)$) to $\widetilde{\alpha}$, contradicting the hypothesis. So, $U \cap \Gamma_e = \emptyset$, $\eta \in \Gamma_i$. Since Γ is a minimal Thurston obstruction, $\Gamma_i = \Gamma_e$.

If $\eta \in P_e(\gamma_1) \cap P_e(\gamma_2)$ with $\gamma_1, \gamma_2 \in \Gamma_i$ different, and for $j \in \{1, 2\}$, $\widetilde{\gamma}_j$ is a connected component of $f^{-1}(\gamma_j)$ homotopic (rel $PC(f)$) to η, it would be $f\big(PC(f) \cap D(\eta)\big) \subset D(\gamma_1) \cap D(\gamma_2)$, in contradiction with $D(\gamma_1) \cap D(\gamma_2) = \emptyset$ since $D(\gamma_1), D(\gamma_2) \in \Gamma_i$. Therefore, if γ_1, γ_2 are Jordan curves distinct from Γ_i, then $P_e(\gamma_1)$, $P_e(\gamma_2)$ are nonempty disjoint sets of Jordan curves, and as P_e is a function from the finite set Γ_e to $\mathscr{P}(\Gamma_e)$, the elements of the range of P_e are sets with only one element. Thus, there is one and only one connected component of $f^{-1}(\gamma) \in \Gamma_e$, and P_e is a permutation of Γ_e which, as Γ is a minimal Thurston obstruction, is a cyclic permutation.

Let $p = \#\Gamma_e$, $\gamma_0 \in \Gamma_e$, and define inductively $\gamma_{j-1}(P_e)^{-1}(\gamma_j)$ with $\gamma_p = \gamma_0$. For every $j \in \{0, \ldots, p-1\}$, the curve γ_j is the only element of Γ_e homotopic to a connected component of $f^{-1}\big(\gamma_{j+1}\big)$. Since $f^{-1}\big(D(\gamma_{j+1})\big)$ is a union of sets homeomorphic to disjoint open disks not containing ∞, there is a unique connected component $\widetilde{\gamma}_j$ of $f^{-1}\big(\gamma_{j+1}\big)$ homotopic to γ_j, denoted $d_j = \deg(f_{|\widetilde{\gamma}_j})$. Order the elements of Γ $\gamma_0, \ldots, \gamma_{p-1}$ followed by curves in Γ_n. The corresponding transition matrix is of the form

$$M_\Gamma = \begin{bmatrix} A & 0 \\ C & B \end{bmatrix}, \qquad A = \begin{bmatrix} 0 & \frac{1}{d_0} & 0 & \cdots & 0 \\ \vdots & \ddots & \ddots & \ddots & \vdots \\ \vdots & & \ddots & \ddots & 0 \\ 0 & & & \ddots & \frac{1}{d_{p-2}} \\ \frac{1}{d_{p-1}} & 0 & \cdots & \cdots & 0 \end{bmatrix},$$

with $\lim B^m = 0$. The eigenvalues of B have absolute value < 1, and those of M_Γ not eigenvalues of B are the zeros of the polynomial

$$\det(A - \lambda I_p) = (-1)^n \left(\lambda^p - \prod_{j=0}^{p-1} \frac{1}{d_j}\right).$$

Since there is a Thurston obstruction, this polynomial has a zero ≥ 1. As $d_j \geq 1$, such a zero is 1 with multiplicity p. Thus, $d_j = 1$ for $j \in \{0, \ldots, p-1\}$, and $\{\gamma_0, \ldots, \gamma_p\}$ is a Levy cycle. □

(8.18) *Every postcritically finite topologic polynomial f with all periodic orbits contained in positive orbits with some critical point does not have Thurston obstructions.*

Proof If there is a Thurston obstruction, by the preceding result, there exists a Levy cycle $\Lambda = \{\gamma_0, \ldots, \gamma_{p-1}\}$; denote $\gamma_p = \gamma_0$.

As $PC(f)$ is finite, the orbit of each critical point is eventually periodic. If $x \in PC(f) \cap D(\gamma_j)$, where $D(\gamma_j)$ is the connected component of $\mathbb{C}_\infty \backslash \{\gamma_j\}$ which does not contain ∞, then $f(x) \in D(\gamma_j)$ and $\cup_{j=0}^{p-1} D(\gamma_j)$ contain the positive orbit of x. If for some $\gamma_j \in \Lambda$ the point y is not periodic, but $z = f(y) \in D(\gamma_j)$ is, the components of $f^{-1}\big(D(\gamma_j)\big)$ are homeomorphic to open disks, and one and only one has boundary $\widetilde{\gamma}_j$ homotopic (rel $PC(f)$) to some element of Λ.

As in the proof of the preceding result, Λ_e is the set of curves $\gamma \in \Lambda$ such that $\lim_{j \to +\infty} T^j([\gamma]) \neq 0$, with T the function defined in the set of homotopy classes of non-peripheral Jordan curves to $\mathbb{C}_\infty \backslash PC(f)$ by the transition matrix M_Λ of the multicurve Λ, and Λ_i the set of curves $\gamma \in \Lambda_e$ such that $D(\gamma) \cap \Lambda_e = \emptyset$.

If $\Lambda \subset \Lambda_i$, exactly one of the curves $\gamma_j \in \Lambda$ contains the periodic point $y \in f^{-1}(z)$ and γ_j is homotopic to $\widetilde{\gamma}_j$. As $f_{|D(\gamma_j)} : D(\gamma_j) \to D\big(\widetilde{\gamma}_j\big)$ is injective, $y \notin D\big(\widetilde{\gamma}\big)$ and, therefore, also $y \notin \cup_{j=0}^{p-1} D(\gamma_j)$. So, none of the nonperiodic critical values of f belongs to $\cup_{j=0}^{p-1} D(\gamma_j)$, i.e., the critical values in sets with boundary γ_j, for $j \in \{0, \ldots, p-1\}$, homeomorphic to disks without ∞ are periodic points. As each periodic cycle has a critical point, the connected component of $f^{-1}(\gamma_{j+1})$ homotopic to γ_j has a critical point for some $j \in \{0, \ldots, p-1\}$. As $f_{|D(\gamma_j)} : D(\gamma_j) \to D\big(\widetilde{\gamma}_j\big)$ is a proper function, $\deg\big(f_{|D(\gamma_j)}\big) = \deg\big(f_{|\partial D(\gamma_j)}\big) = 1$, contradicting that γ_j contains a critical point.

In conclusion, there are no Thurston obstructions. □

Thurston Theorem

We prove versions of Thurston theorem of characterization of postcritically finite rational functions a bit more general than the original.

(8.19) *Let $\mathscr{R} = \big((\mathbb{C}_\infty, f), \Omega\big)$, with $\Omega \subset \mathbb{C}_\infty$ finite, $\#\Omega \geq 3$, $\Omega \supset PC(f)$ and $f(\Omega) \subset \Omega$, be a marked branched covering of $\mathbb{C}_\infty$ such that φ_0 is a homeomorphism preserving orientation of $\mathbb{C}_\infty$ onto $\mathbb{C}_\infty$ and $\{(\varphi_n, f_n)\}$ be a sequence with $\varphi_n : \mathbb{C}_\infty \to \mathbb{C}_\infty$ coincident with φ_0 in $\Omega' \subset \Omega$ with $\#\Omega' = 3$ and*

(continued)

$f_n = \varphi_{n-1} \circ f \circ \varphi_n^{-1}$ a rational function. If the signature of the orbifold $\mathcal{O}_{\mathcal{R}}$ is not $(2, 2, 2, 2)$ and (f, Ω) does not have Thurston obstructions:

1. *The Thurston pullback $\widetilde{T}_f$ from* $\mathrm{Teich}(\mathbb{C}_\infty \backslash \Omega)$ *to* $\mathrm{Teich}(\mathbb{C}_\infty \backslash f^{-1}(\Omega))$ *has a unique fixed point τ.*
2. $\lim[\varphi_n] = \tau$ *in* $\mathrm{Teich}(\mathbb{C}_\infty \backslash \Omega)$.
3. *The sequence $\{f_n\}$ is uniformly convergent in $\mathbb{C}_\infty$, and the limit is a rational function F.*
4. *For each $z \in \Omega$, the sequence $\{\varphi_n(z)\}$ converges, and if*

$$X = \{\lim \varphi_n(z) : z \in \Omega\},$$

the set of $(\varphi, \psi) \in (f, \Omega) \times (F, X)$ such that $[\varphi] = [\psi] = \tau$ is an equivalence relation.

Proof For $n \in \mathbb{N}$, $\tau_0 \in \mathrm{Teich}(\mathbb{C}_\infty \backslash \Omega)$ consider the sequence $\tau_n = (\widetilde{T}_f)^n(\tau_0)$, and the geodesic δ in $\mathrm{Teich}(\mathbb{C}_\infty \backslash \Omega)$ with ends τ_0, τ_1, and $\delta_n = \left(\widetilde{T}_f\right)^n(\delta)$. By (8.8), in the operator norm, $\|D_\tau \widetilde{T}_f\| = \|(D_\tau \widetilde{T}_f)^*\| = \|f_*\| \leq 1$. So, the length of δ_n in the Teichmüller distance D_T is $\leq \delta_{n-1}$. Thus, $D_T(\tau_n, \tau_{n+1}) \leq D_T(\tau_0, \tau_1)$. By (8.14), $|w(\tau_{n+1})| \leq |w(\tau_n)| + 2\, D_T(\tau_0, \tau_1)$. Let $m \in \mathbb{N}$ and $C > 0$ be as in (8.15), and let $C_1 = \max\{C, w(\tau_0)\}$ and

$$j_n = \max\{j \in [0, n] \cap \mathbb{N} : \tau_j \in \mathrm{Teich}_{C_1}(\mathbb{C}_\infty \backslash \Omega)\}, \quad n \in \mathbb{N} \cup \{0\}.$$

If $j_n < n$, with q_n, r_n the quotient and remainder of integer division of $n - j_n - 1$ by m, for $j \in [j_n + 1, n]$, $w(\tau_j) > C_1 \geq C$. By (8.15) and (8.14),

$$\begin{aligned} w(\tau_n) &\leq w(\tau_{j_n+1+r_n}) \leq w(\tau_{j_n+1+r_n}) \\ &\quad + (r_n + 1)\, 2D_T(\tau_0, \tau_1) \leq C_1 + m 2D_T(\tau_0, \tau_1). \end{aligned}$$

Hence, if $j_n < n$, then $\tau_n \in \mathrm{Teich}_{C_1 + m2D_T(\tau_0, \tau_1)}(\mathbb{C}_\infty \backslash \Omega)$, which also holds if $j_n = n$, because $\mathrm{Teich}_{C_1}(\mathbb{C}_\infty \backslash \Omega) \subset \mathrm{Teich}_{C_1 + m2D_T(\tau_0, \tau_1)}(\mathbb{C}_\infty \backslash \Omega)$.

Let $C_2 = C_1 + (m + 1) + 2D_T(\tau_0, \tau_1)$. By (8.14), $\delta_n \in \mathrm{Teich}_{C_2}(\mathbb{C}_\infty \backslash \Omega)$ for $n \in \mathbb{N} \cup \{0\}$. By (8.12), there is $\lambda \in [0, 1[$ such that $\|D_\tau (\widetilde{T}_f)^{\#\Omega}\| < \lambda$ for $\tau \in \mathrm{Teich}_C(\mathbb{C}_\infty \backslash \Omega)$. So, for $n \in \mathbb{N} \cup \{0\}$ and $j \in \{1, \ldots, \#\Omega\}$,

$$D_T(\tau_{n\,\#\Omega}\,, \tau_{n\,\#\Omega + j}) \leq \lambda^n D_T(\tau_0, \tau j) \leq \#\Omega\, D_T(\tau_0, \tau j)\, \lambda^n.$$

Since $\lambda \in [0, 1[$, whatever $\varepsilon > 0$, there is $N \in \mathbb{N}$ such that $n > N$ implies $\#\Omega\, D_T(\tau_0, \tau j)\, \lambda^n < \varepsilon$, and, therefore, $\{\tau_n\}$ is a Cauchy sequence in $\mathrm{Teich}(\mathbb{C}_\infty \backslash \Omega)$. As this metric space is complete, the sequence converges to some $\tau \in \mathrm{Teich}(\mathbb{C}_\infty \backslash \Omega)$. Since $\tau_{n+1} = (\widetilde{T}_f)^n(\tau_n)$ and $(\widetilde{T}_f)^n$ is continuous, the limit when $n \to +\infty$ gives $\tau = (\widetilde{T}_f)(\tau)$. Therefore, $\{\tau_n\}$ converges to a fixed point of $\widetilde{T}_f$.

If $\tau, \tau' \in \mathrm{Teich}(\mathbb{C}_\infty \backslash \Omega)$ were fixed points of different $\widetilde{T}_f$ and δ is a geodesic in $\mathrm{Teich}(\mathbb{C}_\infty \backslash \Omega)$ with end points τ, τ', $(\widetilde{T}_f)^{\#\Omega}(\delta)$ would be a curve in $\mathrm{Teich}(\mathbb{C}_\infty \backslash \Omega)$

with end points τ, τ' and length smaller than δ, in contradiction to δ being a geodesic. Therefore, the fixed point of $\widetilde{\mathcal{T}}_f$ in Teich$(\mathbb{C}_\infty \backslash \Omega)$ is unique and independent of τ_0.

As $\{\tau_n\} \subset$ Teich$(\mathbb{C}_\infty \backslash \Omega)$ converges to the unique fixed point $\tau \in$ Teich$(\mathbb{C}_\infty \backslash \Omega)$ of $\widetilde{\mathcal{T}}_f$, the sequence $\{(g_{\tau_n}, \varphi_{\tau_n}, \psi_{\tau_n})\} \in \mathbb{T}_{f,\Omega}$ whose terms represent the, resp., terms of the sequence $\{\tau_n\}$ converges to $\{(g_\tau, \varphi_\tau, \psi_\tau)\} \in \mathbb{T}_{f,\Omega}$, where

$$\mathbb{T}_{f,\Omega} \stackrel{def}{=} \text{Rat}_d \times (\mathbb{C}_\infty)^{\Omega} \times (\mathbb{C}_\infty)^{f^{-1}(\Omega)}$$

is the space introduced after (8.9). Denote $X = \{\lim \varphi_{\tau_n}(\omega) : \omega \in \Omega\}$. The function φ_τ is a bijection from Ω onto X and coincides with $\psi_\tau : \Omega \to X$. So, $g_\tau(X) \subset X$, $PC(f_\tau) \subset X$, and f_τ are postcritically finite. It is easy to check that (φ, ψ) is an equivalence relation in $(f, \Omega) \times (F, X)$. □

The sequence (φ_n, f_n) is called a **Thurston algorithm** to obtain a rational function under the stated conditions. An immediate corollary is: *If* $\#\Omega > \deg(f)+1$, *then there are Thurston obstructions.*

It is made explicit below another immediate corollary containing the original Thurston theorem of characterization of postcritically finite rational functions as a particular case with $\Omega = PC(f)$.

(8.20) *Let* $\mathscr{R} = \big((\mathbb{C}_\infty, f), \Omega\big)$, *with* $\Omega \subset \mathbb{C}_\infty$ *finite,* $\#\Omega \geq 3$, $\Omega \supset PC(f)$ *and* $f(\Omega) \subset \Omega$, *be a marked branched covering of* $\mathbb{C}_\infty$. *If the signature of the orbifold* $\mathcal{O}_{\mathscr{R}}$ *is not* $(2, 2, 2, 2)$ *and* (f, Ω) *does not have Thurston obstructions, then the combinatorial equivalence class of* (f, Ω) *has a rational function, unique up to conjugacy with a Möbius transformation: if* f_1 *and* f_2 *are rational functions in the same combinatorial equivalence class of* (f, Ω), *there is a unique Möbius transformation* M *such that* $M \circ f_1 = f_2 \circ M$.

Spectral Condition for Branched Covering of a Rational Function

In the preceding section, it was found that a sufficient condition for a postcritically finite branched covering $\mathscr{R} = \big((\mathbb{C}_\infty, f), PC(f)\big)$ to correspond to a rational function of the combinatorial equivalence class of $\big(f, PC(f)\big)$ is that this ordered pair does not have Thurston obstructions (i.e., the signature of the orbifold $\mathcal{O}_{\mathscr{R}}$ is not $(2, 2, 2, 2)$ and the maximum eigenvalue of the transition matrix of any stable multicurve Γ is $\lambda_\Gamma < 1$). C. McMullen proved that for branched coverings associated with rational functions, $\lambda_\Gamma \leq 1$. Hence, for rational functions with branched coverings associated with a Thurston obstruction, it is $\lambda_\Gamma = 1$ (i.e., the Thurston obstruction $\lambda_\Gamma < 1$ fails marginally).[43] To study it, it is convenient to consider that $PC(f)$ may be infinite and the multicurve Γ may not be invariant under f.

[43] On this subject we mainly follow parts of an appendix of McMullen, C., *Complex Dynamics and Renormalization*, Princeton University Press, Princeton, New Jersey, 1994.

The rings of a finite set of disjoint rings $A = \{A_1, \dots, A_n\} \subset \mathbb{C}_\infty$ are said to be **nested** if there is a pair of points separated by each one of the rings. Denote the smallest ring containing $\cup_{j=1}^{n} A_j$ by $\boldsymbol{\mathcal{M}(A)}$.

By the monotonicity of the ring module (1.13), $M\big(\mathcal{M}(A)\big) \geq \sum_{j=1}^{n} M(A_j)$, with equality if and only if $\mathcal{M}(A) \setminus \cup_{j=1}^{n} A_j$ is a finite set of Jordan curves.

A stable multicurve $\Gamma = \{\gamma_1, \dots, \gamma_n\}$ of a marked branched covering of $\mathbb{C}_\infty$ is **Irreducible** if the transition matrix $M_\Gamma = [m_{jk}]_{j,k=1}^{n,n}$ is Irreducible, i.e., for $j, k \in \{1, \dots, n\}$ there is $p \in \mathbb{N}$ such that $m_{jk}^{p} > 0$.

By the Perron-Frobenius theory for nonnegative square matrices, if $\lambda(M_\Gamma) > 0$, there is an irreducible submatrix X of M_Γ with spectral radius $\lambda(\Gamma)$. Thus, if Γ is a stable multicurve of a marked branched covering, there is an irreducible multicurve $\Gamma' \subset \Gamma$ with $\lambda_{\Gamma'} = \lambda_\Gamma$.

If f is a rational function of degree $d \geq 2$ with associated marked branched covering $\big((\mathbb{C}_\infty, f), PC(f)\big)$ of $\mathbb{C}_\infty$, $\Gamma = \{\gamma_1, \dots, \gamma_n\}$ is an irreducible stable multicurve in $\mathbb{C}_\infty \setminus PC(f)$, and, for each $j \in \{1, \dots, n\}$, C_j is a ring conformal to a collar of a geodesic Jordan curve γ_j, consider the set of rings $C' = \{C'_1, \dots, C'_n\}$ **obtained from the rings $\boldsymbol{C_j}$ by *pullback*** $C'_j = \mathcal{M}(D_1, \dots, D_n)$ with D_j the set of connected components of $f^{-1}\big(\cup_{j=1}^{n} C_j\big)$ isotopic to γ_j in $\mathbb{C}_\infty \setminus PC(f)$ (as Γ is Irreducible, $f^{-1}\big(\cup_{j=1}^{n} C_j\big) \neq \emptyset$); denote $\boldsymbol{C' = f^* C}$.

By monotonicity of ring module (1.13), $M(C') \geq \sum_{j,k=1}^{n} m_{jk} M(C_k)$.

(8.21) *If f is a rational function of degree $d \geq 2$ with associated marked branched covering $\big((\mathbb{C}_\infty, f), PC(f)\big)$ of $\mathbb{C}_\infty$ and Γ is a stable multicurve in $\mathbb{C}_\infty \setminus PC(f)$, then $\lambda_\Gamma \leq 1$.*

Proof Without loss of generality, $\Gamma = \{\gamma_1, \dots \gamma_n\}$ is Irreducible. Let C_{γ_j} be rings conformal to collars of geodesic Jordan curves γ_j, $j \in \{1, \dots, n\}$, and $C^0 \stackrel{def}{=} \{C_\gamma\}_{\gamma \in \Gamma}$. The sequence of the modules of the sequence of rings $\{C^n = (f^*)^n C^0\}_{n \in \mathbb{N}}$ is upper bounded. Thus, the irreducible stable multicurve transition matrix of Γ is M_Γ and with $\mathbf{v}_0 = \big(M(C_{\gamma_1}), \dots, M(C_{\gamma_n})\big)$ the sequence of vectors $(M_\Gamma)^n \mathbf{v}_0$ in $\mathbb{R}^n$ is upper bounded. So, M_Γ has no eigenvalues with modulus > 1. □

Meromorphic Quadratic Differentials with Almost All Trajectories Closed

C. McMullen classified the marked branched coverings associated with rational functions with principal eigenvalue of a stable multicurve transition matrix equal to 1, establishing that when they are not Lattès functions they have rotation rings (Siegel disks or Herman rings). To prove this, it is useful to have detailed information about rings where all orbits of a rational function are closed and to use properties of quadratic differentials[44] with almost all trajectories closed.

[44] This subsection follows parts of the K. Strebel book listed in the final bibliography.

Fig. 8.7 Trajectories of quadratic differentials in the neighborhood of (from left to right, top to bottom): Regular point (neither zero nor pole), zero of order 1, pole of order 1, poles of order 2 with coefficient of the order -2 term of the Laurent series $a_{-2} < 0$, $a_{-2} > 0$, $\mathcal{I}m\, a_{-2} \neq 0$, poles of orders 3, 4, 5

A **meromorphic quadratic differential** in a Riemann surface S is a set of elements of a meromorphic function φ_ν defined in coordinate neighborhoods transforming with coordinates changes $z_\mu = h(z_n u)$ as

$$\varphi_\nu(z) = \varphi_\mu\big(h(z)\big)[h'(z)]^2,$$

or, for short, $\varphi_\nu(z_\nu)\, dz_\nu^2 = \varphi_\mu\big(z_\mu\big)\, dz_\mu^2$ (the formula for transforming the square of the derivative of functions). If the function elements are holomorphic, it is said to be a **holomorphic quadratic differential**. Given a meromorphic quadratic differential in a Riemann surface S and removing the poles from S gives a Riemann surface S' where the quadratic differential is holomorphic.

A holomorphic quadratic differential φ in a region $\Omega \subset \mathbb{C}$ defines a **direction field** or **line field** by $\varphi(z)\, dz^2 > 0$, i.e., $\varphi\big(z(t)\big)[z'(t)]^2 > 0$ in Ω, which corresponds to $\operatorname{Arg} z'(t) = -\frac{1}{2}\operatorname{Arg}\varphi\big(z(t)\big)$, where $\operatorname{Arg} w$ denotes the argument of $w \in \mathbb{C}\setminus\{0\}$ in $[0, 2\pi[$. The curves $t \mapsto z(t)$ in Ω tangent at each point $z(t)$ to the line with slope $-\frac{1}{2}\operatorname{Arg}\varphi$ are called **trajectories** of the quadratic differential φ in $\Omega \subset \mathbb{C}$ (Fig. 8.7).

For example, a holomorphic quadratic differential in $\mathbb{C}_\infty$ is given by a $\varphi \in H(\mathbb{C})$ that for conformal homeomorphisms h in $\mathbb{C}$ satisfies $\varphi = (\varphi \circ h)[h']^2$ in $\mathbb{C}$ and in a coordinate neighborhood of ∞ the formula is the same as in a neighborhood of 0 with the change of variables $w = \frac{1}{z}$ giving $\varphi(z) = \varphi\left(\frac{1}{z}\right)\frac{1}{z^4}$, e.g., the constant function $\varphi = 1$ in $\mathbb{C}$ when viewed as defining a quadratic differential has a pole of order 4 at ∞.

The **meromorphic quadratic differential metric** φ is the metric with element of length $\sqrt{|\varphi(z)|}\,|dz|$. The associated element of area is $|\varphi(x, y)|\,dx\,dy$. The total area of the Riemann surface S where φ is defined is $\|\varphi\| = \int_S |\varphi(x, y)|\,dx\,dy$. It is finite if and only if the possibly existing poles of φ in S are simple, in which case $\|\varphi\|$ is a norm of φ.

The differentiable paths $\gamma : T \to S$, with $T \subset \mathbb{R}$ an interval, describing geodesics in the meromorphic quadratic differential metric φ are characterized by $\operatorname{Arg}\varphi(z)\,dz^2$ *a constant θ on γ*, i.e., $\operatorname{Arg}\varphi\big(\gamma(t)\big)[\gamma'(t)]^2 = \theta$, $t \in T$.

Hence, geodesic curves do not pass through zeros or poles of φ and maximal geodesic curves (i.e., not proper subsets of geodesics) not closed are open curves (i.e., described by simple paths defined in open intervals) approaching 0 or a pole of φ at the end points. A geodesic curve is said to be **horizontal** if $\theta = 0$ and **Vertical** if $\theta = \pi$.

If the complement of the union of the closed trajectories of a meromorphic quadratic differential $\varphi \neq 0$ in a Riemann surface S has zero area, we say **it has almost all trajectories closed**. Such a quadratic differential $\varphi \neq 0$ has no poles of order > 2, and the coefficient of the term of order -2 in the Laurent series[45] of φ at each pole of order 2 is $a_{-2} < 0$. The complement in S of the union of maximal open rings equal to unions of closed trajectories, called **characteristic rings** of φ, has zero area. An open ring S_j in the Riemann surface of S **has the homotopic type** of a Jordan curve γ_j if every Jordan curve separating the two connected components of its boundary is homotopic to γ_j.

Removing from a characteristic ring S_j a geodesic α_j in the meromorphic quadratic differential metric φ with end points in each connected component of ∂S_j results in a simply connected region where $\sqrt{\varphi}$ has primitive Φ and the image of $S_j \backslash \alpha_j$ by a branch of Φ is a rectangle R_j with horizontal and vertical sides with lengths, resp.,

$$a_j = \int_{\gamma_j} \sqrt{|\varphi(z)|}\,|dz|, \quad b_j = \int_{\widetilde{\alpha}_j} \sqrt{|\varphi(z)|}\,|dz| = a_j M(R_j) = a_j M(S_j),$$

with $\widetilde{\alpha}_j$ the geodesic curve on $S_j \backslash \alpha_j$ in the metric of φ with end points in each one of the connected components of ∂S_j and $M(R_j)$ the rectangle with module R_j equal to the module of the ring S_j; we call a_j the **Length** and b_j the **height** of the characteristic ring S_j of the quadratic differential φ. The norm of the restriction of φ to S_j is $\|\varphi_{|S_j}\| = \int_{S_j} |\varphi(x, y)|\,dxdy = \int_{R_j} 1 = a_j b_j$. Quadratic differentials with almost all trajectories closed can be characterized as follows.

---

[45] Laurent, Pierre Alphonse (1813–1854).

(8.22) *A meromorphic quadratic differential φ in a Riemann surface S with $\|\varphi\| \in \mathbb{R}$ has almost all trajectories closed if and only if*

$$\|\varphi\| = \sum_{j \in J} a_j b_j = \sum_{j \in J} a_j^2 M(S_j) = \sum_{j \in J} b_j^2 \frac{1}{M(S_j)},$$

where $\{S_j\}_{j \in J}$ is the set of characteristic rings of φ.

A set of disjoint Jordan curves with no pair of them homotopic in a Riemann surface S and none of them homotopic to a point in S is called an **admissible set of curves** in S.

J. Jenkins obtained the following property in 1957.

(8.23) *If $\Gamma = \{\gamma_j\}_{j \in J}$, with $J \neq 0$, is an admissible set of curves in a Riemann surface S, a_j is the length of γ, $\rho \in L^2(S)$ and $a_j \leq \int_\gamma \rho(z)\,|dz|$ for almost all Jordan curves γ in S homotopic to γ_j (i.e., except possibly in a set of these curves with area 0), and $\{S_j\}_{j \in J}$ is a set of disjoint rings in S with S_j of the homotopic type of γ_j, then $\sum_{j \in J} a_j^2 M(S_j) \leq \int_S \rho^2$.*

Proof Without loss of generality, we assume the Rings in S_j are nondegenerate, i.e., they are neither circles nor points. For each S_j, there is a conformal homeomorphism mapping it to an annulus, and there is a conformal homeomorphism of the complement of a radial segment in the annulus onto a rectangle R_j with the images of each circle in the annulus with the same center as the annulus horizontals of R_j with Length a_j and height $b_j = a_j M(S_j)$. So, with $\widetilde{\rho}_j$ in R_j corresponding to ρ for the mentioned mappings, $\sum_j a_j b_j \leq \sum_j \int_{R_j} \widetilde{\rho}_j$. For every $n \in \mathbb{N}$,

$$\langle \zeta, \eta \rangle = \sum_{k=1}^{n} \int_{R_k} \zeta_k \eta_k, \qquad \eta = (\eta_1, \ldots, \eta_n),\ \zeta = (\zeta_1, \ldots, \zeta_n) \in [L^2(R_j)]^n,$$

defines an inner product in $[L^2(R_j)]^n$, and with $\mathbf{1} = (1, \ldots, 1) \in [L^2(R_j)]^n$ and $\widetilde{\rho} = (\widetilde{\rho}_{j_1}, \ldots, \widetilde{\rho}_{j_n})$ for $j_1, \ldots, j_n$ distinct elements of J,

$$\|\mathbf{1}\|^2 = \sum_{k=1}^{n} \int_{R_{j_k}} 1 = \sum_{k=1}^{n} a_{j_k} b_{j_k} \leq \sum_{k=1}^{n} \int_{R_{j_k}} \widetilde{\rho}_{j_k} = \langle \mathbf{1}, \widetilde{\rho} \rangle.$$

By the Cauchy-Schwarz inequality, $\|\mathbf{1}\|^4 \leq |\langle \mathbf{1}, \widetilde{\rho} \rangle|^2 \leq \|\mathbf{1}\|^2 \|\widetilde{\rho}\|^2$, and

$$\sum_{k=1}^{n} a_{j_k}^2 M(R_{j_k}) = \sum_{k=1}^{n} b_{j_k}^2 \frac{1}{M(R_{j_k})} = \sum_{k=1}^{n} a_{j_k} b_{j_k} = \|\mathbf{1}\|^2 \leq \|\widetilde{\rho}\|^2 = \sum_{k=1}^{n} \int_{R_{j_k}} \widetilde{\rho}_{j_k}^2 \leq \int_S \rho^2.$$

As $\rho \in L^2(S)$ and $M(R_j) = M(S_j)$, the proof is finished. □

This result allows to prove that the metric of a meromorphic quadratic Differential φ in a Riemann surface S with almost all trajectories closed is the *unique extremal metric* in the sense of being the only one giving the area of S, denoted $\|\varphi\|$, upper bounded by the area in any metric with element of length $\rho(z)\,|dz|$ which for almost all closed curves gives lengths $\geq$ those in the metric with element of length $\sqrt{|\varphi(z)|}\,|dz|$ of the metric of the quadratic differential φ.

(8.24) *If φ is a meromorphic quadratic differential in a Riemann surface S with almost all trajectories closed and $\rho \in L^2(S)$, $\rho \geq 0$ satisfies $\int_{\gamma_j} \rho(z)\,|dz| \geq a_j$ for almost all closed trajectories γ_j of φ, with a_j the length of γ_j in the metric of φ, then $\|\varphi\| \leq \int_S \rho^2$, with equality if and only if $\rho = \sqrt{|\varphi|}$* a.e. *in S.*

Proof The first statement is a direct application of the preceding result. In case of equality $\|\varphi\| = \int_S \rho^2$, there is equality in the Cauchy-Schwarz inequality in the proof of the preceding result and, with the notation there, $\widetilde{\rho} = c\mathbf{1}$ for some constant c. The composition of $|\varphi|$ with a conformal homeomorphism of S_j onto R_j is 1 a.e. in R_j for $j \in J$. As $\int_S |\varphi| = \int_S \rho^2$ and at each R_j we have $\int_{R_j} 1 \leq \int_{R_j} \widetilde{\rho}_j^2$, also $\int_{S_j} |\varphi| = \int_{S_j} \rho^2$ and, therefore, $\int_{R_j} 1 = \int_{R_j} c^2$. So, $c = 1$ and $\widetilde{\rho} = \mathbf{1}$, implying $\rho = \sqrt{|\varphi|}$ a.e. in S_j. Since $S \backslash \cup_{j \in J} S_j$ has area 0, $\rho = \sqrt{|\varphi|}$ a.e. in S. □

(8.25) *If $\varphi \neq 0$ is a holomorphic quadratic differential in a Riemann surface S with $\|\varphi\| \in \mathbb{R}$, $\Gamma = \{\gamma_1, \ldots, \gamma_n\}$ is an admissible set of curves in S with the maximal ring module of the homotopic type of each finite γ_j, and $a_j = \inf_{\gamma \sim \gamma_j} \int \sqrt{|\varphi(z)|}\,|dz|$, where $\gamma \sim \gamma_j$ is for any rectifiable Jordan curve γ homotopic to γ_j in S, then for every set of rings $\widetilde{S}_j$ of the homotopic type of γ_j, $\sum_j a_j^2 M(\widetilde{S}_j) \leq \|\varphi\|$, with equality if and only if $S \backslash \cup_j \widetilde{S}_j$ has area 0 and is a union of trajectories γ of φ with $\operatorname{Arg} \varphi(\gamma(t))[\gamma'(t)]^2 = \theta$ fixed.*

Proof Since the modules of maximal rings with the homotopic type of each γ_j are finite, $a_j > 0$ for $j \in \{1, \ldots, n\}$ and, by the preceding result, $\sum_{j=1}^n a_j^2 M(\widetilde{S}_j) \leq \|\varphi\|$.

In case of equality, with $\rho = \sqrt{|\varphi|}$ in the proof of the preceding result, there is equality in the Cauchy-Schwarz inequality, and, therefore, $\widetilde{\rho}$ is a multiple of $\mathbf{1} \in [L^2(\widetilde{R}_j)]^n$; so, it is a constant a.e. in R_j. Thus, also ρ and $|\varphi| = c$ are constants a.e. in S, and $S \backslash \cup_{j=1}^n \widetilde{S}_j$ has area 0. Hence, $\varphi_{|\widetilde{R}_j} = c\, e^{i\theta_j}$ for $\theta_j \in]-\pi, \pi]$, $j \in \{1, \ldots, n\}$. It is now proved that the θ_j are all equal. Let $p_1 \in \widetilde{S}_1$, $p_2 \in \widetilde{S}_2$, and α_1 be a geodesic arc in the metric of φ in S from p_1 to p_2 not containing any zero of φ. For the arc α_1 with end points p_1, p_2, let β_1, β_2 be arcs of distinct geodesics in the metric of φ with end points, resp., p_1, p_1' and p_2, p_2' such that the same geodesic arc α_1' contained in a geodesic parallel to that containing the arc α_1 passes through p_1' and p_2'. Each ring $\widetilde{S}_j$ intersecting α_1 also intersects α_1' and the intersection of $\widetilde{S}_j$ with the quadrilateral with vertices p_1, p_2, p_2', p_1' is a parallelogram. The lengths of α_1 and α_1' in the metric of φ are equal, and the lengths of β_1 and β_2 are also equal; so, $\theta_j = \theta_1$. Considering $p_j \in \widetilde{S}_j$ also for $j \in \{3, \ldots, n\}$ and α a polygonal line of segments $\alpha_1, \ldots, \alpha_{n-1}$ arcs of geodesic curves in the metric of φ in S whose vertices are, by order, $p_1, \ldots, p_n$, not containing zeros of φ, with the preceding argument paragraph, $\theta_j = \theta_1$ for all $j \in \{1, \ldots, n\}$. □

(8.26) *If $\varphi \neq 0$ is a holomorphic quadratic differential in a Riemann surface S with almost all trajectories closed and $\|\varphi\| \in \mathbb{R}$, $\Gamma = \{\gamma_1, \ldots, \gamma_n\}$ is an admissible set of curves in S with finite maximal modules of rings of the homotopic type of each γ_j, $\{S_1, \ldots, S_n\}$ is the set of characteristic rings of φ, and $\{\widetilde{S}_1, \ldots, \widetilde{S}_n\}$ is a set of disjoint rings with $\widetilde{S}_j$ of the same homotopic type of γ_j,*

$$\sum_{j=1}^{n} a_j^2 M(\widetilde{S}_j) \leq \sum_{j=1}^{n} a_j^2 M(S_j), \qquad \sum_{j=1}^{n} \frac{b_j^2}{M(\widetilde{S}_j)} \geq \sum_{j=1}^{n} \frac{b_j^2}{M(S_j)},$$

with equality if and only if $\widetilde{S}_j = S_j$ for $j \in \{1, \ldots, n\}$.

Proof As $\|\varphi\| = \sum_{j=1}^{n} a_j^2 M(S_j)$, by the inequality in the first paragraph of the proof of (8.25), $\sum_j a_j^2 M(\widetilde{S}_j) \leq \sum_{j=1}^{n} a_j^2 M(S_j)$. In case of equality, if for some $j \in \{1, \ldots, n\}$ the rings S_j and $\widetilde{S}_j$ are not degenerate, a_j is the length of γ_j in the metric of φ, the argument θ in (8.25) is $\theta = 0$, and $\widetilde{S}_j = S_j$ to $j \in \{1, \ldots, n\}$.

If for all $j \in \{1, \ldots, n\}$ one of the rings S_j, $\widetilde{S}_j$ is degenerate, for S_j, $\widetilde{S}_k$ with $S_j \cap \widetilde{S}_k = \emptyset$ the curves γ_j and γ_k are disjoint nonhomotopic, and if $\widetilde{\alpha}$ is a geodesic Jordan curve in the metric of φ in $\widetilde{S}_k$ and α is a closed trajectory of φ in S_j, as the curves are not parallel, they intersect in at least two points ends of arcs of each one of the curves whose union is the boundary of a simply connected region in S, contradicting the Teichmüller lemma following this proof. So, for every $j \in \{1, \ldots, n\}$, both rings S_j, $\widetilde{S}_j$ are nondegenerate, and $\widetilde{S}_j = S_j$.

If some Ring $\widetilde{S}_j$ is degenerate, the second inequality in the statement is trivial, since $M(\widetilde{S}_j) = 0$ and $b_j > 0$. Consider $M(\widetilde{S}_1), \ldots, M(\widetilde{S}_n) > 0$. We start as in the first period of the proof of (8.23) from each ring $\widetilde{S}_j$, but with each rectangle $\widetilde{R}_j$ of Height b_j equal to the height of the characteristic ring S_j and length $\widetilde{a}_j = \frac{b_j}{M(\widetilde{S}_j)}$, which, without loss of generality, we consider $\widetilde{R}_j =]0, \widetilde{a}_j[\times]0, b_j[$. Since each horizontal α of $\widetilde{R}_j$ corresponds to a closed curve in $\widetilde{S}_j$ of the homotopic type of γ_j, the length a_j of γ_j in the metric of φ satisfies $a_j \leq \int_0^{\widetilde{a}_j} \sqrt{|\widetilde{\varphi}(x)|}\, dx$, with $\widetilde{\varphi}$ in $\widetilde{R}_j$ corresponding to φ in $\widetilde{S}_j$ by the conformal homeomorphisms considered. Thus,

$\sum_{j=1}^{n} a_j b_j \leq \sum_{j=1}^{n} \int_{\widetilde{R}_j} \sqrt{|\widetilde{\varphi}|}$. With the inner product and the notation of that proof, and by the Cauchy-Schwarz inequality,

$$\left(\sum_{j=1}^{n} a_j b_j\right)^2 = \left(\sum_{j=1}^{n} \int_{\widetilde{R}_j} \sqrt{|\widetilde{\varphi}|}\right)^2 = |\langle \mathbf{1}, \sqrt{|\widetilde{\varphi}|}\,\rangle|^2 \leq \|\mathbf{1}\|^2 \, \|\sqrt{|\widetilde{\varphi}|}\,\|^2$$
$$= \left(\sum_{j=1}^{n} \widetilde{a}_j b_j\right)\left(\sum_{j=1}^{n} \int_{\widetilde{R}_j} |\varphi|\right) = \left(\sum_{j=1}^{n} \widetilde{a}_j b_j\right)\left(\sum_{j=1}^{n} a_j b_j\right).$$

Hence,

$$\sum_{j=1}^{n} \frac{b_j^2}{M(S_j)} = \sum_{j=1}^{n} a_j b_j \leq \sum_{j=1}^{n} \widetilde{a}_j b_j = \sum_{j=1}^{n} \frac{b_j^2}{M(\widetilde{S}_j)},$$

with equality if and only if $a_j \leq \int_0^{\widetilde{a}_j} \sqrt{|\widetilde{\varphi}(x)|}\, dx$ holds with equality and as well as Cauchy-Schwarz inequality used, and in this case $\widetilde{\varphi}$ is a constant $c > 0$ in all the $\widetilde{R}_j$. Substituting in the first of these equalities, $a_j = c\widetilde{a}_j$. So,

$$\|\varphi\| = \sum_{j=1}^{n} a_j b_j = \sum_{j=1}^{n} \frac{\widetilde{a}_j}{c} b_j,$$

and, by (8.22), $\frac{\widetilde{a}_j}{\widetilde{c}} = a_j$ for $j \in \{1, \ldots, n\}$ and $\widetilde{S}_1, \ldots, \widetilde{S}_n$ are the characteristic rings of φ in S; thus, $\widetilde{S}_j = S_j$ for $j \in \{1, \ldots, n\}$. □

The following lemma was published by O. Teichmüller in 1939.

(8.27) Teichmüller lemma: *If $D \subset \mathbb{C}$ is a region bounded by a polygonal Jordan curve α concatenation of curves $\alpha_1, \ldots, \alpha_n$ (in this order) which are arcs of geodesics in the metric of a meromorphic quadratic differential φ in* cl *D with zeros or poles of φ in α only at vertices of this polygonal curve, with $\theta_j \in [0, 2\pi[$ denoting the interior angle of the region defined by the tangents to α_j and to α_{j+1} at the common vertex $j \in \{1, \ldots, n-1\}$ and $\theta_n \in [0, 2\pi[$ denoting the angle interior to the region defined by the tangents to α_n and to α_1 at the common vertice x, then*

$$\sum_{j=1}^{p} \left(1 - \theta_j \frac{n_j+2}{2\pi}\right) = \sum_{k=1}^{q} m_k + 2,$$

where p, q are the sum of the number of poles with the number of zeros of φ in, resp., ∂D, D, and $\pm n_j$, $\pm m_k$ are the orders of successive zeros or poles of φ in, resp., ∂D, D, with sign $+/-$ depending on whether it is a zero or a pole.

Proof Along each α_j, $\operatorname{Arg}\varphi(z)\,dz^2$ is constant, and as

$$d\big(\operatorname{Arg}\varphi(z)\,dz^2\big) = d\operatorname{Arg}\varphi(z) + 2\,d(\operatorname{Arg}dz) = 0\,,$$

$\frac{1}{2\pi}\int_\alpha d\operatorname{Arg}\varphi(z) = -\frac{1}{\pi}\int_\alpha d(\operatorname{Arg}dz)$. $\operatorname{Arg}dz$ increases from $2\pi - \sum_{j=1}^{p}(\pi-\theta_j)$ along α ; thus, $\sum_{k=1}^{q} m_k + \sum_{j=1}^{p}\frac{\theta_j}{2\pi}n_j = -2 + \sum_{j=1}^{p}\big(1-\frac{\theta_j}{\pi}\big)$. At the vertices, $1-\frac{\theta_j}{\pi}-\frac{\theta_j}{2\pi}n_j = 1-\theta_j\frac{n_j+2}{2\pi}$, giving the equality in the statement. □

It is necessary to guarantee the existence of a quadratic differential on a Riemann surface S with the elements of a given multicurve closed trajectories and the associated characteristic rings of specified sizes. It is a problem of existence of extrema of functionals that can be specified in three ways: by the lengths of the horizontals, by the modules, or by the lengths of the verticals of rectangles to which they are conformal. The first of these problems was solved by J. Jenkins in 1957 and involves the possibility of rings collapsing to a curve, the second was solved in 1967 by K. Strebel, and although it does not involve the possibility of collapsing rings, the functional to be minimized is more difficult to handle, and the third was solved in 1976 by J. Hubbard and H. Masur,[46] and, independently, by H. Renelt,[47] and it has the advantage of not involving to collapse rings and of relying on an easier functional minimization. It is the H. Renelt solution of the last existence problem with a simplification by E. Reich that is given here, as presented by K. Strebel in the book listed in the final bibliography.

(8.28) *If $\Gamma = \{\gamma_1,\dots,\gamma_n\}$ is an admissible set of curves in a Riemann surface S and the supremum of the modules of the maximal rings S_j of the homotopic type of each γ_j is $M^+ \in \mathbb{R}$, then for every $b_1,\dots,b_n > 0$, there is a unique holomorphic quadratic differential φ on S with almost all trajectories closed and height b_j of S_j for $j\in\{1,\dots,n\}$, and $\|\varphi\| = \sum_{j=1}^{n} b_j^2/M_j(S_j)$.*

Proof Let $\{S_{1m},\dots,S_{nm}\}_{m\in\mathbb{N}}$ be a sequence of sets of rings, each one of them of the homotopic type of γ_j and such that $\sum_{j=1}^{n}\frac{b_j^2}{M(S_{jm})}$ tends to the infimum of these possible sums when $m\to+\infty$, and g_{jm} a conformal homeomorphism of the annulus included in the disk $B_1\subset\mathbb{C}$ with one of the boundary components coincident with the circle ∂B_1 onto S_{jm} . This annulus is $B_1\backslash \operatorname{cl} B_{r_{jm}}$ with $r_{jm} = e^{-2\pi M(S_{jm})}$. As $\Big\{\frac{b_j^2}{M(S_{jm})}\Big\}_{m\in\mathbb{N}}$ is bounded, for each $j\in\{1,\dots,n\}$, the sequence $\big\{M(S_{jm})\big\}_{m\in\mathbb{N}}$ does not tend to 0 . Since $M^+\in\mathbb{R}$, for every $j\in\{1,\dots,n\}$, the sequence $\{g_{jm}\}_{m\in\mathbb{N}}$

[46] Hubbard, J.H., Masur, H., On the existence and uniqueness of Strebel differentials, *Bulletin of the AMS* **82** (1976), 77–79.

[47] Renelt, H., Konstruktion gewisser quadratischer Differentiale mit Hilfe von Dirichletintegralen. *Math. Nachr.* **73** (1976) 125–142. Masur, Howard (1949-).

is a normal family. Thus, it has a subsequence convergent in every compact subset of $B_1\backslash \mathrm{cl}\, B_{r_j}$ to a conformal homeomorphism g_j of $B_1\backslash \mathrm{cl}\, B_{r_j}$ onto S_j, where $r_j = \lim_{m\to+\infty} r_{jm}$. Thus, $M(S_j) = \frac{1}{2\pi}\log\frac{1}{r_j}$, R_j has the homotopic type of γ_j and the set of rings $\{S_1, \ldots, S_n\}$ is extremal, i.e., $\sum_{j=1}^n \frac{b_j^2}{M(S_j)}$ is minimal.

For every quasiconformal mapping of S onto S homotopic to the identity in S mapping rings S_j to $\widetilde{S}_j$, $\sum_{j=1}^n \frac{b_j^2}{M(\widetilde{S}_j)} \ge \sum_{j=1}^n \frac{b_j^2}{M(S_j)}$. Apply conformal homeomorphisms mapping S_j, $\widetilde{S}_j$ to annuli, resp., $C_j = B_1\backslash \mathrm{cl}\, B_{\rho_j}$, $\widetilde{C}_j = B_1\backslash \mathrm{cl}\, B_{\widetilde{\rho}_j} \subset \mathbb{C}$. Cutting C_j along a radius and applying a suitable multiple of a logarithm branch gives a rectangle R_j of height b_j and length $a_j = \frac{b_j}{M(S_j)}$. The cut of the ring $\widetilde{S}_j$ by the quasiconformal mapping and the analogous transformation leads to a quadrilateral of height b_j and length (in the horizontal) $\widetilde{a}_j$. So, $M(\widetilde{s}_j) = M(\widetilde{R}_j) = \frac{b_j}{\widetilde{a}_j}$. f is called the quasiconformal mapping from R_j onto $\widetilde{R}_j$ induced by the initial quasiconformal mapping from S_j onto $\widetilde{S}_j$ and by the conformal transformations applied to get the annuli C_j, $\widetilde{C}_j$ and the quadrilaterals R_j, $\widetilde{R}_j$ (Fig. 8.8).

Let $\zeta^* = f(\zeta)$, with $\zeta = \xi + i\eta$, $\xi, \eta \in \mathbb{R}$. The length of the image of each horizontal α in R_j by f is

$$\widetilde{a}_j \le \int_{f(\alpha)} 1\,|d\zeta^*| = \int_\alpha \left|\frac{\partial f}{\partial \zeta} + \frac{\partial f}{\partial \bar{\zeta}}\right| d\xi\,.$$

With $J = \left|\frac{\partial f}{\partial \zeta}\right|^2 - \left|\frac{\partial f}{\partial \bar{\zeta}}\right|^2$, the absolute value of the Jacobian of f and κ the Beltrami coefficient of f, by the Cauchy-Schwarz inequality in $L^2(R_j)$ and changing the integration variables by f, $\int_{R_j} J\, d\xi d\eta = \int_{\widetilde{R}_j} 1 = \widetilde{a}_j b_j$,

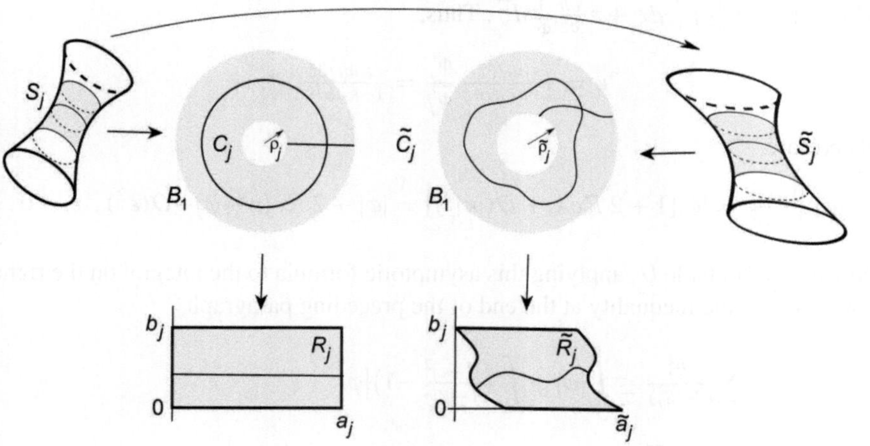

Fig. 8.8 Transformation of quasiconformal rings $S_j, \widetilde{S}_j$ in a Riemann surface S to annuli $C_j, \widetilde{C}_j \subset \mathbb{C}$ and of these to quasiconformal quadrilaterals $R_j, \widetilde{R}_j \subset \mathbb{C}$ with heights equal to a given $b_j > 0$

$$
\begin{aligned}
(\widetilde{a}_j b_j)^2 &= \left(\int_{R_j} J^{1/2}\tfrac{1}{J^{1/2}}\left|\tfrac{\partial f}{\partial \zeta}+\tfrac{\partial f}{\partial \overline{\zeta}}\right| d\xi d\eta\right)^2 \\
&\le \left(\int_{R_j} J\, d\xi d\eta\right)\left(\int_{R_j}\tfrac{1}{J}\left|\tfrac{\partial f}{\partial \zeta}+\tfrac{\partial f}{\partial \overline{\zeta}}\right|^2 d\xi d\eta\right) = \widetilde{a}_j b_j \int_{R_j}\tfrac{1+|\kappa|^2}{1-|\kappa|^2}\, d\xi d\eta\,.
\end{aligned}
$$

Therefore,

$$
\frac{b_j^2}{M(\widetilde{R}_j)} = \widetilde{a}_j b_j \le \int_{R_j}\frac{1+|\kappa|^2}{1-|\kappa|^2}\, d\xi d\eta\,.
$$

Denoting Φ_j the conformal homeomorphism of R_j onto S_j minus a cut along a curve connecting the two connected components of the annulus boundary, and defining $\varphi(z) = [\Phi'_j(x)]^2$ in terms of local coordinates z in coordinate neighborhoods of S_j and $\varphi = 0$ in $S \backslash \cup_{j=1}^n S_j$, with a change of integration variables,

$$
\sum_{j=1}^n \frac{b_j^2}{M(\widetilde{S}_j)} \le \sum_{j=1}^n \int_{S_j}\frac{1+|\kappa|^2}{1-|\kappa|^2}\,|\varphi|\, dxdy = \int_{S_j}\frac{1+|\kappa|^2}{1-|\kappa|^2}\,|\varphi|\, dxdy\,.
$$

What was obtained in the two preceding paragraphs is valid for any quasiconformal mapping f of S_j onto $\widetilde{S}_j$. Now f is chosen appropriately to obtain the result. Let z be a local coordinate in a neighborhood of coordinates U of S and $h : S \to \mathbb{C}$ C^1 with compact support included in U.The function $z \mapsto w = z + \varepsilon h(z)$ is equal to the identity outside of U, and it is a quasiconformal mapping from S to S for every small $\varepsilon \in \mathbb{C}$. This function is homotopic to 1_S and $dw = \left(1+\varepsilon\frac{\partial h}{\partial z}\right)dz + \varepsilon\frac{\partial h}{\partial \overline{z}}\, d\overline{z}$. Taking into account that the conformal homeomorphism Φ_j is from S_j onto R_j, $dw = \left(1+\varepsilon\frac{\partial h}{\partial z}\right)\frac{1}{\Phi'}d\zeta + \varepsilon\,\frac{\partial h}{\partial \overline{z}}\frac{1}{\overline{\Phi'}}d\overline{\zeta}$. Thus,

$$
\kappa = \frac{\varepsilon\,\partial h/\partial \overline{z}}{1+\varepsilon\,\partial h/\partial z}\,\frac{\Phi'_j}{\overline{\Phi'_j}} = \frac{\varepsilon\,\partial h/\partial \overline{z}}{1+\varepsilon\,\partial h/\partial z}\,\frac{\varphi}{|\varphi|}\,.
$$

Therefore,

$$
|\varphi|\,\frac{|1+\kappa|^2}{1-|\kappa|^2} = |\varphi|[1+2\,\mathcal{R}e\,\kappa + O(|\kappa|^2)] = |\varphi| + 2\,\mathcal{R}e\!\left(\varepsilon\frac{\partial h}{\partial \overline{z}}\varphi\right) + O(\varepsilon^2)\,,\ \ \varepsilon \to 0\,.
$$

Since κ is 0 outside U, applying this asymptotic formula to the integral on the right-hand side of the inequality at the end of the preceding paragraph,

$$
\begin{aligned}
\sum_{j=1}^n \frac{b_j^2}{M(\widetilde{S}_j)} &\le \int_S |\varphi| + \int_U \left(\frac{|1+\kappa|^2}{1-|\kappa|^2} - 1\right)|\varphi| \\
&= \sum_{j=1}^n \frac{b_j^2}{M(S_j)} + 2\,\mathcal{R}e\left(\varepsilon\int_U \frac{\partial h}{\partial \overline{z}}\varphi\right) + O(\varepsilon^2)\,, \quad \varepsilon \to 0\,.
\end{aligned}
$$

By the second paragraph of this proof, $\sum_{j=1}^{n} \frac{b_j^2}{M(\tilde{S}_j)} \geq \sum_{j=1}^{n} \frac{b_j^2}{M(S_j)}$ and the argument of $\varepsilon \in \mathbb{C}$ is arbitrary. Thus, $\int_U \frac{\partial h}{\partial \bar{z}} \varphi = 0$ for every C^1 function with compact support h included in U. By the Weyl lemma (1.15), φ is equal to a holomorphic function a.e. in U, and considering a cover of S by coordinate neighborhoods, we get a function φ equal to a holomorphic quadratic differential a.e. in S denoted φ. The set $S \backslash \cup_{j=1}^{n} S_j$ has zero area and $\varphi = (\Phi_j')^2$ in S_j for $j \in \{1, \ldots, n\}$. Each S_j is a ring included in a characteristic ring, as it is the union of closed trajectories of φ and cannot be a proper subset of the characteristic ring because that would contradict that $S \backslash \cup_{j=1}^{n} S_j$ has area 0. So, $\{S_1, \ldots, S_n\}$ is the set of characteristic rings of φ and the height of each S_j is b_j, concluding the proof of existence of a holomorphic quadratic differential with almost all trajectories closed and each characteristic ring S_j with a given height b_j.

If φ_1, φ_2 are quadratic differentials satisfying the conditions in the statement, by (8.24) with $\varphi = \varphi_1$ and $\rho = \sqrt{|\varphi_2|}$, it is $\sqrt{|\varphi_2|} = \sqrt{|\varphi_1|}$ a.e. in S. Thus, $\varphi_1, \varphi_2 \in H(S)$, and $|\varphi_2| = |\varphi_1|$. In an open circle $V \subset S$ without zeros of φ_1, the holomorphic function $\frac{\varphi_2}{\varphi_2}$ has absolute value 1. So, it is constant in V and, therefore, also in S, with a value $e^{i\theta}$ with $\theta \in [0, 2\pi[$. As, for example, γ_1 is a trajectory of both φ_1 and φ_2, the corresponding tangents at each point coincide and, therefore, $\theta = 0$. Consequently, $\varphi_2 = \varphi_1$. □

The following two results were obtained by K. Strebel, first for compact Riemann surfaces in 1967, and in general in 1976.

(8.29) *If $\Gamma = \{\gamma_1, \ldots, \gamma_n\}$ is an admissible set of curves in a Riemann surface S and Φ^1 is the set of holomorphic quadratic differentials φ in S such that $\gamma_1, \ldots, \gamma_n$ are closed trajectories of φ, the complement in S of the union of closed trajectories of φ has zero area and $\|\varphi\| = 1$, then the set of vectors $\mathbf{x} = (b_1^2, \ldots, b_n^2)$ with components the squares of the heights of the characteristic rings S_j of φ of the same homotopic type of γ_j is for $\varphi \in \Phi^1$ a **surface** $\mathscr{H} \subset \mathbb{R}^n$ with boundary the elements of the set boundary of $\mathscr{H}$ with at least 1 component 0, included in the intersection of the half spaces of $\mathbb{R}^n$ without 0 and are bounded by the hyperplanes $\mathscr{E}_{\mathbf{x}}$ passing at each $\mathbf{x} \in \mathscr{H}$ and orthogonal to $\mathbf{n_x} = \left(\frac{1}{M(S_1)}, \ldots, \frac{1}{M(S_n)}\right)$.*

Proof Let $\mathbf{u} = (u_1, \ldots, u_n) \in [0, +\infty[^n$ with $\|\mathbf{u}\| = 1$ in the canonical inner product in $\mathbb{R}^n$.

By (8.28), there is a holomorphic quadratic differential $\widetilde{\varphi}$ in S with characteristic rings $S_1, \ldots, S_n$ and each S_j of the homotopic type of γ_j with Heights $b_j = \sqrt{u_j}$, for $j \in \{1, \ldots, n\}$. Therefore, there is one unique holomorphic quadratic differential with norm 1 and the same height as those characteristic rings and $\varphi = \frac{1}{\|\widetilde{\varphi}\|} \widetilde{\varphi}$, which is a continuous function of $\mathbf{u}$, so that $\mathbf{x} = \frac{1}{\|\varphi\|} \mathbf{u}$ is also a continuous function of $\mathbf{u}$.

The range of this function is a hypersurface $\mathscr{H}$ in $\mathbb{R}^n$ with boundary $\partial\mathscr{H}$ whose elements have at least one component equal to 0. $\mathscr{H}$ is the surface of the vectors in $\mathbb{R}^n$ with components the squares of the heights of the characteristic rings of the associated holomorphic quadratic differentials with norm 1 defined in S of the same homotopic type of, resp., $\gamma_1, \ldots, \gamma_n$.

If $\mathbf{x} = (b_1^2, \ldots, b_n^2)$, $\widetilde{\mathbf{x}} = (\widetilde{b}_1^2, \ldots, \widetilde{b}_n^2) \in \mathscr{H}$, with associated holomorphic quadratic differentials, resp., $\varphi, \widetilde{\varphi}$, and with the n characteristic rings of the homotopic type of $\gamma_1, \ldots, \gamma_n$, resp., $S_1, \ldots, S_n$ and $\widetilde{S}_1, \ldots, \widetilde{S}_n$, by the second inequality in (8.26),

$$\widetilde{\mathbf{x}} \cdot \mathbf{n}_{\mathbf{x}} = \sum_{j=1}^{n} \frac{\widetilde{b}_j^2}{M(S_j)} > \sum_{j=1}^{n} \frac{\widetilde{b}_j^2}{M(\widetilde{S}_j)} = \widetilde{\mathbf{x}} \cdot \mathbf{n}_{\widetilde{\mathbf{x}}} = 1 .$$

Since, by (8.22), $\mathbf{x} \cdot \mathbf{n}_{\mathbf{x}} = \sum_{j=1}^{n} \frac{b_j^2}{M(S_j)} = \|\varphi\| = 1$, $(\widetilde{\mathbf{x}} - \mathbf{x}) \cdot \mathbf{n}_{\mathbf{x}} > 0$. Thus, $\mathscr{H} \setminus \{\mathbf{x}\}$ belongs to the open half space of $\mathbb{R}^n$ not containing 0 and bounded by the plane $\mathscr{E}_{\mathbf{x}}$ orthogonal to $\mathbf{n}_{\mathbf{x}}$ and containing $\mathbf{x}$. □

(8.30) *If $\Gamma = \{\gamma_1, \ldots, \gamma_n\}$ is an admissible set of curves in a hyperbolic Riemann surface S and $\mathbf{m} = (m_1, \ldots, m_n) \in]0, +\infty[^n$ is a unit vector in the canonical norm of $\mathbb{R}^n$, then there is $c > 0$ and 1 unique quadratic differential φ with $\|\varphi\| = 1$ in S and closed trajectories homotopic to γ_j with boundaries of rings C_j with $M(C_j) = c\, m_j$, $j \in \{1, \ldots, n\}$, and in the set of disjoint open rings $C = \{C_1, \ldots, C_n\}$ representing isotopy classes of Γ where the equality holds for some $c > 0$, there is only one maximizing $\sum_{j=1}^{n} M(C_j)$.*

Proof Let $\mathbf{u} \in \mathbb{R}^n$ be a unit vector with the direction and orientation of $\left(\frac{1}{m_1}, \ldots, \frac{1}{m_n}\right)$ and $\mathbf{x}$ the point closest to the origin of the intersection of a plane $\mathscr{E}_{\mathbf{x}}$ orthogonal to $\mathbf{u}$ with the surface $\mathscr{H}$ of the preceding result. $\mathbf{x} \notin \partial\mathscr{H}$, since if it has only 1 component 0, without loss of generality the last, with $\mathbf{x}' \in \mathscr{H}$ and $\mathbf{x}' = \tau\mathbf{x} + x_n'\mathbf{e}_n$, $x_n' > 0$, $\tau \in \mathbb{R}$, where $(\mathbf{e}_1, \ldots, \mathbf{e}_n)$ is the canonical base of $\mathbb{R}^n$, it is $\tau \in]0, 1[$, $\tau \to 1$ when $x_n' \to 0$, and

$$(\mathbf{x}' - \tau\mathbf{x}) \cdot \mathbf{u} = x_n'\mathbf{e}_n - (1-\tau) \sum_{j=1}^{n-1} x_j \mathbf{e}_j < 0 ,$$

for small $x_n' > 0$, implying that the points of $\mathscr{H}$ close to $\mathbf{x}' \in \mathscr{H}$ are in the half space bounded by $\mathscr{E}_{\mathbf{x}}$ and containing 0 and contradicting the definition of $\mathscr{E}_{\mathbf{x}}$. If $n - m > 1$ components of $\mathbf{x}$ are 0, without loss of generality the last ones, we get $\mathbf{x}' \in \mathscr{H}$ in the half space bounded by $\mathscr{E}_{\mathbf{x}}$ and containing 0, contradicting the definition of $\mathscr{E}_{\mathbf{x}}$. The heights of the cylinders of the quadratic differential φ with $\|\varphi\| = 1$ corresponding to $\mathbf{x}$ are $b_j = \sqrt{x_j}$ and a normal to $\mathscr{H}$ at $\mathbf{x}$ is $\left(\frac{1}{M(C_1)}, \ldots, \frac{1}{M(C_n)}\right)$, collinear with $\left(\frac{1}{m_1}, \ldots, \frac{1}{m_n}\right)$, and there is $c > 0$ such that $M(C_j) = c\, m_j$ for $j \in \{1, \ldots, n\}$.

To prove uniqueness, suppose $\widetilde{\varphi}$ is another quadratic differential satisfying the conditions with set of rings $\widetilde{C}=\{\widetilde{C}_1,\ldots,\widetilde{C}_n\}$. There is $\widetilde{c}>0$ such that $M(\widetilde{C}_j)=\widetilde{c}\,m_j$ for $j\in\{1,\ldots,n\}$, $k=\frac{\widetilde{c}}{c}>0$, and $M(\widetilde{C}_j)=kM(C_j)$ for $j\in\{1,\ldots,n\}$. With $a_j=\inf_{\gamma^*}\int_\gamma\sqrt{|\varphi(z)|}\,|dz|$ and γ in the set of rectified Jordan curves in S in the homotopy class of γ_j,

$$\sum_{j=1}^{n} a_j^2 M(\widetilde{C}_j)=k\sum_{j=1}^{n} a_j^2 M(C_j),$$

Thus, $\|\varphi\|=k\,\|\widetilde{\varphi}\|$ and as $\|\varphi\|=\|\widetilde{\varphi}\|=1$, it is $k=1$. C and $\widetilde{C}$ are sets of rings representing the isotopy classes of Γ. Therefore, $M(\widetilde{C}_j)=M(C_j)$, and $\widetilde{C}=C$. □

Dynamics of Rational Functions with Thurston Obstructions

C. McMullen proved that rational functions with Thurston obstructions have domains of rotation or are Lattès functions.

(8.31) *If f is a rational function of degree $d\geq 2$ with associated marked branched covering $((\mathbb{C}_\infty,f),PC(f))$ of $\mathbb{C}_\infty$ and $\Gamma=\{\gamma_1,\ldots,\gamma_n\}$ is a stable multicurve irreducible in $\mathbb{C}_\infty\backslash PC(f)$, $\mathbf{m}=(m_1,\ldots,m_n)\in\,]0,+\infty[^n$ is an eigenvector associated with the eigenvalue 1 of the transition matrix M_Γ of Γ and $C=\{C_1,\ldots,C_n\}$ is the set of rings guaranteed by* (8.30), *then there is $c>0$ such that $M(C_j)=c\,m_j$ for $j\in\{1,\ldots,n\}$ and $f^*C=C$.*

Proof By (8.30), for every connected component of $\mathbb{C}_\infty\backslash PC(f)$, there is $c>0$ such that $M(C_j)=c\,m_j$ for $j\in\{1,\ldots,n\}$. With c_0 the minimum value of the $c>0$ corresponding to these connected components and $v_j=M(C_j)-c_0m_j$ in each of the other connected components, if it were $\mathbf{v}=(v_1,\ldots,v_n)\neq 0$ in one of the components, as M_Γ is Irreducible, for every $j\in\{1,\ldots,n\}$, there would exist $m\in\mathbb{N}$ such that the j component of $M_\Gamma\mathbf{v}$ is >0. So, with $C'=(f^m)^*(C)$, in this connected component, $M(C_j')>M(C_j)$ and $M(C_k')>M(C_k)$ for all $k\in\{1,\ldots,n\}$. By the uniqueness in (8.30), $C_k'=C_k$ for all $k\in\{1,\ldots,n\}$, contradicting $\mathbf{v}\neq 0$. Hence, $M(C_j)=c_0m_j$ for all $j\in\{1,\ldots,n\}$.

Since $M_\Gamma\mathbf{m}=\mathbf{m}$, with $C'=f^*C$, $M(C_j')\geq M(C_j)$ for $j\in\{1,\ldots,n\}$ and, by the uniqueness in (8.30), $C'=C$. Therefore, $f^*C=C$. □

If f is a rational function of degree $d\geq 2$ with associated marked branched covering $((\mathbb{C}_\infty,f),PC(f))$ of $\mathbb{C}_\infty$, $\Gamma=\{\gamma_1,\ldots,\gamma_n\}$ is an irreducible stable multicurve in $\mathbb{C}_\infty\backslash PC(f)$ and, for each $j\in\{1,\ldots,n\}$, C_j is a ring conformal to a collar of the geodesic Jordan curve γ_j, we call a **cluster** of $C=\{C_1,\ldots,C_n\}$ to a ring $A=\mathcal{M}(D_1,\ldots,D_n)$ with $M(A)=\sum_{j=1}^n M(D_j)$ for some set $\{D_1,\ldots,D_n\}$ of connected components of $D=f^{-1}\left(\cup_{j=1}^n C_j\right)$ nested inside each other.

(8.32) *Let f be a rational function of degree $d \geq 2$ with associated marked branched covering $((\mathbb{C}_\infty, f), PC(f))$ of $\mathbb{C}_\infty$ and $\Gamma = \{\gamma_1, \dots, \gamma_n\}$ an irreducible stable multicurve in $\mathbb{C}_\infty \setminus PC(f)$ and $A = \mathcal{M}(D_1, \dots, D_k)$ a cluster of $C = \{C_1, \dots, C_n\}$.*

1. *If A does not contain critical points of f, then $f(A)$ is a cluster and (A, f) is a covering of $f(A)$.*
2. *If A contains critical points of f and B_1, B_2 are distinct connected components of ∂A, then there are clusters $A_1, A_2 \subset A$ without critical points of f and analytic arcs $I_1, I_2 \subset \mathbb{C}_\infty \setminus PC(f)$ with end points in $PC(f)$ such that $\partial f(A_j)$ is a disjoint union of $f(B_j)$ and I_j for $j \in \{1, 2\}$.*

Proof

(1) $f(D_j) = C_\ell$ for some $\ell \in \{1, \dots, n\}$. If there are not critical points of f in the circles separating the sets $D_1, \dots, D_k$, the rings $C_1, \dots, C_n$ are disjoint and are nested into each other; so, (A, f) is a covering of $f(A)$. Consequently, $M\big(f(A)\big) = \sum_{j=1}^{M}(C_j)$ and $f(A)$ are a cluster.

(2) The critical points of f belong to the union of the circles separating the sets $D_1, \dots, D_k$. For $j \in \{1, 2\}$, let A_j be the ring with B_j one of the connected components of ∂A_j containing the union of the sets D_ℓ without critical points of f, and let S_j be the another connected component of ∂A_j. $S_j \cap A_j \neq \emptyset$ and S_j are a connected component of $f^{-1}\big(\mathbb{C}_\infty \setminus \cup_{\ell=1}^{n} C_\ell\big)$; so, it is a branched covering of $f(S_j)$. So, $I_j = f(S_j)$ is an analytic arc with end points critical values. Therefore, (A_j, f) is a covering of $f(A_j)$, $\partial f(A_j) = \partial f(A_j)$, and this set is the disjoint union of $f(B_j)$ with $f(S_j)$. □

(8.33) *Let f be a rational function of degree $d \geq 2$ with associated marked branched covering $((\mathbb{C}_\infty, f), PC(f))$ of $\mathbb{C}_\infty$ and $\Gamma = \{\gamma_1, \dots, \gamma_n\}$ an irreducible stable multicurve in $\mathbb{C}_\infty \setminus PC(f)$ with $\lambda_\Gamma = 1$. Then:*

1. *If f is postcritically finite, the canonical orbifold $\mathcal{O}_f$ has signature $(2, 2, 2, 2)$, and it is doubly covered by a complex torus.*
2. *If f is not postcritically finite, then $\gamma_1, \dots, \gamma_n$ are the geodesic Jordan curves in the Poincaré metric of the, resp., ring of a set of n rings included in Siegel disks or Herman rings permuting when f is applied, each a connected component of $\mathbb{C}_\infty \setminus PC(f)$.*

Proof Firstly, assume there is a ring C_j such that none of the images $A_m = f^m(C_j)$ for $m \in \mathbb{N}$ has critical points of f. Each A_m is a cluster and $M(A_{m+a}) \geq M(A_m)$. As the number of clusters is finite, there are $k > m$ such that $A_k = A_m$ and

$f^{k-m}(A_m) = A_k$ by degree 1. Since A_m is a minimal ring containing the union of some of the rings $C_1, \ldots, C_n$, for some $\ell \in \mathbb{N}$, $j \in \{1, \ldots, n\}$, $f^{\ell}(C_j) = C_j$ by degree 1. Therefore, C_j is contained in a Siegel disk or a Herman ring of f. For each Siegel disk or Herman ring U, $\partial U \subset PC(f)$ and $f\big(PC(f) \cap U\big) \subset U$.

Now, assume that $\cup_{m=0}^{\infty} f^m(C_j)$ contains some critical point of f. There is ∂C_k with some connected component an analytic arc. Let m be the smallest number in $\mathbb{N} \cup \{0\}$ such that $A = f^m(C_k)$ contains some critical point of f. The set A is a cluster and a connected component C_1 of A is an analytic arc. By (8.32.2), there is a cluster $A_1 \subset A \backslash PC(f)$ such that $\partial f(A_1) = f(C_1) \cup I_1$, and it is the union of two analytic arcs. Therefore, $\mathbb{C}_\infty = A' \cup f(C_1) \cup I_1$, with A' a cluster in $f(A_1)$.

The quadratic differential φ guaranteed by (8.30) has extension to a meromorphic quadratic differential in $\mathbb{C}_\infty$ satisfying $f^*\varphi = \deg(f)\,\varphi$. If φ were 0 at a point z, it would also be 0 at all preimages of z, what is impossible since the zeros of φ are discrete. Similarly, if φ had poles of order ≥ 2 in z, it would have poles in all preimages of z, which is impossible. So, φ does not have zeros and can only have simple poles. The number of poles of a meromorphic quadratic differential in $\mathbb{C}_\infty$ is the number of zeros plus 4. So, φ has four simple poles and no zeros. Thus, at each critical value of f the quadratic differential φ has one simple pole. The image by f of such a pole is a pole in $PC(f)$. There are no other poles, since if φ had a pole at $z \notin PC(f)$, the preimages by f of each point of a negative orbit of z under f would be poles, which is impossible. So, the set of poles of φ is $PC(f)$, and the signature of the canonical orbifold $\mathcal{O}_f$ is (2, 2, 2, 2), and[48] f induces in $\mathbb{C}$ by a lift of a universal covering an affine transformation, with two sheets and $|a| = \sqrt{d}$. So, $\mathcal{O}_f$ is doubly covered by a complex torus. □

Application of Thurston Theorem to Maximize the Number of Zeros of Harmonic Polynomials in the Plane

The Thurston Theorem of characterization of postcritically finite rational functions has many other applications in Complex Dynamics and also in other areas. To illustrate this in another context, we present a curious and formative application to obtain the optimal number of zeros of a class of harmonic polynomials.

A **harmonic polynomial** of a complex variable is a harmonic function $f : \mathbb{C} \to \mathbb{C}$ such that the real and complex parts of $f(x + iy)$ are polynomial functions of the real variables x, y. Thus,[49] as for any complex valued harmonic function defined in $\mathbb{C}$, there are functions $h, g \in H(\mathbb{C})$ such that $f = h + \overline{g}$, and since the real and imaginary parts of $f(x + iy)$ are polynomial functions of (x, y), h and g can be chosen to be complex polynomials.

By the fundamental theorem of algebra, nonconstant complex polynomials of degree $n \in \mathbb{N}$ have a finite number of zeros (n counting multiplicities). For harmonic polynomials of complex variable, it is radically different, e.g., $z^n - \overline{z^n}$ with $n \in \mathbb{N}$

[48] See Section 14 of Chapter 13 of *CADOVA*.

[49] See Section 2 of Chapter 9 of *CADOVA*.

has infinitely many #$\mathbb{R}$ solutions, namely, the elements of the union of n lines in $\mathbb{C}$ containing 0 with directions angularly equally spaced and one of them the real axis. Another simple example is $z^2 + c\overline{z}$ with $c \in \mathbb{R}$, which, with $z = x + iy$, $x, y \in \mathbb{R}$, is $(x^2 - y^2 + cx) + iy(2x - c) = 0$. If $c \neq 0$, there are two real simple zeros (0 and $-c$) and two zeros with imaginary part $\neq 0$, namely, $\frac{c}{2}(1 \pm i\sqrt{3})$, also simple, or if $c = 0$, 1 double real zero at 0; thus, there are four simple zeros (two real and two non-real complex conjugates) or 1 double real zero, according to $c \neq 0$ or $c = 0$.

In 1992, T. Sheil-Small[50] posed the problem of finding upper bounds of the number of zeros of harmonic polynomials of complex variable $f = h + \overline{g}$ with h, g polynomials of degrees, resp., $n, m \in \mathbb{N}$ with $n > m$, and conjectured that n^2 would be the optimal upper bound, what was proved in 1998 by A. Wilmshurst for $m = n - 1$, who also proved that if $m = 1$, the optimal upper bound would be $3n{-}2$. This is trivial for $n = 2$, because for $f(x, y) = \big(u(x, y), v(x, y)\big)$, $v(x, y)$ is the product of y and a real polynomial of degree 1 and $u(x, y)$ is a real polynomial of degree 2 in $x, y \in \mathbb{R}$, and, therefore, the real zeros of f are the zeros of the degree 2 real polynomial $u(x, 0)$ and the non-real zeros are (ky, y) with k a constant and y a zero of the second-degree real polynomial $u(ky, y)$. Thus, the maximum number of zeros in f is $4 = 3n - 2$ for $n = 2$, and for optimality, we have the example at the end of the preceding paragraph with $c \neq 0$, which has four zeros.

In 2000, B. Bshouty and A. Lyzzaik[51] proved that the upper bound $3n - 2$ for the number of zeros of harmonic polynomials $f(z) = h(z) - \overline{z}$ with h a polynomial of degree $n \in \mathbb{N}\backslash\{1\}$ is optimal for $n = 4, 5, 6, 8$.

D. Khavinson and G. Świątek proved in an article of 2003 the A. Wilmshurst[52] Conjecture with Complex Dynamics, observing that the zeros of $P(z) - \overline{z}$, with P a complex polynomial, are the fixed points of $\overline{P} \circ \overline{P}$, but they did not prove optimality. At the time, they cited a (unpublished) communication of D. Saranson and R. Crofoot[53] in 1999 with a proof of the upper bound of the number of zeros by $3n - 2$ for $n = 3$. In this case the optimality can be obtained e.g., finding that the zeros of $\frac{1}{2}z(z^2 - 3) + \overline{z}$ are 0, ± 1, $\pm\frac{1}{2}\big(\sqrt{7} + i\big)$, $\pm\frac{1}{2}\big(\sqrt{7} - i\big)$, amounting for $7 = 3n{-}2$ for $n = 3$ (Fig. 8.9). The optimality of the upper bound obtained in 2003 by D. Khavinson and G. Świątek was left open for all $n \geq 9$ and for $n = 7$.

In the same article, D. Khavinson and G. Świątek reported that the communication of D. Saranson and R. Crofoot in 1999 included the following conjecture: *For every $n \in \mathbb{N}\backslash\{1\}$ there is a complex polynomial P of degree n and $n - 1$ distinct points $z_1, \ldots, z_{n-1}$ such that $P'(z_j) = 0$ and $P(z_j) = \overline{z_j}$ for $j \in \{1, \ldots, n-1\}$* and proved the validity of this conjecture implies the optimality of the upper bound $3n - 2$.

Next, we present proofs of these results.[54]

---

[50] Sheil-Small, Terence (1937-).

[51] Bshouty, Daoud (1953-). Lyzzaik, Abdallah.

[52] Wilmshurst, Andrea.

[53] Saranson, Donald (1933–2017). Crofoot, Robert.

[54] Wilmshurst, D., Świątek, G., On the number of zeros of certain harmonic polynomials, *Proc. American Mathematical Society* **131** (2003), 409–414.

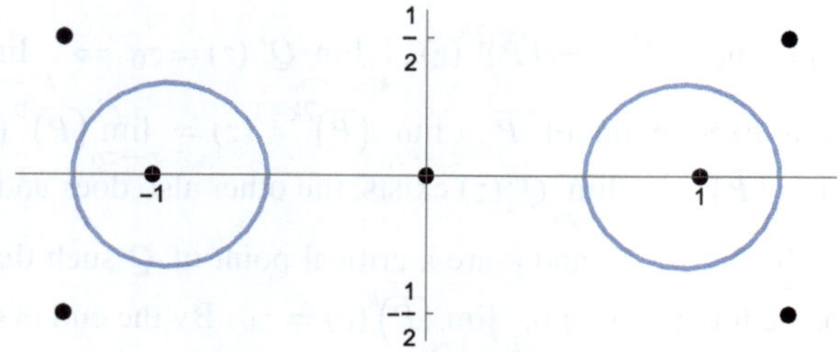

Fig. 8.9 Zeros of $P(z)-\overline{z}$ with $P(z)=\frac{1}{2}z\left(\frac{1}{3}z^2-3\right)$ and curve $|P'(z)|=1$

(8.34) *If P is a complex polynomial of degree $n\in\mathbb{N}\setminus\{1\}$, the number of zeros of $P(z)-\overline{z}$ where $|P'|\le 1$ is upper bounded by $n-1$.*

Proof Let $Q=\overline{P}\circ\overline{P}$ and $P(z_0)-\overline{z_0}=0$. With $P(z)=\sum_{j=0}^{n}c_j z^j$,

$$Q(z)=\overline{\sum_{j=0}^{n}c_j\left(\sum_{k=0}^{n}\overline{c_k z^k}\right)^j}=\sum_{j=0}^{n}\overline{c_j}\left(\sum_{k=0}^{n}c_k z^k\right)^j=\sum_{j=0}^{n}\overline{c_j}\,(P(z))^j\,;$$

Q is a complex polynomial of degree n^2, $Q(z_0)=\sum_{j=0}^{n}\overline{c_j}\,\overline{z_0}^j=\overline{P(z_0)}=z_0$,

$$Q'(z)=\sum_{j=0}^{n}\overline{c_j}\,j\,(P(z))^{j-1}P'(z)=\overline{P'\big(\overline{P(z)}\big)}P'(z)\,,$$

and $Q'(z_0)=\overline{P'(z)}P'(z)=|P'(z)|^2$. Thus, if $|P'(z_0)|\le 1$, then z_0 is an attracting, superattracting, or parabolic fixed point of Q, and, therefore, there is a critical point of Q in the attracting basin[55] of the orbit of z_0 under Q.

If c is a critical point of Q, i.e., $Q'(c)=0$, then c or $\overline{P}(c)$ is a critical point of P, and if one of these points, denoted ζ, is a critical point of P and z is such that $\overline{P}(z)=\zeta$, then $Q'\big(\overline{P}(z)\big)=Q'(\zeta)=0$. Thus, all preimages of ζ by $\overline{P}$ are critical points of Q. The preimages of ζ by $\overline{P}$ are the zeros z of the polynomial $P(z)-\overline{\zeta}$, and the number of these zeros, counting multiplicities, is n. If $\zeta\neq\overline{P}(\zeta)$, then ζ is not a preimage of ζ by $\overline{P}$, and, therefore, as ζ is a Critical Point of Q, the number of critical points of Q in the grand orbit of ζ is $\ge n+1$, counting multiplicities. If $\zeta=\overline{P}(\zeta)$ and the multiplicity of the critical point ζ of P is k, as $Q'(\zeta)=\overline{P'\big(\overline{P}(\zeta)\big)}P'(\zeta)=[P'(\zeta)]^2$, the multiplicity of the critical point ζ of Q is $2k>k$, and, therefore, the number of critical points of Q in the grand orbit of ζ is also $\ge n+1$, counting multiplicities.

If $P(z_0)-\overline{z_0}=0$, as $\overline{P}$ is continuous,

$$\lim_{k\to+\infty}\left(\overline{P}\right)^k(z)=z_0\Longrightarrow\lim_{k\to+\infty}Q^k(z)=z_0\,,$$

[55] See Sections 6 and 10 of Chapter 13 of *CADOVA*.

and since $Q^k(z) = (\overline{P})^{2k}(z)$, $\lim_{k\to+\infty} Q^k(z) = z_0 \Rightarrow \lim_{k\to+\infty} (\overline{P})^k(z) = z_0$ and, as z_0 is a fixed point of $\overline{P}$, $\lim_{z\to z_0} (\overline{P})^{2k+1}(z) = \lim_{z\to z_0} (\overline{P})^{2k}(\overline{P}(z)) = z_0$. So, if one of $\lim_{z\to z_0} (\overline{P})^k(z)$, $\lim_{z\to z_0} Q^k(z)$ exists, the other also does and is equal.

If $P(z_0) - \overline{z_0}$ and c are a critical point of Q such that $\lim_{k\to+\infty} Q^k(c) = z_0$, by the preceding paragraph, $\lim_{k\to+\infty} (\overline{P})^k(c) = z_0$. By the end of the second paragraph of this proof, there is a grand orbit of some critical point c of P under $\overline{P}$ with $\geq n+1$ critical points of Q, and in all of them $\lim_{k\to+\infty} (\overline{P})^k$ is z_0. By the first paragraph of this proof, each zero of $P(z) - \overline{z}$ where $|P'(z)| \leq 1$ attracts some critical point of Q, and, from what we saw, then it attracts $n+1$ critical points, counting multiplicities. Since distinct zeros attract disjoint sets of critical points and the degree of Q is n^2, the number of critical points of Q counting multiplicities is $n^2 - 1 = (n-1)(n+1)$. Therefore, the number of zeros in $P(z) - \overline{z}$ with $|P'| \leq 1$ is upper bounded by $n-1$.

□

The next proof uses a generalization of the argument principle for harmonic functions of complex variable by T.J. Suffridge and J.W. Thompson in 2000, following a version obtained in 1996 for functions without poles by P. Duren, W. Hengartner, and R.S. Laugesen.[56]

If f is a harmonic function in an open set $U \subset \mathbb{C}$ and $f(z_0) = 0$ with $z_0 \in U$, we say that z_0 is a **regular zero** of f if the Jacobian J_f of f at z_0 is $\neq 0$, i.e., with h, g holomorphic functions in a neighborhood of z_0 such that $f = h + \overline{g}$, $J_f(z_0) = |h'(z_0)|^2 - |g'(z_0)|^2 \neq 0$. If $J_f(z_0) > 0$, the **order of a zero** z_0 of f, $\mathcal{O}(z_0)$, is the smallest $k \in \mathbb{N}$ such that $h^{(k)}(z_0) \neq 0$, and if $J_f(z_0) < 0$, it is $-k$ with k the order of zero z_0 of $\overline{f} = g + \overline{h}$, i.e., the smallest $k \in \mathbb{N}$ such that $g^{(k)}(z_0) \neq 0$.

We say that $z_0 \in \mathbb{C}$ is a **pole of a harmonic function** f in the complement of z_0 in one of its neighborhoods if $\lim_{z\to z_0} |f(z)| = +\infty$. The **order of the pole** z_0 of f, $\mathcal{O}(z_0)$, is $-\frac{1}{2\pi}\Delta_C \arg f$ for a small circle C with center z_0, with, for a Jordan curve C, $\boldsymbol{\Delta_C \arg f}$ the increment of a continuous argument of f along a simple path describing it in the positive direction relative to region of $\mathbb{C}$ it bounds.

(8.35) Argument principle for complex harmonic functions:
If $\Omega \subset \mathbb{C}$ is a simply connected region, $p_1, \ldots, p_m$, with $m \in \mathbb{N}$, are distinct points of Ω, $f : \Omega \backslash \{p_1, \ldots, p_m\} \to \mathbb{C}$ is a harmonic function with regular zeros and poles $p_1, \ldots, p_m$, $C \subset \Omega$ is a Jordan curve not containing poles or

(continued)

[56] Suffridge, T.J., Thompson, J.W., Local behavior of harmonic mappings, *Complex Variables Theory* **41** (2000), 63–80, and Duren P., Hengartner W., Laugesen R.S., The argument principle for harmonic functions, *The Amer. Math. Monthly,* **103** (1996), 411–415.

zeros of f, $\Omega_1 \subset \mathbb{C}$ is the region bounded by C, and $N_0(f;\Omega_1)$, $N^p(f;\Omega_1)$ are the sums of the orders of, resp., zeros and poles of f in Ω_1, then

$$N_0(f;\Omega_1) - N^p(f;\Omega_1) = \tfrac{1}{2\pi}\,\Delta_C \arg f\,.$$

Proof Firstly, assume f does not have poles nor zeros in Ω_1. Since $\mathrm{cl}\,\Omega_1$ is a simply connected region, $C = \partial\Omega_1$ is homotopic in $\mathrm{cl}\,\Omega_1$ to a small circle centered on that point. So, $\mathrm{cl}\,\Omega_1$ is homeomorphic to the closed disk bounded by this circle and the restriction of a homeomorphism from $\mathrm{cl}\,\Omega_1$ to $C = \partial\Omega_1$ is a homeomorphism of C onto such disk. The disk and the circle are homeomorphic to, resp., the square $Q = [-1, 1]^2 \subset \mathbb{C}$ and ∂Q. Let h be a homeomorphism of Q onto $\mathrm{cl}\,\Omega_1$, and, therefore, $h_{|\partial\Omega_1}$ is a homeomorphism of C onto ∂Q. Since $0 \notin f(\Omega_1 \cup C)$, $F = f \circ h$ does not have zeros in Q. Since Q is compact, by the Heine-Cantor theorem, F is uniformly continuous in Q and, by the Weierstrass theorem of extrema of continuous functions, $|F|$ has a minimum $m_F > 0$ in Q.

Now assume f has zeros but not poles in Ω_1. As the zeros are regular, they are isolated and in a finite number n, $z_1, \ldots, z_n$. Let $\delta > 0$ be such that $\mathrm{cl}\,B_\delta(z_1), \ldots, \mathrm{cl}\,B_\delta(z_n) \subset \Omega$ are disjoint. Connect each $\partial B_\delta(z_j)$ to C by a piecewise regular curve C_j, so that the set of these curves is disjoint (Fig. 8.10). Let γ be a path describing in the positive direction relative to the region bounded by C each arc of C between consecutive end points of curves $C_1, \ldots, C_n$ in C and when it reaches the end point of C_j in C goes along it up to the other end point, and then goes along $\partial B_\delta(z_j)$ in the negative direction relative to $B_\delta(z_j)$, continuing along C_j from the end point in $\partial B_\delta(z_j)$ to the end point in C, and then the arc from C to the end point of another curve C_k, and so on until reaching the starting point. γ describes a curve γ^* bounding a region $R \subset \Omega$ without zeros of f, with the positive orientation relative to R. So, $\Delta_{\gamma^*}(f) = 0$. Contributions to the change of a continuous argument over γ in each C_j cancel, and the change in $\partial B_\delta(z_j)$ is $-2\pi\,\mathcal{O}(z_j)$. Thus $0 = \Delta_{\gamma^*}(f) = \Delta_C(f) - 2\pi \sum_{j=1}^{n} \mathcal{O}(z_j)$, and $N_0(f;\Omega_1) = \sum_{j=1}^{n} \mathcal{O}(z_j) = \frac{1}{2\pi}\,\Delta_C(f)$. □

(8.36) *If P is a complex polynomial of degree $n \in \mathbb{N}\backslash\{1\}$, the number of zeros of $P(z) - \overline{z}$ is upper bounded by $3n - 2$.*

Fig. 8.10 Auxiliary figure for the proof of 8.35

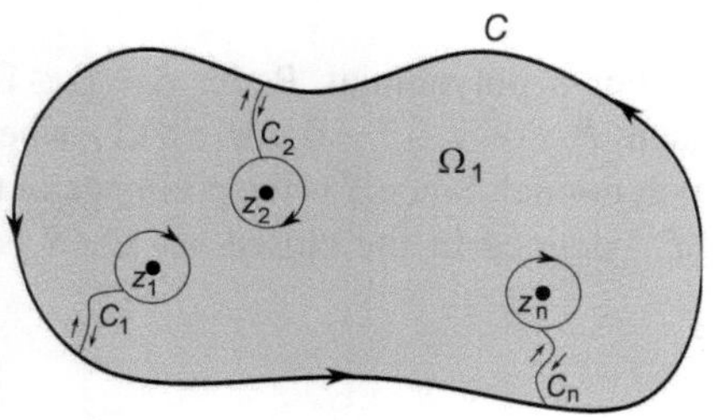

Proof By (8.34), the number of zeros of $f(z) = P(z) - \bar{z}$ with $|P'| \le 1$ upper bounded by $n-1$. It remains to upper bound the number of zeros of f where $|P'| > 1$. Let Ω_+, Ω_- be the sets where f, resp., preserves, inverts orientation. Since the Jacobian of f is $J_f = |P'|^2 - 1$, it is

$$\Omega_- = \{z \in \mathbb{C} : |P'(z)| < 1\}, \quad \Omega_+ = \{z \in \mathbb{C} : |P'(z)| > 1\}.$$

Since $\lim_{|z| \to +\infty} J_f(z) = +\infty$, Ω_- is bounded, and, as $\Gamma = \partial\Omega_+ = \partial\Omega_-$ has Cartesian equation $|P'(z)| = 1 \Leftrightarrow P'(z) = e^{i\theta}$ with $\theta \in [0, 2\pi[$, by the implicit function theorem, Γ is a closed curve parameterized by θ, analytic except possibly at 0 or 2π, corresponding to a z_0 with $P'(z_0) = 1$. Ω_-, Ω_+ are the regions in $\mathbb{C}$ separated by Γ (Fig. 8.9).

P is said to be a regular polynomial if f has no zeros where $|P'| = 1$.

Since $\lim_{|z| \to +\infty} |f(z)| = +\infty$, by the argument principle, for $R > 0$ large enough, $\Delta_{\partial B_R} \arg(f) = 2\pi n$ with ∂B_R described in the positive direction relative to B_R, and take $R > 0$ even larger if necessary so that the open disk B_R) contains $\Omega_- \cup \Gamma$ and all zeros of f. Assuming P is a regular polynomial, the argument principle can be applied considering the increment of a continuous argument of f along Γ, as this does not contain zeros or poles of f. As, by (8.34) and the argument principle, taking into account that f reverses the orientation in Ω_- and the zeros of f in Ω_- have order -1, $\Delta_\Gamma \arg(f) \ge -2\pi(n-1)$, with Γ described in the positive direction relative to Ω_-,

$$\Delta_{\partial B_R(0)} \arg(f) - \Delta_\Gamma \arg(f) \le 2\pi n + 2\pi(n-1) = 2\pi(2n-1).$$

By the mean value principle applied to $\Omega_+^0 = \mathrm{cl}\,\Omega_+ \cap \mathrm{cl}\,B_R$, the sum of the orders of the zeros of f in Ω_+^0, and also in Ω_+,

$$N_0(\Omega_+^0; f) = \tfrac{1}{2\pi}\big[\Delta_{\partial B_R(0)} \arg(f) - \Delta_\Gamma \arg(f)\big] \le 2n-1.$$

As f preserves the orientation in Ω_+^0, the orders of the zeros in Ω_+^0 are ≥ 1, and the number of zeros of f in Ω_+ is upper bounded by $2n-1$. Since, by (8.34), the number of zeros of f in Ω_- is upper bounded by $n-1$, the number of zeros of f in $\mathbb{C}$ is upper bounded by $(2n-1)+(n-1) = 3n-2$. So, for regular polynomials P, the number of zeros in F is upper bounded by $3n-2$.

For small perturbations of a regular polynomial P in the uniform convergence topology in the spherical metric in $\mathbb{C}_\infty$, by the implicit function theorem, the number of zeros of f in Ω_+ does not decrease. Therefore, for P in the closure of the set of regular polynomials, the number of zeros in f is upper bounded by $3n-2$.

For a polynomial P, let $S = \{z \in \mathbb{C} : |P'(z_0)| = 1\}$ and $S' = f(S)$. If $c \notin S'$, then $P(z) - c - \bar{z} \ne 0$ for $z \in \mathbb{C}$, where $|P'(z)| = 1$, and $P(z) - c$ is a regular polynomial. Since S is the preimage of the closed set $\{1\}$ by the continuous function $|P'|$ defined in the closed set $\mathbb{C}$, S is a closed set, and, as P' is a polynomial,

$|P'(z)| \to +\infty$ when $|z| \to +\infty$, and S is bounded. Thus, $S \subset \mathbb{C}$ is a compact set and, as images of compact sets by continuous functions are compact sets, S' is compact. If it were $\operatorname{int} S' \neq \emptyset$, as f is a continuous function, $f^{-1}(\operatorname{int} S')$ would be a nonempty open set and $\operatorname{int} S \neq \emptyset$, which, by the implicit function theorem, is false. Therefore, S' is compact with empty interior and $\{c \in \mathbb{C} : P - c \text{ is a regular polynomial}\}$ is an open dense subset of $\mathbb{C}$. Since every polynomial belongs to the closure of the regular polynomials, the upper bound $3n-2$ holds for the zeros of f.

□

(8.37) *If P is a complex polynomial of degree $n \in \mathbb{N}\backslash\{1\}$ with $n-1$ distinct points $z_1, \ldots, z_{n-1}$ such that $P'(z_j) = 0$ and $P(z_j) = \overline{z_j}$ for $j \in \{1, \ldots, n-1\}$, then the number of zeros of $P(z) - \overline{z}$ is $n-1$, and the existence of P with these properties implies the optimality of the upper bound $3n-2$ of the number of zeros of $P(z) + \overline{z}$ for P complex polynomials of degree n.*

Proof In this case, the upper bound in (8.34) is optimal. With the proof of (8.36), the upper bound in the statement is also optimal. □

The optimality of the upper bound $3n-2$ of the number of zeros of harmonic polynomials of complex variable $f(z) = h(z) - \overline{z}$ with h a polynomial of degree n for all $n \in \mathbb{N}\backslash\{1\}$ was settled in 2008 by L. Geyer[57] using the Thurston theorem of characterization of postcritically finite rational functions (8.20) and the S. Levy result (8.18) for topological polynomials to prove the existence of polynomials satisfying conditions more restrictive than those of the mentioned conjecture of D. Saranson and R. Crofoot, and thus proving the conjecture holds.

(8.38) *For every $n \in \mathbb{N}\backslash\{1\}$, there is a complex polynomial P of degree n with real coefficients and $n-1$ distinct points $z_1, \ldots, z_{n-1}$ such that $P'(z_j) = 0$, $P''(z_j) \neq 0$ and $P(z_j) = \overline{z_j}$ for $j \in \{1, \ldots, n-1\}$, implying that the upper bound $3n-2$ of the number of zeros of harmonic polynomials of complex variable $f(z) = h(z) - \overline{z}$ with h a polynomial of degree $n \in \mathbb{N}\backslash\{1\}$ is optimal.*

Proof We construct a topological polynomial f such that $f(\overline{z}) = \overline{f(z)}$ for $z \in \mathbb{C}$ and $f(c) = \overline{c}$ for every Critical Point c of f. The construction differs for polynomials of even or odd degrees.

If f is a polynomial of odd degree $n = 2m+1$ with $m \in \mathbb{N}$, we consider a polynomial g of degree n with real coefficients and distinct critical points

[57] Geyer, L., Sharp bound for the valence of certain harmonic polynomials, *Proc. Amer. Math. Soc.* **136** (2008), 549–555.

$c_1, \ldots, c_m$ in the upper complex half plane such that the corresponding critical values $v_j = g(c_j)$ are distinct points in the lower complex half plane. For example, for small $\varepsilon > 0$, with $c_{2k-1} = ik$, $c_{2k} = i(k+\varepsilon)$ and $g(z) = -\int_0^z \prod_{k=1}^m (\zeta^2 - c_k^2)\, d\zeta$. Let h be a homeomorphism of the closed lower complex half plane onto itself such that $h(v_j) = \overline{c}_j$. Extend h by reflection on the real axis to a homeomorphism of $\mathbb{C}$ to $\mathbb{C}$, and define $f = h \circ g$, which is a topological polynomial of degree $2m+1$ commuting with the reflection relative to the real axis and mapping each critical point to its complex conjugate. It is $PC(f) = \{c_1, \ldots, c_m, \overline{c_1}, \ldots, \overline{c_m}, \infty\}$. By (8.18) and (8.20), f is postcritically equivalent to a polynomial P, and, therefore, there are homeomorphisms φ, ψ preserving orientation and homotopic (rel $PC(f)$) such that $\varphi \circ f = g \circ \psi$, since $Q(z) = \overline{P(\overline{z})}$ is postcritically equivalent to P, i.e., conjugated to Q, which is a polynomial with real coefficients. It is $\varphi(c_j) = \psi(c_j)$ for $j \in \{1, \ldots, m\}$, $PC(P) = \{\varphi(c_1), \ldots, \varphi(c_m), \overline{\varphi(c_1)}, \ldots, \overline{\varphi(c_m)}, \infty\}$ and $P(z_j) = \overline{z_j}$ for $j \in \{1, \ldots, m\}$.

If f is a polynomial of even degree $n = 2m+2$ with $m \in \mathbb{N} \cup \{0\}$, consider a polynomial g of degree n with real coefficients and different critical points $c_0 \in \mathbb{R}$ and $c_1, \ldots, c_m$ in the upper complex half plane such that the critical values $v_j = g(c_j)$ for $j \in \{1, \ldots, m\}$ are distinct points in the lower complex half plane with coefficients c_k chosen as before, then $g(z) = -\int_0^z (\zeta - 1) \prod_{k=1}^m (\zeta^2 - c_k^2)\, d\zeta$ has the desired properties. Let h be a homeomorphism of the closed lower complex half plane on itself with $h(g(c_0)) = c_0$ and $h(v_j) = \overline{c_j}$ for $j \in \{1, \ldots, m\}$. Then, proceed as above.

There is a complex polynomial P with the properties in the statement and, by (8.38), the number of zeros in P is $3n-2$. So, the upper bound D. Khavinson and G. Świątek got in 2003 is optimal for harmonic polynomials $f(z) = h(z) + \overline{z}$ with h a polynomial of degree $n \in \mathbb{N} \setminus \{1\}$. □

In 2006, D. Khavinson and G. Neumann obtained analogously an upper bound of the number of zeros of harmonic rational functions $f(z) = R(z) - \overline{z}$, with R rational function of degree $n \in \mathbb{N} \setminus \{1\}$.

(8.39) *If R is a complex rational function of degree $n \in \mathbb{N} \setminus \{1\}$, the number of zeros of $R(z) - \overline{z}$ where $|R'(z)| \leq 1$ is upper bounded by $2n-2$.*

Proof Let $R = p/q$ with p, q relatively prime polynomials. As in the proof of (8.34) with $Q = \overline{R} \circ R$, a rational function of degree n^2, and in $\mathbb{C}_\infty$ instead of $\mathbb{C}$. Each zero of $R(z) - \overline{z}$ with $|R'(z)| \leq 1$ attracts $\geq n+1$ critical points of Q, as in the proof of (8.34). The number of critical points of Q is $2(n^2-1)$, counting multiplicities.[58] So, the number of zeros of $R(z) - \overline{z}$ with $|R'(z)| \leq 1$ multiplied by $n+1$ is upper bounded by $2n^2 - 2 = 2(n-1)(n+1)$, and the number of zeros by $2(n-1)$. □

---

58 See Section 10 of Chapter 13 of *CADOVA*.

(8.40) *If R is a complex rational function of degree $n \in \mathbb{N}\setminus\{1\}$, the number of zeros of $R(z)-\overline{z}$ is upper bounded by $5n-5$.*

Proof Let $R=p/q$ with p, q relatively prime polynomials of degrees, resp., n_p, n_q, $f(z)=R(z)-\overline{z}$, and n_+, n_- the numbers of zeros of f that, resp., preserve or invert the orientation.

A rational function R is said here to be regular if f does not have zeros that are critical points of R. It is assumed that R is a regular rational function. In particular, the zeros of f where the orientation is reversed are those considered in the preceding result; thus, $n_- \leq 2n-2$.

Consider B_r with $r>0$ large enough for B_r to contain all poles and zeros of f.

Alternatively, it is $n=n_q \geq n_p \geq 0$ or $n=n_p \geq n_q+1$. These cases are considered separately.

If $n=n_q \geq n_p \geq 0$, the set of critical points of f is bounded, and the restriction of R to B_r is bounded, ∞ is not a zero of f and, with $g(z)=\overline{z}$, for r larger if necessary, then $\Delta_{\partial B_r}(f)=\Delta_{\partial B_r}(g)=-2\pi$, with ∂B_r described once with positive orientation relative to B_r. By the argument principle (8.35),

$$(n_+-n_-)-n=\tfrac{1}{2\pi}\Delta_{\partial B_r}(f)=-1 .$$

So, the number of zeros of f is

$$n_+ + n_+ = 2n_- + n-1 \leq 2(2n-2)+n-1 = 5n-5 .$$

If $n=n_p \geq n_q+1$, as p, q are relatively prime polynomials, they are not both 0 at 0. We consider separately the cases *(a)* $p(z)\neq 0$ and *(b)* $p(z)=0\neq q(z)$.

(a) If $p(z)\neq 0$, then $f(0)\neq 0$ and the function $F(z)=\frac{q(1/z)}{p(1/z)}-\overline{z}$ is of the form $F(z)=\frac{z^{n_p-n_q}\widetilde{q}(z)}{\widetilde{p}(z)}-\overline{z}$, with $\widetilde{p}(z)$ and $z^{n_p-n_q}\widetilde{q}(z)$ prime polynomials of degrees, resp., n_p and $\leq (n_p-n_q)+n_q=n_p$, and, therefore, F satisfies the conditions of f in the preceding alternative, whereby the number of zeros of F is upper bounded by $5n-5$. Since $F(z)+\overline{z}=\frac{1}{R(z)}=\frac{1}{f(1/z)+1/\overline{z}}$, for $z\neq 0$,

$$F(z)=0 \iff \tfrac{1}{\overline{z}}=f\big(\tfrac{1}{z}\big)+\tfrac{1}{\overline{z}} \iff f\big(\tfrac{1}{z}\big)=0 ,$$

and, as $f(0)\neq 0$, the number of zeros in f is that of F; so, it is upper bounded by $5n-5$.

(b) If $p(z)=0\neq q(z)$, as the polynomials p, q are relatively prime, for $c\in\mathbb{C}\setminus\{0\}$ small, the polynomials $p(z)+c$, $q(z)$ are relatively prime and $p(0)+c\neq 0$. The functions f and $f_c(z)=\frac{p(z)+c}{q(z)}-\overline{z}$ have the same poles and $f_c\to f$ when $c\to 0$ uniformly on compact subsets of $\mathbb{C}$ without poles of f. Considering the function F_c defined from f_c as in *(a)* F was defined from f, it follows that the number of zeros of f_c is upper bounded by $5n$–5. If the number of zeros of f in $\mathbb{C}$ were $>5n-5$, since it is finite, there would be small closed disks centered at each zero, all with a radius $\rho>0$, without any zeros of f other than the, resp., center, nor critical points of the regular rational function f. If f_c would

not have any zeros in one of those disks, by the Weierstrass theorem of extrema of continuous functions, $|f_c|>0$ would assume a minimum value $m>0$ at the boundary of each disk, and, for small c, the increment of a continuous argument of $f=f_c\big(1+\frac{f-f_c}{f_c}\big)$ along the circle in the positive direction relative to the disk it bounds would be 0, contradicting the argument principle (8.35) and the center of the circle being a zero of f. Hence, for small c, each disk considered has at least one zero of f_c, and, therefore, the number of zeros of f is larger if f is a regular rational function.

This concludes the proof for regular rational functions.

For non-regular rational functions R, it is proved with the approximation argument used at the end of the proof of (8.36). □

D. Khavinson and G. Neumann gave the example with $n=2$ and five zeros $f(z)=\frac{z^2+z-1/2}{z^2-(3/2)z+1}-\overline{z}$, whose zeros are $\frac{1}{2}$, $1\pm\sqrt{2}$, $\frac{1}{2}(1\pm i\sqrt{11})$, showing that the upper bound $5n-5$ is optimal for $n=2$ (Fig. 8.11[59]), but S.H. Rhie had obtained in 2003 examples of functions $R(z)-\overline{z}$ with R rational function of any degree $n\in\mathbb{N}\setminus\{1\}$ with this number of zeros by a method of small perturbations in the context of the study of gravitational lenses resulting from the curvature of light rays by masses as predicted by General Relativity (Fig. 8.12).

The connection of this type of harmonic rational functions with gravitational lenses results from considering n point masses in a plane between a point source of light and the observer perpendicular to the direction of observation to obtain the position of the source of light in the parallel plane that contains it depending on the positions and, resp., point masses, in complex representation originating from the intersection of the planes with the straight line in the direction of observation passing through the observer and parallel to the real and complex axes, with the gravitational lensing equation $z-\sum_{j=1}^{n}\frac{\sigma_j}{\overline{z}-\overline{z_j}}=0$, where z_j is the position of the

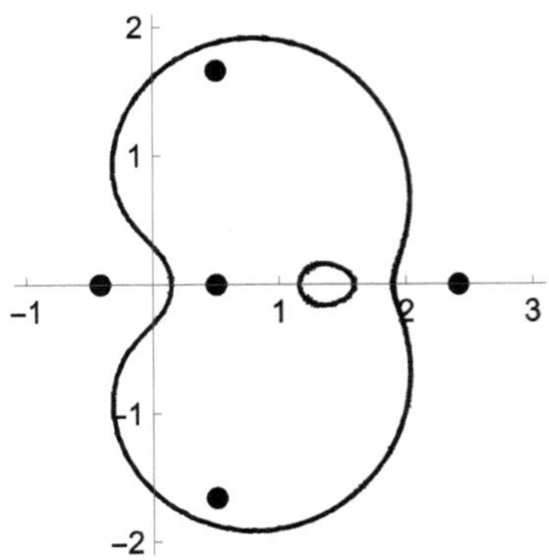

Fig. 8.11 Zeros of $R(z)-\overline{z}$ with $R(z)=\frac{z^2+z-1/2}{z^2-(3/2)z+1}$ and curve $|R'(z)|=1$

[59] Computed with *Mathematica*.

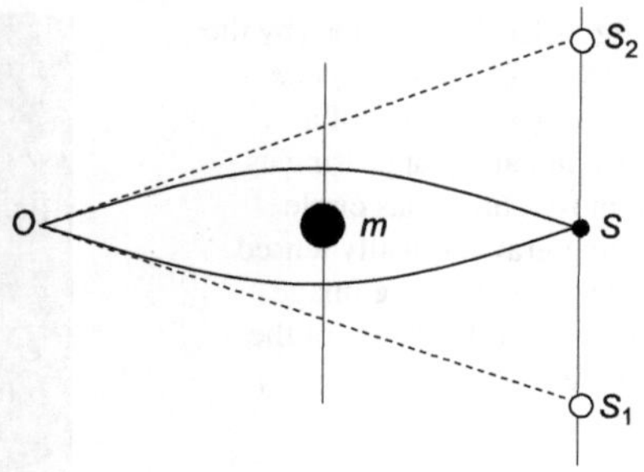

Fig. 8.12 Gravitational lensing by a mass m between point light source S and observer O, in section through a plane showing two images S_1, S_2 of S in that plane

mass j projected to the plane of the gravitational lens and $\sigma_j \in \mathbb{R}$ is a constant proportional to the mass[60] j.

In 1998, A. Wilmshurst obtained the upper bound n^2 for the number of zeros of harmonic polynomials of complex variable $P(z) - \overline{Q(z)}$ with P, Q polynomials of degree, resp., $n, m \in \mathbb{N}$ with $n > m$ and conjectured that the optimal upper bound would be $3n - 2 + m^2 - m$. In 2015, S.-Y. Lee, A. Lerario, and E. Lundberg invalidated the conjecture, obtaining the existence of infinite examples where it does not hold. In 2018, D. Khavinson, S.-Y. Lee and A. Saéz[61] obtained the upper bound $n^2 - n$ for the number of zeros outside the real and complex axes and the existence of harmonic polynomials of the type above with number of zeros $\geq \max\{3n - 2, m^2 + m + n\}$, but so far no optimal upper bound was obtained except $3n - 2$ for $m = 1$ with $Q(z) = z$ of (8.36) and (8.38).

Figure 8.13[62] is of a photograph taken by the *NASA/ESA Hubble Space Telescope* in 2011 with the galaxy LRG 3-757 (the bright spot in the center) acting as gravitational lens of a distant galaxy that appears as almost a circle around the image of LRG 3-757 due to an almost perfect alignment of the two galaxies with the telescope and LRG 3-757 has an almost circular cross section, with a configuration called Einstein ring;[63] LRG 3-757 was discovered in 2007.

Figure 8.14[64] is of a photograph taken by *NASA/ESA Hubble Space Telescope* with the galaxy UZC J224030.2+032131 (the diffuse spot in the middle of the four

---

[60] For an interesting description of the relationship of gravitational lenses with the number of zeros of harmonic rational functions $f(z) = \frac{P(z)}{Q(z)} - \overline{z}$, with P, Q polynomials, see Khavinson, D., Neumann, G., From the fundamental theorem of algebra to astrophysics: A "harmonious" path, *Notices of the Notices of the AMS* **55** (2008), 666–675.

[61] Khavinson, D., Lee, S.-Y., Saéz, A., Zeros of harmonic polynomials, critical lemniscates, and caustics, *Complex Analysis and Its Synergies* **4** (2018), 2, 20 pgs. Lee, Seung-Yeop (1975-). Lerario, Antonio (1984-). Lundberg, Erik. Saéz, Andrés.

[62] Author and date of figure: ESA/Hubble & NASA (19.12.2011), License *Creative Commons Public Domain, World Wide Web link:* https://apod.nasa.gov/apod/image/1112/lensshoe_hubble_3235.jpg.

[63] *Einstein Ring.* The nearly complete 1st Einstein ring (B1938+666) was discovered in 1998 in collaboration of astronomers at *University of Manchester* and the *NASA's Hubble Space Telescope*.

[64] Author and date of the figure: ESA/Hubble & NASA (23.01.2012), under License *Creative Commons Attribution 3.0 Unported, World Wide Web link:* https://esahubble.org/images/potw1204a/.

Fig. 8.13 Photograph (by the *NASA/ESA Hubble Space Telescope* in 2011) of a distant galaxy appearing as almost a luminous circle being gravitationally lensed by a galaxy whose image appears at the center of the circle

brightest points) acting as a gravitational lens for a quasar that appears as four distinct luminous points and not as a circle due to the elongated shape of the galaxy, in an arrangement called the Einstein cross;[65] this quasar is seen in the direction of the constellation Pegasus about 11 billion light-years away, with the galaxy about 10 times closer.

Fig. 8.14 Fotograph (by the *NASA/ESA Hubble Space Telescope* in 2012) of a quasar appearing as four distinct luminous points due to gravitational lensing by a galaxy whose image is the more diffuse luminous spot in the middle of the four points

[65] *Einstein cross*. This name was initially given to the image of quasar Q2237+030 under the effect of gravitational lensing of the galaxy ZW 2237+030 discovered in 1985 and considered evidence of the prediction of the curvature of light rays in the presence of masses predicted by general relativity and applied later on to similar arrangements.

Chapter 9
Teichmüller Spaces of Riemann Surfaces

9.1 Introduction

The last chapter of this book is mainly on the notions of Teichmüller space of holomorphic and rational dynamical systems and the, resp., deformation spaces, introduced in 1998 by C. McMullen and D. Sullivan. It is natural to precede it by the basics of Teichmüller spaces of Riemann surfaces. Teichmüller spaces allow to characterize the possible conformal structures of Riemann surfaces by considering classes of quasiconformal mappings of Riemann surfaces homotopic modulo conformal homeomorphisms. It was for this purpose that O. Teichmüller introduced Teichmüller spaces of compact Riemann surfaces and, in addition to the relationship with quasiconformal mappings, he discovered its strong relationship with quadratic differentials, proving that on a compact Riemann surface of genus $g > 1$ every holomorphic quadratic differential determines a quasiconformal mapping which is the unique extremal in its homotopy class, and that such a Teichmüller space is homeomorphic to $\mathbb{C}^{3(g-1)}$

In an article published in 1954 entitled *On quasiconformal mappings*,[1] L. Ahlfors initiated the work he continued in subsequent years of giving a clear and rigorous foundation of Teichmüller spaces, including the introduction of a geometric definition of quasiconformal mappings, and attracting an attention that led to a systematic development of the Teichmüller spaces theory and applications by many authors.

In 1960–1965, L. Bers started another route to the theory of Teichmüller spaces, here initiated in Chap. 4 on holomorphic motions with a first contact with Teichmüller spaces of regions of $\mathbb{C}_\infty$ covered by $\mathbb{H}$. In this approach, the quadratic differentials are found as Schwarzian derivatives of conformal extensions of quasiconformal mappings considered in the universal covering of the Riemann

[1] Ahlfors, L., On quasiconformal mappings, *J. d'Analyse Math.* **3** (1953/54), 1–58.

L. T. Magalhães, *Quasiconformal Mappings in the Plane and Complex Dynamics*,
https://doi.org/10.1007/978-3-031-80115-0_9

surface and obtained as solutions of the Beltrami equation, allowing to extend the theory of Teichmüller spaces to noncompact Riemann surfaces.

In 1964–1965, L. Bers[2] noticed that the Teichmüller space of $\mathbb{H}$ contains as closed subsets the Teichmüller spaces of all the other Riemann surfaces with universal covering by $\mathbb{H}$ and that some general properties can be obtained by properties of this space, which he called universal Teichmüller space, and is one of the Teichmüller spaces considered in Chap. 4 about holomorphic motions. This chapter has six descriptions of the universal Teichmüller space, from different perspectives, providing a deeper understanding of what the universal Teichmüller space is, and consequently also of what other Teichmüller spaces of Riemann surfaces covered by $\mathbb{H}$ are.

The chapter includes the Teichmüller theorem for compact Riemann surfaces of genus $g > 1$, a particular case for hyperbolic Riemann surfaces with hyperbolic Teichmüller spaces of the more general result established in 1939 by O. Teichmüller, and the Royden theorem of 1971 establishing that in the Teichmüller space of a Riemann surface the Teichmüller metric coincides with the Kobayashi metric, defined on a complex differential manifold M by the largest distance such that the holomorphic functions $f : B_1 \to M$ satisfy $\|f'(0)\| \leq 1$, obtaining for Teichmüller spaces a result similar to the Schwarz lemma for conformal homeomorphisms from B_1 to B_1.

So, this chapter extends the study of Teichmüller spaces initiated in the last section of Chap. 4 and continued in Chap. 8, and which will be further extended in the next chapter.

9.2 Quasiconformal Mappings in Teichmüller Space

Once the relationship of quasiconformal mappings with Teichmüller spaces is known in the case of subsets of $\mathbb{C}_\infty$, considered in Chap. 4 for holomorphic motions, the L. Bers definitions of Teichmüller space of a Riemann surface and of quasiconformal mapping between Riemann surfaces[3] are natural and easy.

A ***K*-quasiconformal mapping from a Riemann surface S_1 to a Riemann surface S_2** is a function $f : S_1 \to S_2$ such that its representation in local coordinates $h_2 \circ f \circ h_1^{-1}$, for each pair of coordinate neighborhoods U_1, U_2 of, resp., S_1, S_2 with $f(U_1) \cap U_2 \neq \emptyset$, and associated coordinate systems, resp., h_1, h_2, is a K-quasiconformal mapping. A **quasiconformal mapping from a Riemann surface S_1 to a Riemann surface S_2** is a K-quasiconformal mapping from S_1 to S_2 for some $K \geq 1$. As before, when convenient, denote by K_f the infimum of the possible dilatations of a quasiconformal mapping f between Riemann surfaces.

---

[2] Bers, L., Automorphic forms and general Teichmiiller spaces. *Proc. Conf. Complex Analysis, Minneapolis 1964*, 109–113, Springer-Verlag (1965).

[3] The opening sections of this chapter follow parts of the O. Lehto book listed in the final bibliography, but by a very different order. For general aspects of Riemann surfaces assumed here, if necessary, see Chapters 11 and 12 of *CADOVA*.

It is necessary to ensure the compatibility of the definition for different coordinate systems in intersecting coordinate neighborhoods, which holds since the coordinate systems of Riemann surfaces are conformal homeomorphisms and, therefore, compositions with them or with their inverses do not change the dilatation of quasiconformal mappings.

Given a Riemann surface S, in the space of the quasiconformal mappings of S to any Riemann surface, we consider the equivalence relation of $f_1 : S \to S_1$ with $f_2 : S \to S_2$ if $f_1(S)$ and $f_2(S)$ are conformal sets.[4] The set of the associated equivalence classes is called the **Riemann space of** S, denoted $\mathbf{Riem}(S)$. For example, if $S = \mathbb{H}$, this equivalence relation is so weak that all quasiconformal mappings from $\mathbb{H}$ to any Riemann surface are equivalent, and, therefore, the corresponding Riemann space has only one point. In 1939, O. Teichmüller observed (though only for compact Riemann surfaces) that a useful space is obtained by strengthening the equivalence relation as follows.

The **Teichmüller space of Riemann surface** S, denoted $\mathbf{Teich}(S)$, is the set of the equivalence classes of quasiconformal mappings from S to any Riemann surface considering quasiformal mappings $f_1 : S \to S_1$, $f_2 : S \to S_2$ equivalent if $f_2 \circ f_1^{-1}$ is **homotopic modulo the boundary**, i.e., with the restrictions of both functions to ∂S coincident along an entire homotopy, if S is a Riemann surface with boundary, to a conformal homeomorphism of $f_1(S)$ onto $f_2(S)$.

The **origin of** Teich(S) is the equivalence class in Teich(S) with the identity in S; so, of all conformal homeomorphisms from S onto S.

It is useful to have results on conformal homeomorphisms between Riemann surfaces homotopic to the identity or homotopic modulo the boundary of the domain that can be obtained with lifts by universal coverings in the case of Riemann surfaces with covering by $\mathbb{H}$, considering properties of the, resp., covering groups.

A **limit point of a group** G acting on a set X is a $p \in X$ such that $p = \lim_{n \to +\infty} g_n(q)$ for some $q \in X$ and some sequence $\{g_n\} \subset G$ with distinct terms. The **limit set of a group** is the set L of its limit points. G is said to be an **elementary group** if $\#L \leq 2$.

(9.1) *If S_1, S_2 are Riemann surfaces with universal coverings, resp., $(\mathbb{H}, \pi_1)$, $(\mathbb{H}, \pi_2)$ and covering groups, resp., G_{π_1}, G_{π_2}, both non-elementary, and $f_1, f_2 : S_1 \to S_2$ are quasiconformal mappings with F_1 a Lift of f_1 by the considered coverings, then f_1 and f_2 induce the same group isomorphism of G_{π_1} to G_{π_2} if and only if there is lift F_2 of f_2 by the coverings that coincides with F_1 in the limit set of the group G_{π_1}.*

[4] The arbitrariness of the codomain Riemann surface of these functions introduces a harmless ambiguity in this framework, often avoided in the literature with the notion of **Riemann surface** S **marked by Riemann surface** S_j, $j \in \{1, 2\}$. The equivalence classes are considered between ordered pairs (S, S') of Riemann surfaces with a quasiconformal mapping of S onto S'.

Proof If F_2 is a lift of f_2 coinciding with F_1 in the limit set L_1 of G_{π_1}, since F_1, F_2 map L_1 onto the limit set L_2 of G_{π_2} and L_1 is invariant under G_{π_1}, then $F_1 \circ g \circ F_1^{-1}{}_{|L_2} = F_2 \circ g \circ F_2^{-1}{}_{|L_2}$ for $g \in G_{\pi_1}$. Both sides of the equality are Möbius transformations, and as they coincide in a set with at least three points, they coincide in the whole $\mathbb{H}$.

Conversely, if the preceding equality holds in $\mathbb{H}$ and $h = F_1^{-1} \circ F_2$, then $g \circ h = h \circ g$ for $g \in G_{\pi_1}$. If z is a fixed point of some g, then $g\big(h(z)\big) = h(z)$, i.e., $h(z)$ is also a fixed point of g. If z is an attracting fixed point of g and $w \in \mathbb{H}$, then

$$\lim_{n\to+\infty} g^{\circ n}\big(h(w)\big) = z, \qquad \lim_{n\to+\infty} g^{\circ n}\big(h(w)\big) = \lim_{n\to+\infty} h\big(g^{\circ n}(w)\big) = h(z).$$

If z is a repelling fixed point of g, it is an attracting fixed point of g^{-1}, which also commutes with h. Thus, $h(z) = z$ for all fixed points z of G_{π_1}, as this set is dense in L_1 and $F_1(z) = F_2(z)$ for $z \in L_1$. □

(9.2) *If S_1, S_2 are Riemann surfaces with universal coverings, resp., $(\mathbb{H}, \pi_1)$, $(\mathbb{H}, \pi_2)$ and covering groups, resp., G_{π_1}, G_{π_2}, homeomorphisms f_1, f_2 of S_1 onto S_2 induce the same group isomorphism $g \mapsto f_j \circ g \circ f_j^{-1}$, $j \in \{1, 2\}$, if and only if they are homotopic.*

Proof Let f_1, f_2 be homeomorphisms and, resp., F_1, F_2 their lifts by the considered coverings.

If f_1, f_2 are homotopic, as the lift of a homotopy of f_1 to f_2 by the coverings is a homotopy $H : \mathbb{H} \times [0, 1] \to \mathbb{H}$ of F_1 to F_2, for $g \in G_{\pi_1}$ and $z \in \mathbb{H}$ the paths $t, \mapsto H\big(g(z), t\big)$ and $t \mapsto F_1 \circ g \circ F_1^{-1}\big(H(z, t)\big)$ have the same initial and final points and the same projection on S_2, $t \mapsto \pi_2\big(H(z, t)\big)$. Therefore, they coincide, and, at $t = 1$, $F_1 \circ g \circ F_1^{-1} = F_2 \circ g \circ F_2^{-1}$ and, consequently, also the projections satisfy $f_1 \circ g \circ f_1^{-1} = f_2 \circ g \circ f_2^{-1}$.

Conversely, if the last equality holds for all $g \in G_{\pi_1}$, define a homotopy H from F_1 to F_2 in $\mathbb{H}$ with $H(z, t)$ equal to the point of the geodesic in the Poincaré metric of $\mathbb{H}$ with end points $F_1(z), F_2(z)$ which divides the length of this geodesic in the ratio $t : (1 - t)$. As the function $\theta(g) = F_1 \circ F_1^{-1}$ keeps invariant Poincaré distances, $[\theta(g)]\big(H(z, t)\big) = [\theta(g)]\big(H(g(z), t)\big)$. So, $\theta(g) \circ H(z, t) = H(g(z), t)$ for $z \in \mathbb{H}$, $t \in [0, 1]$, i.e., the functions $z \mapsto H(z, t)$ determine the same homomorphism in the group G_{π_1}, and, therefore, we can consider a function $\pi_2\big(H(\pi_1^{-1}(z), t)\big)$ defined for $(z, t) \in S_1 \times [0, 1]$, which is an homotopy from f_1 to f_2. □

(9.3) *If S is a Riemann surface with universal covering $(\mathbb{H}, \pi)$, the identity is the only conformal homeomorphism from S to S homotopic to the identity.*

Proof If $f : S \to S$ is a conformal homeomorphism homotopic to the identity, by the preceding result, the covering group isomorphism induced by f coincides with that induced by the identity in S. Since the identity in $\mathbb{H}$ is a lift f by the covering, by (9.1), the projection f of this lift is the identity in S. □

(9.4) *If S_1, S_2 are Riemann surfaces with universal covering, resp., $(\mathbb{H}, \pi_1)$, $(\mathbb{H}, \pi_2)$ and covering groups, resp., G_{π_1}, G_{π_2}, then conformal homeomorphisms f_1, f_2 from S_1 to S_2 are homotopic modulo the boundary (i.e., coincident in ∂S_1 throughout an homotopy) if and only if there are lifts by the covering coinciding on ∂S_1.*

Proof If f_1, f_2 are homotopic modulo the boundary, F_1 is a lift of f_1 and F_2 is a lift of f_2 by the, resp., coverings, with F_2 homotopic to F_1 with homotopy coincident with F_1 in the set B which is the complement on the real axis of the limit set L of G_{π_1}. By (9.2), F_1, F_2 induce the same group isomorphism, and, by (9.1), $F_1 = F_2$ in L. Therefore, $F_1 = F_2$ on the real axis.

Conversely, if F_1, F_2 are lifts of, resp., f_1, f_2 by the, resp., coverings such that $F_1 = F_2$ on the real axis, construct a homotopy from F_1 to F_2 in $\mathbb{H}$ as in the proof of (9.2) which, projected by the coverings, gives a homotopy from f_1 to f_2 with the values in ∂S_1 fixed throughout the homotopy. □

For compact Riemann surfaces, the following holds.

(9.5) *If S_1, S_2 are homeomorphic compact Riemann surfaces, every homotopy class of homeomorphisms from S_1 onto S_2 preserving orientation contains quasiconformal mappings.*

Proof Let f be the a orientation preserving homeomorphism of S_1 onto S_2. By the compactness of S_1, there is a finite cover $\{U_1, \dots, U_N\}$ of S by open sets conformal to B_1 with ∂U_j analytic curves. We consider the sequence of functions defined recursively by $f_0 = f$ and for $k \in \{1, \dots, N\}$ by $f_k = f_{k-1}$ in $S \backslash U_k$ and in U_k coinciding with the extension obtained by applying the Beurling-Ahlfors extension (3.3) of the function defined in the boundary of U_k by f_{k-1}, i.e., applied after transforming U_k and $f_{k-1}(U_k)$ by conformal homeomorphisms onto $\mathbb{H}$ normalized to a function induced from $\mathbb{H}$ to $\mathbb{H}$ with ∞ a fixed point and applying the inverses of these conformal homeomorphisms. Each $f_{k|U_k}$ is a diffeomorphism, and, therefore, it is locally quasiconformal. By the proof of (3.3), if f_{k-1} is quasiconformal at a point $z, \in \partial U_k$, for some neighborhood $V \subset S_1$ of z, the function $f_{k|U_k \cap V}$ is quasiconformal and, as ∂U_k has zero Lebesgue measure and, therefore, it has a removable singularity, also f_k is quasiconformal at z. Hence, f_N is a quasiconformal mapping in S_1. The function $(z, t) \mapsto t f_k(z) + (1 - t) f_{k-1}(z)$ defines a homotopy from f_{k-1} to f_k, for $k \in \{1, \dots, N\}$. Therefore, the quasiconformal mapping f_N in S is homotopic to f. □

For non-compact Riemann surfaces, this may fail, e.g., B_1 and $\mathbb{C}$ are homotopic, and they are not quasiconformally equivalent.

Analogously to the particular case considered in Chap. 4, we call **Teichmüller distance** in $\mathrm{Teich}(S)$ to

$$D_{T,S}(\tau,\sigma) = \tfrac{1}{2}\inf\bigl\{\log K_{g\circ f^{-1}} : f\in\tau,\ g\in\sigma\bigr\}.$$

In checking the conditions of the definition of distance, it only is not trivial to prove that $D_{T,S}(\tau,\sigma)=0$ implies $\tau=\sigma$, which results from the existence of each equivalence class of an **extremal quasiconformal mapping**, i.e., with dilatation not greater than the dilatation of any other element of the equivalence class, as proved next. Thus,

$$D_{T,S}(p,q)=\tfrac{1}{2}\inf\{\log K_f : f\in F\},$$

where F is the set of quasiconformal mappings from $f(S)$ onto $g(S)$ homotopic to $g\circ f^{-1}$, for fixed $f\in p$, $g\in q$, and also the infimum can be replaced by minimum in the above formulas for Teichmüller distance.

(9.6) *For every Riemann surface S in each equivalence class of* $\mathrm{Teich}(S)$, *there is an extremal quasiconformal mapping.*

Proof By the uniformization theorem of Riemann surfaces,[5] each Riemann surface has a universal covering by $\mathbb{C}_\infty$, $\mathbb{H}$ or $\mathbb{C}$. The result is trivial for universal covering by $\mathbb{C}_\infty$ and also for non-compact Riemann surfaces with universal covering by $\mathbb{C}$. For a Riemann surface with covering by $\mathbb{H}$, we can prove for lifts of the functions associated with universal coverings of the surface, as in (4.9), and through the covering projection, we get a quasiconformal mapping in each element of $\mathrm{Teich}(S)$. For a compact Riemann surface with covering $\mathbb{C}$, if the covering group is trivial or cyclic, all images of the Riemann surface by quasiconformal mappings are conformal. It remains to consider toric Riemann surfaces, for which there are extremal quasiconformal mappings as a consequence of the following result. □

(9.7) *In each equivalence class of homeomorphisms preserving orientation from a torus $S_1=\mathbb{C}/G_1$ to another torus $S_2=\mathbb{C}/G_2$, with universal coverings, resp., $(\mathbb{C},\pi_1)$, $(\mathbb{C},\pi_2)$ and covering groups, resp., G_{π_1}, G_{π_2}, there are affine extremal quasiconformal mappings whose lifts by the coverings are affine functions, and they are unique modulo compositions with conformal homeomorphisms homotopic to the identity.*

[5] See Chapter 12 of *CADOVA*.

Proof For $j \in \{1, 2\}$, the elements of G_{π_j} are translations $z \mapsto z + n\omega_j + n\sigma_j$ for any $m, n \in \mathbb{Z}$, where $\omega_j, \sigma_j \in \mathbb{C}\backslash\{0\}$ such that $\mathcal{I}m\, \frac{\omega_j}{\sigma_j} \neq 0$. We call (ω_j, σ_j) a **basis of the group** G_{π_j} and say that it is a **normalized basis** if $\mathcal{I}m\, \frac{\omega_j}{\sigma_j} > 0$. For $j \in \{1, 2\}$, given $p_j, q_j \in S_j$ and $z_j \in \pi_j^{-1}(p_j)$, $w_j \in \pi_j^{-1}(q_j)$, as translations commute with each element of G_{π_j}, for each $t \in [0, 1]$, the function

$$\zeta \mapsto \zeta + t(w_j + z_j)$$

has projection by π_j in a conformal homeomorphism $f_{j,t} : S_j \to S_j$, giving a homotopy of the identity 1_{S_j} to a conformal homeomorphism[6] $f_{j,1}$, and $f_{j,1}(p_j) = q_j$. Therefore, with fixed points $p_j \in S_j$, we can restrict to functions $f : S_1 \to S_2$ normalized by the condition $f(p_1) = p_2$.

Choosing $p_j = \pi_j(0)$, there is one single lift by the coverings $F : \mathbb{C} \to \mathbb{C}$ such that $F(0) = 0$, which is called the **normalized lift** of f and induces a group isomorphism of G_{π_1} onto G_{π_2} mapping

$$z \mapsto z + m\omega_1 + \sigma_1 \quad \text{to} \quad z \mapsto z + mF(\omega_1) + nF(\sigma_1).$$

Every group isomorphism θ of G_{π_1} onto G_{π_2} is of this type, because if (ω_1, σ_1) is the basis of G_{π_1} the affine transformation with 0 fixed and mapping ω_1, σ_1 to, resp., $\theta(\omega_1), \theta(\sigma_1)$ is projected by the coverings onto a homeomorphism of S_1 onto S_2. For this reason and because the conformal homeomorphisms from $\mathbb{C}$ to $\mathbb{C}$ are affine, we just consider functions with normalized lifts. $\mathcal{L}$ denotes the set of these normalized lifts.

Given a fixed normalized basis (ω_1, σ_1) of G_{π_1}, there is a normalized basis (ω_2, σ_2) of G_{π_2} such that $\mathcal{F}$ is the set of homeomorphisms $F : \mathbb{C} \to \mathbb{C}$ with $F(0) = 0$ and $F(z + m\omega_1 + n\sigma_1) = F(z) + m\omega_2 + n\sigma_2$ for $z \in \mathbb{C}$, $m, n \in \mathbb{Z}$, since the identity is the only automorphism of G_{π_2}. It is clear that in $\mathcal{F}$, there is only one affine transformation A.

For a K-quasiconformal mapping $F \in \mathcal{F}$, consider $F_k(z) = \frac{1}{k}F(kz)$, for $k \in \mathbb{N}$, which are also K-quasiconformal mappings. By the equality in the preceding paragraph, $\lim F_k = A$ uniformly in the Euclidean metric. Therefore, the dilatation of A is also upper bounded by K, and the projection of A by the coverings is an extremal quasiconformal mapping in the homotopy class considered. rotating the real and imaginary axes of $\mathbb{C}$ so that the direction of the real axis coincides with that of maximal elongation under application of A. Denoting the function in the new coordinates also by A, it is $A(x + iy) = Kx + iy$ for $x, y \in \mathbb{R}$. Denote by F any extremal quasiconformal mapping in the homotopic class containing A.

It remains to prove the uniqueness of the extremal quasiconformal mapping modulo compositions with conformal homeomorphisms. As the function $F - A$ has periods ω_1, σ_1, by the Weierstrass theorem of extrema of continuous functions, $|F - A|$ has a maximum $M \geq 0$. Since F is an absolutely continuous function on lines, for any $r > 0$,

[6] Thus, for tori, it is very different for Riemann surfaces S covered by $\mathbb{H}$ with non-elementary covering groups, for which, by (9.3), the only conformal deformation of S onto S is the identity.

$$\int_0^r \left|\tfrac{\partial}{\partial x}F(x+iy)\right|dx \geq \left|\int_0^r \tfrac{\partial}{\partial x}F(x+iy)\,dx\right| \geq Kr-2M, \quad \text{a.e. for } y\in[0,r].$$

So, with $Q=[0,r]^2$, it is $\iint_Q \left|\frac{\partial}{\partial x}F(x+iy)\right| dxdy \geq Kr^2-2Mr$. The number of period parallelograms congruent to a parallelogram P intersecting Q is asymptotically $\frac{1}{\text{area}(P)}\,r^2[1+o(1)]$, $r\to+\infty$. Therefore,

$$\iint_P \left|\tfrac{\partial}{\partial x}F(x+iy)\right| dxdy \geq \tfrac{\text{area}(P)}{r^2[1+o(1)]}\,(Kr^2-2Mr)\,, \qquad r\to+\infty.$$

Consequently, $\iint_P \left|\frac{\partial}{\partial x}F(x+iy)\right| dxdy \geq K\,\text{area}(P)$. Since F is a K-quasiconformal mapping, if JF denotes the Jacobian of F,

$$\left|\tfrac{\partial}{\partial x}F(x+iy)\right|^2 \leq KJF(x+iy)\,, \qquad \text{a.e. for } x+y\in\mathbb{C}.$$

By the Cauchy-Schwarz inequality,

$$K^2[\text{area}(P)]^2 \leq \text{area}(P)\iint_P \left|\tfrac{\partial}{\partial x}F(x+iy)\right|^2 dxdy \leq K\text{area}(P)\,\text{area}\big(F(P)\big).$$

Since $F(P)$ and $A(P)$ are fundamental domains of G_{π_2} with zero measure boundaries, $\text{area}\big(F(P)\big)=\text{area}\big(A(P)\big)=K\,\text{area}(P)$, and we have equality in the preceding inequality. Thus,

$$\left|\tfrac{\partial}{\partial x}F(x+iy)\right|^2 = KJF(x+iy)\,, \qquad \text{a.e. for } x+y\in\mathbb{C}.$$

Since a.e. $\frac{\partial}{\partial x}F(x+iy)=\frac{\partial}{\partial z}F(z)+\frac{\partial}{\partial \bar z}F(z)$ for $x+iy+z$, and it is

$$JF = \left|\tfrac{\partial}{\partial z}F\right|^2 + \left|\tfrac{\partial}{\partial \bar z}F\right|^2,$$

we have $\frac{\partial F/\partial \bar z}{\partial F/\partial z}=\frac{K-1}{K+1}$ a.e.. Hence, F is an affine transformation, and $F=A$. Consequently, the extremal quasiconformal mapping is affine, and it is unique modulo compositions with conformal homeomorphisms homotopic to the identity.

□

A simple consequence of the definition of Teichmüller space of Riemann surfaces by quasiconformal mappings is that the Teichmüller spaces of quasiconformal Riemann surfaces are isometric.

(9.8) *The Teichmüller spaces of quasiconformal Riemann surfaces are isometric. If S_1, S_2 are Riemann surfaces and $h: S_1\to S_2$ is a quasiconformal mapping onto S_2, then the function mapping $[f_1]\in\text{Teich}(S_1)$ to $[f_1\circ h^{-1}]\in\text{Teich}(S_2)$ is an isometric bijection of* Teich(S_1) *onto* Teich(S_2) *with the, resp., Teichmüller distances and maps $[h]\in\text{Teich}(S_1)$ to the origin of* Teich(S_2).

Proof The function $f \mapsto f \circ h^{-1}$ is a bijection of the set of quasiconformal mappings of S_1 onto S_1 onto the set of quasiconformal mappings of S_2 onto S_2. Quasiconformal mappings f_1, f_2 defined in S_1 belong to a same element of Teich(S_1) if and only if $f_1 \circ h^{-1}$, $f_2 \circ h^{-1}$ belong to a same element of Teich(S_2). The function mapping $[f_1] \in \text{Teich}(S_1)$ to $[f_1 \circ h^{-1}] \in \text{Teich}(S_2)$ is a bijection of Teich(S_1) onto Teich(S_2), and with the, resp., Teichmüller distances, it is an isometry. It is obvious that this function maps $[h] \in \text{Teich}(S_1)$ to the origin of Teich(S_2). □

Consequently, since compact Riemann surfaces of the same genus are homeomorphic and, by (9.5), they also are quasiconformal mappings: *All Teichmüller spaces of compact Riemann surfaces of the same genus with the, resp., Teichmüller distances are isometric.*

9.3 Modular Group of Riemann Surface

By the preceding section, if S is a Riemann surface and $h : S \to S$ is a quasiconformal mapping, $[f] \mapsto [f \circ h^{-1}]$ is an isometry of Teich(S) onto itself. The set of these isometries with the operation of equivalence classes of quasiconformal mappings induced by the composition of functions is a group, called **modular group** of Teich(S), denoted **Mod(*S*)**.

If $S = \mathbb{H}$, Mod(S) coincides with the universal modular group of two sections above.

Denote $\boldsymbol{QC(S)}$ the group of quasiconformal mappings from S to S and $\boldsymbol{QC_0(S)}$ the normal subgroup of quasiconformal mappings of $QC(S)$ homotopic in S to the identity.

For $h \in QC(S)$, consider the function $[f] \mapsto [f \circ h^{-1}]$, which is an element of Mod(S). As for $h_1, h_2 \in QC(S)$ such that $h_2^{-1} \circ h_1 \in QC_0(S)$, it is

$$[f \circ h_1^{-1}] = [f \circ h_2^{-1}],$$

h defines a function from $QC(S)/QC_0(S)$ onto Mod(S) that is a group homomorphism. If S is such that the only conformal homeomorphism from S to S is the identity, the function so defined is a bijection of $QC(S)/QC_0(S)$ onto Mod(S) and a group isomorphism, which also holds for Riemann surfaces quasiconformal to such a Riemann surface S.

The Riemann space Riem(S) of a Riemann surface S was mentioned before the definition of Teichmüller space of Riemann surface. The study of this space is called the **Riemann moduli problem**.

(9.9) *For Riemann surfaces* S, Riem(S) = Teich(S)/Mod(S).

Proof If $[f], [g] \in \text{Teich}(S)/\text{Mod}(S)$, there is a quasiconformal mapping $h : S \to S$ such that $[f \circ h^{-1}] = [g]$. So, $f(S)$ and $g(S)$ are conformal homeomorphisms, i.e., f, g belong to the same point of $\text{Riem}(S)$.

Conversely, if f, g belong to the same point of $\text{Riem}(S)$, there is a conformal homeomorphism $T : f(S) \to g(S)$ and $h = g^{-1} \circ T \circ f$ is a quasiconformal mapping of S into S. Since $g = T \circ (f \circ h^{-1})$, g and $f \circ h^{-1}$ belong to the same point of $\text{Riem}(S)$. □

9.4 Differential in a Riemann Surface

If $f : S \to \mathbb{C}$ is a holomorphic function in a Riemann surface S, and h_1, h_2 are coordinate systems of two intersecting coordinate neighborhoods of S, the coordinates representations of f are, resp., $f_1 = f \circ h_1^{-1}$ and $f_2 = f \circ h_2^{-1}$, and the derivative of $f_1 \circ h_1 = f_2 \circ h_2$ is $(f_1' \circ h_1)\, h_1' = (f_2' \circ h_1)\, h_2'$, i.e., we have the invariance $(f_1' \circ h_1) \frac{h_1'}{h_2'} = (f_2' \circ h_1)$.

A **differential** in a Riemann surface S is a set of functions $\varphi = \{\varphi_j\}$ with φ_j defined in a coordinate neighborhood U_j with coordinate system h_j of S, and the set of coordinate neighborhoods $\{U_j\}$ considered is a cover of S such that

$$(\varphi_j \circ h_j) \frac{h_j'}{h_k'} = (\varphi_k \circ h_k), \quad U_j \cap U_k \neq \emptyset, \text{ for all } j, k.$$

Generalizing, for $m, n \in \mathbb{Z}$, a **(m, n)-differential** in S is a set $\varphi = \{\varphi_j\}$ satisfying the invariance relation

$$(\varphi_j \circ h_j) \left(\frac{h_j'}{h_k'} \right)^m \left(\overline{\frac{h_j'}{h_k'}} \right)^n = (\varphi_k \circ h_k) \text{ for all } j, k \text{ in each } U_j \cap U_k \neq \emptyset.$$

A (m,n)-differential $\varphi = \{\varphi_j\}$ in S is called a **holomorphic (m,n)-differential** or a **meromorphic (m,n)-differential** in S if all functions φ_j are, resp., holomorphic or meromorphic. It is clear that a differential is a $(1, 0)$-differential and a quadratic differential is a $(2, 0)$-differential.

For each pair $m, n \in \mathbb{Z}$, the set of (m, n)-differentials in a Riemann surface S is a complex linear space with the usual operations of addition and multiplication by scalars, and the set of (m, n)-differentials holomorphic in S with the same operations is one of its linear subspaces.

A **Beltrami differential** in a Riemann surface S is a $(-1, 1)$-differential in S with all elements Lebesgue measurable with L^∞ norms < 1.

If (R, p) is a universal covering of a Riemann surface S_1, with $R = B_1$ or $R = \mathbb{C}$ (thus, excluding the trivial case $R = \mathbb{C}_\infty$), and f is a quasiconformal mapping from S_1 to a Riemann surface S_2, then the Beltrami coefficient or complex dilation κ_f of f is naturally defined in R. If G is the covering group, then $\kappa_f = (\kappa_f \circ g) \frac{\overline{g'}}{g'}$ for $g \in G$; thus, κ_f is a Beltrami differential of f in the Riemann surface S_1.

Another way to obtain the Beltrami differential of f is to consider universal coverings of S_1 and S_2 (as f is a quasiconformal mapping of S_1 to S_2, the base spaces of these coverings are conformal). The function between the covering spaces of the universal coverings obtained lifting f is a quasiconformal mapping, and the Beltrami differential of f is the projection of the Beltrami coefficient of this quasiconformal mapping.

The Beltrami differentials κ_1, κ_2 in a Riemann surface S are said to be **equivalent Beltrami differentials** if the corresponding quasiconformal mappings are equivalent. Therefore, each point of Teich(S) can be seen as an equivalence class of Beltrami differentials in S. With no ceremony with notation, we will consider points of Teichmüller spaces as equivalence classes of quasiconformal mappings or Beltrami differentials as appropriate to context.

The Teichmüller distance with Beltrami differentials is

$$D_T(p,q) = \tfrac{1}{2} \inf \left\{ \log \tfrac{1+\|(\mu-\nu)/(1-\overline{\mu}\nu)\|_{L^\infty}}{1-\|(\mu-\nu)/(1-\overline{\mu}\nu)\|_{L^\infty}} : \mu \in p\,,\ \nu \in q \right\}.$$

Since

$$\tanh \log \tfrac{1+x}{1-x} = \frac{(1+x)/(1-x)-(1-x)/(1+x)}{(1+x)/(1-x)+(1-x)/(1+x)} = \frac{(1+x)^2-(1-x)^2}{(1+x)^2+(1-x)^2} = \frac{2x}{1+x^2}\,, \quad x \in [0,1[\,.$$

With $\beta_{T,S} = \tanh D_{T,S}$, as tanh and log are strictly increasing,

$$\beta_{T,S}(p,q) = \inf\{ \| \tfrac{\mu-\nu}{1-\overline{\mu}\nu)} \|_{L^\infty} : \mu \in p\,,\ \nu \in q \}.$$

$\beta_{T,S}$ is also a distance in Teich(S) and, since for $x \in [0,1[$, $x \le \frac{2x}{1+x^2} \le 2x$, it is equivalent to the Teichmüller distance $D_{T,S}$.

In this framework, we have the following generalization of the measurable Riemann mapping theorem (2.9): *If κ is a Beltrami differential in a Riemann surface S_1, then there is a quasiconformal mapping from S_1 to a Riemann surface S_2 with Beltrami coefficient κ, unique modulo left composition by conformal homeomorphisms.*

9.5 Conformal Structures in Riemann Surfaces

By the end of the preceding section, if κ is a Beltrami differential in a Riemann surface S and U is a coordinate neighborhood of S with coordinate system h, there is a quasiconformal mapping $g : h(U) \to \mathbb{C}$ with Beltrami coefficient $\kappa \circ h^{-1}$. Then, $f = g \circ h$ is a quasiconformal mapping from U to $\mathbb{C}$ with Beltrami coefficient κ. If f_1, f_2 are functions obtained in this way for coordinate neighborhoods U_1, U_2 with $U_1 \cap U_2 \neq \emptyset$ with coordinate systems, resp., h_1, h_2, the function $f_2 \circ f_1^{-1}$ is a conformal homeomorphism defined in $f_1(U_1 \cap U_2)$. Therefore: *Every Beltrami differential in a Riemann surface S defines a conformal structure in S.*

If $\mathscr{E}$ is the initial conformal structure of a Riemann surface S and $\mathscr{E}_\kappa$ is the one induced by a Beltrami differential κ, this conformal structure is determined by quasiconformal mappings of open sets of S with the initial conformal structure in $\mathbb{C}$ with Beltrami coefficients restrictions of κ to such sets. These functions are conformal homeomorphisms in the conformal structure $\mathscr{E}_\kappa$.

Conformal structures $\mathscr{E}_1, \mathscr{E}_2$ of a Riemann surface are said to be **equivalent conformal structures by deformation** if they can be **conformally deformed** one to the other, i.e., if there is a conformal homeomorphism of S with conformal structure $\mathscr{E}_1$ onto S with conformal structure $\mathscr{E}_2$ homotopic to the identity.

(9.10) *Conformal structures $\mathscr{E}_1, \mathscr{E}_2$ induced by Beltrami coefficients κ_1, κ_2 in a Riemann surface S, such that the identity in S preserves the orientation in both structures, are equivalent by deformations if and only if the quasiconformal mappings in S with Beltrami coefficient, resp., κ_1, κ_2 belong to the same point of* Teich(S).

Proof If $\varphi : S \to S$ is a conformal homeomorphism in S with a conformal structure $\mathscr{E}_1$ in S with conformal structure $\mathscr{E}_2$ homotopic to the identity and f_1, f_2 are quasiconformal mappings in S with Beltrami coefficients, resp., κ_1, κ_2, then $hg = f_2 \circ \varphi \circ f_1^{-1} g : f_1(S) \to f_2(S)$ is a conformal homeomorphism and $f_2 \circ f_1^{-1}$ is homotopic to h. So, κ_1, κ_2 are equivalent Beltrami differentials.

If κ_1, κ_2 are equivalent Beltrami differentials, there are conformal mappings $h : f_1(S) \to f_2(S)$ such that $\varphi = f_2^{-1} \circ h \circ f_1 : S \to S$ is homotopic to the identity, and $\varphi : S \to S$ is a conformal homeomorphism of S with a conformal structure $\mathscr{E}_1$ in S with conformal structure $\mathscr{E}_2$. Therefore, $\mathscr{E}_1, \mathscr{E}_2$ are equivalent by deformation. □

This has the following very important consequence: *The Teichmüller space of a Riemann surface can be seen as the set of equivalence classes by deformation of conformal structures in S.*

For compact Riemann surfaces, the following property holds.

(9.11) *On a compact Riemann surface, every conformal structure is equivalent by deformation to a conformal structure induced by a Beltrami differential in S, with the identity in S preserving the orientation in both structures.*

Proof Let $\mathscr{E}_1, \mathscr{E}_2$ be conformal structures in S. By (9.5), there is a quasiconformal mapping f of S with the conformal structure $\mathscr{E}_1$ and the conformal structure $\mathscr{E}_2$ homotopic to the identity. If κ_f is the Beltrami coefficient of f, the coordinate systems of $\mathscr{E}_2$ are $h \circ f^{-1}$ such that h is the coordinate system of $\mathscr{E}_1$. Therefore, the two conformal structures are equivalent by deformation in S. □

As a result, with (9.5), we have the following characterization of Teichmüller spaces of compact Riemann surfaces.

(9.12) *The Teichmüller space of a compact Riemann surface S is isomorphic to the set of equivalence classes by deformation of conformal structures in S.*

9.6 Universal Teichmüller Space

In 1965, L. Bers noticed that every Teichmüller space Teich(S) of a Riemann surface S with universal covering by $\mathbb{H}$ is a closed subset of the Teichmüller space of the open upper complex half plane Teich$(\mathbb{H})$. L. Bers called Teich$(\mathbb{H})$ the **universal Teichmüller space**, because the identity from Teich(S) to Teich$(\mathbb{H})$ is a homeomorphism and, therefore, some properties obtained in Chap. 4 for Teichmüller spaces of regions of $\mathbb{C}_\infty$ with covering $(\mathbb{H}, \pi)$ can be extended to any Teichmüller space of Riemann surface with covering by $\mathbb{H}$, as the universal Teichmüller space is the particular case of those spaces with π the identity.

The Teichmüller spaces of Riemann surfaces with covering by $\mathbb{H}$ are ordered by inclusion, since if S_1, S_2 are Riemann surfaces with universal covering, resp., $(\mathbb{H}, \pi_1)$, $(\mathbb{H}, \pi_2)$ and covering group, resp., G_{π_1}, G_{π_2}, **Teich$(S_1) \subset$ Teich(S_2)** if and only if π_2 is a subgroup of π_1.

Since the covering $(\mathbb{H}, \pi_0)$ of $\mathbb{H}$ is with $\pi_0 = 1_{\mathbb{H}}$, the covering group G_{π_0} is the trivial group, with just the identity, which is a subgroup of the covering group G_π of any Riemann surface S with universal covering $(\mathbb{H}, \pi)$. Therefore, *every Teichmüller space* Teich(S) *of a Riemann surface with covering by* $\mathbb{H}$ *is a subset of the universal Teichmüller space* Teich$(\mathbb{H})$.

If S is a Riemann surface with covering by $\mathbb{H}$ *and* $D_{T,\mathbb{H}}$, $D_{T,S}$ *are the Teichmüller distances of, resp.,* Teich$(\mathbb{H})$, Teich$(\mathbb{S})$, *the restriction of* $D_{T,\mathbb{H}}$ *to* $[\text{Teich}(S)]^2$ *is a distance in* Teich(S) *and* $D_{T,\mathbb{H}|\text{Teich}(\mathbb{S})} \leq D_{T,S}$.

It can be proved, as in Chap. 4 for Teich(Ω) with $\Omega \subset \mathbb{C}_\infty$ with covering by $\mathbb{H}$, that *if S is a Riemann surface with covering by* $\mathbb{H}$, *the metric space* $\big(\text{Teich}(S), D_{T,S}\big)$ *is complete.*

For sharpening the idea made of Teichmüller spaces, it is useful to view the universal Teichmüller space from diverse perspectives. Next, we consider six different perspectives, each one giving a different view of this space and, therefore, also of any Teichmüller space covered by $\mathbb{H}$.

Directly from the definition and the preceding section, we have three ways of looking at the universal Teichmüller space:

(1) *Set of the equivalence classes of quasiconformal mappings in* $\mathbb{H}$ *normalized by having extensions by continuity to the real axis with* $0,\ 1,\ \infty$ *fixed points*

(2) *Set of the equivalence classes of the open ball of radius* 1 *and center* 0 *of* $L^{\infty}(\mathbb{H})$ *(these are Beltrami coefficients of quasiconformal mappings).*

(3) *Set of the equivalence classes by deformation of conformal structures in* $\mathbb{H}$.

By the Beurling-Ahlfors extension (3.3), there is a bijection from the set of equivalence classes of the quasisymmetric functions in $\mathbb{H}$ considered in (1) to the set of the quasisymmetric functions defined in the real axis with fixed points $0, 1, \infty$. So, another way to see the universal Teichmüller space is:

(4) *Set of the quasisymmetric functions in* $\mathbb{R}$ *with fixed points* $0, 1, \infty$.

Consequently: *Each point of the universal Teichmüller space can be represented by an analytic function of real variables with complex values quasiconformal as a complex function, or by an analytic function of real variable with complex values and* L^{∞} norm <1.

If f is an element of the set $\mathscr{F}$ of quasiconformal mappings in $\mathbb{H}$ normalized by having extensions by continuity to the real axis with $0, 1, \infty$ fixed points, also $f^{-1} \in \mathscr{F}$, and if $f, g \in \mathscr{F}$, also $f \circ g \in \mathscr{F}$. So, $\mathscr{F}$ with operation the composition of functions is a group. $f_1, f_2 \in \mathscr{F}$ are said to be **equivalent quasiconformal functions** if $f_2 \circ f_1{}^{-1}_{|\mathbb{R}}$ is the identity. If $g: \mathbb{H} \to \mathbb{H}$ is a quasiconformal mapping and $f \in \mathscr{F}$, there is a Möbius transformation $h_g: \mathbb{H} \to \mathbb{H}$ such that

$$h_g \circ f \circ g^{-1} \in \mathscr{F}. \qquad \omega_{[g]}: \mathrm{Teich}(\mathbb{H}) \to \mathrm{Teich}(\mathbb{H})$$

is defined by $\omega_{[g]}([f]) = [h_g \circ f \circ g^{-1}]$, a bijection with $\omega_0 = \omega_{[1_{\mathbb{H}}]} = 1_{\mathrm{Teich}(\mathbb{H})}$,

$$\left(\omega_{[g_1]} \circ \omega_{[g_2]}\right)([f]) = [\, h_{g_1} \circ h_{g_2} \circ f \circ g_2^{-1} \circ g_1^{-1}\,] = \omega_{[g_1]\circ[g_2]}([f]).$$

$\omega_{[g]}: g: \mathbb{H} \to \mathbb{H}$ is a quasiconformal mapping$\}$ with the composition of functions is a group, called **universal modular group**. Therefore, another way of seeing the universal Teichmüller space is:

(5) *Quotient group* $\mathscr{F}/N$ *with* N *the normal subgroup*[7] *of the group of functions equivalent to the identity, with the composition of functions.*

The operation in this subgroup is $[f] \circ [g] = [f \circ g]$ with identity the equivalence class containing the identity function (or the equivalence class of Beltrami coefficients containing the zero function). Therefore, also:

The set of quasimetric functions with $0, 1, \infty$ *fixed points and operation the composition of functions is a group, isomorphic to the group* $\mathrm{Teich}(\mathbb{H}) = \mathscr{F}/N$.

A quasiconformal mapping in $\mathbb{H}$, $f \in \mathscr{F}$, can be extended to a quasiconformal mapping in $\mathbb{C}$ by reflection relative to the real axis, but this is not fertile. In 1960,

[7] A **normal subgroup** of a group G is a subgroup N of G invariant under **conjugation**, i.e., such that $gng^{-1} \in N$ for $n \in N$, $g \in G$.

L. Bers noted that it is much more fertile to extend Beltrami coefficients from $\mathbb{H}$ to $\mathbb{C}$, so that they be conformal in $-\mathbb{H}$, an idea precursor of quasiconformal surgery by two decades.

Thus, denoting f^κ a quasiconformal mapping from $\mathbb{C}$ to $\mathbb{C}$ extension of the quasiconformal mapping f_κ from $\mathbb{H}$ to $\mathbb{H}$ belonging to $\mathscr{F}$ and with Beltrami coefficient κ, obtained as a solution of the Beltrami equation in $\mathbb{C}$ with Beltrami coefficient equal to κ in $\mathbb{H}$ and equal to 0 in $-\mathbb{H}$ with $0, 1, \infty$ fixed points, the following holds.

(9.13) *Beltrami coefficients κ_1, κ_2 are equivalent if and only if the conformal homeomorphisms $(f^{\kappa_1})_{|-\mathbb{H}}$ and $(f^{\kappa_2})_{|-\mathbb{H}}$ coincide.*

Proof If $(f^{\kappa_1})_{|-\mathbb{H}}$ and $(f^{\kappa_2})_{|-\mathbb{H}}$ coincide, $f^{\kappa_1} \circ (f_{\kappa_1})^{-1}$ and $f^{\kappa_2} \circ (f_{\kappa_2})^{-1}$ are conformal homeomorphisms in $\mathbb{H}$, and their images of $\mathbb{H}$ are the same quasidisk. Since $0, 1, \infty$ are fixed points of both, they coincide in $\mathbb{H}$ and, therefore, also in the real axis. As f^{κ_1} and f^{κ_2} coincide in the real axis, so f_{κ_1} and f_{κ_2} coincide in this axis, and κ_1, κ_2 are equivalent Beltrami coefficients.

Conversely, if f_{κ_1} and f_{κ_2} coincide in the real axis, the function g equal to $f^{\kappa_1} \circ (f_{\kappa_2})^{-1}$ in $f_{\kappa_2}(\mathrm{cl}\,(-\mathbb{H}))$ and to $f^{\kappa_1} \circ (f_{\kappa_1})^{-1} \circ f_{\kappa_2} \circ (f^{\kappa_2})^{-1}$ in $f_{\kappa_2}(\mathbb{H})$ is a homeomorphism defined in $\mathbb{C}$ which is a conformal homeomorphism in $f^{\kappa_2}(-\mathbb{H})$. It also is a conformal homeomorphism in $f^{\kappa_2}(\mathbb{H})$, because $f^{\kappa_1} \circ (f_{\kappa_1})^{-1}$ and $f_{\kappa_2} \circ (f^{\kappa_2})^{-1}$ are conformal homeomorphisms in $f^{\kappa_2}(\mathbb{H})$. Since $f^{\kappa_2}(\mathbb{R})$ is a quasiconformal curve, has zero measure, and, by the measurable Riemann mapping theorem, differs from the identity in $\mathbb{C}$ by composition at the left by a Möbius transformation, g is a Möbius transformation. As $0, 1, \infty$ are fixed points of this Möbius transformation, it is the identity in $\mathbb{C}$ and, therefore, $(f^{\kappa_1})_{|-\mathbb{H}}$ and $(f^{\kappa_2})_{|-\mathbb{H}}$ coincide. □

Since every equivalence class $[f_\kappa]$ has analytic functions of real variables, every equivalence class $[f^\kappa]$ has analytic functions defined in $\mathbb{H}$, but f^κ can be very irregular on the real axis (e.g., there are quasiregular curves with Hausdorff dimension any value of $[1, 2[$), even if they are analytic as functions of real variables in $\mathbb{H}$ and conformal homeomorphisms in $-\mathbb{H}$.

So, we have another way of seeing the universal Teichmüller space:

(6) *Set of the quasiconformal mappings from $\mathbb{C}$ to $\mathbb{C}$ conformal in $-\mathbb{H}$ with $0, 1, \infty$ fixed.*

$\mathbb{H}$ is trivially a subset of $\mathbb{C}_\infty$ with covering $(\mathbb{H}, G_{1_\mathbb{H}})$. So, the results in Chap. 4 for Teichmüller spaces of subsets of $\mathbb{C}_\infty$ covered by $\mathbb{H}$ apply to the universal Teichmüller space $\mathrm{Teich}(\mathbb{H})$. In particular:

1. $(\mathrm{Teich}(\mathbb{H}), D_{T,\mathbb{H}})$ *is a connected complete metric space;*
2. *There is a Bers embedding,, i.e., a homeomorphism h from* $\mathrm{Teich}(\mathbb{H})$ *with distance $D_{T,\mathbb{H}}$ onto $h(\mathrm{Teich}(\mathbb{H})) \subset B_{3/2}(0) \subset B(-\mathbb{H}, G_{1_\mathbb{H}})$, with the latter the*

Banach space of the quadratic differentials in $\mathbb{H}$ *with the Nehari norm* $\| \ \|_{\mathcal{N}_{-\mathbb{H}}}$ *and* $h(0)=0$, $h(\mathrm{Teich}(\mathbb{H}))$ *is a noncomplete metric subspace of the Banach space* $(B(-\mathbb{H}, G_{\{1_{\mathbb{H}}\}}), \| \ \|_{\mathcal{N}_{-\mathbb{H}}})$ *with elements Schwarzian derivatives of quasiconformal mappings and the largest open ball* $B_r(0) \subset B(-\mathbb{H}, G_{1_{\mathbb{H}}})$ *it contains is with* $r = \frac{1}{2}$.

The geodesics in the universal Teichmüller space with Teichmüller distance can be described with extremal Beltrami coefficients of points in this space. An **extremal Beltrami coefficient of a point in the universal Teichmüller space** p is $\kappa \in B_1 \subset L^\infty(\mathbb{C})$ such that $\|\kappa\|_{L^\infty}$ is the minimum of the possible L^∞ norms of Beltrami coefficients of elements of p.

(9.14) *If* κ *is a extremal Beltrami coefficient of a point* p *in the universal Teichmüller space, then*

$$\kappa_t = \frac{(1+\kappa)^t-(1-|\kappa|)^t}{(1+\kappa)^t+(1-|\kappa|)^t}\,\frac{\kappa}{|\kappa|}, \qquad t\in[0, 1],$$

is extremal Beltrami coefficient of $p_t \in \mathrm{Teich}(\mathbb{H})$ *corresponding to the class of complex dilatations equivalent to* κ_t. *The curve described by* $t \mapsto p_t$ *is a geodesic with end points* 0 *and* p *in the universal Teichmüller space with the Teichmüller distance, and* $D_{T,\mathbb{H}}(p_t, 0) = t\, D_{T,\mathbb{H}}(p, 0)$.

Proof $\kappa_t(z)$ divides the length in the Poincaré metric of B_1 of the line segment with end points 0 and $\kappa(z)$ in the ratio $t:(1-t)$. If f_κ has dilatation K, the function f_{κ_t} has dilatation K^t and $f_\kappa \circ (f_{\kappa_t})^{-1}$ has dilatation K^{1-t}. If $g \in p_t$, then

$$\varphi = f_\kappa \circ (f_{\kappa_t})^{-1} \circ g \in p, \text{ and } K \le K_\varphi \le K^{1-t} K_g .$$

So, $K_g \ge K^t$ and, therefore, κ_t is an extremal Beltrami coefficient. It is $D_{T,\mathbb{H}}(p_t, 0) = \frac{1}{2}\log K^t = t\, D_{T,\mathbb{H}}(p, 0)$. Since $f_\kappa \circ (f_{\kappa_t})^{-1}$ has dilatation K^{1-t}, $D_{T,\mathbb{H}}(p_t, p) \le (1-t)\, D_{T,\mathbb{H}}(p, 0)$ and

$$D_{T,\mathbb{H}}(0, p_t) + D_{T,\mathbb{H}}(p_t, p) = D_{T,\mathbb{H}}(0, p), \quad t\in[0, 1].$$

Repeating the argument for any subarc of $t \mapsto p_t$, we conclude that $t \mapsto p_t$ describes a geodesic. □

9.7 Geodesics in Teichmüller Space of Riemann Surface Covered by $\mathbb{H}$

If S is a Riemann surface with covering $\mathbb{H}$, *the geodesics in* $\mathrm{Teich}(S)$ *with Teichmüller distance can be described as at the end of the preceding section for the universal Teichmüller space* $\mathrm{Teich}(\mathbb{H})$.

The **extremal Beltrami coefficient of a point of** Teich(S) is defined as for points in universal Teichmüller space. *We get the analogue of* (9.14) using the following result analogous to (9.13) established for the universal Teichmüller space, and noting that, with the notation of (9.14), each κ_t , $t \in [0, 1]$, represents a point of Teich(S) , since κ has this property and it is a Beltrami differential in S, i.e., if $(\mathbb{H}, \pi)$ is a universal covering of S and G_π is the covering group, $\kappa = (\kappa \circ g)\, \frac{\overline{g'}}{g'}$ and from the formula for κ_t, it is immediate that κ_t also satisfies the condition defining a Beltrami differential in S. So, κ_t represents a point of Teich(S) .

(9.15) *If S is a Riemann surface with covering by $\mathbb{H}$, Beltrami differentials κ_1, κ_2 are equivalent if and only if the conformal homeomorphisms $(F^{\kappa_1})_{|\mathbb{H}}$ and $(F^{\kappa_2})_{|\mathbb{H}}$ coincide, or if and only if $(F_{\kappa_1})_{|\mathbb{R}}$ and $(F_{\kappa_2})_{|\mathbb{R}}$ coincide, where F_κ is the quasiconformal mapping from $-\mathbb{H}$ to $-\mathbb{H}$ with Beltrami coefficient κ and $0, 1, \infty$ fixed, and F^κ is the quasiconformal mapping in $\mathbb{C}$ with Beltrami coefficient κ in $-\mathbb{H}$ and conformal in $\mathbb{H}$.*

Proof Let $(\mathbb{H}, \pi)$ be a universal covering of S.

If κ_1, κ_2 are equivalent, and $f_{\kappa_1}, f_{\kappa_2}$ with lifts by the considered universal covering of S, resp., $F_{\kappa_1}, F_{\kappa_2}$, there exists a conformal homeomorphism $h : f_{\kappa_1}(S) \to f_{\kappa_2}(S)$ such that $h \circ f_{\kappa_1}$ is homotopic to f_{κ_2} in S. Therefore, by (9.4), if H is the lift of h by the universal covering, $F_{\kappa_2} = H \circ F_{\kappa_1}$ in $\mathbb{R}$. Since H is the Möbius transformation of $\mathbb{H}$ whose extension by continuity to $\mathbb{R}$ has fixed points $0, 1, \infty$, H is the identity and, therefore, $(F^{\kappa_1})_{|\mathbb{H}}$ and $(F^{\kappa_2})_{|\mathbb{H}}$ coincide.

Conversely, if $(F_{\kappa_1})_{|\mathbb{R}}$ and $(F_{\kappa_2})_{|\mathbb{R}}$ coincide, F_{κ_1} and F_{κ_2} induce the same group isomorphism in the universal covering group of S by $-\mathbb{H}$ on a group G',

$$(\pi \circ F_{\kappa_1})(S) = (\pi \circ F_{\kappa_2})(S) = (-\mathbb{H})/G' .$$

By (9.4), $\pi \circ F_{\kappa_1}, \pi \circ F_{\kappa_2}$ are homotopic and, therefore, the Beltrami differentials κ_1, κ_2 are equivalent.

As in the proof of (9.13), we get that $(F_{\kappa_1})_{|\mathbb{R}}$ and $(F_{\kappa_2})_{|\mathbb{R}}$ coincide if and only if $(F^{\kappa_1})_{|\mathbb{H}}$ and $(F^{\kappa_2})_{|\mathbb{H}}$ do. □

(9.16) *Teichmüller spaces of hyperbolic Riemann surfaces with the Teichmüller distance are path connected.*

Proof By the generalization of (9.14) to Teichmüller spaces of hyperbolic Riemann surfaces, with the extremal Beltrami coefficient κ of $p \in$ Teich(S) , define the path $t \mapsto \kappa_t$, $t \in [0, 1]$, of Beltrami coefficients of points $p_t \in$ Teich(S) by the formula in (9.14), leading to a path $t \mapsto p_t$, $t \in [0, 1]$, in Teich(S) with end points 0 and p . Since every $p \in$ Teich(S) can be path connected to 0 in Teich(S) , this space is path connected.

Another path in $\text{Teich}(S)$ connecting 0 to an arbitrary point $p \in \text{Teich}(S)$ with extremal Beltrami coefficient κ is $t \mapsto p_t$, $t \in [0, 1]$, where p_t is the point of $\text{Teich}(S)$ containing a quasiconformal mapping with maximal Beltrami coefficient $t\kappa$. It suffices to check that $t \mapsto p_t$ is continuous in $[0, 1]$, which follows from

$$D_{T,S}(p_{t_1}, p_{t_2}) \leq \frac{1}{2}\left\{\log \frac{1+|t_1-t_2|\,\|\kappa/(1-t_1t_2|\kappa|^2)\|_{L^\infty}}{1-|t_1-t_2|\,\|\kappa/(1-t_1t_2|\kappa|^2)\|_{L^\infty}}\right\}, \qquad t_1, t_2 \in [0, 1],$$

and in the equivalent distance introduced three sections above

$$\beta_{T,S}(p_{t_1}, p_{t_2}) \leq |t_1-t_2| \left\| \frac{\kappa}{1-t_1t_2|\kappa|^2} \right\|_{L^\infty} \leq |t_1-t_2|\, \|\kappa\|_{L^\infty}.$$

□

9.8 Schwarzian Derivatives and Quadratic Differentials

If S is a Riemann surface with covering by $\mathbb{H}$, G is the covering group, and κ is a Beltrami differential in S for G, we know that, with F_κ, the quasiconformal mapping from $\mathbb{H}$ to $\mathbb{H}$ with Beltrami coefficient κ and fixed points $0, 1, \infty$ and F^κ the quasiconformal mapping in $\mathbb{C}$ with Beltrami coefficient κ in $\mathbb{H}$ which is conformal in $-\mathbb{H}$, $\text{Teich}(S)$ can be characterized as the set of conformal homeomorphisms $(F^\kappa)_{|-\mathbb{H}}$. For $g \in G$, $F^\kappa \circ g \circ (F^\kappa)^{-1}$ is a Möbius transformation, and the Schwarzian derivative of $F^\kappa_{|-\mathbb{H}}$ is

$$\mathscr{S}(F^\kappa)_{|-\mathbb{H}} = \mathscr{S}\big((F^\kappa \circ g \circ (F^\kappa)^{-1}) \circ (F^\kappa)_{|-\mathbb{H}}\big) = \mathscr{S}F^\kappa \circ g_{|-\mathbb{H}} = (\mathscr{S}(F^\kappa)_{|-\mathbb{H}} \circ g)(g')^2 .$$

Therefore, the Schwarzian derivative $\mathscr{S}(F^\kappa)_{|-\mathbb{H}}$ is a quadratic differential in S for the group G acting on $\mathbb{H}$.

(9.17) *If S is a Riemann surface with covering by $\mathbb{H}$, G is the covering group, κ is a Beltrami differential in S for G, F_κ is the quasiconformal mapping from $\mathbb{H}$ to $\mathbb{H}$ with Beltrami coefficient κ and fixed points $0, 1, \infty$ and F^κ is the quasiconformal mapping in $\mathbb{C}$ with Beltrami coefficient κ in $\mathbb{H}$ and conformal in $-\mathbb{H}$, then:*

1. $\mathscr{S}(F^\kappa)_{|-\mathbb{H}}$ is a quadratic differential in S for G.
2. $F^\kappa \circ g \circ (F^\kappa)^{-1}$ is a Möbius transformation in $f^\kappa(-\mathbb{H})$ to $g \in G$.

$2 \iff 3$. $F_\kappa \circ g \circ (F_\kappa)^{-1}$ is a Möbius transformation in $\mathbb{R}$ for $g \in G$.

Proof 1 and the first condition in 2 follow directly from the equalities preceding the statement.

If this condition holds, as $h = F^\kappa \circ g \circ (F^\kappa)^{-1}$ is a Möbius transformation in $f^\kappa(-\mathbb{H})$ for $g \in G$, $H = F_\kappa \circ (F^\kappa)^{-1} \circ h \circ F^\kappa \circ (F_\kappa)^{-1}$ is a Möbius transformation from $\mathbb{H}$ to $\mathbb{H}$ which coincides with $F_\kappa \circ g \circ (F_\kappa)^{-1}$ in $\mathbb{R}$. Therefore, the last condition in 2 also holds. Conversely, if this condition holds, $F_\kappa \circ g \circ (F_\kappa)^{-1}$ is a Möbius transformation in $\mathbb{R}$, and $h = F^\kappa \circ g \circ (F^\kappa)^{-1}$ is a homeomorphism of $\mathbb{C}$ onto $\mathbb{C}$ conformal in $F^\kappa(-\mathbb{H})$ and also in $F_\kappa(\mathbb{H})$. Thus, h is a Möbius transformation and the first condition in 2 holds. □

Let G be a group of transformations acting on $\mathbb{H}$.

Analogously to the quadratic differentials in Chap. 4, consider the set of functions f holomorphic in $-\mathbb{H}$ quadratic differentials for G such that

$$\|f\|_{\mathcal{N}_{-\mathbb{H}}} \stackrel{def}{=} \sup_{z \in -\mathbb{H}} (\mathcal{I}m z)^2 |f(z)| \in \mathbb{R}.$$

With the usual operations, this is a normed complex space with this norm, also called **Nehari norm**, defined by considering the element of length of the Poincaré metric of $-\mathbb{H}$, $\frac{|dz|}{|\mathcal{I}m\, z|}$ and so as to be invariant under conformal mappings from $-\mathbb{H}$ to $-\mathbb{H}$, denoted $\boldsymbol{Q(G)}$. The relationship with the notation for quadratic differentials in Chap. 4 is $Q(1_{\mathbb{H}}) = B(-\mathbb{H}, G)$.

If G_1, G_2 are groups of transformations acting on $\mathbb{H}$ and G_1 is a subgroup of G_2, then $Q(G_2)$ is a linear subspace of $Q(G_1)$. The spaces $Q(G)$ with G a group of transformations acting on $\mathbb{H}$ are linear subspaces of the Banach space $Q(1_{\mathbb{H}})$. If $\{f_n\}$ is a Cauchy sequence in $Q(G)$, since $Q(1_{\mathbb{H}})$ is a Banach space, it has limit $f \in Q(1_{\mathbb{H}})$. For $g \in G$, $\lim_{n\to+\infty} f_n \circ g = f \circ g$ uniformly in compact subsets of $\mathbb{H}$. Thus,

$$(f \circ g)(g')^2 = \lim_{n\to+\infty} (f_n \circ g)(g')^2 = \lim_{n\to+\infty} f_n = f,$$

and, therefore, $f \in Q(G)$. Consequently, every space $Q(G)$ is complete, and, therefore, it is a Banach subspace of $Q(1_{\mathbb{H}})$.

The elements of $Q(1_{\mathbb{H}})$ and, therefore, also those of $Q(G)$ with G a group of transformations acting on $\mathbb{H}$, are locally injective Schwarzian derivatives of meromorphic functions in $\mathbb{H}$. Consider the set

$$U(G) = \{\mathscr{S} f \in Q(G) \colon f \text{ is Univalent in } -\mathbb{H}\}.$$

It is $U(G) = U(1_{\mathbb{H}}) \cap Q(G)$. By the preceding paragraph, $Q(G)$ is a closed subspace of the Banach space $Q(1_{\mathbb{H}})$, and from what was seen in Chap. 4, $U(1_{\mathbb{H}})$ is a closed subset of $Q(1_{\mathbb{H}})$. So, *$U(G)$ is a closed subset of $Q(1_{\mathbb{H}})$*.

Also from Chap. 4, for f a univalent function in $-\mathbb{H}$, $\|\mathscr{S} f\|_{\mathcal{N}_{-\mathbb{H}}} \le \frac{3}{2}$. So, by (9.17), the elements of $U(G)$ are the Schwarzian derivatives of the univalent functions f in $-\mathbb{H}$ such that $f \circ g \circ f^{-1}$ is a Möbius transformation in $f(-\mathbb{H})$ for $g \in G$, i.e., every univalent function on $-\mathbb{H}$ with $\mathscr{S} f \in U(G)$ induces an isomorphism of G onto the group $G(f) = \{f \circ g \circ f^{-1} : g \in G\}$ of the Möbius transformations acting on $f(-\mathbb{H})$.

Denote $\boldsymbol{T(G)}$ the subset of $U(G)$ of the Schwarzian derivatives $\mathscr{S} f \in U(G)$ of functions f that can be extended to a quasiconformal mapping in $\mathbb{C}$ with Beltrami coefficient a Beltrami differential in the Riemann surface S for G.

$T(G)$ is the image of Teich(S) *by the function mapping the point of* Teich(S) *corresponding to the Beltrami differentials in S equivalent to a Beltrami differential* κ *in S in the Schwarzian derivative* $\mathscr{S}(f^{\kappa})_{|-\mathbb{H}}$, where f^{κ} is the quasiconformal mapping from $\mathbb{C}$ to $\mathbb{C}$ with Beltrami coefficient κ in $\mathbb{H}$ and conformal in $-\mathbb{H}$.

Since the restriction to $[\text{Teich}(S)]^2$ of the Teichmüller distance in the universal Teichmüller space Teich$(\mathbb{H})$ is upper bounded by the Teichmüller distance in Teich(S), i.e., $D_{T,\mathbb{H}} \leq D_{T,S}$ in S^2, *the function of* Teich(S) in $Q(G)$ *considered in the preceding period is continuous.*

(9.18) *If G is a group of transformations acting on* $\mathbb{H}$ *and* $B_{\frac{1}{2}}(0) \subset Q(G)$,

$$T(G) \supset B_{\frac{1}{2}}(0).$$

Proof For $\varphi \in B_{1/2}(0) \subset Q(G)$, let $\kappa_{\varphi}(z) = -2(\mathcal{I}m\, z)^2 \varphi(\overline{z})$. By the Ahlfors-Weill section theorem (4.14), $\varphi \mapsto [\kappa_{\varphi}]$ is the inverse function in $B_{1/2}(0) \subset Q(G)$ of $[\kappa] \mapsto \mathscr{S}(f^{\kappa})_{|-\mathbb{H}}$. Thus, $B_{1/2}(0) \subset T(G)$. □

9.9 Equivalence of Teichmüller Spaces Metrics

For Riemann surfaces with covering by $\mathbb{H}$ the Teichmüller distance of each Teichmüller space is uniformly equivalent to the restriction to that space of the Teichmüller distance of the universal Teichmüller space, as shown below, narrowing the relationship of Teichmüller spaces of different Riemann surfaces and allowing the generalization of several properties obtained for the universal Teichmüller space.

(9.19) *If S is a hyperbolic Riemann surface, the Teichmüller distance* $D_{T,S}$ *in* Teich(S) *is uniformly equivalent to the restriction to this space of the Teichmüller distance* $D_{T,\mathbb{H}}$ *in the universal Teichmüller space* Teich$(\mathbb{H})$, *i.e., the identity from the metric space* $\big(\text{Teich}(S), D_{T,S}\big)$ *to the metric space* $\big(\text{Teich}(S), D_{T,\mathbb{H}\,|\text{Teich}(S)}\big)$ *is uniformly continuous as well as its inverse. For* $p \in \text{Teich}(S)$,

$$\overline{\lim_{q\to p}} \frac{D_{T,S}(p,q)}{D_{T,\mathbb{H}}(p,q)} \leq 3.$$

For each bounded $A \subset \text{Teich}(S)$,

$$D_{T,S}(p,q) \leq 3\big[1 + \text{diam}_{D_{T,S}}(A)\big] D_{T,\mathbb{H}}(p.q),$$

where $\text{diam}_{D_{T,S}}(A)$ *is the diameter of A with the distance* $D_{T,S}$.

Proof Consider a covering of S by $\mathbb{H}$ with covering group G. If $p \in \text{Teich}(S)$, κ is a Beltrami coefficient in S such that $p = [\kappa]$ and f^κ is the quasiconformal mapping in $\mathbb{C}$ with Beltrami coefficient κ in $\mathbb{H}$ and extension by continuity to $\mathbb{R}$ with fixed points $0, 1, \infty$, as $\text{Teich}(S)$ is a subset of the universal Teichmüller space $\text{Teich}(\mathbb{H})$, by (4.10.8), $\|\mathscr{S}(f^\kappa)_{|-\mathbb{H}}\|_{\mathcal{N}_{-\mathbb{H}}} \leq \sigma_0(-\mathbb{H})\,\|\kappa\|_{L^\infty}$. As $\beta_{T,\mathbb{H}}(p,0) = \|\kappa\|_{L^\infty}$ and, analogously to (4.13), we obtain for $g : -\mathbb{H} \to \mathbb{C}_\infty$ univalent $\|\mathscr{S} g\|_{\mathcal{N}_{-\mathbb{H}}} \leq \frac{3}{2}$, and $\sigma_0(-\mathbb{H}) \leq \frac{3}{2}$, $\|\mathscr{S}(f^\kappa)_{|-\mathbb{H}}\|_{\mathcal{N}_{-\mathbb{H}}} \leq \frac{3}{2}\,\beta_{T,\mathbb{H}}(p,0)$. On the other hand, if $\varphi \in Q(G)$ with $\|\varphi\|_{\mathcal{N}_{-\mathbb{H}}} \geq \frac{1}{2}$ and $\kappa_\varphi(z) = -2(\mathcal{I}m\, z)^2 \varphi(\overline{z})$, since $(\varphi \circ g)(g')^2 = \varphi$, and $\mathcal{I}m\, z = \frac{\mathcal{I}m\, g(z)}{|g'(z)|}$ for $g \in G$, κ_φ is a Beltrami differential in S for G, and therefore $[\kappa_\varphi] \in \text{Teich}(S)$. By the Ahlfors-Weill section theorem (4.14), $\varphi \mapsto [\kappa_\varphi]$ is the inverse function in $B_{1/2} \subset Q(G)$ of $[\kappa] \mapsto \mathscr{S}(f^\kappa)_{|-\mathbb{H}}$, and $\|\mathscr{S}(f^\kappa)_{|\mathbb{H}}\|_{\mathcal{N}_{-\mathbb{H}}} \geq \frac{1}{2}\,\beta_{T,S}(p,0)$, $\beta_{T,S}(p,0) \leq 3\,\beta_{T,\mathbb{H}}(p,0)$, which, since $\beta_{T,X} = \tanh D_{T,X}$ and tanh are strictly increasing, is equivalent to the first inequality in the statement for $q = 0$.

To get this inequality with $q \neq 0$, fix $p = [\kappa] \in \text{Teich}(S)$ and consider the function $\widetilde{\alpha}_\kappa$ mapping a Beltrami coefficient μ in S to a Beltrami coefficient $\widetilde{\alpha}_\kappa(\mu)$ in S such that $f_{\widetilde{\alpha}_\kappa(\mu)} = f_\mu \circ (f_\kappa)^{-1}$, where f_κ is the quasiconformal mapping from $\mathbb{H}$ to $\mathbb{H}$ obtained as a solution of the Beltrami equation in $\mathbb{C}$ with Beltrami coefficient equal to κ in $\mathbb{H}$ and to 0 in $-\mathbb{H}$, and with $0, 1, \infty$ fixed points. The function $\widetilde{\alpha}_\kappa$ induces the function $[\mu] \mapsto \alpha_\kappa([\mu]) = [\widetilde{\alpha}_\kappa(\mu)]$, which is an isometry of the universal Teichmüller space onto itself, and it is also an isometry of $\text{Teich}(S)$ onto $\text{Teich}(S^\kappa)$ with the, resp., Teichmüller distances, where $S^\kappa = \mathbb{H}/G^\kappa$, with $G^\kappa = \{f_\kappa \circ g \circ (f_\kappa)^{-1} : g \in G\}$.

The function α_κ maps $p = [\kappa]$, $q = [\mu] \in \text{Teich}(S)$ to, resp., 0 and $[\nu] = \alpha_\kappa([\mu]) \in \text{Teich}(S^\kappa)$. So, for $p, q \in \text{Teich}(S)$,

$$\beta_{T,\mathbb{H}}(p,q) \leq \beta_{T,S}(p,q) = \beta_{T,S^\kappa}(0,[\nu]) \leq 3\,\beta_{T,\mathbb{H}}(0,[\nu]) = 3\,\beta_{T,\mathbb{H}}(p,q)\,.$$

The distances $\beta_{T,S}$ and $\beta_{T,\mathbb{H}}$ are uniformly equivalent, and since they are compositions of the function tanh with the corresponding Teichmüller distances, the Teichmüller distance $D_{T,s}$ of $\text{Teich}(S)$ and the restriction to this space of the universal Teichmüller space distance are uniformly equivalent. As the range of the real function tanh is $]-1, 1[$, $\tanh^{-1} t = \frac{1}{2}\log\frac{1+t}{1-t}$, $t \mapsto \frac{1+t}{1-t}$ is strictly increasing for $t \in [0, 1[$, and $\tanh x \leq x$ for $x \geq 0$,

$$D_{T,S} = \tanh^{-1}\beta_{T,S} = \frac{1}{2}\log\frac{1+\beta_{T,S}}{1-\beta_{T,S}} \leq \frac{1}{2}\log\frac{1+3\,\beta_{T,\mathbb{H}}}{1-3\,\beta_{T,\mathbb{H}}} \leq \frac{1}{2}\log\frac{1+3\,D_{T,\mathbb{H}}}{1-3\,D_{T,\mathbb{H}}}\,.$$

Since $\log x \leq x - 1$ for $x > 0$, $D_{T,S} \leq \frac{1}{2}\left(\frac{1+3\,D_{T,\mathbb{H}}}{1-3\,D_{T,\mathbb{H}}} - 1\right) = \frac{3\,D_{T,\mathbb{H}}}{1-3\,D_{T,\mathbb{H}}}$, as $D_{T,\mathbb{H}} \in \left[0, \frac{1}{3}\right[$, and $\frac{D_{T,S}}{D_{T,\mathbb{H}}} \leq 3$. Thus, we have the upper bound of the limit in the statement.

As $t \mapsto \frac{3}{1-3t}$ is strictly increasing for $t \in [0, \frac{1}{3}[$, by the inequality in the preceding paragraph, $D_{T,S} \leq \frac{3\,D_{T,\mathbb{H}}}{1-3\,D_{T,\mathbb{H}}}$, for $D_{T,\mathbb{H}} \leq a < \frac{1}{3}$, $D_{T,S} \leq \frac{3}{1-3a}\,D_{T,\mathbb{H}}$.

If $A \subset \text{Teich}(S)$ is bounded and $p, q \in A$, then for $D_{T,\mathbb{H}} \leq a < \frac{1}{3}$, the last inequality in the preceding paragraph holds, and for $D_{T,\mathbb{H}} > a$, with $a = \frac{\text{diam}_{D_{T,S}}(A)}{3[1+\text{diam}_{D_{T,S}}(A)]}$,

$$D_{T,S}(p,q) \le \mathrm{diam}_{D_{T,S}}(A) < \tfrac{1}{a}\mathrm{diam}_{D_{T,S}}(A) D_{T,S}(p,q)$$
$$\le 3\big[1+\mathrm{diam}_{D_{T,S}}(A)\big] D_{T,\mathbb{H}}(p,q).$$

□

As a result, since the metric space $\big(\mathrm{Teich}(S), D_{T,S}\big)$ is complete, for Riemann surfaces S with covering by $\mathbb{H}$: Teich(S) *is a closed subset of the universal Teichmüller space* $\big(\mathrm{Teich}(\mathbb{H}), D_{T,\mathbb{H}}\big)$.

As homeomorphisms preserve the property of a set being closed: $T(G)$ *is a closed subset of* $T(1_{\mathbb{H}})$.

9.10 Bers Embedding

What was proved in the preceding section allows to easily generalize to Teichmüller spaces of hyperbolic Riemann surfaces the Bers embedding obtained in Chap. 4 for the particular case considered there.

(9.20) *If S is a Riemann surface with covering by $\mathbb{H}$ with covering group G, then $[\kappa] \mapsto \mathscr{S}(f^{\kappa})_{|-\mathbb{H}}$, with f^{κ} the quasiconformal mapping in $\mathbb{C}$ with Beltrami coefficient κ in $\mathbb{H}$ and extension by continuity to $\mathbb{R}$ with fixed points $0, 1, \infty$, is a homeomorphism from $\big(\mathrm{Teich}(S), D_{T,S}\big)$ onto the metric space $T(G)$ with the distance of the Nehari norm in $-\mathbb{H}$.*

Proof By the Bers embedding (4.16) for the particular case considered in Chap. 4, $[\kappa] \mapsto \mathscr{S}(f^{\kappa})_{|-\mathbb{H}}$ is a homeomorphism from Teich(S) with a restriction to this space of the Teichmüller distance of the universal Teichmüller space over $T(G)$. As established in the preceding section, this distance is equivalent to the Teichmüller distance $D_{T,S}$ in Teich(S). □

By (9.18), for G a group of transformations acting on $\mathbb{H}$, $T(G)$ contains the ball of the normed space $Q(G)$ with radius $\frac{1}{2}$ and center 0. With the Bers embedding, we can prove that for any point of $T(G)$, there is an open ball centered at the point and contained in $T(G)$ in that normed space. So, $T(G)$ is an open subset of the normed space $Q(G)$. The first proof of this property was in 1965 by L. Bers.

(9.21) *If G is a group of transformations acting on $\mathbb{H}$, then $T(G)$ is an open subset of normed space $Q(G)$, an open and closed subset of $Q(G) \cap T(1_{\mathbb{H}})$, and a connected component of this space containing 0.*

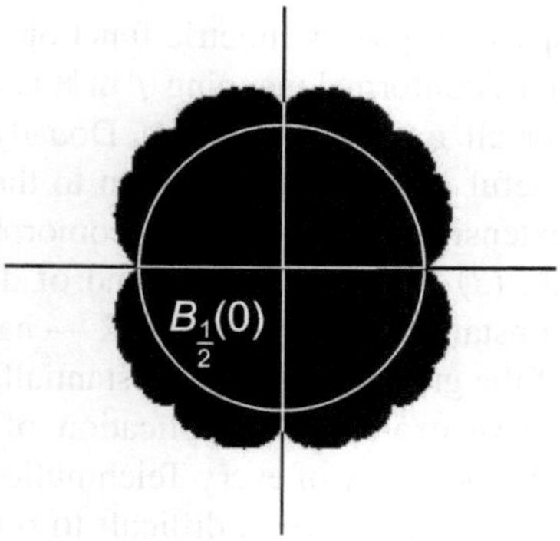

Fig. 9.1 Image of the Teichmüller space $T(G)$ of a torus with 1 puncture by the Bers embedding (in black), $\partial B_{\frac{1}{2}}(0) \subset T(G)$

Proof Let $\psi \in T(1_{\mathbb{H}})$ be such that $\psi\big(\mathscr{S}(f^\kappa)_{|-\mathbb{H}}\big) = \mathscr{S}\widetilde{\alpha}_\kappa(\mu)$, where f^κ is as in the hypothesis of the preceding result, $\widetilde{\alpha}_\kappa$, α_κ are as in the second paragraph of the proof of (9.19), and h is the Bers embedding of the universal Teichmüller space onto $T(1_{\mathbb{H}})$; thus, $\psi = h \circ \alpha_\kappa \circ h^{-1}$. So, ψ is a homeomorphism of $T(1_{\mathbb{H}})$ onto itself, $\psi\big(T(G)\big) = T(G^\kappa)$ and $\psi\big(Q(G) \cap T(1_{\mathbb{H}})\big) = Q(G^\kappa) \cap T(1_{\mathbb{H}})$. If $B_r(0)$ is an open ball of $Q(G^\kappa)$ such that $T(G^\kappa) \supset B_r$, for some neighborhood of 0 in $T(1_{\mathbb{H}})$, it is $B_r(0) = Q(G^\kappa) \cap T(1_{\mathbb{H}}) \cap V_0$. Since $T(1_{\mathbb{H}})$ is an open subset of $Q(1_{\mathbb{H}})$, the set $\psi^{-1}(V) = Q(G) \cap \psi^{-1}(V_0)$ is a neighborhood of $\mathscr{S}(f^\kappa)_{|-\mathbb{H}}$ in $Q(G)$, included in $\psi^{-1}\big(T(G^\kappa)\big)_{|f(\mathbb{H})} = T(G)$. So, $T(G)$ is an open subset of $Q(G)$.

Since $T(G)$ is an open subset of $Q(G)$, and, as seen at the end of the preceding section, it is a closed subset of $T(1_{\mathbb{H}})$, it is an open and closed subset of $Q(G) \cap T(1_{\mathbb{H}})$. Since $T(G)$ is connected, it is a connected component of the set $Q(G) \cap T(1_{\mathbb{H}})$, and it is clear that $0 \in T(G)$. □

In 1972, L. Bers observed that given a Riemann surface S with covering $\mathbb{H}$ and a quadratic differential in $Q(G)$, it was not known whether it belonged or not to the image $T(G)$ of the Teichmüller space of S. In 1979, R.M. Porter published a first algorithm for computing numerically points of the boundary of $T(G)$ for S a torus with 1 puncture, but a good algorithm for this appeared only in 2004, developed by Y. Komori and T. Sugawa. In 2006, Y. Komori, T. Sugawa, M. Wada, and Y. Yamashita[8] presented a more direct algorithm that for S a torus with 1 puncture gives the image $T(G)$ of Teich(S) by the Bers embedding in Fig. 9.1.[9]

In 1985, following partial contributions by I. Kra in 1969 and O. Lehto in 1974, P. Tukia proved with a long and deep proof the property with remarkable consequences that every quasisymmetric function h with fixed points $0, 1, \infty$ such that $h \circ g \circ h^{-1}$ is the restriction of a Möbius transformation for all $g \in G$, where G is a group of transformations acting on $\mathbb{H}$ with the metric inherited from the

[8] Komori, Y., Sugawa, T., Wada, M., Yamashita, Y., Drawing Bers embeddings of the Teichmüller space of once-punctured tori, *Experimental Mathematics* **15** (2006), 51–60. Porter, R. Michael. Komori, Yohei. Sugawa, Toshiyuki (1965-). Wada, Masaaki. Yamashita, Yasushi.

[9] Author and date of the base figure: Café Bene (31.03.2015), under license *Creative Commons Attribution-Share Alike 3.0 Unported* (https://commons.wikimedia.org/wiki/File:Bers-Einbettung.png), in which colors were eliminated, passing $T(G)$ for black and adding the circle $\partial B_{1/2}$.

space of quasisymmetric functions with fixed points $0, 1, \infty$, has an extension to a quasiconformal mapping f in $\mathbb{H} \cup \mathbb{R}$ with $f \circ g \circ f^{-1}$ a conformal homeomorphism for all $g \in G$. In 1986, A. Douady and C. Earle[10] published a more explicit and useful proof that in addition to the three basic properties of the Beurling-Ahlfors extension (3.3) — *(1)* Diffeomorphism, *(2)* Lipschitzian in the Poincaré metric of $\mathbb{H}$, *(3)* With an upper bound of dilatation depending only on the quasisymmetry constant of the function in $\mathbb{R}$ — has the property: *(4)* Compatibility with the action of the group G, which substantially widens the usefulness of the extension.

An example of application of the **Douady-Earle extension** is the proof of contractibility of every Teichmüller space, i.e., that it can be continuously deformed to a point, which is difficult to prove because if κ and μ are equivalent Beltrami differentials $t\kappa$ and $t\mu$, for $t \in]0, 1]$, may not be equivalent, as shown in 1970 by F. Gehring, though the contractibility of the universal Teichmüller space, proved in 1967 by C. Earle and J. Eells,[11] can be proved with the Beurling-Ahlfors extension without using the Douady-Earle extension. But the Douady-Earle extension also allows to prove $T(G) = Q(G) \cap T(1_{\mathbb{H}})$, establishing a closer relationship of any Teichmüller space with the universal Teichmüller space immediately giving properties such as $T(G)$ being an open subset of $T(1_{\mathbb{H}})$ and a closed subset of $Q(G)$ that were proved above with an alternative, but much longer, argument.

9.11 Foliations

In what follows, we use the notion of foliation of a one-dimensional complex differential manifold M, as its decomposition in real one dimensional analytic manifolds—the foliation leaves—locally diffeomorphic to intervals in $\mathbb{R}$ such that there are holomorphic functions defined in some neighborhood of every point of M to sets $I \times J$ with $I, J \subset \mathbb{R}$ open intervals with the image of each leaf a fixed set $I \times \{y\}$ for $y \in J$.

A **foliation** of a one-dimensional complex differential manifold M is a set $\mathcal{F}$ of coordinate systems $\{(U_a, \varphi_a)\}_{a \in A}$ of M with $\cup_{a \in A} U_a = M$, $\varphi_a(U_a) = U_{a1} \times U_{a2}$ with $U_{a1}, U_{a2} \subset \mathbb{R}$ open sets, and if $U_a \cap U_b \neq \emptyset$, the change of coordinates $\varphi_b \circ \varphi_a^{-1} : \varphi_a(U_a \cap U_b) \to \varphi_b(U_a \cap U_b)$ is a holomorphic function of the form

$$\varphi_b \circ \varphi_a^{-1}(x, y) = \big(h_1(x, y), h_2(y)\big).$$

$\{(U_a, \varphi_a)\}_{A \in A}$ is a set of coordinate systems of one-dimensional real analytic manifolds, called the **leaves** of M.

Given foliations $\mathcal{F}$ and $\mathcal{F}'$ of a Riemann surface X, the existence of a homeomorphism of X onto X isotopic to the identity mapping $\mathcal{F}$ to $\mathcal{F}'$ defines an equivalence relation $\mathcal{F} \sim \mathcal{F}'$.

On a Riemann surface S, consider the **holomorphic quadratic differentials** $\varphi(z)\, dz^2$ in S, with $\varphi \in H(S)$. Such a holomorphic quadratic differential in S

[10] Douady, A., Earle, C., Conformally natural extension of homeomorphisms of the circle, *Acta Mathematica,* **157** (1986), 23–48.

[11] Eells, James (1926–2007).

determines a foliation with leaves tangent to the vectors $\mathbf{v}$ such that $\varphi(\mathbf{v}) > 0$. A holomorphic quadratic differential in S is said to be **integrable** if $\|\varphi\| = \int_S |\varphi|$ is finite.

The set of integrable holomorphic quadratic differentials in S is a Banach space with the norm $\|\varphi\|$ (equal to the area of S in the metric with element of length $\sqrt{|\varphi(z)|\,dz}$), denoted $\boldsymbol{Q(S)}$.

A Riemann surface S is said to be of **finite type** (g, n) if there is a compact Riemann surface $\widetilde{S}$ of genus g and a set E, called the set of **punctures** of S, with $\#E = n \in \mathbb{N}$ and such that $S = \widetilde{S} \setminus E$.

(9.22) *Let S be a Riemann surface of finite type (g, n). If $\varphi(z)\,dz^2$ is an integrable quadratic differential in S, the singularity of φ at each puncture of S is removable or a simple pole.*

Proof With coordinates z in a small neighborhood V of a puncture p with the origin at p and $z \in B_\varepsilon$ with $\varepsilon > 0$ small and $m \in \mathbb{Z}$, the integral

$$\int_{B_\varepsilon} \frac{1}{|x+iy|^m}\,dx\,dy = \int_0^{2\pi}\int_0^{\varepsilon} \frac{1}{r^m}\,r\,dr\,d\theta$$

is finite if and only if $m \leq 1$. □

As in (4.15): *If S is a Riemann surface of finite type (g, n), the complex linear space $Q(S)$ of quadratic differentials in S has finite dimension* $3(g-1)+n$ *except if* $(g=0, n \in \{0, 1, 2\})$ *or* $(g=1)$, *when it is, resp.,* 0 *or* $1+n$.

(9.23) *If S is an oriented compact complex manifold,* $\mathrm{Teich}(S)$ *is homeomorphic to a closed ball of* $\mathbb{C}^n$ *with* $n = \dim Q(S)$.

Proof If S is a compact Riemann surface of genus $g \geq 2$, it can be decomposed by $3(g-1)$ Jordan curves in $2(g-1)$ "pants" (Fig. 9.2[12]). These Jordan curves can be chosen as geodesics whose lengths determine "pants" up to isometry. To reconstruct S from the "pants," they have to be "glued" at the "waist" and at the "legs"; so, 1 parameter is needed for reversing orientation. Therefore, $\mathrm{Teich}(S)$ is parameterized by $\mathbb{R}^{2[3(g-1)]}$ or by $\mathbb{C}^{3(g-1)}$ and, by what was seen in the paragraph preceding the statement of this result, $\dim \mathbb{C}^{3(g-1)} = 3(g-1) = \dim Q(S)$.

[12] Author and date of figure: Oleg Alexandrov (07.09.2007), under License *Public Domain CC* (https://commons.wikimedia.org/wiki/File:Triple_torus_illustration.png) to which curves have been added.

Fig. 9.2 Decomposition of compact Riemann surface of genus $g=3$ by $3(g-1)=6$ geodesic Jordan curves in $2(g-1)=4$ "pants"

It is similar for manifolds with boundary or with a smaller genus. □

A **measurable foliation** of a one-dimensional complex manifold M is a foliation $\mathcal{F}$ with an associated measure α on the transversal curves to $\mathcal{F}$ preserved by the natural functions defined by transversals to the leaves.

Given a measurable foliation $\mathcal{F}$ of a Riemann surface S and a conformal metric with element of length $\rho(z)\,dz$ in S, the **length of the foliation** $\mathcal{F}$ is

$$\ell_\rho(\mathcal{F}) = \int_{\mathcal{F}} \rho(z)\,|dz| = \iint_S \rho(z)\,\alpha(w)\,|dz|\,|dw|\,.$$

The **extremal length of a foliation** $\mathcal{F}$ of S is

$$\lambda_S(\mathcal{F}) = \frac{\inf_{\mathcal{F}'\sim\mathcal{F}} \ell_\rho^2(\mathcal{F}')}{\int_S \rho^2(x+iy)\,dx dy}\,.$$

A measurable foliation $\mathcal{F}$ is said to be a **geodesic foliation** or an **extremal foliation** for the metric with element of length $\rho(z)\,|dz|$ if it minimizes, resp., $\ell_\rho(\mathcal{F}')$ or $\lambda_S(\mathcal{F}')$, in the class of foliations $\mathcal{F}'\sim\mathcal{F}$.

(9.24) *If $\varphi \in Q(S)\setminus\{0\}$ is a holomorphic quadratic differential in a Riemann surface S of finite type and $\rho=\sqrt{|\varphi|}$, then for every foliation of S and metric with element of length $\widetilde{\rho}(z)\,|dz|$ such that the areas of S in the two metrics defined by ρ and $\widetilde{\rho}$ are equal, $\ell_{\widetilde{\rho}}\big(\mathcal{F}(\varphi)\big) \le \ell_{\widetilde{\rho}}\big(\mathcal{F}(\varphi)\big)$, with equality if and only if $\widetilde{\rho}=c\rho$ for some $c>0$.*

Proof By the Cauchy-Schwarz inequality,

$$\ell_{\widetilde{\rho}}^2(\mathcal{F}) = \left(\int_{\mathcal{F}} \widetilde{\rho}\right)^2 = \left(\int_S \widetilde{\rho}\,\rho\right)^2 \le \left(\int_S \widetilde{\rho}\right)\left(\int_S \rho\right) = \mathrm{Area}_\rho^2(S)\,,$$

with equality if and only if there is $c>0$ such that $\widetilde{\rho} = c\rho$. □

(9.25) *If $\varphi \in Q(S)\backslash\{0\}$ is a holomorphic quadratic differential in a Riemann surface S of finite type, $\rho = \sqrt{|\varphi|}$ and $\mathcal{F}(\varphi)$ are the foliation determined by φ, then $\lambda_S(\mathcal{F}(\varphi)) = \|\varphi\|$ and only $c\rho$ with $c > 0$ define extremal lengths of $\mathcal{F}(\varphi)$.*

Proof $\mathcal{F} = \mathcal{F}(\varphi)$ is the unique geodesic foliation, because ρ defines a metric with curvature ≤ 0, since if $\mathcal{F}' = f(\mathcal{F})$ with f isotopic to the identity, defining $g: S \to S$ mapping each leaf L' of $\mathcal{F}'$ to the corresponding leaf $L = f^{-1}(L')$ of $\mathcal{F}$ by the projection on the nearest point in the metric with element of length $\rho(z)\,|dz|$, and since the metric with this element of length has curvature ≤ 0, the projection does not increase distances. Thus,

$$\lambda_S(\mathcal{F}) \geq \frac{\inf_{\mathcal{F}' \sim \mathcal{F}} \ell_\rho^2(\mathcal{F}')}{\text{Area}_\rho(S)} \geq \frac{\ell_\rho^2(\mathcal{F})}{\text{Area}_\rho(S)} = \text{Area}_\rho(S) = \|\varphi\|\,.$$

By the preceding result,

$$\lambda_S(\mathcal{F}) = \sup_{\widehat{\rho}} \frac{\inf_{\mathcal{F}' \sim \mathcal{F}} \ell_{\widehat{\rho}}^2(\mathcal{F}')}{\text{Area}_\rho(S)} \leq \sup_{\widehat{\rho}} \frac{\ell_{\widehat{\rho}}^2(\mathcal{F})}{\text{Area}_\rho(S)} \leq \frac{\ell_\rho^2(\mathcal{F})}{\text{Area}_\rho(S)} = \|\varphi\|\,,$$

and $\lambda_S(\mathcal{F}) = \|\varphi\|$, and the condition holds with equality. □

Geodesic foliations may not be unique, e.g., for a torus $S = \mathbb{Z} \oplus \tau\mathbb{Z}$ with $\varphi = |dz|^2 \in Q(S)$, $\mathcal{F} = \mathcal{F}(\varphi)$ and $\rho = \sqrt{|\varphi|}$, the projection on the second component $\pi: S \to S^1 = \mathbb{R}/(\mathcal{I}m\,\tau\mathbb{Z})$, whose fibers are the leaves of $\mathcal{F}$, the transversal measure of $\mathcal{F}$ induces a measure α in S^1, and, conversely, each measure α' in S^1 determines a measurable foliation $\mathcal{F}'$ with the same leaves of $\mathcal{F}$. If the total measure of α and α' is the same, then $\mathcal{F}' \sim \mathcal{F}$ and the two foliations are geodesic for the metric defined by ρ.

A measurable foliation is said to be **uniquely ergodic** if it admits a unique transversal invariant measure modulo multiplication by positive constants. *If $\mathcal{F}(\varphi)$, with $\varphi \in Q(S)$, is uniquely ergodic, e.g., an "irrational" foliation of S, then it is the unique geodesic measurable foliation in the metric with element of length $\sqrt{|\varphi(z)|}\,dz$.*

A **Teichmüller mapping** is a quasiconformal mapping $f: S_1 \to S_2$ between Riemann surfaces of finite type for which there are $c \in [0, 1[$ and $\varphi \in Q(S)$ such that the Beltrami coefficient of f is $\kappa_f = c\,\frac{\overline{\varphi}}{|\varphi|}$.

(9.26) Teichmüller theorem: *If S_1 is a Riemann surface of finite type, in every homotopy class of orientation preserving homeomorphisms of S_1 onto a Riemann surface S_2, there is 1 unique Teichmüller mapping $f: S_1 \to S_2$, which is an extremal quasiconformal mapping in its homotopy class.*

Proof Consider the function $\pi : Q(S_1)_1 \to \text{Teich}(S_1)$, where $Q(S_1)_1$ is the open ball with radius 1 and center 0 in $Q(S_1)$, such that $\pi(0) = [1_{s_1}]$ and for $\varphi \neq 0$, $\pi(\varphi) = f$ with $f : S_1 \to S_2$ the quasiconformal mapping with Beltrami coefficient $\kappa_f = \frac{\|\varphi\|}{|\varphi|}\,\overline{\varphi}$ and S_2 the result of an expansion along the leaves of the foliation $\mathcal{F}(\varphi)$ determined for each $\varphi \in Q(S_1)_1$ by the factor $\frac{1+\|\varphi\|}{1-\|\varphi\|}$. This function $f : S_1 \to S_2$ is a Teichmüller function.

By the preceding result, f is an extremal quasiconformal mapping in its homotopy class and $D_T\big(S_1, \pi(S_1)\big) = \frac{1}{2}\log K_f$. By uniqueness and as f determines φ, we can get φ from π, and π is injective. As π is continuous and, by (9.23), the dimensions of the domain and range of π are equal, π is a local homeomorphism and a proper function. Thus, $\big(Q(S_1)_1, \pi\big)$ is a covering of $\pi\big(Q(S_1)_1\big)$, and π is a homeomorphism from $Q(S_1)_1$ to $\pi\big(Q(S_1)_1\big)$. Therefore, every Riemann surface of finite type S_2 is the image of a Teichmüller mapping defined in S_1. □

If S is a Riemann surface, any complex structure in S can be specified by a Beltrami coefficient $\kappa \in M(S)_1$, with $\boldsymbol{M(S)_1}$ the open ball with radius 1 and center 0 of $L^\infty(S)$. So, there is a surjective holomorphic function $\pi : M(S)_1 \to \text{Teich}(S)$ with $\pi(0) = S$. S is conformal to $\mathbb{H}/\Gamma$, with Γ a discrete subgroup of the special linear group $SL(2,\mathbb{R})$ of nonsingular 2×2 matrices with real components. The lift from $\kappa \in M(S)_1$ to $\mathbb{H}$ can be extended by reflection to $-\mathbb{H}$ giving a Beltrami coefficient $\widetilde{\kappa} \in M(\mathbb{C}_\infty)_1$. Denote F the solution of the Beltrami equation with coefficient $\widetilde{\kappa}$ and fixed points $0, 1, \infty$. F maps $\mathbb{H}$ to itself. As $\widetilde{\kappa}$ is invariant under Γ, F conjugates the group Γ to another discrete subgroup Γ' of the special linear group $SL(2,\mathbb{R})$, and, with $S' = \mathbb{H}/\Gamma'$, the function obtained from F by projection is a quasiconformal mapping $f : S \to S'$ such that $\pi(\kappa) = (f, S')$.

One can be prove Teich(S) is a differential manifold. The tangent space $T_S\text{Teich}(S)$ is the quotient space of the tangent space of $M(S)_1$ at 0, denoted $\boldsymbol{M(S)}$, by the space of deformations of S determining complex structures coincident with the original complex structure modulo an isotopy, i.e., $\kappa = \frac{\partial v}{\partial \bar{z}}(z)$ for some quasiconformal mapping v in S (the space of such v can be seen as the tangent space at the group identity of the quasiconformal mappings of S to itself). Denote $\boldsymbol{M_0(S)}$ the preimage by π of the equivalence class in Teich(S) with the identity, which is a closed subspace of $M(S)$, and P the subset of $Q(S)$ such that $M_0(S) = \big\{\kappa \in M(S) : \int_S \kappa p = 0 \text{ to } p \in Q(S)\big\}$. For every $p \in P$,

$$\int_S p\,\frac{\partial v}{\partial \bar{z}} = -\int_S \frac{\partial p}{\partial \bar{z}}\,v = 0,$$

for all C^∞ vector fields v. So, $\frac{\partial p}{\partial \bar{z}} = 0$ in the sense of distributions and $M_0(S) \subset Q(S)$. As the inverse inclusion holds, $M_0(S) = Q(S)$.

If $\kappa \in M(S)$ and $f_\varepsilon : S \to S'$ are a quasiconformal mapping with Beltrami coefficient $\varepsilon\kappa$ for $\varepsilon > 0$ small,

$$d(S, S') = \tfrac{1}{2}\log K_f = \tfrac{1}{2}\log\frac{1+\varepsilon\|\kappa\|}{1-\varepsilon\|\kappa\|} = \varepsilon\|\kappa\| + O(\varepsilon^2), \quad \text{when } \varepsilon \to 0.$$

Adding to κ elements of $M_0(S)$ gives all functions isotopic to f in the limit when $\varepsilon \to 0$. Therefore, the definition of the Teichmüller distance with the factor $\frac{1}{2}$ in $D_T = \frac{1}{2} \inf \log K_f$ is compatible with the natural norms L^1 and L^∞ in the tangent and cotangent spaces of Teich(S).

Summarizing the three preceding paragraphs, we have the following:

(9.27) *If S is a Riemann surface of genus $g > 1$, for each $p \in$ Teich(S), the tangent space T_pTeich(S) is isomorphic to $M(S)/M_0(S)$, and the cotangent space T_p^* Teich(S) is isomorphic to $Q(S)$. The Teichmüller distance corresponds to the distance defined by the norm of L^∞ in $M(S)/M_0(S)$ and by the norm of L^1 in $Q(S)$.*

One additional reason for the factor $\frac{1}{2}$ in the definition of the Teichmüller metric is that it coincides with the Kobayashi metric introduced with a natural motivation by the Schwarz lemma.

The **Kobayashi metric** on a complex manifold M is the largest pseudometric[13] δ_M such that for every Holomorphic function from B_1 to M, $\|f'(0)\| \leq 1$. The next result was obtained in 1971 by H. Royden.[14] To establish this Royden theorem of equality of the Teichmüller and Kobayashi metrics, it is convenient to begin by establishing the following result obtained in 1994 by C. Earle, I. Kra, and S. Krushkal.[15]

(9.28) *Under the conditions and with the notation of the paragraph that follows the proof of the Teichmüller theorem* (9.26), *if $f : B_1 \to$ Teich(S) is holomorphic, there is a holomorphic function $g : B_1 \to M(S)_1$ such that $\pi \circ g = f$, which can be chosen to satisfy $g(0) = \kappa_0$ with $\pi(\kappa_0) = f(0)$.*

Proof By the Bers embedding (4.16), $f(B_1) \subset B(-\mathbb{H}, G_p)$, where $B(-\mathbb{H}, G_p)$ is the complex linear space of quadratic differentials in $p(\mathbb{H})$ with $(\mathbb{H}, p)$ a universal

---

[13] A **pseudometric** or **pseudodistance** is defined as distance except that it may be degenerate, i.e., besides $d(x, x) = 0$, there may exist $x \neq y$ with $d(x, y) = 0$. The Kobayashi pseudometric was introduced in 1967 by Shoshichi Kobayashi (1932–2012). An important property of this pseudometric is that all holomorphic functions of $\mathbb{C}$ onto a compact one-dimensional complex manifold M are constant if and only if M is **Kobayshi-Hyperbolic**, i.e., the Kobayshi pseudometric is a metric in M.

[14] Royden, H.L., Automorphisms and isometries of Teichmüller space. *Advances in the Theory of Riemann Surfaces (Proc. Conf., Stony Brook, N.Y., 1969)*, 369–383, Ann. of Math. Stud., No. 66 Princeton University Press, Princeton, NJ, 1971. Royden, H.L., Remarks on the Kobayashi metric. *Several complex variables, II (Proc. Internat. Conf., Univ. Maryland, College Park, Md., 1970)*, 125–137, Lecture Notes in Math., Vol. 185, Springer-Verlag, Berlin-New York, 1971.

[15] In their article cited at the end of this chapter introduction.

covering of $p(\mathbb{H}) \subset \mathbb{C}_\infty$ and G_p the group of this covering. Without loss of generality, $f(0)=0$, as can be achieved with translations and adjustment of the group G_p. The set $E=(-\mathbb{H}) \cup \mathbb{R} \cup \{\infty\}$ is invariant under G_p. For each $\tau \in B_1$, let ϕ_τ be a meromorphic function in E with Schwarzian derivative the quadratic differential $f(\tau)$. Each ϕ_τ is a *Schlicht* function in $-\mathbb{H}$ (i.e., a univalent function with $\phi_\tau(0)=0$ and $\phi'_\tau(0)=1$), which can be extended to a *Schlicht* function in E and uniquely specified by requiring 0, 1, ∞ to be fixed points. In particular, we have $\phi_0(z)=z$ for $z \in E$. Thus, $\{\phi_\tau\}_{\tau \in B_1}$ is a holomorphic motion of E. Since $f(B_1) \subset B(-\mathbb{H}, G_p)$, for each $\tau \in B_1$ and $\gamma \in G_p$, there is a Möbius transformation $T_{\tau,\gamma}$ such that $\phi_\tau\big(\gamma(z)\big)=T_{\tau,\gamma}\big(\phi_\tau(z)\big)$ for $z \in E$. By the Slodkowski equivariant theorem (4.8), this holomorphic motion of E has extension to a G_p-equivariant holomorphic motion of $\mathbb{C}_\infty$. For each $\tau \in B_1$, denote $g(\tau)$ the Beltrami coefficient of ϕ_τ in $\mathbb{H}$. The equivariance of the holomorphic motion implies $g(\tau) \in M(S)_1$ for $\tau \in B_1$. By (4.2), obtained in 1986 by L. Bers and H. Royden, the function $g: B_1 \to M(S)_1$ is holomorphic and, by the Bers embedding, $\pi\big(g(\tau)\big)$ is the Schwarzian derivative of the *Schlicht* function ϕ_τ in $-\mathbb{H}$ and, therefore, $\pi \circ g = f$. It is $g(0)=0$, but if $\kappa_0 \in M(S)_1$, with a translation, we get $\widetilde{g}$ such that $\pi \circ \widetilde{g} = f$, $\widetilde{g}(0)=\kappa_0$ and $\pi(\kappa_0)=0$. □

The equality of the Teichmüller and Kobayashi metrics was proved in 1971 by H. Royden for Riemann surfaces of finite type $(g, 0)$, i.e., compact Riemann surfaces without punctures. In 1987, F. Gardiner proved it for all finite type of Riemann surfaces. The following simple proof was given by C. Earle, I. Kra, and S. Krushkal in 1994 using the previous result they obtained at the time and a particular case of a result obtained in 1979 by L. Harris[16] giving equality of the metrics in $M(S)_1$, where the Teichmüller metric is

$$D_{T,M(S)_1}(\kappa_1, \kappa_2) = \tfrac{1}{2} \log \tfrac{1+N(\kappa_1,\kappa_2)}{1-N(\kappa_1,\kappa_2)}, \qquad N(\kappa_1, \kappa_2) = \Big\| \tfrac{\kappa_1-\kappa_2}{1-\overline{\kappa_1}\kappa_2} \Big\|_{L^\infty}.$$

(9.29) Royden theorem: *If S is a compact Riemann surface of genus $g>1$, in* Teich(S) *the Teichmüller and the Kobayashi metrics coincide.*

Proof If $\kappa_1 \neq 0$, $\kappa_2=0$, and $f: B_1 \to M(S)_1$ are a holomorphic function such that $f(0)=0$ and $f(\tau)=\kappa_1$ with $\tau \in B_1$, by the Schwarz lemma, $\|\kappa_1\|_{L^\infty} \le |\tau|$, and $D_{T,M(S)_1}(\kappa_1, 0) \le \delta_{M(S)_1}(\varphi_1, 0)$. On the other hand, for $f: B_1 \to M(S)_1$ such that $f(\tau) = \frac{\tau\kappa_1}{\|\kappa_1\|_{L^\infty}}$, it is $f(\|\kappa_1\|_{L^\infty})=\kappa_1$. Therefore, $\delta_{M(S)_1}(\kappa_1, 0) \le D_{T,M(S)_1}(\kappa_1, 0)$ and $D_{T,M(S)_1}(\kappa_1, 0)=\delta_{M(S)_1}(\varphi_1, 0)$ if $\kappa_1 \neq 0$.

If $0 \neq \kappa_1 \neq \kappa_2 \neq 0$, the function defined in $M(S)_1$ by $f(\kappa)=\frac{\kappa_2-\kappa}{1-\overline{\kappa_2}\kappa}$ is a conformal homeomorphism of $M(S)_1$ onto itself and

$$D_{T,M(S)_1}(\kappa_1, \kappa_2) = D_{T,M(S)_1}\big(f(\kappa_1), 0\big) = \delta_{M(S)_1}\big(f(\kappa_1), f(\kappa_2)\big) = \delta_{M(S)_1}(\kappa_1, \kappa_2).$$

[16] Harris, Lawrence.

By the symmetric property of distances and the triviality of equality for $\kappa_1 = \kappa_2 = 0$, it is $D_{T,M(S)_1}(\kappa_1, \kappa_2) = \delta_{M(S)_1}(\kappa_1, \kappa_2)$ for all $\kappa_1, \kappa_2 \in M(S)_1$.

By the definition of the Kobayashi metric, if $\varphi_1, \varphi_2 \in \mathrm{Teich}(S)$,

$$\delta_{\mathrm{Teich}(S)}(\varphi_1, \varphi_2) \leq \inf\big\{\delta_{M(S)_1}(\kappa_1, \kappa_2) : \kappa_j \in M(S)_1, \quad \pi(\kappa_j) = \varphi_j, \, j \in \{1, 2\}\big\}.$$

Therefore, $\delta_{\mathrm{Teich}(S)}(\varphi_1, \varphi_2) \leq D_T(\varphi_1, \varphi_2)$.

For $f : B_1 \to \mathrm{Teich}(S)$ such that $f(0) = \varphi_1$ and $f(\tau) = \varphi_2$ with $\tau \in B_1$, by the preceding result, there is a holomorphic function $g : B_1 \to M(S)_1$ such that $f = \pi \circ g$ and, applying again the conclusion of two paragraphs above,

$$D_T(\varphi_1, \varphi_2) = D_T\big(\pi(g(0)), \pi(g(\tau))\big) \leq D_{T,M(S)_1}\big(g(0), g(\tau)\big) \leq d_{B_1}(0, \tau),$$

where d_{B_1} is the Poincaré metric of B_1. The infimum for all f satisfying the mentioned conditions gives $D_T(\varphi_1, \varphi_2) \leq \delta_{\mathrm{Teich}(S)}(\varphi_1, \varphi_2)$.

By the 2 preceding paragraphs, $D_T(\varphi_1, \varphi_2) = \delta_{\mathrm{Teich}(S)}(\varphi_1, \varphi_2)$, $\varphi_1, \varphi_2 \in \mathrm{Teich}(S)$.

□

The Royden theorem has the following corollary.

(9.30) *If S is a compact Riemann surface of genus >1, every holomorphic function $f : \mathrm{Teich}(S) \to \mathrm{Teich}(S)$ is a contraction in the Teichmüller distance, i.e., $D_T\big(f(\varphi), f(\psi)\big) \leq D_T(\varphi, \psi)$.*

Chapter 10
Teichmüller Spaces of Dynamical Systems

10.1 Introduction

In 1998, C. McMullen and D. Sullivan published an important contribution to the foundations of Complex Dynamics in the notable article already cited in the introduction of Chap. 5 and used in that chapter. In addition to the issues related to structural stability and bifurcation described in that chapter, they defined the Teichmüller space of a holomorphic dynamical system and the corresponding space of deformations, which can be identified with the open ball of radius 1 of Beltrami coefficients defined in the domain of the holomorphic function giving the dynamics by iteration and described the Teichmüller space of a complex rational function f as a product of three components:

- The open ball of radius 1 of the Beltrami coefficients on the Julia set of f, which is a polydisk of dimension the number of ergodic components of the maximal subset of the Julia set supporting an invariant line field;
- A Teichmüller space associated with foliated sets, which is a polydisk of dimension equal to the sum of the number of cycles of Herman rings with the number of foliated equivalence classes of non-periodic critical points with positive orbits eventually cycles of Siegel disks, Herman rings, or attracting basins of superattracting periodic orbits;
- A Teichmüller space associated with orbits eventually entering attracting basins of attracting or parabolic periodic orbits involving a finite union of Riemann surfaces, one for each cycle of attracting or parabolic connected components, with each attracting basin of attracting periodic orbits and each parabolic basin contributing with a surface of a punctured torus at as many points as the number of grand orbits eventually entering the attracting or parabolic basin, in the latter case plus 2.

They considered conjugacy classes of rational functions by quasiconformal mappings, conformal and quasiconformal deformations of rational functions, and obstructions to deformations.

L. T. Magalhães, *Quasiconformal Mappings in the Plane and Complex Dynamics*,
https://doi.org/10.1007/978-3-031-80115-0_10

The main objective of this chapter is to present these contributions of C. McMullen and D. Sullivan.[1] So, this chapter extends the study of Teichmüller spaces initiated in the last section of Chap. 4 and continued in Chap. 8 and Chap. 9.

The chapter includes a section dedicated to characterizations obtained in 1988 by C. Earle and C. McMullen of quasiconformal mappings in a hyperbolic Riemann surface, which are isotopic relative to the ideal boundary through quasiconformal conjugacies, as they are used to prove properties of Teichmüller spaces of rational dynamical systems obtained by C. McMullen and D. Sullivan. This section is preceded by one on the notions of conformal barycenter and conformal barycentric extension, introduced in 1986 by A. Douady and C. Earle for extending homeomorphisms in ∂B_1 to homeomorphisms in B_1, as they are used to establish the already-mentioned characterizations due to C. Earle and C. McMullen.

10.2 Teichmüller Space of Holomorphic Dynamical System

Although the aim of this chapter is the notion of Teichmüller space of a rational dynamical system, in this section, we consider the more general notion of Teichmüller space of a holomorphic dynamical system.

A **holomorphic dynamical system** is (S, f) with S a one-dimensional complex differential manifold[2] and $f : S \to S$ a Holomorphic function.

An **isomorphism of holomorphic dynamical systems** $(S_1, f_1), (S_2, f_2)$ is a conformal homeomorphism α of S_1 onto S_2 such that $\alpha \circ f_1 = f_2 \circ \alpha$.

Given holomorphic dynamical systems $(S_1, f_1), (S_2, f_2)$, a quasiconformal conjugacy $\varphi : S_1 \to S_2$ of (S_1, f_1) with (S_2, f_2) defines a **marking** of (S_1, f_1) by (S_2, f_2).

Holomorphic dynamical systems (S_j, f_j), with $j \in \{1, 2\}$, marked by quasiconformal conjugacies $\varphi_j : S_j \to Y_j$ of the holomorphic dynamical system (S_j, f_j) with the holomorphic dynamical system (Y_j, g_j) are said to be **isomorphic marked dynamical systems** if there is an isomorphism α from (Y_1, g_1) to (Y_2, g_2) respecting the marking, i.e., $\varphi_2 = \alpha \circ \varphi_1$.

The **deformation space** of a holomorphic dynamical system (S, f), denoted **Def(S, f)**, is the space of isomorphism classes of marked holomorphic dynamical systems by (S, f).

Since quasiconformal mappings in S are solutions of the Beltrami equation with coefficients $\kappa \in L^\infty(S)$ with $\|\kappa\|_{L^\infty} < 1$, Def$(S, f)$ can be identified with the open ball of radius 1 and center 0 of the $L^\infty(S)$ subspace of the invariant Beltrami coefficients under f, denoted $M(S)_1^f$, via the bijection $\varphi \mapsto \frac{\partial\varphi/\partial\bar{z}}{\partial\varphi/\partial z}$. As $M(S)_1^f$ is

[1] We follow parts of the 1998 C. McMullen and D. Sullivan article cited above and of the notes McMullen, C., *Riemann surfaces, dynamics and geometry*, Course Notes, October 7, 2020, made available by C. McMullen in the Internet.

[2] Possibly disconnected, so it may not be a Riemann surface

an open subset of a complex Banach space, Def(S, f) *is a complex differential manifold.*

If S is a hyperbolic Riemann surface, it is conformal to B_1 or $B_1 \setminus \Gamma$, with Γ a discrete subgroup of the group $\operatorname{Aut} B_1$ of automorphisms of B_1. The **ideal boundary of** S, denoted $\partial_I S$, is the boundary $\Omega \setminus \Gamma$ of the orbifold $(B_1 \cup \Omega)/\Gamma$, where Ω is the complement of the limiting set of the group Γ in S^1. If S is a one-dimensional complex differential manifold, possibly disconnected and with connected hyperbolic, elliptic, or parabolic components, the **ideal boundary of** S, denoted $\partial_I S$, is the union of the ideal boundaries of the hyperbolic connected components of S.

The quasiconformal conjugacies φ of a holomorphic dynamical system (S, f) with itself form a group $\boldsymbol{QC(S, f)}$, which acts on Def(S, f) modifying the marking.The normal subgroup $\boldsymbol{QC_0(S, f)}$ of the group $QC(S, f)$ consists of the isotopic quasiconformal conjugacies of (S, f) with itself isotopic (rel $\partial_I S$) to the identity by uniformly quasiconformal conjugacies.

The **Teichmüller space** of a holomorphic dynamical system (S, f) is the quotient space of the marked dynamical system up to isotopy (rel $\partial_I S$), denoted **Teich(S, f)** $=$ Def$(S, f)/QC_0(S, f)$.

The **modular group** is **Mod(S, f)** $= QC(S, f)/QC_0(f)$, and the **modular space** is the set $\boldsymbol{\mathcal{M}(S, f)} =$ Teich(S, f)/Mod(S, f) of equivalence classes of isomorphisms of holomorphic dynamical systems conjugated with f.

The subgroup of $QC(S, f)$ of the conformal conjugacies is denoted **Aut(S, f)**. Its image in Mod(S, f) is the stabilizer[3] of $[(S, f)] \in$ Teich(S, f).

(10.1) Examples:

1. **Irrational rotation in an annulus:** Consider the holomorphic dynamical system (S, f) with $S_r = B_r \setminus \operatorname{cl} B_1 \subset \mathbb{C}$, where $r > 1$, and $f(z) = e^{i2\pi\alpha} z$ with $2\pi\alpha \in \mathbb{R}\setminus\mathbb{Q}$. Aut$(S_r, f)$ consists of the rotations around 0. So, it is isomorphic to S^1 with the multiplication of $\mathbb{C}$.

   The deformation space Def(S_r, f) is identified with the open ball of the Beltrami coefficients in $L^\infty(S_r)$ invariant under f with center 0 and radius 1, which is identified with the open ball of $L^\infty([1, r])$ with center 0 and radius 1, since the invariance is under rotations around 0 and the Beltrami coefficient values of the in S_r are determined by the values in a radius. As every Riemann surface Y quasiconformal to S_r is conformal to S_s for some $s > 1$, the rotation number of f is a topological invariant and conjugacies φ can be normalized to $\varphi(1) = 1$ by composition with automorphisms of S_s, Def(S_r, f) can be identified as a marked dynamical system to the set of the (S_s, f, φ) such that $\varphi : S_r \to S_s$ is a conformal homeomorphism invariant under rotations with center at 0 and $\varphi_{|S^1}(1) = 1_{S^1}$.

[3] The **stabilizer** of a group G for a $g_0 \in G$ is the subgroup of G whose elements leave g_0 fixed, $\{g \in G : gg_0 = g_0\}$.

The group $QC(S_r, f)$ consists of quasiconformal mappings from S_r to S_r commuting with rotations around 0. Considering orbits of rotations around 0 in S_r as circles centered at 0, with the metric with element of length $\frac{|dz|}{|z|}$, the conjugacies $\varphi \in QC(S_r, f)$ preserve the foliation of S_r by these circles, mapping leaves to leaves by isometries, and are Lipschitzian functions determined by the values they have in $]1, r[$. They are $\varphi(z) = e^{i2\pi\theta(|z|)}z$, with $\theta : [1, r] \to \mathbb{C}$ Lipschitzian and $\theta(1), \theta(r) \in \mathbb{R}$. The condition that φ is a quasiconformal mapping is equivalent to $\mathcal{R}e\,\theta$, $\mathcal{I}m\,\theta$ to be Lipschitzian and the impossibility of $\frac{1}{r} = 2\pi\,(\mathcal{I}m\,\theta)'(r)$, i.e., $\inf_{r\in[1,r]}\left[\frac{1}{r} - 2\pi\,\beta'(r)\right] > 0$; thus, $r \mapsto |\varphi(r)|$ is strictly increasing, with secants to its graph with slopes lower bounded by a positive number.

The group $QC_0(S_r, f)$ consists of the quasiconformal mappings $\varphi \in QC(S_r, f)$ homotopic in S_r to the identity through homotopies keeping fixed the values in the images of 1 and r of the restriction of the rotation f to S^1, i.e., $\varphi_{|\partial S_r}$ is the identity, and there is no torsion along the homotopy; so, the function θ of the preceding period satisfies $\theta(0) = \theta(r) \in \mathbb{Z}$.

Since points $(S_s, f, \varphi) \in \mathrm{Def}(S_r, f)$ can be normalized so that $\varphi(1) = 1$, i.e., $\theta(1) = 0$, φ is determined modulo an isotopy (rel $\partial_I S_r$) by the value of $\theta(r) \in \mathbb{R}$. The function $(S_s, f, \varphi) \mapsto -\frac{1}{i2\pi}\log\varphi(r) = \theta(r) + i \bmod S_s$, which is holomorphic and has range, the open upper complex half plane $\mathbb{H}$ induces an isomorphism of $\mathrm{Teich}(S_r, f)$ onto $\mathbb{H}$.

The modular group $\mathrm{Mod}(S_r, f)$ is isomorphic to $\mathbb{R}^2/\{(n, n) : n \in \mathbb{Z}\}$ with an isomorphism, the function $\varphi \mapsto (\theta(1), \theta(r))$.

The modular group $\mathrm{Mod}(S_r, f)$ acts on $\mathrm{Teich}(S_r, f)$ modifying the values of $\varphi \in QC(S_r, f)$ in ∂S_r, with the action in $\mathbb{H}$ given by $(s, t)(z) = z + (t - s)$.

The modular space $\mathcal{M}(S_r, f) = \mathrm{Teich}(S_r, f)/\mathrm{Mod}(S_r, f)$ is isomorphic to $\mathbb{H}/G$, with G the group of translations of $\mathbb{H}$. The function $(Y, g) \mapsto \bmod Y$, equivalent to $z \mapsto \mathcal{I}m\,z$, is an isomorphism of groups of $\mathbb{H}/G$ to $]0, +\infty[$ with the multiplication of real numbers. The irrational rotations in an annulus are classified in this way, modulo conjugacies.

2. **Linear transformation with attracting fixed point:** Consider the holomorphic dynamical system $(\mathbb{C}, f)$ with $f(z) = \lambda z$ with $0 < |\lambda| < 1$.

   $\mathrm{Aut}(\mathbb{C}, f)$ is the group $\mathbb{C}\setminus\{0\}$ with the multiplication of $\mathbb{C}$.

   Consider the torus T obtained by identifying in $\mathbb{C}\setminus\{0\}$ points of each grand orbit of the dynamical system.

   The deformation space $\mathrm{Def}(\mathbb{C}\setminus\{0\}, f)$ is identified with the open ball of the Beltrami coefficients in $L^\infty(\mathbb{C})$ invariant under f with center 0 and radius 1, which is identified with the open ball of $L^\infty(B_1\setminus \mathrm{cl}\, B_{|\lambda|})$ with center 0 and radius 1 because the invariance is under multiplications by $\lambda \in B_1\setminus\{0\}$ and, therefore, the Beltrami coefficient values in $\mathbb{C}$ are determined by the values in the fundamental domain $B_1\setminus \mathrm{cl}\, B_{|\lambda|}$. The conjugacy φ can be normalized so that $\varphi(1) = 1$ and $\mathrm{Def}(\mathbb{C}\setminus\{0\}, f)$ can be identified as a marked dynamical system with the set $(\mathbb{C}\setminus\{0\}, g, \varphi)$ such that $g(z) = \mu z$ for some $\mu \in B_1\setminus\{0\}$ and $\varphi : \mathbb{C}\setminus\{0\} \to \mathbb{C}\setminus\{0\}$ is a conformal homeomorphism invariant under multiplication by λ.

Teich($\mathbb{C}\backslash\{0\}$, f) is isomorphic to Teich(T) and, therefore, also to $\mathbb{H}$. An isomorphism is induced by the function $(\mathbb{C}\backslash\{0\}, g, \varphi) \mapsto \frac{\log \mu}{i2\pi}$, where $\log \mu$ is well defined by the holomorphic marked dynamical system $(\mathbb{C}\backslash\{0\}, g)$ by $(\mathbb{C}\backslash\{0\}, f)$, choosing a curve $\gamma \subset \mathbb{C}\backslash\{0\}$ connecting 1 to λ, so that the curve $\varphi(\gamma) \subset \mathbb{C}\backslash\{0\}$ connects 1 to μ and there is a continuous branch of the logarithm along this curve with $\log(1) = 0$.

With a covering $(\mathbb{C}\backslash\{0\}, p)$ of T, the modular group Mod(T) is isomorphic to the special linear group $SL(2, \mathbb{Z})$ of 2×2 nonsingular matrices with integer components with the product of matrices, and the group Mod($\mathbb{C}\backslash\{0\}$, f) is isomorphic to the subgroup of $SL(2, \mathbb{Z})$ of the matrices $\begin{bmatrix} 1 & n \\ 0 & 1 \end{bmatrix}$ with $n \in \mathbb{Z}$. So, it is also isomorphic to the additive group $\mathbb{Z}$.

The modular group Mod($\mathbb{C}\backslash\{0\}$, f) acts on Teich($\mathbb{C}\backslash\{0\}$, f) with an action given in $\mathbb{H}$ by $z \mapsto z + n$.

The modular space $\mathcal{M}(\mathbb{C}\backslash\{0\}, f) = \text{Teich}(\mathbb{C}\backslash\{0\}, f)/\text{Mod}(\mathbb{C}\backslash\{0\}, f)$ is isomorphic to $\mathbb{H}/\mathbb{Z}$ and $B_1\backslash\{0\}$, and an isomorphism onto the latter is simply induced by $\big(\mathbb{C}\backslash\{0\}, \mu z\big) \mapsto \mu$.

10.3 Conformal Barycenter and Barycentric Extension

The following section is mainly devoted to prove the very useful conditions equivalent to a quasiconformal mapping $g : S \to S$ in a hyperbolic Riemann surface S to be isotopic (rel $\partial_I S$) to the identity through uniformly quasiconformal conjugacies, as considered in the preceding section in the definition of $QC_0(S, g)$. These conditions were obtained in 1988 by C. Earle and C. McMullen using the barycentric extension of homeomorphisms in S^1 to cl B_1 introduced in 1986 by A. Douady and C. Earle.

Let $\mathbf{\Lambda}_{S^1}$ be the set of continuous functions $\lambda : S^1 \to [0, +\infty[$, $\int_{S^1} \lambda(z)\, dz = 1$. The conformal barycenter[4] of $\lambda \in \Lambda_{S^1}$ corresponds to the center of mass of a mass distribution in S^1 with density λ in the Poincaré metric of B_1.

(10.2) $\xi_\lambda : B_1 \to \mathbb{C}$ *such that* $\xi_\lambda(z) = (1 - |z|^2) \int_{S^1} T_z(w)\, \lambda(w)\, |dw|$, *with* $\lambda \in \Lambda_{S^1}$ *and* T_z *the Möbius transformation* $T_z(w) = \frac{w - z}{1 - \bar{z} w}$ *has a unique zero* **CB(λ)**, *called the* ***conformal barycenter of*** λ.

[4] For conformal barycenter and barycentric extension, we follow parts of the A. Douady and C. Earle article cited in the preceding chapter section on Bers embedding. The article by Cantarella, J., Schumacher, H., Computing the conformal barycenter, *SIAM J. Applied Algebra and Geometry,* **6** (2022), 503–530, gives algorithms to compute conformal barycenters and barycentric extensions. The computation of barycentric extensions was previously addressed by W. Abykoff and T. Ye in 1997 and by V. Jaćimović in 2019. Abykoff, William (1944-). Ye, Taiping. Jaćimović, Vladimir (1974-). Cantarella, Jason (1972-). Schumacher, Henrik.

Proof For $w \in S^1$, $|\overline{z}w| = |z| < 1$, $\frac{1}{1-\overline{w}z} = \sum_{j=0}^{\infty}(\overline{w}z)^j = 1 + \overline{w}z + o(z)$, $z \to 0$,

$$\xi_\lambda(z) = (1-|z|^2)\int_{S^1}(w-z)(1+\overline{z}w)\,\lambda(w)\,|dw| + o(z)$$

$$= \xi_\lambda(0) - z + \overline{z}\int_{S^1} w^2\lambda(w)\,|dw| + o(z)\,.$$

The Jacobian of ξ_λ at 0 is

$$J\xi_\lambda(0) = \left|\tfrac{\partial \xi_\lambda}{\partial z}(0)\right|^2 - \left|\tfrac{\partial \xi_\lambda}{\partial \overline{z}}(0)\right|^2 = 1 - \int_{S^1\times S^1} s^2\overline{w}^2\lambda(s)\,\lambda(w)\,|ds|\,|dw|\,,$$

and, as for $s, w \in S^1$,

$$|s^2-w^2|^2 = (s^2-w^2)(\overline{s}^2-\overline{w}^2) = |s|^4 + |w|^4 - s^2\overline{w}^2 - w^2\overline{s}^2 = 2 - s^2\overline{w}^2 - w^2\overline{s}^2,$$

$$J\xi_\lambda(0) = \tfrac{1}{2}\int_{S^1\times S^1}|s^2-w^2|^2\lambda(s)\,\lambda(w)\,|ds|\,|dw| > 0\,.$$

Thus, if $\xi_\lambda(0) = 0$, 0 is an isolated zero of ξ_λ with Index 1.

For $T \in \mathrm{Aut}\, B_1$, $\xi_{T\cdot\lambda} = T\cdot\xi_\lambda$, with $T\cdot\lambda = (\lambda\circ T^{-1})(T^{-1})'$ and

$$T\cdot\xi_\lambda = (\xi_\lambda T')\circ T^{-1}\,.$$

So, any zero of ξ_λ in B_1 is an isolated zero with Index 1. To complete the proof, it is enough to verify that ξ_λ does not have zeros in disks with center 0 and radii $r \in]0, 1[$ close to 1 and the vector field associated with ξ_λ on each of these circles C_r with radius r points to the interior of the disk bounded by C_r.

Let $\ell > 0$ be such that every arc of circle $C \subset S^1$ with length upper bounded by ℓ is $\int_C \lambda(w)\,|dw| \le \frac{1}{3}$ and let $x_0 \in]-1, 1[$ be such that the length of the arc $T_{x_0}(C_\ell)$ of the circle $C_\ell \subset S^1$ with length ℓ centered at 1 is $\frac{3\pi}{2}$. For $w \in B_1$ with $|w| = r \ge x_0$, let $T \in \mathrm{Aut}\, B_1$ preserve the orientation and satisfying $T(w) = 0$, $T\big(-\frac{w}{|w|}\big) = 1$. Denote $\nu = \lambda\circ T^{-1}$. If $\gamma \in S^1$ is the arc of a circle with end points with arguments $\pm\frac{\pi}{4}$ and $\int_\gamma \lambda(w)\,|dw| \ge \frac{2}{3}$, then

$$\mathcal{R}e\big(\xi_\lambda(0)\big) = \int_{S^1}\mathcal{R}e(w)\,\lambda(w)\,|dw| \ge -\tfrac{1}{3} + \tfrac{2}{3\sqrt{2}} > 0\,.$$

Hence, the vector $\xi_\nu(0)$ points to the interior of the disk bounded by $T(C_r)$ and $\xi_\lambda(w)$ points to the interior of the disk bounded by C_r. Consequently, ξ_λ has only one zero, which is the conformal barycenter of λ (Fig. 10.1). □

The **barycentric extension** to $\mathrm{cl}\, B_1$ of a homeomorphism h of S^1 onto S^1 is the function $\mathbf{BE}(h) : \mathrm{cl}\, B_1 \to \mathrm{cl}\, B_1$ such that $[\mathrm{BE}(h)](z) = h(z)$ for $z \in S^1$ and $\mathrm{BE}(z) = w$ for $z \in B_1$, with w the unique element of B_1 such that

$$F_h(z, w) = \tfrac{1}{2\pi}\int_{S^1}\tfrac{h(s)-w}{1-\overline{w}\,h(s)}\,\tfrac{1-|z|^2}{|z-s|^2}\,|ds| = 0\,.$$

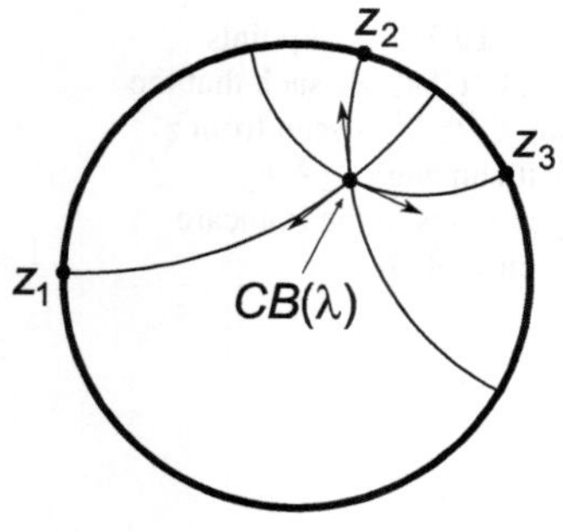

Fig. 10.1 Conformal barycenter $\mathrm{CB}(\lambda)$ of $\lambda: S^1 \to [0, +\infty[$ with support in the union of small neighborhoods of 3 points $z_1, z_2, z_3 \in S^1$ and integral $\frac{1}{3}$ in each one of these neighborhoods

(10.3) *The barycentric extension* $\mathrm{BE}(h): \mathrm{cl}\, B_1 \to \mathrm{cl}\, B_1$ *of a homeomorphism* h *of* S^1 *to* S^1 *is* ***conformally natural****, i.e., for all* $Q, T \in \mathrm{Aut}\, B_1$,

$$\mathrm{BE}(Q \circ h \circ T) = Q \circ \mathrm{BE}(h) \circ T$$

is a homeomorphism of $\mathrm{cl}\, B_1$ *to* $\mathrm{cl}\, B_1$ *and the restriction of* $\mathrm{BE}(h)$ *to* B_1 *is an holomorphic diffeomorphism of* B_1 *to* B_1 *as a function of real variables.*

Proof It is left as an exercise to check that BE is conformally natural.

First, we prove that $\mathrm{BE}(h)$ is continuous at every point of S^1. For each arc of a circle $C \subset S^1$, let $V(C)$ be the set of points $z \in B_1$ such that C is seen from z by an angle $\geq \frac{\pi}{2}$ in the Poincaré metric of B_1. $\partial V(C)$ is an arc of a circle Γ intersecting S^1 with resp. tangents at the point of intersection making an angle $\frac{\pi}{4}$ (Fig. 10.2). For every $w \in \Gamma$, there is $T \in \mathrm{Aut}\, B_1$ such that $T(w) = 0$, $T(C) \subset S^1$ is the arc of circle centered at 1 with end points in S^1 with arguments $\pm\frac{\pi}{4}$ and $T\big(V(C)\big) = B_1 \cap B_{1/\sqrt{2}}(1/\sqrt{2})$. From what was seen in the paragraph before the final paragraph of the proof of the preceding result and the definition of conformally natural, if $\int_C \lambda(w)\,|dw| \geq \frac{2}{3}$, ξ_λ in Γ points to the interior of $V(C)$, and therefore $\mathrm{CB}(\lambda) \in V(C)$. With $U(C) = \big\{z \in B_1 : \frac{1}{2\pi}\int_C \frac{1-|z|^2}{|z-w|^2}\,|dw| \geq \frac{2}{3}\big\}$, we have $[\mathrm{CB}(h)]\big(U(C)\big) \subset V\big(h(C)\big)$. So, if $w \in S^1$, for $C \subset S^1$ in a neighborhood of w, $C \cup U(C)$ is a neighborhood of w in $\mathrm{cl}\, B_1$ and the sets $h(C) \cup V\big(h(C)\big)$ generate a fundamental system of neighborhoods of $h(w)$ in $\mathrm{cl}\, B_1$. Thus, $\mathrm{BE}(h)$ is continuous at w, and as $w \in S^1$ is arbitrary, $\mathrm{BE}(h)$ is continuous in S^1.

Now, it suffices to prove that $\mathrm{BE}(h)$ is analytic as a function of real variables with Jacobian without zeros in B_1. As BE is conformally natural, we can assume without loss of generality that $[\mathrm{BE}(h)](0) = 0$ and $h: S^1 \to S^1$ has degree 1. For $z \in B_1$, it is $[\mathrm{BE}(h)](z) = w$ with w such that

$$F_h(z, w) = \frac{1}{2\pi}\int_{S^1} \frac{h(s)-w}{1-\overline{w}\,h(s)}\,\frac{1-|z|^2}{|z-s|^2}\,|ds| = 0\,.$$

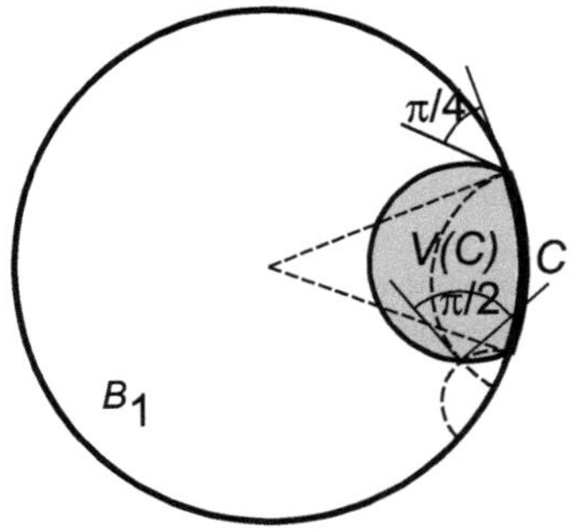

Fig. 10.2 Set of points $z \in V(C) \subset B_1$ such that the arc $C \subset S^1$ is seen from z with an angle $\geq \frac{\pi}{2}$ by geodesics in the Poincaré metric of B_1

The function F_h is analytic in $B_1 \times B_1$ as a function of real variables and

$$\frac{\partial F_h}{\partial z}(0,0) = \frac{1}{2\pi}\int_{S^1} \bar{s}\, h(s)\, |ds|\,, \qquad \frac{\partial F_h}{\partial \bar{z}}(0,0) = \frac{1}{2\pi}\int_{S^1} s\, h(s)\, |ds|\,,$$

$$\frac{\partial F_h}{\partial w}(0,0) = -1\,, \qquad \frac{\partial F_h}{\partial \bar{w}}(0,0) = \frac{1}{2\pi}\int_{S^1} h^2(s)\, |ds|\,,$$

$$\Big|\frac{\partial F_h}{\partial w}(0,0)\Big|^2 - \Big|\frac{\partial F_h}{\partial \bar{w}}(0,0)\Big|^2 = 1 - \int_{S^1\times S^1} \frac{h^2(s)}{2\pi}\,\frac{\bar{h}^2(t)}{2\pi}\, |ds||dt|\,.$$

At the end of the first paragraph of the proof of the preceding result, we got

$$J F_h(0,0) = \Big|\frac{\partial F_h}{\partial w}(0,0)\Big|^2 - \Big|\frac{\partial F_h}{\partial \bar{w}})(0,0)\Big|^2 = \frac{1}{2}\frac{1}{4\pi^2}\int_{S^1\times S^1} |h^2(s) - h^2(t)|^2\, |ds||dt| > 0\,.$$

By the implicit function theorem, [BE](h) is analytic as a function of real variables, and the Jacobian of BE(h) at 0 is

$$J[\mathrm{BE}(h)](0) = \Big|\frac{\partial[\mathrm{BE}(h)]}{\partial z}(0)\Big|^2 - \Big|\frac{\partial[\mathrm{BE}(h)]}{\partial \bar{z}}(0)\Big|^2 = \frac{|(\partial F_h/\partial z)(0,0)|^2 - |(\partial F_h/\partial \bar{z})(0,0)|^2}{|(\partial F_h/\partial w)(0,0)|^2 - |(\partial F_h/\partial \bar{w})(0,0)|^2}\,.$$

The homeomorphism $h : S^1 \to S^1$ has Fourier series $h(z) = \sum_{n\in\mathbb{Z}} c_n z^n$. The Jacobian of the harmonic function u in B_1 solution of the Dirichlet problem with boundary values given by h satisfies[5]

$$J u(0) = |c_1|^2 - |c_{-1}|^2 = \frac{1}{4\pi^2}\int_{S^1\times S^1} \mathcal{R}e\Big(h(\zeta)\,\overline{h(z)}\,(\zeta\bar{\zeta} - \bar{z}\zeta)\Big)\, |d\zeta||dz|\,.$$

With $z = e^{is}$, $\zeta = e^{it}$, and $h(e^{iu}) = e^{i\psi(u)}$, where $\psi : \mathbb{R} \to \mathbb{R}$ is continuous, strictly increasing, and satisfies $\psi(u + 2\pi) = \psi(u) + 2\pi$,

[5] This has been known since at least 1926, in lessons of Hellmuth Kneser (1898–1973).

$$\begin{aligned} Ju(0) &= |c_1|^2 - |c_{-1}|^2 = \tfrac{2}{4\pi^2} \int_0^{2\pi}\!\!\int_0^{2\pi} \sin(s-t)\, \sin\big(\psi(s)-\psi(t)\big)\, ds\, dt \\ &= \tfrac{2}{4\pi^2} \int_0^{\pi} \sin(u) \int_0^{2\pi} \sin\big(\psi(t+u)-\psi(t)\big)\, dt\, du \\ &= \tfrac{2}{4\pi^2} \int_0^{\pi} \sin(u) \int_0^{2\pi} H(t,u)\, dt\, du, \end{aligned}$$

$$\begin{aligned} H(t,u) \overset{def}{=} {} & \sin\big(\psi(t+u)-\psi(t)\big) + \sin\big(\psi(t+2\pi)-\psi(t+u+\pi)\big) \\ & + \sin\big(\psi(t+u+\pi)-\psi(t+\pi)\big) + \sin\big(\psi(t+\pi)-\psi(t+u)\big). \end{aligned}$$

The last integral in the formula for $Ju(0)$ is positive because the sum of 4 sines of positive angles with sum 2π is always positive. □

Denote the Banach space of continuous functions of S^1 and of $\mathrm{cl}\, B_1$ in $\mathbb{C}$ with the supremum norm by, resp., $\boldsymbol{\mathscr{C}(S^1)}$ and $\boldsymbol{\mathscr{C}(\mathrm{cl}\, B_1)}$, the resp. subspaces of homeomorphisms, resp., of S^1 onto S^1 and of $\mathrm{cl}\, B_1$ onto $\mathrm{cl}\, B_1$ by, resp., $\boldsymbol{\mathscr{H}(S^1)}$ and $\boldsymbol{\mathscr{H}(\mathrm{cl}\, B_1)}$, the space of the C^∞ functions from B_1 to $\mathbb{C}$ with the topology of uniform convergence in compact subsets of both the function and its derivatives of any order by $\boldsymbol{C^\infty(B_1)}$, and the subspace of this space of the diffeomorphisms of B_1 onto B_1 by $\mathbf{Diff}(\boldsymbol{B_1})$.

(10.4) *The barycentric extension of homeomorphisms of S^1 onto S^1,*

$$\mathrm{BE}(h) : \mathscr{H}(S^1) \to \mathrm{Diff}(B_1) \cap \mathscr{H}(\mathrm{cl}\, B_1),$$

is continuous. The functions $h \mapsto \frac{\partial[\mathrm{BE}(h)]}{\partial z}(0)$, $h \mapsto \frac{\partial[\mathrm{BE}(h)]}{\partial \bar{z}}(0)$ defined in $\mathscr{H}(S^1)$ are continuous.

Proof We begin proving that $(z,h) \mapsto [\mathrm{BE}(h)](z)$ from $S^1 \times \mathscr{H}(S^1)$ to $\mathrm{cl}\, B_1$ is continuous. Let $z_0 \in S^1$, $h_0 \in \mathscr{H}(S^1)$, and V_1 be a neighborhood of $h_0(z_0)$ in $\mathrm{cl}\, B_1$. Taking into account the argument in the second paragraph of the preceding result, there is an arc of circle $C_1 \subset S^1$ centered at $h_0(z_0)$ such that $\mathrm{cl}\, V(C_1) \subset V_1$, an arc of circle $C_0 \subset S^1$ centered at z_0, and a neighborhood W of h_0 in $\mathscr{C}(S^1)$ such that $h(C_0) \subset C_1$ for all $h \in W$. Therefore, $\mathrm{cl}\, U(C_0)$ is a neighborhood of z_0 in $\mathrm{cl}\, B_1$ and $[\mathrm{BE}(h)]\big(\mathrm{cl}\, U(C_0)\big) \subset \mathrm{cl}\, V(C_1)$ for all $h \in W$, proving the continuity of $(z,h) \mapsto [\mathrm{BE}(h)](z)$ from $S^1 \times \mathscr{H}(S^1)$ to $\mathrm{cl}\, B_1$.

Let $\Omega = \{(z,w,h) \in B_1 \times \mathbb{C} \times \mathscr{C}(S^1) : |w|\, \|h\| < 1\}$, F_h defined as in the proof of (10.3), $G : \Omega \to \mathbb{C}$ with $G(z,w,h) = F_h(z,w)$ and $h_0 \in \mathscr{H}(S^1)$, $z_0 \in B_1$, $w_0 = [\mathrm{BE}(h_0)](z_0)$. It is $F(z_0, w_0, h_0) = 0$ and $\big[\,\big|\frac{\partial F}{\partial w}\big|^2 - \big|\frac{\partial F}{\partial \bar{w}}\big|^2\big](z_0, w_0, h_0) > 0$ because it is a positive Jacobian multiple of ξ_λ at its single zero w_0, where

$\lambda(s) = \frac{1}{2\pi} \frac{1-|z_0|^2}{|z_0 - h^{-1}(s)|^2} (h^{-1})'(s)$ for $s \in S^1$. By the implicit function theorem, in a neighborhood of (z_0, w_0, h_0), the zeros of F are the points on the graph of an analytic function as a function of real variables $w = K(z, h)$ with K defined in a neighborhood of (z_0, h_0) in $B_1 \times \mathscr{C}(S^1)$. Therefore, $[\mathrm{BE}(h)](z) = K(z, h)$ for (z, h) in a neighborhood of (z_0, h_0) in $B_1 \times \mathscr{H}(S^1)$. Consequently, the functions $h \mapsto \frac{\partial[\mathrm{BE}(h)]}{\partial z}(0)$ and $h \mapsto \frac{\partial \mathrm{BE}(h)}{\partial \bar{z}}(0)$ are defined and continuous in $\mathscr{H}(S^1)$. □

(10.5) *If a homeomorphism h of* S^1 *to* S^1 *has extension to a quasiconformal mapping in* cl B_1, *then the barycentric extension of h to* cl B_1 *is a quasiconformal mapping.*

Proof Let $\mathscr{H}_1(S^1)$ be the set of homeomorphisms from S^1 to S^1 of degree 1. For $h \in \mathscr{H}_1(S^1)$, consider the functions

$$(z) = \frac{1-|z|^2}{1-|[\mathrm{BE}(h)](z)|^2} \left(\left| \frac{\partial[\mathrm{BE}(h)]}{\partial z}(z) \right| + (-1)^j \left| \frac{\partial[\mathrm{BE}(h)]}{\partial \bar{z}}(z) \right| \right), \quad j \in \{1, 2\}.$$

Since BE is conformally natural, for $Q, T \in \mathrm{Aut}\, B_1$ preserving orientation, $\alpha_j(Q \circ h \circ T) = \alpha_j(h) \circ T$ for $j \in \{1, 2\}$. $h \mapsto [\alpha_j(h)](0)$, $j \in \{1, 2\}$, by (10.3), have values > 0 for $h \in \mathscr{H}_1(S^1)$, and by (10.4), they are continuous. By (3.22), the set $\mathscr{H}^K(S^1) \subset \mathscr{H}_1(S^1)$ of functions with extension to a K-quasiconformal mapping in cl B_1 with fixed points $1, i, -1$ is compact, and

$$\inf\left\{[\alpha_1(h)](0) : h \in \mathscr{H}^K(S^1)\right\},\ \sup\left\{[\alpha_2(h)](0) : h \in \mathscr{H}^K(S^1)\right\} \in \,]0, +\infty[\,.$$

□

10.4 Quasiconformal Isotopies

This section is mainly for proving the very useful conditions obtained in 1988 by C. Earle and C. McMullen[6] equivalent to a quasiconformal mapping $g : S \to S$ in a hyperbolic Riemann surface S to be isotopic (rel $\partial_I S$) to the identity through uniformly quasiconformal conjugacies, as considered in the preceding section in the definition of $QC_0(S, g)$.

(10.6) *If S is a hyperbolic orbifold, i.e, conformal to* B_1/Γ *for* Γ *a discrete subgroup of* $\mathrm{Aut}(B_1)$, *and* $g : S \to S$ *is a K-quasiconformal mapping, the following conditions are equivalent:*

(continued)

[6] This follows Earle, C., McMullen, C., Quasiconformal isotopies, in *Holomorphic Functions and Moduli I,* Vol. 10, 143–154, MSRI Publ., Springer-Verlag, Berlin, 1988.

1. *There is a lift of g to a continuous function from B_1 to B_1 with continuous extension to* cl B_1 *that restricted to S^1 is the identity.*
2. *g is homotopic* (rel $\partial_I S$) *to the identity.*
3. *g is isotopic* (rel $\partial_I S$) *to the identity through K'-quasiconformal mappings, where K' depends only on K.*

Proof $3 \Rightarrow 2 \Rightarrow 1$ are easy.

($1 \Rightarrow 3$*)* Let g commute with the elements of Γ and κ be the Beltrami coefficient of γ. By the measurable Riemann mapping theorem (2.9), for each $t \in [0, 1]$, there is 1 unique quasiconformal mapping from B_1 to B_1 with Beltrami coefficient $t\kappa$ and fixed points $-1, 1, \infty$. By (2.2), γ_t is an isotopy in cl B_1 (but possibly not (rel S^1)). Since κ is invariant under Γ, $\Gamma_t = \gamma_t^{-1} \circ \Gamma \circ \gamma_t$ is a family of discrete subgroups of Aut B_1 isomorphic to Γ. As $\gamma_{0|S^1} = \gamma_{1|S^1} = 1_{S_1}$, it is $\Gamma_0 = \Gamma_1 = \Gamma$.

Let $\beta_t = \mathrm{BE}(\gamma_t^{-1})$ be the barycentric extension of the values on the boundary of γ_t^{-1}. By (10.5), these extensions are K'-quasiconformal with K' function of K. As they are continuous functions of t, β_t is an isotopy in cl B_1, and it is a conjugacy of γ_t with Γ in B_1. β_0, β_1 are the identity, as they are barycentric extensions of the identity. Since $\psi_t = \beta_t \circ \gamma_t$ is Γ-equivariant, the projection of the covering considered for the lift from g to γ projects ψ_t to an isotopy $g_t : S \to S$ from 1_S to g. Since β_t and α_t^{-1} coincide in S^1, ψ_t is an isotopy (rel S^1), and g_t is an isotopy (rel $\partial_I S$). □

Teich(S) can be seen as the set of ordered pairs (Y, α) such that $\alpha : S \to Y$ is a quasiconformal mapping of orbifolds modulo the equivalence relation from (Y_1, α_1) to (Y_2, α_2) defined by the existence of a conformal homeomorphism $\gamma : Y_1 \to Y_2$ such that a restriction of $\varphi = \alpha_2^{-1} \circ \gamma \circ \alpha_1$ to S^1 is the identity. By the preceding result, we get the same equivalence relation by replacing this condition by *φ is quasiconformally isotopic* (rel $\partial_I S$) *to the identity*, which is a more intrinsic property of S.

With the preceding result, we have the following simple proof of the theorem obtained in 1971 by L. Bers and L. Greenberg.[7]

(10.7) Bers-Greenberg theorem: *If S is a hyperbolic orbifold with branching points set B,* Teich(S) *and* Teich($S \backslash B$) *are canonically isomorphic.*

Proof The restriction of quasiconformal mappings from S to $S \backslash B$ is a canonical bijection from the homeomorphisms defined in S to those defined in $S \backslash B$. An

[7] Bers, L., Greenberg, L., Isomorphisms between Teichmüller spaces. *Advances in the Theory of Riemann Surfaces (Proc. Conf., Stony Brook, N.Y., 1969),* 53–79, Ann. of Math. Stud., No. 66, Princeton University Press, Princeton, NJ, 1971. Greenberg, Leon (1931-)

isotopy to the identity uniformly quasiconformal in $S\backslash B$ can be extended to an isotopy in S leaving the points of B fixed, which is an isotopy in the orbifold. Conversely, the restriction to $S\backslash B$ of an isotopy to the identity in S keeping the points of B fixed is an isotopy in $S\backslash B$. The Riemann surface with boundary $(S\backslash B) \cup \partial_I(S\backslash B)$ can be canonically identified with $(S \cup \partial_I S)\backslash B$, since a universal covering of the Riemann surface $S\backslash B$ can be factored by the universal covering (B_1, p) of S. Therefore, one of the isotopies considered is an isotopy relative to the ideal boundary if and only if the other one is.

With the quasiconformal isotopy (rel $\partial_I S$) referred as an alternative for the definition of Teich(S), functions α, β belong to the same equivalence class as Teich(S) if and only if $\alpha_{|S\backslash B}$, $\beta_{|S\backslash B}$ belong to the same equivalence class as Teich$(S\backslash B)$. □

With (10.6) we also have a simple proof of the contractibility of the group of diffeomorphisms in a compact Riemann surface S of genus $g \geq 2$ isotopic to the identity with the uniform topology C^∞, given in 1969 by C. Earle and J. Eells.

A topological space S is **contractible** if the identity in S is homotopic in S to a constant function, i.e., if S can be continuously deformed in S to a point p, and such homotopy is said to be a **contraction** of S to p.

(10.8) *If S is a compact Riemann surface with genus $g \geq 2$, the group* Diff$_0(S)$ *of diffeomorphisms defined in S isotopic to the identity with the uniform C^∞ topology is contractible.*

Proof By (10.6), Diff$_0(S)$ is isomorphic to the group of diffeomorphisms in S with restrictions to S^1 equal to the identity. Since $\partial_I S = \emptyset$, every diffeomorphism in S is a quasiconformal mapping. By the isotopy g_t obtained at the end of the proof of (10.6), there is a contraction of Diff$_0(S)$ to the identity in S. □

Next, we consider consequences of (10.6) for regions included in $\mathbb{C}_\infty$ obtained in 1988 by C. Earle and C. McMullen in the same article where they established the preceding result.

(10.9) *If S is a region in $\mathbb{C}_\infty$ with $\#(\mathbb{C}_\infty\backslash S) \geq 3$, every homotopy* (rel ∂S) *is a homotopy* (rel $\partial_I S$).

Proof We begin by proving there is a function $\beta : [0, +\infty[\to [0, +\infty[$ continuous at 0 such that every path γ in S including its end points and every one of its lifts $\widehat{\gamma}$ satisfy $\operatorname{diam}(\widehat{\gamma}) \leq \beta(\operatorname{diam}(\gamma))$. Without loss of generality, $S \subset \mathbb{C}_\infty\backslash\{0, 1, \infty\}$, as can be ensured with a Möbius transformation modifying spherical diameters of subsets of $\mathbb{C}_\infty$ by multiplication with a bounded factor. Let (B_1, π) be a universal covering of S, $w=\pi(0)$ and (B_1,λ) be a universal covering of $\mathbb{C}_\infty\setminus\{0,1,\infty\}$ with $\lambda(0)=w$. It is $\pi=\lambda\circ\alpha$, where $\alpha: B_1\to B_1$ with $\alpha(0)=0$. Since all branches of π^{-1} have a bounded derivative on any closed ball $B\subset S$ centered at w, there is $M>0$ such that if $\gamma\subset B$, it is $\operatorname{diam}(\widehat{\gamma})\le M\operatorname{diam}(\gamma)$, and the result follows simply with $\beta(t)=Mt$. In the general case, let $\gamma_0=\alpha(\widehat{\gamma})$ be the lift of γ by the covering (B_1,λ). For a closed set $E\subset\operatorname{cl}B_1$, let ω_E be the harmonic measure of E, i.e., the only harmonic function in $B_1\setminus E$ with value 1 in ∂E and 0 in ∂B_1 ($\omega_E(z)$ can be described like for diffusion processes as the probability that a random path starting at $z\in B_1\setminus E$ reaches ∂E before ∂B_1).

By the maximum principle, comparing the values of $\omega_{\widehat{\gamma}}(z)$ and $\omega_{\widehat{\gamma}_0}(\alpha(z))$ in $B_1\setminus\alpha^{-1}(\gamma_0)$ leads to $\omega_{\widehat{\gamma}}(0)\le\omega_{\widehat{\gamma}_0}(0)$.

The derivative of any branch of λ^{-1} with the spherical metric in the domain and the Euclidean metric in the codomain in $\mathbb{C}_\infty$ with 3 punctures is of the order $O\big(\frac{1}{d}\big)$ when the distance d to the nearest of these punctures tends to 0. Therefore, if γ is not included in a ball of radius $\sqrt{\operatorname{diam}(\gamma)}$ centered at one of the 3 punctures, then $\operatorname{diam}(\gamma_0)=O\big(\sqrt{\operatorname{diam}(\gamma)}\big)$ when $\operatorname{diam}(\gamma)\to 0$. Since for every path in B_1 including the end points and not passing through 0, $\omega_{\gamma_0}(0)=O\big(-\frac{1}{\log\operatorname{diam}(\gamma_0)}\big)$ when $\operatorname{diam}(\gamma_0)\to 0$, it is $\omega_{\gamma_0}(0)=O\big(-\frac{1}{\log\operatorname{diam}(\gamma)}\big)$ when $\operatorname{diam}(\gamma)\to 0$. If γ is included in the ball with radius $\sqrt{\operatorname{diam}(\gamma)}$ centered at one of the 3 punctures, its lift γ_0 is included in a ball bounded by a horocycle[8] of diameter $O\big(-\frac{1}{\log\operatorname{diam}(\gamma)}\big)$ and, as the harmonic measure of a ball bounded by a horocycle is comparable to the diameter of that horocycle, $\omega_{\gamma_0}(0)=O\big(-\frac{1}{\log\operatorname{diam}(\gamma)}\big)$ when $\operatorname{diam}(\gamma)\to 0$.

From the probabilistic description of the harmonic measure $\omega_{\widehat{\gamma}}$, it is to be expected that $\operatorname{diam}(\widehat{\gamma})=O\big(\omega_{\widehat{\gamma}}(0)\big)$ when $\operatorname{diam}(\widehat{\gamma})\to 0$, as the probability of a random path beginning at 0 to hit a set with a given diameter before hitting ∂B_1 is smaller if the set is closer to ∂B_1, and for such a set $\widehat{\gamma}$, $\operatorname{diam}(\widehat{\gamma})=O\big(\omega_{\widehat{\gamma}}(0)\big)$ when $\operatorname{diam}(\widehat{\gamma})\to 0$. The proof of this property (and of the last assertion in the preceding sentence) is given after the current proof.

The inequalities in the last three paragraphs above imply there exists a function $\beta:[0,+\infty[\,\to[0,+\infty[$ continuous at 0 such that for every path γ in S including its end points and every one of its lifts $\widehat{\gamma}$, it is $\operatorname{diam}(\widehat{\gamma})\le\beta\big(\operatorname{diam}(\gamma)\big)$.

Let $\phi_t:\operatorname{cl}S\to\operatorname{cl}S$ be an isotopy from φ_0 to the identity in $\operatorname{cl}S$ keeping fixed the points of ∂S, with $t\in[0,1]$, and $\widehat{\varphi}_t$ its lift by the universal covering (B_1,π) of S to an isotopy in B_1 of the lift of ϕ_0 to the identity in B_1. We wish to prove that $\widehat{\varphi}_t$ can be extended to an isotopy in $\operatorname{cl}B_1$ keeping the points of S^1 fixed. For this, construct for each $\varepsilon>0$ a neighborhood U of S^1 such that $\operatorname{diam}\big(\{\widehat{\varphi}_t(z): t\in[0,1], z\in U\}\big)<\varepsilon$. Let $\delta>0$ be such that $\beta(\delta)<\varepsilon$ and V a neighborhood of ∂S with

$$\operatorname{diam}\big(\{\widehat{\varphi}_t(x): t\in[0,1], x\in V\}\big)<\delta,$$

which exists because the isotopy keeps the points of ∂S fixed. If $z\in B_1$ is such that $\pi(z)\in V$, then $\operatorname{diam}\big(\{\widehat{\varphi}_t(z): t\in[0,1]\}\big)<\varepsilon$. As $\operatorname{cl}S\setminus V$ is compact, if $\pi(z)\in S\setminus V$,

[8] See the definition and some properties in an exercise at the end of Chapter 10 of *CADOVA*.

the diameter of $\{\widehat{\varphi}_t(z) : t \in [0,1]\}$ in the Poincaré metric of B_1 is upper bounded by a constant depending only of V. As the quotient of distances in the Poincaré and the Euclidean metrics tends to ∞ when the points approach S^1, there is a neighborhood U of S^1 such that $\operatorname{diam}\big(\{\widehat{\varphi}_t(z) : t \in [0,1]\}\big) < \varepsilon$ if $\pi(z) \in S \backslash V$. Thus, $\widehat{\varphi}_t \to 1_{S^1}$ uniformly when $|z| \to 1$. So, $\widehat{\varphi}_t$ can be extended to an isotopy in cl B_1 keeping the points of S^1 fixed. The corresponding extension of φ_t is an isotopy (rel $\partial_I S$). □

In 1985 by C. Fitzgerald, B. Rodin and S. Warschawski[9] proved:

(10.10) *If E is a continuum included in cl B_1 and ω_E is the harmonic measure of E in $B_1 \subset E$, then $\omega_E(0) \geq \frac{1}{\pi} \arcsin \frac{\operatorname{diam}(E)}{2}$.*

Proof Let $z_1, z_2 \in E$ be such that $\operatorname{diam}(E) = |z_1 - z_2|$ with $|z_2| \leq |z_1|$. Assume, without loss of generality, $z_1 = r > 0$, as this can be achieved with a rotation. It is $E \subset \mathrm{cl}\, B_1 \backslash]r, 1]$. Denote F_r the Riemann mapping of $B_1 \backslash [r, 1[$ onto B_1 extended by continuity to the prime ends[10] of $B_1 \backslash [r, 1[$ and normalized so that $F_r(0) = 0$ and $F_r(r) = 1$. $F_r(E)$ is a *continuum* in cl B_1 containing the points 1 and $F_r(z_2)$. By the maximum principle for harmonic functions, $\omega_E(0) \geq \omega_{E,r}(0) = \omega_{F_r(E)}(0)$, where $\omega_{E,r}$ is the harmonic measure of E in $B_1 \backslash [r, 1[$.

Let $z \in S^1 \cap \mathbb{H}$ and $\alpha = \{e^{it} : t \in [\operatorname{Arg} z, \pi]\}$. By the maximum principle for harmonic functions and invariance by conformal homeomorphisms, $\omega_E \alpha \geq \omega_{\alpha_r}(0) = \omega_{F_r(\alpha)}(0)$. Therefore, by the Löwner lemma (10.11) proved following the current proof, the length of the arc $F(\alpha)$ is upper bounded by the length of the arc α. Since the left end point -1 of α is fixed under F_r, $\operatorname{Arg} F_r(z) \in [\operatorname{Arg} z, \pi]$. With the harmonic function in $B_1 \backslash [r, 1[$ such that $h(z) = \operatorname{Arg} \frac{F_r(z)}{z}$, where Arg is the function argument in $\mathbb{C} \backslash \{0\}$ with values in $]-\pi, \pi]$, it is $h(z) \geq 0$ for $z \in S^1 \cap \mathbb{H}$. As for points in $z \in B_1 \cap \mathbb{H}$ close to $]r, 1]$, it is $\operatorname{Arg} z$, $\operatorname{Arg} F_r(z) \geq 0$, the limit of $h_{B_1 \cap \mathbb{H}}$ at points $x \in]r, 1]$ is ≥ 0. Due to the symmetry of the region, $F_r(\overline{z}) = \overline{F_r(z)}$ and $h(\overline{z}) = -h(z)$ for $z \in B_1$. So, $h(x) = 0$ for $x \in [-1, r]$. By the maximum principle for harmonic functions, $h(z) \geq 0$ for $z \in \big(B_1 \backslash [r, 1[\big) \cap \mathbb{H}$. Therefore, if $z = |z| e^{i\theta}$ with $|z| \in]0, 1]$, $\theta \in]0, \pi]$, then $F_r(z) = |F_r(z)|\, e^{i\Theta}$ with $\Theta \in]\theta, \pi]$, and by the Schwarz lemma,[11] $|z| \leq |F_r(z)|$.

Adapting to *continua* instead of curves the result of D. Gaier (10.12) proved below, $\omega_{F_r(E)}(0) \geq \frac{1}{\pi} \arcsin \frac{|F_r(z_2) - 1|}{2}$. By the preceding paragraph,

$$\omega_E(0) \geq \omega_{F_r(E)}(0) \geq \frac{1}{\pi} \arcsin \frac{|F_r(z_2)-1|}{2} \geq \frac{1}{\pi} \arcsin \frac{|z_2 - r|}{2} = \frac{1}{\pi} \arcsin \frac{\operatorname{diam}(E)}{2}.$$

□

[9] Fitzgerald, Carl (1942–2021). Rodin, Burton (1933-). Warschawski, Stefan (1904–1989)

[10] See Sections 5 and 6 of Chapter 10 of *CADOVA*.

[11] See Section 3 of Chapter 10 of *CADOVA*.

Next, we give proofs of the two results used in the proof above not proved yet. The first was given in 1923 by K. Löwner.

(10.11) Löwner lemma: *If f is a holomorphic function of B_1 to B_1, $f(0)=0$, $\gamma \subset S^1$ is an arc without the end points and $|f(z)| \to 1$ when $z \to \gamma$, the length of γ is upper bounded by that of $f(\gamma)$.*

Proof By the Schwarz symmetry principle,[12] f has a holomorphic extension to γ, and since $|f(z)|$ is strictly increasing when $z \to \gamma$, $f'(z) \neq 0$. So, f is a conformal homeomorphism in a neighborhood of γ and if $\operatorname{Arg} z$ increases, as conformal homeomorphisms preserve orientation, so does $\operatorname{Arg} f(z)$.

For $w = e^{i\theta} \in \gamma$, $\theta \in \mathbb{R}$ fixed and $g_w : B_1 \to B_1$ such that $g_w(z) = \frac{f(wz)}{f(w)}$, $\left[\frac{\partial}{\partial r} g_w(re^{i\theta})\right]_{r=1} = \frac{wf'(w)}{f(w)}$. For f to be a conformal homeomorphism in a neighborhood of γ, $\operatorname{Arg} f'(w) = \frac{\operatorname{Arg} f(w)}{w}$ and $\left|\left[\frac{\partial}{\partial r} g_w(re^{i\theta})\right]_{r=1}\right| = |f'(w)|$. By the Schwarz lemma,

$$|1 - g_w(re^{i\theta})| \geq 1 - |g_w(re^{i\theta})| \geq 1 - R, |f'(w)| = \left| \lim_{r \to 1} \frac{1 - g_w(re^{i\theta})}{1 - r} \right| \geq 1.$$

Integrating $|f'|$ on γ, we get that the length of $f(\gamma)$ is greater than that of γ. □

The following result was obtained in 1972 by D. Gaier.[13]

(10.12) *If γ is a simple curve in B_1 with end points 1 and $\zeta \in B_1$, $\widetilde{\gamma}$ is the arc of the circle S^1 from 1 to the point of the intersection of S^1 with the arc of a circle K through ζ and orthogonal to S^1 at the points where they intersect nearest to ζ, $\widetilde{\widetilde{\gamma}}$ is the arc of a circle with end points ζ and 1 orthogonal to S^1 at 1, and ξ_E is the harmonic measure of E in $B_1 \backslash \gamma$ for $E \subset S^1 \cup \gamma$* (Fig. 10.3), *then*

$$\xi_\gamma(0) > \tfrac{1}{2}\, \xi_{\widetilde{\widetilde{\gamma}}}(0) > \xi_{\widetilde{\gamma}}(0) > \tfrac{1}{\pi} \arcsin \tfrac{|\zeta - 1|}{2}.$$

Proof We begin by proving the first inequality in the statement. Assume, without loss of generality, $0 \notin \gamma$. Consider the functions $f_1 = T_{\overline{\zeta}}(1)\, T_\zeta$ and $f_2 = \sqrt{f_1}$, with any of the two square roots, where T_a is, as usual, the Möbius transformation $T_a(z) = \frac{z-a}{1-\overline{a}z}$. The function f_2 maps $B_1 \backslash \gamma$ to the part of B_1 bounded by the

[12] See one of the exercises at the end of Chapter 10 of *CADOVA*.

[13] Gaier, D., Estimates of conformal mappings near the boundary, *Indiana U. Math. J.* **21** (1972), 581–595. Gaier, Dieter (1928–2002).

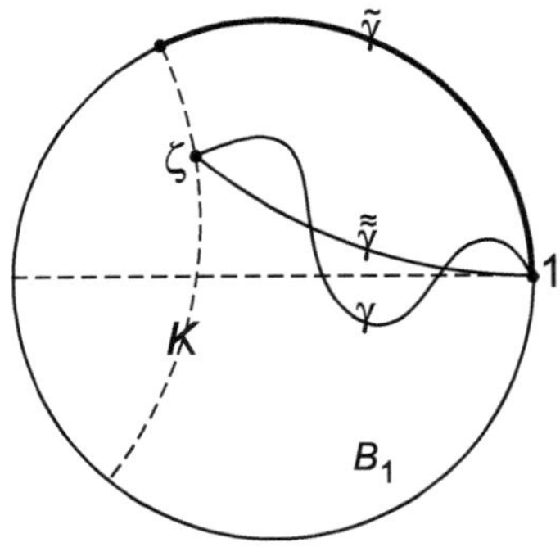

Fig. 10.3 Auxiliary drawing for the statement and the proof of (10.12)

semicircle $S^1 \cap \mathbb{H}$ and the curve $f_2(\gamma)$ passing through 0 and ending at ± 1. It is $\xi_\gamma(0) = \xi_\gamma\big(f_2(\gamma)\big)\big(f_2(0)\big) > \xi_\alpha\big(f_2(0)\big)$, where $\alpha = S^1 \cap (-\mathbb{H})$. If $\mathcal{I}m\ f_2(0) \le 0$, then $\xi_\alpha\big(f_2(0)\big) \ge \frac{1}{2}$, and therefore, $\xi_\gamma(0) > \frac{1}{2}$. Thus, $\mathcal{I}m\ f_2(0) > 0$, and taking into account that $f_2^{-1}\big([-1,\,1]\big) = \widetilde{\gamma}$,

$$\xi_\alpha\big(f_2(0)\big) = \tfrac{1}{2}\,\xi_{[-1,1]\,,\,B_1\cap\mathbb{H}}\big(f_2(0)\big) = \tfrac{1}{2}\,\xi_{\widetilde{\gamma}}(0) = \tfrac{1}{2}\,\xi_{\widetilde{\widetilde{\gamma}}}(0)\,,$$

where $\xi_{S,Y}$ denotes the harmonic measure of S in Y. In any case, $\xi_\gamma(0) > \frac{1}{2}\,\xi_{\widetilde{\widetilde{\gamma}}}(0)$, which is the first inequality in the statement.

Now, we prove the second inequality in the statement. Without loss of generality, $\mathcal{I}m\,\zeta \ge 0$ and $0 \notin \widetilde{\widetilde{\gamma}}$. $G(z) = \frac{1-\bar{z}}{1+\bar{z}}$ maps B_1 to $\mathbb{H}$ and K to a semicircle $\big(\partial B_r\big) \cap \mathbb{H}$ with $r \in]0,\,1[$, γ in $]0,\,ir[$ and $\widetilde{\widetilde{\gamma}}$ to an arc $G\big(\widetilde{\widetilde{\gamma}}\big)$ with end points 0 and $re^{i\varphi}$, $\varphi \in \big[0,\,\frac{\pi}{2}\big[$. It is left as an exercise to check that if L is a straight line segment or a ray in the real axis and $z \in \mathbb{H}$, then $\xi_{L\,,\,\mathbb{H}}(z) = \frac{\omega}{\pi}$, where ω is the angle of view of L from z. Therefore,

$$\xi_\gamma(0) = \xi_{G(\widetilde{\widetilde{\gamma}})}(0) = \tfrac{1}{\pi}\arctan r\,.$$

With $G_1(re^{i\varphi}) = \frac{1}{\cos\varphi}\big(\frac{r}{z} + i\,\sin\varphi\big)$, $G_1 \circ G$ maps $B_1 \backslash \widetilde{\widetilde{\gamma}}$ to the complex half plane to the right of the vertical line $z = 1$, i.e., $(-i\mathbb{H})\backslash[1,+\infty[$. The function,

$$G_2(z) = \tfrac{1}{2}\big(\sqrt{z} + \tfrac{1}{\sqrt{z}}\big),$$

where $\sqrt{z}$ is any of the two square roots, maps $(-i\mathbb{H})\backslash[1,+\infty[$ to $\mathbb{H}$ and $G_2\big([1,+\infty[\big) = [0,+\infty[$. It is left as an exercise to check that

$$\operatorname{Arg} G_2 \circ G_1 \circ G(0) = 2\theta\,,$$

with θ solution of the equation

$$\frac{r^2}{(\cos^2\theta)\cos^2\varphi} - \frac{\tan^2\varphi}{\sin^2\theta} = 1\,, \qquad \varphi \in \big[0,\,\tfrac{\pi}{2}\big[\,,\ \theta \in \big]0,\,\tfrac{\pi}{2}\big[\,.$$

By the angle of view property used above,

$$\xi_{\widetilde{\widetilde{\gamma}}}(0) = \xi_{[0,+\infty[\,,\,\mathbb{H}}\big(G_2 \circ G_1 \circ G(0)\big) = 1 - \tfrac{2}{\pi}\theta\,.$$

With $x = \cos^2\varphi\,,\ y = \frac{1}{\cos^2\theta}\,,$ the preceding equation is $r^2y^2 - y(1+r^2) + x = 0\,,$ which, for fixed r, is the equation of a parabola through $(0, 0)$ and $(1, 1)$. So, for $(x, y) \in\,]0, 1]\times\,]1, +\infty[\,,\ y < y(0) = \frac{1+r^2}{r^2}$. Thus, $\cos^2\theta > \frac{r^2}{1+r^2}$, i.e., $\operatorname{cotan}\theta > r$, and, therefore,

$$\tan\big(\tfrac{\pi}{2}\,\xi_{\widetilde{\gamma}}(0)\big) = \operatorname{cotan}\theta > r\,, \qquad \xi_{\widetilde{\gamma}}(0) > \tfrac{2}{\pi}\arctan r = 2\,\xi_{\widetilde{\widetilde{\gamma}}}(0)\,,$$

what includes the second inequality in the statement.

The last inequality is geometric, and it is also left as an exercise. □

The following example shows that the reciprocal of (10.9) does not hold.

(10.13) Example: Consider $S = \mathbb{H}\backslash\big([\{\frac{1}{n} : n \in \mathbb{N}\}] \cup \{0\}\big) \times [0, 1]\big)$ and an isotopy (rel $\partial_I S$) starting at identity and displacing each point $\big(\frac{2n+1}{2n(n+1)}, \frac{3}{4}\big)$ to $\big(\frac{2n+1}{2n(n+1)}, \frac{1}{4}\big)$. The diameters of the preimages of each rectangle $]\frac{1}{n}, \frac{1}{n+1}[$ by the projection of the universal covering of S defined in B_1 converge to 0 in the Euclidean metric. As there is no continuous extension of the isotopy close to $\{0\} \times [0, 1]$, the isotopy (rel $\partial_I S$) is not an isotopy (rel ∂S).

A **bounded isotopy** in a region $S \subset \mathbb{C}_\infty$ with $\#\mathbb{C}_\infty\backslash S \geq 2$ is an isotopy in S through paths with lifts to B_1 with uniformly bounded Poincaré diameters, i.e., moving points in a uniformly bounded set in the Poincaré metric of S.

(10.14) *Let S be a region in $\mathbb{C}_\infty$ with $\#(\mathbb{C}_\infty\backslash S) \geq 3$.*

1. *Every isotopy* (rel S^1) *in B_1 K-quasiconformal is bounded.*
2. *Every bounded isotopy in S is an isotopy* (rel ∂S) *in S.*
3. *If $\varphi : S \to S$ is a quasiconformal mapping, the following are equivalent:*
   (a) *$\varphi_{|S^1}$ is the identity.*
   (b) *φ is homotopic* (rel ∂S) *to the identity.*
   (c) *There is an isotopy h of $\mathbb{C}_\infty$ by K-quasiconformal mappings with the points of $\mathbb{C}_\infty\backslash S$ fixed and $h_{|S}$ an isotopy of φ to 1_S.*

Proof

(1) A K-quasiconformal mapping in $\operatorname{cl} B_1$ with the points of S^1 fixed can only move points in B_1 a bounded distance in the Poincaré metric of B_1.

(2) As the ratio of the Poincaré metric of S to the spherical metric tends to ∞ close to ∂S, the isotopy has a continuous extension to $1_{\partial S}$.

(3) Clearly (c) ⇒ (b). By (10.9), (b) ⇒ (a). If (a) holds, by (10.6), there is an isotopy K'-quasiconformal from φ to the identity, which, by 1 and 2,

has extension to an isotopy in $\mathbb{C}_\infty$ defined outside S by the identity. The continuous extension of a K'-quasiconformal mapping by the identity is a K'-quasiconformal mapping, and (c) holds. Therefore, (a) $\Rightarrow$ (c). □

(10.15) *If S is a region in $\mathbb{C}_\infty$ with $\#(\mathbb{C}_\infty \backslash S) \geq 2$, φ is a homotopy* (rel $\partial_I S$) *in S if and only if φ is a homotopy* (rel S^1) *in S.*

Proof Let $\widetilde{S}$ be the Orbifold $S \cup \partial_I S$, Γ the discrete subgroup of the group $\operatorname{Aut} B_1$ such that $\widetilde{S} = S/\Gamma$, $\Omega \subset S^1$ the complement of the limit set of γ, $\{K_j\}_{j\in\mathbb{N}}$ a sequence of connected compact sets such that $S_j \subset S_{j+1}$ for $j \in \mathbb{N}$ and $\widetilde{S} = \cup_{j\in\mathbb{N}} K_j$, γ a curve in S and $\widehat{\gamma}$ its lift by the universal covering (B_1, π) of S. The union of Ω with the hyperbolic fixed points of Γ is a dense subset of S^1. Let $\varepsilon > 0$ and F be a finite subset of this set such that the union of circles with radius $\varepsilon > 0$ and centers at the points of F covers S^1 and L is the convex hull of F. L/Γ is a compact subset of $\widetilde{S}$. So, for $N \in \mathbb{N}$ large, $L/\Gamma \subset K_N$. The connected components of $\operatorname{cl} B_1 \backslash L$ have diameters $< \varepsilon$. If $\gamma \cap K_N = \emptyset$, then $\widehat{\gamma} \cap L = \emptyset$ and $\operatorname{diam}(\widehat{\gamma}) < \varepsilon$. Consequently, there is a sequence $\{\alpha_n\}$ convergent to 0 such that $\operatorname{diam}(\widehat{\gamma}) \leq \alpha_{N(\gamma)}$, with $N(\gamma)$ the largest natural number such that $\gamma \cap K_N = \emptyset$.

Similarly to the proof of (10.9), we consider a homotopy φ_t (rel $\partial_I S$) and $\widehat{\varphi}_t$, $t \in [0, 1]$, and its lift to a homotopy in $B_1 \cup \Omega$ such that $\widehat{\varphi}_0$ is the identity. It is necessary to consider a neighborhood U of S^1 such that $\operatorname{diam}(\widehat{\varphi}_{[0,1]}(z)) < \varepsilon$ for all $z \in U$. For this, let $n \in \mathbb{N}$ be such that $\alpha_N < \varepsilon$ and let $M \in \mathbb{N}$ be such that $K_M \supset \varphi_{[0,1]}(K_N)$. So, $\operatorname{diam}(\widehat{\varphi}_{[0,1]}(z)) < \varepsilon$ if $\pi(z) \notin K_m$. If $L \subset B_1 \cup \Omega$ is a compact set such that $\pi(L) = K_M$, then $\widehat{\varphi}_{[0,1]}(\gamma L) = \gamma(\widehat{\varphi}_{[0,1]}(L))$ for $\gamma \in \Gamma$. Since $\widehat{\varphi}_{[0,1]}(L)$ is a compact subset of the set of points of discontinuity, the Euclidean diameters of the translations of this set by Γ converge to 0. Hence, $\operatorname{diam}(\widehat{\varphi}_{[0,1]}(z)) < \varepsilon$ except possibly for z in one of the finite translations of L, and, as these form a compact set, the set of the isotopy fixed points is $L \cap S^1$. So, for z in some neighborhood of S^1, $\operatorname{diam}(\widehat{\varphi}_{[0,1]}(z)) < \varepsilon$. □

10.5 Teichmüller Space of a Rational Dynamical System

A rational dynamical system is a holomorphic dynamical system $(\mathbb{C}_\infty, f)$ with f a rational function. Denote $\boldsymbol{J}$ the Julia set of f, $\widehat{\boldsymbol{J}}$ the closure of the union of the grand orbits of the periodic points with the grand orbits of the critical points of f, and $\widehat{\boldsymbol{F}} = \mathbb{C}_\infty \backslash \widehat{J}$. Denote $\boldsymbol{\Omega}^{dis}$ and $\boldsymbol{\Omega}^{fol}$ the unions of the open sets of points with grand orbits, resp., discrete and not discrete; so, $\{\Omega^{dis}, \Omega^{fol}\}$ is a partition of $\widehat{F}$. For $S \subset \mathbb{C}_\infty$ open and invariant under f denote $\boldsymbol{S/f}$ the quotient of S by the equivalence relation of grand orbits.

By the Sullivan classification theorem,[14] $\widehat{J}$ is the union of the Julia set, grand orbits of attracting or superattracting periodic points, centers of Siegel disks (a countable set), and leaves of the foliations invariant under f, which contain grand orbits of critical points (a countable union of one-dimensional *continua*). $(\widehat{F}, f)$ is a covering of $\widehat{F}$ without periodic points.

(10.16) *The Teichmüller space* $\mathrm{Teich}(\mathbb{C}_\infty, f)$ *of a rational function* f *of degree* $d \geq 2$ *is isomorphic to*

$$M(J)_1^f \times \mathrm{Teich}(\Omega^{fol}, f) \times \mathrm{Teich}(\Omega^{dis}/f),$$

where Ω^{dis}/f *is a complex manifold and* $M(J)_1^f$ *is the intersection of* J *with the open ball of the Banach subspace of* $L^\infty(\mathbb{C}_\infty)$ *of the invariant Beltrami coefficients under* f *with center at* 0 *and radius* 1, $M(\mathbb{C}_\infty)_1^f$.

Proof Since $\widehat{J}$ contains a countable subset dense in the union of $\widehat{J}$ with the grand orbits of f, for every $\omega \in QC_0(\mathbb{C}_\infty, f)$, $\omega_{|\widehat{J}} = 1_{\widehat{J}}$, and, therefore,

$$QC_0(\mathbb{C}_\infty, f) = QC_0(\Omega^{fol}, f) \cup QC_0(\Omega^{dis}, f).$$

As $\widehat{J} \backslash J$ has measure 0, there is a factorization in Cartesian products as in the statement, with $\mathrm{Teich}(\Omega^{dis}, f)$ instead of $\mathrm{Teich}(\Omega^{dis}/f)$. Ω^{dis} is a countable union of disjoint open sets Ω_j^{dis} totally invariant under f such that Ω_j^{dis}/f are connected. Hence,

$$\mathrm{Teich}(\Omega^{dis}, f) = \textstyle\prod'_j \mathrm{Teich}(\Omega_j^{dis}, f) = \prod'_j \mathrm{Teich}(\Omega_j^{dis}/f) = \mathrm{Teich}(\Omega^{dis}/f),$$

where $\prod'_j T_j$ is the quotient of the restricted product of the corresponding deformation spaces by the restricted product of the corresponding groups QC_0 and the **restricted product** of topological groups is the set of ${t_j}_j \subset \cup_j T_j$ such that $t_j \in T_j$ except possibly for j in a finite subset of indices.

Since Ω^{dis} does not contain periodic points of f, the orbifold Ω^{dis}/f has no singularities, and it is a complex manifold. □

A **polydisk** of dimension n is a set homeomorphic to an open ball of $\mathbb{C}^n$.

An **ergodic component** of a rational function $f : \mathbb{C}_\infty \to \mathbb{C}_\infty$ is a maximal connected set $A \subset \mathbb{C}_\infty$ such that $f^{-1}(A) = A$ and $\mathbb{C}_\infty \backslash A$ or A has measure 0.

If f is a complex rational function, $x, y \in \mathbb{C}_\infty \backslash J(f)$ are said to be in the same **foliated equivalence class** if the closures of its grand orbits coincide.

[14] See Section 12 of Chapter 13 of *CADOVA*.

(10.17) *In the preceding result:*

1. $M(J)_1^f$ *is a polydisk of dimension* $\leq 2(d-1)$ *equal to the number of ergodic components of the maximal measurable subset of the Julia set of* f *supporting a line field.*
2. $\mathrm{Teich}(\Omega^{fol}, f)$ *is a polydisk of dimension equal to the number of cycles of Herman rings plus the number of foliated equivalence classes of non-cyclic critical points with orbits eventually entering Siegel disks, Herman rings, or superattracting basins.*
3. Ω^{dis}/f *is a finite union of Riemann surfaces, one for each cycle of attracting or parabolic components of the Fatou set of* f. *Each attracting basin contributes with a torus with* n *punctures and each parabolic basin with a sphere with* $n+2$ *punctures, where* n *is the number of grand orbits of critical points of* f *eventually entering into the basin.*
4. $\dim \mathrm{Teich}(\mathbb{C}_\infty, f) = n_{AC} + n_{HR} + n_{LF} - n_P$, *resp., the numbers of foliated equivalence classes of non-cyclic critical points in the Fatou set of* f, *cycles of Herman rings, ergodic components supporting line fields in the Julia set of* f, *cycles of parabolic basins.*

Proof

(1) If $\kappa \in M(J)_1^f$, the extension to $\mathbb{C}_\infty$ with $\kappa = 0$ in $\mathbb{C}_\infty \backslash J$ is an injection of $M(J)_1^f$ into $\mathrm{Def}(\mathbb{C}_\infty, f)$. Let $\eta : \mathrm{Def}(\mathbb{C}_\infty, f) \to \mathrm{Rat}_d/\mathrm{Aut}(\mathbb{C}_\infty)$ be the holomorphic function mapping each $(\mathbb{C}_\infty, g, \varphi)$ identified with a point of $\mathrm{Def}(\mathbb{C}_\infty, g)$ with g a rational function determined up to conjugacy with conformal transformations φ. These rational mappings g form an orbifold of dimension $2(d-1) = \dim \mathrm{Rat}_d$.

If it were $n > 2(d-1)$, the preimages of elements of $\mathrm{Rat}_d/\mathrm{Aut}(\mathbb{C}_\infty)$ by $\eta_{|B_1}$ would be analytic manifolds of dimension ≥ 1. Therefore, for each g_0 in an orbifold there is an analytic arc from φ_t to g_0 that projects onto g_0, and with t varying, the marking of g_0 changes by a quasiconformal isotopy, in contradiction with the existence of an injection of B_1^n in $\mathrm{Teich}(\mathbb{C}_\infty, f)$.

Consequently, $M(J)_1^f$ is a subset of a finite dimensional complex linear space, and therefore, it is a polydisk. Line fields[15] supported by ergodic components of J form a basis of $M(J)^f$.

(2) Ω^{fol} is a disjoint union of totally invariant open sets under f, $\Omega^{fol} = \cup_j \Omega_j^{fol}$, with Ω_j^{fol}/f connected for each j, and

$$\mathrm{Teich}(\Omega^{fol}, f) = \textstyle\prod_j' \mathrm{Teich}(\Omega_j^{fol}, f),$$

[15] See Section 3 of Chapter 5 of *CADOVA*.

with the notation for restricted product of the proof of (10.16). Each factor in the right-hand side of the equation is conformal to a disk or it is a point. Each factor conformal to a disk has a lift in $\mathrm{Def}(\mathbb{C}_\infty, f)$, and, as $\dim \mathrm{Rat}_d$ is finite, the number of factors conformal to disks is finite, and they correspond to Herman rings, Siegel disks, or superattracting basins of periodic orbits. A cycle of foliated regions with n leaves with critical points gives n periodic rings if they are Siegel disks or superattracting orbits, and it gives $n+1$ periodic rings if they are Herman rings. If two critical points belong to a same leaf, they belong to the same foliated equivalence class.

(3) By the Sullivan nonwandering domains theorem (6.3), each connected component of Ω^{dis} is preperiodic. Hence, each connected component S of Ω^{dis}/f can be represented by Y/f^p with Y obtained from an attracting basin or a parabolic basin U of period p removing the periodic points and the grand orbits of critical points.

If U is the attracting basin of a periodic point x with period p, which, without loss of generality, is assumed to be $\neq\infty$, and if $\lambda=(f^p)'(x)$ is its multiplier, the linearization at x is a holomorphic function[16]

$$\psi(z) = \lim_{n\to+\infty} \lambda^{-n}[f^{np}(z)-x],$$

mapping U to $\mathbb{C}$ injectively and satisfying $\psi\circ f^p = \lambda\psi$. If U' is the complement in U of the grand orbit of x, the space of the grand orbits in U' is conformal to $(\mathbb{C}\setminus\{0\})/T_\lambda$, where $T_\lambda(z)=\lambda z$, which is a complex torus, and eliminating from U' the points of orbits of critical points, we get Y/f^p.

If U is a parabolic basin, there is[17] $\psi: U\to\mathbb{C}$ such that $\psi\circ f^p(z)=z+1$. The space of grand orbits in U is a cylindrical surface, and eliminating the points of orbits of critical points, we also get Y/f^p.

In both cases, it is necessary to eliminate $n\geq 1$ orbits of critical points so that the number of connected components of Ω^{dis}/f is upper bounded by the number of critical points of f, which is $2(d-1)$.

(4) The dimension of the Teichmüller space of a torus with n punctures is n and of a sphere with $n+2$ punctures is $n-1$. Therefore, $\dim \mathrm{Teich}(\Omega^{dis}/f)$ is equal to the number of grand orbits of acyclic critical points in Ω^{dis} minus n_P. By 2, the number of the remaining orbits of critical points (modulo foliations equivalence) plus n_{HR} is equal to $\dim \mathrm{Teich}(\Omega^{fol}, f)$. On the other hand, $\dim M(J)_1^f = n_{LF}$. Therefore, by (10.16), we have the formula in 4 for $\dim \mathrm{Teich}(\mathbb{C}_\infty, f)$. □

An immediate consequence is an alternative proof that the number of cycles of stable regions of a complex rational function f of degree $d\in\mathbb{N}\setminus\{1\}$ is finite, since every superattracting, attracting, or parabolic orbit attracts at least one

---

[16] See Section 6 of Chapter 13 of *CADOVA*.

[17] See Section 9 of Chapter 13 of *CADOVA*.

critical point, and also, with a proof of P. Fatou,[18] that with arbitrarily small perturbations of f at least half of the neutral cycles can be made attracting, and since $\dim(\mathbb{C}_\infty, f) \leq 2(d-1)$ and this is also the number of critical points of f counting multiplicities, by the preceding result, f has $\leq 2(d-1)$ Herman rings and $\leq 4(d-1)$ Siegel disks. Consequently, the number of cycles of stable regions is upper bounded by $8(d-1)$ (although, as shown in Chap. 7, this upper bound can be improved to $2(d-1)$ with the M. Shishikura contribution published in 1998).

These results also give an alternative proof of the Sullivan nonwandering domains theorem (6.3), beginning with the first two paragraphs of the proof given in Chap. 6, guaranteeing that if there would exist a wandering domain Ω_0, for large m, $f^m(\Omega_0) \subsetneqq \mathbb{C}$ would be a simply connected nonwandering region, and therefore, it would be conformal to B_1, and pursuing with the following simple and short argument: for large m, $f^m(\Omega_0)$ does not contain critical points. Thus, f is a bijection of $f^m(\Omega_0)$ onto $f^{m+1}(\Omega_0)$ and the Riemann surface Ω^{dis}/f has a connected component S conformal to B_1 with a finite number of punctures, one for each grand orbit containing a critical point and intersecting Ω_0. So, the Teichmüller space of S would be infinite dimensional, what is false for Teichmüller spaces of rational functions.

It can be proved as in (8.7) that *the action of the modular group* $\mathrm{Mod}(\mathbb{C}_\infty, f)$ *of a rational function* f *of degree* $d \geq 2$ *in* $\mathrm{Teich}(\mathbb{C}_\infty, f)$ *is properly discontinuous.*

[18] See the argument in the proof of the second result of Section 10 of Chapter 13 of *CADOVA*.

Appendix A
Elements of Lebesgue Integral and Measure

A.1 Introduction

This appendix is included to facilitate access to certain aspects of Lebesgue integral and measure to readers who might not know them yet, as this book is intended for second-university-year students with a good background in Mathematical Analysis studying the author book *CADOVA* or an equivalent source, e.g., one of those mentioned in the Preface, with minimal additional basic bibliography.

The first section can be ignored by readers who master the basic aspects of Lebesgue integral and measure, including those in Appendix B of *CADOVA* following the simplified approach for $\mathbb{R}^n$ given by F. Riesz in 1924 which begins by defining Lebesgue integral with sequences of step functions and defines Lebesgue measure through the integral, but most likely some readers cannot spare the last section, on relationships of functions of completely additive set functions with indefinite integrals and derivatives, volume derivative, and integrability of homeomorphism Jacobians.

The notion of Lebesgue integral is due to H. Lebesgue in 1902 in his doctoral thesis with the elegant title *Intégrale, Longueur, Aire*, where he created Measure Theory and defined the integral of functions approximated by simple functions constant in measurable sets except possibly in subsets with measure 0, through series of the values of these functions in each set where they are constant multiplied by the set measure.

Geometrically, the idea of the Lebesgue integral of real valued functions of real variable compared to that of the Riemann integral, given by B. Riemann in 1854, can be described as the Lebesgue integral resulting from approximating the area bounded by the function graph and the abscissa axis by the sum of the areas of sets defined by constant functions in subsets of the domain obtained by finite partitions in subintervals in the function range, instead of the areas of rectangles obtained by subdivision of the domain in a finite number of subintervals

L. T. Magalhães, *Quasiconformal Mappings in the Plane and Complex Dynamics*,
https://doi.org/10.1007/978-3-031-80115-0

corresponding to integrals of step functions or Darboux[1] sums. This geometric idea of the Lebesgue integral, replacing finite partitions of the domain by finite partitions of bounded intervals in the codomain, requires to consider measure of sets preimages of intervals with extremes the constant values considered in the codomain for the approximations. This is why the need to consider the measure of sets that are not intervals arises. So, the introduction of the Lebesgue integral for real functions of real variables based on the original idea of H. Lebesgue starts by defining the measure of sets and obtaining several properties of this measure.

In 1920 F. Riesz presented a direct construction of the Lebesgue integral in $\mathbb{R}^n$ following the line of definition initiated with the Riemann integral through approximations by step functions, considering nondecreasing sequences of step functions with bounded integrals converging except possibly in a measure 0 set to a function, called and **upper function** (instead of also considering approximations from above), whereby the need to consider the measure of sets is reduced to measure 0 sets. These are simply sets that can be covered by a countable set of intervals with sum of lengths, in the sense of series when infinite, that can be made arbitrarily small.[2] With this idea, the construction of the Lebesgue integral is simplified to the point of following the line of definition of the Riemann integral and greatly reducing the preliminary properties needed to justify the definition, since properties of the measure of nonzero measure sets are not needed, with the advantage of being based, from the definition itself, on switching limits with integrals and, therefore, being directly linked to the practical issues of evaluating integrals. With this definition of Lebesgue integral, which is the one adopted in this appendix, definitions of measurable set and Lebesgue measure are obtained from the definition of integral and not the reverse.

Chapter 1 requires for certain properties of quasiconformal mappings, the notion of absolutely continuous functions, a particular case of functions of bounded variation. The latter were introduced in 1881 by C. Jordan who characterized them as differences of nondecreasing functions and noted that the indefinite integral of a Riemann integrable function is a function of bounded variation. The existence of derivative of a monotone function almost everywhere, i.e., except possibly in a measure 0 set and, therefore, also of a function of bounded variation, and the integrability of that derivative were established by H. Lebesgue in 1904 (though at the time with the additional unnecessary condition of continuity); this result is called Lebesgue differentiation theorem. To simplify the analysis of this question, it is useful to consider upper- or lower-right/left derivatives of a function at a point, introduced by U. Dini in 1878, to study in detail the fundamental theorem of calculus with the Riemann integral, as well as the rising Sun lemma established by F. Riesz in 1932, according to which the "shadow" set of the domain of a continuous function is an open set, imagining a function graph as a landscape profile illuminated

[1] Darboux, Gaston, 1842–1917

[2] The Magistral book of F. Riesz and B. Sz.-Nagy, *Functional Analysis*, Frederick Ungar Pub. Co., New York, 1978, whose reading is still very useful today, contains an exposition of this route to the Lebesgue integral. Szőkefalvi-Nagy, Béla (1913–1998).

horizontally by the "Sun" considered at $+\infty$. To prove that the derivative of a monotone function is integrable, we use the Fatou lemma, obtained by P. Fatou in 1806, according to which if a sequence of non-negative functions with uniformly bounded integrals in the same set converges almost everywhere to a function in that set, this function is integrable and has an integral bounded by the lower limit of the sequence of the integrals of the functions of the initially considered sequence.[3]

One consequence of the Lebesgue differentiation theorem is the Fubini theorem of term-by-term differentiation of convergent series of monotone functions, established by G. Fubini in 1915. This result allows short proofs of two fundamental results: *(i)* the Lebesgue density theorem, obtained by H. Lebesgue in 1903, establishing that, except possibly on zero measure sets, points of any subset of real numbers are points of density, in the sense of the exterior measure of the intersection of the set with an interval centered at the point divided by the length of the interval to converge to 1 when the length of the interval tends to 0, and *(ii)* the part of the fundamental theorem of calculus for Lebesgue integral concerning derivatives of indefinite integrals of integrable functions, which are continuous functions of bounded variation.

The second part of the fundamental theorem of calculus for the Lebesgue integral, corresponding to the Barrow rule that gives the integral of a function in a bounded interval by the difference of an antiderivative of the function in the extreme points of the interval, concerns to functions with absolutely continuous antiderivative. It was announced in 1904 by H. Lebesgue without proof and was proved in 1905 by G. Vitali, who created the name absolutely continuous function, though this property had been identified by A. Harnack[4] in 1882 for indefinite integrals in the context of improper Riemann integrals of unbounded functions. In 1907 H. Lebesgue gave a much simpler proof of this part of the fundamental theorem of calculus.

In order to establish the equivalence of the geometric and analytic definitions of quasiconformal mappings in Chap. 1, the integrability of a homeomorphism Jacobian is used in a region of the plane where it has finite partial derivatives almost everywhere (a.e.), *i.e.*, except in a measure 0 set. This is why this appendix includes a final section for that purpose, considering completely additive set functions and, in particular, absolutely continuous set functions, and the notion of their derivative, establishing that it is measurable almost everywhere with integral on Borel sets bounded by the value of the function on the set of integration, a result for completely additive set functions analogous to the part of the fundamental theorem of calculus for integrals of derivatives of real functions of real variables. This result was established by H. Lebesgue in 1910 and later simplified by others. It is applied to the volume derivative of homeomorphisms to obtain the relationship with a homeomorphism Jacobian in a region of the plane where it has finite partial derivatives almost everywhere and to prove integrability of the Jacobian.

---

[3] Dini, Ulisse (1845–1918).

[4] Vitali, Giuseppe (1875–1932). Harnack, Carl Gustav Alex (1851–1888).

A.2 Bounded Variation and Absolutely Continuous Functions and Fundamental Theorem of Calculus

For a short proof of the Lebesgue differentiation theorem, stating that monotone real valued functions of one real variable have finite derivative a.e., the following version of the F. Riesz rising Sun lemma of 1932 is useful.

(A.1) Rising Sun lemma:
If $g:[a,b]\to\mathbb{R}$ is continuous and $S\subset[a,b]$ is an open set, then the set

$$S_g=\{x\in S:\exists_{y>x}\;]x,y[\subset S,\;g(x)>g(y)\}$$

is open, and if $]c,d[$ is one of its connected components, $g(c)\geq g(d)$.

Proof The proof that S_g is an open set is left as an exercise (Fig. A.1[5]). Let $x\in]c,d[$ and $s=\max\{r\in[x,d]:g(x)\geq g(r)\}$. If it were $s<d$, it would be $g(x)<g(d)$, $s\in S_g$, and there would exist $t>s$ with $]s,t[\subset S$ and $g(s)>g(t)$. If it were $t\leq d$, it would be $g(x)\geq g(s)>g(t)$ contradicting the definition of s. If it were $t>d$, it would be $g(d)>g(x)\geq g(s)>g(t)$ and $d\in S_g$, contradicting that $]c,d[$ is a connected component of S_g. Hence, $s=d$ and $g(x)\geq g(d)$ for every $x\in]c,d[$. Since g is continuous, $g(c)\geq g(d)$. □

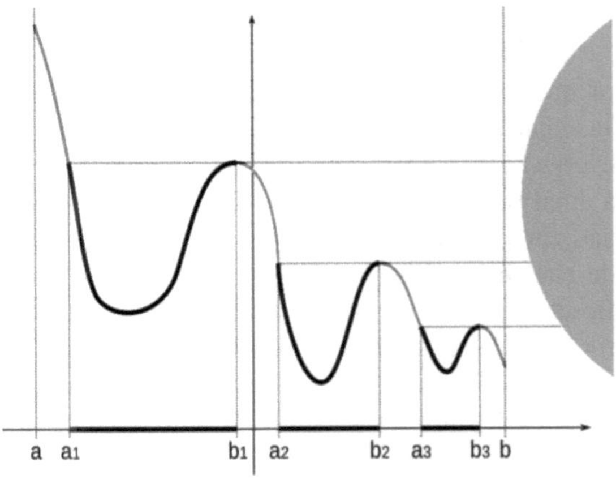

Fig. A.1 "Shadow" set S_g in the rising Sun lemma for a continuous function $g:[a,b]\to\mathbb{R}$

[5] Author and date of the figure: Claudio Rocchini (24.01.2014), under License *Creative Commons Attribution-Share Alike 3.0 Unported*. https://commons.wikimedia.org/wiki/File:Rising_sun_lemma.svg

(A.2) Lebesgue differentiation theorem: *Every monotone function $f:[a,b]\to\mathbb{R}$ is differentiable* a.e.. *Discontinuities of f, if they exist, are countable jump discontinuities, and $f'\in L^1([a,b])$, $\int_a^b|f'|\le|f(b)-f(a)|$.*

Proof The result is first established with the additional hypothesis of continuity of f, and without loss of generality, f is assumed to be nondecreasing.

The upper-right and lower-right derivative of f, denoted, resp., $D^+f(t)$ and $D_+f(t)$, are defined at a point t, by, resp., the upper and lower derivative of the incremental ratio $\frac{1}{h}[f(t+h)-f(t)]$ when $h\to 0+$. Similarly, the upper-left and lower-left derivatives at a point t, denoted $D^+f(t)$ and $D_-f(t)$, are defined by the, resp., upper and lower limit of the incremental ratio when $h\to 0-$. It is

$$D^+f(t)\ge D_+f(t),\, D^-f(t)\ge D_-f(t)$$

and f is differentiable at t if and only if

$$D^+f(t)=D_+f(t)=D^-f(t)=D_-f(t)\ne\pm\infty\,.$$

It suffices that *(i)* $D^+f\ne\infty$ and *(ii)* $D^+f\le D_-f$ a.e., since applying *(ii)* to $t\mapsto -f(-t)$ gives $D^-f\le D_+f$ a.e. and, therefore,

$$-\infty<D^+f\le D_-f\le D^-f\le D_+f\le D^+f<+\infty, \quad \text{a.e.};$$

thus, once *(i)* and *(ii)* are proven, the inequalities $\le$ are equalities, and the result is valid for continuous functions.

(i) If $E_\infty=\{t\in[a,b]:D^+f(t)=\infty\}$, $E_C=\{t\in[a,b]:D^+f(t)>C\}$ for $C\in\mathbb{R}$ and $y\in E_C$, there is $x>y$ such that $\frac{f(x)-f(y)}{x-y}>C$, i.e., $g(x)>g(y)$ where $g(t)=f(t)-Ct$. By the rising Sun lemma, $E_C\subset S_g=\cup_n\,]c_n,d_n[$ and $f(c_n)-Cc_n=g(c_n)\ge g(d_n)=f(d_n)-Cd_n$. Therefore,

$$C\sum_n(d_n-c_n)\le\sum_n[f(d_n)-f(c_n)]\le f(b)-f(a)\,.$$

Since $E_\infty\subset E_C$ for every $C>0$, E_∞ is covered by a countable set of intervals with sum of lengths $\le\frac{f(b)-f(a)}{C}$ for $C>0$ and, therefore, E_∞ has measure 0.

(ii) If $0<c<C$, $E_{c,C}=\{t\in[a,b]:D^+f(t)>C>c>D_-f(t)\}$ and $E=\{t\in[a,b]:D^+f(t)>D_-f(t)\}$, then $E\subset\cup_{c,C\in\mathbb{Q},\,c<C}\,E_{c,C}$. Thus, to prove that E has measure 0 it suffices to prove that $E_{c,C}$ has measure 0 for each pair of rational numbers $c<C$. By the rising Sun lemma with $g(t)E=f(-t)E+ct$,

$$E_{c,C}\subset -S_g=-\cup_n\,]c_n,d_n[=\cup_n\,]-d_n,-c_n[\,,\quad f(-c_n)-f(-d_n)\le c(d_n-c_n)\,,$$

and, also by the rising Sun lemma, but with $g(t)=f(t)-Ct$,

$$E_{c,C}\subset S_g=\cup_m\,]c'_m,d'_m[\,,\qquad Cf(d'_m)-f(c'_m)\le(d'_m-c'_m)\,.$$

Hence, $E_{c,C}$ is a subset of the union of the intervals that are intersections of the intervals in the two considered unions, which are a countable set denoted $\{]c_{n,k}, d_{n,k}[\}$, with $]c_{n,k}, d_{n,k}[\subset]c_n, d_n[$; therefore,

$$Cf(d_{n,m}) - f(c_{n,m}) \leq (d_{n,m} - c_{n,m}).$$

Denoting

$$\Sigma_1 = \sum_n (d_n - c_n), \qquad \Sigma_2 = \sum_{n,m} (d_{n,m} - c_{n,m}),$$

$$V_1 = \sum_n [f(d_n) - f(c_n)], \qquad V_2 = \sum_{n,m} [f(d_{n,m}) - f(c_{n,m})],$$

it is $C\Sigma_2 \leq V_2 \leq V_1 \leq c\Sigma_1$, and therefore, $\Sigma_2 \leq \frac{c}{C}\Sigma_1$. Repeating and alternating this process gives sets of intervals with elements included in the preceding intervals and sums of the intervals lengths satisfying $\Sigma_{2j} \leq \frac{c}{C}\Sigma_{2j-1}$ for $j \in \mathbb{N}$; hence, $\Sigma_{2j} \leq \left(\frac{c}{C}\right)\Sigma_1 \to 0$ when $j \to +\infty$, and therefore, $E_{c,C}$ has measure 0.

With the differentiability a.e. of continuous nondecreasing functions proved, to extend it to discontinuous functions, it suffices to modify the argument replacing the values of the functions g to which the rising Sun lemma was applied by the corresponding lateral limits at points of discontinuity, which exist since f is a monotone function.

A monotone function in an interval $[a, b]$ can only be discontinuous in a countable set of points, as each point of discontinuity corresponds to a jump of the function values from the left to the right lateral limits, and each interval with these limits as extreme points can be labeled by one rational number it contains. Thus, the set of discontinuity points of f is countable, and therefore, it has measure 0.

Consequently, $g(t) = \lim\limits_{h \to 0} \frac{1}{h}[f(t+h) - f(t)]$ defines a real valued function a.e. in $[a, b]$; thus, f is differentiable a.e. in $[a, b]$, with $f' = g$. Defining $g_m(t) = m\left[f\left(t + \frac{1}{m}\right) - f(t)\right]$, where it is considered that $f(t) = f(b)$ for $t \geq b$, gives $g_m \geq 0$, since f is nondecreasing, and $g_m(t) \to g(t) = f'(t)$ a.e. for $t \in [a, b]$; therefore, f' is measurable, and by the Fatou lemma,[6]

$$\int_a^b f' \leq \varliminf_{m \to +\infty} \int_a^b g_m = \varliminf_{m \to +\infty} m \int_a^b \left[f\left(t + \tfrac{1}{m}\right) - f(t)\right] dt$$

$$= \varliminf_{m \to +\infty} \left(m \int_b^{b+\frac{1}{m}} f - m \int_a^{a+\frac{1}{m}} f\right) = \varliminf_{m \to +\infty} \left(f(b) - m \int_a^{a+\frac{1}{m}} f\right) \leq f(b) - f(a).$$

Therefore, $f' \in L^1([a, b])$ and $\int_a^b f' \leq f(b) - f(a)$. □

As a consequence, the following result of G. Fubini in 1915 can be proved.

---

[6] The Fatou lemma ensures that the integral of the limit a.e. of a sequence of measurable non-negative functions in a measurable set is bounded by the lower limit of the sequence of integrals of the terms of the sequence of functions considered. It can be proved analogously to proving the Lebesgue dominated convergence theorem.

(A.3) Fubini theorem for derivative of series of monotone functions: *If $\{f_n\}$ is a sequence of monotone functions in $[a, b]$ of the same type such that $\sum_{n=1}^{\infty} f_n$ converges, $\left(\sum_{n=1}^{\infty} f_n\right)' = \sum_{n=1}^{\infty} f_n'$ a.e. in $[a, b]$.*

Proof Without loss of generality, $\{f_n\}$ is nondecreasing and $f_n(a) = 0$ for all $n \in \mathbb{N}$, since the value of each function at a can be subtracted from each one of the functions and the subtracted values added to the series partial sums. Denote $f(t) = \sum_{n=1}^{\infty} f_n(t)$. As each f_n is nondecreasing and ≥ 0, so is f. By the Lebesgue differentiation theorem, the functions f_n, for $n \in \mathbb{N}$, and f are differentiable a.e. in $[a, b]$, and the union E of the sets where each of these functions is not differentiable is a set of measure 0 such that the functions $s_m = \sum_{n=1}^{m} f_n$ are nondecreasing and $s_m'(t) \leq s_{m+1}'(t) \leq f'(t)$ for $t \in [a, b] \backslash E$; so, $\sum_{n=1}^{\infty} f_n'$ is defined and converges a.e. in $[a, b]$.

As $\lim s_m(b) = f(b)$, there is a subsequence $\{s_{m_k}\}$ of $\{s_m\}$ with $f(b) - s_{m_k}(b) < \frac{1}{2^k}$ for $k \in \mathbb{N}$, and $f(t) - s_{m_k}(t) < f(b) - s_{m_k}(b) < \frac{1}{2^k}$ for $k \in \mathbb{N}$, $t \in [a, b]$. The sequence $\{f - s_{m_k}\}$ is monotone nondecreasing with terms ≥ 0 in $[a, b]$ and, for the same reason as in the preceding paragraph, $\sum_{k=1}^{\infty} (f' - s_{m_k}')$ converges a.e. in $[a, b]$; thus, $f' - s_{m_k}' \to 0$ when $k \to +\infty$ a.e. in $[a, b]$. As $\{s_m'(t)\}$ is a nondecreasing sequence at each point t where it is defined, $\lim s_m' = f'$ a.e. in $[a, b]$. □

A useful consequence is the Lebesgue density theorem in $\mathbb{R}$.

If $A \subset \mathbb{R}^n$, the **density of x in** A is defined to be $d_A(\mathbf{x}) = \lim\limits_{\varepsilon \to 0} \frac{\mu^*(A \cap B_\varepsilon(\mathbf{x}))}{\mu(B_\varepsilon(\mathbf{x}))}$, where μ is the Lebesgue measure and μ^* is the exterior Lebesgue measure $\mathbb{R}^n$. A $\mathbf{x} \in \mathbb{R}^n$ is said to be a **point of density of** A if $d_A(\mathbf{x}) = 1$.

For example, for a square in $\mathbb{R}^2$ the density of a point is 1, 0, $\frac{1}{2}$ or $\frac{1}{4}$ according to the point is, resp., interior, exterior, in the boundary but not a verticex, or a verticex of the square.

(A.4) Lebesgue density theorem in $\mathbb{R}$: *If $A \subset \mathbb{R}$ and $D = \{p \in A : p$ is a point of density of $A\}$, then $N \overset{def}{=} A \backslash D$ has measure 0.*

Proof For $k \in \mathbb{N}$ fixed let $f(t) = \mu^*\left(A \cap \,]-k, \min\{t, k\}[\right)$, $\{\Sigma_n\}$ be a sequence of open subsets of $[-k, k]$ containing $A \cap \,]-k, k[$ with $\Sigma_{n+1} \subset \Sigma_n$ and lengths tending to $\mu^*\left(A \cap \,]-k, \min\{t, k\}[\right)$, and $f_n(t) = \mu^*\left(\Sigma_n \cap \,]-k, \min\{t, k\}[\right)$. As each Σ_n is a countable union of open intervals and, therefore, also $\Sigma_n \cap \,]-k, \min\{t, k\}[$ is, and $f_n(t)$ is the sum of the lengths of disjoint open intervals with union $\Sigma_n \cap \,]-k, \min\{t, k\}[$, for $t \in A$ with $t \leq k$, it is $f_n'(t) = 1$. Applying the Fubini theorem for the derivative of series of monotone functions to $\{f_n - f\}$ and considering the restrictions of these functions to $[-k, k]$, it is $f_n' - f' \to 0$ a.e. in $[-k, k]$;

so, $f' = 1$ a.e. in $A \cap [-k, k]$. Since $k \in \mathbb{N}$ is arbitrary, $f' = 1$ a.e. in A and the density of t in A is $f'(t)$. □

$f : [a, b] \to \mathbb{R}$ is said to be a **function of bounded variation** if there is $M > 0$ such that for every finite partition of $[a, b]$ by points $a = t_0 < t_1 < \cdots < t_m = b$, then $\sum_{k=1}^{m} |f(t_k) - f(t_{k-1})| \le M$, and in this case, the supremum of the sums $\sum_{k=1}^{m} |f(t_k) - f(t_{k-1})|$ for all possible finite partitions of $[a, b]$ is called the **total variation** of f in $[a, b]$, denoted $\boldsymbol{T_a^b(f)}$. Denoting the positive and negative parts of f by, resp., $[f]^+ = \max\{f, 0\}$ and $[f]^- = -\min\{f, 0\}$, the **positive variation** and **negative variation** of f are defined to be the supremum of, resp.,

$$\sum_{k=1}^{m} [f(t_k) - f(t_{k-1})]^+, \qquad \sum_{k=1}^{m} [f(t_k) - f(t_{k-1})]^-,$$

denoted, resp., $\boldsymbol{P_a^b(f)}$ and $\boldsymbol{N_a^b(f)}$.

It is immediate from the definition that:

1. *Monotone functions f in $[a, b]$ are functions of bounded variation, with total variation bounded by $|f(b) - f(a)|$.*
2. *Differentiable functions with bounded derivative in $[a, b]$ are functions of bounded variation, with total variation bounded by $(b - a)\sup_{[a,b]} |f'|$.*
3. *The elements of $C^1[a, b]$ are functions of bounded variation.*
4. *The sum of a finite number of functions of bounded variation in $[a, b]$ is a function of bounded variation, with the total variation bounded by the sum of the functions total variations.*
5. *The multiplication of a constant by a function of bounded variation is a function of bounded variation with total variation equal to the product of the constant absolute value by the function total variation.*
6. *The functions of bounded variation in $[a, b]$ with the usual addition and multiplication by scalars are a linear space, denoted $\boldsymbol{BV([a, b])}$.*
7. If f is a function of bounded variation in $[a, b]$, then

$$T_a^b(f) = P_a^b(f) + N_a^b(f), \qquad f(b) - f(a) = P_a^b(f) - N_a^b(f).$$

(A.5) *$f \in BV([a, b])$ if and only if it is the difference of two nondecreasing functions in $[a, b]$, e.g., $f(t) = \big[P_a^t(f) + \frac{f(a)}{2}\big] - \big[N_a^t(f) - \frac{f(a)}{2}\big]$.*
If $f \in BV([a, b])$, then it is differentiable a.e. *in $[a, b]$ and*

$$f' \in L^1([a, b]), \qquad \int_a^b |f'| \le 2\,T_a^b(f), \qquad \Big|\int_a^b f'\Big| \le T_a^b(f).$$

Proof If $f \in BV([a, b])$, by the preceding property 7,

$$f(t) = P_a^t(f) - N_a^t(f) + f(a) = \big[P_a^t(f) + \tfrac{f(a)}{2}\big] - \big[N_a^t(f) - \tfrac{f(a)}{2}\big],$$

and the first statement follows since the functions $t \mapsto P_a^t(f)$ and $t \mapsto N_a^t(f)$ are nondecreasing.

As $t \mapsto T_a^t([a,b])$ is nondecreasing, by (A.2), the function of bounded variation f is differentiable a.e. in $[a,b]$, $f'(t) = \frac{d}{dt}[T_a^t(f)] - \frac{d}{dt}[T_a^t(f) - f(t)]$ and f' is measurable. Hence, $\frac{d}{dt}[T_a^t(f)]$, $\frac{d}{dt}[T_a^t(f) - f(t)] \geq 0$, and, therefore,

$$|f'(t)| \leq \tfrac{d}{dt}[T_a^t(f)] + \tfrac{d}{dt}[T_a^t(f) - f(t)] .$$

It is known that $|f(b) - f(a)| \leq T_a^b$. Therefore,

$$\int_a^b |f'(t)|\,dt \leq \int_a^b \tfrac{d}{dt}[T_a^t(f)]\,dt + \int_a^b \tfrac{d}{dt}[T_a^t(f) - f(t)]\,dt$$
$$\leq T_a^b(f) + [T_a^b(f) - (f(b) - f(a))] \leq 2\,T_a^b(f) ,$$

$$\left|\int_a^b f'(\tau)\,d\tau\right| = \left|\int_a^b \tfrac{d}{dt}[T_a^t(f)]\,dt - \int_a^b \tfrac{d}{dt}[T_a^t(f) - f(t)]\,dt\right|$$

$$\leq \max\left\{\int_a^b \tfrac{d}{dt}[T_a^t(f)]\,dt, \int_a^b \tfrac{d}{dt}[T_a^t(f) - f(t)]\,dt\right\} \leq \max\{T_a^b(f), T_a^b(f)\} = T_a^b(f).$$

□

The following result establishes part of the fundamental theorem of calculus for Lebesgue integrals, namely, that the derivative of the indefinite integral of an integrable function is the integrand function a.e. and that indefinite integrals of integrable functions are functions of bounded variation.

(A.6) Fundamental Theorem of Calculus—Derivative of indefinite integral: *If $f \in L^1([a,b])$, then $F(t) = \int_a^t f$ is a continuous function of bounded variation with total variation in $[a,b]$ $T_a^b = \int_a^b |f|$, and $F' = f$ a.e. in $[a,b]$.*

Proof If $a = t_0 < t_1 < \cdots < t_m = b$, then

$$\sum_{k=1}^m |F(t_k) - F(t_{k-1})| = \sum_{k=1}^m \left|\int_{t_{k-1}}^{t_k} f\right| \leq \sum_{k=1}^m \int_{t_{k-1}}^{t_k} |f| = \int_a^b |f| .$$

Thus, $T_a^b(F) \leq \int_a^b |f| < +\infty$, and F is a function of bounded variation.

To prove $F' = f$ a.e. in $[a,b]$, assume, without loss of generality, that f is an upper function in $[a,b]$, since an integrable function is a difference of upper limit functions. Let $\{s_n\}$ be a sequence of step functions in $[a,b]$ such that $s_n \nearrow f$ a.e. in $[a,b]$ and $\{\int_a^b s_n\}$ is upper bounded. With $S_n(t) = \int_a^t s_n$, it is

$$F = \lim_{n\to+\infty} S_n = S_1 + \textstyle\sum_{n=1}^{\infty}(S_{n+1} - S_n) .$$

Functions S_1 and $S_{n+1} - S_n$ are nondecreasing in $[a, b]$. By the Fubini theorem for derivative of series of monotone functions, the series can be differentiated term by term. Hence, $F' = S_1' + \sum_{n=1}^{\infty}(S_{n+1}' - S_n')$ a.e.. in $[a, b]$. Since $S_n' = s_n$ a.e. in $[a, b]$ for $n \in \mathbb{N} \cup \{0\}$, it is $F' = s_1 + \sum_{n=1}^{\infty} s_n = \lim_{n \to +\infty} s_n = f$ a.e. in $[a, b]$. □

$f : [a, b] \to \mathbb{R}$ is said to be an **absolutely continuous function** in $[a, b]$ if for every $\varepsilon > 0$ there is $\delta > 0$ such that if $\{]t_j, t_j^*[\}$ is a finite family of disjoint subintervals of $[a, b]$ with sum of lengths $< \delta$, then $\sum_{k=1}^{m} |f(t_j^*) - f(t_j)| \leq \varepsilon$.

Every absolutely continuous function in $[a, b]$ *is uniformly continuous and of bounded variation.* The first statement follows directly from the definition and the second from the existence of $\delta > 0$ such that the total variation of f in each subinterval of $[a, b]$ with length $<\delta$ is < 1, and since $[a, b]$ can be subdivided in a finite number of subintervals with lengths $< \delta$, the total variation of f is finite.

A function $f : D \to \mathbb{R}$ with $D \subset \mathbb{R}$ is said to be a **Lipschitz function** if there is $L > 0$ such that $|f(y) - f(x)| \leq L|y - x|$ for $x, y \in D$.

It follows directly from the definition that *every Lipschitz function in an interval* $[a, b]$ *is absolutely continuous in* $[a, b]$.

Denoting $L([a, b])$, $AC([a, b])$, $UC([a, b])$, $DAE([a, b])$ the linear spaces of functions, resp., Lipschitz, absolutely continuous, uniformly continuous, differentiable a.e. in $[a, b]$,

$$C^1([a, b]) \subsetneqq L([a, b]) \subsetneqq AC([a, b]) \subsetneqq UC([a, b]) \subsetneqq C^0([a, b]),$$
$$AC([a, b]) \subsetneqq BV([a, b]) \subsetneqq DAE([a, b]).$$

It is left as exercise to find functions to prove these inclusions are strict.

Since the sum of two absolutely continuous functions in $[a, b]$ and the product of a constant by an absolutely continuous function in $[a, b]$ are absolutely continuous functions in $[a, b]$, *the absolutely continuous functions in* $[a, b]$ *with the usual addition and multiplication operations by scalars form a linear subspace of the intersection of the linear spaces of the continuous functions and the functions of bounded variation in* $[a, b]$.

(A.7) *A function* f *is absolutely continuous in* $[a, b]$ *if and only if it is a difference of absolutely continuous nondecreasing functions.*

Proof Sufficiency is immediate. For necessity, observe that an absolutely continuous function f has bounded variation and, therefore, it is a difference of two nondecreasing functions; in particular, $f(t) = T_a^t(f) - [T_a^t(f) - f(t)]$ as observed in the proof of (A.5). Since absolutely continuous functions form a linear space, to prove that these two functions are absolutely continuous, it suffices to prove that one of them is, e.g., that $v(t) = T_a^t(f)$ is absolutely continuous. Let $\varepsilon > 0$ and choose δ so that the condition in the definition of absolute continuity for f holds.

Considering a family of open intervals $]a_j, b_j[\subset[a, b]$ with sum of lengths $<\delta$, $\sum_j[v(b_j)-v(a_j)]$ is the supremum of the sums $\sum_k\sum_m|f(x_{k,m})-f(x_{k,m-1})|$ for all possible finite partitions of the intervals $]a_j, b_j[$, with $j=1,\ldots,N$. These sums are upper bounded by ε and, therefore, $\sum_j|v(b_j)-v(a_j)|\leq\varepsilon$. □

The absolutely continuous functions in $[a, b]$ are precisely the functions satisfying the part of the fundamental theorem of calculus for Lebesgue integral corresponding to the Barrow rule. More explicitly, this rule is valid for the integral of a function in $[a, b]$ if and only if this function has an antiderivative a.e. in $[a, b]$, which is absolutely continuous in $[a, b]$.

(A.8) Fundamental theorem of calculus – Barrow rule:
A function F is absolutely continuous in $[a, b]$ if and only if there is $f\in L^1([a, b])$ such that $F(t)-F(a)=\int_a^t f$ for $t\in[a, b]$; if yes, $\int_a^t F'=F(t)-F(a)$.

Proof Let $F(t)-F(a)=\int_a^t f$ for $t\in[a, b]$ with $f\in L^1([a, b])$. If, additionally, f is bounded, i.e., $|f(t)|\leq M$ for some $M>0$, then

$$|F(y)-F(x)|=\left|\int_x^y f\right|\leq\int_x^y|f|\leq M|y-x|,\quad x, y\in[a, b];$$

thus, F is a Lipschitz function, and therefore, it is absolutely continuous. If f is unbounded, denoting $f_n(t)=f(t)$ if $|f(t)|\leq n$, $f_n(t)=n$ if $|f(t)|>n$ and $f_n(t)=-n$ if $|f(t)|<-n$, then $f(t)=f_n(t)+[f(t)-f_n(t)]$ for $t\in[a, b]$. If $\{]t_j, t_j^*[\}$ is a finite set of disjoint subintervals of $[a, b]$ with sum of lengths $<\delta$, then

$$\sum_{k=1}^{m}\left|\int_{t_j}^{t_j^*}f_n\right|\leq\sum_{k=1}^{m}\int_{t_j}^{t_j^*}|f_n|\leq\sum_{k=1}^{m}\int_{t_j}^{t_j^*}n\leq n\delta.$$

Since $\lim_{n\to+\infty}(f-f_n)=0$ and $|f-f_n|\leq|f|+|f_n|\leq 2|f|$, by the Lebesgue dominated convergence theorem, $\lim_{n\to+\infty}\int_a^b|f-f_n|$. Hence, for every $\varepsilon>0$, there is $N\in\mathbb{N}$ such that $n\geq N$ implies $\int_a^b|f-f_n|<\varepsilon$ and

$$\sum_{k=1}^{m}\left|\int_{t_j}^{t_j^*}(f-f_n)\right|\leq\sum_{k=1}^{m}\int_{t_j}^{t_j^*}|f-f_n|\leq\int_a^b|f-f_n|<\varepsilon.$$

From the two preceding paragraphs, for every $\varepsilon>0$, if $\{]t_j, t_j^*[\}$ is a finite set of disjoint subintervals of $[a, b]$ with sum of lengths $<\frac{\varepsilon}{N}$, then $\sum_{k=1}^{m}\left|\int_{t_j}^{t_j^*}f\right|\leq 2\varepsilon$, and, therefore, $f\in AC([a, b])$.

Conversely, if $f \in AC([a,b])$, for every $\varepsilon > 0$, there is $\delta > 0$ such that if $\{]t_j, t_j^*[\}$ is a finite set of disjoint subintervals of $[a,b]$ with sum of lengths $< \delta$, then $\sum_{k=1}^{m} |F(t_j^*) - F(t_j)| \leq \varepsilon$. This condition implies that functions $t \mapsto T_a^t(F)$, $t \mapsto P_a^t(F)$, and $t \mapsto N_a^t(F)$ are also absolutely continuous in $[a,b]$; thus, as $F(t) = \left[P_a^t(F) + \frac{F(a)}{2}\right] - \left[N_a^t(F) - \frac{F(a)}{2}\right]$ for $t \in [a,b]$, it can be assumed, without loss of generality, that F is increasing. The incremental ratio of F in t with $h \neq 0$ satisfies $0 \leq \frac{1}{h}[F(t+h) - F(t)] \to F'(t)$ when $h \to 0$ a.e. for $t \in [a,b]$, and the integral of this incremental ratio in $]\alpha, \beta[\subset [a,b]$, with $h > 0$ small, and, taking into account that F is continuous,

$$\frac{1}{h}\int_{\beta}^{\beta+h} F - \frac{1}{h}\int_{\alpha}^{\alpha+h} F \to F(\beta) - F(\alpha), \quad h \searrow 0.$$

Hence, by the Fatou lemma[6], $F' \in L^1([a,b])$ and $\int_\alpha^\beta F' \leq F(\beta) - F(\alpha)$. With $G(t) = \int_a^t F'$,

$$[F(\beta)-G(\beta)]-[F(\alpha)-G(\alpha)] = [F(\beta)-F(\alpha)]-[G(\beta)-G(\alpha)] \geq \int_\alpha^\beta F' - \int_\alpha^\beta F' = 0;$$

thus, $H = F - G$ is nondecreasing in $[a,b]$. As $F \in AC([a,b])$ and, by the preceding paragraph, $G \in AC([a,b])$, also $H \in AC([a,b])$. By the other part of the fundamental theorem of calculus (A.6), $H' = F' - G' = F' - F' = 0$ a.e. in $[a,b]$. It now suffices to prove that H is constant, by proving that every nondecreasing function $H \in AC([a,b])$ with $H' = 0$ a.e. in $[a,b]$ is constant.

Let $H \in AC([a,b])$ be nondecreasing with $H' = 0$ a.e. in $[a,b]$, and

$$E = \left\{t \in [a,b] : H'(t) \neq 0 \text{ or } H'(t) \text{ does not exist}\right\}.$$

Since E has measure 0, for every $\delta > 0$ E can be covered by a countable set of intervals with sum of lengths $< \delta$ and, as H is absolutely continuous, for every $\varepsilon > 0$ the sum of the lengths of the intervals images of the intervals of the cover of E is $\leq \varepsilon$ if δ is small; so, $H(E)$ has measure 0. Since $H' = 0$ in $]a,b[\setminus E$, for every $\varepsilon > 0$ and $t \in]a,b[\setminus E$, there is $x > t$ such that $H(x) - H(t) < \varepsilon(x - t)$, i.e., for $\varepsilon > 0$ fixed, $]a,b[\setminus E$ is a subset of the "shadow" set S_{g_ε} of the rising Sun lemma (A.1) for the function $g_\varepsilon(t) = \varepsilon t - H(t)$. Hence, S_{g_ε} is an open set, and, therefore, it is a countable union of open intervals $]a_n, b_n[$ and $g_\varepsilon(a_n) \leq g_\varepsilon(b_n)$, i.e., $H(b_n) - H(a_n) \leq \varepsilon(b_n - a_n)$. Therefore, the sum of the lengths of the intervals $]H(a_n), H(b_n)[$ with union $H(S_{g_\varepsilon})$ is $\leq \varepsilon(b - a)$, and since $H(]a,b[\setminus E) \subset S_{g_\varepsilon}$ for all $\varepsilon > 0$, the set $H(]a,b[\setminus E)$ and, therefore, also $H([a,b]\setminus E)$ have measure 0. As $H(E)$ and $H([a,b]\setminus E)$ have measure 0, so does $H([a,b])$. Since H is continuous, $H([a,b])$ is a connected subset of $\mathbb{R}$ with measure 0. As the connected subsets of $\mathbb{R}$ are the intervals or the sets with just one point and, as any interval of $\mathbb{R}$ has measure > 0, $H([a,b])$ has only 1 point, i.e., H is constant in $[a,b]$.

The last statement follows directly from (A.6). □

An immediate consequence is: *the absolutely continuous functions with derivative* 0 a.e. *in* $[a, b]$ *are the constant functions,* since $F(t) - F(a) = \int_a^t f$ with $f \in L^1([a, b])$ implies $F' = f$ a.e. in $[a, b]$ and if $F' = 0$ a.e. in $[a, b]$, it is $f = 0$ a.e. and $F(t) - F(a) = 0$ for $t \in [a, b]$.

Another important consequence follows.

(A.9) Lebesgue decomposition of continuous monotone or of bounded variation functions: *Every continuous monotone or of bounded variation function f in $[a, b]$ is $f = g + h$ with g absolutely continuous and $h' = 0$* a.e. *in $[a, b]$; if f is monotone, then g and h also are. g and f are called, resp., the* ***absolutely continuous part*** *and the* ***singular part*** *of f.*

Proof If f is continuous and nondecreasing, by the Fatou lemma[6] applied as in the proof of (A.8), it is $f' \in L^1([a, b])$ and with $g(t) = \int_a^t f'$, $h = f - g$ is nondecreasing, and $f' \geq 0$, also g is nondecreasing and, by (A.6), $h' = f' - g' = 0$ a.e. in $[a, b]$. If f is continuous and nonincreasing, so is $-f$. If f is a continuous function of bounded variation, it is the difference of continuous nondecreasing functions and this result holds for these functions. □

An example of a continuous monotone function with zero absolutely continuous part is the **Cantor function**,[7] which, for each $t \in [0, 1]$ with ternary representation $t = \sum_{n=1}^{\infty} \frac{a_n}{3^n}$ with $a_n t \in \{0, 1, 2\}$, is the number with binary representation $f(t)t = \sum_{n=1}^{N-1} \frac{a_n/2}{2^n} t + \frac{1}{2^N}$, where $N = \min\{n : a_n = 1\}$ if any $a_n = 1$ and otherwise $N = \infty$. It is left as and exercise to verify that f is continuous and nondecreasing, with $f' = 0$ a.e. in $[0, 1]$ with range $[0, 1]$, not differentiable in an uncountable subset of $[0, 1]$, and uniformly continuous but not absolutely continuous in $[0, 1]$.

A.3 Volume Derivative and Integrability of Homeomorphism Jacobians

By the change of integration variables theorem for integrals, if $\Omega \subset \mathbb{R}^n$ is an open connected set, $B \subset \Omega$ is measurable, $\mathbf{f}: \Omega \to \mathbb{R}^n$ is C^1, and $J_\mathbf{f}$ is the **Jacobian** of **f**, $J_\mathbf{f} = \det D\mathbf{f}$, then there is $\mu(\mathbf{f}(B)) = \int_{\mathbf{f}(B)} 1 = \int_B |J_\mathbf{f}|$; so, $J_\mathbf{f} \in L(B)$. It is proved below that if **f** is a homeomorphism with finite partial derivatives a.e., then also $J_\mathbf{f} \in L(B)$, and if, additionally, **f** is locally absolutely continuous in Ω, then also $\mu(\mathbf{f}(B)) = \int_B |J_\mathbf{f}|$. To do this, it is useful to consider completely additive set

[7] It is also called the **devil staircase function**. It was introduced in 1884 by G. Cantor, who mentioned at the time that Ludwig Sheeffer (1859–1885) had given it as a counterexample to the attempted extension of the fundamental theorem of calculus presented by A. Harnack.

functions and their Lebesgue decomposition as a sum of two completely additive set functions, one absolutely continuous and the other singular, analogous to the Lebesgue decomposition of continuous monotone or bounded variation functions in (A.9).

The condition for a function to be an indefinite integral, expressed in terms of absolute continuity, was considered for additive set functions of Borel sets by J. Radon[8] in 1913.

The set of Borel subsets of $\mathbb{R}^n$ is denoted $\mathscr{B}$. If $\emptyset \neq S \subset \mathbb{R}^n$, a function $F:\{B\in\mathscr{B}: B\subset S\}\to\mathbb{R}\cup\{-\infty, +\infty\}$ such that $F(\cup_j B_j)=\sum_j F(B_j)$ for every countable family $\{B_j\}$ of disjoint subsets of the domain is called a **completely additive set function** in S.

A completely additive set function F in $S\subset\mathbb{R}^n$ is said to be **absolutely continuous** in S if it is not identically $+\infty$ and for every $\varepsilon>0$ there is $\delta>0$ such that $B\subset S$, $B\in\mathscr{B}$, $\mu(B)<\delta$ implies $|F(B)|<\varepsilon$; it is said to be **locally absolutely continuous** in S if it is absolutely continuous in every compact subset of S.

A completely additive set function F in $S\subset\mathbb{R}^n$ is said to be **singular** in S if it is zero except in a Borel set of measure 0.

(A.10) Characterization of absolutely continuous completely additive set functions: *A completely additive set function $\emptyset\neq S\subset\mathbb{R}^n$ is absolutely continuous in S if and only if it is 0 in measure 0 subsets of S.*

Proof Directly from the definition, a completely additive absolutely continuous set function in $S\subset\mathbb{R}^n$, $S\neq\emptyset$ vanishes in measure 0 subsets of S. Conversely, if F is a completely additive set function not absolutely continuous in $S\subset\mathbb{R}^n$, $S\neq\emptyset$, there are $\varepsilon>0$, a decreasing sequence $\{\delta_j\}\subset\,]0,+\infty[$ such that $\delta_j\to 0$, and a sequence $\{B_j\}\subset\mathscr{B}$ with $B_j\subset S$, $\mu(B_j)<\delta_j$, $|F(B_j)|\geq\varepsilon$. Without loss of generality, $\{B_j\}$ is a sequence of sets such that $B_{j+1}\subset B_j$ for $j\in\mathbb{N}$, since if not, considering a subsequence of $\{\delta_j\}$ such that $\delta_{j+1}\leq\frac{\delta_j}{2}$ for $j\in\mathbb{N}$ and the corresponding subsequence of $\{B_j\}$ and, if necessary, replacing this subsequence by $\{B'_m\}$ with $B'_m=\cup_{j=m}^{\infty}B_j$, it is $\mu(B'_j)<\delta_j$, where $\{\delta_j\}$ is the initial sequence. It is $\mu\left(\cap_{j=m}^{\infty}B'_j\right)=0$ and $F\left(\cap_{j=m}^{\infty}B'_j\right)\geq\varepsilon$. Thus, if F is not absolutely continuous, there are Borel subsets of S with measure 0 where F is not 0. □

It is left as an exercise to prove that *set functions defined in the nonempty Borel subsets of $S\subset\mathbb{R}^n$ by integrals of an integrable function in S are completely additive and absolutely continuous in S.*

If $\emptyset\neq X\subset S$ and F is a bounded completely additive set function in S, the supremum and the infimum of F in the Borel subsets of X, denoted, resp., $\overline{T}_F(X)$ and $\underline{T}_F(X)$ are called, resp., **upper variation** and **lower variation** of F in X. $T_F(X)=\overline{T}_F(X)-\underline{T}_F(X)$ is called **total variation** of F in X.

[8] Radon, Johann (1887–1956).

Since under the conditions of the preceding paragraph, for each $B \in \mathscr{B}$ with $B \subset X$ it is $F(B) \in \mathbb{R}$ because F is bounded, and $B = B \cup \emptyset$ implies $F(B) = F(B) + F(\emptyset)$, *if* $\emptyset \neq X \subset S$ *and* F *is a bounded completely additive set function in* S, *then* $F(\emptyset) = 0$ *and* $\underline{T}_F(X) \leq 0 \leq \overline{T}_F(X)$.

(A.11) *Every bounded completely additive set function* F *in a set* $S \in \mathscr{B}$ *is the sum of two completely additive set functions, one nondecreasing and another nonincreasing relative to set inclusion, e.g.,* $F = \overline{T}_F + \underline{T}_F$ *in* S.

Proof $\overline{T}_F$ and $\underline{T}_F$ are bounded completely additive set functions, resp., nondecreasing and nonincreasing relative to set inclusion. For $Y \subset X \subset S$ it is $Y = X \setminus (X \setminus Y)$; thus, $F(Y) = F(X) - F(X \setminus Y) \leq F(X) - \underline{T}_F(X)$ and $\overline{T}_F(X) \leq F(X) - \underline{T}_F(X)$, and for $-F$ it is $\underline{T}_F(X) \geq F(X) - \overline{T}_F(X)$. With the two inequalities together, $F = \overline{T}_F + \underline{T}_F$ in S. □

The following result was published by H. Hahn[9] in 1921.

(A.12) *If* F *is a bounded completely additive set function in a set* $S \in \mathscr{B}$, *then there is* $P \in \mathscr{B}$ *such that* $F(X) \geq 0$ *for every* $X \in \mathscr{B}$ *such that* $X \subset P$, *and* $F(X) \leq 0$ *for all* $X \in \mathscr{B}$ *such that* $X \subset S \setminus P$.

Proof For $m \in \mathbb{N}$ let $B_m \subset S$ be a Borel set such that $\overline{T}_F(S) - \frac{1}{2^m} \leq F(B_m)$. Since $F(B_m) = \overline{T}_F(B_m) + \underline{T}_F(B_m)$,

$$\underline{T}_F(B_m) \geq \overline{T}_F(S) - \overline{T}_F(B_m) - \tfrac{1}{2^m} \geq -\tfrac{1}{2^m}.$$

With $P = \underline{\lim}\, B_m$, it is $0 \geq \underline{T}_F(P) \geq \underline{\lim}\, \underline{T}_F(B_m) \geq -\lim \frac{1}{2^m} = 0$ and, therefore, $\underline{T}_F(P) = 0$, and $F(X) \geq 0$ for all $X \in \mathscr{B}$ with $X \subset P$.

For every Borel set $B \subset S \setminus B_m$, it is $B \cup B_m \subset S$, and therefore,

$$\overline{T}_F(S) \geq F(B \cup B_m) = F(B) + F(B_m), \qquad F(B) \leq \overline{T}_F(S) - F(B_m) \leq \tfrac{1}{2^m}.$$

Therefore, $\overline{T}_F(S \setminus B_m) \leq \frac{1}{2^m}$.

Since $S \setminus P = \overline{\lim}\, S \setminus B_m \subset \sum_{m=k} S \setminus B_m$ for all $k \in \mathbb{N}$,

$$\overline{T}_F(S \setminus P) \leq \sum_{m=k} \overline{T}_F(S \setminus B_m) \leq \tfrac{1}{2^{k-1}}, \quad k \in \mathbb{N}.$$

Hence, $\overline{T}_F(S \setminus P) = 0$; thus, $F(X) \leq 0$ for all $X \in \mathscr{B}$ with $X \subset S \setminus P$. □

[9] Hahn, Hans (1879–1934).

(A.13) *If $F \geq 0$ is a bounded completely additive set function in a set $S \in \mathscr{B}$, for every $a > 0$, there is a sequence of disjoint measurable sets $\{E_m\}_{m \in \mathbb{N} \cup \{0\}}$ such that $\mu(E_0) = 0$ and for every $X \in \mathscr{B}, X \subset E_m$ it is*

$$a\,(m-1)\,\mu(X) \leq F(X) \leq a\,m\,\mu(X), \quad m \in \mathbb{N}.$$

Proof By the preceding result, for every $a > 0, m \in \mathbb{N}$ there is $P_m \subset \mathscr{B}$ such that $F(X) - a\,m\,\mu(X) \geq 0$ for all $X \subset P_m$ and $F(X) - a\,m\,\mu(X) \leq 0$ for all $X \in \mathscr{B}$ such that $X \subset S \backslash P_m$. Let $B_m = \cup_{k=m}^{\infty} P_k$. Every measurable $X \subset B_m$ is $X = \cup_{k=m}^{\infty} X_k$ with $X_k \subset P_k$ measurable and $X_k \cap X_j = \emptyset$ for $j \neq k$ with $j, k = m, m+1, \ldots$, and

$$F(X) = \sum_{k=m}^{\infty} F(X_k) \geq \sum_{k=m}^{\infty} a\,k\,\mu(X_k) \geq \sum_{k=m}^{\infty} a\,m\,\mu(X),\ m \in \mathbb{N}, X \in \mathscr{B}, X \subset S \backslash P_m.$$

Analogously, $F(X) \leq \sum_{k=m}^{\infty} a\,m\,\mu(X)$ for $m \in \mathbb{N}, X \in \mathscr{B}, X \subset S \backslash P_m$. With $E_1 = S \backslash B_1$, $E_{m+1} = B_m \backslash B_{m+1}$ for $m \in \mathbb{N}$ and $B_0 = \cap_{m=1} B_m$, the sets B_m with $m \in \mathbb{N} \cup \{0\}$ are measurable, disjoint and $S = \cup_{m=0} B_m$. The inequality in the statement holds for all $m \in \mathbb{N}$, and since $B_0 \subset B_m$ for $m \in \mathbb{N}$, it is $F(B_0) \geq a\,m\,\mu(B_0)$ for $m \in \mathbb{N}$, and, therefore, $\mu(B_0) = 0$. □

(A.14) Lebesgue decomposition (singular + absolutely continuous) of completely additive function:
If $S \subset \mathbb{R}^n$ is a countable union of finite measure Borel sets, every bounded completely additive function F in S is a unique sum of two completely additive set functions, one singular, σ, and the other absolutely continuous, α, in S, which in every $X \in \mathscr{B}$ is $\int_X f$ with $\alpha(X) = f \in L(S)$. If $F \geq 0$, then $\alpha, \sigma \geq 0$.

Proof By (A.11), $F = \overline{T}_F + \underline{T}_F$ with $\overline{T}_F \geq 0$ and $\underline{T}_F \leq 0$ bounded completely additive set functions in S; so, without loss of generality, $F \geq 0$.

Considering first the case of S with finite measure and $a > 0$ arbitrary, by the previous result, for each $k \in \mathbb{N}$ there is a sequence $\{B_m^k\}_{m \in \mathbb{N} \cup \{0\}}$ of disjoint sets such that for each $k \in \mathbb{N}$, $S = \cup_{m=0} B_m^k$, $\mu(B_0^k) = 0$ and

$$\tfrac{m-1}{2^k}\,\mu(X) \leq F(X) \leq \tfrac{m}{2^k}\,\mu(X), \quad X \in \mathscr{B}, X \subset B_m^k.$$

Therefore, for $k, m, j \in \mathbb{N}$,

$$\tfrac{m}{2^k}\,\mu(B_m^k \cap B_j^{m+1}) \leq F(B_m^k \cap B_j^{m+1}) \leq \tfrac{j-1}{2^{k+1}}\,\mu(B_m^k \cap B_j^{m+1}),$$

$$\tfrac{j}{2^{k-1}}\,\mu(B_m^k \cap B_j^{m+1}) \leq F(B_m^k \cap B_j^{m+1}) \leq \tfrac{m-1}{2^k}\,\mu(B_m^k \cap B_j^{m+1});$$

thus,

$$(2m-j+1)\,\mu(B_m^k\cap B_j^{m+1})\geq 0\,,\qquad (j-2m+2)\,\mu(B_m^k\cap B_j^{m+1})\geq 0\,,$$

$\mu(B_m^k\cap B_j^{m+1})=0$ if $j>2m+1$ or $j<2m-2$, and $B_m^k\subset\big(\cup_{j=2m-2}^{2m+1}B_j^{k+1}\big)\cup A_m^k$ with $\mu(A_m^k)=0$. With $B_0=\big(\cup_{k=1}^{\infty}B_0^k\big)\cup\big(\cup_{k,m=1}^{\infty}A_m^k\big)$ and $g_k(\mathbf{x})=\frac{m-1}{2^k}$ if $\mathbf{x}\in B_m^k\backslash B_0$ for $k,m\in\mathbb{N}$ and $g_k(\mathbf{x})=0$ for $\mathbf{x}\in B_0$, then $\{g_k\}$ is a sequence of Borel functions with $g_k\geq 0$ in S for $k\in\mathbb{N}$. Since $|g_{k+1}-g_k|\leq\frac{1}{2^k}$ in S, there is a measurable $g:S\to[0,+\infty[$ such that $g_k\to g$ uniformly for S.

For every measurable $X\subset S$ and $k\in\mathbb{N}$,

$$F(X)\geq F(X\cap B_0)+\sum_{m=1}^{\infty}\tfrac{m-1}{2^{-k}}\mu(X\cap B_m^k)\;=\;F(X\cap B_0)+\int_X g_k\,,$$

$$F(X)\leq F(X\cap B_0)+\sum_{m=1}^{\infty}\tfrac{m}{2^{-k}}\mu(X\cap B_m^k)\;=\;F(X\cap B_0)+\int_X g_k+\tfrac{1}{2^k}\mu(X)\,.$$

The limit when $k\to+\infty$ gives the decomposition $F(X)=F(X\cap B_0)+\int_X g$.

If S is a countable union of disjoint finite measure Borel sets, $S=\cup_{m=1}^{\infty}S_m$, by the preceding paragraph, for $m\in\mathbb{N}$ there is an integrable function $f_m\colon S_m\to[0,+\infty[$ and a set of measure 0, $B_{m,0}\subset S_m$, such that $F(X\cap S_m)=F(X\cap B_{m,0})+\int_{X\cap S_m}f_m$ for $m\in\mathbb{N}$. With $B_0=\cup_{m=1}^{\infty}B_{m,0}$ and $f=f_m$ in each S_m, $m\in\mathbb{N}$, f is integrable in S, $B_0\subset S$ has measure 0 and

$$F(X)=\sum_{m=1}^{\infty}F(X\cap B_{m,0})+\sum_{m=1}^{\infty}\int_{X\cap S_m}f\;=\;F(X\cap B_0)+\int_X f\,.$$

For each $X\subset S\backslash B_0$, it is $F(X\cap B_0)=F(\emptyset)=0$; so, $\sigma(X)=F(X\cap B_0)$ defines a completely additive singular set function on S. The integral in the right-hand side vanishes for every X of measure 0, and, by (A.10), $\alpha(X)=\int_X f$ is a completely additive absolutely continuous set function. These functions are the Lebesgue decomposition. Of course, if $F\geq 0$, also $\sigma,\alpha\geq 0$. It remains to prove uniqueness. If σ_1,α_1 give a singular-absolutely continuous decomposition of F in S, then $\sigma_1+\alpha_1=\sigma+\alpha$, and $\sigma_1-\sigma=\alpha_1-\alpha$ in S is a completely additive set function both singular and absolutely continuous; so, it is 0 in S except possibly in a subset of measure 0. Since it is continuous, it is 0 in S, i.e., $\sigma_1-\sigma=\alpha_1-\alpha=0$ in S, what proves uniqueness. □

O. Nikodym[10] obtained the following in 1930.

[10] Nikodym, Otto (1887–1974).

(A.15) Radon-Nikodym theorem: *If $S \subset \mathbb{R}^n$ is a countable union of finite measure sets, for a completely additive set function F to be absolutely continuous in S, it is necessary and sufficient that $F(E) = \int_E f$ for some integrable function f in each measurable set $E \subset S$.*

Proof This follows directly from the Lebesgue decomposition. □

If $B, B' \subset \mathbb{R}^n$ are Borel sets and $\mathbf{f}$ is a homeomorphism of B to B', since homeomorphisms map open sets to open sets, Borel sets included in B are mapped by $\mathbf{f}$ to Borel sets included in B', and the function can be associated with the homeomorphism $\mathbf{f}$ of sets $V_{\mathbf{f}}$ defined in $\mathscr{B}$ by $V_{\mathbf{f}}(A) = \mu\big(\mathbf{f}(A \cap B)\big)$. If $V_{\mathbf{f}}$ is (locally) absolutely continuous in B, it is said that $\mathbf{f}$ is a **(locally) absolutely continuous homeomorphism.**

(A.16) *An absolutely continuous homeomorphism defined from a Borel subset of $\mathbb{R}^n$ to $\mathbb{R}^n$ maps sets of measure 0 to sets of measure 0 and countable unions of sets with finite exterior measure to measurable sets.*

Proof Directly from the definitions, each set B of measure 0 is contained in a Borel set A of measure 0, and as $V_{\mathbf{f}}(A) = 0$ and, therefore, $\mathbf{f}(A)$ has measure 0, also $\mathbf{f}(B)$ has measure 0.

If A is a countable union of subsets of the domain of $\mathbf{f}$ with finite exterior measure, since there is a Borel set $B \supset A$ of the domain of $\mathbf{f}$ such that $\mu(B \backslash A) = 0$ and $\mathbf{f}$ is an absolutely continuous homeomorphism, it is $\mu\big(\mathbf{f}(B) \backslash \mathbf{f}(A)\big) = \mu\big(\mathbf{f}(B \backslash A)\big) = 0$, and therefore, $\mathbf{f}(B) \backslash \mathbf{f}(A)$ is measurable. As $\mathbf{f}(B) \in \mathscr{B}$ is measurable, so is $\mathbf{f}(A)$. □

The following definition is natural, given the definition of point of density in the preceding section.

If $\emptyset \neq S \subset \mathbb{R}^n$ and $F \geq 0$ is a completely additive set function in S, the limit, upper limit, and lower limit of $\frac{F(E_m)}{\mu(E_m)}$, where $\{E_m\} \subset \Omega$ is a sequence of closed n cubes centered at $\mathbf{z}$ decreasing with set inclusion such that $\mu(E_m) \to 0$, is called, resp., **derivative, upper derivative**, and **lower derivative** of F in S at $\mathbf{z} \in S$, denoted, resp., $F'(\mathbf{z})$, $\overline{D}F(\mathbf{z})$, and $\underline{D}F(\mathbf{z})$.

If $\emptyset \neq S \subset \mathbb{R}^n$, $F \geq 0$ is a completely additive set function in S and $\mathbf{z} \in S$, then $F'(\mathbf{z})$ exists if and only if $\overline{D}F(\mathbf{z}) = \underline{D}F(\mathbf{z})$; if yes, they are the same.

(A.17) *If σ is a bounded completely additive singular set function in a nonempty set $S \subset \mathbb{R}^n$, then σ' exists and $\sigma' = 0$ a.e. in S.*

Proof There is $B_0 \in \mathscr{B}$ with $B_0 \subset S$ of measure 0 such that $\sigma = 0$ in $S \backslash B_0$ and $\overline{T}_\sigma$, $\underline{T}_\sigma$ are singular with $\overline{T}_\sigma = \underline{T}_\sigma = 0$ in $S \backslash B_0$. Denoting $Q_c = \{\mathbf{x} \in B \backslash B_0 :|, \overline{T}'_\sigma(\mathbf{x}) > c\}$, if it were $\mu(Q_0) > 0$, for some $N \in \mathbb{N}$ it would be $\mu\big(Q_{\frac{1}{N}}\big) > 0$, and for some bounded interval $I \subset \mathbb{R}^n$, it would be $\mu\big(Q_{\frac{1}{N}} \cap I\big) > 0$. Since $Q_{\frac{1}{N}} \subset Q_0$, it is $\overline{T}_\sigma\big(Q_{\frac{1}{N}} \cap I\big) \le \overline{T}_\sigma(Q_0 \cap I)$. For $\mathbf{x} \in Q_{\frac{1}{N}}$, it is $\overline{T}'_\sigma(\mathbf{x}) > \frac{1}{N}$, and for some $N \in \mathbb{N}$, it is $\overline{T}_\sigma\big(Q_{\frac{1}{N}} \cap I\big)/\mu\big(Q_{\frac{1}{N}} \cap I\big) \ge \frac{1}{N}$. Hence, if it were $\mu(Q_0) > 0$, it would be $\overline{T}_\sigma\big(Q_{\frac{1}{N}} \cap I\big) > 0$, contradicting $\overline{T}_\sigma = 0$. Thus, $\mu(Q_0) > 0$, i.e., $\overline{T}'_\sigma = 0$ a.e. in S, and $-\sigma\ \underline{T}'_\sigma = -\overline{T}_{-\sigma}' = 0$ a.e. in S. By (A.11), $\sigma = \overline{T}_\sigma + \underline{T}_\sigma$, and, therefore, $\sigma' = \overline{T}'_\sigma + \underline{T}'_\sigma = 0$ a.e. in S. □

A set of closed subsets $\mathscr{E}$ of $\mathbb{R}^n$ is said to be a **Vitali cover** of $B \subset \mathbb{R}^n$ if for every $\mathbf{x} \in B$ there is a sequence of elements of $\mathscr{E}$ with union containing $\mathbf{x}$ with diameters $\neq 0$ converging to 0.

To prove the Lebesgue theorem for integrals of derivatives of completely additive set functions, the following two results are useful. The first — Vitali lemma — was proved for $n = 1$ by G. Vitali in 1908.

(A.18) Vitali lemma:
Every Vitali cover $\mathscr{E}$ of a subset A of $\mathbb{R}^n$ by closed balls has a sequence (infinite or finite) $\{E_m\}$ with disjoint terms such that $A \backslash \cup_m E_m$ has measure 0.

Proof Let $U \supset A$ be an open subset of $\mathbb{R}^n$ with finite measure $\mu(U)$. Without loss of generality, the elements of $\mathscr{E}$ are assumed to be closed subsets of U. The sequence of elements of $\mathscr{E}$ is constructed with an arbitrary $E_1 \in \mathscr{E}$, and once $E_1, \ldots, E_N$ are chosen, since for $k_N = \sup\{\mu(E) : E \in \mathscr{E},\ E \cap \cup_{m=1}^N E_m = \emptyset\}$, it is $k_N \le \mu(U)$, and if $A \not\subseteq \cup_{m=1}^N E_m$, choose $E_{N+1} \in \mathscr{E}$ with $E_{N+1} \cap \cup_{m=1}^N E_m = \emptyset$ and $\mu(E_{N+1}) > \frac{k_N}{2}$. The sets $E_1, \ldots, E_N \subset U$ are disjoint and $\sum_m \mu(E_m) \le \mu(U)$. Hence, for every $\varepsilon > 0$, there is $N \in \mathbb{N}$ such that $\sum_{m \ge N+1} \mu(E_m) < \varepsilon$.

If $\mathbf{x} \in A \backslash \cup_{m=1}^N E_m$, as $\mathbf{x} \notin \cup_{m=1}^N E_m$ and this set is closed, there is $E \in \mathscr{E}$ containing $\mathbf{x}$ such that $E \cap \cup_{m=1}^N E_m = \emptyset$. If $E \cap E_j = \emptyset$ for $j \le m$, then $\mu(E) \le k_m < 2\mu(E_{m+1})$. Since $\mu(E_m) \to 0$, $E \cap E_m \neq \emptyset$ for some m; denote m the smallest natural number for which this holds. It is $m > N$ and $\mu(E) \le k_{m-1} \le 2\mu(E_m)$. Since the n-volume of a ball B with radius $r(B)$ is $\mu(B) = [r(B)]^n \mu(B_1)$, it is $\left[\frac{r(E)}{r(E_m)}\right]^n = \frac{\mu(E)}{\mu(E_m)} \le 2$, and the distance of $\mathbf{x} \in E$ to the ball E_m center is upper bounded by

$$2r(E) + r(E_m) \le (2\sqrt[n]{2} + 1)\, r(E_m) \le 5\, r(E_m).$$

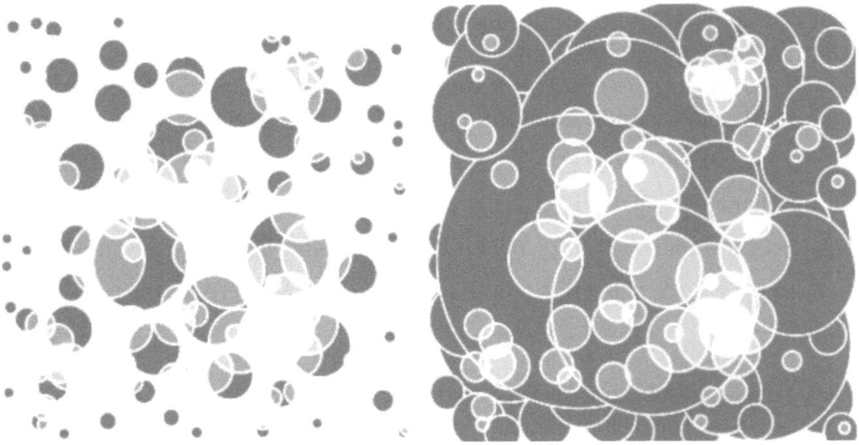

Fig. A.2 Illustration of the Vitali lemma: At left a set of balls in two gray levels, the darker disjoint, and at right the balls with the same centers of the darker ball and radii the triple of the resp. ball at left, covering the union of all balls at left

Therefore, $\mathbf{x}$ belongs to the ball J_m with the same center and 5 times larger radius than that of E_m (Fig. A.2;[11] if $\mathscr{E}$ is finite, 3 times larger is sufficient). Then, $A\backslash\cup_{m=1}^{N} E_m \subset \cup_{m=N+1}^{\infty} J_m$ and

$$\mu^*\big(A\backslash \cup_{m=1}^{N} E_m\big) \leq \sum_{m\geq N+1} \mu(J_m) = 5\sum_{m\geq N+1} \mu(E_m) < 5\,\varepsilon\,.$$

Since $\varepsilon > 0$ is arbitrary, $A\backslash\cup_{m=1}^{N} E_m$ has measure 0. □

(A.19) *The derivatives $\overline{D}F$, $\underline{D}F$ of a completely additive set function F in a nonempty measurable set $S\subset\mathbb{R}^n$ are measurable.*

Proof If $a\in\mathbb{R}$, then

$$DFt^{-1}\big(]a,+\infty[\big) = \cup_{k=1}^{\infty}[\overline{D}F]^{-1}\big([a+\tfrac{1}{k},+\infty[\big).$$

Since unions of intervals are measurable, $\overline{D}F$ is measurable if and only if for every $k\in\mathbb{N}$ the set $[\overline{D}F]^{-1}\big([a+\frac{1}{k},+\infty[\big)$ is measurable. A point $\mathbf{z}\in S$ belongs to this set if and only if there is a decreasing sequence (in set inclusion) $\{E_m\}$ of closed-n cubes centered at $\mathbf{z}$ with $\mu(E_m)\to 0$ such that $\frac{F(E_m)}{\mu(E_m)}\geq a+\frac{1}{k}$ for $m\in\mathbb{N}$. Therefore,

[11] Author and date of figure: Claudio Rocchini (24.01.2014), under License *Creative Commons Attribution-Share Alike 3.0 Unported.* https://commons.wikimedia.org/wiki/File:Rising_sun_lemma.svg

if A_j is the union of n-cubes E such that $\mu(E) \leq \frac{1}{j}$ and $\frac{F(E)}{\mu(E)} \geq a + \frac{1}{k}$, then it is measurable, and, therefore, also $[\overline{D}F]^{-1}\big([a+\frac{1}{k}, +\infty[\big) = \cap_{j=1}^{\infty} A_j$ is measurable. Consequently, $\overline{D}F$ is measurable, and as $\underline{D}F = -\overline{D}(-F)$, also $\underline{D}F$ is. □

(A.20) *If $\Omega \subset \mathbb{R}^n$ is an open connected set and $F \geq 0$ is a completely additive set function in Ω such that $\overline{D}F \geq \delta$ in $A \subset \Omega$, then $F(X) \geq \delta\, \mu(A)$ for all bounded Borel sets X with $A \subset X \subset S$.*

Proof If $\varepsilon > 0$, there is a bounded open set $U \subset \mathbb{R}^n$ such that $X \subset U$ and $F(X) \geq F(U) - \varepsilon$. For $\sigma < \delta$ let $\mathscr{E}_\sigma$ be the set of closed balls $E \subset U$ such that $F(E) \geq \sigma\, \mu(E)$. As $\overline{D}F \geq \delta > \sigma$ in A, $\mathscr{E}_\sigma$ is a Vitali cover of A, and by the Vitali lemma, there is a sequence $\{E_m\} \subset \mathscr{E}_\sigma$ of disjoint terms such that $A \backslash \cup_m E_m$ has measure 0. Thus,

$$F(U) \geq \sum_m F(E_m) \geq \sigma \sum_m \mu(E_m) \geq \sigma\, \mu(A),$$

and $F(X) \geq F(U) - \varepsilon \geq \sigma\, \mu(A) - \varepsilon$. As $\varepsilon > 0$ and $\sigma < \delta$ are arbitrary, $F(X) \geq \delta\, \mu(A)$. □

In 1910 H. Lebesgue gave a proof of the following result on existence of finite derivative, later on simplified and extended by several persons.

(A.21) Lebesgue theorem for integrals of derivatives of completely additive set functions:
If $\Omega \subset \mathbb{R}^n$ is an open connected set and $F \geq 0$ is a bounded completely additive set function, then F has finite derivative a.e. *in Ω and F' is measurable.*

Proof If it were $\overline{D}F > \underline{D}F$ in $A \subset \Omega$ with $\mu(A) > 0$, considering for $h, k \in \mathbb{N}$, $A_{h,k} = \{\mathbf{x} \in A : \overline{D}F > \frac{h+1}{k} > \frac{h}{k} > \underline{D}F\}$, as $A = \cup_{h,k} A_{h,k}$, there would exist $h_0, k_0 \in \mathbb{N}$ such that $\mu(A_{h_0,k_0}) > 0$. With arbitrary $\varepsilon > 0$, there would exist a bounded set $B \subset A_{h_0,k_0}$ with $\mu^*(B) > 0$ and an open bounded set $U \supset B$ such that $\mu(U) \leq \mu(B) + \varepsilon$. The set $\mathscr{E}$ of closed balls $E \subset U$ with $F(E) \leq \frac{h_0}{k_0}\, \mu(E)$ would be a Vitali cover of B. By the Vitali lemma A.18, there would exist a sequence of disjoint sets $\{E_m\}$ covering B except for a measure 0 set,

$$F\big(\cup_m E_m\big) = \textstyle\sum_m F(E_m) \leq \frac{h_0}{k_0} \sum_m \mu(E_m) \leq \frac{h_0}{k_0}\mu(U) \leq \frac{h_0}{k_0}\big(\mu(B) + \varepsilon\big).$$

Since $\overline{D}F > \frac{h_0+1}{k_0}$ in B and $\mu\big(B \backslash \cup_m E_m\big) = 0$, by (A.20), $F\big(\cup_m E_m\big) \geq \frac{h_0}{k_0}\mu(B)$ and, therefore, $(h_0 + 1)\, \mu(B) \leq h_0\, [\, \mu(B) + \varepsilon\,]$, i.e., $\mu(B) \leq \varepsilon$, contradicting $\mu^*(B) > 0$, since $\varepsilon > 0$ is arbitrary. Hence, $\overline{D}F = \underline{D}F$, i.e., F' exists, a.e. in Ω. If F' were not finite a.e. in Ω, there would exist $r > 0$ such that $F' = +\infty$ in a subset

S dof $B_r(\mathbf{a}) \subset \Omega$ with $\mu(S) > 0$, and, once more by (A.20), it would be $F' = +\infty$ in a subset S of $F\big(B_r(\mathbf{a})\big)$, contradicting that F is bounded a.e. in Ω, by hypothesis. Therefore, F has finite derivative a.e. in Ω. By (A.19), F' is measurable. □

The following result is analogous to the Fubini theorem for derivative of series of monotone functions (A.3), but for completely additive set functions.

(A.22) *If $\{F_m\}$ is a uniformly bounded monotone (with set inclusion) sequence of completely additive set functions in a set $\emptyset \neq S \subset \mathbb{R}^n$ and $F(B) = \lim F_m(B)$ for $B \in \mathscr{B}$ with $B \subset S$, then $F' = \lim F'_m$ a.e. in S.*

Proof Without loss of generality, assume $\{F_m\}$ is nondecreasing. Since a sequence with terms $G_m = F \backslash F_m$ is nonincreasing, also $\{\overline{D}G_m\}$ is. It suffices to prove $\overline{D}G_m \to 0$ a.e. in S.

Let $X_0 \subset S$ be the set where $\overline{D}G_m$ does not converge to 0 when $m \to +\infty$. If it were $\mu(X_0) > 0$, since $X_0 = \cup_{k=1}^{\infty} X_k$, where $X_k = \left\{\mathbf{x} \in S : \lim \overline{D}G_m(\mathbf{x}) \geq \frac{1}{k}\right\}$, there would exist $j \in \mathbb{N}$ such that $\mu(X_j) > 0$ and a bounded measurable set $A \subset X_j$ with $\mu(A) > 0$. As $\{\overline{D}G_m\}$ is nonincreasing, it would be $\overline{D}G_m \geq \frac{1}{m}$ in A and, by (A.20), for every bounded interval $I \supset A$, $G_m(I) \geq \frac{1}{j}\mu(A) > 0$ for $m \in \mathbb{N}$, contradicting $\lim G_m(I) = F(I) - \lim F_m(I) = 0$. Thus, $\mu(X_0) = 0$ and $\overline{D}G_m \to 0$ a.e. in S. □

(A.23) *If $\emptyset \neq S \subset \mathbb{R}^n$ and $V_S(X) = \mu(X \cap S)$ for measurable $X \subset \mathbb{R}^n$, then $V'_S = \chi_S$ a.e. in S and if S is measurable, then $V'_S = \chi_S$ a.e. in $\mathbb{R}^n$.*

Proof There is a nonincreasing sequence (with set inclusion) of open sets $\{U_m\} \subset \mathbb{R}^n$ such that S is a subset of $U = \cap_{m=1}^{\infty} U_m$ with $\mu(S) = \mu(U)$. It is $V'_{U_m} = 1$ in U_m for $m \in \mathbb{N}$ and, therefore, also $V'_{U_m} = 1$ in S for $m \in \mathbb{N}$. The sequence $\{V_{U_m}\}$ is nonincreasing and converges to $V_U = V_S$. By the preceding result,

$$V'_S = \lim V'_{U_m} = 1 = \chi_S \text{ a.e. in } S.$$

If S is measurable, also $S^c = \mathbb{R}^n \setminus S$ is, and $V_S + V_{S^c} = \mu(X)$ for all measurable $X \subset \mathbb{R}^n$. Therefore, $V'_S(\mathbf{x}) + V'_{S^c}(\mathbf{x}) = 1$ at each point $\mathbf{x}$ where V'_S and V'_{S^c} exist, which, by (A.21), is a.e. in S. As $V'_{S^c} = 1$ a.e. in S^c, it is $V'_S = 0 = \chi_S$ a.e. in S^c. Hence, $V'_S = \chi_S$ a.e. in $\mathbb{R}^n$. □

(A.24) *If $\emptyset \neq S \subset \mathbb{R}^n$ is measurable, $f \in L(S)$ and $F(B) = \int_B f$ for each $B \in \mathcal{B}$ such that $B \subset S$, then $F' = f$ a.e. in S.*

Proof Without loss of generality, take $f \geq 0$. By the preceding result, $F' = f$ a.e. in S for $f = \chi_S$ if S is measurable and, therefore, also if f is a linear combination of characteristic functions of measurable sets. Since every function $f \in L(S)$ with $f \geq 0$ in a measurable set S is the limit of a nondecreasing sequence $\{f_m\}$ of functions linear combination of characteristic functions of measurable sets, with $F_m(B) = \int_B f_m$ for each $B \in \mathcal{B}$ with $B \subset S$, it is $F' = \lim F'_m = \lim f_m = f$ a.e. em S. □

(A.25) *If $\Omega \subset \mathbb{R}^n$ is an open connected set and $F \geq 0$ is a bounded completely additive set function, then for every $B \in \mathscr{B}$ such that $B \subset \Omega$, it is $F' \in L(B)$ with $\int_B F' \leq F(B)$, with equality for all $B \in \mathscr{B}$, and $B \subset \Omega$ if and only if F is locally absolutely continuous in Ω.*

Proof By the Lebesgue decomposition (A.14), $F = \sigma + \alpha$, with $\sigma, \alpha \geq 0$ bounded completely additive set functions, resp., singular and absolutely continuous, and $\alpha(B) = \int_B f$ for $B \in \mathscr{B}$ with $B \subset \Omega$ and $f \in L(\Omega)$. By (A.17), $\sigma' = 0$ a.e. in Ω, and, by (A.24), $\alpha' = f$ a.e. em Ω; thus, $F' = \alpha' = f$ a.e. in Ω and $\alpha(B) = \int_B F'$ for $B \in \mathscr{B}$ with $B \subset \Omega$. Hence, $F(B) = \sigma(B) + \int_B F'$ and, as $\sigma(B) \geq 0$, it is $\int_B F' \leq F(B)$. Therefore, $F' \in L(B)$ for $B \in \mathscr{B}$ with $B \subset \Omega$.

It is $\int_B F' = F(B)$ for $B \in \mathscr{B}$ with $B \subset \Omega$ if and only if $\sigma(B) = 0$ for $B \in \mathscr{B}$, $B \subset \Omega$, i.e., if and only if $F = \alpha$, what is equivalent to F being absolutely continuous. □

This provides an alternative proof of the Barrow rule, considering $n = 1$, with $f \geq 0$ and integrable function in $[a, b]$ and $F([a, t]) = \int_a^t f$ for $t \in [a, b]$, because the set function F satisfies the conditions of the hypothesis, has finite measurable derivative a.e. in $[a, b]$ and $\int_a^t F' = F([a, t]) = \int_a^t f$. Thus, $\int_a^t (F' - f) = 0$ for $t \in [a, b]$, and $\int_s^t (F' - f) = 0$ for $s, t \in [a, b]$, and $F' = f$ a.e. in $[a, b]$. Defining $\widetilde{F}(t) = \widetilde{F}(a) + F([a, t])$, $\widetilde{F}' = F' = f$ a.e. in $[a, b]$ and $\int_a^t f = F([a, t]) = \widetilde{F}(t) - \widetilde{F}(a)$. The proof in the preceding section simplifies this proof for real functions of a real variable with the rising Sun lemma instead of the Vitali lemma.

If $\mathbf{f}$ is a homeomorphism of an open connected set $\Omega \subset \mathbb{R}^n$ to $\mathbb{R}^n$ and the derivative $V'_{\mathbf{f}}$ of $V_{\mathbf{f}}(B) = \mu(\mathbf{f}(B \cap \Omega))$, with $B \in \mathscr{B}$, exists a.e. in Ω, it is called the n-**volume derivative** of $\mathbf{f}$. The following statement holds.

(A.26) Lebesgue theorem for a homeomorphism volume derivative:
If $\mathbf{f}$ is a homeomorphism of an open connected set $\Omega \subset \mathbb{R}^n$, then the n-volume derivative $V'_{\mathbf{f}}$ of $\mathbf{f}$ is locally integrable in Ω (i.e., integrable in compact subsets of Ω) with $\int_B V'_{\mathbf{f}} \leq \mu(B)$ and is locally absolutely continuous in Ω.

A.4 Banach Spaces L^p

For $p \in [1, +\infty[$ and a nonempty measurable set $S \subset \mathbb{R}^n$ the real (resp., complex) linear space of measurable functions $f: S \to \mathbb{K}$ with $\mathbb{K}=\mathbb{R}$ (resp., $\mathbb{K}=\mathbb{C}$) such that $|f|^p \in L(S)$, identifying equal functions a.e. in S, with the norm $\|f\|_{L^p} = \left(\int_S |f|^p\right)^{1/p}$, is denoted $\boldsymbol{L^p(S)}$.

The particular cases $p=1$ and $p=2$ are considered in the section 2 of Appendix B of *CADOVA*, where it is proved they are complete normed spaces, i.e., **Banach spaces**, and for $p=2$, it is a complete Euclidean space, i.e., a **Hilbert space**.

Checking the conditions of the normed linear space definition for $L^p(S)$ with $p \in [1, +\infty[$ and $\emptyset \neq S \subset \mathbb{R}^n$ is very easy with the exception of the triangular inequality, proved next.

The function defined in $[0, +\infty[$ by $y = x^{p-1}$, with $p > 1$, has inverse defined in $[0, +\infty[$ by $x = y^{\frac{1}{p-1}} = y^{q-1}$, with $q = 1 + \frac{1}{p-1} = \frac{p}{p-1} = \frac{1}{1-1/p}$, i.e., with $\frac{1}{p} + \frac{1}{q} = 1$, and these functions have integrals $\int_0^a x^{p-1} dx = \frac{a^p}{p}$ and $\int_0^b y^{q-1} dy = \frac{b^q}{q}$, for $a, b > 0$. Therefore, $ab \leq \frac{a^p}{p} + \frac{b^q}{q}$ for $a, b > 0$, $p, q > 1$ with $\frac{1}{p} + \frac{1}{q} = 1$, and there is equality if and only if $a^p = b^q$ (Fig. A.3). For $f, g \in L^p(S) \setminus \{0\}$, with $p > 1$, $\frac{1}{p} + \frac{1}{q} = 1$, $a = \frac{|u_j|}{\|\mathbf{u}\|_{L^p}}$ and $b = \frac{|v_j|}{\|\mathbf{v}\|_{L^q}}$ in inequality $ab \leq \frac{a^p}{p} + \frac{b^q}{q}$,

$$\frac{f(\mathbf{x})\, g(\mathbf{x})|}{\|f\|_{L^p} \|g\|_{L^q}} \leq \frac{|f(\mathbf{x})|^p}{p\, \|f\|_{L^p}^p} + \frac{|g(\mathbf{x})|^q}{q\, \|g\|_{L^q}^q} .$$

Integrating both sides,

$$\frac{\int_S |fg|}{\|f\|_{L^p} \|g\|_{L^q}} \leq \frac{1}{p} + \frac{1}{q} = 1 ,$$

we get the **Hölder inequality**, which for $p=2$ is the Cauchy-Schwarz inequality,

$$\|fg\|_{L^1} \leq \|f\|_{L^p} \|g\|_{L^q} , \quad \text{for } p \geq 1 ,\ \frac{1}{p} + \frac{1}{q} = 1 ,$$

with equality for $p > 1$ with $f, g \neq 0$ if and only if $\frac{|f|^p}{\|f\|_{L^p}^p} = \frac{|g|^q}{\|g\|_{L^q}^q}$ a.e., i.e., if and only if $|f|^p = c|f|^q$ a.e., with $c = \frac{\|f\|_{L^p}^p}{\|g\|_{L^q}^q}$. If $|f|^p = c|g|^q$ a.e. in S with $c > 0$, integrating gives $\|f\|^p = c\|g\|^q$. On the other hand, if $f = 0$ or $g = 0$ a.e. in S, the inequality holds with an equality. So, *there is equality in Hölder inequality with $p > 1$ if and only if $|f|^p$ and $|g|^q$ are linearly dependent in the linear space $L^1(S)$.*

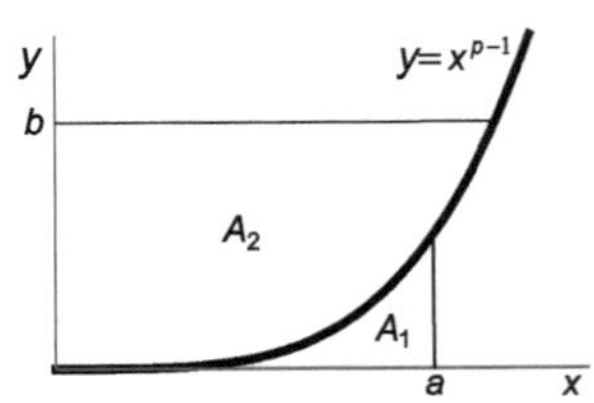

Fig. A.3 $ab \leq \int_0^a x^{p-1} dx + \int_0^b y^{q-1} dy = \frac{a^p}{p} + \frac{b^q}{q}$, $a, b > 0$, $\frac{1}{p} + \frac{1}{q} = 1$, $p, q > 1$

Since

$$(x+y)^p = (x+y)(x+y)^{p-1} = x(x+y)^{p-1} + y(x+y)^{p-1}, \quad x, y \geq 0, \ p \geq 1,$$

with $f, g \in L^p(S)$, $h = |f| + |g|$ and $k = h^{p-1}$, by Hölder inequality, as $(p-1)q = p$, it is $\|k\|_{L^q} = \|h\|_{L^p}^{p/q}$, and

$$\|h\|_p^p = \int_S |h|^p = \int_S |f|\,|h|^{p-1} + \int_S |g|\,|h|^{p-1}$$

$$\leq \|f\|_{L^p}\|k\|_{L^q} + \|g\|_{L^p}\|k\|_{L^q} = \big(\|f\|_{L^p} + \|g\|_{L^p}\big)\|k\|_{L^q} = \big(f\|_{L^p} + \|g\|_{L^p}\big)\|h\|_{L^p}^{\frac{p}{q}}.$$

Dividing by $\|h\|_{L^p}^{p/q}$ for $\|h\|_{L^p} \neq 0$ gives $\|h\|_{L^p} \leq \|f\|_{L^p} + \|g\|_{L^p}$. With the triangular inequality in $\mathbb{R}$,

$$\|f+g\|_{L^p}^p = \int_S |f+g|^p \leq \int_S (|f|+|g|)^p = \|h\|_{L^p}^p \leq (\|f\|_{L^p} + \|g\|_{L^p})^p,$$

giving the **Minkowski inequality**[12]

$$\|f+g\|_{L^p} \leq \|f\|_{L^p} + \|g\|_{L^p}, \quad \text{for } p > 1,$$

also holding trivially for $f = 0$ or $g = 0$, which is the triangular inequality in $L^p(S)$ for $p > 1$.

The proof of the Riesz-Fisher[13] theorem in the section 2 of Appendix B of *CADOVA* for $L^1(S)$ and $L^2(S)$, establishing they are complete normed spaces, also proves that *if* $\emptyset \neq S \subset \mathbb{R}^n$ *is measurable and* $p \in [1, +\infty[$, *then* $L^p(S)$ *is a Banach space, and also a Hilbert space if* $p = 2$.

If $S \subset \mathbb{R}^n$ is a nonempty measurable set with finite measure, since for $1 \leq p < q$ and $f : S \to \mathbb{K}$, if $|f(\mathbf{x})| > 1$, then $|f(\mathbf{x})|^q < |f(\mathbf{x})|^p$, and if $|f(\mathbf{x})| \leq 1$ then $|f(\mathbf{x})|^q \leq 1$. Applying the Lebesgue dominated convergence theorem, *if* $S \subset \mathbb{R}^n$ *is a nonempty measurable set with finite measure and* $1 \leq p < q$, *then* $L^q(S) \subset L^p(S)$.

Other very useful real (resp., complex) Banach spaces of measurable functions defined from a nonempty measurable set $S \subset \mathbb{R}^n$ to $\mathbb{K}$, with $\mathbb{K} = \mathbb{R}$ or $\mathbb{K} = \mathbb{C}$, are the bounded functions in the complement of a set of measure 0 in S, identifying equal functions a.e. in S, with the norm $\|f\|_{L^\infty} = \inf\{C \geq 0 : |f| \leq C \text{ a.e.}\}$, denoted $\boldsymbol{L^\infty(S)}$.

It is easy to verify that if a nonempty set $S \subset \mathbb{R}^n$ is measurable, then $L^\infty(S)$ satisfies the conditions of the definition of normed linear space. To prove it is a Banach space, it remains to prove it is complete. If $\{f_n\} \subset L^\infty(S)$ is a Cauchy sequence, for every $\varepsilon > 0$, there is $N \in \mathbb{N}$ such that $n \geq N$ and $m \in \mathbb{N}$ imply

[12] Minkowski, Hermann (1864–1909).

[13] Fisher, Ernst (1875–1950).

$\|f_n - f_{n+m}\|_{L^\infty} < \varepsilon$, and, therefore, $|f_n(\mathbf{x}) - f_{n+m}(\mathbf{x})| < \varepsilon$ for $\mathbf{x} \in R$ with $R \subset S$ such that $S \backslash R$ has measure 0. Hence, for $\mathbf{x} \in R$, $\{f_n(\mathbf{x})\}$ is a Cauchy sequence in $\mathbb{K}$, and as this is a complete space, this sequence converges to some point in $\mathbb{K}$, denoted $f(\mathbf{x})$. Define the function $f: S \to \mathbb{K}$ to be equal to $f(\mathbf{x})$ for $\mathbf{x} \in R$ and to 0 in $S \backslash R$. For every $\mathbf{x} \in R$, $|f(\mathbf{x})| \leq \varepsilon + |f_N(\mathbf{x})|$, and, therefore, $\|f\|_{L^\infty} \leq \varepsilon + \|f_N\|_{L^\infty}$; thus, $f \in L^\infty$. For $n \geq N$ and $\mathbf{x} \in R$, $|f_n(\mathbf{x}) - f(\mathbf{x})| < \varepsilon$, implying that $\|f_n - f\|_{L^\infty} < \varepsilon$, and that $\{f_n\}$ has limit f in L^∞. Hence, *if $S \subset \mathbb{R}^n$ is a nonempty measurable set, then $L^\infty(S)$ is a Banach space.*

If $S \subset \mathbb{R}^n$ is a nonempty measurable set with finite measure, and $f: S \to \mathbb{K}$ is continuous, by the Weierstrass theorem of extrema of continuous functions, in each compact set, $|f|$ is bounded, and since int S is a countable union of compact subsets and ∂S has measure 0, it is $f \in L^\infty(S)$; thus, denoting $C^0(S)$ the set of continuous functions from S to $\mathbb{K}$, it is $C^0(S) \subset L^\infty(S)$. Furthermore, in each interval $I \subset S$ the function f is the upper limit of step functions a.e. in I, and, as each $g \in L^1(S)$ is also upper limit of step functions in each I, $C^0(S)$ is a dense subset of $L^1(S)$, and, therefore, also of $L^q(S)$ for $q \in [1, +\infty[\, \cup \{\infty\}$. Hence, *if $S \subset \mathbb{R}^n$ is a nonempty measurable set with finite measure, $q \in [1, +\infty[\, \cup \{\infty\}$ and $1 \leq p < q$, then $C^0(S)$ and $L^q(S)$ are dense subsets of $L^p(S)$.*

It is proved next that for $p \in [1, +\infty[$ the dual space of L^p is L^q with $\frac{1}{p} + \frac{1}{q} = 1$. It was proved for $p > 1$ by F. Riesz in 1910 after his announcement in 1907 of its validity for $p = 2$, a particular case called **Riesz representation theorem**. For $p = 1$ it was proved by H. Steinhaus[14] in 1918.

(A.27) *If $S \subset \mathbb{R}^n$ is a nonempty measurable set and $p \in [1, +\infty[$, then a linear transformation $T: L^p(S) \to \mathbb{R}$ is continuous if and only if there is $g \in L^q(S)$, with $\frac{1}{p} + \frac{1}{q} = 1$ (if $p = 1$, then $q = \infty$), such that $T(f) = \int_S fg$; if yes, g is unique, and the operator norm of T is $\|T\| = \|g\|_{L^q}$.*

Proof If $g \in L^q(S)$, T defined as indicated is a linear transformation of $L^p(S)$ to $\mathbb{R}$. By the Hölder inequality, $|T(f)| \leq \|f\|_{L^p} \|g\|_{L^q}$, and T is continuous.

If $g, \widetilde{g} \in L^q(S)$ were such that $T(f) = \int_S fg = \int_S f\widetilde{g}$, it would be $\int_S f(g - \widetilde{g}) = 0$ for $f \in L^p(S)$, and, therefore, for every measurable set with finite measure $E \subset S$ with f the characteristic function of E (1 in E and 0 outside E), it is $g - \widetilde{g} = 0$, i.e., $g = \widetilde{g}$, a.e. in E, proving that if g exists, it is unique.

If $\|T\| = 0$, the result is valid with $g = 0$. Thus, it suffices to prove the existence of g for T a continuous linear transformation with $\|T\| > 0$.

Firstly, consider S with finite measure. For each measurable set $E \subset S$, let χ_E be the characteristic function of E and $G(E) = T(\chi_E)$. If $E_1, E_2 \subset S$ are disjoint measurable sets, then $G(E_1 \cup E_2) = G(E_1) + G(E_2)$. If $\{E_m\} \subset S$ is a sequence of disjoint intervals and $A_k = \cup_{m=0}^{k} E_m$, by the Lebesgue dominated convergence theorem and as $p \in [1, +\infty[$, it is $\lim \|\chi_E - \chi_{A_k}\|_{L^p(S)} = 0$, and, by the continuity

[14] Steinhaus, Hugo (1887–1972).

of T, $\lim G(A_k) = G\big(\cup_{m=0}^{\infty} E_m\big)$. Hence, G is a completely additive set function in S, and if $E \subset S$ has measure 0, since $\|\chi_E\|_{L^p(S)} = 0$, it is $G(E) = 0$. Therefore, by (A.10), G is absolutely continuous in S. By the Radon-Nikodym theorem (A.15), there is a function g integrable in each measurable set $E \subset S$ such that $G(E) = \int_E g$. Therefore, $T(\chi_E) = \int_E g = \int_S \chi_E g$ for every measurable $E \subset S$.

If $f \in L^\infty(S)$, then f is the limit of a uniformly convergent sequence $\{f_m\}$ of functions linear combinations of characteristic functions $\chi_{E_{m,k}}$ of measurable sets $E_{m,k} \subset S$, and $T(f_m) = \int_S f_m g$. As T is continuous in $L^p(S)$,

$$\lim \| f_m - f \|_{L^p(S)} = 0 \Rightarrow T(f) = \lim T(f_m) = \int_S fg\,,$$

the last equality by the Lebesgue dominated convergence theorem; therefore, $T(f) = \int_S fg$, for $f \in L^\infty(S)$ and g defined as above.

We want to show that $g \in L^q(S)$ and $\|g\|_{L^q} = \|T\|$.

If $p = 1$, for every measurable set $E \subset S$

$$\Big| \int_E g \Big| = |G(E)| = T(\chi_E) \le \|T\|\, \|\chi_E\|_{L^1(S)} = \|T\| \int_S \chi_E\,;$$

Hence, $|g| \le \|T\|$ a.e. in S; therefore, $g \in L^\infty(S)$, and $\|g\|_{L^\infty(S)} \le \|T\|$.

If $p \in]1, +, \infty[$, with h the measurable function with $h(\mathbf{x})$ 1 or -1 according to $g(\mathbf{x}) \ge 0$ or $g(\mathbf{x}) < 0$, it is $hg = |g|$. Consider the measurable set

$$E_m = \{\mathbf{x} \in S : |g(\mathbf{x})| \le m\}.$$

The function defined by $f = h|g|^{q-1}\chi_{E_m}$ belongs to $L^\infty(S)$ since h and g do, and

$$|f|^p = |g|^{pq-p} = |g|^q\,, \quad |f| = |g|^{q/p}\,, \quad \text{in } E_m\,, \qquad fg = |g|^q \chi_{E_m}\,, \quad \text{in } S\,.$$

Therefore,

$$\int_{E_m} |g|^q = \int_S fg = T(f) \le \|T\|\, \|f\|_{L^p(S)} = \|T\| \Big[\int_{E_m} |g|^q\Big]^{\frac{1}{p}}.$$

where the second equality follows from three paragraphs above since $f \in L^\infty(S)$. Hence, $\big(\int_{E_m} |g|^q\big)^{1-1/p} \le \|T\|$, and, therefore, $\big(\int_S \chi_{E_m} |g|^q\big)^{1/q} \le \|T\|$ for $m \in \mathbb{N}$; by the monotone convergence theorem, $\|g\|_{L^q(S)} \le \|T\|$.

By the 5 preceding paragraphs, if S has finite measure, $p \in [1, +\infty[$ and $f \in L^\infty(S)$, then $T(f) = \int_S fg$ with $g \in L^q(S)$ and $\|T\| = \|g\|_{L^q(S)}$. Since $L^\infty(S)$ is a dense subset of $L^p(S)$, by the continuity of both sides of the equality $T(f) = \int_S fg$, this equality is valid for $f \in L^p(S)$, ending the proof if S has finite measure.

It remains to prove for measurable S with infinite measure. The function of variable in $\mathbb{R}^n$ $\varphi(\mathbf{x}) = \frac{e^{-\|\mathbf{x}\|}}{\|\mathbf{x}\|}\,\mathbf{x}$, with $\|\mathbf{x}\|$ the norm of the canonical inner product in $\mathbb{R}^n$, is a C^∞ bijection of $\mathbb{R}^n \backslash \{0\}$ onto $B_1 \backslash \{0\}$, where B_1 is the ball of $\mathbb{R}^n$ with center 0 and radius 1. $T : L^p(S) \to \mathbb{R}$ is a continuous linear transformation if and only if $\widetilde{T}: L^p(\varphi(S)) \to \mathbb{R}$ such that $\widetilde{T}(\widetilde{f}) = T(\widetilde{f} \circ \varphi^{-1})$ is a continuous linear transformation. The set $\varphi(S) \subset \mathbb{R}^n$ is bounded and measurable, and, therefore, $\varphi(S)$ has finite measure. Hence, by what was established in the preceding paragraphs of this proof, for all $f \in L^p(S)$, with $\widetilde{f} = f \circ \varphi$ there is $\widetilde{g} \in L^q(\varphi(S))$ such that

$$\widetilde{T}(\widetilde{f}) = \int_{\varphi(S)} \widetilde{f}\,\widetilde{g} = \int_{\varphi(S)} \big(f\,(\widetilde{g} \circ \varphi^{-1})\big) \circ \varphi = \int_S \big(f\,(\widetilde{g} \circ \varphi^{-1})\big)\,\left|\det D\varphi^{-1}\right|,$$

the last equality by the change of integration variable theorem with $\mathbf{y} = \varphi(\mathbf{x})$. So, the result holds with $g = \widetilde{g} \circ \varphi^{-1} |\det D\varphi^{-1}|$, as clearly $g \in L^q(S)$. □

If $S \subset \mathbb{R}^n$ is a nonempty measurable set, the linear transformation $T: L^\infty(S) \to \mathbb{R}$ with $T(f) = f(0)$ is continuous, and there is not a $g \in L^1(S)$ such that $T(f) = \int_S fg$. Thus, the result cannot be extended to $p = \infty$.

A.5 Looman-Menchoff Theorem

In 1935 D. Menchoff completed a proof initiated by H. Looman in 1923 that the Cauchy-Riemann equations are sufficient for differentiability of continuous complex functions in an open subset of $\mathbb{C}$. The proof uses Morera theorem, Baire theorem, and Lebesgue measure, including the Barrow rule for Lebesgue integral proved in Section 2 of this appendix. Although the Looman-Menchoff theorem is not needed for this book chapters, due to its general interest, we present now a proof that also uses the simplification of R. Narasimhan and Y. Nievergelt in their 2001 book listed in the final bibliography of *CADOVA*.

(A.28) Looman-Menchoff theorem: *If $f: \Omega \to \mathbb{C}$ is continuous and satisfies the Cauchy-Riemann equations in an open set $\Omega \subset \mathbb{C}$, then $f \in H(\Omega)$.*

Proof Let $\Omega' = \{z \in \Omega : \exists_{U \text{neighborhood of } z} : f_{|U} \in H(U)\}$ and $E = \Omega \backslash \Omega'$. To prove the result amounts to show that $E = \emptyset$.

Suppose $E \neq \emptyset$. For $k \in \mathbb{N}$ denote

$$\Omega_k^j = \left\{(x, y) \in \Omega : \left|f\big((x, y) \pm \delta \mathbf{e}_j\big) - f(x, y)\right| \leq k\delta \text{ for } 0 < \delta \leq \tfrac{1}{k}\right\}, \quad j = 1, 2,$$

where $\mathbf{e}_1 = (1, 0)$, $\mathbf{e}_2 = (0, 1)$, and $\Omega_k = \Omega_k^1 \cap \Omega_k^2$. Since f is continuous, Ω_k is closed relative to Ω, and since $\frac{\partial f}{\partial x}, \frac{\partial f}{\partial y}$ exist in Ω, $\Omega = \cup_{k=1}^\infty \Omega_k$ and $\cup_{k=1}^\infty (\Omega_k \cap E) = E$. By the Baire category theorem,[15] for some $k_0 \in \mathbb{N}$, $\operatorname{int}(\Omega_{k_0} \cap E) \neq \emptyset$. There are $r > 0$, $z_0 \in \Omega$ such that $\operatorname{cl} B_r(z_0) \subset \Omega$ and

[15] See section 3 of Appendix A of *CADOVA*.

$\emptyset \neq B_r(z_0) \cap E \subset \Omega_{k_0} \cap E$. By the Weierstrass theorem of extrema of continuous functions in compact sets, there is $L>0$ such that $|f| < \frac{L}{2}$ in $\operatorname{cl} B_r(z_0)$; thus, for $(x, y) \in E \cap B_r(z_0)$, $\frac{1}{k_0} < \delta < r$, we have $\left|f\big((x, y) \pm \delta \mathbf{e}_j\big) - f(x, y)\right| \leq L k_0 \delta$. Therefore, with $M = \max\{k_0, Lk_0\}$,

$$\left|f\big((x, y) \pm \delta \mathbf{e}_j\big) - f(x, y)\right| \leq M\delta, \qquad (x, y) \in E \cap B_r(z_0),\ 0<\delta<r,\ j=1,2.$$

Let $R=[a, b] \times [c, d] \subset B_r(z_0)$, $a<b$, $c<d$ and $A>0$ be such that $\frac{1}{A} \leq \frac{d-c}{b-a} \leq A$. Let $\varepsilon < 0$ and $U \supset E$ be an open set with $\mu_2(U \setminus E) < \varepsilon$, where μ_2 is Lebesgue measure in $\mathbb{R}^2$; such a set U exists since any closed set is measurable. By successive N subdivisions of the sides the rectangle R in equal length segments in each step passing to the double of segments, we have a partition of R in a set of subrectangles R_ℓ, $\ell \in \{1, \ldots, 4^N\}$. For N large, if $R_\ell \cap E \neq \emptyset$, then $R_\ell \subset U$. If $R_\ell \subset B_r(z_0) \setminus E$, then $f \in H(\operatorname{cl} R_\ell)$, and, by the Cauchy theorem, $\int_{\delta R_\ell} f(z)\,dz = 0$. Therefore,

$$\int_{\delta R} f(z)\,dz = \sum_{\ell=1}^{4^N} \int_{\delta R_\ell} f(z)\,dz = \sum_{\{\ell : R_\ell \cap E \neq \emptyset\}} \int_{\delta R_\ell} f(z)\,dz.$$

For any rectangle $S \subset R$ let $\widetilde{S}$ be the intersection of all closed rectangles containing $S \cap E$. Due to the Cauchy theorem and the uniform continuity of a continuous function in closed rectangles (by the compactness of closed bounded subsets of $\mathbb{C}$ and the Heine-Cantor theorem[15], $\int_{\delta R_\ell} f(z)\,dz = \int_{\delta \widetilde{R_\ell}} f(z)\,dz$. For each $\ell \in \{1, \ldots, 4^N\}$ such that $R_\ell \cap E \neq \emptyset$,

$$\left|\int_{\delta R} f(z)\,dz\right| \leq \sum_{\{\ell : R_\ell \cap E \neq \emptyset\}} \left|\int_{\delta R_\ell} f(z)\,dz\right| = \sum_{\{\ell : R_\ell \cap E \neq \emptyset\}} \left|\int_{\delta \widetilde{R_\ell}} f(z)\,dz\right|.$$

Denote $\widetilde{R} = I \times J$, $I = \left[\widetilde{a}, \widetilde{b}\right]$, $J = \left[\widetilde{c}, \widetilde{d}\right]$, and for $x \in I$ denote $E_x = \{y \in J : (x, y) \in E\}$. If $E_x \neq \emptyset$, by the auxiliary result (A.29),

$$\left|f(x, \widetilde{d}) - f(x, \widetilde{c}) - \int_{E_x} \tfrac{\partial f}{\partial y}\,dy\right| \leq M \mu_1(J \backslash E_x) \leq 4A\,\mu_1(J \backslash E_x) \leq 4A\big(d-c-\mu_1(E_x)\big),$$

where μ_1 denotes Lebesgue measure in $\mathbb{R}$. If $E_x = \emptyset$, as there are $\xi, \xi' \in I$ such that $\big(\xi, \widetilde{c}\big), \big(\xi', \widetilde{d}\big) \in R \cap E$, then

$$\begin{aligned}
\left|f\big(x, \widetilde{d}\big) - f\big(x, \widetilde{c}\big)\right| &\leq \left|f\big(x, \widetilde{d}\big) - f\big(\xi', \widetilde{d}\big)\right| + \left|f\big(\xi', \widetilde{d}\big) - f\big(\xi', \widetilde{c}\big)\right| \\
&\quad + \left|f\big(\xi', \widetilde{c}\big) - f\big(\xi, \widetilde{c}\big)\right| + \left|f\big(\xi, \widetilde{c}\big) - f\big(x, \widetilde{c}\big)\right| \\
&\leq M\big(|\widetilde{d} - \widetilde{c}| + |x - \xi'| + |\xi' - \xi| + |\xi - x|\big) \\
&\leq M[(d-c) + 3(b-a)] \leq M(1+3A)(d-c) \leq 4AM(d-c);
\end{aligned}$$

so, the inequality between the first and the last terms in the preceding period also holds if $E_x = \emptyset$. Integrating both sides of this inequality,

$$\left|\int_{\widetilde{a}}^{\widetilde{b}}[f(x,\widetilde{d})-f(x,\widetilde{c})]dx-\int_{R\cap E}\tfrac{\partial f}{\partial y}\,dxdy\right|$$
$$\leq 4AM\big[(\widetilde{b}-\widetilde{a})(d-c)-\mu(\widetilde{R}\cap E)\big]\leq 4AM\mu\big(R\backslash(R\cap E)\big).$$

Switching in the preceding argument the roles of x and y,

$$\left|\int_{\widetilde{c}}^{\widetilde{d}}[f(\widetilde{b},y)-f(\widetilde{a},y)]dy-\int_{R\cap E}\tfrac{\partial f}{\partial x}\,dxdy\right|\leq 4AM\mu\big(R\backslash(R\cap E)\big).$$

Since

$$\int_{\partial\widetilde{R}}f(z)\,dz=i\int_{\widetilde{c}}^{\widetilde{d}}[f(\widetilde{b},y)-f(\widetilde{a},y)]dy+\int_{\widetilde{a}}^{\widetilde{b}}[f(x,\widetilde{d})-f(x,\widetilde{c})]dx\,,$$

and the validity of the Cauchy-Riemann equations for $f=u+iv$ with u,v real valued implies $i\frac{\partial f}{\partial x}=\frac{\partial f}{\partial y}=i\frac{\partial(u+iv)}{\partial x}-\frac{\partial(u+iv)}{\partial y}=0$,

$$\left|\int_{\delta\widetilde{R_\ell}}f(z)\,dz\right|\leq 8AM\,\mu\big(R_\ell\backslash(R_\ell\cap E)\big)\,.$$

As $R_\ell\cap E\neq\emptyset\Rightarrow R_\ell\subset U$, it is $\sum_{\{\ell\,:\,R_\ell\cap E\neq\emptyset\}}\mu\big(R_\ell\backslash(R_\ell\cap E)\big)\leq\mu\big(U\backslash(U\cap E)\big)<\varepsilon$, and, from the two preceding paragraphs, $\left|\int_{\delta R}f(z)\,dz\right|<8AM\varepsilon$. Since $\varepsilon>0$ is arbitrary, it is $\int_{\delta R}f(z)\,dz=0$ for all closed rectangles $R\subset B_r(z_0)$, and, therefore, also $\int_{\delta\Delta}f(z)\,dz=0$ for all closed triangle $\Delta\subset B_r(z_0)$. Hence, by Morera theorem,[16] $f_{|B_r}\in H(B_r(z_0))$, contradicting the assumption $E\neq\emptyset$. Consequently, $f\in H(\Omega)$. □

(A.29) *If $I=[a_0,b_0]\subset\mathbb{R}$, $a_0<b_0$, $f:I\to\mathbb{C}$ is differentiable, $\emptyset\neq E\subset I$ is closed, and $M>0$ are such that $|f(x)-f(y)|\leq M|x-y|$ for $x\in E$, $y\in I$,*

$$\left|f(b_0)-f(a_0)-\int_E f'(x)\,dx\right|\leq M\mu_1(I\backslash E)\,.$$

Proof Let $J=[a,b]\subset I$ and $f_j:\mathbb{R}\to\mathbb{C}$ be the affine function with values at a,b coincident with those of f. It is $|f_J(x)-f_j(y)|\leq\frac{1}{b-a}|f(b)-f(a)||x-y|$ for $x,y\in\mathbb{R}$. Let $E_0=E\cup\{a\}\cup\{b\}$ and $g:I\to\mathbb{C}$ be such that $g_{|E_0}=f_{|E_0}$ and if J is the closure of a connected component of $I\setminus E_0$, then $g_{|J}=f_{J\,|J}$.

We now prove that $|g(x)-g(y)|\leq L|x-y|$ for $x,y\in I$, with $L=M$. If $x<y$ belong to the same interval J closure of a connected component of $I\setminus E_0$, the inequality holds with $L=\frac{|f(b)-f(a)|}{b-a}\leq M$; so, the inequality also holds with $L=M$. If $x<y$ do not belong to the same such interval J, there is $\xi\in E$ such that $x<\xi<y$;

[16] See the beginning of Chapter 3 of *CADOVA*.

if $x \in E_0$, then $|g(x)-g(\xi)| = |f(x)-f(\xi)| \leq M(\xi - x)$; if $x \notin E_0$, J is the closure of the connected component of $I \setminus E_0$ containing x and x' is the right end point of the interval J, then

$$|g(x)-g(\xi)| \leq |g(x)-g(x')| + |g(x')-g(\xi)| \leq M(x'-x) + M(\xi - x') = M(\xi - x),$$

where for the first term of the second inequality, we use the preceding period, and for the other term of the same inequality, we note that $\xi \in E, x' \in E_0$. We get analogously, $|g(\xi) - g(y)| \leq M(y - \xi)$. Therefore, adding the two inequalities gives $|g(x) - g(y)| \leq M(x - y)$ for $x, y \in I$.

Thus, the function g is absolutely continuous in the interval I. By the Barrow rule for Lebesgue integral (A.8), $g(b_0) - g(a_0) = \int_I g' = \int_E g' + \int_{I \setminus E} g'$. As f is differentiable, at every non-isolated point x of E, $g'(x) = f'(x)$. As sets of isolated points of subsets of $\mathbb{R}$ are countable, $g' = f'$ a.e. in E. Since then, $|g'| \leq M$ a.e. in I and $g(a_0) = f(a_0)$, $g(b_0) = f(b_0)$,

$$\left| f(b_0) - f(a_0) - \int_E f'(x)\, dx \right| = \left| \int_{I \setminus E} g'(x)\, dx \right| \leq M\mu_1(I \setminus E).$$

□

An example of a function that satisfies the Cauchy-Riemann equations in $\mathbb{C}$ and is not holomorphic in any neighborhood of 0 is $f(z) = e^{-z^4}$ if $z \neq 0$ and $f(0) = 0$; since $\lim_{r \to 0+} \frac{f(re^{i\pi/4})}{re^{i\pi/4}} = \infty$, f is not holomorphic at 0 (it has an essential singularity at 0).

The continuity of a function f in $\mathbb{C}$ and the validity of the Cauchy-Riemann equations at a point are not sufficient for differentiability at the point, as the example $f(z) = \frac{z^5}{|z|^4}$ for $z \neq 0$ and $f(z) = 0$ shows at the point 0.

A much weaker condition for sufficiency of the validity of the Cauchy-Riemann equations for holomorphy of functions is the Weyl lemma (1.26) in Chap. 1, namely, that the function be locally Lebesgue integrable and the Cauchy-Riemann equations are satisfied with derivatives in the sense of distributions.

Appendix B
Elements of Distributions

B.1 Introduction

Differential equations express relations of values and derivatives of functions at each point, but it is not possible to experimentaly measure a quantity at a point (itself impossible to materialize), and measurements are averages in neighborhoods of the point, with appropriate weights, corresponding to weighted integrals of functions. Differential equations are often limits of integral relations at points when regions of integration shrink to points. It may even happen that in the limit, the function does not have derivative in the classical sense or is discontinuous. It is therefore natural to extend the notions of function and derivative to correspond to relations valid under integrals, allowing to define derivatives of non-differentiable continuous functions and even of more general functions. This is the idea of distributions and why they play a central role in the study of Partial Differential Equations and in the Calculus of Variations and their applications.[1]

Another case concerns problems in Mechanics, Electromagnetism, or other areas for which it is of interest to consider distributions of mass density, electric charge, or densities of scalar quantities by length, area, or volume and also point masses, charges, or similar quantities totally concentrated at a point. For example, if in a straight line we consider intervals with a continuous density of mass per length and isolated points with concentrated mass $\neq 0$, it is useful to be able to obtain the total mass and establish other relations with integrals that allow to consider "densities" corresponding to masses concentrated at isolated points.

Other cases are of impulses concentrated at a point seen as limits of integrable functions with integrals with limit $\neq 0$ when the supports shrink to the point and of making sense of differentiating functions with jump discontinuities, as in circuits, systems, or signal analysis.

[1] This appendix is mainly based on the first chapter of the book Hörmander, L., *Linear Partial Differential Operators, Springer-Verlag, Berlin, 1976.* Lars Hörmander (1931–2012) was awarded the *Fields Medal* in 1962 for contributions to the theory of linear differential operators.

L. T. Magalhães, *Quasiconformal Mappings in the Plane and Complex Dynamics*,
https://doi.org/10.1007/978-3-031-80115-0

For example, the linear partial differential equation $\frac{\partial^2 u}{\partial x\,\partial y}(x, y) = f(x)$ in $\mathbb{R}^2$, with f continuous, has solutions $u(x, y) = yF(x) + g(y) + h(x)$, where F is the antiderivative of f, g is differentiable in $\mathbb{R}$, and h is a function defined in $\mathbb{R}$ possibly with discontinuities, but many of these functions are not solutions of the equation $\frac{\partial^2 u}{\partial y\,\partial x}(x, y) = f(x)$, obtained by changing the order of derivatives. The first derivative to be applied to solutions of the equation, $\frac{\partial u}{\partial x}$, does not even exist at points where h is not differentiable. It is natural to establish a framework for the solutions of partial differential equations to agree for different derivative orders. The domain of definition of derivatives can be extended through the linear transformation defined by the derivatives under integrals, as weighted averages, and the resp. adjoint, in the case of the partial differential equation considered,

$$\iint \frac{\partial^2 u}{\partial x\,\partial y}\,(x, y)\,\varphi(x, y)\,dx\,dy = \iint f(x)\,\varphi(x, y)\,dx\,dy\,,$$

where weights φ, called **test functions**, are C^2 functions with bounded supports and can be integrated by iterated integrals and by parts passing the derivatives of u to φ, leading to

$$\iint u(x, y)\,\frac{\partial^2 \varphi}{\partial y\,\partial x}\,(x, y)\,dx\,dy = \iint \big[\,yF(x) + g(y) + h(x)\,\big]\,\frac{\partial^2 \varphi}{\partial y\,\partial x}\,(x, y)\,dx\,dy\,,$$

which is also equivalent to

$$\iint \frac{\partial^2 u}{\partial y\,\partial x}\,(y, x)\,\varphi(x, y)\,dx\,dy = \iint u(x, y)\,\frac{\partial^2 \varphi}{\partial x\,\partial y}\,(x, y)\,dx\,dy\,.$$

Since the above integral equations can only be satisfied by unique continuous functions, resp., $\frac{\partial^2 u}{\partial x\,\partial y}$, u, $\frac{\partial^2 u}{\partial y\,\partial x}$, it is natural to define $\frac{\partial^2 u}{\partial x\,\partial y} = f$ in the **weak sense**, even for $u(x, y) = y\,F(x) + g(y) + h(x)$ with h not differentiable and possibly discontinuous, but integrable in a neighborhood of each point, and to consider this function a solution of the differential equations $\frac{\partial^2 u}{\partial x\,\partial y} = f$ and $\frac{\partial^2 u}{\partial y\,\partial x} = f$, which, in this weak sense, are equivalent.

The basics of distributions were introduced by S. Sobolev in 1936 to solve a particular type of differential equations. He also introduced for this particular type of equations the appropriate topology of the space of test functions, the definition as continuous linear functionals, the identification of locally integrable functions with distributions, the product of C^∞ functions by distributions, and the derivatives of any order of distributions, but he did not consider applications for other differential equations. In fact, the important notion of distribution was not considered in general before the work of L. Schwartz from 1944 onward.

In 1893, O. Heaviside had used the "delta function" in his operational calculus, which only finds a solid basis of definition in the context of distributions, considering it non-rigorously as derivative of the unit step function, 0 at points < 0 and 1 at points > 0, today usually called Heaviside function. G. Kirchoff also considered the "delta function" in 1882 when dealing with the fundamental solution of the wave equation, and S.-D. Poisson had considered it in 1815, J. Fourier in 1822, and A.-L. Cauchy in 1823 and 1827, in connection with linear differential equations and Fourier series. Much later, the "delta function" was also called "Dirac delta function," following its consideration by P. Dirac[2] in a 1927 article entitled *The physical interpretation of the quantum dynamics*, though P.Dirac came rather late to consider it.

In 1927, S. Bochner extended the notion of Fourier transform to functions f with $\frac{f(x)}{1+|x|^k}$ integrable for some $k \in \mathbb{N}$, which may fail to be functions but are distributions, without giving an overall picture of them.

In 1944, L. Schwartz considered distributions as linear functionals continuous in a space of test functions C^∞ with compact support and dedicated close to 15 years developing and making widely known the Theory of Distributions and their applications, having been largely responsible for the widespread adoption of distributions as a natural framework for the study of partial differential equations and other problems. The term "distribution" is due to L. Schwartz in 1945. Distributions began to have widespread use in Partial Differential Equations mainly after the publication in 1950–51 of the L.Schwartz book *Théorie des Distributions*.[3]

The regularization of functions by regularizing kernels or mollifiers is useful for building test functions of distributions. It was created by K.O. Friedrichs[4] in 1944, when he also introduced the term *mollifier,* for establishing equivalence of weak and strong solutions of linear partial differential equations in spaces such as L^2 and L^1.

In 1960 L. Hörmander contributed decisively to the Theory of Linear Partial Differential Equations with variable coefficients giving necessary conditions for existence of solution for any independent term $f \in C^\infty$, following an example given by H. Lewy[5] of a partial differential equation with variable coefficients that, in contrast to linear partial differential equations with constant coefficients, has no solution (even in the sense of distributions), and establishing that these conditions properly strengthened are sufficient for existence of local solutions for any independent term $f \in C^\infty$.

In this chapter, the notion of derivative is used in the weak sense of distributions exemplified in the fourth paragraph of this introduction, and the regularization of functions by mollification is also used. This appendix is included to facilitate the access to these concepts to readers who did not encounter them before. Most of the last two sections are not needed for this book chapter, but they are included to provide simple access to basic properties of distributions and their limits, derivatives, and products.

B.2 Space of Test Functions

If $S \neq \emptyset$ is an open subset of $\mathbb{R}^n$, denote $C_0^k(S)$, $k \in \mathbb{N} \cup \{0, \infty\}$, the set of C^k function with bounded support in S with values in $\mathbb{K}$, where $\mathbb{K} = \mathbb{R}$ or $\mathbb{K} = \mathbb{C}$) (the

[2] Heaviside, Oliver (1850–1925). Paul Dirac (1902–1984) received the Nobel Prize of Physics in 1933 by the discovery of new forms of atomic theory.

[3] Poisson, Siméon Denis (1781–1840). Kirchoff, Gustav (1824–1887). Bochner, Solomon (1889–1982).

[4] Friedrichs, Kurt Otto (1901–1982).

[5] Lewy, Hans (1904–1988).

case $S \subset \mathbb{C}$ is as that of $S \subset \mathbb{R}^2$). In the context of distributions, the elements of $C_0^\infty(S)$ are called **test functions** in S, since they are used as possible weights in integrals as in the introduction to verify if an identity, e.g., a differential equation holds in the weak sense.

$C_0^k(S)$ for $k \in \mathbb{N} \cup \{0, \infty\}$ are linear subspaces of $C^\infty(S)$. A test function ≥ 0 in $\mathbb{R}^2$ or $\mathbb{C}$ with support $B_{0,1}$, where $B_{\mathbf{a},r} = \{\mathbf{x} \in \mathbb{R}^n : \|\mathbf{x}\| \leq r\}$ is the closed ball of $\mathbb{R}^n$ with center $\mathbf{a}$ and radius r, is $\varphi(\mathbf{x}) = \psi\left(\|\mathbf{x}\|^2 - 1\right)$, with $\psi(t) = e^{-1/t^2}$ if $t < 0$ and 0 if $t \geq 0$. Dividing φ by its integral gives a test function $\Psi \geq 0$ with support $B_{0,1}$ and integral 1 in $\mathbb{R}^n$.

The function

$$\Psi_{\mathbf{a},\varepsilon}(\mathbf{x}) = \frac{1}{\varepsilon^2}\,\Psi\left(\frac{\mathbf{x}-\mathbf{a}}{\varepsilon}\right), \qquad \varepsilon > 0,$$

is a test function ≥ 0 *in* $\mathbb{R}^n$ *with support* $B_{\mathbf{a},\varepsilon}$ *and integral* 1 *in* $\mathbb{R}^n$ (Fig. B.1). *Denote* $\Psi_\varepsilon = \Psi_{0,\varepsilon}$.

The convolution of these function with a function of $L^1(S)$ with bounded support in an open set S smoothens it[6] leading for $\varepsilon > 0$ small to infinitely differentiable functions arbitrarily close to the initial function in $L^1(S)$.

(B.1) Smoothening of functions in L^1 or L^2 by mollifying:
If $u \in L^p(S)$, $p = 1, 2$, $S \neq \emptyset$ *is an open subset of* $\mathbb{R}^n$ *with* $\operatorname{supp} u \subset S$ *bounded, then* $u_\varepsilon = u * \Psi_\varepsilon \in C_0^\infty(S)$ *is a test function in S with support included in the set of points distant* $\leq \varepsilon$ *from* $\operatorname{supp} u$, $\|u_\varepsilon\|_{L^1} \leq \|u\|_{L^1}$, $u_\varepsilon \to u$ *in* L^p *when* $\varepsilon \to 0$ *and* $C_0^\infty(S)$ *is dense in* $L^p(S)$ *for* $p = 1, 2$ (Fig. B.2).

If, additionally, u is continuous, $u_\varepsilon \to u$ *uniformly when* $\varepsilon \to 0$.

$M_\varepsilon(u) = u_\varepsilon = u * \Psi_\varepsilon$ *is a continuous linear transformation from* $L^p(S)$ *to* $C_0^\infty(S)$, *with operator norm* $\|M_\varepsilon\| = 1$ *for* $p = 1, 2$.

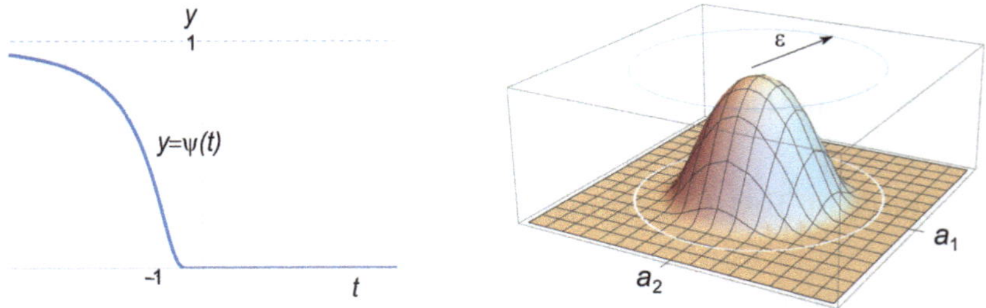

Fig. B.1 $\psi(t) = e^{-\frac{1}{t^2}}$ and test function $\Psi_{\mathbf{s},\varepsilon}$ with support $B_{\mathbf{a},\varepsilon} \subset \mathbb{R}^2$, $\mathbf{a} = (a_1, a_2)$

[6] This technique of smoothening functions by convolution with a smooth function is called ***mollifying***, and smoothening functions, like $\Psi_{\mathbf{a},\varepsilon}$, are called **mollifiers**.

Proof If $u\in L^p(S)$, with the functions in this space with values in $\mathbb{K}$, has compact support, then the convolution $u*\Psi_\varepsilon$, i.e., the function

$$u_\varepsilon(\mathbf{x}) = \big(u*\Psi_\varepsilon\big)(\mathbf{x}) = \int_{\mathbb{R}^n} u(\mathbf{y})\,\Psi_\varepsilon(\mathbf{x}-\mathbf{y})\,d\mathbf{y}$$

is defined in $\mathbb{R}^n$, since if $p=1$, for every $\mathbf{x}\in\mathbb{R}^n$, $|u(\mathbf{y})\,\psi_\varepsilon(\mathbf{x}-\mathbf{y})|\le|u(\mathbf{y})|$, and as $u\in L^1(S)$ and $\operatorname{supp}u\subset S$, also $\mathbf{y}\mapsto u(\mathbf{y})\,\psi_\varepsilon(\mathbf{x}-\mathbf{y})$ is in $L^1(\mathbb{R}^n)$, and

$$\|u_\varepsilon\|_{L^1} = \int_{\mathbb{R}^n}|u_\varepsilon| \le \int_{\mathbb{R}^n}|u(\mathbf{y})|\,d\mathbf{y} = \|u\|_{L^1}\,.$$

If $p=2$, as $\Psi_\varepsilon\in L^2(\mathbb{R}^n)$, by the Cauchy-Schwarz inequality,

$$\big|u_\varepsilon(\mathbf{x})\big| = \left|\int_{\mathbb{R}^n} u(\mathbf{y})\,\Psi_\varepsilon(\mathbf{x}-\mathbf{y})\,d\mathbf{y}\right| \le \|u\|_{L^2}\|\Psi_\varepsilon\|_{L^2}\,,$$

and u_ε is bounded, and it is C^∞, since by the general properties of integral partial derivatives relative to $\mathbf{x}$ can be switched with an integral, and with estimates similar to the preceding ones with the corresponding derivative instead of Ψ_ε, we get integrable functions. u_ε has bounded support because it is 0 at points distant $>\varepsilon$ of $\operatorname{supp}u$; thus, $u_\varepsilon\in C_0^\infty(S)$ (Fig. B.2).

The change of integration variables $\mathbf{x}-\mathbf{y}=\mathbf{z}$ gives

$$|u_\varepsilon(\mathbf{x})-u(\mathbf{x})| = \left|\int_{\mathbb{R}^n}\big[u(\mathbf{x}-\mathbf{z})-u(\mathbf{x})\big]\Psi_\varepsilon(\mathbf{z})\,d\mathbf{z}\right| \le \int_{\mathbb{R}^n}\big|u(\mathbf{x}-\mathbf{z})-u(\mathbf{x})\big|\Psi_\varepsilon(\mathbf{z})\,d\mathbf{z}\,.$$

If u is continuous, as it has bounded support, by the Heine-Cantor theorem, it is uniformly continuous. For every $\delta>0$, there is $\varepsilon>0$ such that $\|\mathbf{x}-\mathbf{z}\|\le\varepsilon$ implies $\big|u(\mathbf{x}-\mathbf{z})-u(\mathbf{x})\big|\le\delta$; $|u_\varepsilon(\mathbf{x})-u(\mathbf{x})|\le\delta\int_{\mathbb{R}^n}\Psi_\varepsilon(\mathbf{z})\,d\mathbf{z}=\delta$, since $\Psi_\varepsilon(\mathbf{z})=0$ for $\|\mathbf{z}\|>\varepsilon$. Hence, if u is continuous, $u_\varepsilon\to u$ uniformly when $\varepsilon\to 0$.

For every $\delta>0$ there is $v\in C_0^0(S)$ such that $\|u-v\|<\delta$ and, therefore,

$$\|u_\varepsilon-v_\varepsilon\|_{L^1} = \|(u-v)_\varepsilon\|_{L^1} \le \|(u-v)\|_{L^1} \le \delta\,.$$

By the Heine-Cantor theorem, for $\varepsilon>0$ small $\|v_\varepsilon-v\|_{L^p}\le\delta V_n(I)$, $p=1,2$, where I is a bounded interval containing $\operatorname{supp}v$. By the triangular inequality,

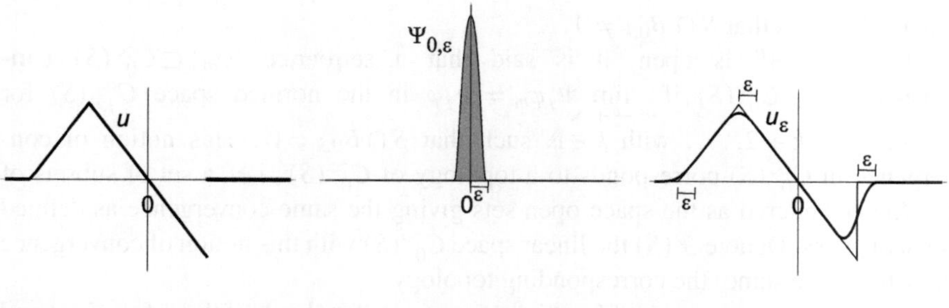

Fig. B.2 Test function $u_\varepsilon = u * \Psi_{0,\varepsilon}$ with support in an open set $S\subset\mathbb{R}^n$, arbitrarily close to an integrable function with bounded support u in S, exemplified with $n=1$ (with the gray area 1)

$$\|u_\varepsilon - u\|_{L^p} \le \|u_\varepsilon - v_\varepsilon\|_{L^p} + \|v_\varepsilon - v\|_{L^p} + \|v - u\|_{L^p} \le \delta\big(2 + V_n(I)\big), \quad p = 1, 2.$$

Hence, $u_\varepsilon \to u$ in $L^p(S)$ when $\varepsilon \to 0$, for $p = 1, 2$.

The last statement follows from

$$\max_S |u_\varepsilon| \le \int_{\mathbb{R}^n} |u| = \|u\|_{L^1} \le \|u\|_{L^2},$$

the last inequality obtained applying the Cauchy-Schwarz inequality; thus, in both cases, $p = 1, 2$, it is $\|M_\varepsilon\| \le 1$, and the reverse inequality is obtained considering the constant function 1 if S is bounded or a sequence of functions equal to 1 in an expansive sequence of compact subsets of S. □

If K is a compact subset of an open set $S \ne \emptyset$ of $\mathbb{R}^n$ and $\varepsilon < \varepsilon' < \varepsilon + \varepsilon'$ with $\varepsilon + \varepsilon'$ smaller than the distance of points in the exterior of S to points in K, considering a function u with $u = 1$ in the compact set K'_ε of the points distant $\le \varepsilon'$ from K and $u = 0$ outside K'_ε. The function $u_\varepsilon = u * \Psi_\varepsilon$ has support in $K_{\varepsilon+\varepsilon'}$ and $u_\varepsilon = 1$ in $K_{\varepsilon-\varepsilon'}$. Therefore, *if K is a compact subset of an open subset $S \ne \emptyset$ of $\mathbb{R}^n$, there is $\psi \in C_0^\infty(S)$ such that $0 \le \psi \le 1$ and $\psi = 1$ in an open subset of $\mathbb{R}^n$ containing K.*

The following condensed notation is adopted for partial derivatives,

$$D^{\mathbf{k}}\varphi(\mathbf{x}) = \frac{\partial^{k_1+\cdots+k_n}\varphi}{\partial^{k_1}x_1\cdots\partial^{k_n}x_n}(\mathbf{x}), \quad \mathbf{k} = (k_1, \ldots, k_n),\ |\mathbf{k}| = \sum_{j=1}^n k_j,\ k_1, \ldots, k_n \in \mathbb{N}\cup\{0\}.$$

For $\emptyset \ne S \subset \mathbb{R}^n$ open and $r > 0$, denote $C_{r,0}^\infty(S)$ the set of functions of $C_0^\infty(S)$ with support $B_{0,r}$. If $S \ne \emptyset$, $C_{r,0}^\infty(S)$ is a linear subspace of $C_0^\infty(S)$. For every $m \in \mathbb{N}$, it is a normed linear space with norm

$$\|\varphi\|_m = \sup\big\{\big|D^{\mathbf{k}}\varphi(\mathbf{x})\big| : \ |\mathbf{k}| = k,\ \mathbf{k} = (k_1, \ldots, k_n),\ k \in \{0, \ldots, m\},\ \|\mathbf{x}\| \le r\big\}.$$

In this space, a sequence converges to $\varphi \in C_{r,0}^\infty(S)$ if and only if the function and all its partial derivatives of order $\le m$ converge uniformly in compact subsets of S, resp., to φ and to its corresponding partial derivatives. It is

$$C_{k,0}^\infty(S) \subsetneqq C_{k+1,0}^\infty(S) \subsetneqq C_{k+2,0}^\infty(S) \subsetneqq \cdots, \qquad C_0^\infty(S) = \cup_{j=k}^\infty C_{j,0}^\infty(S), \quad k \in \mathbb{N},$$

where S is such that $S \cap B_{0,k} \ne \emptyset$.

If $\emptyset \ne S \subset \mathbb{R}^n$ is open, it is said that a **sequence** $\{\varphi_m\} \subset C_0^\infty(S)$ **converges** to $\varphi \in C_0^\infty(S)$ if $\lim_{m\to+\infty} \Psi_j\varphi_m = \Psi_j\varphi$ in the normed space $C_{j,0}^\infty(S)$ for $j = k, k+1, k+2, \ldots$, with $k \in \mathbb{N}$ such that $S \cap B_{0,k} \ne \emptyset$. This notion of convergence in $C_0^\infty(S)$ corresponds to a topology of $C_0^\infty(S)$, i.e., a set of subsets of $C_0^\infty(S)$ considered as the space open sets giving the same convergence as defined for sequences. Denote $\mathscr{D}(S)$ the linear space $C_0^\infty(S)$ with this notion of convergence or (what is the same) the corresponding topology.

In this space, a sequence of functions converges if and only if these functions and all their partial derivatives of order $\le m$ converge uniformly in compact subsets of S, resp., to a function in the space and to its corresponding partial derivatives.

B.3 Definition and Basic Properties of Distributions

A linear functional $T:\mathscr{D}(S)\to\mathbb{K}$ ($\mathbb{K}=\mathbb{R}$ or $\mathbb{C}$), with $S\neq\emptyset$ open subset of $\mathbb{R}^n$, is said to be **continuous** or **bounded** if for sequences $\{\varphi_m\}\subset\mathscr{D}(S)$ such that $\varphi_n\to\varphi$ in $\mathscr{D}(S)$ implies $f(\varphi_m)\to f(\varphi)$ in $\mathbb{R}$, when $m\to+\infty$.

A **distribution** in S, with $S\neq\emptyset$ open subset of $\mathbb{R}^n$ or $\mathbb{C}$, is defined as a continuous linear functional $T:\mathscr{D}(S)\to\mathbb{K}$. The **set of distributions in S is denoted $\mathscr{D}^*(S)$**. It is a subspace of the linear space of the functions from $\mathscr{D}(S)$ to $\mathbb{K}$.

Analogously to linear transformations between normed spaces, a linear functional $T:C_0^\infty(S)\to\mathbb{K}$ is continuous if for every compact $\emptyset\neq K\subset S$, there is $M>0$ such that

$$|T(\varphi)|\leq M\sum_{|\mathbf{k}|\leq k}\sup|D^{\mathbf{k}}\varphi|,\quad \varphi\in C_0^\infty(S);$$

thus, *a linear functional T in $C_0^\infty(S)$ is a Distribution in S if and only if it satisfies the preceding condition*. This condition on linear functionals defined in $\mathscr{D}(S)$ can be adopted as an alternative definition of distribution in S.

A **finite-order distribution** in S is a distribution T such that the preceding condition holds with a same k for all compact subsets K of S, and the smallest integer k with this property is called the **order of the distribution** T. The set of the finite-order distributions in S is denoted $\mathscr{D}_F^*(S)$.

The sets of distributions of finite-order k in S and of distributions of any finite-order $\mathscr{D}_F^*(S)$ are subspaces of the linear space of distributions $\mathscr{D}^*(S)$ in S with scalars in $\mathbb{K}$.

A function $f:S\to\mathbb{K}$ integrable in every compact subset of an open subset $S\neq\emptyset$ of $\mathbb{R}^n$ or $\mathbb{C}$ is said to be **locally integrable** in S. The set of these functions is denoted $L^1_{\text{loc}}(S)$, and the set of measurable functions $f:S\to\mathbb{K}$ such that $f^2\in L^1_{\text{loc}}(S)$ is denoted $L^2_{\text{loc}}(S)$. If S is a bounded set, $L^2_{\text{loc}}(S)\subset L^1_{\text{loc}}(S)$.

$L^1_{\text{loc}}(S)$ and $L^2_{\text{loc}}(S)$, with $S\neq\emptyset$ an open subset of $\mathbb{R}^n$ or $\mathbb{C}$, are subspaces of the linear space of the functions defined in S with scalar values.

If $f\in L^1_{\text{loc}}(S)$, the function defined in $\mathscr{D}(S)$ for all $\mathbf{k}\in(\mathbb{N}\cup\{0\})^2$ by $T(\varphi)=\int_{\mathbb{R}^n}f\,D^{\mathbf{k}}\varphi$ is a distribution of finite-order $|\mathbf{k}|$. Distributions mentioned in the motivation at the beginning of this appendix in connection with partial differential equations are of this type.

If $f_1, f_2\in L^1_{\text{loc}}(S)$, they define distributions of order 0 as in the preceding paragraph, resp., $T_j(\varphi)=\int_{\mathbb{R}^n}f_j\varphi$, $j=1,2$. If $T_1=T_2$, then $\int_{\mathbb{R}^n}(f_1-f_2)\varphi=0$ for $\varphi\in C_0^\infty$. As in the preceding section, define

$$(f_1-f_2)_\varepsilon=(f_1-f_2)*\Psi_{0,\varepsilon}=\int_{\mathbb{R}^n}(f_1-f_2)\,\Phi_{\mathbf{y},\varepsilon},$$

which is 0 since $\Phi_{\mathbf{y},\varepsilon}\in C_0^\infty$, and, by the result on convolutions of the preceding section, $(f_1-f_2)_\varepsilon\to f_1-f_2$ in $L^1(S)$ when $\varepsilon\to0$; thus, $f_1-f_2=0$ in $L^1(S)$, i.e., $f_1=f_2$ *a.e.* in S. Therefore, *the relation $f\mapsto\int_{\mathbb{R}^n}f\varphi$ defines a one-to-one*

Fig. B.3 Graphical representations of the Dirac distribution (mass/charge 1 concentrated at 0) or sequence of functions ≥ 0 C^∞ with integral 1 in $\mathbb{R}$ and supports intervals centered at 0 with lengths tending to 0

correspondence of $L^1_{\text{loc}}(S)$ *with the set of distributions of order* 0 *in* S *defined by* $T_g(\varphi) = \int_{\mathbb{R}^n} g\varphi$ *for some* $g \in L^1_{\text{loc}}(S)$.

Based on this, **each function** $f \in L^1_{\text{loc}}(S)$ **is identified with the distribution** T_f of order 0 in an open subset $S \neq \emptyset$ of $\mathbb{R}^n$ or $\mathbb{C}^n$ defined by $T_f(\varphi) = \int_S f\varphi$ and **the distributions in** S **are considered as extensions of the locally integrable functions in** S.

The function defined in $\mathscr{D}(\,]0,1[\,)$ by $T(\varphi) = \sum_{j=1}^{\infty} \varphi\big(\frac{1}{j}\big)$ is a distribution in $]0,1[$, and it is not of finite order.

The function defined in $\mathscr{D}(\mathbb{R}^n)$ by $T(\varphi) = \varphi(0)$ is a distribution in $\mathbb{R}^n$ that is not of finite order. It is denoted δ and called **Dirac distribution** or **Dirac δ-function**[7] in $\mathbb{R}^n$. The distribution defined by $\delta_{\mathbf{a}}(\varphi) = \varphi(a)$ is called the **Dirac distribution centered at a** or **Dirac δ-function centered at a** (Fig. B.3).

It is also common to write $\int_{\mathbb{R}^n} \delta\varphi$ and $\int_{\mathbb{R}^n} \delta_{\mathbf{a}}\varphi$ for, resp., $\varphi(0)$ and $\varphi(\mathbf{a})$, in spite of the nonexistence of a δ-function and a $\delta_{\mathbf{a}}$-function defined in $\mathbb{R}^n$ with this property, since it suggests that δ and $\delta_{\mathbf{a}}$ are, e.g., the distributions of mass density by volume corresponding to a mass 1 concentrated at the point, resp., 0 and $\mathbf{a}$. It is also common to write $\delta_{\mathbf{a}}(\mathbf{x}) = \delta(\mathbf{x}-\mathbf{a})$. These should be seen just as symbolic notations without analytic meaning.

The distribution $m\,\delta_{\mathbf{a}}$, with $m > 0$, corresponds to a density in $\mathbb{R}^n$ with total quantity (e.g., mass) m concentrated at the point $\mathbf{a}$; therefore, it solves one of the objectives mentioned in the introduction to this appendix of considering densities of scalar quantities by volume compatible with their concentration at points.

With $\emptyset \neq S \subset \mathbb{R}^n$ an open set, we say that a **sequence converges to** 0 **in** $\mathscr{D}(S)$ if it is a sequence $\{\varphi_m\} \subset \mathscr{D}(S)$ such that there is a compact set $K \subset S$ containing the supports of φ_m for m large with $D^{\mathbf{k}}\varphi_m \to 0$ uniformly in K when $m \to +\infty$, for all $\mathbf{k} = (k_1, \ldots, k_n) \in (\mathbb{N} \cup \{0\})^n$.

If T is a distribution in S and $\{\varphi_m\} \subset \mathscr{D}(S)$ is a sequence converging to 0 in $\mathscr{D}(S)$, since there is $M > 0$ such that

[7] The distribution δ is not a function defined in $\mathbb{R}^n$ in the sense of possible identification with a function $f_\delta \in L^1_{\text{loc}}(\mathbb{R}^n)$ such that $\delta(\varphi) = \int_{\mathbb{R}^n} f_\delta\varphi$, since this would require f_δ to be 0 outside 0 and to have integral 1 in $\mathbb{R}$, what is contradictory.

$$|T(\varphi)| \leq M \sum_{|\mathbf{k}|\leq k} \sup |D^{\mathbf{k}}\varphi|, \quad \varphi \in \mathscr{D}(S),$$

$T(\varphi_m) \to 0$ when $m \to +\infty$. Conversely, if T is a linear functional of $C_0^\infty(S)$ and there is a compact subset K of S such that the preceding inequality fails for all $M > 0$ and $k \in \mathbb{N} \cup \{0\}$, then for every $m \in \mathbb{N} \cup \{0\}$, the condition fails with $M = k = m$, and, as T is a linear transformation, $\psi_m = \frac{\varphi_m}{T(\varphi_m)} \in C_0^\infty(K)$ is such that $T(\psi_m) = 1$ and $\sup |D^{\mathbf{k}}\psi_j| \leq \frac{1}{m}$ if $|\mathbf{k}| \leq m$; therefore, the sequence $\{T(\psi_m)\}$ does not converge to 0 when $m \to +\infty$. Consequently, *a linear functional of $C_0^\infty(S)$, with $\emptyset \neq S \subset \mathbb{R}^n$ an open set, is a distribution in S if and only if $T(\varphi_m) \to 0$ for every sequence $\{\varphi_m\}$ converging to 0 in $\mathscr{D}(S)$*. This condition for linear functionals of $C_0^\infty(S)$ can be adopted as an alternative definition of distribution in S.

The **restriction** of a distribution T in S to an open set S' with $\emptyset \neq S' \subset S$ is defined simply by restricting the domain of the linear functional T from $\mathscr{D}(S)$ to $\mathscr{D}(S')$. $T_1, T_2 \in \mathscr{D}^*(S)$ are said to be **distributions equal in a neighborhood of** $\mathbf{a} \in S$ if there is a neighborhood of $\mathbf{a}$ where their restrictions coincide.

It is now seen that the local behavior of a distribution in S in a neighborhood of each point of S completely determines a distribution. If T_1, T_2 are distributions in S such that every point $\mathbf{a} \in S$ has a neighborhood where T_1 and T_2 are equal, for each $\varphi \in C_0^\infty(S)$ and $\mathbf{a} \in K = \operatorname{supp} \varphi$ such neighborhoods form an open cover of the compact set K; therefore, there is a finite subcover $U_1, \ldots, U_k$ of K. If $\varphi_j \in C_0^\infty(U_j)$, $j \in [\mathbf{k}]$, are a partition of unity in K subordinated to that cover,[8] and $\varphi = \sum_{j=1}^k \varphi_j \varphi$, then

$$T_1(\varphi) = \sum_{j=1}^k T_1(\varphi_j \varphi) = \sum_{j=1}^k T_2(\varphi_j \varphi) = T_2(\varphi).$$

Therefore, *if T_1, T_2 are distributions in S such that in a neighborhood of each point $\mathbf{a} \in S$ T_1 and T_2 are equal, then $T_1 = T_2$ in S*.

The **support of a distribution** T in an open set $\emptyset \neq S \subset \mathbb{R}^n$ is defined to be the set of points of S without any neighborhood where T is equal to 0, and it is denoted $\operatorname{supp} T$.

It is immediate from the definition that the support of a distribution in S is a *subset of S closed relatively to S*, i.e., it is the intersection of S with a closed subset of $\mathbb{R}^n$. By two paragraphs above, $T = 0$ in the complement of $\operatorname{supp} T$ in S, i.e., $T(\varphi) = 0$ for $\varphi \in C_0^\infty(S)$ with $\operatorname{supp} T \cap \operatorname{supp} \varphi = \emptyset$. Therefore, the definition of support of a distribution agrees with that of continuous functions seen as distributions.

If $f \in L^1(S)$, with $\emptyset \neq S \subset \mathbb{R}^n$ an open set, the set $C_{\operatorname{supp} f}^\infty(S)$ of the functions $\varphi \in C^\infty(S)$ such that $\operatorname{supp} \varphi \cap \operatorname{supp} f$ *is a compact subset of S is a subspace of the linear space $C^\infty(S)$*, and $T_f(\varphi) = \int_{\mathbb{R}^n} f\varphi$ defines a continuous linear functional

[8] i.e., $\sum_{j=1}^k \varphi_j = 1$ in K and $\operatorname{supp} \varphi_j \subset U_j$ for $j \in [\mathbf{k}]$, which exists always. Here and in what follows, for $k \in \mathbb{N}$, $[\mathbf{k}]$ denotes the set $\{1, 2, \ldots, k\}$

in $C^{\infty}_{\operatorname{supp} f}(S)$, with convergence of sequences in this space defined as in $\mathscr{D}(\mathbb{R}^n)$; $T_f(\varphi)=0$ if $\operatorname{supp}\varphi\cap\operatorname{supp} f=\emptyset$. It is seen next that the domain of each distribution can be extended analogously.

If T is a distribution in S, the set $C^{\infty}_{\operatorname{supp} T}(S)$ of the functions $\varphi\in C^{\infty}(S)$ such that $\operatorname{supp}\varphi\cap\operatorname{supp} T$ *is a compact subset of S is a subspace of the linear space $C^{\infty}(S)$*. If $\varphi\in C^{\infty}_{\operatorname{supp} T}(S)$, by a result of the preceding section, there is a test function $\psi\in C_0^{\infty}(S)$ with $\psi=1$ in an open subset of S containing a compact set that contains $\operatorname{supp} T\cap\operatorname{supp}\varphi$. Thus, with $\varphi_1=\psi\varphi\in C_0^{\infty}(S)$ and $\varphi_2=(1-\psi)\varphi$, it is $\varphi=\varphi_1+\varphi_2$ and $\operatorname{supp} T\cap\operatorname{supp}\varphi_2=\emptyset$. If $\widetilde{T}$ is a linear functional in $C^{\infty}_{\operatorname{supp} T}(S)$ that coincides with T in $C_0^{\infty}(S)$ and is 0 in $\varphi\in C^{\infty}$ with $\operatorname{supp} T\cap\operatorname{supp}\varphi=\emptyset$, then

$$\widetilde{T}(\varphi)=\widetilde{T}(\varphi_1)+\widetilde{T}(\varphi_2)=\widetilde{T}(\varphi_1)=T(\varphi_1).$$

If also $\varphi=\widetilde{\varphi}_1+\widetilde{\varphi}_2$ with $\widetilde{\varphi}_1\in C_0^{\infty}$ and $\operatorname{supp} T\cap\operatorname{supp}\widetilde{\varphi}_2=\emptyset$, then

$$\varphi_1-\widetilde{\varphi}_1=\widetilde{\varphi}_2-\varphi_2\,,\qquad \varphi_1-\widetilde{\varphi}_1=\psi\varphi\in C_0^{\infty}(S)\,,$$
$$F\cap\operatorname{supp}(\varphi_1-\widetilde{\varphi}_1)=F\cap\operatorname{supp}(\varphi_2-\widetilde{\varphi}_2)=\emptyset\,,$$

and $0=T(\varphi_1-\widetilde{\varphi}_1)=T(\varphi_1)-T(\widetilde{\varphi}_1)$. Thus, $\widetilde{T}(\varphi)=T(\varphi_1)$ defines a linear functional in $C^{\infty}_{\operatorname{supp} T}(S)$ that coincides with T in C_0^{∞} and is 0 at $\varphi\in C^{\infty}$ with $\operatorname{supp} T\cap\operatorname{supp}\varphi=\emptyset$. Therefore, *if T is a distribution in an open set $\emptyset\neq S\subset\mathbb{R}^n$, then there is 1 unique linear functional $\widetilde{T}$ in $C^{\infty}_{\operatorname{supp} T}(S)$ extending T from $C_0^{\infty}(S)$ to $C^{\infty}_{\operatorname{supp} T}(S)$ such that $\widetilde{T}(\varphi)=0$ for $\varphi\in C^{\infty}(S)$ with* $\operatorname{supp} T\cap\operatorname{supp}\varphi=\emptyset$, *where $C^{\infty}_{\operatorname{supp} T}(S)$ is the linear space of the functions $\varphi\in C^{\infty}(S)$ with* $\operatorname{supp}\varphi\cap\operatorname{supp} T$ *a compact subset of S.*

The resulting extension is not usually distinguished by notation and it is assumed that a distribution T in S is not only defined as a continuous linear functional in $\mathscr{D}(S)$ but also as a continuous linear functional in the linear space $C^{\infty}_{\operatorname{supp} T}(S)$ of functions $\varphi\in C^{\infty}(S)$ such that $\operatorname{supp}\varphi\cap\operatorname{supp} T$ is a compact subset of S.

If the support of a distribution T in S is compact, the preceding extension of T is a linear functional in $C^{\infty}(S)$ since in this case $C^{\infty}(S)=C^{\infty}_{\operatorname{supp} T}(S)$. Considering in $C^{\infty}(S)$ the convergence of sequences

$$\varphi_m\to\varphi\iff\sum_{|\mathbf{k}|\in\mathbb{N}\cup\{0\}}\ \sup_{\text{compacts}}\left|D^{\mathbf{k}}\varphi_m-D^{\mathbf{k}}\varphi\right|\to 0\,,\quad\text{when } m\to+\infty\,,$$

where the supremum is over all compact subsets of S, it can be verified that the extension of a distribution T in S with compact support to the space $C^{\infty}(S)$ with this convergence of sequences in this space is a continuous linear functional. The space of test functions $C^{\infty}(S)$ with the considered convergence of sequences is usually denoted $\mathscr{E}(S)$, and **the space of distributions in S with compact support extended to $C^{\infty}(S)$ is denoted $\mathscr{E}^*(S)$**.

Distributions are defined as continuous linear functionals in a space of test functions $\mathscr{D}(S)$ chosen as a reasonable compromise of two opposite objectives for test functions:

1. To be sufficiently restrictive (in this case C^∞ with compact support) with a relatively weak topology for the space of distributions to be ample.
2. To have sufficient diversity to distinguish different locally integrable functions, allowing the distributions to be seen as an extension of those functions.

B.4 Limit, Derivatives, and Product of Distributions

We say that a **sequence** $\{T_m\}$ **of distributions in** S **converges** to a distribution T in S if the sequence of numbers $\{T_m(\varphi)\}\subset\mathbb{K}$ converges to $T(\varphi)$, for $\varphi\in\mathscr{D}(S)$.

The set of the convergent sequences of distributions in S **is a subspace** V **of the linear space** $\left[\mathscr{D}^*(S)\right]^{\mathbb{N}}$ **of the sequences of distributions in** S, since it is $\neq\emptyset$ (e.g., contains the zero distribution) and satisfies the addition and multiplication by scalars of $\mathbb{K}$ closure axioms. The function assigning to each convergent sequence of distributions in S its limit is a linear transformation from V to $\mathscr{D}^*(S)$.

A **continuous function in the space of distributions** is by definition a function $f:\mathscr{D}^*(S)\to\mathscr{D}^*(S)$ such that $f(T_m)\to f(T)$ for every sequence $\{T_m\}\subset\mathscr{D}^*(S)$ such that $T_m\to T\in\mathscr{D}^*(S)$, when $m\to+\infty$.

The derivative of distributions is to be defined so as to agree with the "weak sense" considered for differential equations in one of the motivations of distributions in the introduction to this appendix. Since by integration by parts,

$$\int_{\mathbb{R}^n}\frac{\partial f}{\partial x_k}\,\varphi=-\int_{\mathbb{R}^n}f\,\frac{\partial\varphi}{\partial x_k}\,,\qquad \varphi\in C_0^\infty(S)\,,$$

for f integrable in $S\subset\mathbb{R}^n$, it is natural to adopt the following definition.

The **derivative of a distribution** T in an open set $\emptyset\neq S\subset\mathbb{R}$ is defined to be the distribution in S such that

$$DT(\varphi)=-T(D\varphi)\,,\qquad \varphi\in C_0^\infty(S)\,.$$

If T is a distribution in an open set $\emptyset\neq S\subset\mathbb{R}^n$, since partial derivatives of C^∞ functions in $\mathbb{R}^n$ are invariant with permutation of partial derivatives relative to different variables,[9]

$$\begin{aligned}D^{(k,j)}T(\varphi)&=-D^jT(D^k\varphi)=T(D^jD^k\varphi)=T(D^kD^j\varphi)\\&=-D^kT(D^j\varphi)=D^{(j,k)}T(\varphi)\,,\qquad\qquad k,j\in[\mathbf{n}]\,,\ \varphi\in C_0^\infty\,.\end{aligned}$$

Thus, partial derivatives of any order of distributions in an open set $\emptyset\neq S\subset\mathbb{R}^n$ are invariant with permutation of single variables relative to which the derivatives are taken, and the **k-derivative of a distribution** T in an open set $\emptyset\neq S\subset\mathbb{R}^n$ is

$$D^{\mathbf{k}}T(\varphi)=(-1)^{|\mathbf{k}|}\,T(D^{\mathbf{k}}\varphi)\,,\quad \mathbf{k}=(k_1,\ldots,k_n)\in(\mathbb{N}\cup\{0\})^n\,,\ \ \varphi\in C_0^\infty\,.$$

[9] By the Schwarz lemma in $\mathbb{R}^n$ of invariance of continuous partial derivatives relative to single variables under permutation of variables

Therefore, *distributions are infinitely differentiable, and as they are considered as an extension of locally integrable functions, every such function is infinitely differentiable in the sense of distributions.* In particular, continuous functions not differentiable in the classical sense have derivatives of arbitrarily high order in the sense of distributions.

As an example, consider the continuous ramp function defined in $\mathbb{R}$ by $G(x)=x$ for $x\geq 0$ and $G(x)=0$ for $x<0$, identified with the distribution in $\mathbb{R}$ $T_G(\varphi)=\int_{\mathbb{R}} f\varphi$, $\varphi\in\mathscr{D}(\mathbb{R})$. This function does not have derivative at 0, but its version as a distribution T_G has derivative given by

$$\begin{aligned} T'_G(\varphi) &= -\int_{\mathbb{R}} G\varphi' = -\int_0^{+\infty} G(x)\,\varphi'(x)\,dx = \int_0^{+\infty} G'(x)\,\varphi(x)\,dx \\ &= \int_0^{+\infty}\varphi(x)\,dx = \int_{\mathbb{R}} H\varphi = T_H(\varphi)\,, \end{aligned}$$

for $\varphi\in\mathscr{D}(\mathbb{R})$, where $H:\mathbb{R}\to\{0,1\}$ is 1 in $[0,+\infty[$ and 0 in $]-\infty,0[$, called **Heaviside function** or **unit step function**, and in the sense of distributions $G'=H$ (Figs. B.4 and B.5). Besides,

$$T''_G(\varphi) = T'_H(\varphi) = \int_{\mathbb{R}} H\varphi' = -\int_0^{+\infty}\varphi'(x)\,dx = \varphi(0) = \delta(\varphi)\,, \quad \varphi\in\mathscr{D}(\mathbb{R})\,,$$

and $H'=\delta$ in the sense of distributions (Figs. B.4 and B.5). Therefore,

$$T^k_G(\varphi) = \delta^{(k-2)}(\varphi) = (-1)^k\delta(\varphi^{(k-2)}) = (-1)^k\varphi^{(k-2)}(0)\,, \qquad k\in\mathbb{N}\backslash\{1\}.$$

Hence, the successive derivatives in the sense of distributions of the ramp function G (not differentiable at 0 in the classical sense) are

$$G'=H\,,\ G''=H'=\delta\,,\ G'''=H''=-\delta',\ \cdots\,,\ G^{(k+2)}=H^{(k+1)}=(-1)^k\delta^{(k)},\ \cdots\,,$$

and the derivatives of the Dirac distribution are

$$\delta^{(k)}(\varphi) = (-1)^k\varphi^{(k)}(0)\,, \quad \varphi\in\mathscr{D}(\mathbb{R})\,,\ k\in\mathbb{N}\,.$$

A function $g\in L^1_{\mathrm{loc}}(S)$ is identified with the distribution in S given by $T_g(\varphi)=\int_{\mathbb{R}^n} g\varphi$, $\varphi\in\mathscr{D}(S)$, and the product of a function $f\in C^\infty(S)$ by $g\in L^1_{\mathrm{loc}}(S)$ is identified with the distribution $T_{fg}=\int_{\mathbb{R}^n} fg\varphi=\int_{\mathbb{R}^n} gf\varphi$, where $f\varphi\in C^\infty_0(S)$ for $\varphi\in\mathscr{D}(S)$. So, it is natural to adopt the following definition of product of C^∞ functions by distributions.

The **product of a function** $f\in C^\infty(S)$ **by a Distribution** T in S is defined to be the distribution in S such that

$$(fT)(\varphi) = (Tf)(\varphi) = T(f\varphi)\,, \quad \varphi\in\mathscr{D}(S)\,.$$

It is always $\operatorname{supp} fT\subset\operatorname{supp} f\cap\operatorname{supp} T$.

The product so defined is a bilinear function from $C^\infty(S)\times\mathscr{D}^*(S)$ to $\mathscr{D}^*(S)$ with left neutral element $f\in C^\infty(S)$ such that $f=1$ in S and right neutral element $E\in\mathscr{D}^*(S)$ with $E(\varphi)=\int_{\mathbb{R}^n}\varphi$, which is the distribution $E=1$ identified with the

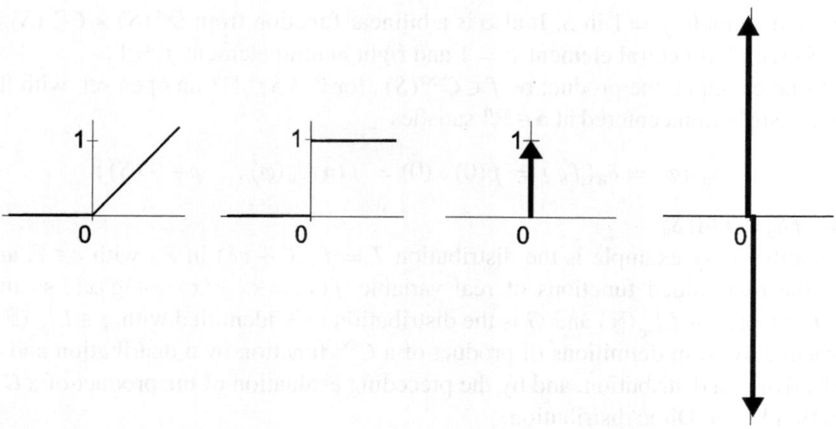

Fig. B.4 Graphical representation of derivatives in the sense of distributions of a ramp function, the Heaviside function, and the Dirac distribution, as a function or concentrated charges (for δ a positive charge at 0 and for δ' a pair of charges one positive an another negative (Fig. B.5)

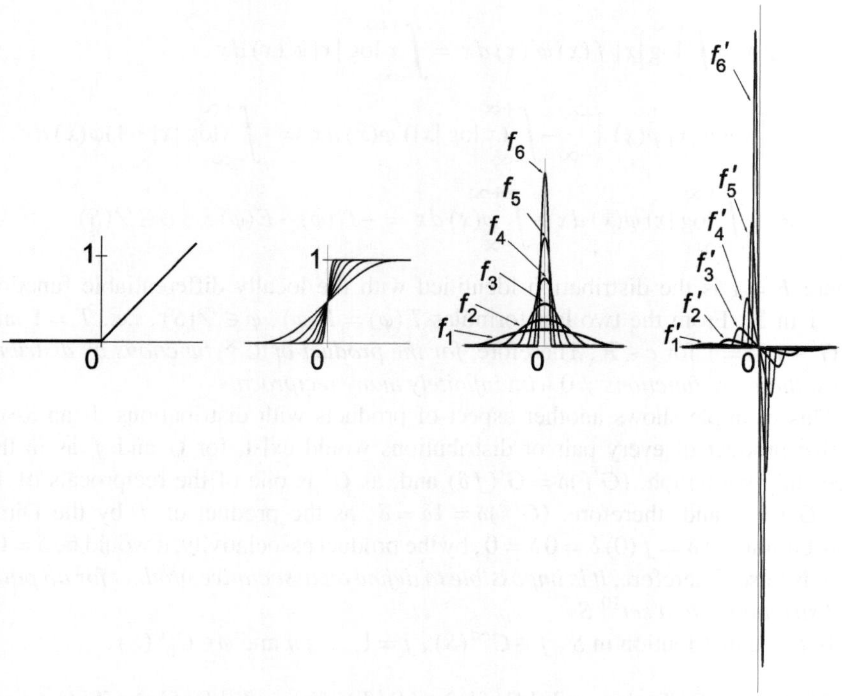

Fig. B.5 Graphical representation of the distributions in the preceding figure as sequences of C^∞ functions ≥ 0 with bounded integrals and supports intervals centered at 0 with lengths $\to 0$

locally integrable $f=1$ in S. It also is a bilinear function from $\mathscr{D}^*(S)\times C^\infty(S)$ to $\mathscr{D}^*(S)$ with left neutral element $E=1$ and right neutral element $f=1$.

As an example, the product of $f\in C^\infty(S)$, for $\emptyset\neq S\subset\mathbb{R}^n$ an open set, with the Dirac distribution centered at $\mathbf{a}\in\mathbb{R}^n$ satisfies

$$(f\delta_{\mathbf{a}})(\varphi) = \delta_{\mathbf{a}}(f\varphi) = f(0)\,\varphi(0) = f(\mathbf{a})\,\delta_{\mathbf{a}}(\varphi)\,, \quad \varphi\in\mathscr{D}(S)\,;$$

thus, $f\delta_{\mathbf{a}}=f(\mathbf{a})\,\delta_{\mathbf{a}}$.

Another easy example is the distribution $T=f(G'+c\delta)$ in $\mathbb{R}$, with $c\in\mathbb{R}$ and f,g the real valued functions of real variable $f(x)=x$, $g(x)=\log|x|$, so that $f\in C^\infty(\mathbb{R})$, $g\in L^1_{\mathrm{loc}}(\mathbb{R})$ and G is the distribution in $\mathbb{R}$ identified with $g\in L^1_{\mathrm{loc}}(\mathbb{R})$. Immediately from definitions of product of a C^∞ function by a distribution and of derivative of a distribution, and by the preceding evaluation of the product of a C^∞ function by the Dirac distribution,

$$\begin{aligned} T(\varphi) &= \big[f\big(G'+c\delta\big)\big](\varphi) = \big[fG'\big](\varphi) + \big[c\delta\big](\varphi) = G'\big(f\varphi\big) + cf(0)\,\delta \\ &= -G\big((f\varphi)'\big) = -G\big(f'\varphi+f\varphi'\big) = -G\big(\varphi\big) - G\big(f\varphi'\big)\,, \quad \varphi\in\mathscr{D}(S)\,. \end{aligned}$$

Since, with integration by parts,

$$\begin{aligned} G(f\varphi') &= \int_{-\infty}^{+\infty} \log|x|\,f(x)\,\varphi'(x)\,dx = \int_{-\infty}^{+\infty} x\log|x|\,\varphi'(x)\,dx \\ &= \Big[x\log|x|\,\varphi(x)\Big]_{-\infty}^{-\infty} - \int_{-\infty}^{+\infty} (x\log|x|)'\varphi(x)\,dx = -\int_{-\infty}^{+\infty} (\log|x|+1)\varphi(x)\,dx \\ &= -\int_{-\infty}^{+\infty} \log|x|\,\varphi(x)\,dx - \int_{-\infty}^{+\infty} \varphi(x)\,dx = -G(\varphi) - E(\varphi)\,, \quad \varphi\in\mathscr{D}(S)\,, \end{aligned}$$

where $E=1$ is the distribution identified with the locally differentiable function $f=1$ in $\mathbb{R}$. From the two last formulas $T(\varphi)=E(\varphi)$, $\varphi\in\mathscr{D}(S)$, i.e., $T=1$ and $f(G'+c\delta)=1$ for $c\in\mathbb{R}$. Therefore, *for the product of C^∞ functions by distributions, there are functions $\neq 0$ with infinitely many reciprocals.*

This example shows another aspect of products with distributions. If an associative product of every pair of distributions would exist, for G and f as in the preceding paragraph, $(G'f)\delta=G'(f\delta)$ and, as G' is one of the reciprocals of f, i.e., $G'f=1$ and, therefore, $(G'f)\delta=1\delta=\delta$, as the product of f by the Dirac distribution is $f\delta=f(0)\,\delta=0\,\delta=0$, by the product associativity, it would be $\delta=0$, what is false. Therefore, *it is impossible to define an associative product for all pairs of distributions in a set*[10] S.

If T is a distribution in S, $f\in C^\infty(S)$, $j=1,\dots,n$ and $\varphi\in C_0^\infty(S)$,

$$\begin{aligned} &\big[(D^jf)T+f(D^jT)\big](\varphi) = T([D^jf]\varphi)+(D^jT)(f\varphi) = T([D^jf]\varphi)-T\big(D^j(f\varphi)\big) \\ &= T\big([D^jf]\varphi-[D^jf]\varphi-f\,D^j\varphi\big) = -T(f\,D^j\varphi) = -(fT)(D^j\varphi = D^j(fT)(\varphi)\,; \end{aligned}$$

[10] This impossibility was first shown by L.Schwartz in an article of 1954.

thus, $D^j(fT) = (D^j f)T + f(D^j T)$, $j \in [\mathbf{n}]$, i.e., *the rule of derivative of products of functions holds also for the product of* C^∞ *functions by distributions.*

We prove next the rule for derivative of the composition of a locally integrable function with partial derivatives in the sense of distributions with a C^2 homeomorphism, applied in Chap. 1 to prove that the geometric and analytic definitions of quasiconformal mappings are equivalent.

(B.2) *If* $U \subset \mathbb{R}^2$ *is an open set,* $\mathbf{g} = (g_1, g_2) : U \to \mathbb{R}^2$ *is a* C^2 *homeomorphism from* U *onto* S*, and* $\mathbf{f} \in L^1_{\mathrm{loc}}(S)$ *has partial derivatives in the sense of distributions in* $L^1_{\mathrm{loc}}(S)$*, then* $\mathbf{f} \circ \mathbf{g}$ *also does and*

$$D_1(\mathbf{f}\circ\mathbf{g}) = (D_1\mathbf{f}\circ\mathbf{g})D_1\mathbf{g}_1 + (D_2\mathbf{f}\circ\mathbf{g})D_1\mathbf{g}_2\,,$$

$$D_2(\mathbf{f}\circ\mathbf{g}) = (D_1\mathbf{f}\circ\mathbf{g})D_2\mathbf{g}_1 + (D_2\mathbf{f}\circ\mathbf{g})D_2\mathbf{g}_2\,.$$

Proof With $J = \det D\mathbf{g}$,

$$D\mathbf{g}^{-1} = [D\mathbf{g}]^{-1}\, J^{-1} \begin{bmatrix} D_2(\mathbf{g}^{-1})_2 & -D_2(\mathbf{g}^{-1})_1 \\ -D_1(\mathbf{g}^{-1})_2 & D_1(\mathbf{g}^{-1})_1 \end{bmatrix},$$

in corresponding points. Hence, for every test function $\varphi \circ \mathbf{g} \in C_0^\infty(U)$, with changes of integration variables, integration by parts, rules of derivatives of operations of functions and the Schwarz lemma for higher order continuous derivatives of functions,

$$\begin{aligned}
&\iint [(D_1\mathbf{f}\circ\mathbf{g})D_1\mathbf{g}_1 + (D_2\mathbf{f}\circ\mathbf{g})D_1\mathbf{g}_2](\varphi\circ\mathbf{g}) \\
&\quad = \iint [(D_1\mathbf{f}\circ\mathbf{g})J^{-1}D_1\mathbf{g}_1 + (D_2\mathbf{f}\circ\mathbf{g})J^{-1}D_1\mathbf{g}_2](\varphi\circ\mathbf{g})J \\
&\quad = \iint \big[D_1\mathbf{f}D_2(\mathbf{g}^{-1})_2 - D_2\mathbf{f}D_1(\mathbf{g}^{-1})_1\big]\varphi \\
&\quad = \iint \mathbf{f}\big(-D_1[\varphi D_2(\mathbf{g}^{-1})_2] + D_2[\varphi D_1(\mathbf{g}^{-1})_2]\big) \\
&\quad = \iint \mathbf{f}\big(-(D_1\varphi)D_2(\mathbf{g}^{-1})_2 + (D_2\varphi)D_1(\mathbf{g}^{-1})_2]\big) \\
&\quad = \iint (\mathbf{f}\circ\mathbf{g})\big(-[(D_1\varphi)\circ\mathbf{g}]J^{-1}D_1\mathbf{g}_1 - [(D_2\varphi)\circ\mathbf{g}]J^{-1}D_1\mathbf{g}_2\big)J \\
&\quad = \iint (\mathbf{f}\circ\mathbf{g})\big(-[(D_1\varphi)\circ\mathbf{g}]D_1\mathbf{g}_1 - [(D_2\varphi)\circ\mathbf{g}]D_1\mathbf{g}_2\big) = -\iint (\mathbf{f}\circ\mathbf{g})D_1(\varphi\circ\mathbf{g}),
\end{aligned}$$

giving the first formula. The second follows by symmetry. □

Appendix C
Fourier Transform in the Plane

The Fourier transform was introduced by J.Fourier, S.D. Poisson, and A.-L. Cauchy. J. Fourier's first contribution was in a paper on heat propagation presented to the Paris Academy of Sciences in 1811 and published only in 1824 due to the controversy generated by his ideas about Fourier series implying that very general functions can be represented as a countable superposition of simple trigonometric functions, what was very difficult to accept at the time, though it was also mentioned in his book *La Théorie Analytique de la Chaleur* published in 1822. The paper submitted by J. Fourier in 1811 contains the Fourier transform inversion formula, but it was not established rigorously there. S.D. Poisson introduced the Fourier integral in an article entitled *Mémoire sur la théorie des ondes* published in 1816. A.L. Cauchy considered the Fourier integral in a 1816 paper entitled *Théorie de la propagation des ondes*, which was the first systematic work on surface waves in fluids. In 1823 and 1827, he considered the Fourier transform inversion theorem and applied integrals on paths in the complex plane to compute inverse Fourier transforms for solving partial differential equations. In 1825 and 1845, A.L. Cauchy also used them in an "operational calculus" to solve differential equations, though this had already been used by other authors.

In 1836, J.Liouville[1] related Fourier series to Fourier integrals by a limit process. However, the Fourier transform was only made rigorous with the work of several mathematicians at the end of the nineteenth and the beginning of the twentieth centuries, including the adoption of the Lebesgue integral.

The extension of Fourier transform and isometry in L^2 — Fourier-Plancherel transform — was in 1910 by M. Plancherel. The development of Fourier analysis led to the area of mathematics called Harmonic Analysis.

Denote $L^j = L^j(\mathbb{C})$ for $j = 1, 2$, $C^0(\mathbb{C}) = C^0$, $C_0^\infty = C_0^\infty(\mathbb{C})$ and, for a complex function f and $z, w \in \mathbb{C}$, $f_w(z) = f(z+w)$.

[1] Liouville, Joseph (1809–1882).

L. T. Magalhães, *Quasiconformal Mappings in the Plane and Complex Dynamics*,
https://doi.org/10.1007/978-3-031-80115-0

The **Fourier transform** of $f \in L^1$, denoted $\widehat{f}$ or $f\widehat{\ }$, is

$$\widehat{f}(\xi, \eta) = \iint_{\mathbb{R}^2} f(x, y)\, e^{-i2\pi(x\xi + y\eta)} dxdy\,, \qquad \xi, \eta \in \mathbb{R}\,.$$

The **convolution** of $f, g \in L^1$ is defined by

$$f * g\,(x, y) = \iint_{\mathbb{R}^2} f(s, t)\, g(x-s, y-t)\, dsdt\,.$$

By the Fubini theorem,

$$\begin{aligned}\iint_{\mathbb{R}^2} |f * g\,(x, y)|\, dxdy &= \iint_{\mathbb{R}^2}\iint_{\mathbb{R}^2} |f(s, t)\, g(x-s, y-t)|\, dsdt\, dxdy \\ &= \iint_{\mathbb{R}^2} |f(s, t)| \iint_{\mathbb{R}^2} |g(x-s, y-t)|\, dxdy\, dsdt \\ &= \iint_{\mathbb{R}^2} |f(s, t)|\, \|g\|_{L^1} dsdt = \|f\|_{L^1} \|g\|_{L^1}.\end{aligned}$$

Hence[2] $f * g \in L^1$ and $\|f * g\|_{L^1} \leq \|f\|_{L^1} \|f\|_{L^1}$; therefore, $f, g \in L^1 \Rightarrow f * g \in L^1$.

(C.1) *The Fourier transform satisfies for all $f, g \in L^1$:*

1. $f \mapsto \widehat{f}$ *is a linear transformation from* L^1 *to* $L^\infty \cap C^0$ *and* $\|\widehat{f}\|_{L^\infty} \leq \|f\|_{L^1}$.
2. $\widehat{f}(\xi, \eta) \to 0$ *when* $|(\xi, \eta)| \to +\infty$ ***(Riemann-Lebesgue lemma).***
3. $\widehat{f_\lambda}(\zeta) = \frac{1}{\lambda^2} \widehat{f}\left(\frac{\zeta}{\lambda}\right)$, *if* $f_\lambda(z) = f(\lambda z)$ *and* $\lambda \in \mathbb{R} \setminus \{0\}$.
4. $(f * g)\widehat{\ } = \widehat{f}\, \widehat{g}$.
5. $\widehat{\widehat{f}}(z) = f(-z)$ a.e. *for* $z \in \mathbb{C}$, *if, in addition,* $\widehat{f} \in L^1$.
6. $\widehat{fg} = \widehat{f} * \widehat{g}$, *if, in addition,* $\widehat{f}, \widehat{g}, \widehat{fg} \in L^1$.
7. $\widehat{f} = \widehat{g} \Rightarrow f = g$ $(f(z) = g(z)$ a.e. *for* $z \in \mathbb{C}$*)*.
8. *If* $f \in L^1 \cap L^2$, *then* $\widehat{f} \in L^2$ *and* $\|\widehat{f}\|_{L^2} = \|f\|_{L^2}$.
9. *If, in addition,* $\frac{\partial f}{\partial x}(x, y)$, $\frac{\partial f}{\partial y}(x, y) \in L^1$, *then*
   $\widehat{\frac{\partial f}{\partial x}}(\xi, \eta) = i2\pi\xi \widehat{f}$, $\quad \widehat{\frac{\partial f}{\partial y}}(\xi, \eta) = i2\pi\eta \widehat{f}(\xi, \eta)$,
   $\widehat{f_z}(z) = i\pi \bar{z} \widehat{f}(z)$, $\quad \widehat{f_{\bar{z}}}(z) = i\pi z \widehat{f}(z)$.

Proof

1. By the Lebesgue dominated convergence theorem, the integral defining $\widehat{f}$ exists, and since the integrand function is continuous in (ξ, η), $\widehat{f} \in C^0$ and as for $\omega \in \mathbb{C}$, it is $|\widehat{f}(\omega)| \leq \iint_{\mathbb{R}^2} |f(x, y)|\, dxdy = \|f\|_{L^1}$, $\widehat{f}$ is bounded, and $\|\widehat{f}\|_{L^\infty} \leq \|f\|_{L^1}$.

[2] The Fubini theorem guarantees that the integral in the definition of $f * g$ exists a.e. in $\mathbb{R}^2$.

Thus, the Fourier transform $f \mapsto \widehat{f}$ is a linear transformation from L^1 to $L^\infty \cap C^0$.

2. If f is the characteristic function of an interval $[a, b]\times[c, d]$, then

$$\widehat{f}(\xi, \eta) = \int_a^b e^{-i2\pi x\xi} dx \int_a^b e^{-i2\pi y\eta} dy,$$

and if $|\xi| \to +\infty$, then

$$|\widehat{f}(\xi, \eta)| \leq \left|\frac{e^{-i2\pi a\xi} - e^{-i2\pi b\xi}}{i2\pi\xi}\right| \left|\int_c^d e^{-i2\pi y\eta} dy\right| \leq \frac{1}{\pi|\xi|}\int_c^d |e^{-i2\pi y\eta}|\, dy = \frac{d-c}{\pi|\xi|} \to 0,$$

and analogously if $|\eta| \to +\infty$; thus, the result holds for characteristic functions of bounded intervals, and by the linearity of the integral, it also holds for step functions.

If $f \in L^1$ for $\varepsilon > 0$, there is a step function s such that $\| f - s\|_{L^1} < \varepsilon$, and by the preceding period, there is $M > 0$ such that $|(\xi, \eta)| > M \Rightarrow |\widehat{s}(\xi, \eta)| < \varepsilon$. With 1,

$$|\widehat{f}(\xi, \eta)| \leq |\widehat{f}(\xi, \eta) - \widehat{s}(\xi, \eta)| + |\widehat{s}(\xi, \eta)| \leq \|\widehat{f} - \widehat{s}\|_{L^\infty} + \varepsilon \leq \|\widehat{f} - \widehat{s}\|_{L^1} + \varepsilon \leq 2\varepsilon.$$

Hence, for every $\varepsilon > 0$, there is $M > 0$ such that $|\widehat{f}(\xi, \eta)| \leq 2\varepsilon$; therefore, $\widehat{f}(\xi, \eta) \to 0$ when $|(\xi, \eta)| \to +\infty$.

3.
$$\begin{aligned}\widehat{f_\lambda}(\xi, \eta) &= \iint_{\mathbb{R}^2} f(\lambda x, \lambda y) e^{-i2\pi(x\xi + y\eta)} dx\, dy \\ &= \frac{1}{\lambda^2}\iint_{\mathbb{R}^2} f(s, t) e^{-i2\pi\frac{1}{\lambda}(s\xi + t\eta)} ds\, dt = \frac{1}{\lambda^2}\widehat{f}\left(\frac{\xi}{\lambda}, \frac{\eta}{\lambda}\right).\end{aligned}$$

4. With the Fubini theorem,

$$\begin{aligned}(f * g)\widehat{\ }(\xi, \eta) &= \iint_{\mathbb{R}^2}\iint_{\mathbb{R}^2} f(s, t)\, g(x - s, y - t)\, dsdt\, e^{-i2\pi(x\xi + y\eta)} dxdy \\ &= \iint_{\mathbb{R}^2} f(s, t)\, e^{-i2\pi(s\xi + t\eta)} \iint_{\mathbb{R}^2} g(x - s, y - t)\, e^{-i2\pi((x-s)\xi + (y-t)\eta)} dxdy\, dsdt \\ &= \iint_{\mathbb{R}^2} f(s, t)\, e^{-i2\pi(s\xi + t\eta)}\, \widehat{g}(\xi, \eta)\, dsdt = \widehat{f}(\xi, \eta)\, \widehat{g}(\xi, \eta),\end{aligned}$$

i.e., $(f * g)\widehat{\ } = \widehat{f}\,\widehat{g}$.

5. For $s, t, x, y \in \mathbb{R}$ and $\lambda > 0$ let $K_\lambda(t) = e^{-\lambda|t|}$, $k_\lambda(x) = \int_{\mathbb{R}} K_\lambda(t)\, e^{i2\pi tx} dt$, $\widetilde{k}_\lambda(x, y) = k_\lambda(x)\, k_\lambda(y)$ and $\widetilde{K}_\lambda(s, t) = K_\lambda(t)\, K_\lambda(s)$. It is $k_\lambda(x) = \frac{2\lambda}{\lambda^2 + (2\pi x)^2}$, $\int_{\mathbb{R}} k_\lambda = 1$, $0 < K_\lambda(t) \leq 1$, $K_\lambda(t) \nearrow 1$ when $\lambda \to 0$. Hence, $\iint_{\mathbb{R}^2} \widetilde{k}_\lambda(x, y)\, dxdy = 1$, $0 < \widetilde{K}_\lambda(t) \leq 1$ and $\widetilde{K}_\lambda(t, s) \nearrow 1$ when $\lambda \to 0$.

If $f \in L^1$ and $(x, y) \in \mathbb{C}$, the function $(s, t) \mapsto f(s, t)\, \widetilde{k}_\lambda(x-s, y-t)$ belongs to L^1 since it is measurable and $\widetilde{k}_\lambda$ is bounded. By the Fubini theorem,

$$\begin{aligned} f * \widetilde{k}_\lambda(x, y) &= \iint_{\mathbb{R}^2} f(s, t)\, \widetilde{k}_\lambda(x-s, y-t)\, ds\, dt \\ &= \iint_{\mathbb{R}^2} f(s, t) \iint_{\mathbb{R}^2} \widetilde{K}_\lambda(s', t')\, e^{i2\pi[s'(x-s)+t'(y-t)]}\, ds' dt' ds dt \\ &= \iint_{\mathbb{R}^2} \iint_{\mathbb{R}^2} f(s, t)\, e^{-i2\pi(s's+t't)}\, ds\, dt\, \widetilde{K}_\lambda(s', t')\, e^{i2\pi(s'x+t'y)} ds' dt' \\ &= \iint_{\mathbb{R}^2} \widehat{f}(s', t')\, \widetilde{K}_\lambda(s', t')\, e^{i2\pi(s'x+t'y)} ds' dt'. \end{aligned}$$

If, in addition, $\widehat{f} \in L^1$, the integrand in the last integral is a measurable function, and its absolute value is $\leq |\widehat{f}| \in L^1$; thus, by the Lebesgue dominated convergence theorem,

$$f * \widetilde{k}_\lambda(x, y) \to \iint_{\mathbb{R}^2} \widehat{f}(s, t)\, e^{i2\pi(sx+ty)} ds dt = \widehat{\widehat{f}}\big(-(x, y)\big), \quad \text{for } (x, y) \in \mathbb{C}, \text{ when } \lambda \to 0,$$

On the other hand, if $f \in L^1$, for all $(x, y) \in \mathbb{C}$,

$$\begin{aligned} \big| f * \widetilde{k}_\lambda(x, y) - f(x, y) \big| &= \left| \iint_{\mathbb{R}^2} \big[f(s, t) - f(x, y) \big] \widetilde{k}_\lambda(x-s, y-t)\, ds dt \right| \\ &= \left| \iint_{\mathbb{R}^2} \big[f(x-s', y-t') - f(x, y) \big] \widetilde{k}_\lambda(s', t')\, ds' dt' \right| \\ &\leq \iint_{\mathbb{R}^2} \big| f(x-s, y-t) - f(x, y) \big|\, \widetilde{k}_\lambda(s, t)\, ds dt\,. \end{aligned}$$

Integrating and applying the Fubini theorem, with $h(s, t) = \| f_{-(s,t)} - f \|_{L^1}$,

$$\begin{aligned} \| f * \widetilde{k}_\lambda - f \|_{L^1} &= \iint_{\mathbb{R}^2} \big| f * \widetilde{k}_\lambda(x, y) - f(x, y) \big| dx dy \\ &\leq \iint_{\mathbb{R}^2} \| f_{-(s,t)} - f \|_{L^1} \widetilde{k}_\lambda(s, t)\, ds dt = h * \widetilde{k}_\lambda(0)\,. \end{aligned}$$

By the triangular inequality,

$$|h(s, t)| \leq \| f_{-(s,t)} \|_{L^1} + \| f \|_{L^1} = 2 \| f \|_{L^1}, \quad (s, t) \in \mathbb{C},$$

and, therefore, h is a bounded function. By the Lebesgue dominated convergence theorem, h is continuous since it is given by an integral of an integrable function that depends continuously on (s, t).

If h is a continuous bounded function in $\mathbb{C}$, changing integration variables,

$$\begin{aligned} h*\widetilde{k}_\lambda(0)-h(0) &= \iint_{\mathbb{R}^2}[h(s,t)-h(0)]\,\widetilde{k}_\lambda(-s,-t)\,ds\,dt \\ &= \iint_{\mathbb{R}^2}[h(s,t)-h(0)]\,\tfrac{1}{\lambda^2}\,\widetilde{k}_\lambda\big(\tfrac{-s}{\lambda},\tfrac{-t}{\lambda}\big)\,ds\,dt \\ &= \iint_{\mathbb{R}^2}\big[h\big(-\lambda(s',t')\big)-h(0)\big]\,\widetilde{k}_\lambda(s',t')\,ds'dt'. \end{aligned}$$

For $0<\lambda\le 1$, it is $0<\widetilde{k}_\lambda(x,y)\le \widetilde{k}_1(x,y)=\frac{4}{[1+(2\pi x)^2][1+(2\pi y)^2]}$, and the integrand in the last integral is bounded by $2\|h\|_{L^\infty}\widetilde{k}_1\in L^1$, is continuous in $\mathbb{C}$, and converges to 0 when $\lambda\to 0$; by the Lebesgue dominated convergence theorem, $h*\widetilde{k}_\lambda(0)\to h(0)$ when $\lambda\to 0$.

By the three preceding paragraphs,

$$f*\widetilde{k}_\lambda(z)-\widehat{\widehat{f}}\,(-z)\to 0 \ \text{ for } z\in\mathbb{C}, \qquad \big\|f*\widetilde{k}_\lambda - f\big\|_{L^1}\to 0, \qquad \text{when } \lambda\to 0.$$

As $\big\|f*\widetilde{k}_\lambda - f\big\|_{L^1}\to 0$, it is[3] $f*\widetilde{k}_\lambda(z)\to f(z)$ a.e. for $z\in\mathbb{C}$, and, as

$$f*\widetilde{k}_\lambda(z)-\widehat{\widehat{f}}\,(-z)\to 0, \quad z\in\mathbb{C},$$

it follows that $f(z)=\widehat{\widehat{f}}\,(-z)$ a.e. for $z\in\mathbb{C}$.

6. By 4, $\big(\widehat{f}*\widehat{g}\big)^\wedge=\widehat{\widehat{f}}\,\widehat{\widehat{g}}$. By 5, $\widehat{f}*\widehat{g}=\widehat{fg}$ a.e. in $\mathbb{C}$. By 1, the functions in this equality are continuous, and, therefore, the equality holds in all $\mathbb{C}$.
7. If $\widehat{f}=\widehat{g}$, with $h=f-g$, then $\widehat{h}=0$ and $\widehat{\widehat{h}}=0$; thus, by 4, $h=0$ a.e. in $\mathbb{C}$.
8. If $f\in L^1\cap L^2, \widetilde{f}(z)=\overline{f(-z)}$, then $h=f*\widetilde{f}\in L^1$ and, with the Cauchy-Schwarz inequality,

$$\big|f*\widetilde{f}(x,y)\big|=\left|\iint_{\mathbb{R}^2}f(s,t)\,\widetilde{f}(x-s,y-t)\,ds\,dt\right|=\big|\langle f,f_{-(x,y)}\rangle\big|\le\|f\|_{L^2}^2;$$

therefore, h is bounded. Proceeding analogously with $f*\widetilde{f}(x,y)-f*\widetilde{f}(x',y')$,

$$|h(x,y)|=\big|f*\widetilde{f}(x,y)-f*\widetilde{f}(x',y')\big|\le\|f\|_{L^2}\big\|f_{-(x,y)}-f_{-(x',y')}\big\|_{L^2}.$$

Since $(x,y)\mapsto f_{-(x,y)}$ is continuous, by the Lebesgue dominated convergence theorem, $h=f*\widetilde{f}$ is continuous in $\mathbb{C}$. With the functions $\widetilde{k}_\lambda$ and $\widetilde{K}_\lambda$ introduced at the beginning of the proof of 4 and by the Fubini theorem,

[3] See the Riesz-Fisher Theorem in Appendix A.

$$\begin{aligned}
h*\widetilde{k}_\lambda(0) &= \iint_{\mathbb{R}^2} h(s,t)\,\widetilde{k}_\lambda(-s,-t)\,ds\,dt\\
&= \iint_{\mathbb{R}^2} h(s,t)\iint_{\mathbb{R}^2} \widetilde{K}_\lambda(s',t')\,e^{-i2\pi(s's+t't)}ds'dt'ds\,dt\\
&= \iint_{\mathbb{R}^2}\iint_{\mathbb{R}^2} h(s,t)\,e^{-i2\pi(s's+t't)}ds\,dt\,\widetilde{K}_\lambda(s',t')\,ds'dt'\\
&= \iint_{\mathbb{R}^2} \widehat{h}(s',t')\,\widetilde{K}_\lambda(s',t')\,ds'dt'.
\end{aligned}$$

As $\widehat{h}=(f*\widetilde{f})\widehat{\ }=\widehat{f}\,\widehat{\widetilde{f}}=\widehat{f}\,\overline{\widehat{f}}=|\widehat{f}\,|^2\geq 0$ and $\widetilde{K}_\lambda(s',t')\nearrow 1$ when $\lambda\to 0$, by the Levi monotone convergence theorem,[4] $h*\widetilde{k}_\lambda(0)\to \|\widehat{f}\|_{L^2}$. By the paragraph preceding the last one in the proof of 4, $h*\widetilde{k}_\lambda(0)\to h(0)=\langle f,f\rangle=\|f\|^2_{L^2}$, when $\lambda\to 0$. By the last two periods, $\|\widehat{f}\|_{L^2}=\|f\|^2_{L^2}$ for $f\in L^1\cap L^2$.

9. By 2 and integration by parts,

$$\begin{aligned}
\widehat{\frac{\partial f}{\partial x}}(\xi,\eta) &= -\iint_{\mathbb{R}^2} f(x,y)\,(-i2\pi\xi)\,e^{-i2\pi(x\xi+y\eta)}\,dx\,dy\\
&= i2\pi\xi\,\widehat{f}(\xi,\eta),\quad (\xi,\eta)\in\mathbb{C},
\end{aligned}$$

and, analogously, $\widehat{\frac{\partial f}{\partial y}}=i2\pi\eta\widehat{f}$. Since[5] $f_z=\frac{1}{2}\big(\frac{\partial f}{\partial x}-i\frac{\partial f}{\partial y}\big)$ and $f_{\overline{z}}=\frac{1}{2}\big(\frac{\partial f}{\partial x}+i\frac{\partial f}{\partial y}\big)$, $\widehat{f_z}(z)=i\pi\overline{z}\widehat{f}(z)$ e $\widehat{f_{\overline{z}}}(z)=i\pi z\widehat{f}(z)$, for $z\in\mathbb{C}$. □

C_0^∞ is dense in L^j, $j=1,2$, but $\widehat{\varphi}\notin C_0^\infty$ for $\varphi\in C_0^\infty\setminus\{0\}$. Therefore, it is convenient to extend C_0^∞ to a space of C^∞ functions with Fourier transforms in that space.

If $\varphi\in C^\infty$, for the conditions allowing to permute partial derivatives in $D^{\mathbf{j}}$, $\mathbf{j}\in(\mathbb{N}\cup\{0\})^2$, (with the notation introduced in the preceding appendix) to hold with the integral in $\widehat{\varphi}(\xi,\eta)=\iint_{\mathbb{R}^2}\varphi(x,y)\,e^{-i2\pi(x\xi+y\eta)}dx\,dy$, it is necessary and sufficient that the function $(\xi,\eta)\mapsto D^{\mathbf{j}}_{(\xi,\eta)}\big[\varphi(x,y)\,e^{-i2\pi(x\xi+y\eta)}\big]$ belong to $L^1(\mathbb{R}^n)$, where $D^{\mathbf{j}}_{(\xi,\eta)}$ means that the partial derivatives are relative to the variables (ξ,η), and, as

$$\begin{aligned}
\big|D^{\mathbf{j}}_{(\xi,\eta)}\big(\varphi(x,y)e^{-i2\pi(x\xi+y\eta)}\big)\big| &= \big|\varphi(x,y)(-i2\pi)^{|\mathbf{j}|}(x,y)^{\mathbf{j}}e^{-i2\pi(x\xi+y\eta)}\big|\\
&= (2\pi)^{|\mathbf{j}|}\big|(x,y)^{\mathbf{j}}\varphi(x,y)\big|,
\end{aligned}$$

where for $\mathbf{j}=(j_1,j_2)\in(\mathbb{N}\cup\{0\})^2$, it is $(x,\eta)^{\mathbf{j}}=x^{j_1}y^{j_2}$. In order that $\widehat{\varphi}\in C^\infty$ for all $\varphi\in C^\infty$, it is necessary and sufficient that the function $(x,y)\mapsto(x,y)^{\mathbf{j}}\varphi(x,y)$ belong to L^1 for all $\mathbf{j}\in(\mathbb{N}\cup\{0\})^2$ and $\varphi\in C^\infty$.

[4] Levi, Beppo (1875–1961).

[5] See the beginning of the second section of Chap. 1.

It is desired that derivatives $D^{\mathbf{j}}\varphi$ and $\mathbf{j} \in (\mathbb{N} \cup \{0\})^2$ have Fourier transform in C^∞, which, by the preceding argument, is equivalent to the function $(x, y) \mapsto (x, y)^{\mathbf{j}} D^{\mathbf{k}}\varphi(x, y)$ to belong to L^1 for all $\mathbf{j}, \mathbf{k} \in (\mathbb{N} \cup \{0\})^2$ and $\varphi \in C^\infty$. Since functions $(x, y) \mapsto \frac{1}{(1+|(x,y)|)^\alpha}$ are integrable in $\mathbb{R}^2$ if and only if $\alpha > 2$, the condition holds if and only if for all $\mathbf{j}, \mathbf{k} \in (\mathbb{N} \cup \{0\})^2$, the function

$$(x, y) \mapsto (1+|(x, y)|)^\alpha |(x, y)^{\mathbf{j}} D^{\mathbf{k}}\varphi(x, y)|$$

is bounded for some $\alpha > 2$, i.e., if and only if functions $\varphi \in C^\infty$ as well as their partial derivatives of all orders decay to 0 when $|(x, y)| \to +\infty$ faster than the product of all integer powers of the independent variables x, y. Thus, it is natural to extend the space C_0^∞ as follows.

We call **Schwartz space** the set of function $\varphi \in C^\infty$ such that

$$\sup_{\substack{(x,y)\in\mathbb{C} \\ \mathbf{j},\mathbf{k}\in(\mathbb{N}\cup\{0\})^2}} \left|(x, y)^{\mathbf{j}} D^{\mathbf{k}}\varphi(x, y)\right| < \infty,$$

with the notion of convergence of sequences of functions $\{\varphi_m\}$ defined by $\varphi_m \to \varphi$ when $m \to +\infty$ if

$$\sup_{\substack{(x,y)\in\mathbb{C} \\ \mathbf{j},\mathbf{k}\in(\mathbb{N}\cup\{0\})^2}} \left|(x, y)^{\mathbf{j}} D^{\mathbf{k}}(\varphi_m - \varphi)(x, y)\right| \to 0, \quad m \to +\infty.$$

The Schwartz space with this notion of convergence is **denoted** $\mathscr{S}$.

$\mathscr{S}$ is a subspace of the linear space C^∞. By an argument similar to that preceding the definition of the Schwartz space, *$\mathscr{S}$ is the largest subspace of the linear space L^1 invariant under partial derivatives and products by integer powers of the independent variables, and $\mathscr{S} \subset L^2$*, since $\mathscr{S}$ is a space of integrable C^∞ functions. A simple example of a $\varphi \in \mathscr{S} \setminus C_0^\infty$ is $\varphi(z) = e^{-a|z|^2}$, with $a > 0$; so, $C_0^\infty \subsetneqq \mathscr{S} \subsetneqq C^\infty$. From what was seen in the preceding appendix on distributions, $C_0^\infty \subset L^1 \cap L^2$ is dense in L^1 and in L^2, and, therefore, also *$\mathscr{S} \subset L^1 \cap L^2$ is dense in L^1 and in L^2*.

With the argument preceding the definition of the Schwartz space, *the Fourier transform $f \mapsto \widehat{f}$ is a continuous linear transformation from $\mathscr{S}$ to $\mathscr{S}$ such that*

$$\widehat{\ }(\xi, \eta) = (\xi, \eta)^{\mathbf{j}}\, \widehat{\varphi}(\xi, \eta), \qquad \left[(x, y)^{\mathbf{j}}\varphi(x, y)\right]^{\widehat{\ }}(\xi, \eta) = (-1)^{|\mathbf{j}|} D^{\mathbf{j}}\widehat{\varphi}(\xi, \eta),$$

for $(x, y), (\xi, \eta) \in \mathbb{C}$, $\mathbf{j} = (j_1, j_2) \in (\mathbb{N} \cup \{0\})^2$, $\varphi \in \mathscr{S}$.

The inversion of the Fourier transform $\varphi(z) = \widehat{\widehat{\varphi}}(-z)$ a.e. for $z \in \mathbb{C}$ leads to

$$\begin{aligned} \varphi(x, y) &= \iint_{\mathbb{R}^2} \widehat{\varphi}(\xi, \eta)\, e^{i2\pi(x\xi+y\eta)} d\xi d\eta \\ &= \iint_{\mathbb{R}^2} e^{i2\pi(x\xi+y\eta)} \iint_{\mathbb{R}^2} \varphi(x', y')\, e^{-i2\pi(x'\xi+y'\eta)} dx'dy'd\xi d\eta, \end{aligned}$$

but the existence of the corresponding integral in $\mathbb{R}^4$ is not guaranteed for $\varphi \in \mathscr{S}$. Therefore, the order of integration cannot be changed in general, but multiplying the function by $\psi(\xi,\eta)$, with $\psi \in \mathscr{S}$, we have integrability, and the integration order can be changed, leading to

$$\iint_{\mathbb{R}^2} \widehat{\varphi}(\xi,\eta)\,\psi(\xi,\eta)\,e^{i2\pi(x\xi+y\eta)}d\xi d\eta$$
$$= \iint_{\mathbb{R}^2} \varphi(x',y') \iint_{\mathbb{R}^2} \psi(\xi,\eta)\,e^{-i2\pi[(x'-x)\xi+(y'-y)\eta]}\,d\xi d\eta\,dx'dy'$$
$$= \iint_{\mathbb{R}^2} \varphi(x',y')\,\widehat{\psi}(x'-x,y'-y)\,dx'dy' = \iint_{\mathbb{R}^2} \varphi(\xi+x,\eta+y)\,\widehat{\psi}(\xi,\eta)\,d\xi d\eta\,.$$

Replacing $\psi(\xi,\eta)$ by $\psi\big(\varepsilon(\xi,\eta)\big)$, with $\varepsilon>0$, which has Fourier transform $\frac{1}{\varepsilon^2}\widehat{\psi}\big(\frac{1}{\varepsilon}(\xi,\eta)\big)$,

$$\iint_{\mathbb{R}^2} \widehat{\varphi}(\xi,\eta)\,\psi\big(\varepsilon(\xi,\eta)\big)\,e^{i2\pi(x\xi+y\eta)}d\xi d\eta = \iint_{\mathbb{R}^2} \varphi(\varepsilon\xi+x,\varepsilon\eta+y)\,\widehat{\psi}(\xi,\eta)\,d\xi d\eta\,.$$

Since $\widehat{\varphi}, \widehat{\psi} \in \mathscr{S} \subset L^1$ and $\varphi, \psi \in C^0$ are bounded functions, the limit when $\varepsilon \to 0$ gives

$$\psi(0)\iint_{\mathbb{R}^2} \widehat{\varphi}(\xi,\eta)\,e^{i2\pi(x\xi+y\eta)}d\xi d\eta = \varphi(x,y)\iint_{\mathbb{R}^2} \widehat{\psi}(\xi,\eta)\,d\xi d\eta = \varphi(x,y)\|\widehat{\psi}\|_{L^1}.$$

The function $\psi(z)=e^{-\pi|z|^2}$ belongs to $\mathscr{S}$, and it is a fixed point of the Fourier transform, i.e., $\widehat{\psi}=\psi$, and $\|\widehat{\psi}\|_{L^1}=1$. With this function in the preceding formula, we conclude: *if $f \in \mathscr{S}$, then $\widehat{\widehat{f}}\,(z)=f(-z)$ for $z \in \mathbb{C}$, and the Fourier transform is a linear space isomorphism of $\mathscr{S}$ onto $\mathscr{S}$.*

If $f, g \in \mathscr{S}$, as $\mathscr{S} \subset L^1$, then $f*g \in L^1$, and by (C.1.4), $(f*g)\widehat{\ } = \widehat{f}\,\widehat{g}$. Since $\widehat{f}, \widehat{g} \in \mathscr{S}$, also $\widehat{f}\,\widehat{g} \in \mathscr{S}$, and, therefore, $f*g \in \mathscr{S}$. Consequently, *the convolution of functions of $\mathscr{S}$ belongs to $\mathscr{S}$.*

By (C.1.8), the Fourier transform defined in L^1 satisfies for all $f \in L^1 \cap L^2$, $\widehat{f} \in L^2$, and $\|\widehat{f}\|_{L^2} = \|f\|_{L^2}$. Since $\mathscr{S} \subset L^1 \cap L^2$, *the Fourier transform is an isometry of $\mathscr{S}$ onto $\mathscr{S}$.* As $\mathscr{S}$ is dense in L^1 and in L^2, by continuity, there is a unique extension of the Fourier transform to an isometry of L^2 onto L^2, called **Fourier-Plancherel transform**, still denoted $f \mapsto \widehat{f}$.

(C.2) *The Fourier-Plancherel transform satisfies for all $f, g \in L^2$:*

1. $\langle f, g\rangle_{L^2} = \langle \widehat{f}, \widehat{g}\rangle_{L^2}$.
2. $\widehat{f_\lambda}(\zeta) = \frac{1}{\lambda^2}\widehat{f}\big(\frac{\zeta}{\lambda}\big)$, *if $f_\lambda(z)=f(\lambda z)$ and $\lambda \in \mathbb{R}\backslash\{0\}$.*
3. $(f*g)\widehat{\ } = \widehat{f}\,\widehat{g}$.

(continued)

4. $\widehat{fg} = \widehat{f} * \widehat{g}$.
5. $\widehat{\widehat{\widehat{f}}}(z) = f(-z)$ a.e. *for* $z \in \mathbb{C}$.
6. $\|f\|_{L^\infty} \leq \|\widehat{f}\|_{L^1}$.
7. *If* $f \in L^1_{loc} \cap L^2$, $\frac{\partial f}{\partial x}(x, y)$, $\frac{\partial f}{\partial y}(x, y) \in L^2$, *then*

$$\widehat{\frac{\partial f}{\partial x}}(\xi, \eta) = i2\pi\xi \widehat{f}, \qquad \widehat{\frac{\partial f}{\partial y}}(\xi, \eta) = i2\pi\eta \widehat{f}(\xi, \eta),$$

$$\widehat{f_z}(z) = i\pi\overline{z}\widehat{f}(z), \qquad \widehat{f_{\overline{z}}}(z) = i\pi z \widehat{f}(z).$$

Proof 1 follows from the polarization formula

$$\begin{aligned}\|f+g\|_{L^2} - \|f-g\|_{L^2} + i\|f+ig\|_{L^2} - i\|f-ig\|_{L^2} \\ = 2[\langle f, g\rangle_{L^2} + \langle g, f\rangle_{L^2} + i\langle f, ig\rangle_{L^2} + i\langle ig, f\rangle_{L^2}] \\ = 2[\langle f, g\rangle_{L^2} + \langle g, f\rangle_{L^2} + \langle f, g\rangle_{L^2} - \langle g, f\rangle_{L^2}] = 4\langle f, g\rangle_{L^2},\end{aligned}$$

and from the isometry $f \mapsto \widehat{f}$ of L^2 onto L^2, since replacing in the preceding formula f by $\widehat{f}$ and g by $\widehat{g}$, the first term remains invariant, and, therefore, the resp. last terms are equal; $\langle f, g\rangle_{L^2} = \langle \widehat{f}, \widehat{g}\rangle_{L^2}$.

The other properties follow from those of Fourier transform in (C.1), the definition and properties of $\mathscr{S}$ and the density of $\mathscr{S}$ in L^2. □

We call **Hilbert-Beurling transform** to $\mathcal{L}: L^2 \to L^2$ with $\widehat{\mathcal{L}(f)}(\zeta) = \frac{\overline{\zeta}}{\zeta} \widehat{f}(\zeta)$. By 6 in the preceding result,

$$\widehat{g_z}(\zeta) = i\pi\overline{\zeta}\,\widehat{g}(\zeta) = \frac{\overline{\zeta}}{\zeta}\, i\pi\zeta\,\widehat{g}(\zeta) = \frac{\overline{\zeta}}{\zeta}\,\widehat{g_{\overline{z}}} = \frac{\overline{\zeta}}{\zeta}\,\widehat{h}(\zeta) = \widehat{\mathcal{L}(g_{\overline{z}})}(\zeta).$$

Hence, $g_z = \mathcal{L}(g_{\overline{z}})$, i.e., the Hilbert-Beurling transform maps $g_{\overline{z}}$ to g_z.

By 4 in (C.2), $\widehat{\widehat{\widehat{\widehat{f}}}} = f$ for every $f \in L^2$, i.e., the 4th power of composition of the Fourier-Plancherel transform is the identity in L^2. Consequently, the eigenvalues of the Fourier-Plancherel transform belong to the set of the 4 complex roots of unity $\{\pm 1, \pm i\}$. It was seen three paragraphs before (C.2) that $\psi(z) = e^{-\pi|z|^2}$ belongs to $\mathscr{S} \subset L^2$ and is a fixed point of the Fourier transform. Hence, it also is a fixed point of the Fourier-Plancherel transform, i.e., an eigenvector associated with eigenvalue 1 of this transform. In the next example, we get an eigenvector associated with the eigenvalue $-i$.

(C.3) Example: The Fourier-Plancherel transform of the complex function $\frac{1}{z}$, with $z = x + iy$, $x, y \in \mathbb{R}$, is

$$\widehat{\frac{1}{x+iy}}(\xi,\eta)=\iint_{\mathbb{R}^2}\frac{1}{x+iy}\,e^{-i2\pi(x\xi+y\eta)}\,dxdy=\int_{\mathbb{R}}e^{-i2\pi y\eta}\left(\int_{\mathbb{R}}\frac{1}{x+iy}\,e^{-i2\pi x\xi}dx\right)dy$$

$$=\int_0^{+\infty}e^{-i2\pi y\eta}\left(\int_{\mathbb{R}}\frac{1}{x+iy}\,e^{-i2\pi x\xi}dx\right)dy+\int_{-\infty}^{0}e^{-i2\pi y\eta}\left(\int_{\mathbb{R}}\frac{1}{x+iy}\,e^{-i2\pi x\xi}dx\right)dy\,.$$

The integrands in the integrals in x in these equalities last term do not belong to $L^1(\mathbb{R})$, but it can be proved with the residues theorem that the integrals exist as improper integrals[6], and therefore, they can be evaluated as a Cauchy principal value, i.e., the limit of the integrals on symmetric intervals $[-R,R]\subset\mathbb{R}$ relative to 0 when $R\to+\infty$, simplifying the argument since the Jordan lemma can be used. The complex function $h(z)=\frac{1}{z+iy}\,e^{-i2\pi z\xi}$ has a simple pole at $-iy$, in the imaginary axis and in the lower or upper open complex halfplane according to $y>0$ or $y<0$. Consider a concatenation of a regular path along the real axis from $-R$ a R, with $R>|y|$, and a regular path along the semicircle centered at 0 with radius R in the lower or upper complex halfplane so that the integral on the semicircle tends to 0 when $r\to+\infty$, i.e., so that in such semicircle $-i2\pi z\xi$ would have negative real part, i.e., $\xi\,\mathcal{I}m\,z<0$. Consider the lower or upper semicircle according to $\xi>0$ or $\xi<0$. The resulting path is closed around a pole if and only if $-\xi y<0$, and in the positive or negative direction of polar angles according to this pole lies in the upper or lower complex halfplane, i.e., according to $y<0$ or $y>0$. By the residues theorem and the Cauchy theorem,

$$\int_{\mathbb{R}}\frac{1}{x+iy}\,e^{-i2\pi x\xi}\,dx=\begin{cases}-\frac{y}{|y|}i2\pi\ \mathrm{Res}(h;-iy)-\frac{y}{|y|}i2\pi\,e^{-2\pi y\xi} & \text{, if } \xi y>0\,,\\ \int_{\mathbb{R}}\frac{1}{x+iy}\,e^{-i2\pi x\xi}\,dx & \text{, if } \xi y<0\,.\end{cases}$$

In any case, with the Heaviside function $H(t)$,

$$\widehat{\frac{1}{x+iy}}(\xi,\eta)=H(\xi)\int_0^{+\infty}-i2\pi\,e^{-2\pi(\xi+i\eta)y}dy+H(-\xi)\int_{-\infty}^{0}i2\pi\,e^{-2\pi(\xi+i\eta)y}dy$$

$$=H(\xi)\,\frac{-i}{\xi+i\eta}+H(-\xi)\,\frac{-i}{\xi+i\eta}=\frac{-i}{\xi+i\eta}\,;$$

thus, $\widehat{\frac{1}{z}}(\zeta)=-i\frac{1}{\zeta}\,,\ \zeta\in\mathbb{C}$.

The complex conjugate of this equality is $\widehat{\frac{1}{\bar z}}(\zeta)=i\,\frac{1}{\bar\zeta}\,,\ \zeta\in\mathbb{C}$.

Therefore, *complex functions* $\frac{1}{z}$ *and* $\frac{1}{\bar z}$ *are eigenvectors of the Fourier-Plancherel transform associated with the eigenvalues, resp.,* $-i$ *and* i .

[6] See Example (8.21.2) in Chapter 8 of *CADOVA*.

Appendix D
Partially Absolutely Continuous Functions with Derivatives in L^2_{loc}

Quasiconformal mappings in a plane region are functions partially absolutely continuous (p.a.c.)[1] with partial derivatives in L^2_{loc} as functions of two real variables. We consider here some important properties of these real variables functions used for quasiconformal mappings in this book chapters, beginning with two characterizations, one by equalities of line integrals on the boundary of rectangles with horizontal and vertical sides to double integrals of $L^p_{loc}(\Omega)$ functions, with $p = 1, 2$, and another of having uniform approximations by sequences of functions in C_0^∞ with convergent partial derivatives in L^p to the partial derivatives of the considered functions, also for $p = 1\,, 2$. Afterward, we have a version of Green theorem with conditions on the integrands weaker than usual, which is applied to get that homeomorphisms p.a.c. with derivatives in L^2_{loc} in a region are absolutely continuous.

(D.1) *If $\Omega \subset \mathbb{R}^2$ is a region and $\mathbf{f} : \Omega \to \mathbb{R}^2$ is continuous, then $\mathbf{f}$ is a* p.a.c. *function with partial derivatives in $L^p_{loc}(\Omega)$ ($p = 1$ or 2) if and only if there are $\mathbf{g}, \mathbf{h} \in L^p_{loc}(\Omega)$ such that for every rectangle $R = [a, b] \times [c, d] \subset \Omega$,*

$$\int_{\partial R} \mathbf{f}(x, y)\, dx = -\iint_R \mathbf{g}\,, \qquad \int_{\partial R} \mathbf{f}(x, y)\, dy = \iint_R \mathbf{h}\,,$$

and, if yes, $\mathbf{g} = D_2\mathbf{f}$ and $\mathbf{h} = D_1\mathbf{f}$ a.e. *in Ω.*

Proof If $\mathbf{f}$ is a p.a.c. function with partial derivatives in $L^p_{loc}(\Omega)$ ($p = 1$ or 2), as the restriction of $\mathbf{f}(x, y)$ to each horizontal or vertical edge of R are absolutely

[1] The definition of p.a.c. function is given in Chap. 1 as a function $f = (u, v) : \Omega \to \mathbb{C}$, with $\Omega \subset \mathbb{C}$ a region, such that for each closed rectangle $[a, b] \times [c, d] \subset \Omega$ with sides parallel to the real and imaginary axes the functions $x \mapsto u(x + iy)$, $y \mapsto u(x + iy)$, $x \mapsto v(x + iy)$, $y \mapsto v(x + iy)$ are absolutely continuous a.e. on, resp., $[a, b]$ and $[c, d]$.

L. T. Magalhães, *Quasiconformal Mappings in the Plane and Complex Dynamics*,
https://doi.org/10.1007/978-3-031-80115-0

continuous functions of, resp., x or y,

$$\int_{\partial R} \mathbf{f}(x, y)\, dx = \int_a^b [\mathbf{f}(x, c) - \mathbf{f}(x, d)]\, dx = -\int_a^b \int_c^d D_2\mathbf{f}(x, y)\, dy\, dx = \iint_R D_2\mathbf{f}.$$

the last equality because $D_2\mathbf{f} \in L^2(R) \subset L^1(R)$ and by the Fubini theorem, and the other equality switching x, y.

Conversely, if $\mathbf{g}, \mathbf{h}$ satisfy the stated conditions, by the equalities of integrals in the statement, for $R_{xy} = [a, x] \times [c, y]$,

$$\int_a^x [\mathbf{f}(t, c) - \mathbf{f}(t, y)]\, dt = -\iint_{R_{xy}} \mathbf{g} = \int_a^x \int_c^y \mathbf{g}(t, s)\, ds\, dt\,.$$

If $\{y_n\}$ is an enumeration of $[c, d] \cap \mathbb{Q}$, with derivative of the first and the last terms of this equality in order to x with $y = y_n$ fixed, since the fundamental theorem of calculus for derivative of indefinite integrals holds for absolutely continuous functions a.e., $[\mathbf{f}(x, y_n) - \mathbf{f}(x, c)] = \int_c^{y_n} \mathbf{g}(x, s)\, ds$ for $x \in [a, b] \backslash E_n$ with $E_n \subset [a, b]$ a measure 0 set. If $x \notin E = \cup_{n=1}^{\infty} E_n$, this equality holds for all $y_n \in [c, d] \cap \mathbb{Q}$, and as both sides are continuous in y_n, the equality holds with y_n replaced by any other point of $[c, d]$ except possibly for $x \in E$. Hence, $\mathbf{f}$ is the indefinite integral of an integrable function, and by the fundamental theorem of calculus, the function $y \mapsto \mathbf{f}(x, y)$ is absolutely continuous a.e. for $x \in [a, b]$. Again by the fundamental theorem of calculus, $D_2\mathbf{f}(x, y) = \mathbf{g}(x, y)$ a.e. for $y \in [c, d]$, if $x \in [a, b] \backslash E$. Since $(x, y) \mapsto D_2\mathbf{f}(x, y)$ and $(x, y) \mapsto \mathbf{g}(x, y)$ are measurable, by the Fubini theorem, $D_2\mathbf{f} = \mathbf{g}$ a.e. in R. Switching x, y, $x \mapsto \mathbf{f}(x, y)$ is absolutely continuous a.e. for $y \in [c, d]$ and $D_1\mathbf{f} = \mathbf{h}$ a.e. in R. Since R is an arbitrary closed rectangle with all edges in Ω, the conclusions hold in Ω. □

Next we have another characterization of p.a.c. functions with partial derivatives in L^p_{loc} in a region, now in terms of being uniformly approximated by functions of C_0^∞ with partial derivatives converging in L^p to the partial derivatives of the considered derivatives, for $p = 1, 2$.

(D.2) *If $\Omega \subset \mathbb{R}^2$ is a region and $\mathbf{f} : \Omega \to \mathbb{R}^2$, then $\mathbf{f}$ is a* p.a.c. *function with partial derivatives in $L^p_{loc}(\Omega)$ ($p = 1$ or 2) if and only if there is a sequence of functions $\{\mathbf{f}_n\} \subset C_0^\infty(\Omega)$ and functions $\mathbf{g}_j \in L^p_{loc}$, $j = 1, 2$, such that for every compact set $K \subset \Omega$,*

$$\mathbf{f}_n \to \mathbf{f} \text{ uniformly in } K \text{and } \|D_j\mathbf{f}_n - \mathbf{g}_j\|_{L^p(K)} \to 0\,,\ j = 1, 2\,, n \to +\infty\,,$$

and, if yes, $\mathbf{g}_j = D_j\mathbf{f}$, $j = 1, 2$ a.e. *in Ω. If* $\operatorname{supp}\mathbf{f} \subset \Omega$ *is a compact set, there is $\{\mathbf{f}_n\}$ satisfying the conditions with Ω instead of K.*

Proof If there is a sequence of function $\{\mathbf{f}_n\} \subset C_0^\infty(\Omega)$ and functions $\mathbf{g}_j \in L^p_{loc}$ ($p = 1$ or 2), $j = 1, 2$, such that for every compact set $K \subset \Omega$ the limits in the statement hold, $\mathbf{f}$ is continuous and if $R \subset \Omega$ is a compact rectangle with horizontal and vertical edges,

$$\int_{\partial R} \mathbf{f}(x, y)\, dx = \lim_{n\to+\infty} \int_{\partial R} \mathbf{f}_n(x, y)\, dx = -\lim_{n\to+\infty} \iint_R D_2\mathbf{f}_n = -\iint_R D_2\mathbf{g},$$

where the last inequality when $p=1$ results from

$$\left|\iint_R (D_2\mathbf{f}_n - D_2\mathbf{g})\right| \leq \iint_R |D_2\mathbf{f}_n - D_2\mathbf{g}| = \|D_j\mathbf{f}_n - \mathbf{g}_j\|_{L^1(K)} \to 0,$$

and when $p=2$ results from the Cauchy-Schwarz inequality

$$\left|\iint_R (D_2\mathbf{f}_n - D_2\mathbf{g})\right| = \left|\langle D_2\mathbf{f}_n - D_2\mathbf{g}, 1\rangle_{L^2(K)}\right| \leq \|D_2\mathbf{f}_n - D_2\mathbf{g}\|_{L^2(K)} \|1\|_{L^2(K)} \to 0.$$

Switching x, y, $\int_{\partial R} \mathbf{f}(x, y)\, dy = -\iint_R D_1\mathbf{g}$. By the preceding result, $\mathbf{f}$ is a p.a.c. function with partial derivatives in $L^p_{loc}(\Omega)$ and $\mathbf{g}_j = D_j\mathbf{f}$, $j = 1, 2$ a.e. in Ω, for $p = 1, 2$.

Conversely, if $\mathbf{f}$ is a p.a.c. function with partial derivatives in $L^p_{loc}(\Omega)$ ($p = 1$ or 2), it is possible to construct in $K \subset \Omega$ a sequence with the properties in the statement and $\mathbf{g}_j = D_j\mathbf{f}$, $j = 1, 2$ a.e. in Ω, mollifying as in (B.1), $\mathbf{f}_n = \widetilde{\mathbf{f}} * \Psi_{1/n}$, $n \in \mathbb{N}$, with $\widetilde{\mathbf{f}} = f$ in a compact set $K' \subset \Omega, K' \supset K$ and $\widetilde{\mathbf{f}} = 0$ outside K'. Since every open set is a countable union of expanding compact sets, considering such a sequence for Ω and in each of the compact sets in the sequence a sequence of approximations with the properties in the statement, with the Cantor diagonal argument, we get a sequence with those properties in Ω. It now suffices to prove $D_j\widetilde{\mathbf{f}} * \Psi_{1/n} = D_j(\widetilde{\mathbf{f}} * \Psi_{1/n})$, $j = 1, 2$, in K for $n \in \mathbb{N}$ large.

With $Q_n =]-\frac{1}{n}, \frac{1}{n}[^2$, for $n \in \mathbb{N}$ large, it is $\widetilde{\mathbf{f}}(\mathbf{x} - \mathbf{y}) = \mathbf{f}(\mathbf{x} - \mathbf{y})$ for $\mathbf{x} \in K, \mathbf{y} \in Q_n$, and for $j = 1, 2$,

$$\begin{aligned} D_j\widetilde{\mathbf{f}} * \Psi_{\frac{1}{n}}(x, y) &= \iint_{Q_n} D_j\mathbf{f}\big((x, y) - (s, t)\big)\Psi_{\frac{1}{n}}(s, t)\, ds dt \\ &= \int_{-\frac{1}{n}}^{\frac{1}{n}} \left(\int_{-\frac{1}{n}}^{\frac{1}{n}} D_j\mathbf{f}(x - s, y - t)\Psi_{\frac{1}{n}}(s, t)\, ds\right) dt. \end{aligned}$$

Since $(s, t) \mapsto \mathbf{f}((x, y) - (s, t))$ is absolutely continuous in $I_t =]-\frac{1}{n}, \frac{1}{n}[\times\{t\}$ a.e. for $t \in]-\frac{1}{n}, \frac{1}{n}[$. For such t, the last integral in the preceding formula can be integrated by parts giving

$$\int_{-\frac{1}{n}}^{\frac{1}{n}} D_j\mathbf{f}(x - s, y - t)\, \Psi_{\frac{1}{n}}(s, t)\, ds = \int_{-\frac{1}{n}}^{\frac{1}{n}} \mathbf{f}(x - s, y - t)\, D_j\Psi_{\frac{1}{n}}(s, t)\, ds\,;$$

therefore,

$$D_j\widetilde{\mathbf{f}}*\Psi_{\frac{1}{n}}(x,y)=\int_{-\frac{1}{n}}^{\frac{1}{n}}\left(\int_{-\frac{1}{n}}^{\frac{1}{n}}\mathbf{f}(x-s,\,y-t)\,D_j\Psi_{\frac{1}{n}}(s,t)\,ds\right)dt$$

$$=\iint_{Q_n}\mathbf{f}\big((x,y)-(s,t)\big)\,D_j\Psi_{\frac{1}{n}}(s,t)\,ds\,dt=\widetilde{\mathbf{f}}*D_j\Psi_{\frac{1}{n}}(x,y)\,.$$

Since $\widetilde{\mathbf{f}}*D_j\Psi_{\frac{1}{n}}=D_j\big(\widetilde{\mathbf{f}}*\Psi_{\frac{1}{n}}\big)$, it is $D_j\widetilde{\mathbf{f}}*\Psi_{\frac{1}{n}}=D_j\big(\widetilde{\mathbf{f}}*\Psi_{\frac{1}{n}}\big)$, for $j=1,2$. □

(D.3) Green theorem:
If $\Omega\subset\mathbb{R}^2$ is a region, $\mathbf{f},\mathbf{g}:\Omega\to\mathbb{R}^2$ are p.a.c. *functions with partial derivatives in $L^2_{loc}(\Omega)$, $D\subset\mathbb{R}^2$ is an open set bounded by a rectifiable Jordan curve with* cl $D\subset\Omega$, *and* $\mathbf{g}$ *is a function of bounded variation, then*

$$\int_{\partial D}\mathbf{f}\,D_1\mathbf{g}\,dx+\mathbf{f}\,D_2\mathbf{g}\,dy=\iint_D\big(D_1\mathbf{f}D_2\mathbf{g}-D_2\mathbf{f}D_1\mathbf{g}\big)\,,$$

also holding if $\mathbf{f}$ *is* p.a.c. *with partial derivatives in $L^1_{loc}(\Omega)$ and $\mathbf{g}\in C^1(\Omega)$.*

Proof If $\mathbf{f},\mathbf{g}$ are p.a.c. functions with partial derivatives in $L^2_{loc}(\Omega)$, they can be approximated by sequences, resp., $\{\mathbf{f}_n\}$, $\{\mathbf{g}_n\}\subset C_0^\infty(\Omega)$ with the properties in the preceding result. By the Green theorem for C_0^1 functions,

$$\iint_D\sum_{j=1}^{2}(-1)^{j-1}D_j\mathbf{f}_m D_{3-j}\mathbf{g}_n=\int_{\partial D}\big(\mathbf{f}_m D_1\mathbf{g}_n dx+\mathbf{f}_m D_2\mathbf{g}_n dy\big).$$

When $n\to+\infty$, $\mathbf{g}_n\to\mathbf{g}$ uniformly in ∂D and, as $\mathbf{f}_m,\mathbf{g}_n,\mathbf{g}$ are functions of bounded variation, integrating by parts,

$$\lim_{n\to+\infty}\int_{\partial D}\big(\mathbf{f}_m D_1\mathbf{g}_n\,dx+\mathbf{f}_m D_2\mathbf{g}_n\,dy\big)=-\lim_{n\to+\infty}\int_{\partial D}\big(\mathbf{g}_n D_1\mathbf{f}_m dx+\mathbf{g}_n D_2\mathbf{f}_m dy\big)$$

$$=-\int_{\partial D}\big(\mathbf{g}D_1\mathbf{f}_m dx+\mathbf{g}D_2\mathbf{f}_m dy\big)=\int_{\partial D}\big(\mathbf{f}_m D_1\mathbf{g}\,dx+\mathbf{f}_m D_2\mathbf{g}\,dy\big)\,.$$

If $\mathbf{f}$ is a p.a.c. function with partial derivatives in $L^1_{loc}(\Omega)$ and $\mathbf{g}\in C_0^1(\Omega)$, the formulas hold with $\mathbf{g}_n=\mathbf{g}$ for all $n\in\mathbb{N}$.

As $D_j\mathbf{f}_m\to D_j\mathbf{f}$ in $L^2(D)$, $j=1,2$, when $m\to+\infty$, the double integrals $\iint_D\|D_j\mathbf{f}_m\|^2$, $j=1,2$, are uniformly bounded for $m\in\mathbb{N}$ large, and, by the Cauchy-Schwarz inequality,

$$\left\| \iint_D \sum_{j=1}^{2} (-1)^{j+1} D_j \mathbf{f}_m D_{3-j}(\mathbf{g}_n - \mathbf{g}) \right\| \le \sum_{j=1}^{2} \|D_j \mathbf{f}_m D_{3-j}(\mathbf{g}_n - \mathbf{g})\|_{L^1(D)}$$

$$\le \sum_{j=1}^{2} \|D_j \mathbf{f}_m\|_{L^2(D)} \|D_j(\mathbf{g}_n - \mathbf{g})\|_{L^2(D)} \to 0, \qquad n \to +\infty.$$

Therefore,

$$\iint_D \sum_{j=1}^{2} (-1)^{j+1} D_j \mathbf{f}_m D_{3-j}\mathbf{g} = \int_{\partial D} \big(\mathbf{f}_m D_1 \mathbf{g}\, dx + \mathbf{f}_m D_2 \mathbf{g}\, dy\big).$$

Since $\mathbf{f}_m \to \mathbf{f}$ uniformly when $m \to +\infty$,

$$\int_{\partial D} \big(\mathbf{f}_m D_1 \mathbf{g}\, dx + \mathbf{f}_m D_2 \mathbf{g}\, dy\big) \to \int_{\partial D} \big(\mathbf{f} D_1 \mathbf{g}\, dx + \mathbf{f} D_2 \mathbf{g}\, dy\big), \quad m \to +\infty,$$

and by the Cauchy-Schwarz inequality as above,

$$\iint_D \sum_{j=1}^{2} (-1)^{j+1} D_j(\mathbf{f}_m - \mathbf{f}) D_{3-j}\mathbf{g} \to 0, \quad m \longrightarrow +\infty,$$

implying the equality in the statement.

If $\mathbf{f}$ is a p.a.c. function with partial derivatives in $L^2_{loc}(\Omega)$ and $\mathbf{g} \in C^1_0(\Omega)$,

$$\left\| \iint_D \sum_{j=1}^{2} (-1)^{j+1} D_j(\mathbf{f}_m - \mathbf{f}) D_{3-j}\mathbf{g} \right\|$$

$$\le \sum_{j=1}^{2} \max_{\text{cl}\, D} \|D_{3-j}\mathbf{g}\| \|D_j(\mathbf{f}_m - \mathbf{f})\|_{L^2(D)} \to 0, \quad m \to +\infty,$$

and the equality in the statement also holds in this case. □

We have the following consequence.

(D.4) *Every homeomorphism in a region* $\Omega \subset \mathbb{R}^2$ p.a.c. *with partial derivatives in* $L^2_{loc}(\Omega)$ *is absolutely continuous in* Ω.

Proof By (E.4) proved in Appendix E, $\iint_B |J_{\mathbf{f}}| \le \mu\big(\mathbf{f}(B)\big)$ for $B \in \mathscr{B}$ with $B \subset \Omega$, and equality if and only if $\mathbf{f}$ is locally absolutely continuous in Ω. Without loss of generality, $J_{\mathbf{f}} > 0$. Let $R \subset \Omega$ be a closed rectangle with horizontal and vertical edges. With $R' = \mathbf{f}(R)$, $\mathbf{f}$ is p.a.c. and a function of bounded variation in ∂R, and $\partial R' = \mathbf{f}(\partial R)$ is rectifiable. Applying (D.3) in R and R' with $\mathbf{f} = (u, v)$,

$$\iint_R J = \int_{\partial R} u(D_1 v\, dx + D_2 v\, dy)\,, \qquad\qquad \iint_{R'} 1 = \int_{\partial R'} u\, dv\,.$$

Since $\mathbf{f}$ is a homeomorphism, the line integrals in the right-hand side are related by changes of integration variables, and they are equal. Thus, $\iint_B J_{\mathbf{f}} = \mu\big(\mathbf{f}(B)\big)$ for every B equal to a closed rectangle with horizontal and vertical edges subset of Ω, and the equality also holds for every open set $B \subset \Omega$ since it is a countable union of rectangles with horizontal and vertical edges, each pair of rectangles intersecting at most in subsets of their edges. If $B \in \mathscr{B}$ with $\mathrm{cl}\, B \subset \Omega$, for every $\varepsilon > 0$, there is an open set $U_\varepsilon \supset B$ such that $\mathrm{cl}\, U \subset \Omega$ and $\mu(U_\varepsilon \backslash B) < \varepsilon$. Since $B \mapsto \iint_B J_{\mathbf{f}}$ is a function of sets completely additive and absolutely continuous in Ω,

$$\iint_B J_{\mathbf{f}} = \lim_{\varepsilon \to 0} \iint_{U_\varepsilon} J_{\mathbf{f}} = \lim_{\varepsilon \to 0} \mu\big(\mathbf{f}(U_\varepsilon)\big) \geq \mu(B)\,.$$

Since $\iint_B |J_{\mathbf{f}}| \leq \mu\big(\mathbf{f}(B)\big)$, equality holds for all $B \in \mathscr{B}$ with $\mathrm{cl}\, B \subset \Omega$. So, they also hold for all $B \in \mathscr{B}$ with $B \subset \Omega$, and $\mathbf{f}$ is locally absolutely continuous in Ω. □

Appendix E
Existence and Differentiability of Partial Derivatives

The proof of equivalence of the geometric and analytic definitions of quasi-conformal mappings in Chap. 1 used that existence of partial derivatives of a homeomorphism implies its differentiability *a.e.*. This is false in general, but holds if the partial derivatives are continuous.

This property was established for open functions in 1959 by F. Gehring and O. Lehto with the Egorov theorem[1] (proved independently by C. Severini[2] in 1910 and D. Egorov in 1911) and the Lebesgue density theorem on the real line considered in Appendix A.

The Egorov theorem establishes that a sequence of measurable functions converging in a measurable set A with finite Lebesgue measure converges approximately uniformly, in the sense that for every $\varepsilon > 0$, there is an open subset of a closed subset whose complement in A has Lebesgue measure $< \varepsilon$.

(E.1) Egorov theorem:
If $\{f_k\}$ *is a sequence of measurable functions converging* a.e. *in a measurable set* $A \subset \mathbb{R}^n$ *with* $\mu(A) < \infty$ *to* f *and* $\varepsilon > 0$, *there is a closed set* $B \subset A$ *such that* $\mu(A \setminus B) < \varepsilon$ *and* $f_k \to f$ *uniformly in* B *when* $k \to +\infty$.

---

[1] Egorov theorem is a particular case of the **Littlewood three principles** John Edensor Littlewood (1885–1977) masterfully expressed in *Lectures on the Theory of Functions*, Oxford University Press, 1944, as:

"The extent of knowledge required [in the theory of real functions] *is nothing like so great as is sometimes supposed. There are three principles, roughly expressible in the following terms: Every (measurable) set is nearly a finite union of intervals; every function (of class* L^λ*) is nearly continuous; and every convergent sequence of functions is nearly uniformly convergent. Most of the results of the present section are fairly intuitive applications of these ideas, and the student armed with them should be equal to most occasions when real variable theory is called for. If one of the principles would be the obvious means to settle a problem if it were 'quite' true, it is natural to ask if the 'nearly' is near enough, and for a problem that is actually soluble it generally is."*

[2] Egorov, Dmitri (1869–1931). Severini, Carlo (1872–1951).

L. T. Magalhães, *Quasiconformal Mappings in the Plane and Complex Dynamics*,
https://doi.org/10.1007/978-3-031-80115-0

Proof For $k, m \in \mathbb{N}$ denote $E_{k,m} = \bigcup_{j \geq k} \left\{x \in A : |f_j(x) - f(x)| > \frac{1}{m}\right\}$. For all $m \in \mathbb{N}$, the set $\cap_{k \in \mathbb{N}} E_{k,m}$ has measure 0. As $\mu(A) < \infty$, for every $m \in \mathbb{N}$ there is $k_m \in \mathbb{N}$ such that $\mu(E_{k_m,m}) < \frac{\varepsilon}{2^k}$. If $B = A \setminus \cup_{m \in \mathbb{N}} E_{k_m,m}$, $f_k \to f$ uniformly in B when $k \to +\infty$ and $\mu(A \setminus B) \leq \sum_{m \in \mathbb{N}} \frac{1}{2^m} \varepsilon = \varepsilon$. □

A point $\mathbf{x} \in \mathbb{R}^n$ is said to be a **point of density of A in the direction of $\mathbf{v} \in \mathbb{R}^n \setminus \{0\}$** if 0 is a point of density of $\{t \in \mathbb{R} : \mathbf{x} + t\mathbf{v} \in A\}$.

(E.2) Density of a set in the canonical basis of $\mathbb{R}^n$ directions:
For every set $A \subset \mathbb{R}^n$ with nonzero Lebesgue measure, all points of A are points of density of A in the directions of the vectors of the canonical basis of $\mathbb{R}^n$.

Proof The result is a consequence of the Lebesgue density theorem in $\mathbb{R}$ (A.4), since for $(\mathbf{e}_1, \ldots, \mathbf{e}_n)$ the canonical basis of $\mathbb{R}^n$ and every $\mathbf{x}_0 \in \mathbb{R}^n$ the set of real numbers not points of density of $S_k(\mathbf{x}_0) = \{s \in \mathbb{R} : \mathbf{x}_0 + s\mathbf{e}_k \in A\}$ has measure 0 in $\mathbb{R}$, and $S'_k(\mathbf{x}_0) = \mathbb{R}^{k-1} \times S_k(\mathbf{x}_0) \times \mathbb{R}^{n-k}$ has measure 0 in $\mathbb{R}^n$. It is proved by induction in the successive canonical basis directions. Denote A_k the set of points of A of density of A in the directions of the first $k \in \{1, \ldots, n-1\}$ canonical basis vectors. As $A_{k+1} = A_k \cap L_{k+1}$, with L_{k+1} the set of points of density of A in the direction of $\mathbf{e}_{k+1}$, it is $A \setminus A_{k+1} = (A \setminus A_k) \cup (A \setminus L_{k+1})$. By the Lebesgue density theorem in $\mathbb{R}$ (A.4), $A \setminus L_{k+1}$ has measure 0; thus, if $A \setminus A_k$ has measure 0, also $A \setminus A_{k+1}$ has. Since $A_1 = L_1$ and $A \setminus A_1$ has measure 0, this proves that $A \setminus A_n$ has measure 0, i.e., points of A are points of density of A in the directions of the canonical basis vectors, modulo measure 0 sets. □

(E.3) Gehring-Lehto theorem:
Except for measure 0 sets, open functions from a region of $\mathbb{R}^n$ to $\mathbb{R}^n$ are differentiable if and only if they have partial derivatives.

Proof Necessity of existence of partial derivatives for differentiability is immediate from the definitions, and it is enough to prove sufficiency in compact subsets K of $\Omega \subset \mathbb{R}^n$ where an open function $\mathbf{f} : \Omega \to \mathbb{R}^n$ is defined.

For $\mathbf{x} \in K$ such that the partial derivatives $D_j\mathbf{f}(\mathbf{x})$ exist and $h > 0$ is such that $\mathbf{x} + h\mathbf{e_j} \in \Omega$ for all $j = 1, \ldots, n$, define

$$F_h(\mathbf{x}) = \sum_{j=1}^{n} \left\| \frac{\mathbf{f}(\mathbf{x}+h\mathbf{e_j}) - f(\mathbf{x})}{h} - D_j\mathbf{f}(\mathbf{x}) \right\|, \quad j = 1, \ldots, n.$$

For $|h|$ small, F_h is a Borel function defined a.e. in K. There is $N \in \mathbb{N}$ such that, for $n > N$, functions

$$g_n = \sup_{0<h<\frac{1}{n}} F_h = \sup_{0<h<\frac{1}{n},\, h\in\mathbb{Q}} F_h$$

are Borel functions. As the partial derivatives of f exist a.e. in K, $g_n \to 0$ a.e. in K. By the Egorov theorem, for every $\varepsilon > 0$, there is a closed set $B \subset K$ such that the Lebesgue measure of $\Omega \backslash B$ is $< \varepsilon$ and $\frac{\mathbf{f}(\mathbf{x}+h\mathbf{e_j})-f(\mathbf{x})}{h} \to D_j\mathbf{f}(\mathbf{x})$, when $h \to 0$, uniformly for $\mathbf{x} \in B$. Restrictions of $D_j\mathbf{f}$ to B are continuous functions. If $\delta \in]0, 1[$, then there is $\eta > 0$ such that

$$\left|D_j\mathbf{f}(\mathbf{x}_0+\mathbf{v}) - D_j\mathbf{f}(\mathbf{x}_0)\right|, |F_h(\mathbf{x}_0+\mathbf{v})| < \delta\,,\ \|\mathbf{v}\| < \eta\,,\ \text{if } \mathbf{x}_0, \mathbf{x}_0+\mathbf{v} \in B,\ j = 1, \ldots, n\,.$$

If $\mathbf{x}_0 \in B$, with $\mathbf{v} = (v_1, \ldots, v_n)$, $\mathbf{v}_0 = 0$ and $\mathbf{v}_j = (v_1, \ldots, v_j, 0, \ldots, 0)$ for $j = 1, \ldots, n$,

$$\begin{aligned}
\big\|\mathbf{f}(\mathbf{x}_0+\mathbf{v}) - \mathbf{f}(\mathbf{x}_0) - [D\mathbf{f}(\mathbf{x}_0)]\mathbf{v}\big\| &= \Big\| \sum_{j=1}^{n} \big[\mathbf{f}(\mathbf{x}_0+\mathbf{v}_j) - \mathbf{f}(\mathbf{x}_0+\mathbf{v}_{j-1}) \\
&\quad -[D\mathbf{f}(\mathbf{x}_0+\mathbf{v}_{j-1})](\mathbf{v}_j - \mathbf{v}_{j-1})\big] + \big[D\mathbf{f}(\mathbf{x}_0+\mathbf{v}_{j-1}) - D\mathbf{f}(\mathbf{x}_0)\big](\mathbf{v}_j - \mathbf{v}_{j-1}) \Big\| \\
&\le \sum_{j=1}^{n} \big\|\mathbf{f}(\mathbf{x}_0+\mathbf{v}_{j-1}+v_j\mathbf{e_j}) - \mathbf{f}(\mathbf{x}_0+\mathbf{v}_{j-1}) - [D\mathbf{f}(\mathbf{x}_0+\mathbf{v}_{j-1})]v_j\mathbf{e_j}\big\| \\
&\quad + \Big\| \sum_{j=1}^{n} \big[D\mathbf{f}(\mathbf{x}_0+\mathbf{v}_{j-1}) - D\mathbf{f}(\mathbf{x}_0)\big]v_j\mathbf{e_j} \Big\| \le \max_{j=1,\ldots,n} F_{v_j}(\mathbf{x}_0+\mathbf{v}_{j-1})\|\mathbf{v}\| \\
&\quad + \Big\| \sum_{j=1}^{n} \big[D_j\mathbf{f}(\mathbf{x}_0+\mathbf{v}_{j-1}) - D_j\mathbf{f}(\mathbf{x}_0)\big]v_j \Big\| \le (1+n)\delta\|\mathbf{v}\|\,.
\end{aligned}$$

if $\|\mathbf{v}\| < \eta$ and $\mathbf{x}_0+\mathbf{v}_j \in B$ for $j = 0, 1, \ldots, n$.

If $\mathbf{x}_0 \in B$ is a point of density of B in the canonical basis of $\mathbb{R}^n$ directions, there is a n-dimensional cube centered at $\mathbf{x}_0$, $Q_d = \{\mathbf{x}_0\} + [-d, d]^n$, with edges of length $2d < \eta$ such that for every line segment L contained in $\{\mathbf{x}_0\} + \cup_{j=1}^{n} [-d, d]\,\mathbf{e}_j$, $\mu(L \cap B) > \frac{1}{\delta}\mu(L)$. If $\mathbf{x} = (x_1, \ldots, x_n) \in Q_{\frac{d}{2}} \backslash \{0\}$, then every line segment $\{\mathbf{x}\} +]1-\delta, 1[\, x_j$, $\{\mathbf{x}\} +]1, 1+\delta[\, x_j$, $j = 1, \ldots, n$, intersects B. Thus, there are $2n$ points $\mathbf{x}+a_j\mathbf{e}_j$, $\mathbf{x}+b_j\mathbf{e}_j \in B$, with $a_j < 0 < b_j$, $b_j - a_j < 2\delta|x_j|$, $j = 1, \ldots, n$. The two points of intersection of the interval $R = \times_{j=1}^{n}[x_j + a_j, x_j + b_j]$ boundary with the line through $\mathbf{x}$ with the direction of $\mathbf{e}_j$ belong to B; thus, the inequality in the preceding paragraph implies

$$\big\|\mathbf{f}(\mathbf{z}) - \mathbf{f}(\mathbf{x}) - [D\mathbf{f}(\mathbf{x})](\mathbf{z}-\mathbf{x})\big\| \le (1+n)\,\delta\,\|\mathbf{z}-\mathbf{x}\|\,, \quad \mathbf{z} \in \partial R\,.$$

Since $\mathbf{f}$ is an open function, it satisfies the maximum principle in R, i.e., $\mathbf{z} \mapsto \|f(\zeta) - \mathbf{f}(\mathbf{x}) - [D\mathbf{f}(\mathbf{x})](\mathbf{z}-\mathbf{x})\|$ assumes a maximum value in R at a point $\mathbf{z}^* \in \partial R$ and, therefore,

$$\begin{aligned}\left\|\mathbf{f}(\mathbf{z})-\mathbf{f}(\mathbf{x})-[D\mathbf{f}(\mathbf{x})](\mathbf{z}-\mathbf{x})\right\| &\leq \left\|\mathbf{f}(\mathbf{z}^*)-\mathbf{f}(\mathbf{x})-[D\mathbf{f}(\mathbf{x})](\mathbf{z}-\mathbf{x})\right\| \\ &\leq \left\|\mathbf{f}(\mathbf{z}^*)-\mathbf{f}(\mathbf{x})-[D\mathbf{f}(\mathbf{x})](\mathbf{z}^*-\mathbf{x})\right\|+\left\|[D\mathbf{f}(\mathbf{x})](\mathbf{z}^*-\mathbf{z})\right\| \\ &\leq (1+n)\,\delta\,\|\mathbf{z}^*-\mathbf{x}\|+\|D\mathbf{f}(\mathbf{x})\|\,\|\mathbf{z}^*-\mathbf{z}\|\,, \qquad \mathbf{z}\in Q_{\frac{d}{2}}\,.\end{aligned}$$

As $\|\mathbf{z}^*-\mathbf{x}\|\leq\|\mathbf{z}^*-\mathbf{z}\|+\|\mathbf{z}-\mathbf{x}\|$ and $\|\mathbf{z}^*-\mathbf{z}\|<2\delta\|\mathbf{z}-\mathbf{x}\|$,

$$\begin{aligned}\left\|\mathbf{f}(\mathbf{z})-\mathbf{f}(\mathbf{x})-[D\mathbf{f}(\mathbf{x})](\mathbf{z}-\mathbf{x})\right\| &\leq (1+n)\,\delta\,\big(\|\mathbf{z}^*-\mathbf{z}\|+\|\mathbf{z}-\mathbf{x}\|\big)+\|D\mathbf{f}(\mathbf{x})\|\,\|\mathbf{z}^*-\mathbf{z}\| \\ &\leq \big[2\delta\big((1+n)\delta+\|D\mathbf{f}(\mathbf{x})\|\big)+(1+n)\delta\big]\|\mathbf{z}-\mathbf{x}\|\,,\end{aligned}$$

and, therefore, $\mathbf{f}$ is differentiable in $Q_{d/2}$ and, in particular, at $\mathbf{x}_0$. Hence, $\mathbf{f}$ is differentiable in every point $\mathbf{x}_0$ of density of B in the directions of the canonical basis of $\mathbb{R}^n$. By (E.2), $\mathbf{f}$ is differentiable a.e. in every compact set $K\subset\Omega$ and, therefore, also a.e. in Ω. □

By this result and the Lebesgue theorem for a homeomorphism volume derivative (A.26) at the end of Appendix A, we get an extension of the classical volume transformation formula with C^1 changes of variables for locally absolutely continuous homeomorphisms in a region where they have partial derivatives a.e..

(E.4) *If* $\mathbf{f}$ *is a homeomorphism from a region* $\Omega\subset\mathbb{R}^n$ *to* $\mathbb{R}^n$ *with finite partial derivatives* a.e. *in* Ω, $\int_B|J_\mathbf{f}|\leq\mu\big(\mathbf{f}(B)\big)$ *for* $B\in\mathscr{B}$ *with* $B\subset\Omega$, *and equality if and only if* $\mathbf{f}$ *is locally absolutely continuous in* Ω*; if yes, equality holds for all measurable* $B\subset\Omega$.

Proof By the Gehring-Lehto theorem, $\mathbf{f}$ is differentiable a.e. in Ω.

For every point of differentiability $\mathbf{x}_0$, without loss of generality taken to be $\mathbf{x}_0=0$ with $\mathbf{f}(\mathbf{x}_0)=0$, and for every n-cube $Q_r\overset{def}{=}\left[\frac{r}{2},\frac{r}{2}\right]^n\subset\Omega$ with $r>0$, $D\mathbf{f}(0)\,Q_r$ is a k-parallelogram with $k\leq n$ and $\mu\big(D\mathbf{f}(0)\,Q_r\big)=r^n|J_\mathbf{f}(0)|$. As $\mathbf{f}$ is differentiable at 0, for $r>0$ small,

$$E_r\overset{def}{=}\sup_{\mathbf{y}\in\mathrm{cl}\,B_{r\sqrt{n}\frac{r}{2}}}\frac{1}{\|\mathbf{y}\|}\|\mathbf{f}(\mathbf{y})-D\mathbf{f}(0)\mathbf{y}\|\to 0\,,\quad\text{when } r\to 0\,,$$

and $\|\mathbf{f}(\mathbf{x})-D\mathbf{f}(0)\mathbf{x}\|\leq\|\mathbf{x}\|E_r<r\sqrt{n}\,E_r$ for $\mathbf{x}\in Q_r$. Thus, $\mathbf{f}(Q_r)$ is contained in and contains parallelograms with edges with the same directions as each one of the edges of $D\mathbf{f}(0)\,Q_r$ with lengths, resp., $(1+2r\sqrt{n}\,E_r)$ and $(1-2r\sqrt{n}\,E_r)$ times the length of the corresponding edge of $D\mathbf{f}(0)\,Q_r$, and

$$(1+2\sqrt{n}\,E_r)^n r^n|J_\mathbf{f}(0)|\leq\mu\big(\mathbf{f}(Q_r)\big)\leq(1+2\sqrt{n}\,E_r)^n r^n|J_\mathbf{f}(0)|\,.$$

Hence, the n-volume derivative of $\mathbf{f}$ at 0 is $V'_\mathbf{f}(0)=\lim_{r\searrow 0}\frac{\mu(\mathbf{f}(Q_r))}{r^n}=|J_\mathbf{f}(0)|$.

It is $V'_\mathbf{f}=|J_\mathbf{f}|$ a.e. in Ω, and the Lebesgue theorem for volume derivative of a homeomorphism (A.26) can be applied. □

Appendix F
Invertible Continuous Functions and Global Inverse Function Theorem

This appendix proves that every invertible continuous function from an open subset of the plane to the plane is a homeomorphism onto its range, used in Chap. 3, and the global inverse function theorem, established in 1906 by J. Hadamard for functions from $\mathbb{R}^n$ to $\mathbb{R}^n$, used in Chap. 2. Both results are mainly topological. The first is proved with the invariance of domain theorem,[1] (i.e., images of open connected sets by continuous functions are open and connected), established for $\mathbb{R}^n$ in 1912 by L.E. Brouwer and can be proved with the Brouwer fixed point theorem.[2]

The inverse function theorem of differential calculus for C^k, $k \in \mathbb{N} \cup |, \{\infty\}$, functions of open subsets of $\mathbb{R}^n$ to $\mathbb{R}^n$ establishes the existence of a local C^k inverse in the neighborhood of every point where the derivative is invertible and continuous, but a function can satisfy this condition at all points without having an inverse on the whole set. The final results in this appendix are instances of a result obtained

[1] The name invariance of domain theorem implies that **domains** (i.e., connected open sets, also called **regions**) remain domains under transformations by continuous functions, since connectedness is preserved by such functions. A consequence of the invariance of domain theorem in $\mathbb{R}^n$ is the invariance of dimension theorem, establishing that subsets of $\mathbb{R}^n$ or $\mathbb{R}^m$ with nonempty interiors can be homeomorphic if and only if $m = n$.

[2] This theorem is equivalent to the impossibility of a retraction of a closed ball onto its boundary (i.e., of a continuous function of the ball onto its boundary with the boundary points fixed), and it is usually proved with algebraic topology. It was first proved for $n = 3$ in 1904 by Piers Bohl (1865–1921) and in 1909 by Luitzen Egbertus Brouwer (1881–1966). In 1910, it was proved for $n \in \mathbb{N}$ by J. Hadamard and by L.E. Brouwer. In 1978, J. Milnor gave an elementary proof with the hairy ball theorem, which in everyday language is the impossibility of "continuously combing" a ball in $\mathbb{R}^n$ for n odd (if n is even such a vector field exists, e.g., for $n = 2$, $\mathbf{f}(x, y) = (-y, x)$ for $(x, y) \in \mathbb{R}^2$ with $x^2 + y^2 = 1$), which had been first proved for $n = 3$ by L.E. Brouwer in 1912, but was proved by J. Milnor with elementary differential and integral calculus that a ball in $\mathbb{R}^n$ can be "C^1 combed" using the volume relation given by the change of integration variables theorem for multiple integrals with "continuous combing" obtained with polynomial approximations by the Stone-Weierstrass theorem in Milnor, J., Analytic proofs of the "Hairy Ball Theorem" and the Brouwer Fixed Point Theorem, *The Amer. Math. Monthly* **85** (1978), 521–524. Hadamard, Jacques (1865–1963). Stone, M. Harvey (1903–1989)

L. T. Magalhães, *Quasiconformal Mappings in the Plane and Complex Dynamics*,
https://doi.org/10.1007/978-3-031-80115-0

in 1906 by J. Hadamard in $\mathbb{R}^n$. They establish that for a C^k, $k \in \mathbb{N} \cup \{\infty\}$, from $\mathbb{R}^n$ to $\mathbb{R}^n$ with invertible derivative at each point to have a global C^k inverse, it is necessary and sufficient that it be a proper function and have a local C^k inverse in a neighborhood of each point (a simpler proof is given for $k \geq 2$, followed by another for $k \geq 1$ with topological arguments, namely, homotopy or coverings).

(F.1) Invariance of domain theorem in the plane: *Injective continuous functions from subsets of $\mathbb{R}^2$ to $\mathbb{R}^2$ map domains onto domains.*

Proof Let A be a domain in $\mathbb{R}^2$, $\mathbf{a} \in A$ and $f : A \to \mathbb{R}^2$ be injective and continuous. There is an open disk $B_r(\mathbf{a})$ with radius $r > 0$ and center $\mathbf{a}$ such that $\mathrm{cl}\, B_r(\mathbf{a}) \subset A$. As $J = \mathbf{f}(\partial B_r(\mathbf{a}))$ is a Jordan curve, by the Jordan curve theorem,[3] J separates $\mathbb{R}^2$ in two nonempty domains. As continuous functions map connected sets to connected sets,[4] $\mathbf{f}(B_r(\mathbf{a}))$ is a subset of one of these domains. If it were not the whole such domain containing it, $\mathbb{R}^2 \setminus \mathbf{f}(\mathrm{cl}\, B_r(\mathbf{a}))$ would intersect the two domains, and as it is connected, it would also intersect[4] J, what does not happen. Hence, $\mathbf{f}(B_r(\mathbf{a}))$ is one of the domains of $\mathbb{R}^2$ separated by J, which is a nonempty open subset of $\mathbf{f}(A)$ containing $f(\mathbf{a})$. Therefore, the set $\mathbf{f}(A)$ is open. As continuous functions map connected sets to connected sets, $\mathbf{f}(A)$ is connected. □

(F.2) *An injective continuous function from a domain $A \subset \mathbb{R}^2$ to $\mathbb{R}^2$ is a homeomorphism of A onto $\mathbf{f}(A)$.*

Proof By the preceding result, $\mathbf{f}$ maps open sets to open sets; thus, preimages of open sets by $\mathbf{f}^{-1}$ are open sets. Therefore, $\mathbf{f}^{-1}$ is continuous from $\mathbf{f}(A)$ onto A. □

A mapping $\mathbf{f}: \mathbb{R}^n \to \mathbb{R}^n$ is a **proper function** if it is continuous and preimages of compact sets are compact.

Since preimages of closed sets by a continuous function $\mathbf{f}: \mathbb{R}^n \to \mathbb{R}^n$ are closed sets and the compact subsets of $\mathbb{R}^n$ are the bounded closed sets, a necessary and sufficient condition for a function $\mathbf{f} : \mathbb{R}^n \to \mathbb{R}^n$ to be proper is that preimages of bounded sets be bounded, i.e., *a continuous function $\mathbf{f} : \mathbb{R}^n \to \mathbb{R}^n$ is proper if and only if $\|\mathbf{f}(\mathbf{x})\| \to +\infty$ when $\|\mathbf{x}\| \to +\infty$.*

(F.3) Global inverse function theorem for C^2 functions in $\mathbb{R}^n$:
$\mathbf{f}$: $\mathbb{R}^n \to \mathbb{R}^n$ C^k, $k \in \mathbb{N} \setminus \{1\}$, *is a C^k diffeomorphism onto $\mathbb{R}^n$ if and only if it is proper and a local diffeomorphism in a neighborhood of each point of $\mathbb{R}^n$.*

---

[3] See Section 8 of Appendix A of *CADOVA*.

[4] see Section 5 of Appendix A of *CADOVA*.

Proof The necessity is immediate.

Assume $\mathbf{f}:\mathbb{R}^n \to \mathbb{R}^n$ is proper and a local diffeomorphism. If it is known that $\mathbf{f}$ is bijective, by the inverse function theorem, $\mathbf{f}^{-1}$ is C^k in a neighborhood of every point, and $\mathbf{f}$ is a diffeomorphism of $\mathbb{R}^n$ onto $\mathbb{R}^n$.

To prove $\mathbf{f}$ is surjective, i.e., it assumes every value $\mathbf{b}\in\mathbb{R}^n$, without loss of generality, take $\mathbf{b}=0$ (it suffices to translate the variable). Consider the function $F(\mathbf{x})=\frac{1}{2}\|\mathbf{f}(\mathbf{x})\|^2$, with an inner product norm. If $M>F(0)$, as F is also a proper function, $F^{-1}([0,M])\neq\emptyset$ is compact. By the Weierstrass theorem of extrema of continuous functions, F assumes a minimum at some point $\mathbf{a}\in F^{-1}\big([0,M]\big)$, and $F(\mathbf{a})$ is a minimum of F in $\mathbb{R}^n$, i.e., $F(\mathbf{a})=0$; therefore, $\mathbf{f}(\mathbf{a})=0$. Hence, $\mathbf{f}$ is surjective.

To prove $\mathbf{f}=(f_1,\ldots,f_n)$ is injective, without loss of generality, take $\mathbf{f}(0)=0$ (it suffices translations in the domain and range), and it suffices to prove $\#\mathbf{f}^{-1}(\{0\})=1$. The set $\mathbf{f}^{-1}(\{0\})$ is bounded since $\mathbf{f}$ is a proper function, and it is closed since preimages of closed sets by continuous functions are closed relative to the domain, which in this case is a closed set; thus, the set $\mathbf{f}^{-1}(\{0\})$ is compact. Therefore, every sequence of points of $\mathbf{f}^{-1}(\{0\})$ has a subsequence converging to some point of $\mathbf{a}\in\mathbf{f}^{-1}(\{0\})$. By the continuity of $\mathbf{f}$, $\mathbf{f}(\mathbf{a})=0$ and, as $\mathbf{f}$ is locally injective, the sequence terms eventually coincide with $\mathbf{a}$; thus, the set $\mathbf{f}^{-1}(\{0\})$ is finite. Every point $\mathbf{p}$ of this set has a neighborhood U such that every solution of the differential equation $\mathbf{x}'(t)=-\nabla F\big(\mathbf{x}(t)\big)$, with t real, entering U remains in U and converges to $\mathbf{p}$ when $t\to+\infty$, since if U is such that $\mathbf{p}$ is its only point in $\mathbf{f}^{-1}(\{0\})$ and $\mathbf{x}$ is a solution, which exists since initial value problems for this differential equation have a unique solution because ∇F is C^1, then

$$(F\circ\mathbf{x})'=\big((\nabla F)\circ\mathbf{x}\big)\cdot\big(-(\nabla F)\circ\mathbf{x}\big)=-\|(\nabla F)\circ\mathbf{x}\|^2\leq 0;$$

as $\nabla F(\mathbf{y})=\sum_{k=1}^{n}\sum_{j=1}^{n}\frac{\partial f_j}{\partial y_k}f_k\,\mathbf{e}_k=[D\mathbf{f}]\mathbf{f}$ and the Jacobian matrix $D\mathbf{f}$ of $\mathbf{f}$ is nonsingular at all points, the above inequality is strict except at points where $\mathbf{f}$ vanishes; thus, $F\circ\mathbf{x}$ strictly decreases and converges to $\mathbf{p}$. Every solution of the differential equation converges to a point of $\mathbf{f}^{-1}(\{0\})$. If $A\subset\mathbb{R}^n$ is the set of initial values at $t=0$ of the differential equation solutions converging to $\mathbf{p}$, by the continuous dependence on initial conditions of the solutions of a differential equation $\mathbf{x}'(t)=\mathbf{g}(t)$ with $\mathbf{g}$ C^1 in $\mathbb{R}^n$, W is an open set. $\mathbb{R}^n$ is the union of the open sets of initial values at $t=0$ of the differential equation solutions converging to a point of $\mathbf{f}^{-1}(\{0\})$, and, as this is a finite set, $\mathbb{R}^n$ is a finite union of disjoint open sets; as it is a connected set, only one set of the union is nonempty, and $\#\mathbf{f}^{-1}(\{0\})=1$. □

(F.4) Global inverse function theorem for C^1 functions in $\mathbb{R}^n$:
$\mathbf{f}:\mathbb{R}^n\to\mathbb{R}^n$ C^k, $k\in\mathbb{N}$, *is a* C^k *diffeomorphism onto* $\mathbb{R}^n$ *if and only if it is a proper local diffeomorphism in a neighborhood of each point of* $\mathbb{R}^n$.

Proof We begin as in the proof of the preceding result for the proof of necessity and for surjectivity in the proof of sufficiency.

To prove injectivity take, without loss of generality, $\mathbf{f}(0)=0$, and consider a homotopy from $D\mathbf{f}(0)$ to $\mathbf{f}$ defined by $\mathbf{h}:[0,1]\times\mathbb{R}^n$ with $\mathbf{h}(\lambda,\mathbf{x})=\frac{1}{\lambda}\mathbf{f}(\lambda\mathbf{x})$ for $t\in\,]0,1]$, which extended by continuity to $\lambda=0$, by the 2nd order Taylor formula at 0, gives $\mathbf{h}(0,\mathbf{x})=[D\mathbf{f}(0)]\mathbf{x}$. The function $\mathbf{h}$ is C^1 in $]0,1]\times\mathbb{R}^2$ with partial derivatives $\frac{\partial\mathbf{h}}{\partial x_j}(\lambda,\mathbf{x})=\frac{\partial\mathbf{f}}{\partial x_j}(\lambda,\mathbf{x})$ and, by the 2nd order Taylor formula at 0, denoting $H_{\mathbf{f}}(0)$ the Hessian matrix of $\mathbf{f}$ at 0,

$$
\begin{aligned}
\frac{\partial\mathbf{h}}{\partial\lambda}(\lambda,\mathbf{x}) &= \frac{1}{\lambda}\,[D\mathbf{f}(\lambda\mathbf{x})]\mathbf{x}-\frac{1}{\lambda^2}\mathbf{f}(\lambda\mathbf{x})\\
&= \frac{1}{\lambda^2}\Big([D\mathbf{f}(\lambda\mathbf{x})]\lambda\mathbf{x}-[D\mathbf{f}(0)]\lambda\mathbf{x}+\frac{1}{2}(\lambda\mathbf{x})^t H_{\mathbf{f}}(0)(\lambda\mathbf{x})+o(\lambda^2)\Big)\\
&= \frac{1}{\lambda}\Big([D\mathbf{f}(\lambda\mathbf{x})]-[D\mathbf{f}(0)]\Big)\mathbf{x}+\frac{1}{2}\,\mathbf{x}^t H_{\mathbf{f}}(0)\,\mathbf{x}+o(1)\\
&= \frac{1}{\lambda}\,o(\lambda^2)+\frac{1}{2}\,\mathbf{x}^t H_{\mathbf{f}}(0)\,\mathbf{x}+o(1) = \frac{1}{2}\,\mathbf{x}^t H_{\mathbf{f}}(0)\,\mathbf{x}+o(1)\,,\quad \lambda\to 0;
\end{aligned}
$$

thus, $\mathbf{h}$ is C^1, with $\frac{\partial\mathbf{h}}{\partial\lambda}(0,\mathbf{x})=\frac{1}{2}\,\mathbf{x}^t H_{\mathbf{f}}(0)\,\mathbf{x}$.

If $\mathbf{f}(\mathbf{x}_1)=\mathbf{f}(\mathbf{x}_2)\overset{def}{=}\mathbf{y}$, as $\frac{\partial\mathbf{h}}{\partial\mathbf{x}}(\lambda,\mathbf{x})=D\mathbf{f}(\lambda,\mathbf{x})$ is a nonsingular matrix for all $(\lambda,\mathbf{x})$, the equation $\mathbf{h}(\lambda,\mathbf{x})=\mathbf{y}$ implicitly defines $\mathbf{x}$ as C^1 function of λ in a neighborhood of each point $(\lambda_0,\mathbf{x}_0)$ satisfying the equation. Considering a cover of the compact interval $[0,1]$ with open intervals where these functions are defined, as there is a finite subcover of $[0,1]$, the equation defines implicitly $\mathbf{x}=\widetilde{\mathbf{x}}_j(\lambda)$ as a function of $\lambda\in[0,1]$ with $\widetilde{\mathbf{x}}_j(1)=\mathbf{x}_j$, $j=1,2$. If $\mathbf{x}_1\neq\mathbf{x}_2$, as $\mathbf{h}(0,\mathbf{x}_j)=[D\mathbf{f}(0)]\mathbf{x}_j$ for $j=1,2$, and $D\mathbf{f}(0)$ is nonsingular, it would be $\mathbf{h}(0,\mathbf{x}_1)\neq\mathbf{h}(0,\mathbf{x}_2)$. The curves would intersect contradicting the inverse function theorem applied at the intersection point with smallest value of λ. So, $\mathbf{f}(\mathbf{x}_1)=\mathbf{f}(\mathbf{x}_2)\Rightarrow\mathbf{x}_1=\mathbf{x}_2$, and $\mathbf{f}$ is injective. □

With coverings[5] we have an alternative shorter proof of injectivity, which also holds for functions between finite dimensional differential manifolds and for local homeomorphisms. So, we prove it in this case.

(F.5) *Every proper function $\mathbf{f}:\mathbb{R}^n\to\mathbb{R}^n$ local homeomorphism in a neighborhood of each point with $\mathbf{f}(\mathbb{R}^n)$ simply connected is injective.*

Proof For every $\mathbf{b}\in\mathbf{f}(\mathbb{R}^n)$, $\mathbf{f}^{-1}(\{\mathbf{b}\})$ is compact, and if it were an infinite set, it would have an accumulation point, contradicting the injectivity of $\mathbf{f}$ in a neighborhood of each point. Thus, $\mathbf{f}^{-1}(\{\mathbf{b}\})=\{\mathbf{x}_1,\ldots,\mathbf{x}_N\}$ with $N\in\mathbb{N}$. Let $K\subset\mathbb{R}^n$ be a compact set with $\mathbf{b}$ in its interior. $\mathbf{f}^{-1}(K)$ is compact and contains disjoint neighborhoods U_j of $\mathbf{x}_j$ such that the restriction of $\mathbf{f}$ to U_j is a homeomorphism of U_j onto $\mathbf{f}(U_j)$, which is a neighborhood of $\mathbf{b}$ for $j=1,\ldots,N$. $V=\big[\cap_{j=1}^N\mathbf{f}(U_j)\big]\backslash\mathbf{f}\big(\mathbf{f}^{-1}(K)\backslash\cup_{j=1}^N U_j\big)$ is a neighborhood of $\mathbf{b}$, $\mathbf{f}^{-1}(V)$ is a disjoint union of open sets, and the restriction of $\mathbf{f}$ to each one of them is a homeomorphism onto V. Therefore, $\mathbf{f}$ is a covering of $\mathbf{f}(\mathbb{R}^n)$, and as this is a simply connected set, $(\mathbb{R}^n,\mathbf{f})$ is a universal covering of $\mathbf{f}(\mathbb{R}^n)$, and $\mathbf{f}$ is injective. □

[5] See Section 2 of Chapter 12 of *CADOVA* for Riemann surfaces.

Appendix G
Projections and Separating Hyperplanes of Convex Sets

This appendix considers basic aspects of convex sets in order to establish a property of these sets used in one of the last sections of Chap. 3, as part of the proof that δ-monotone functions preserving orientation and nonconstant in a region of $\mathbb{C}$ are quasiconformal mappings.

A **convex set** in $\mathbb{R}^n$ is $\emptyset \neq S \subset \mathbb{R}^n$ with every line segment with end points $\mathbf{u}, \mathbf{v} \in S$ is included in S, i.e., $(1-s)\mathbf{u}+s\mathbf{v} \in S$ for $s \in [0, 1]$ (Fig. G.1).

$\mathbb{R}^n$ is considered in this appendix a Euclidean space with an arbitrary inner product. If $\emptyset \neq F \subset \mathbb{R}^n$ is a closed convex set and $\mathbf{a} \in \mathbb{R}^n$, the **distance of the point a to the set** F is $d_F(\mathbf{a}) = \inf\{\|\mathbf{v}-\mathbf{a}\| : \mathbf{v} \in \mathbb{R}^n\}$.

(G.1) *If* $\emptyset \neq F \subset \mathbb{R}^n$ *is a closed convex set and* $\mathbf{a} \in \mathbb{R}^n$*, there is a unique point* $\mathbf{f} \in F$ *at a minimal distance of a point* $\mathbf{a}$.

Proof If the distance of a point $\mathbf{a} \in \mathbb{R}^n$ to a closed convex subset F of $\mathbb{R}^n$ is assumed at points $\mathbf{x}_1$ and $\mathbf{x}_2$ of F, i.e., if $\|\mathbf{x}_1 - \mathbf{a}\| = \|\mathbf{x}_2 - \mathbf{a}\| = d_F(\mathbf{a})$, then $\frac{1}{2}(\mathbf{x}_1 + \mathbf{x}_2) \in F$ since it is the middle point of the line segment from $\mathbf{x}_1$ to $\mathbf{x}_2$ and F is convex. By the parallelogram law,

$$\|\mathbf{x}_1 - \mathbf{x}_2\|^2 + \|\mathbf{x}_1 + \mathbf{x}_2 - 2\mathbf{a}\|^2 = 2\|\mathbf{x}_1 - \mathbf{a}\|^2 + 2\|\mathbf{x}_2 - \mathbf{a}\|^2 = 4\, d_F^2(\mathbf{a})\,;$$

thus,

$$\|\mathbf{x}_1 - \mathbf{x}_2\|^2 \leq 4 d_F^2(\mathbf{a}) - 4\|\tfrac{1}{2}(\mathbf{x}_1 + \mathbf{x}_2) - \mathbf{a}\|^2 \leq 4\, d_F^2(\mathbf{a}) - 4\, d_F^2(\mathbf{a}) = 0\,,$$

and, therefore, $\mathbf{x}_1 = \mathbf{x}_2$. Hence, there cannot exist more than one point of F at the minimal distance of $\mathbf{a}$. As $d_F(\mathbf{a}) = \inf\{\|\mathbf{v}-\mathbf{a}\| : \mathbf{v} \in \mathbb{R}^n\}$, there is a sequence $\{\mathbf{v}_n\} \subset F$ such that $\|\mathbf{v}_n - \mathbf{a}\| \to d_F(\mathbf{a})$, and, by the triangular inequality and the parallelogram law,

$$\|\mathbf{v}_m - \mathbf{v}_n\|^2 + 4\, d_F^2 \leq \|\mathbf{v}_m - \mathbf{v}_n\|^2 + 4\|\tfrac{1}{2}(\mathbf{v}_m + \mathbf{v}_n) - \mathbf{a}\|^2 = 2\|\mathbf{v}_m - \mathbf{a}\|^2 + 2\|\mathbf{v}_n - \mathbf{a}\|^2.$$

L. T. Magalhães, *Quasiconformal Mappings in the Plane and Complex Dynamics*,
https://doi.org/10.1007/978-3-031-80115-0

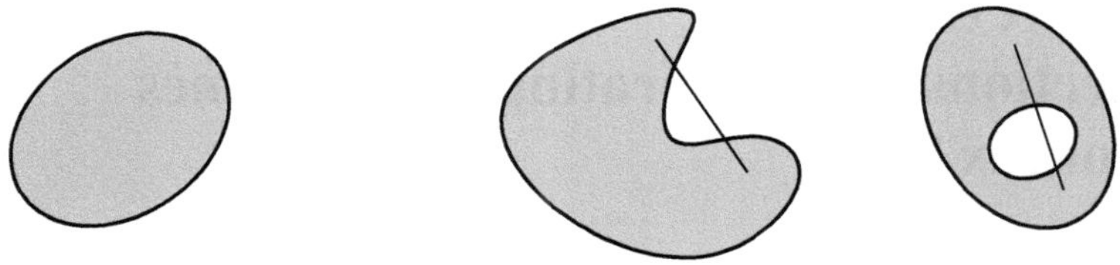

Fig. G.1 Examples of a convex set and of nonconvex sets in a plane

When $m > n \to 0$ the right-hand side of this inequality converges to $4d_F^2$; thus, $\|\mathbf{v}_m - \mathbf{v}_n\| \to 0$ and $\{\mathbf{v}_n\} \subset F$ is a Cauchy sequence. Since $\mathbb{R}^n$ with the usual distance is a complete space, this sequence converges to a point in $\mathbf{f} \in \mathbb{R}^n$, and as F is a closed set, $\mathbf{f} \in F$. As the norm is continuous, $\|\mathbf{f} - \mathbf{a}\| = \lim \|\mathbf{v}_n - \mathbf{a}\| = d_F(\mathbf{a})$. □

The **projection on a closed convex set** $\emptyset \neq F \subset \mathbb{R}^n$ is defined by $P_F : \mathbb{R}^n \to F$ such that $P_F(\mathbf{v})$ is a point of F at minimal distance from $\mathbf{v}$ (Fig. G.2).

(G.2) *If $\emptyset \subset F \subset \mathbb{R}^n$ is a closed convex set, then $\langle \mathbf{a} - P_F(\mathbf{a}), \mathbf{f} - P_F(\mathbf{a})\rangle \leq 0$ for $\mathbf{a} \in \mathbb{R}^n, \mathbf{f} \in F$, and $P_F(\mathbf{a})$ is the unique $\mathbf{u} \in F$ with $\langle \mathbf{a} - \mathbf{u}, \mathbf{f} - \mathbf{u}\rangle \leq 0, \mathbf{f} \in F$.*

Proof If $\mathbf{a}, \mathbf{f} \neq P_F(\mathbf{a})$, then $s = \frac{\langle \mathbf{a} - P_F(\mathbf{a}), \mathbf{f} - P_F(\mathbf{a})\rangle}{\|\mathbf{f} - P_F(\mathbf{a})\|^2} > 0$, and, by the Cauchy-Schwarz inequality, $s \leq 1$, since

$$\langle \mathbf{a} - P_F(\mathbf{a}), \mathbf{f} - P_F(\mathbf{a})\rangle \leq \|\mathbf{a} - P_F(\mathbf{a})\| \, \|\mathbf{f} - P_F(\mathbf{a})\| \leq \|\mathbf{f} - P_F(\mathbf{a})\|^2.$$

Hence, $s\mathbf{f} + (1-s)P_F(\mathbf{a})$ belongs to the convex set F and

$$\|s[\mathbf{f} - P_F(\mathbf{a})] + (P_F(\mathbf{a}) - \mathbf{a})\|^2 = \|[s\mathbf{f} + (1-s)P_F(\mathbf{a})] - \mathbf{a}\|^2 \geq \|P_F(\mathbf{a}) - \mathbf{a}\|^2.$$

Evaluating the norm in the left-hand side of this inequality with the inner product gives the equivalent inequality

$$s^2\|\mathbf{f} - P_F(\mathbf{a})\|^2 + 2s\langle \mathbf{f} - P_F(\mathbf{a}), P_F(\mathbf{a}) - \mathbf{a}\rangle + \|P_F(\mathbf{a}) - \mathbf{a}\|^2 \geq \|P_F(\mathbf{a}) - \mathbf{a}\|^2.$$

Since $\langle \mathbf{f} - P_F(\mathbf{a}), P_F(\mathbf{a}) - \mathbf{a}\rangle = -s\|\mathbf{f} - P_F(\mathbf{a})\|^2$ and $\|\mathbf{f} - P_F(\mathbf{a})\|^2 \neq 0$, if $\mathbf{f} \neq P_F(\mathbf{a})$, the last inequality is equivalent to $-s^2 \geq 0$. Hence, $\langle \mathbf{a} - P_F(\mathbf{a}), \mathbf{f} - P_F(\mathbf{a})\rangle \leq 0$, for $\mathbf{a} \in \mathbb{R}^n, \mathbf{f} \in F$. If $\mathbf{a} \in \mathbb{R}^n$ and $\mathbf{u} \in F$, then $\langle \mathbf{a} - P_F(\mathbf{a}), \mathbf{u} - P_F(\mathbf{a})\rangle \leq 0$, which is equivalent to

$$\langle \mathbf{a}, \mathbf{u}\rangle - \langle \mathbf{a}, P_F(\mathbf{a})\rangle - \langle P_F(\mathbf{a}), \mathbf{u}\rangle + \|P_F(\mathbf{a})\|^2 \leq 0,$$

Fig. G.2 Projection on a closed convex set

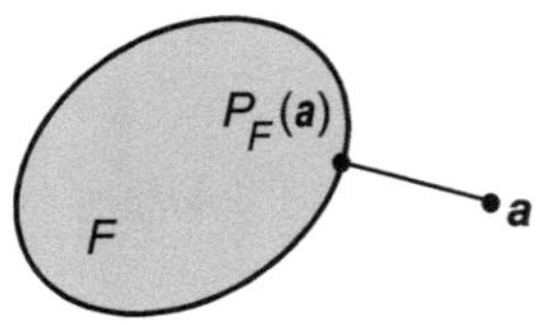

and, therefore,

$$\langle \mathbf{a}-\mathbf{u},\, P_F(\mathbf{a})-\mathbf{u}\rangle = \langle \mathbf{a},\, P_F(\mathbf{a})\rangle - \langle \mathbf{a},\mathbf{u}\rangle - \langle \mathbf{u},\, P_F(\mathbf{a})\rangle + \|\mathbf{u}\|^2$$
$$\geq -\langle P_F(\mathbf{a}),\mathbf{u}\rangle + \|P_F(\mathbf{a})\|^2 - \langle \mathbf{u},\, P_F(\mathbf{a})\rangle + \|\mathbf{u}\|^2 = \|\mathbf{u}-P_F(\mathbf{a})\|^2.$$

Since $P_F(\mathbf{a}) \in F$, if $\mathbf{u} \neq P_F(\mathbf{a})$, it is not $\langle \mathbf{a}-\mathbf{u}, \mathbf{f}-\mathbf{u}\rangle \leq 0$, with $\mathbf{f} \in F$. □

Thus, the condition defining the projection on a closed convex set $F \subset \mathbb{R}^n$, $\|P_F(\mathbf{a})-\mathbf{a}\| \leq \|\mathbf{f}-\mathbf{a}\|$ for $\mathbf{f} \in F$, is equivalent to the following: for every $\mathbf{a} \in \mathbb{R}^n$, $P_F(\mathbf{a})$ is the unique solution $\mathbf{u}$ of the **variational inequality** $\langle \mathbf{a}-\mathbf{u}, \mathbf{f}-\mathbf{u}\rangle \leq 0$, for $\mathbf{f} \in F$. Geometrically, this condition is as follows: the angles of $\mathbf{a}-P_F(\mathbf{a})$ and $\mathbf{f}-P_F(\mathbf{a})$ cannot be acute (*i.e.*, they must be $\geq \frac{\pi}{2}$), for $\mathbf{a} \in \mathbb{R}^n \backslash F$, $\mathbf{f} \in F$ (Fig. G.3).

A **hyperspace** in $\mathbb{R}^n$ is a linear subspace of $\mathbb{R}^n$ orthogonal complement of a vector $\mathbf{n} \in \mathbb{R}^n \backslash \{0\}$, and a **hyperplane** in $\mathbb{R}^n$ is a set $H_c = \{\mathbf{x} \in \mathbb{R}^n : \langle \mathbf{n}, \mathbf{x}\rangle = c\}$ for some fixed $c \in \mathbb{R}$, *i.e.*, a subset of $\mathbb{R}^n$ with Cartesian equation $\langle \mathbf{n}, \mathbf{x}\rangle = c$. It is said that it is the hyperplane orthogonal to $\mathbf{n}$ and also that it is an hyperplane parallel to the considered hyperspace.

If $n \in \mathbb{N}$, a hyperspace in $\mathbb{R}^n$ has dimension $n-1$, and it is said that it has **codimension 1**.

A set $\{\mathbf{x} \in \mathbb{R}^n : \langle \mathbf{n}, \mathbf{x}\rangle \geq c\}$ or $\{\mathbf{x} \in \mathbb{R}^n : \langle \mathbf{n}, \mathbf{x}\rangle \leq c\}$, for some fixed $c \in \mathbb{R}$, is called **closed halfspace** of $\mathbb{R}^n$ defined by a hyperplane $H = \{\mathbf{x} \in \mathbb{R}^n \text{: } \langle \mathbf{n}, \mathbf{x}\rangle = c\}$. For every hyperplane $H \subset \mathbb{R}^n$, $\mathbb{R}^n$ is the union of the two closed half spaces separated by H and with intersection H.

If $\emptyset \neq F \subset \mathbb{R}^n$, we say that a **hyperplane** H **supports** F or it is a **supporting hyperplane** of F if $H \cap F \neq \emptyset$ and F is contained in the same of the two closed halfspaces defined by H (Fig. G.4).

The supporting hyperplanes are for the boundaries of closed convex sets analogous to tangent planes for graphs of differentiable functions.

(G.3) *Every closed convex set* $\emptyset \neq F \subset \mathbb{R}^n$ *has a supporting plane; for every* $\mathbf{a} \in \mathbb{R}^n \backslash F$, $H = \{\mathbf{x} \in \mathbb{R}^n : \langle \mathbf{n}, \mathbf{x}\rangle = \langle \mathbf{n}, P_F(\mathbf{a})\rangle\}$ *with* $\mathbf{n} = \mathbf{a} - P_F(\mathbf{a})$ *is a supporting hyperplane of* F *containing* $P_F(\mathbf{a})$.

Proof If $\mathbf{a} \in \mathbb{R}^n \setminus F$, with the above variational inequality and $\mathbf{n} = \mathbf{a} - P_F(\mathbf{a})$, it is $\langle \mathbf{n}, \mathbf{f}\rangle \leq \langle \mathbf{n}, P_F(\mathbf{a})\rangle$ for $\mathbf{f} \in F$ (Fig. G.4). □

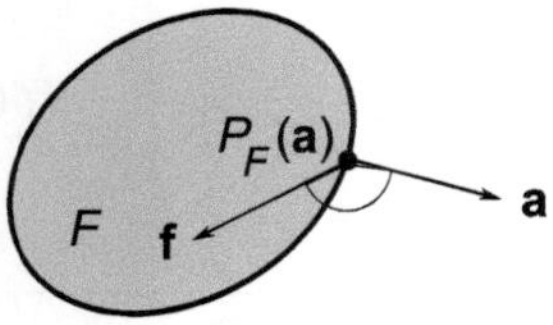

Fig. G.3 Nonacute angles of $\mathbf{a}-P_F(\mathbf{a})$ and $\mathbf{f}-P_F(\mathbf{a})$, with $\mathbf{a} \in \mathbb{R}^n \setminus F$, $\mathbf{f} \in F$ and $\emptyset \neq F \subset \mathbb{R}^n$ a closed convex set

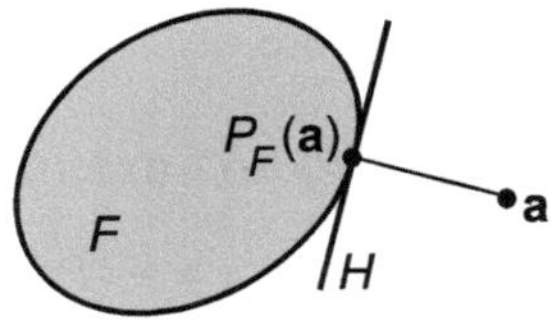

Fig. G.4 Hyperplane H supporting the closed convex subset $F \neq \emptyset$ of $\mathbb{R}^n$ containing the projection of $\mathbf{a} \in \mathbb{R}^n \setminus F$ on the closed convex set F

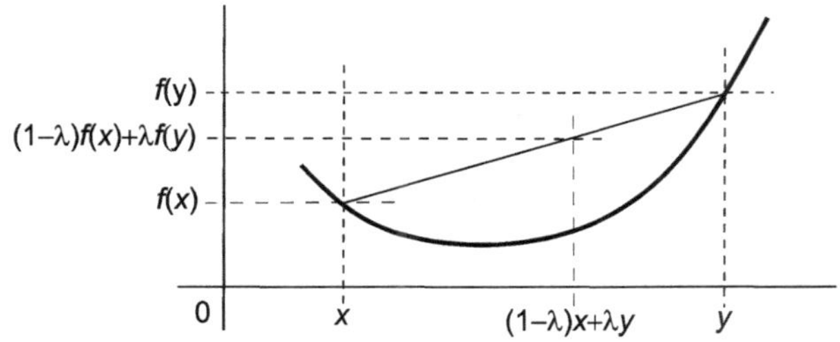

Fig. G.5 Convex function

A **convex function** in $D \subset \mathbb{R}^n$ is a function $f : D \to \mathbb{R}$ such that

$$f\big((1-\lambda)\mathbf{x}+\lambda\mathbf{y}\big) \leq (1-\lambda)\, f(\mathbf{x})+\lambda f(\mathbf{y})\,, \quad \mathbf{x}, \mathbf{y} \in D\,,\ \lambda \in [0,\, 1]\,.$$

Geometrically, convex functions are those with graphs over every striaght line segment in the domain lying below the straight line segment connecting the points in the graph corresponding to images of the end points of the line segment in the domain (Fig. G.5).

(G.4) *Convex functions* $f : D \to \mathbb{R}$, *with* $D \subset \mathbb{R}$, $\operatorname{int} D \neq \emptyset$, *are continuous in* $\operatorname{int} D$, *differentiable except possibly at a countable subset of* $\operatorname{int} D$, *and have lateral derivatives at left and right, resp.,* f_e', f_d', *in* $\operatorname{int} D$, *which are laterally continuous functions, resp., at left and right, such that* $f_e' \leq f_d'$, *and these functions are nondecreasing in* $\operatorname{int} D$. f' *exists at points where* f_e' *and* f_d' *are both continuous. If* f' *exists in an open set* $S \subset \operatorname{int} D$, f *is* C^1 *in* S. *If* f'' *exists at* $c \in \operatorname{int} D$, *then* $f''(c) \geq 0$.

Proof If $f : D \to \mathbb{R}$, with $D \subset \mathbb{R}^n$ and $\operatorname{int} D \neq \emptyset$, is a convex function in D, S is an n-simplex with vertices $\mathbf{s}_0, \ldots, \mathbf{s}_n$ in[1] $\mathbb{R}^n$ such that $\mathbf{a} \in \operatorname{int} S \subset S \subset \operatorname{int} D$ and $B_\delta(\mathbf{a}) \subset S$, with $\delta > 0$, then for every point $\mathbf{x} \in S$, there are $\lambda_0, \ldots, \lambda_n \geq 0$ with $\sum_{j=0}^{n} \lambda_j = 1$ such that $\mathbf{x} = \sum_{j=1}^{n+1} \lambda_j \mathbf{s}_j$, and by the convexity of f,

$$f(\mathbf{x}) \leq \sum_{j=0}^{n} \lambda_j\, f(\mathbf{s}_j) \leq \max_{j \in \{0,\ldots,n\}} f(\mathbf{s}_j)\,.$$

[1] A p-**simplex** with vertices $\mathbf{a}_0, \ldots, \mathbf{a}_p$ in $\mathbb{R}^n$ and $p \in [\mathbf{n}] \cup \{0\}$ is a set $S = \big\{ \sum_{j=0}^{p} c_j \mathbf{a}_j : \sum_{j=0}^{p} c_j = 1\,,\, c_j \in [0,\, 1] \big\}$ with $\{\mathbf{a}_j - \mathbf{a}_0 : j \in [\mathbf{p}]\}$ linearly independent.

With $\mathbf{y}=\mathbf{a}+t\mathbf{v}$, where $t=\frac{\|\mathbf{y}-\mathbf{a}\|}{\|\mathbf{v}\|}\in[0,1]$ and $\|\mathbf{v}\|=\delta$, by the convexity of f,

$$\begin{aligned} f(\mathbf{y}) &= f\big((1-t)\mathbf{a}+t(\mathbf{a}+\mathbf{v})\big) \le (1-t)f(\mathbf{a})+tf(\mathbf{a}+\mathbf{v}) \\ &\le (1-t)f(\mathbf{a})+t\max_{j\in\{0,\dots,n\}} f(\mathbf{s}_j), \end{aligned}$$

and, therefore,

$$f(\mathbf{y})-f(\mathbf{a}) \le t\Big(\max_{j\in\{0,\dots,n\}} f(\mathbf{s}_j)-f(\mathbf{a})\Big).$$

From the convexity of f, it also follows that

$$\begin{aligned} f(\mathbf{a}) = f\big(\tfrac{1}{1+t}\mathbf{y}+\tfrac{t}{1+t}(\mathbf{a}-\mathbf{v})\big) &\le \tfrac{1}{1+t}f(\mathbf{y})+\tfrac{t}{1+t}f(\mathbf{a}-\mathbf{v}) \\ &\le \tfrac{1}{1+t}f(\mathbf{y})+\tfrac{t}{1+t}\max_{j\in\{0,\dots,n\}} f(\mathbf{s}_j); \end{aligned}$$

thus,

$$-t\Big(\max_{j\in\{0,\dots,n\}} f(\mathbf{s}_j)-f(\mathbf{a})\Big) \le f(\mathbf{y})-f(\mathbf{a}).$$

The inequalities of the two preceding periods give, for $\mathbf{y}\in B_\delta(\mathbf{a})$,

$$|f(\mathbf{y})-f(\mathbf{a})| \le t\Big(\max_{j\in\{0,\dots,n\}} f(\mathbf{s}_j)-f(\mathbf{a})\Big) = \tfrac{\|\mathbf{y}-\mathbf{a}\|}{\delta}\Big(\max_{j\in\{0,\dots,n\}} f(\mathbf{s}_j)-f(\mathbf{a})\Big);$$

thus, f is continuous, $\mathbf{a}\in\operatorname{int}D$, and therefore, it is continuous in $\operatorname{int}D$.

If $f: D\to\mathbb{R}$ with $D\subset\mathbb{R}$ a convex function in D and $a\in\operatorname{int}D$, for $0<\lambda<\mu$ with $\mu>0$ small,

$$f(a+\lambda) = f\big(\tfrac{\mu-\lambda}{\mu}a+\tfrac{\lambda}{\mu}(a+\mu)\big) \le \tfrac{\mu-\lambda}{\mu}f(a)+\tfrac{\lambda}{\mu}f(a+\mu),$$

$$\frac{f(a+\lambda)-f(a)}{\lambda} \le \frac{f(a+\mu)-f(a)}{\mu}.$$

For $0<\lambda<\mu$ and $\mu>0$ small, also

$$f(a) = f\big(\tfrac{\lambda}{\lambda+\mu}(a-\mu)+\tfrac{\mu}{\lambda+\mu}(a+\lambda)\big) \le \tfrac{\lambda}{\lambda+\mu}f(a-\mu)+\tfrac{\mu}{\lambda+\mu}f(a+\lambda),$$

and, therefore,

$$\frac{f(a)-f(a-\mu)}{\mu} \le \frac{f(a+\lambda)-f(a)}{\lambda}.$$

Hence, for $\lambda>0$ small the function $\lambda\mapsto\frac{f(a+\lambda)-f(a)}{\lambda}$ is defined, is nondecreasing, and is lower bounded; therefore, it converges when $\lambda\to0^+$ to some value, and the right derivative $f'_d(a)$ of f at a exists. Analogously, we get existence of the left derivative $f_e'(a)$ at a, and from the last inequality above, $f_e'(a) \le f_d'(a)$. Therefore, for $a, b \in \operatorname{int} D$ with $a < b$,

$$f_e'(a) \le f_d'(a) \le \frac{f(b)-f(a)}{b-a} \le f_e'(b) \le f_d'(b);$$

thus, f_e' and f_d' are nondecreasing functions in int D. At every point of discontinuity a of one of these functions, f has a jump discontinuity, and it is possible to choose a rational number strictly between the lateral limits of f at a. Thus, there is a bijection of the set of points of discontinuity of f to a subset of the rational numbers (which is countable); therefore, the sets of points of discontinuity of f_e' as well as those of f_d' are countable. Hence, f_e', f_d' are continuous in int D except possibly in a countable subset. By the last inequality above, at points a where f_e' and f_d' are both continuous it is $f_e'(a) = f_d'(a)$, i.e., f is differentiable and $f'(a) = f_e'(a) = f_d'(a)$. By this inequality and the continuity of convex functions at points of the interior of their domain established in the preceding paragraph, for $a < c < b$,

$$\lim_{c\to a^+} f_d'(c) \le \lim_{c\to a^+} \frac{f(b)-f(c)}{b-c} = \frac{f(b)-f(a)}{b-a};$$

thus, $\lim_{c\to a^+} f_d'(c) \le f_d'(a)$, and, as $f_d'(a) \le f_d'(c)$ for $c > a$, $f_d'(a) \le \lim_{c\to a^+} f_d'(c)$. Hence, $\lim_{c\to a^+} f_d'(c) = f_d'(a)$, and, similarly, $\lim_{c\to a^-} f_e'(c) = f_e'(a)$. If $f''(c)$ exists, as f' exists and is nondecreasing in an interval with c, then $f''(c) \ge 0$. □

It is now proved a statement used by the end of Chap. 3.

(G.5) *If $\emptyset \ne Y \subsetneq \mathbb{R}^n$ is a closed convex set with* $\operatorname{int} Y \ne \emptyset$ *and* $\mathbf{u}, \mathbf{v} \in \mathbb{R}^n$ *are linearly independent vectors, then there is an affine function $A: \mathbb{R}^n \to \mathbb{R}^n$ such that $A(0) \notin Y$ and $A(t\mathbf{u})$, $A(t\mathbf{v}) \in Y$ for some $t > 0$. If $\emptyset \ne \Omega \subset \mathbb{R}^n$ is a region and $\emptyset \ne Y \subsetneq \mathbb{R}^n$ is the intersection of a convex set with Ω, then there is an affine function satisfying those conditions with $A(0) \in \Omega$.*

Proof Let $\mathbf{a} \in \partial Y$ and Y' be the intersection of Y with the plane parallel to $\mathbf{u}$ and $\mathbf{v}$ containing $\mathbf{a}$. Y' is a convex subset of that plane and $\mathbf{a} \in \partial Y'$. So, it suffices to prove the result for $\mathbb{R}^2$, as considered below.

For every $\mathbf{a} \in \partial Y$, there is a supporting straight line of the closed convex set Y containing $\mathbf{a}$. Without loss of generality, $\mathbf{a} = 0$, and the supporting straight line through $\mathbf{a}$ is the abscissa axis of $\mathbb{R}^2$, since this can be achieved with a translation of $\mathbf{a}$ to $0 \in \mathbb{R}^2$ followed by a rotation around 0, with the same rotation applied to $\mathbf{u}$ and $\mathbf{v}$. Since $\operatorname{int} Y \ne \emptyset$, there is an open disk $B_r(\mathbf{y}) \subset Y$ with $r > 0$ and $\mathbf{y} \in Y$ and, as Y is convex, the part of the sector $\{t\mathbf{z} : t \in]0, 1], \mathbf{z} \in B_r(\mathbf{y})\}$ is included in Y; thus, there is an interval $]a, b[\subset \mathbb{R}$ whose closure contains 0 such that for all $x \in]a, b[$ the vertical line containing $(x, 0)$ intersects ∂Y. Denoting (x, y) the point of ∂Y with the smallest ordinate, we get a convex function $f :]a, b[\to \mathbb{R}$ such that $f(x) = y$. By

(G.4), f is differentiable except possibly in a countable set. Let $x_0 \in]a, b[$ be a point of differentiability of f. The straight line tangent to the graph of f at $\big(x_0, f(x_0)\big)$ is the unique supporting line of Y containing this point. Apply a translation moving this point to 0, followed by a rotation around 0 so that this supporting line Y is orthogonal to $\mathbf{u}+\mathbf{v}$ and this vector points to the interior of Y. Due to the tangency, $\{t > 0 : t\mathbf{u}, t\mathbf{v} \in Y\}$ contains an interval $]0, T]$ with $T > 0$. With a small translation displace the origin in the direction opposite to that of the vector $\mathbf{u}+\mathbf{v}$.

The used affine mappings allow to define an invertible affine mapping A^{-1} in $\mathbb{R}^n$ with inverse A such that $A(0) \notin Y$. If the last translation of the origin considered is sufficiently small, for t in some interval $[a, b]$ with $0 < a < b$ and $b - a < T$ it is $A(t\mathbf{u}), A(t\mathbf{v}) \in Y$ (Fig. G.6).

Clearly, $\emptyset \neq Y \subsetneqq \mathbb{R}^n$ is the intersection of the convex set Y with $\emptyset \neq \Omega \subset \mathbb{R}^n$, and A can be defined with a arbitrarily close to 0, so that, additionally, $A(0) \in \Omega$. □

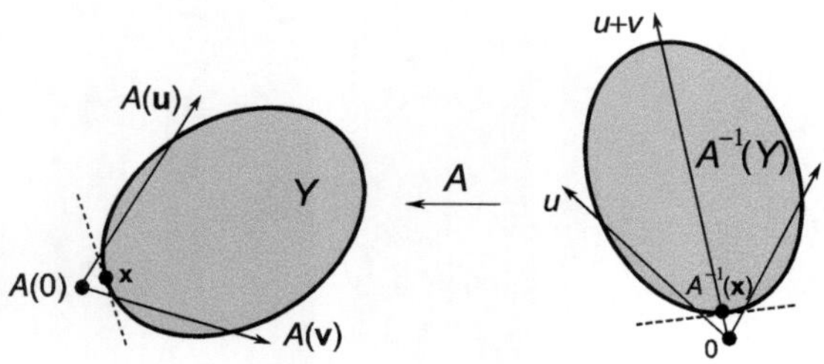

Fig. G.6 Affine function A such that for $\emptyset \neq Y \subsetneqq \mathbb{R}^2$ a convex set and $\mathbf{u}, \mathbf{v} \in \mathbb{R}^2$ linearly independent, $A(0) \notin Y$ and $A(t\mathbf{u}), A(t\mathbf{v}) \in Y$ for $t \in [a, b]$ with $0 < a < b$

Appendix H
Extension of Schwarz Lemma

The proof of the harmonic λ-lemma (4.19) in Chap. 4 uses an extension of the Schwarz lemma for holomorphic functions of $\mathbb{C}$ to a finite dimensional complex Banach space, which is a particular case of the extension of the Schwarz lemma to differentiable functions from complex Banach spaces to finite dimensional spaces.

The simplest classic version of the Schwarz lemma is:[1] *If* $B_1 = B_1(0) \subset \mathbb{C}$, $f : B_1 \to B_1$ *is holomorphic and* $f(0) = 0$, *then* $|f(z)| \leq |z|$ *for* $z \in B_1$.

For Y a Banach space, denote Y^* its **dual space**, i.e., the linear space of the bounded linear transformations from Y to $\mathbb{C}$ with the operator norm

$$\|h\|_{Y^*} = \sup_{y \in B_Y} |h(y)|, \quad h \in Y^*.$$

(H.1) *If* X, Y *are complex Banach spaces,* B_X, B_Y *are the resp. open balls centered at* 0 *with radius* 1 *and* $f : B_X \to B_Y$ *is differentiable with* $f(0) = 0$, *then* $\|f(z)\|_Y \leq \|z\|_X$ *for* $z \in B_X$.

Proof Let $z \in B_X \backslash \{0\}$ be such that $f(z) \neq 0$, and for $b = \frac{f(z)}{\|f(z)\|_Y}$, denote

$$\mathcal{H}_b = \left\{ H \in Y^* : H(b) = \|b\|_Y, \ \|H\|_{Y^*} = 1 \right\}.$$

By the result following this proof, $\mathcal{H}_b \neq \emptyset$. For each $h \in \mathcal{H}_b$, it is $h(0) = 0$ as h is a linear transformation, and since $|h(y)| \leq \|h\|_{Y^*} \|y\|_Y$ for $y \in Y$, if $\|y\|_Y < 1$, then $|h(y)| < 1$. The function $F : B_1 \to \mathbb{C}$ with $F(s) = h\big(f\big(\frac{s}{\|z\|_X} z\big)\big)$ is holomorphic, $F(0) = h\big(f(0)\big) = h(0) = 0$, and $\|\frac{s}{\|z\|_X} z\|_X < 1$ for $s \in B_1$. Thus, $\big\| f\big(\frac{s}{\|z\|_X} z\big)\big\|_Y < 1$, and $F(B_1) \subset B_1$. By the classical Schwarz lemma in $\mathbb{C}$, $|F(s)| \leq |s|$ for $s \in B_1$. With $s = \|z\|_X$, it is $|F(\|z\|_X)| \leq \|z\|_X$, and, as $|F(\|z\|_X)| = h\big(f(z)\big)$, it is

[1] See Section 3 of Chapter 10 of *CADOVA*.

L. T. Magalhães, *Quasiconformal Mappings in the Plane and Complex Dynamics*,
https://doi.org/10.1007/978-3-031-80115-0

$|h(f(z))| \le \|z\|_X$. Since $h(b) = \frac{h(f(z))}{\|f(z)\|_Y}$, it is $h(b)\|f(z)\|_Y = h(f(z))$. Therefore, as $h(b) = \|b\|_Y = 1$,

$$\|f(z)\|_Y = h(b)\, \|f(z)\|_Y = \big|h(f(z))\big| \le \|z\|_X .$$

Hence, $\|f(z)\|_Y \le \|z\|_X$ for $z \in B_X \backslash \{0\}$, and the proof ends observing the inequality holds trivially for $z = 0$. □

(H.2) *If Y is a real or complex Banach space with* $\dim Y = n \in \mathbb{N}$ *and* $b \in Y$, *then there is* $H \in Y^*$ *such that* $H(b) = \|b\|_Y$ *and* $\|H\|_{Y^*} = 1$.

Proof It is first proved for real Banach spaces.

There is a basis of the real linear space Y containing b, $(u_1 = b, u_2, \ldots, u_n)$. Define recursively a linear transformation $H : Y \to \mathbb{R}$ at each one of the basis vectors, by $H(b) = \|b\|_Y$, implying $|H(cb)| = |c|\|b\|_Y$ for $c \in \mathbb{R}$ and, therefore, $|H(w)| = \|w\|_Y$ for w in the linear space generated by $\{b\}$, and assuming H defined at $u_1, \ldots, u_{k-1}$ so that $|H(w)| \le \|w\|_Y$ holds for w in the linear space W spanned by $\{u_1, \ldots, u_{k-1}\}$, define $H(u_k)$ so that $|H(w + cu_k)| \le \|w + cu_k\|_Y$ for $w \in W$, $c \in \mathbb{R}$, which is equivalent to the two following inequalities to hold for all $w \in W$,

$$H(w) + tH(u_k) \le \|w + tu_k\|_Y,\ t > 0, \qquad H(w) - sH(u_k) \le \|w - su_k\|_Y,\ s > 0,$$

which are equivalent to

$$\tfrac{1}{s}\big[H(w) - \|w - su_k\|_Y\big] \le H(u_k) \le \tfrac{1}{t}\big[\|w + tu_k\|_Y - H(w)\big], \quad s, t > 0,\ w \in W.$$

It is possible to define $H(u_k)$ for these conditions to hold if and only if

$$s\big[\|w + tu_k\|_Y - H(w)\big] - t\big[H(w) - \|w - su_k\|_Y\big] \ge 0, \quad s, t > 0,\ w \in W,$$

which, as H is a linear transformation, is equivalent to

$$H\big((s+t)w\big) \le \|sw + stu_k\|_Y + \|tw - stu_k\|_Y , \quad s, t > 0,\ w \in W ,$$

which holds, since

$$\big|H\big((s+t)w\big)\big| \le \|(s+t)w\|_Y \le \|sw + stu_k\|_Y + \|tw - stu_k\|_Y .$$

Thus, $H(u_k)$ can be defined so that $|H(w)| \le \|w\|_Y$ for w in the linear space generated by $\{u_1, \ldots, u_k\}$ for all $k \in \{1, \ldots, n\}$, leading to $H \in Y^*$ such that $H(b) = \|b\|_Y$ and $\|H\|_{Y^*} = 1$.

If Y is a complex Banach space, define

$$H(y) = \mathcal{R}e H(y) + i\, \mathcal{R}e H(iy) = \mathcal{R}e H(y) + i\, \mathcal{R}e\big(i H(y)\big)\,,$$

where $\mathcal{R}e\, H$ is a linear transformation from Y to $\mathbb{R}$ with $\mathcal{R}e\, H(b) = \|b\|_Y$ and $\|\mathcal{R}e\, H\|_Y^* = 1$, whose existence follows from the preceding paragraph. It is

$$H(b) = \|b\|_Y + i\, \mathcal{R}e\big(i\, \|b\|_Y\big) = \|b\|_Y\,,$$

and, with $\theta(y)$ an angular argument of $H(y)$, $|H(y)| = e^{-i\theta(y)} H(y) = H\big(e^{-i\theta(y)} y\big)$; as $\theta(y) \in \mathbb{R}$,

$$|H(y)| = \mathcal{R}e\, H\big(e^{-i\theta(y)} y\big) \le \|\mathcal{R}e\, H\|\, \|e^{-i\theta(y)} y\|_Y = \|y\|_Y\,,$$

and, therefore, $\|H\|_{Y^*} = 1$. □

Note that (H.1) and (H.2) also hold for Y infinite dimensional, with the same proof for the first and proving the second with the Hahn-Banach theorem, which guarantees the existence of an extension of every continuous linear transformation with scalar values in a linear subspace of a Banach space to a continuous linear transformation in the whole space with operator norm with value equal to that of the initial linear transformation, whose proof depends on the assumption of validity of the axiom of choice or of the equivalent Zorn[2] Lemma: *if every totally ordered subset of a partially ordered set is upper bounded, then the set has a maximum.*

[2] Zorn, Max August (1906–1993).

Bibliography

The bibliography on the subjects of this text is immense. The option here is for a relatively short list (in addition to the about 60 publications cited in the book chapters), with some alternative books on the themes considered and including a few books on subjects touched on but not discussed in this text. The main intention is to provide possible references for further study. These books have useful additional bibliographic references.

1. Abikoff, W. *The Real Analytic Theory of Teichmüller Space*. Lecture Notes in Mathematics (Vol. 820). Springer, 1980.
2. Ahlfors, L. V. *Complex Analysis, an Introduction to the Theory of Analytic Functions of One Complex Variable* (3rd ed.). McGraw-Hill Book Company, 1978.
3. Ahlfors, L. V. *Lectures on Quasiconformal Mappings,* (2nd ed.) with additional chapters by C. J. Earle, I. Kra, M. Shishikura & J.H. Hubbard, University lecture series (Vol. 38). American Mathematical Society, 2006.
4. Ahlfors, L. V. *Conformal Invariants—Topics in Geometric Function Theory*. AMS Chelsea Publishing, 2010.
5. Astala, K., Iwaniec, T., & Martin, G. *Elliptic Partial Differential Equations and Quasiconformal Mappings in the Plane*. Princeton Mathematical Series (Vol. 48). Princeton University Press, 2009.
6. Bojarski, B., Gutlyanskii, V., Martio, O.,& Ryazanov, V. *Infinitesimal Geometry of Quasiconformal and Bi-Lipschitz Mappings in the Plane*. EMS tracts in mathematics. European Mathematical Society, 2013.
7. Branner, B., & Fagella, N. *Quasiconformal Surgery in Holomorphic Dynamics*. Cambridge University Press, 2014.
8. Carleson, L., & Gamelin, T. W. *Complex Dynamics*. Springer, 1993.
9. Chabat, B. *Introduction à L'Analyse Complexe, Tome 1: Fonctions d'Une Variable,* Tome 2: Fonctions de Plusieurs Variables. Éditions MIR, Moscou, 1990.
10. Conway, J. B. *Functions of One Complex Variable* (Vols. I, II, 2nd ed.). Springer, 1995.
11. de Faria, E., & de Melo, W. *Mathematical Tools for One-dimensional Dynamics*. Cambridge University Press, 2008.
12. Gardiner, F. P. *Teichmüller Theory and Quadratic Differentials*. John Wiley & Sons, 1987.
13. Gardiner, F. P., & Lakic, N. *Quasiconformal Teichmüller Theory*. American Mathematical Society, 2000.

L. T. Magalhães, *Quasiconformal Mappings in the Plane and Complex Dynamics*,
https://doi.org/10.1007/978-3-031-80115-0

14. Gehring, F. W., Martin, G. J., & Palka, B. P. *An Introduction to the Theory of Higher-dimensional Quasiconformal Mappings*. American Mathematica Society, 2017.
15. Imayoshi Y., & Taniguchi. M. *An Introduction to Teichmüller Spaces*. Springer, 1992.
16. Lehto, O. *Univalent Functions and Teichmüller Spaces*. Springer, 1987.
17. Lehto, O., & Virtanen, K. I. *Quasiconformal Mappings in the Plane*. Springer, 1973.
18. Magalhães, L. T. *Integrais Múltiplos*. Texto Editora, 1993.
19. Magalhães, L. T. *Complex Analysis and Dynamics in One Variable with Applications*. Springer, 2025.
20. McMullen, C. *Complex Dynamics and Renormalization*. Princeton University Press, 1994.
21. Milnor, J. W. *Dynamics in One Complex Variable*. Princeton University Press, 2006.
22. Nag, S. *The Complex Analytic Theory of Teichmüller Spaces*. Canadian Mathematical Society, Wiley-Interscience, 1987.
23. Remmert, R. *Classical Topics in Complex Function Theory*. Springer, 1998.
24. Rudin, W. *Real and Complex Analysis*. McGraw-Hill Book Company, 1987.
25. Stanislav, H., & Koskela, P.[1] *Lectures on Mappings of Finite Distortion*. Lecture notes in mathematics (Vol. 2096). Springer, 2014.
26. Stein, E. M., Shakarchi, R. *Complex Analysis*. Princeton Lectures in Analysis. Princeton University Press, 2003.
27. Strebel, K. *Quadratic Differentials*. Springer, 1984.

[1] Abikoff, William (1944–). Bojarski, Bogdan (1931–2018). Gutlyanskii, Vladimir (1941–). Martio, Olli (1941-). Ryazanov, Vladimir. Lakic, Nicola (1966–). Palka, Bruce (1943–). Imayoshi, Yoichi. Taniguchi, Masahiko. Nag, Subhashis (1955–1998). Stanislav, Hencl (1976–). Koskela, Pekka (1960–)-.

Authors Index

L. T. Magalhães, *Quasiconformal Mappings in the Plane and Complex Dynamics*,
https://doi.org/10.1007/978-3-031-80115-0

Index

L. T. Magalhães, *Quasiconformal Mappings in the Plane and Complex Dynamics*,
https://doi.org/10.1007/978-3-031-80115-0

N

Q

MIX
Papier aus verantwortungsvollen Quellen
Paper from responsible sources
FSC® C105338
FSC
www.fsc.org

If you have any concerns about our products,
you can contact us on
ProductSafety@springernature.com

In case Publisher is established outside the EU,
the EU authorized representative is:
Springer Nature Customer Service Center GmbH
Europaplatz 3, 69115 Heidelberg, Germany

Printed by Libri Plureos GmbH
in Hamburg, Germany